Math Study Skills

Your overall success in mastering the material in your course depends on you. You must be **committed** to doing your best in this course. This commitment means dedicating the time needed to study math and to do your homework.

In order to succeed in math, you must know how to study it. The goal is to study math so that you understand and not just memorize it. The following tips and strategies will help you develop good study habits.

General Tips

ATTEND EVERY CLASS Be on time. If you must miss class, be sure to talk with your instructor or a classmate about what was covered.

MANAGE YOUR TIME School, work, family, and other commitments place a lot of demand on your time. To be successful, you must be able to devote time to study math every day. Writing out a weekly schedule that lists your class schedule, work schedule, and all other commitments with times that are not flexible will help you to determine when you can study. Use the companion resources that accompany the book, such as MyMathLab, so you can study online with tutoring help whenever you have the time.

Instructor Contact Information

Name: _____

Office Hours: _____

Office Location: _____

Phone Number: _____

E-mail Address: _____

Campus Tutoring Center

Location: _____

Hours: _____

DO NOT WAIT TO GET HELP If you are having difficulty, get help immediately. Since the material presented in class usually builds on previous material, it is very easy to fall behind. Ask your instructor if he or she is available during office hours or get help at the tutoring center on campus.

Tests

Tests are a source of anxiety for many students. Being well prepared to take a test can help ease anxiety.

BEFORE A TEST

❐ Review your notes and the sections of the textbook that will be covered on the test.

❐ Read through the Key Concepts and Skills in the textbook and your own summary from your notes.

❐ Do additional practice problems. Select problems from your homework to rework. The textbook also contains Mixed Practice and Review Exercises that provide opportunities to strengthen your skills. You can also review your understanding of concepts by completing the Say Why problems in the text and using the Study Plan in MyMathLab.

❐ Use the Posttest at the end of the chapter as your practice test. While taking the practice test, do not refer to your notes or the textbook for help. Keep track of how long it takes to complete the test. Check your answers with the Chapter Test Prep Videos on YouTube or in MyMathLab for any problems you did not answer correctly. If you cannot complete the practice test within the time you are allotted for the real test, you need additional practice in the tutoring center or in MyMathLab to speed up.

DURING A TEST

❐ Read through the test before starting.

❐ If you find yourself panicking, relax, take a few slow breaths, and try to do some of the problems that seem easy.

❐ Do the problems you know how to do first, and then go back to the ones that are more difficult.

❐ Watch your time. Do not spend too much time on any one problem. If you get stuck while working on a problem, skip it and move on to the next problem.

❐ Check your work, if there is time. Correct any errors you find.

AFTER A TEST

❐ When you get your test back, look through all of the problems.

❐ On a separate sheet of paper, do any problems that you missed.
Use your notes and textbook, if necessary.

❐ Get help from your instructor or a tutor if you cannot figure out how to do a problem, or set up a meeting with your study group to go over the test together. Make sure you understand your errors.

❐ Attach the corrections to the test and place it in your notebook.

Best wishes in your studies!
— Geoffrey Akst   Sadie Bragg

Class Time

BEFORE CLASS

❏ Review your notes from the previous class session.

❏ Read the section(s) of your textbook that will be covered in class to get familiar with the material. Read these sections carefully. Skimming may result in your not understanding some of the material and your inability to do the homework. If you do not understand something in the text, reread it more thoroughly or seek assistance.

DURING CLASS

❏ Pay attention and try to understand every question your instructor asks.

❏ Take good notes.

❏ Ask questions during class if you do not understand something. Chances are that someone else has the same question, but is not comfortable asking it. If you feel that way also, then write your question in your notebook and ask your instructor after class or see the instructor during office hours.

AFTER CLASS

❏ Review your notes as soon as possible after class. Insert additional steps and comments to help clarify the material.

❏ Reread the section(s) of your textbook, focusing on the Example and Practice side by sides and the Tips. After reading through an example, cover it up and try to do it on your own. Do the practice problem that is paired with the example. (The answers to the practice problems are given in the back of the textbook.) The Tips will help you avoid common errors and provide other suggestions to foster understanding.

❏ Work in MyMathLab and use the personalized study plan.

Homework

The best way to learn math is by doing it. Homework is designed to help you learn and apply concepts and to master certain skills. Some tips for doing homework are

❏ Do your homework the same day that you have class. Keeping up with the class requires you to do homework regularly rather than cramming right before tests.

❏ Review the section of the textbook that corresponds to the homework.

❏ Review your notes.

❏ If you get stuck on a problem, look for a similar example in your textbook or notes or use Help Me Solve This in MyMathLab.

❏ Write a question mark next to any problems that you just cannot figure out. Get help from your instructor or the tutoring center, or call someone from your study group.

❏ Check your answer after each problem. The answers to the odd-numbered problems are in the back of the textbook. If you are assigned even-numbered problems, try the odd-numbered problem first and check the answer. If your answer is correct, then you should be able to do the even-numbered problem correctly. If doing your homework in MyMathLab, keep your work and repeat problems until you can work through them successfully.

FORM A STUDY GROUP A study group provides an opportunity to discuss class material and homework problems. Find at least two other people in your class who are committed to being successful. Exchange contact information and plan to meet or work together regularly throughout the semester either in person or via e-mail, MyMathLab, or phone.

USE YOUR BOOK'S STUDY RESOURCES There are additional resources and support materials available with this book to help you succeed. See the list below and in the preface.

Notebook and Note Taking

Taking good notes and keeping a neat, well-organized notebook are important factors in being successful. If you do your homework online through MyMathLab, you should still keep a notebook to stay organized.

YOUR NOTEBOOK Use a loose-leaf binder divided into four sections: (1) notes, (2) homework, (3) graded tests (and quizzes), and (4) handouts. Or combine the resources in MyMathLab with the MyWorkBook with Chapter Summaries.

TAKING NOTES

❏ Copy all important information. Also, write all points that are not clear to you so that you can discuss them with your instructor, a tutor, or your study group.

❏ Write explanations of what you are doing in your own words next to each step of a practice problem.

❏ Listen carefully to what your instructor emphasizes and make note of it.

The following resources are available in MyMathLab, through your college bookstore, and at **www.pearsonhighered.com**:

- Student's Solutions Manual
- Video Resources with Chapter Test Prep Videos
- MyMathLab
- MyWorkBook with Chapter Summaries

Full descriptions are available in the preface.

Introductory Algebra
through Applications

Third Edition

GEOFFREY AKST • SADIE BRAGG
Borough of Manhattan Community College, The City University of New York

PEARSON

Boston   Columbus   Indianapolis   New York   San Francisco   Upper Saddle River
Amsterdam   Cape Town   Dubai   London   Madrid   Milan   Munich   Paris   Montréal   Toronto
Delhi   Mexico City   São Paulo   Sydney   Hong Kong   Seoul   Singapore   Taipei   Tokyo

Editorial Director: Christine Hoag
Editor in Chief: Maureen O'Connor
Executive Content Editor: Kari Heen
Content Editor: Katie DePasquale
Editorial Assistant: Rachel Haskell
Senior Managing Editor: Karen Wernholm
Senior Production Supervisor: Ron Hampton
Senior Cover Designer: Barbara T. Atkinson
Senior Technology/Author Support Specialist: Joe Vetere
Text Design: Leslie Haimes
Composition: PreMediaGlobal
Image Research Manager: Rachel Youdelman
Media Producer: Aimee Thorne
Software Development: TestGen: Mary Durnwald; MathXL: Jozef Kubit
Executive Marketing Manager: Michelle Renda
Marketing Manager: Rachel Ross
Marketing Assistant: Ashley Bryan
Procurement Manager/Boston: Evelyn Beaton
Procurement Specialist: Debbie Rossi
Media Procurement Specialist: Ginny Michaud
Cover Photo: Bamboo on white © Subotina Anna/Shutterstock

Library of Congress Cataloging-in-Publication Data

Akst, Geoffrey.
  Introductory algebra through applications / Geoffrey Akst, Sadie Bragg.—3rd ed.
    p. cm
  Includes index.
  ISBN-13: 978-0-321-74669-6   ISBN-10: 0-321-74669-4 (student ed.: alk. paper)
    1. Mathematics—Textbooks.   I. Bragg, Sadie   II. Title.
  QA152.3.A47 2012
  510—dc22   2011005563

1 2 3 4 5 6 7 8 9 10—CRK—15 14 13 12 11

pearsonhighered.com

To the memory of
Maxine Jefferson
and
Mag Dora Chavis

Contents

Preface

FROM THE AUTHORS

Our goal in writing *Introductory Algebra through Applications* was to create a text that would help students progress and succeed in their college developmental math course. Throughout, we emphasize an applied approach, which has two advantages. First of all, it can help students prepare to meet their future mathematical demands—across disciplines, in subsequent coursework, in everyday life, and on the job. Secondly, this approach can be motivating, convincing students that mathematics is worth learning and more than just a school subject.

We have attempted to make the text readable, with understandable explanations and exercises for honing skills. We have also put together a set of easy-to-grasp features, consistent across sections and chapters.

To address many of the issues raised by national professional organizations, including AMATYC, NCTM, and NADE, we have been careful to stress connections to other disciplines; to incorporate the appropriate use of technology; to integrate quantitative reasoning skills; to include problem sets that facilitate student writing, critical thinking, and collaborative activities; and to emphasize real-world data in examples and exercises.

Above all, we have tried to develop a flexible text that can meet the needs of students in both traditional and redesigned developmental courses.

This text is part of the *through Applications* series that includes the following:

WHAT'S NEW IN THE THIRD EDITION?

Say Why Exercises   New fill-in-the-blank problems, located at the beginning of each chapter review, providing practice in reasoning and communicating mathematical ideas (see page 277).

Updated and Expanded Section Exercise Sets   Additional practice in mastering skills.

Lengthening of Cumulative Review Exercise Sets   Twice as many review exercises in response to user demand (see pages 287–288).

Greater Emphasis on Learning Objectives   End-of-section exercises closely aligned with the learning objectives in order to encourage and facilitate review (see pages 30 and 38–39).

More Examples and Exercises Based on Real Data   Additional and more varied applied problems that are useful, realistic, and authentic (see page 36).

Parallel Paired Exercises   Odd/even pairs of problems that more closely reflect the same learning objective (see page 129).

Easy-to-Locate Features   Color borders added for back-of-book answer, glossary, and index pages.

Highlighting of Quantitative Literacy Skills   Additional exercises that provide practice in number sense, proportional reasoning, and the interpretation of tables and graphs (see page 159).

Increased Attention to Photos and Graphics   Carefully selected photos to make problems seem more realistic, and relevant graphics to better meet the needs of visual learners (see pages 82 and 332).

Newly Expanded and Robust MyMathLab Coverage!   One of *every* problem type is now assignable in MyMathLab.

Now Two MyMathLab Course Options

1. **Standard MyMathLab** allows you to build *your* course *your* way, offering maximum flexibility and complete control over all aspects of assignment creation. Starting with a clean slate lets you choose the exact quantity and type of problems you want to include for your students. You can also select from prebuilt assignments to give you a starting point.

2. **Ready-to-Go MyMathLab** comes with assignments prebuilt and preassigned, reducing start-up time. You can always edit individual assignments, as needed, through the semester.

KEY FEATURES

Math Study Skills Foldout   A full-color foldout with tips on organization, test preparation, time management, and more (see inside front cover).

Chapter Openers   Extended real-world applications at the beginning of each chapter to motivate student interest and demonstrate how mathematics is used (see page 182).

Pretests and Posttests   Chapter tests, which are particularly useful in a self-paced, lab, or digital environment (see pages 113 and 180).

Section Objectives   Clearly stated learning objectives at the beginning of each section to identify topics to be covered (see page 30).

Side-by-Side Example/Practice Format   Distinctive side-by-side format that pairs each example with a corresponding practice exercise and gets students actively involved from the start (see page 191).

Tips   Helpful suggestions and cautions for avoiding mistakes (see page 164).

Journal Entries   Writing assignments in response to probing questions interspersed throughout the text (see page 292).

Calculator Inserts   Optional calculator and computer software instruction to solve section problems (see pages 47 and 249).

Cultural Notes   Glimpses of how mathematics has evolved across cultures and throughout history (see page 123).

For Extra Help   Boxes at the beginning of every section's exercise set that direct students to helpful resources (see page 129).

Mathematically Speaking Exercises   Vocabulary exercises in each section to help students understand and use standard mathematical terminology (see page 83).

Mixed Practice Exercises   Problems in synthesizing section material (see page 101).

Application Exercises   End-of-section problems to apply the topic at hand in a wide range of contexts (see pages 173–174).

Mindstretcher Exercises   Nonstandard section problems in critical thinking, mathematical reasoning, pattern recognition, historical connections, writing, and group work to deepen understanding and provide enrichment (see page 144).

Key Concepts and Skills Summary   With a focus on descriptions and examples, the main points of the chapter organized into a practical and comprehensive chart (see pages 103–105).

Chapter Review Exercises   Problems for reviewing chapter content, arranged by section (see pages 277–281).

Chapter Mixed Application Exercises   Practice in applying topics across the chapter (see pages 281–283).

Cumulative Review Exercises   Problems to maintain and build on the mathematical content covered in previous chapters (see pages 430–431).

Appendixes   Three brief appendixes—Table of Symbols, Factoring the Sum of Cubes and the Difference of Cubes, and Introduction to Graphing Calculators.

U.S. and Metric Unit Tables   Located opposite the inside back cover for quick reference.

Geometric Formulas   A reference on the inside back cover of the text displaying standard formulas for perimeter, circumference, area, and volume.

Coherent Development   Texts with consistent content and style across the developmental math curriculum.

WHAT SUPPLEMENTS ARE AVAILABLE?

For a complete list of the supplements and study aids that accompany *Introductory Algebra through Applications*, Third Edition, see pp. xi.

ACKNOWLEDGMENTS

We are grateful to everyone who has helped to shape this textbook by responding to questionnaires, participating in telephone surveys and focus groups, reviewing the manuscript, and using the text in their classes. We wish to thank Sheila Anderson, *Housatonic Community College;* James J. Ball, *Indiana State University;* Mike Benningfield, *Paris Junior College;* Amanda Bertagnolli-Comstock, *Bishop State Community College;* Susan D. Caire, *Delgado Community College;* Donald M. Carr, *College of the Desert;* Alison Carter, *California State University, Howard;* Kristin Chatas, *Washtenaw Community College;* Karla Childs, *Pittsburgh State University;* O. Pauline Chow, *Harrisburg Area Community College;* Bryan Cockerham, *Front Range Community College;* Vincent Conklin, *Bauder College;* Jonathan Cornick, *Queensborough Community College/CUNY;* Susi Curl, *McCook Community College;* Karena Curtis, *Labette Community College;* Lucy Edwards, *Las Positas College;* Gene Forster, *Southeastern Illinois College;* Mary Ellen Gallegos, *Santa Fe Community College;* Naomi Gibbs, *Pitt Community College;* Tim Hagopian, *Worcester State College;* Anthony Hearn, *Community College of Philadelphia;* Barbara Heim, *Ivy Tech Community College;* Lori Holdren, *Manatee Community College;* Marlene Ignacio, *Pierce College–Puyallup;* Sharon Jackson, *Brookhaven College;* Marilyn Jacobi, *Gateway Community College;* John Jacobs, *Massachusetts Bay Community College;* Jennifer Johnson, *Delgado Community College;* Mary Ann Klicka, *Bucks County Community College;* Sandy

Lanoue, *Tulsa Community College;* Thang Le, *College of the Desert;* Carol Lerch, *Daniel Webster College;* Ken Mead, *Genesee Community College;* Kimberly McHale, *Columbia College;* Debbie Moran, *Greenville Technical College;* Sandra Peskin, *Queensborough Community College/CUNY;* Carol Phillips-Bey, *Cleveland State University;* Sharonda Ragland, *ECPI College of Technology;* Nancy Ressler, *Oakton Community College;* Sylvester Roebuck, *Olive-Harvey College;* Patricia C. Rome, *Delgado Community College;* Dr. Yojana Sharma, *Stark State College;* Mary Pat Sheppard, *Malcolm X College;* Richard Sturgeon, *University of Southern Maine;* Marcia Swope, *Nova Southeastern University;* Jane Tanner, *Onondaga Community College;* Sven Trenholm, *Herkimer County Community College;* Bernadette Turner, *Lincoln University;* Betty Vix Weinberger, *Delgado Community College;* Cora S. West, *Florida Community College at Jacksonville;* Mary Wolyniak, *Broome Community College;* and Jeff Young, *Delaware Valley College.* In addition, we would like to extend our gratitude to our accuracy checkers and to those who helped us perfect the content in many ways: Michael Carlisle; Lisa Collete; Paul Lorczak; Denise Heban; Sharon O'Donnell, *Chicago State University;* Ann Ostberg; Lenore Parens; and Deana Richmond.

Writing a textbook requires the contributions of many individuals. Special thanks go to Greg Tobin, President, Mathematics and Statistics, Pearson Arts and Sciences, for encouraging and supporting us throughout the entire process. We thank Kari Heen and Katie DePasquale for their patience and tact, Michelle Renda, Rachel Ross, and Maureen O'Connor for keeping us abreast of market trends, Rachel Haskell for attending to the endless details connected with the project, Ron Hampton, Laura Osterbrock, Laura Hakala, Marta Johnson, and Rachel Youdelman for their support throughout the production process, Barbara Atkinson for the cover design, and the entire Pearson developmental mathematics team for helping to make this text one of which we are very proud.

Geoffrey Akst

Sadie Bragg

Student Supplements

Student's Solutions Manual
By Deana Richmond
- Provides detailed solutions to the odd-numbered exercises in each exercise set and solutions to all chapter pretests and post-tests, practice exercises, review exercises, and cumulative review exercises

ISBN-10: 0-321-75931-1   ISBN-13: 978-0-321-75931-3

New Video Resources on DVD with Chapter Test Prep Videos
- Complete set of digitized videos on DVD for students to use at home or on campus
- Includes a full lecture for each section of the text
- Covers examples, practice problems, and exercises from the textbook that are marked with the ⊙ icon
- Optional captioning in English is available
- Step-by-step video solutions for each chapter test
- Chapter Test Prep Videos are also available on YouTube (search by using author name and book title) and in MyMathLab

ISBN-10: 0-321-75929-X   ISBN-13: 978-0-321-75929-0

MyWorkBook with Chapter Summaries
By Carrie Green
- Provides one worksheet for each section of the text, organized by section objective, along with the end-of-chapter summaries from the textbook
- Each worksheet lists the associated objectives from the text, provides fill-in-the-blank vocabulary practice, and exercises for each objective

ISBN-10: 0-321-75928-1   ISBN-13: 978-0-321-75928-3

MathXL Online Course (access code required)

InterAct Math Tutorial Website
www.interactmath.com
- Get practice and tutorial help online
- Provides algorithmically generated practice exercises that correlate directly to the textbook exercises
- Retry an exercise as many times as desired with new values each time for unlimited practice and mastery
- Every exercise is accompanied by an interactive guided solution that gives the student helpful feedback when an incorrect answer is entered
- View the steps of a worked-out sample problem similar to the one that has been worked on

Instructor Supplements

Annotated Instructor's Edition
- Provides answers to all text exercises in color next to the corresponding problems
- Includes teaching tips

ISBN-10: 0-321-75709-2   ISBN-13: 978-0-321-75709-8

Instructor's Solutions Manual (download only)
By Deana Richmond
- Provides complete solutions to even-numbered section exercises
- Contains answers to all Mindstretcher problems

ISBN-10: 0-321-75720-3   ISBN-13: 978-0-321-75720-3

Instructor's Resource Manual with Tests and Mini-Lectures (download only)
By Deana Richmond
- Contains three free-response and one multiple-choice test form per chapter, and two final exams
- Includes resources designed to help both new and adjunct faculty with course preparation and classroom management, including sample syllabi, tips for using supplements and technology, and useful external resources
- Offers helpful teaching tips correlated to the sections of the text

ISBN-10: 0-321-75721-1   ISBN-13: 978-0-321-75721-0

PowerPoint Lecture Slides (available online)
- Present key concepts and definitions from the text

TestGen® (available for download from the Instructor's Resource Center)

AVAILABLE FOR STUDENTS AND INSTRUCTORS

MyMathLab® Ready-to-Go Course (access code required)

These new Ready-to-Go courses provide students with all the same great MyMathLab features that you're used to, but make it easier for instructors to get started. Each course includes preassigned homework and quizzes to make creating your course even simpler. Ask your Pearson representative about the details for this particular course or to see a copy of this course.

MyMathLab with Pearson eText—Instant Access for *Introductory Algebra through Applications*

MyMathLab delivers proven results in helping individual students succeed. It provides engaging experiences that personalize, stimulate, and measure learning for each student. And, it comes from a trusted partner with educational expertise and an eye on the future. To learn more about how MyMathLab combines proven learning applications with powerful assessment, visit www.mymathlab.com or contact your Pearson representative.

MathXL—Instant Access for *Introductory Algebra through Applications*

MathXL® is the homework and assessment engine that runs MyMathLab. (MyMathLab is MathXL plus a learning management system.) With MathXL, instructors can

- Create, edit, and assign online homework and tests using algorithmically generated exercises correlated at the objective level to the textbook.
- Create and assign their own online exercises and import TestGen tests for added flexibility.
- Maintain records of all student work tracked in MathXL's online gradebook.

With MathXL, students can

- Take chapter tests in MathXL and receive personalized study plans and/or personalized homework assignments based on their test results.
- Use the study plan and/or the homework to link directly to tutorial exercises for the objectives they need to study.
- Access supplemental animations and video clips directly from selected exercises.

MathXL is available to qualified adopters. For more information, visit www.mathxl.com or contact your Pearson representative.

Index of Applications

INDEX OF APPLICATIONS

Photo Credits

p. 1: Monkey Business Images/Shutterstock; p. 28: Medical Picture/Alamy; p. 50: Timothy A. Clary/AFP/Getty Images; p. 65: Andrew Butterton/Alamy; p. 82: Alexander Raths/Shutterstock; p. 112: Stephen Coburn/Shutterstock; p. 123: Paul Paladin/Alamy; p. 127: Tselichtchev/Shutterstock; p. 160 (l): Keith Levit/Shutterstock; p. 160 (r): Harris & Ewing Collection/Library of Congress Prints and Photographs Division [LC-DIG-hec-04117]; p. 182: Richard Drew/AP Images; p. 198: Beth Anderson/Pearson Education, Inc.; p. 199: Galina Barskaya/Shutterstock; p. 289: Dora Modly-Paris/Shutterstock; p. 329: StockLite/Shutterstock; p. 332: R. Nagy/Shutterstock; p. 334: Corbis RF/Alamy; p. 341 (l): Devi/Shutterstock; p. 341 (r): Irina Tischenko/Shutterstock; p. 347: Mary Evans Picture Library/Alamy; p. 359, 407: Pearson Education, Inc.; p. 366: Gordana Sermek/Shutterstock; p. 372 (tr): Scott Camazine/Alamy; p. 372 (bl): Sebastian Kaulitzki/Shutterstock; p. 372 (br): Scripps Howard Photo Service/Newscom; p. 383: Fotomilan011/Dreamstime; p. 384, 565: DB Images/Alamy; p. 389: Joyce/Dreamstime; p. 392 (l): Egomezta/Dreamstime; p. 392 (r): U.S. Navy/ABACAUSA/Newscom; p. 409: Iorboaz/Dreamstime; p. 428 (l): Jozef Sedmak/Shutterstock; p. 428 (r): Siamionau Pavel/Shutterstock; p. 432: Marny Malin/Frederick News Post/AP Images; p. 454 (l): Lonely Planet Images/Alamy; p. 454 (r): David Lee/Shutterstock; p. 473: Brian McEntire/iStockphoto; p. 477: Gene Krebs/iStockphoto; p. 481 (tl): Sean Pavone Photo/Shutterstock; p. 481 (tr): July Store/Shutterstock; p. 481 (bl): IntraClique, LLC/Shutterstock; p. 486: Jupiter Images/Thinkstock; p. 489: Triff/Shutterstock; p. 490: Brandon Bourdages/Shutterstock; p. 498: Bernhard Classen/Alamy; p. 526: Jason Walton/iStockphoto; p. 532: Vitaly Korovin/Shutterstock; p. 534: Jerry Horbert/Shutterstock; p. 540: Sepavo/Shutterstock; p. 544: Greenland/Shutterstock; p. 566: Leshik/Shutterstock; p. 576: Copestello/Shutterstock; p. 593: Peter Arnold, Inc./Alamy; p. 601 (t): Vitalii Nesterchuk/Shutterstock; p. 601 (b): Oleg Golovnev/Shutterstock; p. 612: Photos 12/Alamy; p. 613: Michael Jung/Shutterstock; p. 624: Bettmann/Corbis; p. 632: Sherri R. Camp/Shutterstock; p. 641: Radu Razvan/Shutterstock

CHAPTER R

Prealgebra Review

Why a Review of Prealgebra Topics Is Important

This chapter is a brief review of the major procedures, concepts, and vocabulary in basic mathematics that you will need to succeed in working through this text. Taking the pretest on page 2 will help you to pinpoint topics that you need to review. For these topics, study the explanations and worked-out examples and then solve the related practice problems and exercises until you have mastered the material.

Knowledge of basic mathematics is important to the study of algebra not only for calculating the value of arithmetic expressions but also for understanding algebraic procedures that are based on arithmetic procedures. For instance, familiarity with adding numerical fractions puts you one step ahead when learning how to add algebraic fractions.

In this chapter, we assume that you are proficient in whole-number arithmetic. Accordingly, we begin with a discussion of exponents and order of operations. Then we consider factors, primes, and least common multiples. These topics are needed for working with fractions—the focus of the following section. Next comes a brief review of decimals and finally percent conversions.

If you find that you need to delve more deeply into any particular arithmetic topic, take the initiative of exploring the resources—books, print material, tutoring, multimedia and online sites—that are available at your college to support your personal arithmetic review. You will benefit from doing this sooner rather than later.

OBJECTIVES

A To review exponents and order of operations

B To review factors, primes, and least common multiples

C To review fractions

D To review decimals

E To review percent conversions

F To review applied problems involving basic mathematics

To see if you have already mastered the topics in this chapter, take this test.

1. Rewrite $7 \cdot 7 \cdot 7$ as a power of 7.

2. Compute: $2^4 \cdot 5^2$

3. Simplify: $16 - 2^3$

4. Find the value of $20 - 2(3 - 1^2)$.

5. What are the factors of 12?

6. Rewrite 20 as a product of prime factors.

7. Write $\dfrac{16}{24}$ in simplest form.

8. Find the sum: $\dfrac{3}{10} + \dfrac{1}{10}$

9. Subtract: $10\dfrac{1}{3} - 2\dfrac{5}{6}$

10. Find the product of $\dfrac{2}{9}$ and $\dfrac{2}{3}$.

11. Calculate: $4 \div 1\dfrac{1}{2}$

12. Write 9.013 in words.

13. Round 3.072 to the nearest tenth.

14. Find the sum: $7 + 4.01 + 9.3003$

15. Find the difference: $8 - 2.34$

16. What is the product of 9.23 and 4.1?

17. Evaluate: 0.235×100

18. What is 0.045 divided by 0.25?

19. Compute: $\dfrac{3.1}{1000}$

20. What is the decimal equivalent of 7%?

21. A college launched a campaign to raise $3 million to build a new science building. If $1.316 million has been raised so far, how much more money, to the nearest million dollars, is needed?

22. In the 111th U.S. Congress, $\dfrac{2}{5}$ of the senators were Republicans, and $\dfrac{1}{10}$ of these Republicans were female. What fraction of all the senators were female Republicans? (*Source: The New York Times Almanac 2010*)

23. A doctor increases a patient's daily dosage of thyroxin from 0.05 mg to 0.1 mg. The new dosage is how many times as great as the previous dosage?

24. Snow weighs 0.1 as much as water. Express this decimal as a percent. (*Source:* nrcs.usda.gov)

25. About 95% of all animal species on Earth are insects. Express this percent as a decimal. (*Source:* elkhornslough.org)

• Check your answers on page A-1.

R.1   Exponents and Order of Operations

Exponents

There are many mathematical situations in which we multiply a number by itself repeatedly. Writing such expressions using *exponents* (or *powers*) provides a shorthand method for representing this repeated multiplication of the same factor:

$$\underbrace{2 \cdot 2 \cdot 2 \cdot 2}_{4 \text{ factors of } 2} = 2\overset{\leftarrow\ \text{exponent}}{\underset{\uparrow\quad\text{base}}{4}}$$

The expression 2^4 is read "2 to the fourth power" or simply "2 to the fourth."

To evaluate 2^4, we multiply 4 factors of 2:

$$\begin{aligned}
2^4 &= \underbrace{2 \cdot 2} \cdot 2 \cdot 2 \\
&= \underbrace{4 \cdot 2} \cdot 2 \\
&= \underbrace{8 \cdot 2} \\
&= 16
\end{aligned}$$

So $2^4 = 16$.

Sometimes we prefer to shorten expressions by using exponents. For instance,

$$\underbrace{3 \cdot 3}_{2 \text{ factors of } 3} \cdot \underbrace{4 \cdot 4 \cdot 4}_{3 \text{ factors of } 4} = 3^2 \cdot 4^3$$

EXAMPLE 1

Express $6 \cdot 6 \cdot 6$ using exponents.

Solution $\underbrace{6 \cdot 6 \cdot 6}_{3 \text{ factors of } 6} = 6^3$

PRACTICE 1

Write $2 \cdot 2 \cdot 2 \cdot 2 \cdot 2$ as a power of 2.

EXAMPLE 2

Calculate: $4^3 \cdot 5^3$

Solution $\begin{aligned}
4^3 \cdot 5^3 &= (4 \cdot 4 \cdot 4) \cdot (5 \cdot 5 \cdot 5) \\
&= 64 \cdot 125 \\
&= 8000
\end{aligned}$

It is especially easy to compute powers of 10:

$$10^2 = \underbrace{10 \cdot 10}_{2 \text{ factors}} = \underbrace{100}_{2 \text{ zeros}}$$

$$10^3 = \underbrace{10 \cdot 10 \cdot 10}_{3 \text{ factors}} = \underbrace{1000}_{3 \text{ zeros}}$$

and so on.

PRACTICE 2

Compute: $7^2 \cdot 2^4$

EXAMPLE 3

The distance from the Sun to the star Alpha-one Crucis is about 1,000,000,000,000,000 mi. Express this distance as a power of 10. (*Source:* infoplease.com)

Solution $\underbrace{1,000,000,000,000,000}_{\text{15 zeros}} = 10^{15}$

So the distance is 10^{15} mi.

PRACTICE 3

In 1850, the world population was approximately 1,000,000,000. Represent this number as a power of 10. (*Source:* census.gov)

Order of Operations

Some mathematical expressions involve more than one operation. For instance, consider the expression $5 + 3 \cdot 2$. This expression seems to have two different values, depending on the order in which we perform the given operations.

Add first	**Multiply first**
$5 + \underline{3 \cdot 2}$	$5 + \underline{3 \cdot 2}$
$= \underline{8 \cdot 2}$	$= \underline{5 + 6}$
$=\ \ 16$	$=\ \ 11$

How do we know which operation to carry out first? By consensus we agree to follow the rule called the **order of operations** so that everyone always gets the same value for an answer.

Order of Operations Rule

To evaluate mathematical expressions, carry out the operations *in the following order:*

1. First, perform the operations within any grouping symbols, such as parentheses ( ) or brackets [ ].
2. Then, raise any number to its power.
3. Next, perform all multiplications and divisions as they appear from left to right.
4. Finally, do all additions and subtractions as they appear from left to right.

Applying the order of operations rule to the previous example gives us the following result:

$$5 + \underline{3 \cdot 2} \qquad \text{Multiply first.}$$
$$= \underline{5 + 6} \qquad \text{Then, add.}$$
$$=\ \ 11$$

So 11 is the correct answer.

EXAMPLE 4

Simplify: $20 - 18 \div 9 \cdot 8$

Solution   The order of operations rule gives us:

$$20 - 18 \div 9 \cdot 8 = 20 - 2 \cdot 8 \qquad \text{Divide first.}$$
$$= 20 - 16 \qquad \text{Then, multiply.}$$
$$= 4 \qquad \text{Finally, subtract.}$$

PRACTICE 4

Evaluate: $8 \div 2 + 4 \cdot 3$

EXAMPLE 5

Find the value of $3 + 2 \cdot (8 + 3^2)$.

Solution

$$
\begin{aligned}
3 + 2 \cdot (8 + 3^2) &= 3 + 2 \cdot (8 + 9) && \text{Perform operations in} \\
&&& \text{parentheses: square the 3.} \\
&= 3 + 2 \cdot \quad 17 && \text{Add 8 and 9.} \\
&= 3 + \quad 34 && \text{Multiply 2 and 17.} \\
&= \quad 37 && \text{Add 3 and 34.}
\end{aligned}
$$

PRACTICE 5

Simplify: $(4 + 1)^2 - 4 \cdot 6$

TIP When a division problem is written as a fraction, parentheses are understood to be around both the numerator and the denominator. For instance,

$$
\frac{10 - 2}{3 - 1} \quad \text{means} \quad (10 - 2) \div (3 - 1)
$$

The following example involves finding an *average*. The **average** (or **mean**) of a set of numbers is the sum of those numbers divided by however many numbers are in the set.

EXAMPLE 6

A student's test scores in a class were 85, 94, 93, 86, and 92. If the average of these scores was 90 or above, she earned an A. Did the student earn an A?

Solution

$$
\begin{aligned}
\text{Average} &= \frac{\text{The sum of the scores}}{\text{The number of scores}} \\
&= \frac{85 + 94 + 93 + 86 + 92}{5} \\
&= (85 + 94 + 93 + 86 + 92) \div 5 \\
&= 450 \div 5 \\
&= 90
\end{aligned}
$$

So the student did earn an A.

PRACTICE 6

The following table shows the number of U.S. Nobel Prize winners in recent years.

Year	Number of U.S. Winners
2010	4
2009	11
2008	5
2007	6
2006	6

Was the yearly average more or less than 6? (*Source:* wikipedia.org)

R.2   Factors, Primes, and Least Common Multiples

Recall that in a multiplication problem involving two or more whole numbers, the whole numbers that are multiplied are called **factors**. For example, since $2 \cdot 4 = 8$, we say that 2 is a factor of 8, or that 8 is divisible by 2. The factors of 8 are 1, 2, 4, and 8.

Now, let's discuss the difference between prime numbers and composite numbers. A **prime number** is a whole number that has exactly two factors, namely itself and 1. A **composite number** is a whole number that has more than two factors. For instance, 5 is prime because its only factors are 1 and 5. However, 9 is a composite because it has more than two factors, namely 1, 3, and 9. Note that both 0 and 1 are considered neither prime nor composite.

Every composite number can be written as the product of prime factors, called its **prime factorization**. For instance, the prime factorization of 12 is $2 \cdot 2 \cdot 3$, or $2^2 \cdot 3$. This factorization is unique.

The **multiples** of a number are the products of that number and the whole numbers. For instance, some of the multiples of 5 are:

$$\underset{0 \times 5}{0} \quad \underset{1 \times 5}{5} \quad \underset{2 \times 5}{10} \quad \underset{3 \times 5}{15}$$

A number that is a multiple of two or more numbers is called a **common multiple** of these numbers. Some of the common multiples of 6 and 8 are 24, 48, and 72.

The **least common multiple** (LCM) of two or more numbers is the smallest nonzero number that is a multiple of each number. For example, the LCM of 6 and 8 is 24.

A good way to find the LCM involves prime factorization.

EXAMPLE 1

Find the LCM of 8 and 12.

Solution   We first find the prime factorization of each number.

$$8 = 2 \cdot 2 \cdot 2 = 2^3 \qquad 12 = 2 \cdot 2 \cdot 3 = 2^2 \cdot 3$$

Since 2 appears *three* times in the factorization of 8 and *twice* in the factorization of 12, it must be included three times in forming the least common multiple. So we use 2^3 (the highest power of 2 in the factorizations). The factor of 3 must also be included. Therefore, the LCM of 8 and 12 is:

$$\text{LCM} = 2^3 \cdot 3 = 8 \cdot 3 = 24$$

PRACTICE 1

What is the LCM of 10 and 25?

Example 1 suggests the following procedure:

To Compute the Least Common Multiple (LCM) of Two or More Numbers

- Find the prime factorization of each number.

- Identify the prime factors that appear in each factorization.

- Multiply these prime factors, using each factor the greatest number of times that it occurs in any of the factorizations.

EXAMPLE 2

What is the LCM of 18, 30, and 45?

Solution   We first find the prime factorization of each number.

$$18 = 2 \cdot 3 \cdot 3 = 2 \cdot 3^2 \qquad 30 = 2 \cdot 3 \cdot 5 \qquad 45 = 3 \cdot 3 \cdot 5 = 3^2 \cdot 5$$

The prime factors that appear in the prime factorizations are 2, 3, and 5. Since 2 and 5 occur at most one time and 3 occurs at most two times in any of the factorizations, we get:

$$LCM = 2 \cdot 3^2 \cdot 5 = 2 \cdot 9 \cdot 5 = 90$$

PRACTICE 2

Find the LCM of 20, 36, and 60.

EXAMPLE 3

A gym that is open every day of the week offers aerobic classes every third day and yoga classes every fourth day. If both classes were offered today, in how many days will the gym again offer both classes on the same day?

Solution   To answer this question, we find the LCM of 3 and 4. Let's begin by finding prime factorizations.

$$3 = 3 \qquad 4 = 2 \cdot 2 = 2^2$$

To find the LCM, we multiply 3 by 2^2.

$$LCM = 2^2 \cdot 3 = 12$$

So both classes will be offered again on the same day 12 days from today.

PRACTICE 3

A patient gets two medications while in the hospital. One medication is given every 6 hours and the other is given every 8 hours. If the patient is given both medications now, in how many hours will both medications again be given at the same time?

R.3   Fractions

A **fraction** can mean a part of a whole. For example, $\frac{2}{3}$ of a class means two of every three students. A fraction can also mean the quotient of two whole numbers.

A fraction has three components:

- the **denominator** (on the bottom), that stands for the number of parts into which the whole is divided,
- the **numerator** (on top) that tells us how many parts of the whole the fraction contains,
- the **fraction line** (or **fraction bar**) that separates the numerator from the denominator and stands for the phrase *out of* or *divided by*.

Note that the denominator of a fraction cannot be zero.

A fraction whose numerator is smaller than its denominator is a **proper fraction**. A **mixed number** consists of a whole number and a proper fraction. A mixed number can also be expressed as an **improper fraction**, which is a fraction whose numerator is larger than or equal to its denominator. For example, the mixed number $1\frac{1}{2}$ can be written as the improper fraction $\frac{3}{2}$.

To Change a Mixed Number to an Improper Fraction

- Multiply the denominator of the fraction by the whole-number part of the mixed number.
- Add the numerator of the fraction to this product.
- Write this sum over the original denominator to form the improper fraction.

EXAMPLE 1

Write $12\frac{1}{4}$ as an improper fraction.

Solution $12\frac{1}{4} = \frac{(4 \times 12) + 1}{4}$

$= \frac{48 + 1}{4} = \frac{49}{4}$

PRACTICE 1

Express $3\frac{2}{9}$ as an improper fraction.

To Change an Improper Fraction to a Mixed Number

- Divide the numerator by the denominator.
- If there is a remainder, write it over the denominator.

EXAMPLE 2

Write $\dfrac{11}{2}$ as a mixed number.

Solution

$$\dfrac{11}{2} = 2\overline{)11} \quad \text{Divide the numerator by the denominator.}$$
$$\text{Remainder} \rightarrow 1$$

$$= 5\dfrac{1}{2} \quad \text{Write the remainder over the denominator.}$$

PRACTICE 2

Express $\dfrac{8}{3}$ as a mixed number.

Two fractions are **equivalent** if they represent the same value. To generate fractions equivalent to a given fraction, say $\dfrac{1}{3}$, multiply both its numerator and denominator by the same nonzero whole number. For instance,

$$\dfrac{1}{3} = \dfrac{1\cdot 2}{3\cdot 2} = \dfrac{2}{6} \qquad \dfrac{1}{3} = \dfrac{1\cdot 3}{3\cdot 3} = \dfrac{3}{9}$$

A fraction is said to be in **simplest form** (or **reduced to lowest terms**) when the only common factor of its numerator and its denominator is 1. To simplify a fraction, we divide its numerator and denominator by the same number, or common factor. To find these common factors, it is often helpful to express both the numerator and denominator as the product of prime factors. We can then divide out (or cancel) all common factors.

EXAMPLE 3

Write $\dfrac{42}{28}$ in lowest terms.

Solution

$$\dfrac{42}{28} = \dfrac{2\cdot 3\cdot 7}{2\cdot 2\cdot 7} \quad \text{Express the numerator and denominator as the product of primes.}$$

$$= \dfrac{\cancel{2}\cdot 3\cdot \cancel{7}}{\cancel{2}\cdot 2\cdot \cancel{7}} \quad \text{Divide out common factors.}$$

$$= \dfrac{3}{2} \quad \text{Multiply the remaining factors.}$$

PRACTICE 3

Reduce $\dfrac{24}{30}$ to lowest terms.

Fractions with the same denominator are said to be **like**; those with different denominators are called **unlike**.

To Add (or Subtract) Like Fractions

- Add (or subtract) the numerators.
- Use the given denominator.
- Write the answer in simplest form.

EXAMPLE 4

Find the sum of $\frac{7}{12}$ and $\frac{2}{12}$.

Solution   Using the rule for adding like fractions, we get:

$$\frac{7}{12} + \frac{2}{12} = \frac{7+2}{12} = \frac{9}{12}, \quad \text{or} \quad \frac{3}{4}$$

Add numerators. Keep the same denominator. Simplest form

○ PRACTICE 4

Add: $\frac{7}{15} + \frac{3}{15}$

EXAMPLE 5

Find the difference between $\frac{11}{12}$ and $\frac{7}{12}$.

Solution

$$\frac{11}{12} - \frac{7}{12} = \frac{11-7}{12} = \frac{4}{12}, \quad \text{or} \quad \frac{1}{3}$$

Subtract numerators. Keep the same denominator. Simplest form

PRACTICE 5

Subtract: $\frac{19}{20} - \frac{11}{20}$

Unlike fractions are more complicated to add (or subtract) than like fractions because we must first change the unlike fractions to equivalent like fractions. Typically, we use their **least common denominator (LCD)**, that is, the least common multiple of their denominators, to find equivalent fractions.

To Add (or Subtract) Unlike Fractions

- Rewrite the fractions as equivalent fractions with a common denominator, usually the LCD.
- Add (or subtract) the numerators, keeping the same denominator.
- Write the answer in simplest form.

EXAMPLE 6

Add: $\dfrac{5}{12} + \dfrac{5}{16}$

Solution First, find the LCD, which is 48.

$$\dfrac{5}{12} + \dfrac{5}{16} = \dfrac{20}{48} + \dfrac{15}{48} \quad \text{Find equivalent fractions.}$$

$$= \dfrac{35}{48} \quad \begin{array}{l}\text{Add the numerators, keeping}\\ \text{the same denominator.}\end{array}$$

The fraction $\dfrac{35}{48}$ is already in lowest terms because 35 and 48 have no common factors other than 1.

EXAMPLE 7

Subtract $\dfrac{1}{12}$ from $\dfrac{1}{3}$.

Solution First, find the LCD, which is 12.

$$\dfrac{1}{3} - \dfrac{1}{12} = \dfrac{4}{12} - \dfrac{1}{12} \quad \text{Write equivalent fractions with a common denominator.}$$

$$= \dfrac{3}{12} \quad \text{Subtract the numerators, keeping the same denominator.}$$

$$= \dfrac{1}{4} \quad \text{Reduce } \dfrac{3}{12} \text{ to lowest terms.}$$

◉ PRACTICE 7

Calculate: $\dfrac{4}{5} - \dfrac{1}{2}$

To Add (or Subtract) Mixed Numbers

- Rewrite the fractions as equivalent fractions with a common denominator, usually the LCD.
- When subtracting, rename (or borrow from) the whole number on top if the fraction on the bottom is larger than the fraction on top.
- Add (or subtract) the fractions.
- Add (or subtract) the whole numbers.
- Write the answer in simplest form.

EXAMPLE 8

Find the sum of $3\frac{3}{5}$ and $7\frac{2}{3}$.

Solution   Since the denominators have no common factor other than 1, the least common denominator is the product of 5 and 3, which is 15.

$$3\frac{3}{5} = 3\frac{9}{15}$$
$$+7\frac{2}{3} = +7\frac{10}{15}$$

    Write the fraction as equivalent fractions.

Now, we use the rule for adding mixed numbers with the same denominator.

$$3\frac{3}{5} = 3\frac{9}{15}$$
$$+7\frac{2}{3} = +7\frac{10}{15}$$
$$\overline{\phantom{+7\frac{2}{3} = +7}10\frac{19}{15}}$$

    Add the fractions and add the whole numbers.

Because $\dfrac{19}{15}$ is an improper fraction, we need to rewrite $10\frac{19}{15}$:

$$10\frac{19}{15} = 10 + \frac{19}{15} = 10 + 1\frac{4}{15} = 11\frac{4}{15}$$

$$\frac{19}{15} = 1\frac{4}{15}$$

So the sum of $3\frac{3}{5}$ and $7\frac{2}{3}$ is $11\frac{4}{15}$.

PRACTICE 8

Add $4\frac{5}{8}$ to $3\frac{1}{2}$.

EXAMPLE 9

Compute: $13\frac{2}{9} - 7\frac{8}{9}$

Solution   First, write the problem, $13\frac{2}{9} - 7\frac{8}{9}$, in a vertical format.

$$13\frac{2}{9}$$
$$-7\frac{8}{9}$$
$$\overline{\phantom{-7\frac{8}{9}}}$$

Since $\dfrac{8}{9}$ is larger than $\dfrac{2}{9}$, we need to rename $13\frac{2}{9}$:

$$13\frac{2}{9} = 12 + 1 + \frac{2}{9} = 12 + \frac{9}{9} + \frac{2}{9} = 12\frac{11}{9}$$

$$13 = 12 + 1$$

PRACTICE 9

Find the difference between $15\frac{1}{12}$ and $9\frac{11}{12}$.

The problem now becomes:

$$13\frac{2}{9} = 12\frac{11}{9}$$
$$-7\frac{8}{9} = -7\frac{8}{9}$$

Finally, we subtract and then write the answer in simplest form.

$$13\frac{2}{9} = 12\frac{11}{9}$$
$$-7\frac{8}{9} = -7\frac{8}{9}$$
$$5\frac{3}{9} = 5\frac{1}{3}$$

Now, let's look at how we multiply fractions.

To Multiply Fractions

- Multiply the numerators.
- Multiply the denominators.
- Write the answer in simplest form.

EXAMPLE 10

Multiply: $\dfrac{2}{3} \cdot \dfrac{4}{5}$

Solution $\quad \dfrac{2}{3} \cdot \dfrac{4}{5} = \dfrac{2 \cdot 4}{3 \cdot 5} = \dfrac{8}{15}$

PRACTICE 10

Compute: $\dfrac{1}{2} \cdot \dfrac{3}{4}$.

In multiplying some fractions, we can first simplify (or cancel) by dividing *any* numerator and *any* denominator by a common factor. Simplifying before multiplying allows us to work with smaller numbers and still gives us the same answer.

EXAMPLE 11

Find the product of $\dfrac{4}{9}$ and $\dfrac{5}{8}$.

Solution

Divide the numerator 4 and the denominator 8 by the same number 4. Then multiply.

$$\frac{4}{9} \cdot \frac{5}{8} = \frac{\overset{1}{4} \cdot 5}{9 \cdot \underset{2}{8}} = \frac{1 \cdot 5}{9 \cdot 2} = \frac{5}{18}$$

PRACTICE 11

Multiply $\dfrac{7}{10}$ by $\dfrac{5}{11}$.

To Multiply Mixed Numbers

- Change each mixed number to its equivalent improper fraction.
- Follow the steps for multiplying fractions.
- Write the answer in simplest form.

EXAMPLE 12

Compute: $2\frac{1}{2} \cdot 1\frac{1}{4}$

Solution

$$2\frac{1}{2} \cdot 1\frac{1}{4} = \frac{5}{2} \cdot \frac{5}{4}$$ Change each mixed number to an improper fraction.

$$= \frac{5 \cdot 5}{2 \cdot 4}$$ Multiply the fractions.

$$= \frac{25}{8}, \text{ or } 3\frac{1}{8}$$ Simplify.

PRACTICE 12

Find the product of $3\frac{3}{4}$ and $2\frac{1}{10}$.

Dividing fractions is equivalent to multiplying by the *reciprocal* of the divisor. The reciprocal is found by *inverting*—switching the position of the numerator and denominator of the divisor.

To Divide Fractions

- Change the divisor to its reciprocal.
- Multiply the resulting fractions.
- Write the answer in simplest form.

EXAMPLE 13

Divide: $\frac{4}{5} \div \frac{3}{10}$

Solution

$$\frac{4}{5} \div \frac{3}{10} = \frac{4}{\underset{1}{5}} \times \frac{\overset{2}{10}}{3} = \frac{4 \times 2}{1 \times 3} = \frac{8}{3}, \text{ or } 2\frac{2}{3}$$

$\frac{3}{10}$ and $\frac{10}{3}$ are reciprocals.

PRACTICE 13

Divide: $\frac{3}{4} \div \frac{1}{8}$

To Divide Mixed Numbers

- Change each mixed number to its equivalent improper fraction.
- Follow the steps for dividing fractions.
- Write the answer in simplest form.

EXAMPLE 14

Calculate: $9 \div 2\frac{7}{10}$

Solution

$$9 \div 2\frac{7}{10} = \frac{9}{1} \div \frac{27}{10}$$ Change the mixed number to an improper fraction.

$$= \frac{\overset{1}{\cancel{9}}}{1} \times \frac{10}{\underset{3}{\cancel{27}}}$$ Invert the divisor and multiply.

$$= \frac{10}{3}, \text{ or } 3\frac{1}{3}$$ Simplify.

PRACTICE 14

Divide: $6 \div 3\frac{3}{4}$

EXAMPLE 15

The world's population is approximately 7,000,000,000. If China has $\frac{1}{5}$ of that population, what is the total population of the rest of the world? (*Source: The World Almanac and Book of Facts 2010*)

Solution Let's break up the question into two parts:

1. First, find $\frac{1}{5}$ of 7,000,000,000.

$$\frac{1}{\cancel{5}} \times \overset{1,400,000,000}{\cancel{7,000,000,000}} = 1,400,000,000$$

2. Then, subtract that result from 7,000,000,000.

7,000,000,000 − 1,400,000,000 = 5,600,000,000

So the total population of the rest of the world is 5,600,000,000.

PRACTICE 15

About one-third of U.S. adults are considered obese. If there are approximately 240 million U.S. adults, about how many of them are not considered obese? (*Sources:* win.niddk.nih.gov and census.gov)

R.4 Decimals

A number written as a *decimal* has

- a whole-number part, which precedes the decimal point, and
- a fractional part, which follows the decimal point.

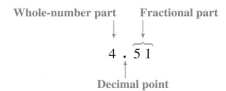

A decimal without a decimal point shown is understood to have the decimal point at the end of the last digit and is the same as a whole number. For instance, 32 and 32. are the same number.

Each digit in a decimal has a place value. The place value system for decimals is an extension of the place value system for whole numbers.

The places to the right of the decimal point are called *decimal places*. For example, the number 64.149 is said to have three decimal places.

For a whole number, the place values are 1, 10, 100, and other powers of 10. By contrast, the place values for the fractional part of a decimal are $\frac{1}{10}$, $\frac{1}{100}$, $\frac{1}{1000}$, and the reciprocals of other powers of 10.

The first decimal place after the decimal point is the ten**th**s place. Working to the right, the next decimal places are the hundred**th**s place, the thousand**th**s place, the ten-thousand**th**s place, and so forth.

The following chart shows the place values in the numbers 0.54 and 513.285.

Hundreds	Tens	Ones	.	Tenths	Hundredths	Thousandths
100	10	1	and	$\frac{1}{10}$	$\frac{1}{100}$	$\frac{1}{1,000}$
		0	.	5	4	
5	1	3	.	2	8	5

Knowing the place value system is the key to changing a decimal to its equivalent fraction and to reading the decimal. For a given decimal, the place value of the rightmost digit is the denominator of the equivalent fraction.

$$0.9 = \frac{9}{10}$$

Read "nine tenths"

$$0.21 = \frac{21}{100}$$

Read "twenty-one hundredths"

Let's look at how to rewrite any decimal as a fraction or mixed number.

To Change a Decimal to the Equivalent Fraction or Mixed Number

- Copy the nonzero whole-number part of the decimal, and drop the decimal point.

- Place the fractional part of the decimal in the numerator of the equivalent fraction.

- Make the denominator of the equivalent fraction 1 followed by as many zeros as the decimal has decimal places.

- Simplify the resulting fraction, if possible.

EXAMPLE 1

Write each decimal as a fraction or mixed number.

a. 0.25   **b.** 1.398

Solution

a. Write 0.25 as $\frac{25}{100}$, which simplifies to $\frac{1}{4}$. So $0.25 = \frac{1}{4}$.

b. The decimal 1.398 is equivalent to a mixed number whose whole-number part is 1. The fractional part of the decimal (398) is the numerator of the equivalent fraction. Since the decimal has three decimal places, the denominator of the fraction has three zeros (that is, it is 1000). So $1.398 = 1\frac{398}{1000} = 1\frac{199}{500}$.

PRACTICE 1

Express each decimal in fractional form.

a. 0.5

b. 2.073

EXAMPLE 2

Write 2.019 in words.

Solution $2.019 = 2\frac{19}{1000}$

We read the original decimal as "two and nineteen thousandths," keeping the whole number unchanged. Note that we use the word *and* to separate the whole number and the fractional part.

PRACTICE 2

Write 4.003 in words.

In computations with decimals, we sometimes *round* the decimal to a certain number of decimal places.

> **To Round a Decimal to a Given Decimal Place**
> - Underline the place to which you are rounding.
> - Look at the digit to the right of the underlined digit—*the critical digit*. If this digit is 5 or more, add 1 to the underlined digit; if it is less than 5, leave the underlined digit unchanged.
> - Drop all digits to the right of the underlined digit.

EXAMPLE 3

Round 94.735 to the nearest tenth.

Solution   First, we underline the digit 7 in the tenths place: 94.7̲35. Since the critical digit 3 is less than 5, we do not add 1 to the underlined digit. Dropping all digits to the right of the 7, we get 94.7. So 94.735 ≈ 94.7 (the symbol ≈ is read "is approximately equal to"). Note that our answer has only one decimal place because we are rounding to the nearest tenth.

PRACTICE 3

Round 748.0772 to the nearest hundredth.

> **To Add Decimals**
> - Rewrite the numbers vertically, lining up the decimal points.
> - Add.
> - Insert a decimal point in the answer below the other decimal points.

EXAMPLE 4

Add: 2.7 + 80.13 + 5.036

Solution   Rewrite the numbers with decimal points lined up vertically so that digits with the same place value are in the same column. Then, add.

$$\begin{array}{r} 2.7 \\ 80.13 \\ + 5.036 \\ \hline 87.866 \end{array}$$

↑ Insert the decimal point in the answer.

PRACTICE 4

Add: 5.92 + 35.872 + 0.3

To Subtract Decimals

- Rewrite the numbers vertically, lining up the decimal points.
- Subtract, inserting extra zeros if necessary for regrouping.
- Insert a decimal point in the answer below the other decimal points.

EXAMPLE 5

Subtract: $5 - 2.14$

Solution

$$\begin{array}{r} 5.00 \\ -\,2.14 \\ \hline 2.86 \end{array}$$

Rewrite 5 as 5.00 and line up decimal points vertically.
Subtract.

Insert the decimal point in the answer.

PRACTICE 5

Find the difference: $3.8 - 2.621$

To Multiply Decimals

- Multiply the factors as if they were whole numbers.
- Find the total number of decimal places in the factors.
- Count that many places from the right end of the product and insert a decimal point.

EXAMPLE 6

Multiply: 6.1×3.7

Solution   First, multiply: $61 \times 37 = 2257$

$$\begin{array}{r} 6\,1 \\ \times\ 3\,7 \\ \hline 4\,2\,7 \\ 1\,8\,3 \\ \hline 2\,2\,5\,7 \end{array}$$

Then, count the total number of decimal places.

$$\begin{array}{r} 6.1 \\ \times\ 3.7 \\ \hline 4\,2\,7 \\ 1\,8\,3 \\ \hline 2\,2.5\,7 \end{array}$$

$\leftarrow$ One decimal place (tenths)
$\leftarrow$ One decimal place (tenths)

$\leftarrow$ Two decimal places (hundredths) in the product

So the answer is 22.57.

PRACTICE 6

Find the product of 2.81 and 3.5.

A shortcut for multiplying a decimal by a power of 10 is to *move the decimal point to the right the same number of places as the power of 10 has zeros.*

EXAMPLE 7

Find the product: $(2.89)(1000)$

Solution   We notice that 1000 is a power of 10 and has three zeros. To multiply 2.89 by 1000, we move the decimal point in 2.89 to the right three places.

$$(2.890)(1000) = 2\,8\,9\,0. = 2890$$

Add a zero to move the decimal point three places.

The product is 2890.

PRACTICE 7
PRACTICE 7

Multiply 32.7 by 10,000.

To Divide Decimals

- If the divisor is not a whole number, move the decimal point in the divisor to the right end of the number.

- Move the decimal point in the dividend the same number of places to the right as we did in the divisor.

- Insert a decimal point in the quotient directly above the decimal point in the dividend.

- Divide the new dividend by the new divisor, inserting zeros at the right end of the dividend as necessary.

EXAMPLE 8

Divide 0.035 by 0.25.

Solution

Move the decimal point to the right end, making the divisor a whole number.

$$0.25\overline{)0.035} \implies 0.25\overline{)0.035}$$

Move the decimal point in the dividend the same number of places.

Finally, we divide 3.5 by 25, which gives us 0.14.

```
        0.1 4
  25)3.5 0
      2 5
      1 0 0
      1 0 0
          0
```

PRACTICE 8

Divide: $2.706 \div 0.15$

A shortcut for dividing a decimal by a power of 10 is to *move the decimal point to the left the same number of places as the power of 10 has zeros.*

EXAMPLE 9

Compute: $\dfrac{7.2}{100}$

Solution   Since we are dividing by the power of 10 with two zeros, we can find this quotient simply by moving the decimal point in 7.2 to the left two places.

$$\frac{7.2}{100} = .072, \quad \text{or} \quad 0.072$$

The quotient is 0.072.

PRACTICE 9

Calculate: $0.86 \div 1000$

EXAMPLE 10

The following graph shows the U.S. outlays for national defense in five consecutive years, expressed in hundreds of billions of current dollars.

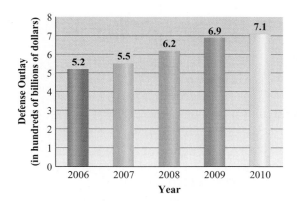

The defense outlay in 2010 was how many times as great as the corresponding outlay 4 years earlier? Round to the nearest tenth. (*Source:* census.gov)

Solution   The defense outlay in 2010 was 7.1 and in 2006 it was 5.2 (both in hundreds of billions of dollars). To find how many times as great 7.1 is as compared to 5.2, we calculate their quotient.

$$5.2\overline{)7.1} = 52.\overline{)71.}$$

$$
\begin{array}{r}
1.36 \\
52\overline{)71.00} \\
\underline{52} \\
19\,0 \\
\underline{15\,6} \\
3\,40 \\
\underline{3\,12} \\
28
\end{array}
$$

Rounding the quotient to the nearest tenth, we conclude that the U.S. defense outlay in 2010 was 1.4 times greater than 4 years earlier.

PRACTICE 10

The table below gives the amount of selected foods consumed per capita in the United States in a recent year.

Food	Annual per Capita Consumption (in pounds)
Red meat	195.2
Poultry	66.5
Fish and shellfish	15.2

The amount of red meat consumed was how many times as great as the amount of poultry, rounded to the nearest tenth? (*Source:* USDA/ Economic Research Service)

R.5 Percent Conversions

Percent means divided by 100. For instance, 18% means 18 divided by 100, or $\frac{18}{100}$, which simplifies to $\frac{9}{50}$. Therefore, the fraction $\frac{9}{50}$ is just another way of writing 18%, suggesting the following rule.

To Change a Percent to the Equivalent Fraction

• Drop the % sign from the given percent and place the number over 100.

• Simplify the resulting fraction, if possible.

EXAMPLE 1

Write 20% as a fraction.

Solution   To change this percent to a fraction, drop the percent sign and write the 20 over 100. Then simplify.

$$20\% = \frac{20}{100} = \frac{1}{5}$$

PRACTICE 1

Write 7% as a fraction.

A percent can also be written as a decimal, since $18\% = \frac{18}{100} = 0.18$. This suggests the following rule:

To Change a Percent to the Equivalent Decimal

• Move the decimal point two places *to the left* and drop the % sign.

EXAMPLE 2

Find the decimal equivalent of 1%.

Solution   The unwritten decimal point lies to the right of the 1. Moving the decimal point two places to the left and dropping the % sign, we get:

$$1\% = 0\underset{\smile}{1}.\% = .01, \text{ or } 0.01$$

PRACTICE 2

Write 5% as a decimal.

To Change a Decimal to the Equivalent Percent

- Move the decimal point two places *to the right* and insert a % sign.

EXAMPLE 3

Write 0.125 as a percent.

Solution   First, move the decimal point two places to the right. Then, insert a % sign.

$$0.1\,2\,5 = 0\,1\,2.5\% = 12.5\%$$

PRACTICE 3

What percent is equivalent to the decimal 0.025?

To Change a Fraction to the Equivalent Percent

- Change the fraction to a decimal.
- Change the decimal to a percent.

EXAMPLE 4

Rewrite $\frac{1}{2}$ as a percent.

Solution   To change the given fraction to a percent, first, find the equivalent decimal. Then, express it as a percent.

$$\frac{1}{2} = 2\overline{)1.0}^{\,0.5} \quad \text{and} \quad 0.5\,0 = 50\%$$
$$\underline{1\,0}$$

PRACTICE 4

Rewrite $\frac{1}{4}$ as a percent.

EXAMPLE 5

A local sales tax rate is 0.0825 of selling prices. Express this tax rate as a percent.

Solution $0.0825 = 0\,0\,8.2\,5\% = 8.25\%$

PRACTICE 5

Suppose that 40% of a student's income goes to paying college expenses. Rewrite this percent as a decimal.

A [R.1] *Rewrite each product using exponents.*

1. $6 \cdot 6 \cdot 6 \cdot 6 \cdot 6$ 2. $7 \cdot 7 \cdot 7 \cdot 7 \cdot 7 \cdot 7 \cdot 7 \cdot 7$ 3. $2 \cdot 2 \cdot 10 \cdot 10 \cdot 10$ 4. $5 \cdot 5 \cdot 5 \cdot 5 \cdot 4 \cdot 4 \cdot 4$

Calculate.

5. $5^2 \cdot 10^3$ 6. $2^4 \cdot 6^2$

Simplify.

7. $24 - 3 \cdot 7$ 8. $20 - 4 \cdot 5$ 9. $2 + 18 \div 3(9 - 7)$

10. $7 + 12 \div 2(7 - 2)$ ◉ 11. $\dfrac{4^2 + 8}{9 - 3}$ 12. $\dfrac{12 + 2^2}{15 - 9 + 2}$

B [R.2] *Find all the factors of each number.*

13. 150   14. 57

Indicate whether each number is prime or composite.

15. 23   16. 47   17. 51   18. 38

Write the prime factorization of each number.

19. 42   20. 54   21. 48   22. 100

Find the LCM.

23. 6 and 8   24. 15 and 20   ◉ 25. 24, 36, and 72   26. 7, 28, and 42

C [R.3] *Write each mixed number as an improper fraction.*

27. $3\dfrac{4}{5}$ 28. $7\dfrac{3}{10}$

Write each fraction as a mixed number.

29. $\dfrac{23}{4}$ 30. $\dfrac{31}{9}$

Simplify.

31. $\dfrac{14}{28}$ 32. $\dfrac{30}{45}$ 33. $5\dfrac{2}{4}$ 34. $6\dfrac{12}{42}$

Calculate. Write answers in lowest terms.

35. $\dfrac{1}{9} + \dfrac{4}{9}$ 36. $\dfrac{3}{10} + \dfrac{7}{10}$ 37. $\dfrac{3}{8} - \dfrac{1}{8}$ 38. $\dfrac{5}{3} - \dfrac{2}{3}$

39. $\dfrac{2}{5} + \dfrac{4}{7}$ 40. $\dfrac{8}{9} + \dfrac{1}{2}$ 41. $\dfrac{3}{10} - \dfrac{1}{20}$ 42. $\dfrac{3}{5} - \dfrac{1}{4}$

43. $1\dfrac{1}{8} + 5\dfrac{3}{8}$ 44. $3\dfrac{1}{5} + 4\dfrac{1}{5}$ 45. $8\dfrac{7}{10} + 1\dfrac{9}{10}$ 46. $5\dfrac{5}{6} + 2\dfrac{1}{6}$

47. $9\frac{11}{12} - 6\frac{7}{12}$

48. $2\frac{5}{9} - 2\frac{4}{9}$

49. $6\frac{1}{10} - 4\frac{3}{10}$

50. $5\frac{1}{4} - 2\frac{3}{4}$

51. $12 - 5\frac{1}{2}$

52. $3 - 1\frac{4}{5}$

53. $7\frac{1}{2} - 4\frac{5}{8}$

54. $5\frac{1}{12} - 4\frac{1}{2}$

55. $\frac{2}{3} \cdot \frac{1}{5}$

56. $\frac{3}{4} \cdot \frac{8}{9}$

57. $1\frac{2}{5} \cdot 10$

58. $20 \cdot 1\frac{5}{6}$

59. $3\frac{1}{4} \cdot 4\frac{2}{3}$

60. $2\frac{1}{2} \cdot 1\frac{1}{5}$

61. $2\frac{5}{6} \div \frac{1}{2}$

62. $1\frac{1}{3} \div \frac{4}{5}$

63. $\frac{2}{3} \div 6$

64. $\frac{1}{10} \div 4$

65. $8 \div 2\frac{1}{3}$

66. $4\frac{1}{2} \div 2\frac{1}{2}$

67. $\left(\frac{3}{4}\right)^2 - \frac{3}{8} \div 6$

68. $8\frac{2}{5} + 2 \div \left(\frac{1}{2} - \frac{1}{3}\right)$

D **[R.4]** *Write each decimal as a fraction or mixed number.*

69. 0.875

70. 2.006

Name the place that the underlined digit occupies.

71. 18.3<u>5</u>9

72. 8024.<u>5</u>

Write each decimal in words.

73. 0.72

74. 0.05

75. 3.009

76. 12.235

Round as indicated.

77. 7.31 to the nearest tenth

78. 9.52 to the nearest tenth

79. 4.3868 to two decimal places

80. 8.6874 to two decimal places

Calculate.

81. $8.2 + 3.91 + 6$

82. $8 + 3.25 + 12.88$

83. $3.8 - 1.927$

84. $2.5 - 1.6$

85. 7.28×0.4

86. 6.24×0.6

87. $2.71 \cdot 1000$

88. 100×5.3

89. $0.006 \div 4$

90. $31.9 \div 10$

91. $12 \div 2.4$

92. $42 \div 2.1$

93. $7.1 + 0.5^2$

94. $8.6 + 0.6^2$

95. $20.8 - 7(4 - 3.1)$

96. $18.6 + 3(9 - 4.2)$

E **[R.5]** *Change each percent to a fraction or mixed number and simplify.*

97. 75%

98. 4%

99. 106%

100. 250%

Change each percent to a decimal.

101. 6%

102. 8%

103. 150%

104. 180%

Change each decimal to a percent.

105. 0.31

106. 0.05

107. 0.0145

108. 0.0148

Change each fraction to a percent.

109. $\dfrac{1}{10}$

110. $\dfrac{3}{8}$

⊙ **111.** $\dfrac{4}{5}$

112. $\dfrac{7}{4}$

Mixed Applications

F *Solve.*

113. In a city, the daily high temperatures for a week in May were 67°F, 72°F, 78°F, 70°F, 65°F, 77°F, and 82°F. What was the average daily high temperature for the week?

114. Two airport shuttle buses arrive at a subway station at 7:30 A.M. If one airport shuttle bus arrives every 8 min and the other airport shuttle bus arrives every 10 min, when is the next time that both airport shuttles will arrive at the subway station together?

115. A trucker drives to a town $\dfrac{1}{2}$ mi away. If he then drives an additional $\dfrac{1}{4}$ mi, how far did he drive in all?

116. A sea otter eats an amount that is about $\dfrac{1}{5}$ of its body weight each day. How much will a 35-lb otter eat in a day?

117. At a florist, a shopper buys a dozen roses for her mother. The regular price of roses is $27 a dozen. Based on the store sign shown, how much does she pay for the roses on sale?

118. Find the missing length in the figure shown:

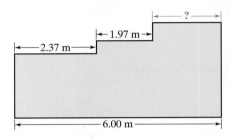

119. A supermarket sells a 4 lb package of 85% lean ground beef for $7.96 and a 3 lb package of 93% lean ground beef for $9.87. What is the difference between the costs per pound of the two types of beef?

120. In May 2010, Arizona voters approved a three-year increase in state sales tax to 6.6%. An item costs $50 before applying the tax. How much change should an Arizona shopper receive after paying for the item with three $20 bills? (*Source:* articles.latimes.com)

121. In survey of ice cream flavors consumed at home, 14% of the ice cream was chocolate. Express this percent as a fraction. (*Source:* usatoday.com)

122. The solar wind streams off the Sun at speeds of about 1,000,000 miles per hour. Express this number as a power of 10. (*Source:* NASA)

• Check your answers on page A-1.

CHAPTER R   Posttest

FOR
EXTRA
HELP

CHAPTER
Test Prep
VIDEOS

The Chapter Test Prep Videos with test solutions are available on DVD, in MyMathLab, and on YouTube (search "AkstIntroductory Alg" and click on "Channels").

To see if you have mastered the topics in this chapter, take this test.

1. Calculate: $8^2 \cdot 2^3$

2. Find the value of $11 \cdot 2 + 5 \cdot 3$.

3. What are the factors of 20?

4. Write $3\frac{1}{4}$ as an improper fraction.

5. Reduce $\frac{10}{36}$ to lowest terms.

6. Find the sum: $\frac{5}{8} + \frac{7}{8}$

7. Add: $7\frac{7}{8} + 4\frac{1}{6}$

8. Calculate: $\frac{4}{9} - \frac{3}{10}$

9. Subtract: $12\frac{1}{4} - 8\frac{3}{10}$

10. Find the product of $\frac{3}{4}$ and $\frac{4}{5}$.

11. Find the quotient: $\frac{2}{3} \div \frac{1}{3}$

12. Calculate: $7 \div 3\frac{1}{5}$

13. Write 2.396 in words.

14. Find the sum: $5.2 + 3 + 8.002$

15. Find the difference: $10 - 3.01$

16. What is the product of 5.02 and 8.9?

17. Evaluate: 2.07×1000

18. Compute: $\frac{0.05}{100}$

19. Express $\frac{1}{8}$ as a decimal and as a percent.

20. Write 0.7 as a percent and as a fraction.

21. A trip to a nearby island takes 3 hours by boat and half an hour by airplane. How many times as fast as the boat is the plane?

22. According to a recent survey, the cost of medical care is approximately 1.94 times what it was a decade ago. Round this decimal to the nearest tenth. (*Source:* kff.org)

23. Find the area (in square meters) of the room pictured, rounded to one decimal place.

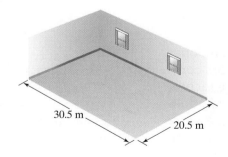

30.5 m

20.5 m

24. The following graph shows the distribution of investments for a retiree. Express as a decimal the percent of investments that are in equities.

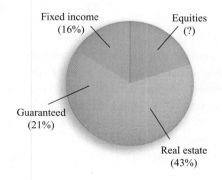

Fixed income
(16%)

Equities
(?)

Guaranteed
(21%)

Real estate
(43%)

25. Typically, the heaviest organ in the human body is the skin, weighing about 9 lb. By contrast, the heart weighs about 0.6 lb. How many times the weight of the heart is the weight of the skin? (*Sources:* faculty.washington.edu and infoplease.com)

• Check your answers on page A-1.

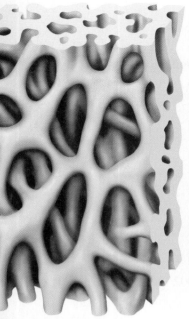

Normal bone structure

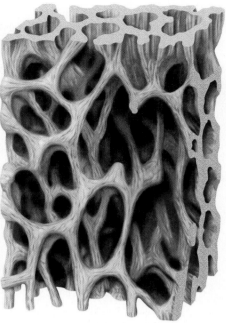

Osteoporotic bone

Introduction to Real Numbers and Algebraic Expressions

Osteoporosis and Real Numbers

Approximately 10 million Americans have osteoporosis, a disease in which bones become weak and are likely to break. To diagnose osteoporosis, doctors commonly employ a bone mineral density (BMD) test. A person's BMD score is compared to a norm based on the optimal density of a healthy 30-year-old adult. Scores below the norm are indicated in negative numbers. For instance, a score of −2 indicates a low bone mass and a score of −2.5 or less is diagnosed as osteoporosis. Generally, a score of −1 is equivalent to a 10% loss of bone density.

The National Osteoporosis Foundation recommends treatment if you have a result that is:

- less than −1.5 with risk factors, or
- less than −2 with no risk factors.

The results of many medical tests are reported in terms of positive and negative numbers.

(*Source:* myhealth.gov)

To see if you have already mastered the topics in this chapter, take this test.

1. Express as a positive or negative integer: a profit of $2000

2. Is 0 a rational number?

3. Graph the number $-\frac{1}{2}$ on the following number line.

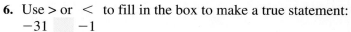

4. What is the opposite of -5?

5. Evaluate $\left| -\frac{2}{3} \right|$.

6. Use $>$ or $<$ to fill in the box to make a true statement:
$-31 \;\rule{1cm}{0pt}\; -1$

7. Subtract: $-6 - 7$

8. Add: $9 + (-4) + 2 + (-9)$

9. Simplify: $3(-7) - 5$

10. What is the reciprocal of $\frac{1}{4}$?

11. Divide: $(-72) \div (-8)$

12. Translate the algebraic expression to words: $3n - 10$

13. Express in exponential form: $-6 \cdot 6 \cdot 6 \cdot 6$

14. Evaluate: $2x - 4y + 8$, if $x = -2$ and $y = 3$

15. Combine like terms: $3n - 7 + n$

16. Simplify: $-5(2 - x) + 9x$

17. Mount Kilimanjaro, Africa's highest point, is 5895 meters (m) above sea level. Lake Assal, its lowest point, is 156 m below sea level. What is the difference in the elevations of these two points? (*Source: National Geographic Family Reference Atlas of the World*, 2002)

18. A wire of length L is to be cut into three pieces of equal length. Write an expression that represents the length of each piece of wire.

19. The formula for the perimeter of a rectangle is $P = 2l + 2w$. Calculate the perimeter of the following rectangle.

20. In golf, scores are given in terms of par; scores above par are positive, and scores below par are negative. The table shows Dustin Johnson's scores for each round of a recent PGA Championship tournament.

Round	First	Second	Third	Fourth
Score	-1	-4	-5	$+1$

(*Source:* espn.com)

In which round did he have the lowest score?

• Check your answers on page A-1.

1.1   Real Numbers

OBJECTIVES

Ⓐ To identify different kinds of real numbers—integers, rational numbers, and irrational numbers

Ⓑ To graph real numbers on the number line

Ⓒ To find the opposite and the absolute value of real numbers

Ⓓ To compare real numbers

Ⓔ To solve applied problems involving the comparison of real numbers

What Algebra Is and Why It Is Important

Algebra is a language that allows us to express the patterns and rules of arithmetic. Consider, for instance, the rule for finding the product of two fractions: Multiply the numerators to get the numerator of the product and multiply the denominators to get the denominator of the product.

$$\frac{3}{4} \cdot \frac{1}{2} = \frac{3 \cdot 1}{4 \cdot 2} = \frac{3}{8}$$

In the language of algebra, we can write the general rule using letters to represent numbers:

$$\frac{a}{b} \cdot \frac{c}{d} = \frac{a \cdot c}{b \cdot d} = \frac{ac}{bd}$$

Can you think of another arithmetic rule that can be expressed algebraically?

Algebra allows us not only to communicate rules like the preceding one but also to solve problems. These problems arise in a variety of disciplines—for example, in chemistry where substances may interact with one another, in medicine where dosage may depend on a patient's age or weight, and in economics where the supply and demand of a product may need to be balanced.

This chapter builds towards a discussion of algebra. But first we consider a more fundamental concept—the real numbers.

Introduction to Real Numbers

Real numbers are numbers that can be represented as points on the number line. They extend the numbers used in arithmetic and allow us to solve problems that we could not otherwise solve.

Let's begin our discussion by looking at different kinds of real numbers.

Integers

In arithmetic, we use **natural numbers** for counting. The natural numbers are 1, 2, 3, 4, 5, 6, 7, 8, 9, ... The three dots mean that these numbers go on forever in the same pattern. The **whole numbers** consist of 0 and the natural numbers: 0, 1, 2, 3, 4, 5, 6, 7, 8, 9 . . . .

We can represent the whole numbers on the number line. A number to the right of another on the number line is the larger number, as shown on the following number line:

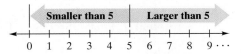

Sometimes we need to consider numbers that are to the left of 0, that is, numbers that are smaller than 0. For instance, we may want to express 2 degrees below zero, as shown on the thermometer to the right.

Numbers to the left of 0 on the number line are said to be *negative*. All negative numbers are smaller than 0. We write -1, -2, -3, -4, and so on. By contrast, the whole numbers 1, 2, 3, 4, and so on are said to be *positive* and can also be written as $+1$, $+2$, $+3$, $+4$, and so on. All positive numbers are larger than 0. The numbers -1, -2, -3, and so on, together with the whole numbers, are called *integers*, as shown on the following number line. Note that the number line extends without end to the right and to the left as the arrows indicate. Also note that numbers become smaller going in the negative direction and become larger going in the positive direction.

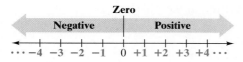

Zero

Negative | Positive

$$\cdots -4 \quad -3 \quad -2 \quad -1 \quad 0 \quad +1 \quad +2 \quad +3 \quad +4 \cdots$$

DEFINITION

The **integers** are the numbers

$$\ldots, -4, -3, -2, -1, 0, +1, +2, +3, +4, \ldots$$

continuing indefinitely in both directions.

EXAMPLE 1

The Dow Jones Industrial Average on the stock market declined 4 points today. Express this situation as an integer.

Solution The number in question represents a decline (or loss), so we write a negative integer, namely, -4.

PRACTICE 1

Represent as an integer: A carnation plant freezes and dies at a temperature of 5° below 0° Fahrenheit (°F).

Rational Numbers

Suppose that we want to represent the following situation: *The New York Giants lost one-half yard on a play.* To express a loss of one-half yard, we need a kind of number other than whole numbers or integers. We need a *rational number*—a number that can be written as the quotient of two integers, where the denominator is not equal to zero.

DEFINITION

Rational numbers are numbers that can be written in the form $\dfrac{a}{b}$, where a and b are integers and $b \neq 0$.

Here are some examples of rational numbers:

- $\dfrac{2}{3}$, since it is the ratio of integers.

- 5, since it can be written in the form $\dfrac{5}{1}$. In general, any integer is also a rational number.

- $7\frac{1}{4}$, since it can be written $\frac{29}{4}$.

- 0.03, since it can be written in the form $\frac{3}{100}$.

- $-\frac{1}{2}$, since it can be written $\frac{-1}{2}$.

A rational number has a decimal representation that either *terminates* or *repeats*. Here are some examples:

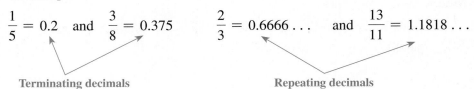

$$\frac{1}{5} = 0.2 \quad \text{and} \quad \frac{3}{8} = 0.375 \qquad \frac{2}{3} = 0.6666\ldots \quad \text{and} \quad \frac{13}{11} = 1.1818\ldots$$

Terminating decimals   Repeating decimals

Just as with integers, we can picture rational numbers as points on a number line. To *graph* a rational number, we locate the point on the number line and mark it as shown.

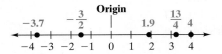

On the above number line, note that

- the point at 0 is called the *origin*,

- numbers to the right of 0 are positive, and numbers to the left of 0 are negative,

- the number 0 is neither positive nor negative.

Until Chapter 8 of this text, most of the numbers with which we work will be rational numbers.

EXAMPLE 2

Graph each number on the same number line.

a. $-\frac{7}{2}$ **b.** 2.1

Solution

a. Because $-\frac{7}{2}$ can be expressed as $-3\frac{1}{2}$, the point $-\frac{7}{2}$ is graphed halfway between -3 and -4.

b. The number 2.1, or $2\frac{1}{10}$, is between 2 and 3 but is closer to 2.

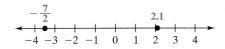

PRACTICE 2

Graph each number on the number line.

a. -1.7 **b.** $\frac{5}{4}$

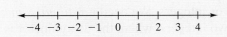

Irrational Numbers

Recall that any rational number can be written as the quotient of two integers. However, there are other real numbers that cannot be written in this form. Such numbers are called *irrational numbers*, and their corresponding decimal representations continue indefinitely and have no repeating pattern. Examples of irrational numbers are

$$\sqrt{2} = 1.4142\ldots \quad \text{The square root of 2}$$
$$-\sqrt{3} = -1.7320\ldots \quad \text{The negative square root of 3}$$
$$\pi = 3.1415\ldots \quad \text{Pi, the ratio of the circumference of a circle to its diameter}$$

In many computations with irrational numbers, we use decimal approximations that are rounded to a certain number of decimal places, for instance:

$$\sqrt{2} \approx 1.41 \qquad -\sqrt{3} \approx -1.73 \qquad \pi \approx 3.14$$

Recall that the symbol $\approx$ means "is approximately equal to."

Irrational numbers, like rational numbers, can be graphed on the number line. The rational numbers and the irrational numbers together make up the *real numbers*.

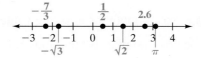

Absolute Value

On the number line, the numbers $+3$ and -3 are *opposites* of each other. Similarly, $-\frac{1}{2}$ and $+\frac{1}{2}$ are opposites. What is the opposite of 0?

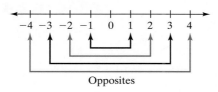

Opposites

> **DEFINITION**
> Two real numbers that are the same distance from 0 on the number line but on opposite sides of 0 are called **opposites**. For any real number n, its opposite is $-n$.

EXAMPLE 3

Find the opposite of each number in the table.

	a.	b.	c.	d.
Number	10	$-\frac{1}{3}$	$10\frac{1}{2}$	-0.3
Opposite				

Solution

	a.	b.	c.	d.
Number	10	$-\frac{1}{3}$	$10\frac{1}{2}$	-0.3
Opposite	-10	$\frac{1}{3}$	$-10\frac{1}{2}$	0.3

PRACTICE 3

Find the opposite of each number.

Number	Opposite
a. -41	
b. $-\frac{8}{9}$	
c. 1.7	
d. $-\frac{2}{5}$	

Because the number -3 is negative, it lies 3 units to the left of 0. The number 3 is positive and lies 3 units to the right of 0.

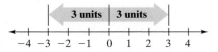

When we locate a number on the number line, the distance that number is from 0 is called its *absolute value*. Thus, the absolute value of $+3$, which we write as $|3|$, is 3. Similarly, the absolute value of -3, or $|-3|$, is 3.

> **DEFINITION**
> The **absolute value** of a number is its distance from 0 on the number line.
> The absolute value of the number n is written $|n|$.

The following properties help us to compute the absolute value of any number:

- The absolute value of a positive number is the number itself.
- The absolute value of a negative number is its opposite.
- The absolute value of a number is always positive or 0.
- Opposites have the same absolute value.

EXAMPLE 4

Evaluate.

a. $|-6|$ **b.** $|4.6|$ **c.** $\left|-\dfrac{1}{3}\right|$ **d.** $-|-2|$

Solution

a. $|-6| = 6$ Because the absolute value of a negative number is its opposite, the absolute value of -6 is 6.

b. $|4.6| = 4.6$ Because the absolute value of a positive number is the number itself, the absolute value of 4.6 is 4.6.

c. $\left|-\dfrac{1}{3}\right| = \dfrac{1}{3}$ Because $-\frac{1}{3}$ is negative, its absolute value is its opposite, namely, $\frac{1}{3}$.

d. $-|-2| = -(2) = -2$ Because $|-2| = 2$, we get $-|-2| = -2$.

PRACTICE 4

Find the value.

a. $\left|\dfrac{1}{2}\right|$ **b.** $|0|$

c. $|-9|$ **d.** $-|-3|$

Comparing Real Numbers

The number line helps us to compare two real numbers, that is, to determine which number is larger.

Given two numbers on the number line, *the number to the right is larger than the number to the left*. Similarly, *the number to the left is smaller than the number to the right*. The *equal sign* $(=)$ and the *inequality symbols* $(\neq, <, \leq, >$ and $\geq)$ are used to compare numbers.

$=$	means *is equal to*	$\dfrac{5}{2} = 2\frac{1}{2}$ is read "$\dfrac{5}{2}$ is equal to $2\frac{1}{2}$."
$\neq$	means *is not equal to*	$3 \neq -3$ is read "3 is not equal to -3."

<	means *is less than*	−1 < 0 is read "−1 is less than 0."
≤	means *is less than or equal to*	4 ≤ 7 is read "4 is less than or equal to 7."
>	means *is greater than*	−2 > −5 is read "−2 is greater than −5."
≥	means *is greater than or equal to*	3 ≥ 1 is read "3 is greater than or equal to 1."

The statements $3 \neq -3$, $-1 < 0$, $4 \leq 7$, $-2 > -5$, and $3 \geq 1$ are *inequalities*.
When comparing real numbers, it is important to remember the following:

- The number 0 is greater than any negative number because all negative numbers on the number line lie to the left of 0. For example, $0 > -5$.

- The number 0 is less than any positive number because all positive numbers lie to the right of 0 on the number line. For example, $0 < 3$.

- Any positive number is greater than any negative number because on the number line all positive numbers lie to the right of all negative numbers. For example, $4 > -1$.

EXAMPLE 5

Indicate whether each inequality is true or false. Explain.

a. $0 > -1$ **b.** $0 \geq \dfrac{1}{2}$ **c.** $-3.5 < 1.5$

d. $-10 \leq -10$ **e.** $-4 > -2$

Solution

a. $0 > -1$ True, because 0 is to the right of −1.

b. $0 \geq \dfrac{1}{2}$ False, because 0 is to the left of $\dfrac{1}{2}$.

c. $-3.5 < 1.5$ True, because −3.5 is to the left of 1.5.

d. $-10 \leq -10$ True, because −10 is equal to −10.

e. $-4 > -2$ False, because −4 is to the left of −2.

PRACTICE 5

Determine whether each inequality is true or false. Explain.

a. $-2 < -1$

b. $0 \leq -5$

c. $\dfrac{10}{4} \geq \dfrac{5}{2}$

d. $0.3 > 0$

e. $-2.4 > 1.6$

We noted in Example 5(a) that $0 > -1$. Is $-1 < 0$? Explain why.

EXAMPLE 6

Graph the numbers $-2, \dfrac{1}{2}, -1$, and $-\dfrac{1}{4}$. Then, list them in order from least to greatest.

Solution

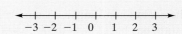

Reading the graph, we see that the numbers in order from the least to the greatest are -2, -1, $-\dfrac{1}{4}$, and $\dfrac{1}{2}$.

PRACTICE 6

Graph the numbers $-2.4, 3, -\dfrac{1}{2}$, and -1.6. Then, write them in order from largest to smallest.

EXAMPLE 7

The average temperature in Fairbanks, Alaska, in the month of December is $-9\,°$F. The average temperature in Barrow, Alaska, in the same month is $-11\,°$F. Which place is warmer in December? Explain. (*Source:* climatetemp.info)

Solution   We need to compare -9 with -11. Because $-9 > -11$, it is warmer in Fairbanks than in Barrow.

PRACTICE 7

The table below shows the elevation of three lakes.

Lakes	Elevation (in feet)
The Caspian Sea (Asia–Europe)	−92
Lake Maracaibo (South America)	0
Lake Eyre (Australia)	−52

(*Source: Geological Survey, U.S. Department of the Interior*)

Which lake has the lowest elevation? Explain.

EXAMPLE 8

Historians use number lines, called *timelines*, to show dates of historical events. On the timeline shown below, the B.C. dates are considered to be negative whereas the A.D. dates are positive.

a. Locate on the following timeline the world history events shown in the table.

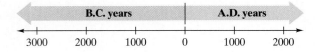

Event	Date
(A) Hieroglyphic writing developed in Egypt.	3200 B.C.
(B) Charlemagne (Charles the Great) was crowned emperor by Pope Leo III in Rome.	A.D. 800
(C) Hun invaders from Asia entered Europe.	A.D. 372
(D) Maya civilization began to develop in Central America.	1500 B.C.
(E) Sweden seceded from the Scandinavian Union.	A.D. 1523
(F) In Greece, the Parthenon was built.	438 B.C.
(G) The city of Rome was founded, according to legend, by Romulus.	753 B.C.
(H) The evolution of England's unique political institutions began with the *Magna Carta*.	A.D. 1215

(*Source: The World Almanac Book of Facts, 2010*)

b. Order the events from the most recent to the earliest event.

Solution B.C. dates are considered to be negative numbers, and A.D. dates positive numbers. So we graph on a timeline similar to the way we graph on a number line. That is, we locate each year on the timeline and mark it as shown below:

a.

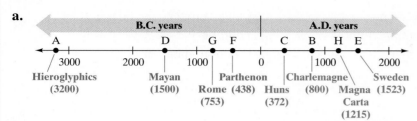

b. The events from the most recent to the earliest: E, H, B, C, F, G, D, and A.

PRACTICE 8

Consider the following table that shows the highlights in the development of algebra.

Event	Date
(A) Babylonian algebra found on cuneiform clay tablets dates back to the reign of King Hammurabi.	1700 B.C.
(B) Greek algebra (as formulated by the Pythagoreans) was geometric.	500 B.C.
(C) The Greek mathematician Diophantus introduced a style of writing equations.	A.D. 250
(D) Greek algebra (as formulated by Euclid) was geometric.	300 B.C.
(E) Bhaskara was one of the most prominent Hindu mathematicians in algebra.	A.D. 1100
(F) Mohammed ibn-Musa al-Khowarizmi wrote the book *Al-jabr* (translated as "algebra" in Latin).	A.D. 825
(G) Modern symbols and notation in algebra emerged.	A.D. 1500

(*Source:* NCTM, *Historical Topics for the Mathematics Classroom*, 1969)

a. Locate on the timeline below the events listed in the table.

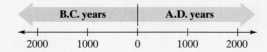

b. Then, order these highlights from the earliest to the most recent event.

Mathematically Speaking

Fill in each blank with the most appropriate term or phrase from the given list.

rational numbers	natural numbers	negative numbers
whole numbers	irrational numbers	real numbers
either terminate or	integers	neither terminate nor
center	origin	

1. The _____ are 1, 2, 3, 4, . . .

2. The _____ consist of 0 and the natural numbers.

3. Numbers to the left of 0 on the number line are _____.

4. The _____ are the numbers . . . , $-4, -3, -2, -1, 0, +1, +2, +3, +4, \ldots$, continuing indefinitely in both directions.

5. The point at 0 on the number line is called the _____.

6. Real numbers that cannot be written as the quotient of two integers are called _____.

7. The corresponding decimal representations of irrational numbers _____ repeat.

8. The rational and the irrational numbers together make up the _____.

A *Express each quantity as a positive or negative number.*

9. 5 km below sea level

10. A profit of $1000

11. A temperature drop of 22.5°C

12. A gain of $6\frac{1}{2}$ lb

13. A withdrawal of $160 from an account

14. A debt of $1500.45

Classify each number by writing a check in the appropriate boxes.

	Whole Numbers	Integers	Rational Numbers	Real Numbers
15. -7				
16. -3				
17. $3\frac{1}{6}$				
18. $4\frac{1}{2}$				
19. 10				
20. 15				

B *Graph each number on the number line.*

21. -3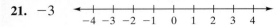

22. -1

23. $-\dfrac{1}{2}$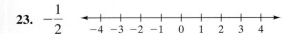

24. $-\dfrac{3}{8}$

25. $3\frac{1}{4}$
$$-4 \; -3 \; -2 \; -1 \; \; 0 \; \; 1 \; \; 2 \; \; 3 \; \; 4$$

26. $2\frac{9}{10}$ $\;$
$$-4 \; -3 \; -2 \; -1 \; \; 0 \; \; 1 \; \; 2 \; \; 3 \; \; 4$$

27. -2.9
$$-4 \; -3 \; -2 \; -1 \; \; 0 \; \; 1 \; \; 2 \; \; 3 \; \; 4$$

28. -3.2
$$-4 \; -3 \; -2 \; -1 \; \; 0 \; \; 1 \; \; 2 \; \; 3 \; \; 4$$

C *Find the opposite of each number.*

29. -3 **30.** -2 **31.** 15 **32.** 36

33. -3.5 **34.** -1.5

Evaluate.

35. $|-4|$ **36.** $|-15|$ **37.** $\left|-\dfrac{2}{3}\right|$ **38.** $\left|-\dfrac{1}{4}\right|$

39. $|-4.6|$ **40.** $|-2.6|$ **41.** $-\left|\dfrac{1}{2}\right|$ **42.** $-\left|\dfrac{6}{5}\right|$

Solve. If impossible, explain why.

43. Name all numbers that have an absolute value of 4.

44. Name all numbers that have an absolute value of 0.4.

45. Name a number whose absolute value is -2.

46. Name three different numbers that have the same absolute value.

D *Indicate whether each inequality is true or false.*

47. $-7 < -5$ **48.** $-3 < -2$ **49.** $-1 > 2.5$

50. $-3 > 4.5$ **51.** $0 \geq -1\frac{1}{4}$ ◉ **52.** $-6 \leq -6$

Replace each ___ with $<$, $>$, or $=$ to make a true statement.

53. 0 ___ -1 **54.** 4 ___ -7 **55.** -1.5 ___ -2 **56.** -1.6 ___ -2

57. 2.5 ___ $2\frac{1}{2}$ **58.** 3.25 ___ $3\frac{1}{4}$ **59.** $|-4|$ ___ $|4|$ **60.** $-|5|$ ___ $|-5|$

61. 6.2 ___ $|-7.1|$ **62.** 7.4 ___ $|-8.6|$

Graph the numbers in each group on the number line. Then, write the numbers from largest to smallest.

63. $3\frac{1}{2}, -1.5, -\dfrac{1}{2}, 0;$
$$-4 \; -3 \; -2 \; -1 \; \; 0 \; \; 1 \; \; 2 \; \; 3 \; \; 4$$

64. $2\frac{1}{2}, -4, 3, -2.5$
$$-4 \; -3 \; -2 \; -1 \; \; 0 \; \; 1 \; \; 2 \; \; 3 \; \; 4$$

65. $-1, 2, -2, -3, 1$
$$-4 \; -3 \; -2 \; -1 \; \; 0 \; \; 1 \; \; 2 \; \; 3 \; \; 4$$

66. $-3, 3, -3.5, 3.5$
$$-4 \; -3 \; -2 \; -1 \; \; 0 \; \; 1 \; \; 2 \; \; 3 \; \; 4$$

Mixed Practice

Solve.

67. Express the quantity as a positive or negative number: a loss of $53.

68. Graph the number $-1\frac{5}{8}$ on the number line.

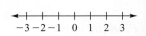

69. Classify the number by writing a check in the appropriate boxes.

	Whole Numbers	Integers	Rational Numbers	Real Numbers
2.6				

70. Find the opposite of the number $-\dfrac{5}{6}$.

71. Evaluate $-|-1.5|$.

72. Name all numbers that have an absolute value of -3. If impossible, explain why.

73. Indicate whether the inequality $-5 \leq -5\frac{1}{3}$ is true or false.

Replace the ▢ *with* $<, >,$ *or* $=$ *to make a true statement.*

74. -7.8 ▢ -8.2

75. $-|3|$ ▢ $|-3|$

76. Graph the numbers $\dfrac{1}{2}, -2\frac{1}{2}, 1\frac{1}{2}, -2$ on a number line. Then, write the numbers from largest to smallest.

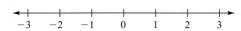

Applications

E *Solve.*

77. Today, a person in debt owes $200. Last week, he owed $2000. Was he better off financially last week, or is he better off today?

78. Will a dieter weigh more if she loses 6 lb or gains 2 lb?

79. Three of the coldest temperature readings ever recorded on Earth were $-90°F$, $-129°F$, and $-87°F$. Of these three temperatures, which was the coldest? (*Source:* ncdc.noaa.gov)

80. Each liquid has its own boiling point—the temperature at which it changes to a gas. Liquid chlorine, for example, boils at $-35°C$, whereas liquid fluorine boils at $-188°C$. Which liquid has the higher boiling point? (*Source: Handbook of Chemistry and Physics*)

81. Astronomers use the term *apparent magnitude* to indicate the brightness of a star as seen from Earth. The following number line shows the apparent magnitude of various stars and other objects. For historical reasons, the brighter a star or object is, the farther to the left it is graphed on the number line.

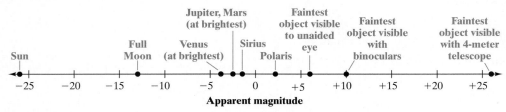

(*Sources:* absolute astronomy.com and topastronomy.com)

a. Which star is brighter as seen from Earth, Polaris or Sirius?

b. Use this number line to estimate the apparent magnitude of the full Moon.

c. The giant star Beta Sagittae lies hundreds of light-years from Earth, with an apparent magnitude of 4.4. Plot this star's apparent magnitude on the number line.

82. The following timeline shows the years of some major technological innovations.

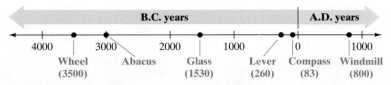

(*Source:* Bill Yenne, *100 Inventions That Shaped World History,* 1993)

a. Which was invented earlier, the lever or the compass?

b. According to the timeline, what was invented between 2000 B.C. and A.D. 1000?

c. The mechanical clock was invented around A.D. 950. Plot a point on the timeline to represent this invention.

83. Below are the birth years for a variety of famous people throughout history.

Leonardo da Vinci (Italian painter)	A.D. 1452	Aristotle (Greek philosopher/scientist)	384 B.C.
Keith Richards (English musician)	A.D. 1943	Wolfgang Amadeus Mozart (Austrian composer)	A.D. 1756
Julius Caesar (Roman statesman)	100 B.C.	Miles Davis (American jazz musician)	A.D. 1926
Charles Dickens (English writer)	A.D. 1812	Marie Antoinette (French queen)	A.D. 1755
William Shakespeare (English playwright)	A.D. 1564	Oliver Cromwell (English statesman)	A.D. 1599
Tiger Woods (American golfer)	A.D. 1976	Socrates (Greek philosopher)	469 B.C.
Attila (Hun king)	A.D. 406		

(*Source: Chambers Biographical Dictionary*)

a. Who was born later, Aristotle or Socrates?

b. Who was born later, Julius Caesar or Attila ?

c. Which of the individuals listed in the table was born the earliest?

d. Which of the individuals listed in the table was born most recently?

84. The following table gives the change from the previous month in the opening stock price per share for Delta Air Lines, Inc. (stock symbol: DAL) for each month of 2009.

Jan.	Feb.	Mar.	Apr.	May	Jun.
$2.86	−$4.22	−$2.24	$0.67	$0.76	−$0.19

Jul.	Aug.	Sept.	Oct.	Nov.	Dec.
−$0.27	$1.22	$0.08	$1.83	−$1.81	$1.17

(*Source:* finance.yahoo.com)

a. What was the greatest increase in the opening stock price?

b. What was the smallest increase in the opening stock price?

c. What was the greatest decrease in the opening stock price?

d. What was the smallest decrease in the opening stock price?

• Check your answers on page A-1.

MIND*Stretchers*

Groupwork

1. Working with a partner, develop a diagram to show the relationship among the real numbers, the rational numbers, the irrational numbers, the integers, the noninteger rational numbers, the whole numbers, the negative integers, the natural numbers, and zero.

Mathematical Reasoning

2. Is there a largest number less than 5 that is
   a. an integer?
   b. a rational number?

Research

3. Using your college library or the Web, investigate the role the Pythagoreans played in discovering irrational numbers. Write a few sentences to summarize your findings.

Cultural Note

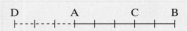

The seventeenth-century English mathematician and cryptographer John Wallis is generally credited with inventing the number line as shown above. Published in his *A Treatise of Algebra* in 1685, the number line gave meaning to negative numbers, the existence of which was controversial. Wallis was the leading English mathematician before Sir Isaac Newton. A professor of geometry at Oxford University, Wallis wrote on a variety of mathematical topics, and introduced the symbol ∞ for infinity. He was the first person to devise a system to teach deaf mutes, and was also one of the founders of the Royal Society, the oldest learned society for science in existence.

(*Sources:* Jan Gullberg, New York: W. W. Norton & Company, *Mathematics, From the Birth of Numbers,*1997; wikipedia.org; newworldencyclopedia.org)

1.2   Addition of Real Numbers

In Chapter R, computations were restricted to zero and positive numbers—whether those positive numbers happened to be whole numbers, fractions, or decimals. Now, we consider computations involving negative numbers as well. Let's start with addition.

Suppose that we want to add two negative numbers: -2 and -3. It is helpful to look at this problem using a real-world situation. If you have a debt of $2 and another debt of $3, altogether you owe $5.

$$-2 + (-3) = -5$$

The number line gives us a picture of adding real numbers. To add -2 and -3, we start at the point corresponding to the first number, -2. The second number, -3, is negative, so we move 3 units to the *left*. We end at -5.

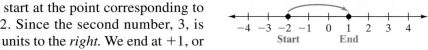

Note that both -2 and -3 are negative and their sum is negative.

Now, suppose that we want to add -2 and 3. On the number line, we start at the point corresponding to the first number, -2. Since the second number, 3, is positive, we move 3 units to the *right*. We end at $+1$, or 1, which is the desired sum.

So $-2 + 3 = 1$.

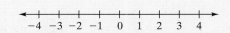

TIP   When using the number line to add two real numbers, move to the left if the second number is negative and to the right if it is positive.

Consider these examples.

EXAMPLE 1

Add -1 and 3 on the number line.

Solution

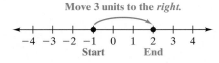

So $-1 + 3 = 2$.

PRACTICE 1

Add 2 and -3 on the number line.

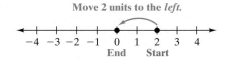

EXAMPLE 2

Add 2 and -2 on the number line.

Solution

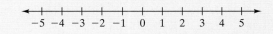

So $2 + (-2) = 0$. Note that 2 and (-2) are opposites, and their sum is 0.

PRACTICE 2

Add -5 and 5 on the number line.

Example 2 suggests that when adding two numbers that are opposites, such as 2 and -2, the sum is 0. We call such numbers *additive inverses.* Every real number has an opposite, or additive inverse, as stated in the following property.

Additive Inverse Property

For any real number a, there is exactly one real number $-a$ such that
$$a + (-a) = 0 \quad \text{and} \quad (-a) + a = 0.$$

In words, this property states that any number added to its opposite is zero.

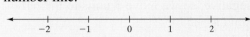

EXAMPLE 3

Using a number line, add $-1\frac{1}{2}$ and 0.

Solution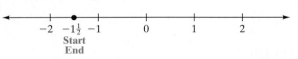

So $-1\frac{1}{2} + 0 = -1\frac{1}{2}$. Note that adding 0 to $-1\frac{1}{2}$ gives us $-1\frac{1}{2}$, the same number we started with.

PRACTICE 3

Add 0 and 0.5 on the following number line.

Example 3 illustrates another important property in adding real numbers—the *additive identity property*, also called the *identity property of 0*.

Additive Identity Property

For any real number a,
$$a + 0 = a \quad \text{and} \quad 0 + a = a.$$

In words, this property states that any number added to zero is the original number.

The previous examples illustrated how to add real numbers on a number line. However, this method is not very efficient, especially for adding large numbers. Instead, let's look at a rule for the addition of real numbers.

To Add Real Numbers

• If the numbers have the same sign, add their absolute values and keep their sign.

• If the numbers have different signs, find the difference between the larger absolute value and the smaller absolute value, and take the sign of the number with the larger absolute value.

Now, let's consider some examples of applying this rule for adding real numbers.

EXAMPLE 4

Add -8 and -19.

Solution Because both numbers are negative, we find their absolute values and then add:

$$|-8| + |-19| = 8 + 19 = 27$$

The sum of two negative numbers is negative.

$$-8 + (-19) = -27$$

Negative numbers ⎯⎯⎯⎯⎯⎯⎯⎯⎯⎯   ⎯⎯ Negative sum

PRACTICE 4

Find the sum: $-13 + (-18)$

TIP When adding numbers with the same sign, the sum has that sign.

EXAMPLE 5

Find the sum of -6.7 and 5.2.

Solution Here we are adding numbers with *different* signs. First, we find the absolute values:

$$|-6.7| = 6.7 \quad \text{and} \quad |5.2| = 5.2$$

Then, we find the difference between the larger absolute value and the smaller one:

$$6.7 - 5.2 = 1.5$$

Because -6.7 has the larger absolute value and its sign is negative, the sum is also negative. So

$$-6.7 + (+5.2) = -6.7 + 5.2 = -1.5$$

Negative ⎯⎯⎯⎯⎯   ⎯⎯ Positive   ⎯⎯ The sum takes the
number   number   sign of the number
                                                with the larger
                                                absolute value.

PRACTICE 5

Add: $10.1 + (-6.6)$

Can you use the number line to explain why the sum in Example 5 is negative?

EXAMPLE 6

Combine: $-\dfrac{2}{9} + \dfrac{2}{9}$

Solution

$$-\frac{2}{9} + \frac{2}{9} = 0$$

Note that $-\dfrac{2}{9}$ and $\dfrac{2}{9}$ are additive inverses. So their sum is 0.

PRACTICE 6

Combine: $-\dfrac{1}{3} + \dfrac{1}{3}$

EXAMPLE 7

Add: $\dfrac{3}{8} + \left(-\dfrac{5}{16}\right)$

Solution

First, we find the LCD of the fractions. The LCD is 16. So

$$\frac{3}{8} + \left(-\frac{5}{16}\right) = \frac{6}{16} + \frac{-5}{16} = \frac{1}{16}.$$

PRACTICE 7

Add: $-\dfrac{7}{18} + \dfrac{1}{9}$

The *commutative property of addition* allows us to add two numbers in any order, getting the same sum. For example, $2 + 6 = 8$ and $6 + 2 = 8$.

> ### Commutative Property of Addition
> For any two real numbers a and b,
> $$a + b = b + a.$$

In words, this property states that the sum of two numbers is the same regardless of order.

By contrast, the *associative property of addition* lets us regroup numbers that are added without affecting the sum. For example, $(2 + 6) + 3 = 8 + 3 = 11$ and $2 + (6 + 3) = 2 + 9 = 11$.

> ### Associative Property of Addition
> For any three real numbers, a, b, and c,
> $$(a + b) + c = a + (b + c).$$

In words, this property states that when adding three numbers, their sum is the same regardless of how they are grouped.

When adding three or more real numbers, it is usually easier to add the positives separately from the negatives. This rearrangement does not affect the sum because of the commutative and associative properties.

EXAMPLE 8

Find the sum: $5 + (-6) + (-9) + 3 + (-5)$

Solution   Let's rearrange the numbers by sign.

$$\underbrace{5 + 3}_{\text{Positives}} \quad + \quad \underbrace{(-6) + (-9) + (-5)}_{\text{Negatives}}$$

$5 + 3 = 8$ Add the positives.

$(-6) + (-9) + (-5) = -20$ Add the negatives.

$8 + (-20) = -12$ Find the sum of the positive and the negative sums.

So $5 + (-6) + (-9) + 3 + (-5) = -12$.

PRACTICE 8

Find the sum:
$-8 + (-4) + 7 + (-8) + 3$

Can you think of another way to get the sum in Example 8?

EXAMPLE 9

In a chemistry class, a student studies the properties of atomic particles, including protons and electrons. She learns that a proton has an electric charge of $+1$, whereas an electron has an electric charge of -1. The charge of a proton cancels out that of an electron. If a charged particle of magnesium has 12 protons and 10 electrons, what is its total charge?

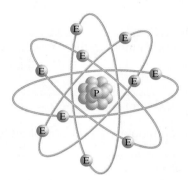

Solution   We can represent the 12 protons and 10 electrons by $+12$ and -10, respectively. To find the total charge of the particle, we add $+12$ and -10.

$$+12 + (-10) = +2$$

So the total charge of the particle of magnesium is $+2$.

PRACTICE 9

The price of a certain stock on Monday was $37.50 per share. On Tuesday, the price of a share went up $2; on Wednesday, it went down $1; and on Thursday, it went down another $2. What was the share price of the stock on Thursday?

Calculators and Real Numbers

To enter a negative number, we need to hit the key that indicates that the sign of the number is negative. Some calculators have a **change-of-sign key,** $\boxed{+/-}$. *Others have a* **negative sign key,** $\boxed{(-)}$. *Do not confuse either of these keys with the* **subtraction key,** $\boxed{-}$. *When entering a negative number, the order in which the change-of-sign key or the negative key must be pressed will depend on the calculator being used. The* $\boxed{+/-}$ *key is usually pressed* **after** *the number is entered, whereas the* $\boxed{(-)}$ *key is usually pressed* **before** *the number is entered.*

EXAMPLE 10

Calculate: $-1.3 + (-5.8)$

Solution

Input

$\boxed{(-)}$ 1.3 $\boxed{+}$ $\boxed{(}$ $\boxed{(-)}$ 5.8 $\boxed{)}$ $\boxed{\text{ENTER}}$

Display

```
- 1.3 + (-5.8)
                    - 7.1
```

PRACTICE 10

Add: $6.002 + (-9.37) + (-0.22)$

Mathematically Speaking

Fill in each blank with the most appropriate term or phrase from the given list.

positive	larger	negative
adding two numbers in any order	regrouping numbers that are added	opposites
identities	additive identity property	smaller
commutative property of addition	associative property of addition	additive inverse property

1. To add real numbers a and b on the number line, start at a and move to the left if b is _____.

2. The _____ states that for any real number a, there is exactly one real number $-a$ such that $a + (-a) = 0$ and $(-a) + a = 0$.

3. Additive inverses are also called _____.

4. The _____ states that for any real number a, $a + 0 = a$ and $0 + a = a$.

5. To add real numbers with different signs, find the difference between their absolute values and take the sign of the number with the _____ absolute value.

6. The commutative property of addition states that _____ results in the same sum.

7. According to the _____, for any two real numbers a and b, $a + b = b + a$.

8. The associative property of addition states that _____ gives us the same sum.

Ⓐ *Find the sum of each pair of numbers using the number line.*

⊙ 9. $4 + (-3)$

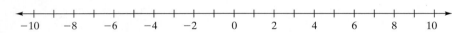

10. $6 + (-2)$

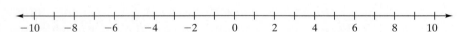

⊙ 11. $8 + (-8)$

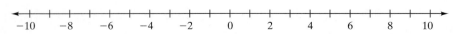

12. $3 + (-3)$

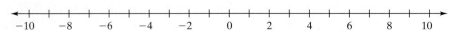

13. $-3 + (-5)$

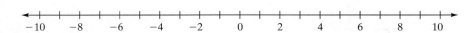

14. $-4 + (-2)$

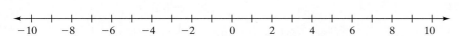

Name the property of addition illustrated.

15. $3 + (-3) = 0$

16. $(-100) + 100 = 0$

17. $5 + (-6) = (-6) + 5$

18. $-1.8 + 2.4 = 2.4 + (-1.8)$

19. $(-4) + 0 = -4$

20. $0 + 2\frac{1}{2} = 2\frac{1}{2}$

21. $(2 + 3) + 6 = 2 + (3 + 6)$

22. $(5 + 6) + (-1) = 5 + [6 + (-1)]$

23. $-a + 0 = -a$

24. $-a + (-b) = -b + (-a)$

Find the sum.

25. $24 + (-1)$

26. $10 + (-6)$

27. $-10 + 5$

28. $-12 + 5$

29. $-6 + (-6)$

30. $-9 + (-4)$

31. $-50 + (-30)$

32. $(-18) + (-18)$

33. $60 + (-90)$

34. $2 + (-10)$

35. $-18 + 18$

36. $-10 + 10$

37. $5.2 + (-0.9)$

38. $6.1 + (-5.9)$

39. $-0.6 + 2$

40. $-0.2 + 8$

41. $-10.5 + 0$

42. $0 + (-0.3)$

◉ **43.** $-9.6 + 3.9$

44. $-7.2 + 2.8$

◉ **45.** $(-9.8) + (-6.5)$

46. $-0.8 + (-0.9)$

47. $-\frac{1}{2} + \left(-\frac{1}{2}\right)$

48. $\left(-\frac{1}{4}\right) + \left(-\frac{3}{4}\right)$

49. $\frac{4}{15} + \left(-\frac{2}{3}\right)$

50. $-\frac{5}{6} + \frac{7}{12}$

51. $-1\frac{3}{5} + 2$

52. $-10 + 3\frac{1}{2}$

53. $2\frac{1}{3} + (-1\frac{1}{2})$

54. $6\frac{1}{4} + (-1\frac{1}{3})$

55. $-24 + (25) + (-89)$

56. $36 + (-17) + (-28)$

57. $15 + (-9) + (-15) + 9$

58. $45 + (-27) + 0 + (-18)$

59. $-0.4 + (-2.6) + (-4)$

60. $(-6.25) + (-0.4) + 3$

61. $(-58) + 10.48 + 58$

62. $-3.7 + 3.7 + (-1.88)$

63. $107 + (-97) + (-45) + 23$

64. $-64 + 7 + (-10) + (-19)$

▦ **65.** $-2.001 + (0.59) + (-8.1) + 10.756$

▦ **66.** $-10 + 6.17 + (-10.005) + (-4.519)$

Mixed Practice

Name the property of addition illustrated.

67. $(-2 + 5) + 8 = -2 + (5 + 8)$

68. $-5.9 + 5.9 = 0$

Solve.

69. Find the sum of 2 and (-9) using the number line.

Find the sum.

70. $37 + 53 + (-38)$

71. $22 + (-15) + (-22)$

72. $-8.5 + 4.8$

73. $2\frac{2}{3} + (-4)$

74. $4.81 + (-0.63) + (-10.002) + 1.05$

Applications

B *Solve. Express your answer as a real number.*

75. The temperature on the top of a mountain was $2°$ below $0°$. If it then got $7°$ warmer, what was the temperature?

76. An elevator goes up 2 floors, down 3 floors, up 1 floor, and finally up 1 floor. What is the overall change in position of the elevator?

77. Last year, a corporation took in $132,000 with expenses of $148,000. How much money did the corporation make or lose?

78. In order to conduct an experiment, a chemist cooled a substance down to $-5°C$. In the course of this experiment, a chemical reaction took place that raised the temperature of the substance by $15°$. What was the final temperature?

79. In the first quarter of Super Bowl XLII, the Pittsburgh Steelers outscored the Arizona Cardinals by three points. In the second quarter, the Cardinals were outscored by seven points. The Steelers scored three points more than the Cardinals in the third quarter. Finally in the fourth quarter, the Cardinals outscored the Steelers by nine points. Who won the game and by how many points? (*Source: Sports Illustrated Almanac 2010*)

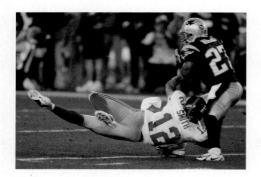

80. Estimate the total annual profits for Dot.com Corporation according to the following chart. (Note that a red number written in parentheses is negative.)

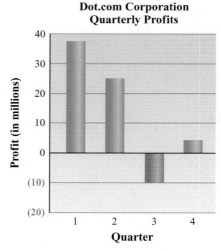

81. With $371.25 in his checking account, an artist writes checks for $71.33 and $51.66. He also deposits $35. After the two checks clear and his deposit is credited, will he still have enough money to cover a check of $250?

82. In January 2005, the average price of a gallon of regular unleaded gasoline was $1.82. In each of the next five Januaries, the average price rose by $0.49, decreased by $0.04, increased by $0.77, decreased by $1.26, and increased by $0.94. What was the average price in January 2010? (*Source:* bls.gov)

83. A plane cruising at an altitude of 32,000 ft hit an air pocket and dropped 700 ft. What was its new altitude?

84. In 2009, the U.S. federal budget allotted $1.795 billion for Supporting Student Success. For the next year, the allotment decreased by $0.254 billion. In 2011, the allotment was $0.245 billion more than for the previous year. How much money was allotted for this program in 2011? (*Source:* whitehouse.gov)

• Check your answers on page A-2.

MINDStretchers

Groupwork

1. Working with a partner, consider the following addition table.

   a. Complete the table.

+	4	−3	−1
6			
−2			
−4			

   b. Explain why the nine numbers that you entered add up to 0.

Mathematical Reasoning

2. Consider the following addition table.

+	x	y	z
x	y	z	x
y	z	x	y
z	x	y	z

   a. Use the table to find $x + y$.

   b. What is the additive identity element in the addition table? Explain how you know.

   c. What is the opposite of x in this table? Explain how you know.

Writing

3. Do you prefer to add real numbers using the number line or using the rule method? Explain why.

1.3 Subtraction of Real Numbers

OBJECTIVES

A To subtract real numbers

B To solve applied problems involving the subtraction of real numbers

The subtraction of real numbers is based on two topics that we have previously covered, namely adding real numbers and finding the opposite of a number.

We already know how to compute the difference between two positive numbers, such as $10 - 2$:

$$10 - (+2) = 8$$

In the previous section, we learned how to add a positive number and a negative number, for example, $10 + (-2)$:

$$10 + (-2) = 8$$

Note that the answers to these two computations are the same. So

$$10 - (+2) = 10 + (-2)$$

More generally, for any real numbers a and b,

$$a - b = a + (-b)$$

In other words, we can change a subtraction problem to an equivalent addition problem by adding the *opposite* of the number being subtracted.

Let's look at the rule for subtracting real numbers.

> **To Subtract Real Numbers**
>
> - Change the operation of subtraction to addition and change the number being subtracted to its opposite.
>
> - Follow the rule for adding real numbers.

Consider these examples.

EXAMPLE 1

Find the difference: $2 - (-5)$

Solution

Change the operation from subtraction to addition.

$$2 - (-5) \quad = \quad 2 + (+5) = 7$$

Change the number being subtracted from -5 to $+5$.

PRACTICE 1

Find the difference: $4 - (-1)$

EXAMPLE 2

Subtract: $-3 - (-9)$

Solution

$$-3 - (-9) = -3 + (+9) = 6$$

Negative 9 · Positive 9

Subtract · Add

PRACTICE 2

Subtract: $-12 - (-15)$

EXAMPLE 3

Find the difference: $2 - 5$

Solution

$$2 - 5 = 2 - (+5) = 2 + (-5) = -3$$

PRACTICE 3

Find the difference: $8 - 12$

EXAMPLE 4

Subtract: $9 - (-13.2)$

Solution

$$9 - (-13.2) = 9 + (+13.2) = 22.2$$

PRACTICE 4

Subtract: $-8.1 - 7.6$

In Chapter R, we discussed the order of operations rule, which states that when a problem involves addition and subtraction, we work from left to right. Let's look at how this rule works with positive and negative numbers.

EXAMPLE 5

Simplify: $-2 - (-6) - (-11)$

Solution

$$
\begin{aligned}
-2 - (-6) - (-11) &= \underline{-2 + (+6)} - (-11) && \text{Subtract } -6 \text{ from } -2. \\
&= \quad\; 4 \quad\;\; - (-11) && \text{Add } -2 \text{ and } +6. \\
&= 4 + 11 && \text{Subtract } -11 \text{ from } 4. \\
&= 15 && \text{Add 4 and 11.}
\end{aligned}
$$

PRACTICE 5

Simplify: $5 - (-8) - (-15)$

EXAMPLE 6

Evaluate: $-7 - (-1) + 5 + (-3)$

Solution

$$-7 - (-1) + 5 + (-3)$$

$$= \underbrace{-7 + (+1)} + 5 + (-3) \qquad \text{Subtract } -1 \text{ from } -7.$$

$$= \underbrace{-6 + 5} + (-3) \qquad\qquad \text{Add } -7 \text{ and } +1.$$

$$= \underbrace{-1 + (-3)} \qquad\qquad\qquad \text{Add } -6 \text{ and } 5.$$

$$= -4 \qquad\qquad\qquad\qquad \text{Add } -1 \text{ and } -3.$$

PRACTICE 6

Simplify: $4 + (-6) - (-11) + 8$

EXAMPLE 7

Simplify: $10 - (5 - 3)$

Solution

$$10 - (5 - 3) = 10 - 2 \qquad \textbf{Perform the operations within parentheses.}$$

$$= 8 \qquad\qquad\qquad \textbf{Subtract 2 from 10.}$$

PRACTICE 7

Evaluate: $5 - (12 - 4)$

EXAMPLE 8

Egypt emerged as a nation in about 3100 B.C. and Ethiopia in about 3000 B.C. How much older is Egypt than Ethiopia? (*Source: The World Book Encyclopedia*)

PRACTICE 8

Paper was invented in China in about 100 B.C., and wood block printing in about A.D. 770. How much older is the invention of paper than that of wood block printing? (*Source: The World Book Encyclopedia*)

Solution Recall that a B.C. year corresponds to a negative integer. So 3100 B.C. and 3000 B.C. are represented by -3100 and -3000, respectively. Because $-3000 > -3100$, we write -3000 first:

$$-3000 - (-3100) = -3000 + (+3100)$$
$$= 100$$

So Egypt is about 100 years older than Ethiopia.

A *Find the difference.*

1. $25 - 8$ **2.** $15 - 9$ **3.** $-24 - 7$ **4.** $-6 - 9$

5. $(-19) - 25$ **6.** $(-49) - 2$ **7.** $52 - (-19)$ **8.** $24 - (-31)$

9. $60 - 95$ **10.** $70 - 92$ **11.** $-34 - (-2)$ **12.** $-30 - (-1)$

13. $16 - (-16)$ **14.** $21 - (-21)$ **15.** $0 - 45$ **16.** $0 - 36$

17. $-31 - 31$ **18.** $-25 - 25$ **19.** $22 - (-22)$ **20.** $8 - (-19)$

21. $200 - (-800)$ **22.** $30 - (-10)$ **23.** $6 - 7.42$ **24.** $10.1 - 11.84$

25. $-7.3 - 0.5$ **26.** $-3 - 0.1$ **27.** $(-5.6) - (-5.6)$ **28.** $(-0.4) - (-0.4)$

29. $8.6 - (-1.7)$ **30.** $9.4 - (-2.5)$ **31.** $-\dfrac{1}{3} - \dfrac{5}{6}$ **32.** $-\dfrac{4}{5} - \dfrac{7}{8}$

33. $-18 - \left(-\dfrac{3}{4}\right)$ **34.** $-12 - \left(-\dfrac{1}{4}\right)$ **35.** $4\frac{3}{5} - (-1\frac{1}{2})$ **36.** $6\frac{1}{2} - (-1\frac{1}{4})$

Simplify.

37. $3 + (-6) - (-15)$ **38.** $12 + (-4) - (-9)$

39. $8 - 10 + (-5)$ **40.** $12 - 16 + (-5)$

41. $-9 + (-4) - 9 + 4$ **42.** $-6 + (-1) - 5 + 8$

43. $9 - 12 - 18$ **44.** $7 - 4 - 12$

45. $-10.722 + (-3.913) - 8.36 - 3.492$ **46.** $1.884 - 0.889 + (-6.12) - (-4.001)$

47. $11 - (8 - 2)$ **48.** $4 - (6 - 1)$

Mixed Practice

Find the difference.

49. $17 - (-31)$ **50.** $-7.6 - 5.8$ **51.** $(-23) - 15$

52. $-28 - (-17)$ **53.** $\dfrac{5}{6} - \left(-\dfrac{3}{4}\right)$

Simplify.

54. $-5 - (-3) + 2 + (-9)$ **55.** $18 + (-13) - (-9)$

56. $-3.712 + (-5.003) - 0.762 - (-4.73)$

Applications

B *Solve. Express your answer as a real number.*

57. The first Olympic Games occurred in 776 B.C. Approximately how many centuries were there between the first Olympic games and the Olympic games held in A.D. 2010?

58. The boiling point of the element radon is $-61.8°C$, whereas its melting point is $-71°C$. How much higher is the boiling point? (*Source:* periodic-table.org.uk)

59. Two airplanes take off from the same airport. One flies north and the other flies south, as pictured. How far apart are the airplanes? (*Hint:* Consider north to be positive and south to be negative.)

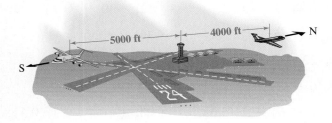

60. The following chart shows the record high and low temperatures (in degrees Fahrenheit) for a number of U.S. states.

State	Record High (°F)	Record Low (°F)
Alabama	114	−27
California	134	−45
Louisiana	114	−16
Minnesota	114	−64
New York	108	−57
Virginia	110	−30

(*Source:* wikipedia.org)

Which state had the greatest difference in record extreme temperatures?

61. A company reported a loss of $281,330 last year and a loss of $5291 this year. By how much money was the company's loss reduced?

62. The value of a computer company's stock rose by $0.50 per share and then dropped by $0.75. What was the overall change in value?

63. The U.S. city with the greatest elevation is Leadville, Colorado. The U.S. city with the lowest elevation is Calipatria, California. If their respective elevations are 10,152 ft above sea level and 184 ft below sea level, what is the difference in the elevations of these cities? (*Source:* wikipedia.org)

64. Two delivery trucks that have been parked in a lot drive off in opposite directions. After driving 20 min, one truck has traveled 10.75 mi while the other has traveled 17.5 mi. At that time, how far apart are the two vehicles?

65. Matter is liquid when its temperature is between its melting point and its boiling point. The following table shows the melting and boiling points (in degrees Celsius) of various elements.

Element	Melting Point (°C)	Boiling Point (°C)
Krypton	−157	−153
Neon	−249	−246
Bromine	−7	59

(*Source: The New York Times Almanac*, 2010)

a. For each of these elements, find the difference between its boiling and melting points.

b. Which of the elements is liquid in the widest range of temperatures?

c. Which of the elements is liquid at 0°C?

66. Superconductors allow for very efficient passage of electric currents. For practical use, a superconductor must work above −196°C, which is the boiling point of nitrogen.

a. In 1986, the first high-temperature superconductor that was able to conduct electricity without resistance at a temperature of −238°C was discovered. How many degrees below the boiling point of nitrogen did this first high-temperature superconductor work?

b. In 1987, the researcher Paul Chu discovered a new class of materials that conduct electricity at −178°C. How many degrees above the boiling point of nitrogen did the new materials conduct electricity?

c. In 1990, other researchers created a miniature transistor that conducts electricity at a temperature that is 48°C above the boiling point of nitrogen. What is this temperature? (*Sources: World Book Encyclopedia* and uh.edu)

• Check your answers on page A-2.

MINDStretchers

Critical Thinking

1. Rearrange the numbers in the square on the left so that it becomes a magic square in which the sum of every row, column, and diagonal is −6.

−3	2	−2
−5	−4	−6
0	−1	1

Mathematical Reasoning

2. Consider the following two problems:
$$8 - (-2) = 8 + 2 = 10$$
$$8 \div \frac{4}{7} = 8 \times \frac{7}{4} = 14$$

   a. Explain in what way the two problems are *similar*.

   b. Explain in what way the two problems are *different*.

Writing

3. Explain whether it is always true that the difference between two numbers is smaller than either of the numbers. Give an example to justify your answer.

1.4   Multiplication of Real Numbers

A To multiply real numbers

B To solve applied problems involving the multiplication of real numbers

In this section we deal with the multiplication of real numbers. From arithmetic, we see that when multiplying two positive numbers, we get a positive number. Now, let's look at multiplying a positive number by a negative number.

Consider, for example, finding the product of 3 and -2. Looking at multiplication as repeated addition, we know that multiplying a number by 3 means the same as adding three of that number:

$$3(-2) = -2 + (-2) + (-2)$$
$$= -6$$

Note that when multiplying a positive number by a negative number, we get a negative answer.

What about multiplying two negative numbers? Let's consider the following pattern:

This number is decreasing by 1 each time.   The product is increasing by 2 each time.

$$3(-2) = -6$$
$$2(-2) = -4$$
$$1(-2) = -2$$
$$0(-2) = \;\;\;0$$
$$-1(-2) = +2 \quad \leftarrow \text{The pattern continues.}$$
$$-2(-2) = +4$$
$$-3(-2) = +6$$

Note that when we multiply two negative numbers, the product is positive. This suggests the following rule:

To Multiply Real Numbers

• Multiply their absolute values.

• If the numbers have the same sign, their product is positive; if they have different signs, their product is negative.

EXAMPLE 1

Multiply -2 by -9.

Solution

$|-2| = 2$ and $|-9| = 9$ Find the absolute values.

$2 \cdot 9 = 18$ Multiply the absolute values.

$-2(-9) = 18$ The product of two negatives is positive.

PRACTICE 1

Find the product of -1 and -100.

TIP

Same Sign

Positive · Positive = Positive   Negative · Negative = Positive

Different Signs

Positive · Negative = Negative   Negative · Positive = Negative

EXAMPLE 2

Multiply: $3(-5)$

Solution

$|3| = 3$ and $|-5| = 5$ Find the absolute values.

$3 \cdot 5 = 15$ Multiply the absolute values.

$3(-5) = -15$ The product of a positive and a negative is negative.

PRACTICE 2

Calculate: $-5 \cdot 3$

Comparing Example 2 and Practice 2, we see that $3(-5) = -5 \cdot 3$, which suggests another property of real numbers, the *commutative property of multiplication.*

Commutative Property of Multiplication

For any two real numbers a and b,

$$a \cdot b = b \cdot a.$$

In words, this property states that the product of two numbers is the same regardless of order.

Other properties involving the multiplication of real numbers are the *multiplicative identity property* and *the multiplication property of zero.*

Multiplicative Identity Property

For any real number a,

$$a \cdot 1 = a \quad \text{and} \quad 1 \cdot a = a.$$

In words, this property states that the product of any number and one is the original number.

Multiplication Property of Zero

For any real number a,

$$a \cdot 0 = 0 \quad \text{and} \quad 0 \cdot a = 0.$$

In words, this property states that the product of any number and zero is zero.

EXAMPLE 3

Find the product.

a. $-9\left(\dfrac{1}{3}\right)$ **b.** $-\dfrac{1}{5}(-25)$ **c.** $-4(0)$

d. $-1.5(-1.5)$ **e.** $0.4(-6)$ **f.** $-7(1)$

Solution

a. $-9\left(\dfrac{1}{3}\right) = \dfrac{\overset{3}{-\cancel{9}}}{1}\cdot\dfrac{1}{\underset{1}{\cancel{3}}} = -3$ Negative · Positive = Negative

b. $-\dfrac{1}{5}(-25) = -\dfrac{1}{\underset{1}{\cancel{5}}}\cdot\dfrac{\overset{5}{-\cancel{25}}}{1} = 5$ Negative · Negative = Positive

c. $-4(0) = 0$ Multiplication property of zero

d. $-1.5(-1.5) = 2.25$ Negative · Negative = Positive

e. $0.4(-6) = -2.4$ Positive · Negative = Negative

f. $-7(1) = -7$ Multiplicative identity property

PRACTICE 3

Multiply.

a. $\left(-\dfrac{2}{3}\right)(-12)$

b. $\left(-\dfrac{1}{3}\right)\left(\dfrac{5}{9}\right)$

c. $(-0.4)(-0.3)$

d. $2.5(-1.9)$

e. $0\cdot(-2.8)$

f. $1\cdot\dfrac{2}{3}$

The next property of real numbers—the *associative property of multiplication*—allows us to *regroup* the product of three numbers. For example, $(2\cdot3)\cdot5 = 2\cdot(3\cdot5)$.

> ### Associative Property of Multiplication
>
> For any three real numbers, a, b, and c,
> $$(a\cdot b)c = a(b\cdot c).$$

In words, this property states that when multiplying three real numbers, the product is the same regardless of how they are grouped.

EXAMPLE 4

Calculate: $-3(-2)(9)$

Solution

$$\begin{aligned} -3(-2)(9) &= 6(9) \quad \text{Multiply } -3 \text{ by } -2.\\ &= 54 \quad \text{Multiply 6 by 9.} \end{aligned}$$

So $(-3)(-2)(9) = 54$.

PRACTICE 4

Multiply: $-8(4)(-2)$

EXAMPLE 5

Multiply: $5(-2)(-1)(3)(-2)$

Solution A good way to calculate this product is to rearrange the numbers by sign.

$$\underbrace{5(3)}_{\text{Positives}} \qquad \underbrace{-2(-1)(-2)}_{\text{Negatives}}$$

$5(3) = 15$ **Find the product of the positives.**

$-2(-1)(-2) = 2(-2) = -4$ **Find the product of the negatives.**

$15(-4) = -60$ **Multiply the two products above.**

So $5(-2)(-1)(3)(-2) = -60$.

PRACTICE 5

Find the product:

$(-6)(-1)(4)(2)(-5)$

In Example 4, the product was positive because there were *two* negative factors. By contrast, in Example 5 the answer was negative because there were *three* negative factors. Can you explain why a product is positive if it has an even number of negative factors, whereas a product is negative if it has an odd number of negative factors?

According to the order of operations rule given in Chapter R, multiplication is performed before either addition or subtraction, working from left to right.

Order of Operations Rule

To evaluate mathematical expressions, carry out the operations *in the following order*:

1. First, perform the operations within any grouping symbols, such as parentheses ( ) or brackets [ ].

2. Then, raise any number to its power.

3. Next, perform all multiplications and divisions as they appear from left to right.

4. Finally, do all additions and subtractions as they appear from left to right.

EXAMPLE 6

Simplify: $-2(24) - 5(-6)$

Solution We use the order of operations rule.

$$\begin{aligned} -2(24) - 5(-6) &= -48 - (-30) \quad &\textbf{Multiply first.} \\ &= -48 + 30 \quad &\textbf{Subtract} -30 \textbf{ from } -48. \\ &= -18 \quad &\textbf{Add} -48 \textbf{ and } 30. \end{aligned}$$

So $-2(24) - 5(-6) = -18$.

PRACTICE 6

Calculate: $4(-25) - (-2)(36)$

Following the order of operations rule, we simplify mathematical expressions by first performing the operations within any grouping symbols such as parentheses ( ) or brackets [ ].

EXAMPLE 7

Simplify: $-3(6 - 10)$

Solution

$$\begin{aligned} -3(6 - 10) &= -3(-4) && \text{Subtract within parentheses.} \\ &= 12 && \text{Multiply.} \end{aligned}$$

So $-3(6 - 10) = 12$.

PRACTICE 7

Calculate: $-5(-9 + 15)$

EXAMPLE 8

Calculate: $5 - 3(2 - 4)$

Solution

$$\begin{aligned} 5 - 3(2 - 4) &= 5 - 3(-2) && \text{Subtract within parentheses.} \\ &= 5 - (-6) && \text{Multiply.} \\ &= 11 && \text{Subtract.} \end{aligned}$$

So $5 - 3(2 - 4) = 11$.

PRACTICE 8

Simplify: $10 - 5(3 + 1)$

EXAMPLE 9

Simplify: $11 - 4(-3 + 1)^2$

Solution

$$\begin{aligned} 11 - 4(-3 + 1)^2 &= 11 - 4(-2)^2 && \text{Add within parentheses.} \\ &= 11 - 4(4) && \text{Square } -2. \\ &= 11 - 16 && \text{Multiply.} \\ &= -5 && \text{Subtract.} \end{aligned}$$

So $11 - 4(-3 + 1)^2 = -5$.

PRACTICE 9

Calculate: $9 - 5(6 - 7)^2$

EXAMPLE 10

Temperatures can be measured in both the Fahrenheit and Celsius systems. The normal *melting point* of the element mercury is about $-37.9°$F. To find the Celsius equivalent of this temperature, we need to compute $\frac{5}{9}(-37.9 - 32)$. Simplify this expression.

(*Source: CRC Handbook of Chemistry and Physics*)

Solution

$$\frac{5}{9}(-37.9 - 32) = \frac{5}{9}(-69.9)$$

$$\approx -38.8$$

So $-37.9°$F is equivalent to $-38.8°$C.

PRACTICE 10

If a rock is thrown upward on the moon with an initial velocity of 10 ft/sec, the rock's velocity after 3 sec will be $[10 - (5.3)(3)]$ ft/sec. Simplify this expression and interpret the result. (Note: Objects moving upward have positive velocity and objects moving downward have negative velocity.) (*Source:* NASA)

Mathematically Speaking

Fill in each blank with the most appropriate term or phrase from the given list.

negative	even	multiply two numbers
regroup the product	any number and 0	in either order
of three numbers	odd	any number and 1
positive	parentheses	brackets

1. The product of two real numbers with the same sign is _____.

2. The product of two real numbers with different signs is _____.

3. The commutative property of multiplication allows us to _____.

4. The multiplicative identity property tells us that the product of _____ is the original number.

5. According to the multiplication property of zero, the product of _____ is zero.

6. The associative property of multiplication allows us to _____.

7. A product is negative if it has a(n) _____ number of negative factors.

8. The grouping symbols [ ] are called _____.

Ⓐ *Name the property of multiplication illustrated.*

9. $2(5) = 5(2)$

10. $1.7(-3) = -3(1.7)$

11. $(-4 \cdot 6) \cdot 3 = -4 \cdot (6 \cdot 3)$

12. $(-8 \cdot 5) \cdot 4 = -8 \cdot (5 \cdot 4)$

13. $-9 \cdot 1 = -9$

14. $1 \cdot (-10) = -10$

15. $-8 \cdot 0 = 0$

16. $0 \cdot \frac{1}{2} = 0$

Find the product.

17. $6(-2)$

18. $7(-4)$

19. $-7(-3)$

20. $-5(-5)$

21. $-12\left(\frac{1}{4}\right)$

22. $-15\left(\frac{2}{3}\right)$

23. $-\frac{1}{3} \cdot \frac{4}{9}$

24. $\left(-\frac{5}{6}\right)\left(\frac{2}{7}\right)$

25. $\left(1\frac{1}{3}\right)\left(-\frac{4}{9}\right)$

26. $\left(2\frac{1}{5}\right)\left(-\frac{2}{7}\right)$

27. $-1.5(-0.6)$

28. $-1.7(-0.4)$

29. $1.2(-50)$

30. $1.5(-60)$

31. $3(-2)(-20)$

32. $-9(-12)(2)$

33. $-15(-3)(0)$

34. $-8.5(0)(2.6)$

35. $-6(1)(-2)(-3)(-4)$

36. $6(-1)(-2)(3)(-4)$

37. $-4(5)(-6)(1)$

38. $10(1)(-10)(-1)$

39. $\left(-\dfrac{1}{3}\right)\left(-\dfrac{1}{3}\right)\left(-\dfrac{1}{3}\right)$

40. $\left(-\dfrac{1}{2}\right)\left(-\dfrac{1}{2}\right)\left(-\dfrac{1}{2}\right)$

Multiply and round to the nearest hundredth.

41. $-6.24(0.08)(-1.97)$

42. $-5.42(-0.19)(-4.8)$

Simplify.

43. $-7 + 3(-2) - 10$

44. $-4 + 2(-5) - 3$

45. $-3 - 5(-6)$

46. $-10 - 2(-8)$

47. $\left(\dfrac{3}{5}\right)(-15) + 6$

48. $\left(\dfrac{3}{4}\right)(-16) + 20$

49. $-5 \cdot (-3 + 4)$

50. $(-10 + 7) \cdot (-3)$

51. $-6 - 3(5 - 9)$

52. $5 - 2(4 - 10)$

53. $7 - 3(-2 + 5)^2$

54. $5(4 - 6)^2 - 9$

Complete each table.

55.

Input	Output
a. -2	$-4(-2) - 3 = ?$
b. -1	$-4(-1) - 3 = ?$
c. 0	$-4(0) - 3 = ?$
d. 1	$-4(1) - 3 = ?$
e. 2	$-4(2) - 3 = ?$

56.

Input	Output
a. 2	$-6(2) + 2 = ?$
b. 1	$-6(1) + 2 = ?$
c. 0	$-6(0) + 2 = ?$
d. -1	$-6(-1) + 2 = ?$
e. -2	$-6(-2) + 2 = ?$

MIXED PRACTICE

Find the product.

57. $-2.8(-1.3)$

58. $\dfrac{2}{5}\left(-\dfrac{3}{4}\right)$

59. $3(-5)(1)(-4)(-2)$

60. Multiply and round to the nearest hundredth: $-3.51(-0.23)(-6.4)$

61. Complete the table.

Input	Output
a. 2	$-5(2) + 4 = ?$
b. 1	$-5(1) + 4 = ?$
c. 0	$-5(0) + 4 = ?$
d. -1	$-5(-1) + 4 = ?$
e. -2	$-5(-2) + 4 = ?$

Solve.

62. Name the property of multiplication that $1.7(-6.3) = -6.3(1.7)$ illustrates.

Simplify.

63. $-5 - 6(-2) + (-3)$

64. $\left(\dfrac{4}{9}\right)(-18) - (-3)$

Applications

B *Solve.*

65. On a double-or-nothing wager, a gambler bets $5 and wins. Express as a signed number the amount of money he won.

66. On January 31, the high temperature in Chicago was 40°F. The high temperature then dropped 3°F per day for 3 days. If 32°F is freezing on the Fahrenheit scale, was it below freezing on February 3?

67. In the 10 games played this season, a team won 3 games by 4 points, won 2 games by 1 point, lost 4 games by 3 points, and tied in the final game. In these games, did the team score more or fewer points than its opposing teams?

68. A start-up company lost $5000 a month for the first 3 months of business. Express this loss as a signed number.

69. During a drought, the water level in a reservoir dropped 3 in. each week for 5 straight weeks. Express the overall change in water level as a signed number.

70. The submarine shown dives to 3 times its current depth of 150 ft below sea level. What is its new depth?

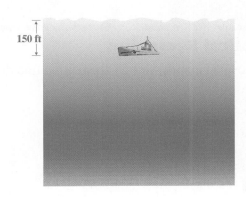

71. The following tables show the number of calories in servings of various foods and the number of calories burned by various activities.

Food	Number of Calories Per Serving
Apple	80
Banana	105
Pretzel, stick	30
Ginger ale, can	125
Donut	210

(*Source: The World Almanac and Book of Facts,* 2000)

Activity (1 hr)	Number of Calories Burned*
Swimming	−288
Bicycling	−612
Football	−498
Basketball	−450
Scrubbing floors	−440

*For a 150-lb person.
(*Source: Exercise & Weight Control,* President's Council on Physical Fitness and Sports, 1986)

Find the net number of calories in each situation.

a. A weight watcher eats 3 servings of pretzel sticks and then plays basketball for $\frac{1}{2}$ hour.

b. A dieter swims for 30 minutes and drinks 2 cans of ginger ale.

c. An athlete eats 3 servings of apples and 2 servings of donuts. Later he goes bicycling for 2 hours and then swims for 1 hour.

72. To discourage guessing on a test, an instructor takes off for wrong answers, grading according to the following scheme:

Performance on a Test Item	Score
Correct	5
Incorrect	−2
Blank	0

What score would the instructor give to each of the following tests?

	Number of Items Correct	Number of Items Incorrect	Number of Items Blank	Test Grade
a.	17	1	2	
b.	19	1	0	
c.	12	7	1	

- Check your answers on page A-2.

MINDStretchers

Mathematical Reasoning

1. Explain how the number line can be used to find the product of two integers.

Historical

2. At a very early age, the eighteenth-century mathematician Carl Friedrich Gauss found the sum of the first 100 positive integers within a few minutes by using the following method. First, he wrote the sum both forward and backward.

$$1 + 2 + 3 + \cdots + 98 + 99 + 100$$
$$100 + 99 + 98 + \cdots + 3 + 2 + 1$$

Then, he added the 100 vertical pairs, getting a sum of 101 for each pair. He concluded that the product 100(101) was twice the correct answer, which turns out to be $\frac{100(101)}{2} = 50(101) = 5050$. Show how to find the sum of the first 1000 *negative* integers using Gauss's method. Explain your work.

Writing

3. A salesman says that he loses a little money on each item sold but makes it up in volume. Explain if this is possible.

1.5   Division of Real Numbers

We now consider division—the last of the four basic operations on real numbers. From arithmetic, we know that every division problem has a related multiplication problem.

$$8 \div 2 = 4 \qquad \text{because} \qquad 2 \cdot 4 = 8$$

Division   Related multiplication

Now, suppose that the division problem involves a negative number, for instance, $-8 \div 2$. From the previous section we know that $(-4) \cdot 2 = -8$. So $(-8) \div 2 = -4$. This problem suggests that when we divide a negative number by a positive number, we get a negative quotient.

Similarly, suppose that we want to calculate $(-8) \div (-2)$. Since $4 \cdot (-2) = -8$, it follows that $(-8) \div (-2) = 4$. This example suggests that the quotient of two negative numbers is a positive number.

We can use the following rule for dividing real numbers:

To Divide Real Numbers

- Divide their absolute values.

- If the numbers have the same sign, their quotient is positive; if the numbers have different signs, their quotient is negative.

EXAMPLE 1

Find the quotient.

a. $(-16) \div (-2)$ **b.** $\dfrac{-24}{6}$ **c.** $\dfrac{-2}{-8}$

d. $\dfrac{9.4}{-2}$ **e.** $\dfrac{-15}{-0.3}$

Solution   In each problem, first we find the absolute values. Next, we divide them. Then, we attach the appropriate sign to this quotient.

a. $(-16) \div (-2)$

$|-16| = 16$ and $|-2| = 2$

$16 \div 2 = 8$

Since the numbers have the *same* signs, their quotient is positive. So $(-16) \div (-2) = 8$.

b. $\dfrac{-24}{6}$ $|-24| = 24$ and $|6| = 6$

$\dfrac{24}{6} = 4$

Since the numbers have *different* signs, the quotient is negative. So $\dfrac{-24}{6} = -4$.

PRACTICE 1

Divide.

a. $40 \div (-5)$

b. $\dfrac{-42}{-6}$

c. $\dfrac{-5}{10}$

d. $\dfrac{-6.3}{9}$

e. $\dfrac{-24}{-0.4}$

EXAMPLE 1 (continued)

c. $\dfrac{-2}{-8}$ $|-2| = 2$ and $|-8| = 8$

$$\frac{2}{8} = \frac{1}{4}$$

Since the numbers have the *same* signs, the quotient is positive.

So $\dfrac{-2}{-8} = \dfrac{1}{4}$.

d. $\dfrac{9.4}{-2}$ $|-9.4| = 9.4$ and $|-2| = 2$

$$\frac{9.4}{2} = 4.7$$

Since the numbers have *different* signs, the quotient is negative.

So $\dfrac{9.4}{-2} = -4.7$.

e. $\dfrac{-15}{-0.3}$ $|-15| = 15$ and $|-0.3| = 0.3$

$$\frac{15}{0.3} = \frac{15\,.0}{0\,.3} = \frac{150}{3} = 50$$

Since the numbers have the *same* signs, the quotient is positive.

So $\dfrac{-15}{-0.3} = 50$.

TIP

Same Sign

$\dfrac{\text{Positive}}{\text{Positive}} = \text{Positive}$ $\dfrac{\text{Negative}}{\text{Negative}} = \text{Positive}$

Different Signs

$\dfrac{\text{Positive}}{\text{Negative}} = \text{Negative}$ $\dfrac{\text{Negative}}{\text{Positive}} = \text{Negative}$

Some division problems involve 0. For instance, $0 \div (-5) = 0$ because $(-5) \cdot 0 = 0$. On the other hand, $(-5) \div 0$ is *undefined* because there is no real number that when multiplied by 0 gives -5.

These two examples lead us to the following conclusion:

Division Involving Zero

For any nonzero real number a,

$$0 \div a = 0.$$

For any nonzero real number a,

$$a \div 0 \text{ is undefined.}$$

In words, these properties state that zero divided by any nonzero number is zero, whereas any number divided by zero is undefined.

Recall that in Chapter R we expressed the rule for dividing fractions in terms of the *reciprocal* of a number. In algebra, we also refer to the reciprocal of a number as its *multiplicative inverse*.

Multiplicative Inverse Property

For any nonzero real number a,

$$a \cdot \frac{1}{a} = 1 \quad \text{and} \quad \frac{1}{a} \cdot a = 1,$$

where a and $\frac{1}{a}$ are **multiplicative inverses** (or **reciprocals**) of each other.

In words, this property states that the product of a number and its multiplicative inverse is one. For example,

- $\frac{1}{3}$ and 3 are multiplicative inverses because $\frac{1}{3} \cdot 3 = 1$

- $-\frac{5}{6}$ and $-\frac{6}{5}$ are multiplicative inverses because $\left(-\frac{5}{6}\right)\left(-\frac{6}{5}\right) = 1$

TIP   A number and its reciprocal have the same sign.

EXAMPLE 2

Complete the following table.

Solution

Number	Reciprocal
a. 4	$\frac{1}{4}$ is the reciprocal of 4 because $4 \cdot \frac{1}{4} = 1$
b. $-\frac{3}{4}$	$-\frac{4}{3}$ is the reciprocal of $-\frac{3}{4}$ because $\left(-\frac{3}{4}\right) \cdot \left(-\frac{4}{3}\right) = 1$
c. -10	$-\frac{1}{10}$ is the reciprocal of -10 because $-10\left(-\frac{1}{10}\right) = 1$
d. $1\frac{1}{2}$	$\frac{2}{3}$ is the reciprocal of $1\frac{1}{2}$ because $1\frac{1}{2} \cdot \frac{2}{3} = \frac{3}{2} \cdot \frac{2}{3} = 1$

PRACTICE 2

Fill in the following table.

Number	Reciprocal
a. -5	
b. $\frac{1}{-8}$	
c. $1\frac{1}{3}$	
d. $-\frac{8}{5}$	

TIP   We usually rewrite a fraction such as $\frac{1}{-10}$ or $\frac{-1}{10}$ as $-\frac{1}{10}$.

Now, let's consider division of real numbers by using reciprocals. Recall that we subtract by adding an opposite. Similarly, we can divide by multiplying by a reciprocal.

Division of Real Numbers

For any real numbers a and b, where b is nonzero,

$$a \div b = \frac{a}{b} = a \cdot \frac{1}{b}.$$

In words, this rule states that the quotient of two numbers is the product of the first number and the reciprocal of the second number.

EXAMPLE 3

Divide.

a. $-\dfrac{1}{3} \div \dfrac{5}{6}$ **b.** $-\dfrac{1}{2} \div 4$

Solution

a. $-\dfrac{1}{3} \div \dfrac{5}{6} = -\dfrac{1}{3} \cdot \dfrac{6}{5} = -\dfrac{1}{{}_1 3} \cdot \dfrac{6^2}{5} = -\dfrac{2}{5}$

b. $-\dfrac{1}{2} \div 4 = -\dfrac{1}{2} \div \dfrac{4}{1} = -\dfrac{1}{2} \cdot \dfrac{1}{4} = -\dfrac{1}{8}$

PRACTICE 3

Divide.

a. $-\dfrac{8}{9} \div \dfrac{2}{3}$

b. $-10 \div \left(-\dfrac{2}{5}\right)$

We use the order of operations rule to simplify the following expressions.

EXAMPLE 4

Simplify.

a. $-8 \div (-2)(-2)$ **b.** $\dfrac{-5 + (-7)}{2}$

Solution

a. $-8 \div (-2)(-2) = 4(-2)$ Perform multiplications and divisions as they occur from left to right. Divide -8 by -2.

$\qquad\qquad\qquad\quad = -8$ Multiply 4 by -2.

b. $\dfrac{-5 + (-7)}{2} = \dfrac{-12}{2}$ Parentheses are understood to be around the numerator. Add -5 and -7.

$\qquad\qquad\quad = -6$ Divide -12 by 2.

PRACTICE 4

Simplify.

a. $(-3)(-4) \div (2)(-2)$

b. $\dfrac{-9 - (-3)}{2}$

EXAMPLE 5

During clinical practice, a student nurse took care of a patient with a fever. He recorded the patient's temperature at the same time every day for five days. The following table shows the change in the patient's temperature each day.

Day	Temperature Change
Monday	Up 2.5°
Tuesday	Down 2°
Wednesday	Down 1.5°
Thursday	Up 1°
Friday	Down 3°

What was the average daily change in the patient's temperature?

Solution To compute the average daily change, we add the five temperature changes and divide the sum by 5, the number of days the temperature was recorded.

$$\frac{2.5 + (-2) + (-1.5) + 1 + (-3)}{5}$$

Since parentheses are understood to be around both the numerator and the denominator, we find the sum in the numerator before dividing by the denominator.

$$\frac{2.5 + (-2) + (-1.5) + 1 + (-3)}{5} = \frac{-3}{5} = -0.6$$

So the average daily change in temperature during the five days was $-0.6°$.

PRACTICE 5

During the past four months, the number of cell phone minutes used changed from the previous month as follows:

Month	Change (in minutes)
1	Down 300
2	Up 200
3	Down 500
4	Up 100

What was the average monthly change in minutes used?

Mathematically Speaking

Fill in each blank with the most appropriate term or phrase from the given list.

zero	different signs	quotient
opposite	absolute values	the same sign
fraction	reciprocal	multiplicative inverse
numbers	undefined	divisor

1. To divide real numbers, first divide the _____.

2. The quotient of two real numbers with _____ is positive.

3. The quotient of two real numbers with _____ is negative.

4. Zero divided by any nonzero real number a is _____.

5. Any nonzero real number a divided by zero is _____.

6. Any nonzero real number a has a multiplicative inverse, or _____, which is written $\frac{1}{a}$.

7. The product of any real number and its _____ is one.

8. To divide two real numbers, multiply the dividend by the reciprocal of the _____.

A *Complete each table.*

9.

	a.	b.	c.	d.	e.
Number	$-\frac{1}{2}$	5	$-\frac{3}{4}$	$3\frac{1}{5}$	-1
Reciprocal					

10.

	a.	b.	c.	d.	e.
Number	-12	$\frac{1}{4}$	7	$-2\frac{1}{3}$	$-\frac{5}{6}$
Reciprocal					

Divide.

11. $-8 \div (-1)$

12. $-12 \div (-1)$

13. $-63 \div 7$

14. $-16 \div 4$

15. $\frac{0}{-9}$

16. $0 \div (-10)$

17. $-250 \div (-10)$

18. $-300 \div (-10)$

19. $-200 \div (-8)$

20. $-400 \div (-5)$

21. $-64 \div (-16)$

22. $-81 \div (-9)$

23. $\frac{-25}{-5}$

24. $\frac{-125}{-5}$

25. $\frac{-2}{16}$

26. $\frac{2}{-10}$

27. $\frac{-10}{-20}$

28. $\frac{-35}{-40}$

29. $\frac{4}{5} \div \left(-\frac{2}{3}\right)$

30. $\frac{7}{12} \div \left(-\frac{1}{6}\right)$

31. $8 \div \left(-\frac{1}{4}\right)$

32. $5 \div \left(-\frac{1}{6}\right)$

33. $2\frac{1}{2} \div (-20)$

34. $3\frac{1}{2} \div (-10)$

35. $(-3.5) \div 7$

36. $(-5.6) \div (8)$

37. $10 \div (-0.5)$

38. $12 \div (-0.3)$

39. $\dfrac{-7.2}{0.9}$

40. $\dfrac{-2.5}{5}$

41. $\dfrac{-3}{-0.3}$

42. $\dfrac{-1.8}{-0.6}$

Divide and round to the nearest hundredth.

43. $(-15.5484) \div (-6.13)$

44. $-6.4516 \div (-3.54)$

45. $-0.8385 \div (0.715)$

46. $0.3102 \div (-0.129)$

Simplify.

47. $-16 \div (-2)(-2)$

48. $-36 \div (-3)(-2)$

49. $(3 - 7) \div (-4)$

50. $(5 - 8) \div (-3)$

51. $\dfrac{2 + (-6)}{-2}$

52. $\dfrac{10 + (-4)}{3}$

53. $(4 - 6) \div (1 - 5)$

54. $(-15 - 3) \div (-2 - 4)$

55. $-56 \div 7 - 4 \cdot (-3)$

56. $32 \div (-8) + (-5) \cdot 6$

57. $(-4)\left(\dfrac{1}{2}\right) - 2 \div \left(-\dfrac{1}{8}\right)$

58. $(-6) \div \left(\dfrac{2}{3}\right) + (-10)\left(\dfrac{2}{5}\right)$

Mixed Practice

Divide.

59. $\left(-4\dfrac{1}{2}\right) \div 3$

60. $\dfrac{5}{6} \div \left(-\dfrac{3}{8}\right)$

61. $(-0.72) \div (-6)$

62. Divide and round to the nearest hundredth: $(-0.7882) \div (2.36)$

63. $-65 \div (-13)$

Solve.

64. Complete the table.

	a.	b.	c.	d.	e.
Number	8	$\dfrac{2}{3}$	$-\dfrac{1}{4}$	-6	$-1\dfrac{1}{3}$
Reciprocal					

Simplify.

65. $-12 \div (5 - 7)$

66. $\dfrac{4 - (-6)}{-2}$

Applications

B *Solve. Express your answer as a signed number.*

67. The following table shows the change from the previous year in the number of customers at a restaurant for each of five years:

Year	Change in the Number of Customers
2007	An increase of 5700
2008	No change
2009	A decrease of 2600
2010	A decrease of 900
2011	A decrease of 1200

For the five years, what was the average change per year in the number of customers?

68. The following bar graph shows the quarterly net income for the company American International Group, Inc. (AIG) for four consecutive quarters. (Note that a red number written in parentheses is negative.)

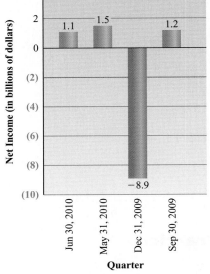

(*Source:* finance.yahoo.com)

To the nearest billion dollars, what was the average net income for a quarter?

69. The change in a stock market index over a 5-day period was −130. What was the average daily change?

70. Two investors bought an equal number of shares of stock in a company. The value of their stock fell by $7000. How much did each of the investors lose?

71. The population of a city decreased by 47,355 people in 10 years. Find the average annual change in population.

72. The federal deficit in 1940 was about $3 billion. Five years later at the end of World War II, it was about $48 billion. How many times the deficit of 1940 was that of 1945? (*Source: Budget of the United States Government, Fiscal Year 2000,* 1999)

73. A football running back lost a total of 24 yd in 6 plays. What was the average number of yards he lost on each play?

74. A client at a weight-loss clinic lost 20 lb in 15 weeks. What was the client's average weekly change in weight?

75. A small company's business expenses for the year totaled $72,000. What were the company's average monthly expenses?

76. Over a 5-year period, the height of a cliff eroded by 4.5 ft. By how many feet did the cliff erode per year?

77. A meteorologist predicted an average daily high temperature of −3°F for a five-day period. During this period, the daily high temperatures (in Fahrenheit degrees) were 2°, 0°, −7° −11°, and 1°. Was the meteorologist's prediction correct?

78. The value of a house decreased from $183,000 to $174,000 during a decade. What was the average amount of depreciation per year?

• Check your answers on page A-2.

MINDStretchers

Patterns

1. Find the missing numbers in the following:
   +1296, +648, −216, −108, +36, +18, −6, _____ , _____, _____

Groupwork

2. Consider the following five integers: −2, 6, −9, 18, −36. Working with a partner, explain which two of the five integers you would choose

   a. to find the smallest quotient.
   b. to find the largest quotient.

Writing

3. Explain the difference between the *opposite* of a number and the *reciprocal* of a number.

1.6 Algebraic Expressions, Translations, and Exponents

OBJECTIVES

A To find the number of terms in an algebraic expression

B To translate algebraic expressions to word phrases and word phrases to algebraic expressions

C To write algebraic expressions using exponents

D To solve applied problems involving algebraic expressions

Algebraic Expressions

In algebra, the word *variable* is used in two ways—as an unknown quantity or as a quantity that can change in value. We can use any letter or symbol to represent a variable. By contrast, a *constant* is a known quantity whose value does not change.

Suppose that each week, you spend $\frac{2}{5}$ of your time on campus working in the student activities area. If n represents the total number of hours that you spend during a particular week on campus, then the *algebraic expression* $\frac{2}{5} \cdot n$ stands for the amount of time you worked in the student activities area that week. So n is a variable, whereas $\frac{2}{5}$ is a constant.

An algebraic expression is an expression in which constants and variables are combined using standard arithmetic operations. When writing an algebraic expression involving a product, we usually omit any multiplication symbol. For instance, we would write $\frac{2}{5}n$ rather than $\frac{2}{5} \cdot n$.

Algebraic expressions consist of one or more *terms*, separated by addition signs. If there are subtraction signs, we can rewrite the expression in an equivalent form using addition. For instance, we can think of the algebraic expression

$$2x + \frac{y}{3} - 4 \quad \text{as} \quad 2x + \frac{y}{3} + (-4).$$

This algebraic expression is made up of three terms.

Terms
$$2x + \frac{y}{3} + (-4)$$

DEFINITION

A **term** is a number, a variable, or the product or quotient of numbers and variables.

EXAMPLE 1

Find the number of terms in each expression.

a. $3y + 1$

b. $\dfrac{a}{b}$

Solution

a. The expression $3y + 1$ has two terms.

b. The expression $\dfrac{a}{b}$ has one term.

PRACTICE 1

Determine how many terms are in each expression.

a. $2a + 3 - b$ **b.** $-4xy$

Translating Algebraic Expressions to Word Phrases and Word Phrases to Algebraic Expressions

In solving word problems, we may need to translate algebraic expressions to word phrases and vice versa. First, let's consider the many ways we can translate algebraic expressions to words.

$x + 5$ **translates to**
• x plus 5
• x increased by 5
• the sum of x and 5
• 5 more than x

$y - 4$ **translates to**
• y minus 4
• y decreased by 4
• the difference between y and 4
• 4 less than y

$\frac{2}{5}n, \frac{2}{5} \cdot n,$ **or** $\left(\frac{2}{5}\right)(n)$ **translates to**
• $\frac{2}{5}$ times n
• the product of $\frac{2}{5}$ and n
• $\frac{2}{5}$ of n

$z \div 3$ **or** $\frac{z}{3}$ **translates to**
• z divided by 3
• the quotient of z and 3
• the ratio of z and 3
• z over 3

Note that there are other possible translations as well.

EXAMPLE 2

Translate each algebraic expression to words.

Solution

Algebraic Expression	Translation
a. $5x$	5 times x
b. $y - (-2)$	the difference between y and -2
c. $-3 + z$	the sum of -3 and z
d. $\dfrac{m}{-4}$	m divided by -4
e. $\dfrac{3}{5}n$	$\dfrac{3}{5}$ of n

PRACTICE 2

Translate each algebraic expression to words.

Algebraic Expression	Translation
a. $\dfrac{1}{3}p$	
b. $9 - x$	
c. $s \div (-8)$	
d. $n + (-6)$	
e. $\dfrac{3}{8}m$	

Note that in Example 2 other translations are also correct.

EXAMPLE 3

Translate each algebraic expression to words.

Solution

Algebraic Expression	Translation
a. $2m + 5$	5 more than twice m
b. $1 - 3y$	the difference between 1 and $3y$
c. $4(x + y)$	4 times the sum of x and y
d. $\dfrac{a + b}{a - b}$	the sum of a and b divided by the difference between a and b

PRACTICE 3

Translate each algebraic expression to words.

Algebraic Expression	Translation
a. $2x - 3y$	
b. $4 + 3m$	
c. $5(a - b)$	
d. $\dfrac{r - s}{r + s}$	

Note that in Example 3(c) the sum $x + y$ is considered to be a single quantity, because it is enclosed in parentheses. Similarly in Example 3(d), the numerator $a + b$ and the denominator $a - b$ are each viewed as a single quantity.

In the previous examples, we discussed translating algebraic expressions to word phrases. Now, let's look at some examples of translating word phrases to algebraic expressions.

EXAMPLE 4

Translate each word phrase to an algebraic expression.

Solution

Word Phrase	Translation
a. twice x	$2x$
b. n decreased by -7	$n - (-7)$
c. the quotient of -6 and z	$(-6) \div z$ or $\dfrac{-6}{z}$
d. $\dfrac{1}{2}$ of n	$\dfrac{1}{2}n$
e. 10 more than y	$y + 10$

PRACTICE 4

Express each word phrase as an algebraic expression.

Word Phrase	Translation
a. $\dfrac{1}{6}$ of n	
b. n increased by -5	
c. the difference between m and -4	
d. the ratio of 100 and x	
e. the product of -2 and y	

EXAMPLE 5

Translate each word phrase to an algebraic expression.

Solution

Word Phrase	Translation
a. the difference between x and the product of 3 and y	$x - 3y$
b. 6 more than 4 times z	$4z + 6$
c. 10 times the quantity r minus s	$10(r - s)$
d. twice q divided by the sum of p and q	$\dfrac{2q}{p + q}$

PRACTICE 5

Express each word phrase as an algebraic expression.

Word Phrase	Translation
a. the sum of m and $-n$	
b. 11 less than the product of 5 and y	
c. the sum of m and n divided by the product of m and n	
d. negative 6 times the sum of x and y	

EXAMPLE 6

If a polygon has n sides, the sum of the measures of its interior angles, in degrees, is 180 times the quantity n minus 2. Write this expression symbolically.

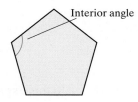
Interior angle

Solution   We translate the word phrase into an algebraic expression as follows:

$$\underset{180}{\underset{\downarrow}{180}} \quad \underset{\cdot}{\underset{\downarrow}{\text{times}}} \quad \underset{(n - 2)}{\underset{\downarrow}{\text{the quantity } n \text{ minus 2}}}$$

So the algebraic expression is $180(n - 2)$, which is in degrees.

PRACTICE 6

To find the area A of a trapezoid, we calculate $\dfrac{1}{2}$ the product of the height h and the sum of the upper base b and the lower base B. Write a formula that expresses this relationship.

Exponents

As discussed in Chapter R, we can use *exponential notation* as a shorthand method for representing repeated multiplication of the same factor. For instance, we can write $5 \cdot 5 \cdot 5 \cdot 5$ in exponential notation as

$$5^4 \quad \longleftarrow \text{ Exponent}$$
$$\uparrow$$
$$\rule{1cm}{0.4pt} \text{ Base}$$

This expression is read "5 to the fourth power."

> **DEFINITION**
>
> For any real number x and any positive integer a,
>
> $$x^a = \underbrace{x \cdot x \cdots x \cdot x,}_{a \text{ factors}}$$
>
> where x is called the **base** and a is called the **exponent** (or **power**).

In exponential notation, the exponent indicates how many times the base is used as a factor.

The expression x^a is read "x to the ath power," or "x to the power a." However, the exponents 2 and 3 are usually read in a special way. For instance, we generally read 5^2 as "5 *squared*" rather than "5 to the second power." Similarly, we read 5^3 as "5 *cubed*" instead of "5 to the third power."

A number raised to the power 1 is that number. For example, $5^1 = 5$ and $x^1 = x$.

In Chapter R, we evaluated 2^4. Now, let's consider the expression $(-2)^4$. To evaluate this expression, we multiply 4 factors of -2:

$$(-2)^4 = \underbrace{(-2)(-2)(-2)(-2)}_{4 \text{ factors of } -2}$$

$$(-2)^4 = \underbrace{(-2)(-2)(-2)(-2)}_{}$$
$$= \underbrace{4(-2)(-2)}_{}$$
$$= \underbrace{(-8)(-2)}_{}$$
$$= 16$$

In short, $(-2)^4 = 16$.

Next, let's consider the expression -2^4. To evaluate this expression, we multiply 4 factors of 2. Then, we take the opposite:

$$-2^4 = -\underbrace{(2)(2)(2)(2)}_{4 \text{ factors of } 2}$$
$$= -16$$

Note that in the expression $(-2)^4$ the base is -2, whereas in -2^4 the base is 2.

EXAMPLE 7

Evaluate.

a. $(-3)^4$ **b.** $-3^4(-2)^2$

Solution

a. $(-3)^4 = (-3)(-3)(-3)(-3) = 81$

b. $-3^4(-2)^2 = -(3)(3)(3)(3)(-2)(-2)$
$$= -(81)(4)$$
$$= -324$$

PRACTICE 7

Evaluate.

a. -6^2

b. $(-6)^2(-3)^2$

Sometimes we put an expression into exponential form. Such expressions may involve more than one base. For instance, the expression $(-4)(-4)(-4)(-3)(-3)$ can be rewritten in terms of powers of -4 and -3:

$$\underbrace{(-4)(-4)(-4)}_{\substack{3 \text{ factors} \\ \text{of } -4}}\underbrace{(-3)(-3)}_{\substack{2 \text{ factors} \\ \text{of } -3}} = (-4)^3(-3)^2$$

Consider the following examples.

EXAMPLE 8

Express in exponential form.

a. $(6)(6)(-10)(-10)(-10)(-10)$

b. $(4)(4)(-3)(-3)(4)$

Solution

a. $\underbrace{(6)(6)}_{\substack{2 \text{ factors} \\ \text{of } 6}}\underbrace{(-10)(-10)(-10)(-10)}_{4 \text{ factors of } -10} = 6^2(-10)^4$

b. $(4)(4)(-3)(-3)(4) = (4)(4)(4)(-3)(-3) = 4^3(-3)^2$

PRACTICE 8

Write using exponents.

a. $(2)(2)(2)(2)(-5)(-5)$

b. $(-6)(-6)(8)(8)(-6)(-6)$

EXAMPLE 9

Rewrite each expression using exponents.

a. $-2n \cdot n$ **b.** $-3x \cdot x \cdot y \cdot y \cdot y \cdot y$

Solution

a. $-2n \cdot n = -2n^2$

b. $-3x \cdot x \cdot y \cdot y \cdot y \cdot y = -3x^2y^4$

PRACTICE 9

Express in exponential form.

a. $-x \cdot x \cdot x \cdot x \cdot x$

b. $2m \cdot m \cdot m \cdot n \cdot n \cdot n \cdot n$

Can you explain the difference between the expressions $2n$ and n^2?

EXAMPLE 10

The population of a small town doubles every 5 yr. If the town's population started with n people, what is its population after 20 yr?

Solution   We know that the town's population started with n people and doubles every 5 yr. To find the population after 20 yr, consider the following table.

Time Passed (in yr)	Number of 5-yr Time Periods	Population of the Town
5	1	$2^1 \cdot n$, or $2n$
10	2	$2^2 \cdot n$, or $4n$
15	3	$2^3 \cdot n$, or $8n$
20	4	$2^4 \cdot n$, or $16n$

↑
Each time period, the population is doubled, that is, multiplied by 2.

So the population after 20 yr is $2^4 n$ or $16n$.

PRACTICE 10

A bacteriologist observes that the population of a bacteria growing in a petri dish triples in size every 2 hr. If x cells were present in the initial population, what was the population after 10 hr? Write the answer using exponents.

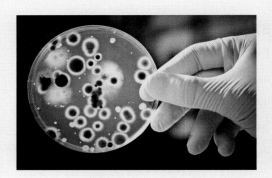

Mathematically Speaking

Fill in each blank with the most appropriate term or phrase from the given list.

base	the ratio of 7 and x	a to the xth power	the quotient of x and 7
7 decreased by x	variable	algebraic expression	
constant	7 less than twice x	the difference between x and 7	x to the ath power
2 times the difference between x and 7	exponent		

1. In algebra, a(n) _____ can be used as an unknown quantity.

2. A(n) _____ is a known quantity whose value does not change.

3. A(n) _____ consists of one or more terms, separated by addition signs.

4. The algebraic expression $x - 7$ can be translated as _____.

5. The algebraic expression $\frac{x}{7}$ can be translated as _____.

6. The word phrase _____ can be translated as $2x - 7$.

7. In the expression x^a, x is called the _____.

8. The expression x^a can be read as _____.

A *Determine the number of terms in each expression.*

9. $-5x$

10. $-7y$

11. $a + b$

12. $10 + 2y$

13. $xy + \frac{x}{y} - z$

14. $x - y + z$

B *Translate each algebraic expression to a word phrase.*

15. $3 + t$

16. $r + 2$

17. $x - 4$

18. $y - 10$

19. $7r$

20. $-4x$

21. $\frac{a}{4}$

22. $x \div 3$

23. $\frac{4}{5}w$

24. $\frac{1}{2}y$

25. $-3 + z$

26. $-5 + m$

27. $2n + 1$

28. $3c + 4$

29. $4(x - y)$

30. $2(m - n)$

31. $1 - 3x$

32. $2 - 5x$

33. $\frac{ab}{a + b}$

34. $\frac{x + y}{x - y}$

35. $2x - 5y$

36. $4a - 7b$

B *Translate each word phrase to an algebraic expression.*

37. 5 more than x

38. 10 more than y

39. 4 less than d

40. 12 less than n

41. the product of -6 and a

42. the product of -7 and x

43. the sum of y and -15

44. the sum of 2 and z

45. $\frac{1}{8}$ of k

46. $\frac{1}{2}$ of m

47. the quotient of m and n

48. n divided by y

49. the difference between a and twice b

50. the difference between three times x and 10

51. 5 more than 4 times z

52. 8 more than 3 times m

53. 12 times the quantity x minus y

54. Two times the quantity m minus n

55. b divided by the difference between a and b

56. x times y divided by the quantity x minus y

Evaluate.

57. -3^2

58. $(-4)^2$

59. $(-3)^3(-4)^2$

60. $-4^2(-3)^2$

C *Write using exponents.*

61. $(-2)(-2)(-2)(4)(4)$

62. $(-5)(-5)(-5)(-5)(2)(2)$

63. $(6)(6)(-3)(-3)(-3)$

64. $(-2)(-2)(4)(4)(4)(4)$

65. $(2)(-1)(-1)(2)(2)(2)$

66. $(-10)(3)(3)(3)(3)(-10)$

67. $3(n)(n)(n)$

68. $2(x)(x)(x)(x)$

69. $-4a \cdot a \cdot a \cdot b \cdot b$

70. $5r \cdot s \cdot s \cdot s \cdot s$

71. $-y \cdot y \cdot y$

72. $(-y)(-y)(-y)$

73. $10a \cdot a \cdot a \cdot b \cdot b \cdot c$

74. $-5x \cdot y \cdot y \cdot z$

75. $-x \cdot x \cdot y \cdot y \cdot y$

76. $(-x)(-x)(-x)(-y)(-y)$

Mixed Practice

Translate each algebraic expression to a word phrase.

77. $8(w - y)$

78. $-5m + 3$

79. $\dfrac{sr}{r - s}$

Write using exponents.

80. $-3p \cdot p \cdot p \cdot q$

81. $a \cdot a(-b)(-b)$

Solve.

82. Determine the number of terms in the expression $a - \dfrac{b}{2}$.

83. Evaluate: $-3^2(-2)^2$

Translate each word phrase to an algebraic expression.

84. $\frac{2}{3}$ of n

85. The sum of x and twice y

86. The sum of x and -4

Applications

D *Solve.*

87. In the triangle shown, write an expression for the sum of the measures of the three angles.

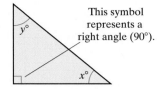

This symbol represents a right angle (90°).

88. Write an expression for the sum of the lengths of the sides in the figure shown.

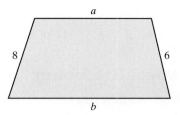

89. Suppose that *p* partners share equally in the profits of an e-business. What is each partner's share, if the profits were $30,000?

90. An investment of *a* dollars doubles in value. What is the new value of this investment?

91. A plane ticket costs *t* dollars before taxes. If the taxes on the plane ticket were *x* dollars, what was the total cost of the ticket?

92. A majority vote needs more than half the voters to accept a proposition. If there are 2*n* voters, how many voters need to accept the proposition for a majority vote?

93. An initial investment of 5000 dollars doubles every 10 years. Write in exponential form the value of the investment after 30 years.

94. A colony of bacteria *E. coli* doubles in size every 20 min when grown in a rich medium. If the colony started with *m* bacteria, how many bacteria were in the colony 2 hr later? Express the answer using exponents. (*Source:* eb.com)

95. The area of a square can be found by squaring the length of a side. Write an expression in exponential form to represent the area of the square shown.

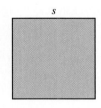

96. The volume of a cube can be found by cubing the length of an edge. Write an expression in exponential form to represent the volume of this cube.

97. A company buys a copier for $10,000. After *n* years, the Internal Revenue Service values the copier at $10,000 times $\frac{1}{20}$ times the quantity 20 minus *n*. Write this value as an algebraic expression.

98. In a math lab, *a* tutors were each assisting *b* students, and *c* tutors were each assisting *d* students. There were an additional *e* students in the lab. How many students were there in the lab altogether?

99. An area rug is placed on the wood floor shown. Find the area of the floor not covered by the rug.

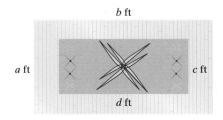

100. In a baseball game, a team scored 2 runs in each of *x* innings and 1 run in each of *y* innings. How many runs did the team score?

• Check your answers on page A-2.

MINDStretchers

Writing

1. Algebra is used in all countries of the world regardless of the language spoken. If you know how to speak a language other than English, translate each of the following algebraic expressions to that language as well as to English.

   a. $\dfrac{x}{2}$ **b.** $2 - x$ **c.** $6 + x$ **d.** $3x$

Mathematical Reasoning

2. Can there be two different numbers a and b for which $a^b = b^a$? Justify your answer.

Groupwork

3. The *algebra tiles* shown below represent the expressions $3x - 2$ and $x^2 + 1$, respectively.

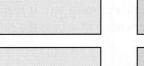   

   Working with a partner, represent each of the following expressions by algebra tiles.

   a. $x - 4$ **b.** $2x + 3$ **c.** $x^2 + x + 1$

1.7 Evaluating Algebraic Expressions and Formulas

In the previous section, we discussed translations involving algebraic expressions. For these expressions to be useful, however, we need to be able to *evaluate* them.

A To evaluate an algebraic expression

Consider the following example. Suppose that you had a temporary job that lasted 75 days. If you were absent from work for d days, then it follows that you were at work for $(75 - d)$ days. To evaluate the expression $75 - d$ for a particular value of d, replace d with that number. For instance, if you were not at work for 4 days, replace d in this expression with 4:

B To evaluate a formula

C To solve applied problems involving algebraic expressions or formulas

$$75 - d = 75 - 4 = 71$$

Replace d with 4.

We can conclude that you were at work 71 days.

The following method is helpful for evaluating expressions:

To Evaluate an Algebraic Expression

- Replace each variable with the given number.

- Carry out the computation using the order of operations rule.

EXAMPLE 1

Evaluate each algebraic expression.

Solution

Algebraic Expression	Value
a. $9 - z$, if $z = -2$	$9 - z = 9 - (-2) = 9 + 2 = 11$
b. $-2cd$, if $c = -1$ and $d = 2$	$-2cd = -2(-1)(2) = 4$

PRACTICE 1

Find the value of each algebraic expression.

Algebraic Expression	Value
a. $25 + m$, if $m = -10$	
b. $-3xy$, if $x = -2$ and $y = 5$	

In evaluating some algebraic expressions, we need to use the order of operations rule.

EXAMPLE 2

Find the value of each expression for $x = -3$, $y = 2$, $w = -5$, and $z = 4$.

a. $2x + 5y$ **b.** $3(w + z)$ **c.** $2w^2 - 3y^3$ **d.** $x^2 - 5x + 1$

Solution

a. $2x + 5y = 2(-3) + 5(2)$ Replace x with -3 and y with 2. Then multiply.
$\quad\quad\quad = -6 + 10$ Add.
$\quad\quad\quad = 4$

b. $3(w + z) = 3(-5 + 4) = 3(-1) = -3$

c. $2w^2 - 3y^3 = 2(-5)^2 - 3(2)^3 = 2(25) - 3(8) = 50 - 24 = 26$

d. $x^2 - 5x + 1 = (-3)^2 - 5(-3) + 1$
$\quad\quad\quad\quad\quad = 9 + 15 + 1$
$\quad\quad\quad\quad\quad = 25$

PRACTICE 2

Evaluate each expression for $a = 2$, $b = -3$, $c = -4$, and $d = 5$.

a. $5a - 2c$

b. $2(d - b)$

c. $2a^3 + 4b^2$

d. $c^2 + 4c - 2$

EXAMPLE 3

Evaluate each expression if $a = 3$, $b = 4$, and $c = -2$.

a. $\dfrac{c}{5 - a}$ **b.** $\dfrac{b + c}{b - c}$ **c.** $(-b)^2$ **d.** $-b^2$

Solution

a. $\dfrac{c}{5 - a} = \dfrac{-2}{5 - 3} = \dfrac{-2}{2} = -1$

b. $\dfrac{b + c}{b - c} = \dfrac{4 + (-2)}{4 - (-2)} = \dfrac{2}{6} = \dfrac{1}{3}$

c. $(-b)^2 = (-4)^2 = (-4)(-4) = 16$

d. $-b^2 = -4^2 = -(4)(4) = -16$

PRACTICE 3

Find the value of each expression when $x = -5$, $y = -3$, and $z = 1$.

a. $\dfrac{x - 2z}{y}$ **b.** $\dfrac{x - z}{x + y}$

c. $(-y)^4$ **d.** $-y^4$

Consider the expressions in Examples 3(c) and 3(d). Explain why they are not equal.

EXAMPLE 4

Physicists have shown that if an object is shot straight upward at a speed of 100 ft/sec, its location after t sec will be

$$-16t^2 + 100t$$

feet above the point of release. How far above or below the point of release will the object be after 5 sec?

Solution   To determine the position of the object after 5 sec, we must substitute 5 for t in the expression $-16t^2 + 100t$.

$$\begin{aligned} -16t^2 + 100t &= -16(5)^2 + 100(5) \\ &= -16(25) + 100(5) \\ &= -400 + 500 \\ &= 100 \end{aligned}$$

So after 5 sec, the object will be 100 ft above the point of release. Note that a position above the point of release is positive, whereas a position below the point of release is negative.

PRACTICE 4

The U.S. per capita consumption of bottled water (in gallons) can be approximated by the expression $2.0x + 15.7$, where x represents the number of years since 2003. Estimate the per capita consumption in 2015. (*Sources: Statistical Abstract of the United States, 2010* and www.census.gov)

Formulas

A *formula* is an equation that indicates how variables are related to one another. Some formulas express geometric relationships; others express physical laws. Just as with algebraic expressions, the letters and mathematical symbols in a formula represent numbers and words.

EXAMPLE 5

To predict the temperature T at a particular altitude a, meteorologists subtract $\frac{1}{200}$ of the altitude from the temperature g on the ground. Here, T and g are in degrees Fahrenheit and a is in feet. Translate this rule to a formula.

Solution Stating the rule briefly in words, the temperature at a particular altitude equals the difference between the temperature on the ground and $\frac{1}{200}$ times the altitude. Now, we translate this rule to mathematical symbols.

$$T = g - \frac{1}{200}a,$$

which is the desired formula.

PRACTICE 5

To convert a temperature C expressed in Celsius degrees to the temperature F expressed in Fahrenheit degrees, we multiply the Celsius temperature by $\frac{9}{5}$ and then add 32. Write a formula that expresses this relationship.

The method of evaluating formulas is similar to that of evaluating algebraic expressions. We substitute all the given numbers for the variables and then carry out the computations using the order of operations rule.

EXAMPLE 6

The formula for finding simple interest is $I = Prt$, where I is the interest in dollars, P is the principal (the amount invested) in dollars, r is the annual rate of interest, and t is the time in years that the principal has been on deposit. Find the amount of interest on a principal of $3000 that has been on deposit for 2 years at a 6% annual rate of interest.

Solution We know that $P = 3000$, $r = 6\%$, and $t = 2$. Converting 6% to its decimal form, we get 0.06. Substituting into the formula gives us:

$$I = Prt$$
$$= 3000(0.06)(2)$$
$$= 360$$

So the interest earned is $360.

PRACTICE 6

Given the distance formula $d = rt$, find the value of d if the rate r is 50 mph and the time t is 1.6 hr.

EXAMPLE 7

The markup M on an item is its selling price S minus its cost C.
a. Express this relationship as a formula.
b. If a digital camera cost a retailer $399.95 and was then sold for $559, how much was the markup on the camera?

Solution
a. We write the formula $M = S - C$.
b. To find the markup, we substitute for S and C.
$$M = S - C$$
$$= 559 - 399.95$$
$$= 159.05$$
So the markup was $159.05.

PRACTICE 7

Kelvin and Celsius temperature scales are commonly used in science. To convert a temperature expressed in Celsius degrees C to degrees Kelvin, K, add 273 to the Celsius temperature.
a. Write this relationship as a formula.
b. Suppose that in a chemistry experiment, C equals -6. What is the value of K?

A *Evaluate each algebraic expression, if $a = 4$, $b = 3$, and $c = -2$.*

1. $-5 + b$
2. $-6 + a$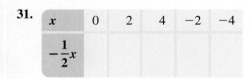
3. $-2ac$
4. $-4cb$

5. $-2a^2$
6. $-4b^2$
7. $2a - 15$
8. $3a - 18$

9. $a + 2c$
10. $4b + 3a$
11. $2(a - c)$
12. $4(a - b)$

13. $-a + b^2$
14. $-b + c^2$
15. $3a^2 - c^3$
16. $c^3 - 2a^2$

17. $\dfrac{a + b}{b - a}$
18. $\dfrac{a - c}{c + a}$
19. $\dfrac{3}{5}(a + b + c)^2$
20. $\dfrac{5}{9}(a - b - c)^2$

Evaluate each algebraic expression if $w = -0.5$, $x = 2$, $y = -3$, and $z = 1.5$.

21. $2w^2 - 3x + y - 4z$
22. $2x - 3w - y^2 + z$
23. $w - 7z - \dfrac{1}{4}(x - 6y)$

24. $5x - \dfrac{2}{5}(8w + 2y) - 4z$
25. $\dfrac{-10xy}{(w - z)^2}$
26. $\dfrac{-(2z - y)^2}{9wx}$

Complete each table.

27.

x	0	1	2	-1	-2
$2x + 5$					

28.

x	0	1	2	-1	-2
$3x + 1$					

29.

y	0	1	2	3	4
$y - 0.5$					

30.

y	0	1	2	3	4
$y - 2.8$					

31.

x	0	2	4	-2	-4
$-\dfrac{1}{2}x$					

32.

x	0	5	10	-5	-10
$-\dfrac{3}{5}x$					

33.

n	2	4	6	-2	-4
$\dfrac{n}{2}$					

34.

n	5	10	-5	-10	-15
$\dfrac{n}{5}$					

35.

g	0	1	2	-1	-2
$-g^2$					

36.

g	0	1	2	-1	-2
g^2					

37.

a	0	1	2	-1	-2
$a^2 + 2a - 2$					

38.

a	0	1	2	-1	-2
$-a^2 - 2a + 2$					

B *Evaluate each formula for the quantity requested.*

Formula	Given	Find
39. $C = \dfrac{5}{9}(F - 32)$	$F = -4°$	C
40. $C = \dfrac{5}{9}(F - 32)$	$F = -8°$	C
41. $A = P(1 + rt)$	$P = \$2000, r = 5\%$, and $t = 2$ yr	A
42. $A = P(1 + rt)$	$P = \$3000, r = 6\%$, and $t = 3$ yr	A
43. $P = 2l + 2w$	$l = 2\frac{1}{2}$ ft and $w = 1\frac{1}{4}$ ft	P
44. $P = 2l + 2w$	$l = 3\frac{1}{2}$ ft and $w = 2\frac{3}{4}$ ft	P
45. $A = \dfrac{a + b + c + d}{4}$	$a = -8, b = -6, c = 4$ and $d = -2$	A
46. $A = \dfrac{a + b + c + d}{4}$	$a = -10, b = -8, c = 6$, and $d = -4$	A
47. $C = \pi d$	$\pi \approx 3.14$ and $d = 100$ m	C
48. $C = \pi d$	$\pi \approx 3.14$ and $d = 120$ m	C
49. $C = A \cdot \dfrac{W}{150\,\text{lb}}$	$A = 100$ mg and $W = 30$ lb	C
50. $C = A \cdot \dfrac{W}{150\,\text{lb}}$	$A = 120$ mg and $W = 40$ lb	C
51. $A = 6e^2$	$e = 1.5$ cm	A
52. $A = 6e^2$	$e = 2.5$ cm	A
53. $V = \pi r^2 h$	$\pi = 3.14, r = 10$ in., and $h = 10$ in.	V
54. $V = \pi r^2 h$	$\pi = 3.14, r = 2$ in., and $h = 25$ in.	V

Mixed Practice

Solve.

55. Evaluate the algebraic expression if $w = 3, x = -1.5, y = -2$, and $z = 0.5$.

$$y^2 - 2z + \frac{1}{3}(2x - w)$$

Complete each table.

56.

x	0	8	12	−8	−12
$-\dfrac{3}{4}x$					

57.

x	0	1	2	−1	−2
$-2x + 4$					

Evaluate each algebraic expression if $a = 3$, $b = -4$, and $c = 2$.

58. $-5(c + b)$

59. $b^2 - 3a^2$

60. $\dfrac{c + a}{b - a}$

Evaluate each formula for the quantity requested.

Formula	Given	Find
61. $A = \dfrac{1}{2}bh$	$b = 7$ in. and $h = 3$ in.	A
62. $F = \dfrac{9}{5}C + 32$	$C = -10°$	F

Applications

C *Write each relationship as a formula.*

63. The average A of three numbers a, b, and c is the sum of the numbers divided by 3.

64. The perimeter P of a rectangle is twice the sum of the length l and the width w.

65. For every right triangle, the sum of the squares of the two legs a and b equals the square of the hypotenuse c.

66. The *aspect ratio* of a hang glider is a measure of how well it can glide and soar. The aspect ratio R is the square of the glider's wingspan s divided by the wing area A.

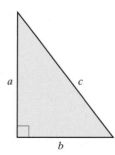

67. The equivalent energy E of a mass equals the product of the mass m and the square of the speed of light c.

68. The weight of an object depends on the gravitational pull of the planet or moon it is on. For instance, the weight E of an astronaut on Earth is 6 times the astronaut's weight m on the moon.

69. The length l of a certain spring in centimeters is 25 more than 0.4 times the weight w in grams of the object hanging from it.

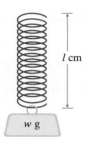

70. In electronics, when two resistors R_1 and R_2 are connected in parallel, the total resistance R between the points X and Y can be founded by dividing the product of the resistances by the sum of the resistances.

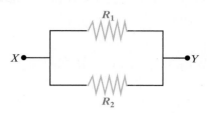

Solve.

71. The distance that a free-falling object drops, ignoring friction, is given by the formula

$$S = \frac{1}{2}gt^2,$$

where S is the distance in feet, g is the acceleration due to gravity, and t is time in seconds. Find the distance that an object falls if $g = 32$ ft/sec^2 and $t = 2$ sec.

72. The volume of a right circular cylinder is given by the formula

$$V = \pi r^2 h,$$

where r is the radius of the base of the cylinder and h is the height of the cylinder. For the cylinder shown, find the volume in cubic centimeters. Use 3.14 for π.

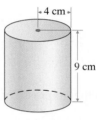

73. The percent markup m of an item based on cost is equal to 100 times the difference of the selling price s of the item and its cost c, divided by the cost.

a. Express this relationship as a formula.

b. If the selling price of a bottle of vitamin E is $8.75 and the cost is $6.25, what is the percent markup on the vitamin E?

74. To calculate the speed f of an object in feet per second, we multiply its speed m in miles per hour by $\frac{22}{15}$.

a. Write this relationship as a formula.

b. If an object is moving at 60 mph, what is its speed in feet per second?

75. The total calories C in a piece of food is the sum of its protein, carbohydrate, and fat calories. A particular piece of food has p grams of protein, f grams of fat, and c grams of carbohydrates. A gram of fat contains 9 calories, and there are 4 calories per gram of carbohydrate or protein. (*Source:* Duyff, *American Dietetic Association Complete Food and Nutrition Guide*)

a. Express C in terms of p, f, and c.

b. How many calories are there in 5 grams of fat, 0 grams of carbohydrates, and 2 grams of protein?

76. A simple approximation of the amount A of water in a human body is 60% of the total weight w.

a. Write this relationship as a formula.

b. If a person weighs 145 lb, approximately how much of the person's weight is water?

• Check your answers on page A-2.

MINDStretchers

Mathematical Reasoning

1. Complete the first three rows of the following table. Make up your own values of a and b in the 4th and 5th rows and complete the table.

a	b	$a + b$	$\lvert a \rvert$	$\lvert b \rvert$	$\lvert a \rvert + \lvert b \rvert$	$\lvert a + b \rvert$	T or F, $\lvert a \rvert + \lvert b \rvert = \lvert a + b \rvert$	T or F, $\lvert a \rvert + \lvert b \rvert \geq \lvert a + b \rvert$
-3	4							
1	-2							
-5	-4							

What general relationship do these examples suggest? Explain why this relationship is true.

Patterns

2. Consider the following table.

Odd number, n	1	3	5	7	...	1999
Counting number, C	1	2	3	4	...	1000

Write a formula that expresses C in terms of n.

Groupwork

3. Some expressions that appear to be different are, in fact, equivalent. Confirm that the expressions $\dfrac{x^2 - 9}{x + 3}$ and $x - 3$ are equal for various values of x.

1.8 Simplifying Algebraic Expressions

In an algebraic expression, the numerical factor of each variable term is said to be its *numerical coefficient* or, simply, its *coefficient*. For instance, the term $3y$ has coefficient 3 and the term $-4y$ has coefficient -4. Note that the coefficient of y is 1 because $y = 1y$. Likewise, the coefficient of $-y$ is -1 because $-y = -1y$.

The terms $3y$, $-4y$, and y are said to be *like* terms because they all contain the variable y raised to the power 1. By contrast, the terms $3x$ and $3x^2$ are *unlike* terms because the variable x is raised to different powers. Similarly, the terms $3x^2$ and $5y^2$ are unlike because they involve different variables.

DEFINITION

Like terms are terms that have the same variables with the same exponents. Terms that are not like are called **unlike terms.**

EXAMPLE 1

For each algebraic expression, identify and name the coefficient of the terms. Then, indicate whether the terms are like or unlike.

a. $3a + 5b$ **b.** $n - 6n$

c. $4x^2 + 2x$ **d.** $2a^2b - 3a^2b$

Solution

a. $3a + 5b$ has terms $3a$ and $5b$; unlike
(coefficient; different variables)

b. $n - 6n$ has terms n and $-6n$; like
(coefficient; same variable)

c. $4x^2 + 2x$ has terms $4x^2$ and $2x$; unlike
(coefficient; same variable but different exponents)

d. $2a^2b - 3a^2b$ has terms $2a^2b$ and $-3a^2b$; like
(coefficient; same variables with same exponents)

PRACTICE 1

Identify and name the coefficients of the terms of each algebraic expression. Then, state whether the terms are like or unlike.

a. $m - 3m$

b. $5x + 7$

c. $2x^2y - 3xy^2$

d. $m + 2m - 4m$

Combining Like Terms

An important property used to simplify algebraic expressions is the *distributive property*.

The Distributive Property

For any real numbers a, b, and c,

$$a \cdot (b + c) = a \cdot b + a \cdot c.$$

This property states that a number times the sum of two quantities is equal to the number times one quantity plus the number times the other quantity.

Another way to express the distributive property is $(b + c)a = ba + ca$. Can you explain why?

The distributive property also holds for $a(b - c)$. Since $b - c = b + (-c)$, it follows that $a(b - c) = ab - ac$ and $(b - c)a = ba - ca$.

EXAMPLE 2

Use the distributive property.

a. $2(x + y)$ **b.** $(3 + 7) \cdot n$

c. $(-5)(a - 2b)$ **d.** $0.6(x - 5)$

Solution

a. $2(x + y) = 2 \cdot x + 2 \cdot y = 2x + 2y$

b. $(3 + 7) \cdot n = 3 \cdot n + 7 \cdot n = 3n + 7n$

c. $(-5)(a - 2b) = (-5) \cdot a + (-5) \cdot (-2b) = -5a + 10b$

d. $0.6(x - 5) = 0.6(x) - 0.6(5) = 0.6x - 3$

PRACTICE 2

Rewrite using the distributive property.

a. $(-10)(4r + s)$

b. $(5 + 1) \cdot w$

c. $3(g - 3h)$

d. $1.5(y + 2)$

We can use the distributive property to simplify algebraic expressions involving like terms. Adding or subtracting like terms using the distributive property is called *combining like terms*. Unlike terms, such as $4x^2$ and $2x$, cannot be combined.

EXAMPLE 3

Simplify.

a. $2x + 6x$ **b.** $a - 8a$

c. $6y - y + 5$ **d.** $3b + 2b - 5b$

Solution

a. $2x + 6x = (2 + 6)x$ Use the distributive property.

$= 8x$ Add 2 and 6.

b. $a - 8a = (1 - 8)a$ Recall that the coefficient of a is 1.

$= -7a$ Use the distributive property.

c. $6y - y + 5 = (6 - 1)y + 5$ Recall that the coefficient of $-y$ is -1. Use the distributive property.

$= 5y + 5$

d. $3b + 2b - 5b = (3 + 2 - 5)b$ Use the distributive property.

$= 0 \cdot b$ Recall that $0 \cdot b = 0$.

$= 0$

PRACTICE 3

Simplify.

a. $5x + x$

b. $-5y - y$

c. $a - 3a + b$

d. $-9t + 3t + 6t$

EXAMPLE 4

Combine like terms if possible.

a. $2x^2 + x$ **b.** $8m^2n^2 + 2m^2n^2$ **c.** $-6a^2b + 5a^2b$

Solution

a. $2x^2 + x$ This algebraic expression cannot be simplified because the terms are unlike.

Unlike terms

b. $8m^2n^2 + 2m^2n^2 = (8 + 2)m^2n^2 = 10m^2n^2$

Like terms

c. $-6a^2b + 5a^2b = (-6 + 5)a^2b = -1a^2b = -a^2b$

Like terms

PRACTICE 4

Combine like terms if possible.

a. $y^2 - 3y^2$

b. $7a^2b + 3ab^2$

c. $4xy^2 - xy^2$

Simplifying Algebraic Expressions Involving Parentheses

Some algebraic expressions involve parentheses. We can use the distributive property to remove the parentheses in order to simplify these expressions.

EXAMPLE 5

Simplify: $\dfrac{1}{2}(x + 6) - 4$

Solution

$$\dfrac{1}{2}(x + 6) - 4 = \dfrac{1}{2}x + 3 - 4 \quad \text{Use the distributive property.}$$

$$= \dfrac{1}{2}x - 1 \quad \text{Combine like terms.}$$

PRACTICE 5

Simplify: $3(y - 4) + 2$

Let's consider simplifying algebraic expressions such as $-(4y - 6)$, in which a negative sign precedes an expression in parentheses.

EXAMPLE 6

Simplify: $-(4y - 6)$

Solution

$$-(4y - 6) = -1(4y - 6) \quad \text{The coefficient of } -(4y - 6) \text{ is } -1.$$

$$= -1 \cdot 4y + (-1)(-6) \quad \text{Use the distributive property.}$$

$$= -4y + 6$$

Because the terms in the expression $-4y + 6$ are unlike, it is not possible to simplify the expression further.

PRACTICE 6

Simplify: $-(2a - 3b)$

EXAMPLE 7

Simplify: $2x - 3 - (x + 4)$

Solution

$$2x - 3 - (x + 4) = 2x - 3 - 1(x + 4)$$ The coefficient of
$-(x + 4)$ is -1.

$$= 2x - 3 - x - 4$$ Use the distributive property.

$$= 2x - x - 3 - 4$$

$$= x - 7$$ Combine like terms.

PRACTICE 7

Simplify: $5y - 6 - (y - 5)$

TIP

• When removing parentheses preceded by a *minus sign*, all the terms in parentheses change to the opposite sign.
• When removing parentheses preceded by a *plus sign*, all the terms in parentheses keep the same sign.

EXAMPLE 8

Simplify: $(14a - 9) - 2(3a + 4)$

Solution

$$(14a - 9) - 2(3a + 4) = 14a - 9 - 6a - 8$$
$$= 8a - 17$$

PRACTICE 8

Simplify: $(y + 3) - 3(y + 7)$

Some algebraic expressions contain not only parentheses but also brackets. When simplifying expressions containing both types of grouping symbols, first remove the innermost grouping symbols by using the distributive property and then continue to work outward.

EXAMPLE 9

Simplify: $5 + 2[-4(x - 3) + 8x]$

Solution

$$5 + 2[-4(x - 3) + 8x] = 5 + 2[-4x + 12 + 8x]$$ Use the distributive property inside the brackets.

$$= 5 + 2[4x + 12]$$ Combine like terms inside the brackets.

$$= 5 + 8x + 24$$ Use the distributive property.

$$= 8x + 29$$ Combine like terms.

PRACTICE 9

Simplify: $10 - [4y + 3(2y - 1)]$

Now, we consider applied problems that involve simplifying expressions. To solve these problems, first we translate word phrases to algebraic expressions, and then simplify.

EXAMPLE 10

When a hospital is filled to capacity, it has p patients in private rooms and 20 more patients in semiprivate rooms than in private rooms. The daily rate for a patient in a private room is $300, and in a semiprivate room, the daily rate is $200. Write an algebraic expression to represent the total amount of money that the hospital takes in per day when all of the rooms are full. Then, simplify the expression.

Solution   We know that p represents the number of patients in private rooms. Since the hospital has 20 more patients in semiprivate rooms than in private rooms, $p + 20$ represents the number of patients in semiprivate rooms. A patient in a private room pays $300 per day, and a patient in a semiprivate room pays $200 per day.

Organizing the information into a table can help us to clarify the relationship between the key quantities in the problem.

Type of Room	Cost per Patient	Number of Patients in Rooms	Total Amount of Money
Private	$300	p	$300p$
Semiprivate	$200	$p + 20$	$200(p + 20)$

So when the hospital is full, the total amount of money that the hospital takes in per day is represented by the algebraic expression $300p + 200(p + 20)$.

Simplifying $300p + 200(p + 20)$, we get:

$$300p + 200(p + 20) = 300p + 200p + 4000$$
$$= 500p + 4000$$

So the total amount of money that the hospital receives per day is $(500p + 4000)$ dollars.

PRACTICE 10

For a concert, a local performing arts center sold c tickets for children and 40 fewer tickets for adults. Write an algebraic expression to represent the total income received by the center if the cost of a ticket is $12 for adults and $5 for children. Then, simplify the expression.

1.8   **Exercises**

FOR
EXTRA
HELP

MyMathLab   *Math XL*   WATCH   READ   REVIEW
     PRACTICE

Mathematically Speaking

Fill in each blank with the most appropriate term or phrase from the given list.

coefficient	unlike	negative
outermost	number	like
positive	innermost	
associative property of addition	distributive property	

1. The _____ of the term $-x$ is -1.

2. Terms that have the same variables with the same exponents are called _____ terms.

3. The _____ states that for any real numbers a, b, and c, $a \cdot (b + c) = a \cdot b + a \cdot c$.

4. We cannot combine _____ terms.

5. When removing parentheses preceded by a(n) _____ sign, change all the terms in parentheses to the opposite sign.

6. To simplify expressions containing brackets and parentheses, begin by removing the _____ grouping symbols.

Ⓐ *Name the coefficient in each term.*

7. $7x$

8. $100a$

9. $-5y$

10. $-18t$

11. ab

12. m

13. $-x^2$

14. $-xy^3$

15. $-0.1n$

16. $2.5s$

17. $\frac{2}{3}a^2b$

18. $-\frac{1}{4}mn$

19. $2x - 5y$

20. $-x + 10y$

Identify the terms, and indicate whether they are like or unlike.

21. $2a - a$

22. $10r + r$

23. $5p + 3$

24. $30 - 2A$

25. $4x^2 - 6x^2$

26. $x^2 + 7x^3$

27. $-20n - 3n$

28. $-15n - 8n$

Ⓑ *Use the distributive property.*

29. $-7(x - y)$

30. $-3(x - y)$

31. $(1 - 10) \cdot a$

32. $(3 - 12) \cdot x$

33. $-0.5(r + 3)$

34. $-0.4(p + 5)$

Ⓒ *Simplify, if possible, by combining like terms.*

35. $3x + 7x$

36. $8p + 3p$

37. $-10n - n$

38. $-y - 5y$

39. $20a - 10a + 4a$

40. $2r - 5r + r$

41. $3y - y + 2$

42. $7y - 2y + 1$

43. $8b^3 + b^3 - 9b^3$

44. $6x^2 + 2x^2 - 14x^2$

45. $-b^2 + ab^2$

46. $-3x^2 - 5x$

47. $3r^2t^2 + r^2t^2$

48. $m^2n^2 + 2m^2n^2$

49. $3x^2y - 5xy^2$

50. $10ab^2 + a^2b$

D *Simplify.*

51. $2(x + 3) - 4$

52. $3(a - 5) - 1$

53. $(7x + 1) + 2(2x - 1)$

54. $(3x - 4) + 2(2x - 18)$

55. $-(3y - 10)$

56. $-(2x + 5)$

57. $5x - 3 - (x + 6)$

58. $5x - 9 - (x + 4)$

59. $-4(n - 9) + 3(n + 1)$

60. $5(2b - 1) + 4(c + b)$

61. $x - 4 - 2(x - 1) + 3(2x + 1)$

62. $a + 1 - 3(a + 1) + 8(4a + 5)$

63. $7 + 3[x - 2(x - 1)]$

64. $2 - [n - 4(n + 5)]$

65. $10 - 3[4(a + 2) - 3a]$

66. $5 + 2[-(z + 7) + 4z]$

Mixed Practice

Simplify, if possible, by combining like terms.

67. $4pq^2 - 6q^2$

68. $12m + 2 - m$

Solve.

69. Identify the terms in the expression $3a + 3a^2$ and indicate whether they are like or unlike.

70. Use the distributive property. $-3(x - 2y)$

71. Name the coefficient in the term xy.

Simplify.

72. $2n - 7 - (3n + 2)$

73. $y - 3 - 4(y - 2) + 2(3y + 1)$

74. $(5a - 6) + 3(2a + 1)$

Applications

E *Write an algebraic expression. Then, simplify.*

75. What is the sum of the angles in the triangle shown?

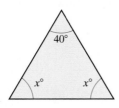

76. According to a will, an estate is to be divided among 2 children and 2 grandchildren. Each grandchild is to receive one-half the amount c that each child receives from the estate. What is the total value of the estate?

77. A baseball fan buys a ticket for a game for d dollars. Two of his friends decide to go to the game at the last minute and purchase tickets for $4 more per ticket. What is the total cost of the 3 tickets?

78. The shape of Colorado is approximately a rectangle. Its length is 100 mi more than its width, w. What is the approximate perimeter of Colorado?

79. If n is the first of 3 consecutive integers, what is the sum of the 3 consecutive integers? (*Hint*: Consecutive integers are integers that differ by 1 unit.)

80. On Tuesday, a company's stock fell by 6% from its value the day before. If v represents the value of the stock on Monday, what was its value on Tuesday?

81. There are 54 questions on the Mathematics section of the Scholastic Aptitude Test (SAT). The raw score for this portion of the test is found as follows: 1 point is awarded to a correct answer, 0 points are given if the question is left unanswered, and $\frac{1}{4}$ of a point is deducted for an incorrect answer. What is the Mathematics raw score of a student who answer c questions correctly and leaves b questions blank?   (*Source:* sat.collegeboard.com)

82. An condo building 100 apartments. Of these, y are 4-room apartments, and the others are 3-room apartments. How many rooms are there in the condo building?

• Check your answers on page A-3.

MIND Stretchers

Groupwork

1. Parentheses have different meanings in different situations. Working with a partner, explain the meaning of parentheses in each context.

   a. $(-2)(-3)$

   b. $5(x + 2)$

   c.

Income	$713,014
Expenditures	$961,882
Profit	($248,868)

   d. I am studying algebra (my favorite subject!).

Research

2. By searching the web or checking in your college's math learning center, investigate how *algebra tiles* can be used to combine like terms. Summarize your findings.

Writing

3. Explain the meaning of *combining like terms* in the following examples.

   a. 5 ft 3 in. +7 ft 6 in.

   b. $16\frac{2}{3} - 5\frac{1}{3}$

   c. $10x - 1 + 4 + 2x$

Key Concepts and Skills

CONCEPT   SKILL

Concept/Skill	Description	Example
[1.1] **Real numbers**	Numbers that can be represented as points on the number line.	$\frac{2}{3}$, $-\frac{1}{2}$, 5, 7.4, and $\sqrt{2}$
[1.1] **Integers**	The numbers $\ldots$, -4, -3, -2, -1, 0, $+1$, $+2$, $+3$, $+4$, $\ldots$ continuing indefinitely in both directions.	-20, -7, 0, $+5$, and $+17$
[1.1] **Rational numbers**	Numbers that can be written in the form $\frac{a}{b}$, where a and b are integers and $b \neq 0$.	-1.6, $\frac{2}{3}$, $-\frac{1}{2}$, 5, and 0.04
[1.1] **Opposites**	Two real numbers that are the same distance from 0 on a number line but on opposite sides. For any real number n, its opposite is $-n$.	3 and -3
[1.1] **Absolute value**	A given number's distance from 0 on a number line. The absolute value of a number n is written $\lvert n \rvert$.	$\lvert 8 \rvert = 8$ $\lvert -8 \rvert = 8$
[1.2] **Additive inverse property**	For any real number a, there is exactly one real number $-a$, such that $$a + (-a) = 0 \quad \text{and} \quad -a + a = 0,$$ where a and $-a$ are **additive inverses (opposites)** of each other.	$7 + (-7) = 0$ and $-7 + 7 = 0$
[1.2] **Additive identity property**	For any real number a, $$a + 0 = a \quad \text{and} \quad 0 + a = a.$$	$7 + 0 = 7$ and $0 + 7 = 7$
[1.2] **To add real numbers**	• If the numbers have the same sign, add the absolute values and keep the sign. • If the numbers have different signs, find the difference between the larger absolute value and the smaller absolute value, and take the sign of the number with the larger absolute value.	$-5 + (-2) = -7$ $+5 + (+2) = +7$, or 7 $-5 + (+2) = -3$ $+5 + (-2) = +3$, or 3
[1.2] **Commutative property of addition**	For any two real numbers a and b, $$a + b = b + a.$$	$6 + 2 = 2 + 6$
[1.2] **Associative property of addition**	For any three real numbers, a, b, and c, $$(a + b) + c = a + (b + c).$$	$(8 + 4) + 1 = 8 + (4 + 1)$
[1.3] **To subtract real numbers**	• Change the operation of subtraction to addition and change the number being subtracted to its opposite. • Follow the rule for adding real numbers.	$2 - (-5)$ $= 2 + (+5) = 7$
[1.4] **To multiply real numbers**	• Multiply their absolute values. • If the numbers have the same sign, their product is positive; if they have different signs, their product is negative.	$(-3)(-8) = 24$ $3(-8) = -24$
[1.4] **Commutative property of multiplication**	For any two real numbers a and b, $$a \cdot b = b \cdot a.$$	$(-2)(-7) = (-7)(-2)$
[1.4] **Multiplicative identity property**	For any real number a, $$a \cdot 1 = a \quad \text{and} \quad 1 \cdot a = a.$$	$3 \cdot 1 = 3$ and $1 \cdot 3 = 3$

continued

Concept/Skill	Description	Example
[1.4] **Multiplication property of zero**	For any real number a, $$a \cdot 0 = 0 \quad \text{and} \quad 0 \cdot a = 0.$$	$2 \cdot 0 = 0$ and $0 \cdot 2 = 0$
[1.4] **Associative property of multiplication**	For any three real numbers, a, b, and c, $$(a \cdot b)c = a(b \cdot c).$$	$(-1 \cdot 2)3 = -1(2 \cdot 3)$
[1.4] **Order of operations rule**	To evaluate mathematical expressions, carry out the operations *in the following order*: **1.** First, perform the operations within any grouping symbols, such as parentheses ( ) or brackets [ ]. **2.** Then, raise any number to its power. **3.** Next, perform all multiplications and divisions as they appear from left to right. **4.** Finally, do all additions and subtractions as they appear from left to right.	$2 + 3 \cdot 5 = 17$ $-3(8 - 4)^2 = -48$
[1.5] **To divide real numbers**	• Divide their absolute values. • If the numbers have the same sign, their quotient is positive; if they have different signs, their quotient is negative.	$-16 \div (-2) = 8$ $-24 \div 3 = -8$
[1.5] **Division involving zero**	For any nonzero real number a, $$0 \div a = 0.$$ For any nonzero real number a, $$a \div 0 \text{ is undefined.}$$	$0 \div 5 = 0$ and $5 \div 0$ is undefined.
[1.5] **Division of real numbers**	For any real numbers a and b where b is nonzero, $$a \div b = \frac{a}{b} = a \cdot \frac{1}{b}.$$	$-\dfrac{3}{5} \div 6 = -\dfrac{3}{5} \div \dfrac{6}{1}$ $= -\dfrac{\overset{1}{3}}{5} \cdot \dfrac{1}{\underset{2}{6}} = -\dfrac{1}{10}$
[1.5] **Multiplicative inverse property**	For any nonzero real number a, $$a \cdot \frac{1}{a} = 1 \text{ and } \frac{1}{a} \cdot a = 1,$$ where a and $\frac{1}{a}$ are **multiplicative inverses (reciprocals)** of each other.	$-\dfrac{3}{4}$ and $-\dfrac{4}{3}$ are multiplicative inverses because $-\dfrac{3}{4} \cdot \left(-\dfrac{4}{3}\right) = 1.$
[1.6] **Variable**	A letter that represents an unknown quantity or one that can change in value.	x, n
[1.6] **Constant**	A known quantity whose value does not change.	$5, -3.2$
[1.6] **Algebraic expression**	An expression in which constants and variables are combined using standard arithmetic operations.	$\dfrac{2}{5}n + 9$
[1.6] **Term**	A number, a variable, or the product or quotient of numbers and variables.	$-11, 3x, \dfrac{n}{4}$

Concept/Skill	Description	Example
[1.6] **Exponential notation**	For any real number x and any positive integer a, $$x^a = \underbrace{x \cdot x \cdots x \cdot x,}_{a \text{ factors}}$$ where x is called the **base** and a is called the **exponent** (or **power**).	Exponent $\downarrow$ $(-2)^3 = (-2)(-2)(-2)$ $\uparrow$ $= -8$ Base
[1.7] **To evaluate an algebraic expression**	• Replace each variable with the given number. • Carry out the computation using the order of operations rule.	If $c = -1$ and $d = 2$, then $-2cd = (-2)(-1)(2) = 4$.
[1.8] **Coefficient**	The numerical factor of a variable term.	Coefficient $\swarrow \quad \downarrow \quad \searrow$ $3x,\ y \text{ or } 1y,\ -2x^2$
[1.8] **Like terms**	Terms that have the same variable and the same exponent.	n and $6n$ $5x^3$ and $-2x^3$
[1.8] **Unlike terms**	Terms that are not like.	x and $3y$ a^2 and a
[1.8] **Distributive property**	For any real numbers a, b, and c, $$a \cdot (b + c) = a \cdot b + a \cdot c$$ and $$(b + c) \cdot a = b \cdot a + c \cdot a.$$	$2(x + y) = 2x + 2y$ and $(3 + 7) \cdot n = 3 \cdot n + 7 \cdot n$
[1.8] **To combine like terms**	• Use the distributive property. • Add or subtract.	$2x + 6x = (2 + 6)x$ $= 8x$

CHAPTER 1   Review Exercises

Say Why
Fill in each blank.

1. The decimal 0.5 _____ a rational number because
   $\overline{\text{is/is not}}$

   _____

   _____ .

2. The additive inverse of 17 _____ −17 because
   $\overline{\text{is/is not}}$

   _____ .

3. The expression $-\dfrac{a}{3}$ and $\dfrac{3}{a}$ _____ multiplicative
   $\overline{\text{are/are not}}$

   inverses because _____ .

4. The expression $-(6a)^2$ _____ equivalent to
   $\overline{\text{is/is not}}$

   $(-6a)(-6a)$ because _____

   _____ .

5. The number 125 _____ a power of 5 because
   $\overline{\text{is/is not}}$

   _____ .

6. The expression $15p^2q^3$ and $-9p^3q^2$ _____ like terms
   $\overline{\text{are/are not}}$

   because _____

   _____ .

[1.1]   *Express each quantity as a signed number.*

7. 3 mi above sea level

8. A withdrawal of $160 from an account

Graph each number on the number line.

9. −1

10. $2\frac{9}{10}$

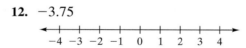

11. 0.5

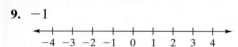

12. −3.75

Find the opposite of each number.

13. −4

14. 6.5

15. $\dfrac{2}{3}$

16. −0.7

Compute.

17. $|-4|$

18. $|0|$

19. $|2.6|$

20. $\left|-\dfrac{5}{9}\right|$

Indicate whether each inequality is true or false.

21. $-7 < -5$

22. $-1 > 3$

[1.2]   *Find the sum of each pair of numbers using the number line.*

23. $-4 + (-1)$

24. $3 + (-7)$

Name the property of addition illustrated by each statement.

25. $5 + (-6) = (-6) + 5$

26. $0 + 25 = 25$

27. $(1 + 3) + 6 = 1 + (3 + 6)$

28. $-10 + 10 = 0$

Find the sum.

29. $9 + (-9)$

30. $4 + (-2)$

31. $-3 + 5$

32. $0 + (-15)$

33. $-3 + (-2)$

34. $-3 + 7 + (-89)$

35. $-0.5 + (-3.6) + (-4)$

36. $-2 + 5.3 + 12$

[1.3] *Find the difference.*

37. $12 - 3$

38. $36 - 47$

39. $-52 - 3$

40. $2 - 5$

41. $-19 - 8$

42. $24 - (-3)$

43. $8 - (-8)$

44. $0 - 5$

45. $6 - 7.42$

46. $-9 - \left(-\dfrac{3}{8}\right)$

Combine.

47. $2 + (-4) - (-7)$

48. $-3 - (-1) + 12$

[1.4] *Name the property of multiplication illustrated by each statement.*

49. $-3(5) = 5(-3)$

50. $(-8 \cdot 6) \cdot 2 = -8 \cdot (6 \cdot 2)$

51. $-9 \cdot 1 = -9$

52. $-7 \cdot 0 = 0$

Find the product.

53. $2(-5)$

54. $-3 \cdot 7$

55. $-60 \cdot 90$

56. $-8(-300)$

57. $(-2.7)(-10)$

58. $\left(\dfrac{3}{4}\right)\left(-\dfrac{1}{3}\right)$

59. $5(-4)(-300)$

60. $(-1)(-12)(3)$

Simplify.

61. $-8 + 3(-2) - 9$

62. $3 - 2(-3) - (-5)$

63. $-9 - 5(-7)$

64. $20 - 3(-6)$

65. $-4(-2 + 5)$

66. $(-12 + 6)(-1)$

[1.5] *Find the reciprocal.*

67. $-\dfrac{2}{3}$

68. 8

Find the quotient. Simplify.

69. $-30 \div (-10)$

70. $6 \div (-1)$

71. $-\dfrac{11}{5}$

72. $\dfrac{4}{5} \div \left(-\dfrac{2}{3}\right)$

Simplify.

73. $-16 \div 2(-4)$

74. $(9 - 23) \div (-13 + 6)$

75. $\dfrac{3 + (-1)}{-2}$

76. $\dfrac{5(7 - 3)}{-8 - 2}$

77. $(-3) + 8 - 2 \cdot (-4)$

78. $10 \div (-2) + (-3) \cdot 5$

[1.6] *Determine the number of terms in each expression.*

79. $-x + y - 3z$

80. $\dfrac{m}{n} + 4$

81. $-9t$

82. $3a - 1 + 7b + c$

Translate each algebraic expression to a word phrase.

83. $-6 + w$

84. $-\dfrac{1}{3}x$

85. $-3n + 6$

86. $5(p - q)$

Translate each word phrase to an algebraic expression.

87. 10 less than x

88. $\dfrac{1}{2}$ of s

89. The quotient of p and q

90. The difference between R and twice V

91. 6 times the quantity 2 less than 4 times n

92. The quantity negative 4 times a divided by the quantity 5 times b plus c.

Write each using exponents.

93. $-3(-3)(-3)(-3)$

94. $-5(-5)(-5)(3)(3)$

95. $4(x)(x)(x)$

96. $-5a \cdot a \cdot b \cdot b \cdot b \cdot c$

[1.7] *Evaluate each algebraic expression if $a = 2$, $b = 5$, and $c = -1$.*

97. $30 + c$

98. $-\dfrac{4}{9}b$

99. $-5a^2$

100. $10(b - c)$

101. $\dfrac{1 - a}{c}$

102. $4a^2 - 4ab + b^2$

[1.8]

103. Apply the distributive property: $-5(x - y)$

104. Combine like terms: $4x + 10x - 2y$

105. Simplify: $3x^2 - x^2 - 4x^2$

106. Combine: $2r^2t^2 - r^2t^2$

107. Simplify: $2(a - 5) + 1$

108. Simplify: $-(3x + 2)$

109. Combine like terms: $-3x - 5 - (x + 10)$

110. Simplify: $(2a - 4) + 2(a - 5) - 3(a + 1)$

Mixed Applications

Solve.

111. Express as a signed number: a gain of $700

112. A company lost $7000 last year. Express this situation as a signed number.

113. The price of a share of stock fell by $0.50 each week for 4 weeks in a row. What was the overall change in the price?

114. In *exothermic* chemical reactions, the surrounding temperature rises; in *endothermic* chemical reactions, the surrounding temperature drops. If a chemical reaction causes the surrounding temperature to change from $-7°C$ to $-4°C$, was it exothermic or endothermic?

115. A patient's temperature drops by 0.5 degrees per hour. If the initial temperature was I degrees, write an algebraic expression for the temperature after h hours.

116. For each 100 people in the United States in a recent year, there were approximately 1.383 births, 0.838 deaths, and a net migration of $+0.425$ people. How many more people per 100 were there at the end of the year than at the beginning? Express as a percent. (*Source:* cia.gov)

117. The following graph shows five countries and their lowest point on land, measured in meters. The countries with negative altitudes have land below sea level.

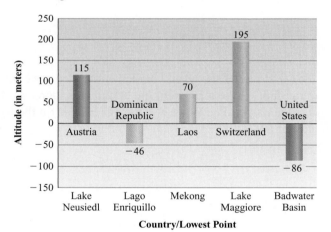

(*Source:* wikipedia.org)

What is the approximate difference in altitude between the country with the highest altitude and the one with the lowest altitude?

118. The bar graph shows the closing price of a share of General Electric Co. stock for a five-day period in November 2010.

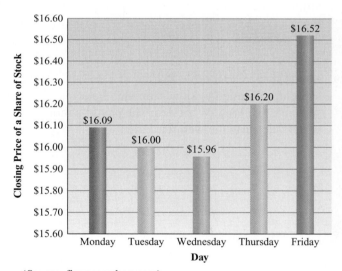

(*Source:* finance.yahoo.com)

Calculate the change in the closing price per share from the previous day for Tuesday through Friday, expressing each change as a signed number.

119. The length of a rectangle is double its width w. Find the rectangle's perimeter.

120. The apparent visual magnitude of a star is a measure of its brightness as viewed from the Earth. The Sun has an apparent visual magnitude of -26.73 and the full Moon, -12.6. About how many times as bright as the full Moon is the Sun? (*Source: The Top 10 of Everything 2010*)

121. A colony of bacteria triples in size every 6 hr. At one point, there are 10 bacteria in the colony. Write in exponential form the number of bacteria in the colony 18 hr later.

122. Firefighters use the formula $S = 0.5N + 26$ to compute the maximum horizontal range S in feet of water from a particular hose, where N is the hose's nozzle pressure in pounds. Calculate S if $N = 90$ lb. (*Source:* firedistrict7. com)

123. On June 1, 2009, the interest rate on a one-year Treasury bill (T-bill) was 0.48%. A year later, the rate dropped to 0.35%. How much less interest, in dollars, would be earned on a one-year T-bill starting at $\$P$ on June 1, 2010 than a one-year T-bill starting at $\$P$ on June 1, 2009? (*Source:* treas.gov)

124. Find the annual interest earned on an investment of $600 if x dollars is invested at an interest rate of 5% per year and the remainder is invested at a rate of 7% per year.

125. The playwright Sophocles was born in 496 B.C. and died approximately 90 yr later. In what year did he die?

126. One year, a company's loss was $60,000. The next year, it was only $20,000. How many times the second loss is the first loss?

127. A homeowner needs to have sufficient current to operate the electrical appliances in the home. Electricians use the formula $I = \dfrac{P}{E}$ to compute the current I in amperes needed in terms of the power P in watts and the energy E in volts. Find I if $P = 2300$ watts and $E = 115$ volts.

128. A checking account has a balance of $410. If $900 is deposited, a check for $720 is written, and two withdrawals of $300 each are made through an ATM, by how much money is the account overdrawn?

129. In addition to a monthly flat fee of x dollars, a college student is charged y dollars for each hour over 20 that she surfs the Web. What does she pay in a month in which she surfs the web for 32 hr?

130. The first day of the term, f students attended the computer lab. As the term went on, attendance increased by s students per day. What was the attendance on the fifth day of the term?

• Check your answers on page A-3.

CHAPTER 1   Posttest

FOR EXTRA HELP

CHAPTER **Test Prep** VIDEOS

The Chapter Test Prep Videos with test solutions are available on DVD, in MyMathLab, and on YouTube™ (search "AkstIntroductory Alg" and click on "Channels").

To see if you have mastered the topics in this chapter, take this test.

1. Express as a positive or negative integer: a loss of 10,000 jobs

2. Is the number $\frac{2}{5}$ rational?

3. Graph the number -2 on the following number line.

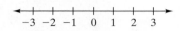

4. Find the additive inverse of 7.

5. Compute: $|-3.5|$

6. True or false? $1 > -4$

7. Add: $10 + (-3)$

8. Combine: $2 + (-3) + (-1) + 5$

9. Simplify: $4 + (-1)(-6)$

10. What is the reciprocal of 12?

11. Divide: $-15 \div 0.3$

12. Translate to an algebraic expression: the sum of x and twice y

13. Write in exponential form: $-5(-5)(-5)$

14. Evaluate $3a + b - c$ if $a = -1, b = 0$, and $c = 2$.

15. Simplify: $4y + 3 - 7y + 10y + 1$

16. Combine: $8t + 1 - 2(3t - 1)$

17. The balance in a checking account was d dollars. Some time later, the balance was 5% higher. What was the new balance?

18. Last year, a company suffered a loss of $20,000. This year, it showed a profit of $50,000. How big an improvement was this?

19. In football, a team has 4 plays, called "downs," to move the football 10 yd or more toward their goal, or else they lose the ball to their opponents. A team had a series of downs in which they gained no ground, lost 10 yd, lost 8 yd, and gained 37 yd. Did the team succeed in keeping the ball?

20. At the beginning of the Deepwater Horizon oil spill of 2010, about 62,000 barrels of oil per day poured into the Gulf of Mexico. If an oil barrel is measured as 42 gallons, how many gallons poured into the Gulf in the first d days of the spill? (*Source:* washingtonpost.com)

• Check your answers on page A-3.

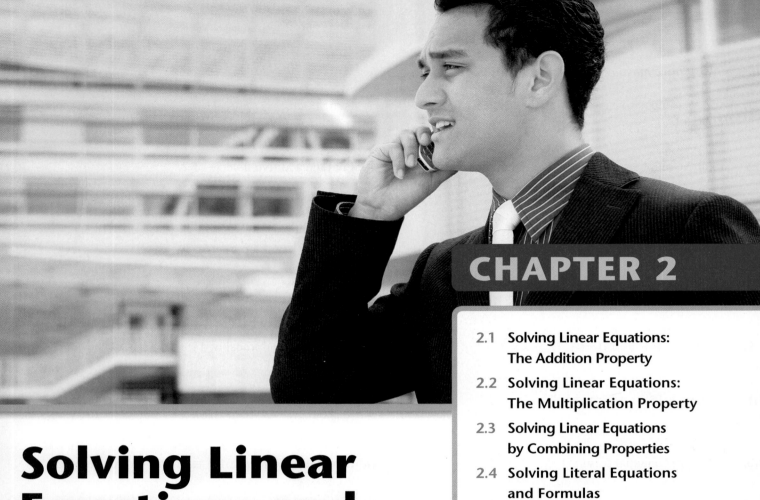

Solving Linear Equations and Inequalities

Cell Phones and Algebra

Today's generation takes wireless communication for granted. The cellular telephone, one of the more important wireless devices, was first introduced to the public in the mid-1980s. Since then, cell phones have shrunk both in size and in price.

Using a cell phone generally entails choosing a calling plan with a monthly service charge. As a rule, this charge consists of two parts: a specified flat monthly fee and a per-minute usage charge varying with the number of minutes of airtime used.

The choice among calling plans can be confusing, especially with the lure of promotional incentives. One nationally advertised calling plan requires a monthly charge of $44.99 for 450 minutes of free local calls and 45 cents per minute for additional local calls. A competing plan is for $64.99 a month for 900 minutes of free local calls plus 40 cents a minute for additional local calls. Under what circumstances is one deal better than the other? The key to answering this question is solving inequalities such as:

$$44.99 + 0.45(x - 450) < 64.99 + 0.4(x - 900)$$

(*Source:* Tom Farley, "The Cell-Phone Revolution," *American Heritage of Invention and Technology*, American Heritage, 2007)

To see if you have already mastered the topics in this chapter, take this test.

1. Is 4 a solution of the equation $7 - 2x = 3x - 11$?

2. What must be done to the equation in order to isolate the variable? $n + 2 = -6$

Solve and check.

3. $\dfrac{y}{-5} = 1$

4. $-n = 8$

5. $\dfrac{2}{3}x - 3 = -9$

6. $4x - 8 = -10$

7. $6 - y = -5$

8. $9x + 13 = 7x + 19$

9. $-2(3n - 1) = -7n$

10. $14x - (8x - 13) = 12x + 3$

11. Solve $v - 5u = w$ for v in terms of u and w.

12. 9 is what percent of 36?

13. 60% of what number is 12?

14. 75% of 80 is what number?

15. Draw the graph of $x \le 2$.

$$\longleftarrow \overset{\;\;}{\underset{-3\;-2\;-1\;\;0\;\;1\;\;2\;\;3}{|\;\;|\;\;|\;\;|\;\;|\;\;|\;\;|}} \longrightarrow$$

16. Solve and graph: $x + 3 > 3$

$$\longleftarrow \overset{\;\;}{\underset{-3\;-2\;-1\;\;0\;\;1\;\;2\;\;3}{|\;\;|\;\;|\;\;|\;\;|\;\;|\;\;|}} \longrightarrow$$

17. An office photocopier makes 30 copies per minute. How long will it take to copy a 360-page document?

18. A florist charges a flat fee of $100 plus $70 for each centerpiece for a wedding. If the total bill for the center-pieces was $1500, how many centerpieces did the florist make?

19. The formula for finding the amount of kinetic energy used is $E = \dfrac{1}{2}mv^2$. Solve this equation for m in terms of E and v.

20. A gym offers two membership options: Option A is $55 per month for unlimited use of the gym and Option B is $10 per month plus $3 for each hour a member uses the gym. For how many hours of use per month will Option A be a better deal?

• Check your answers on page A-3.

2.1 Solving Linear Equations: The Addition Property

What Equations Are and Why They Are Important

In this chapter, we work with one of the most important concepts in algebra, the *equation*. Equations are important because they help us to solve a wide variety of problems.

For instance, suppose that we want to find the measure of $\angle B$ (read "angle B") in triangle ABC.

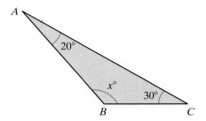

A To determine whether a given number is a solution of a given equation

B To solve linear equations using the addition property

C To solve applied problems using the addition property

We know from geometry that for any triangle, the sum of the measures of the angles is $180°$. To solve this problem, we can write the following equation, where x represents the measure of $\angle B$.

$$20 + 30 + x = 180$$

An equation has two algebraic expressions—one on the left side of the equal sign and one on the right side of the equal sign.

Equal sign

$$\underbrace{20 + 30 + x}_{\text{Left side}} = \underbrace{180}_{\text{Right side}}$$

> **DEFINITION**
> An **equation** is a mathematical statement that two expressions are equal.

Some examples of equations are:

$$1 + 3 + 2 = 6 \qquad y - 6 = 9 \qquad n + 4 = -8 \qquad -5x = 10$$

Solving Equations

An equation may be either true or false. The equation $x + 4 = 9$ is true if 5 is substituted for the variable x.

$$x + 4 = 9$$
$$5 + 4 = 9 \qquad \text{True}$$

However, this equation is false if 6 is substituted for the variable x.

$$x + 4 = 9$$
$$6 + 4 = 9 \qquad \text{False}$$

How many values of x will make $x + 4 = 9$ a true equation? How many values of x will make this equation false?

The number 5 is called a *solution* of the equation $x + 4 = 9$ because we get a true statement when we substitute 5 for x.

DEFINITION

A **solution of an equation** is a value of the variable that makes the equation
a true statement.

EXAMPLE 1

Is 2 a solution of the equation $3x + 1 = 11 - 2x$?

Solution

$$3x + 1 = 11 - 2x$$
$$3(2) + 1 \stackrel{?}{=} 11 - 2(2) \qquad \text{Substitute 2 for } x.$$
$$6 + 1 \stackrel{?}{=} 11 - 4 \qquad \text{Evaluate each side of the equation.}$$
$$7 = 7 \qquad \text{True}$$

Since $7 = 7$ is a true statement, 2 is a solution of $3x + 1 = 11 - 2x$.

PRACTICE 1

Determine whether 4 is a solution of
the equation $5x - 4 = 2x + 5$.

EXAMPLE 2

Determine whether -1 is a solution of the equation
$2x - 4 = 6(x + 2)$.

Solution

$$2x - 4 = 6(x + 2)$$
$$2(-1) - 4 \stackrel{?}{=} 6(-1 + 2) \qquad \text{Substitute } -1 \text{ for } x.$$
$$-2 - 4 \stackrel{?}{=} 6(1) \qquad \text{Evaluate each side of the equation.}$$
$$-6 = 6 \qquad \text{False}$$

Since $-6 = 6$ is a false statement, -1 is *not* a solution of the equation.

PRACTICE 2

Is -8 a solution of the equation
$5(x + 3) = 3x - 1$?

In this chapter, we will work mainly with equations in one variable that have one solution. These equations are called *linear equations* or *first-degree equations* because the exponent of the variable is 1.

DEFINITION

A **linear equation in one variable** is an equation that can be written in the form

$$ax + b = c$$

where a, b, and c are real numbers and $a \neq 0$.

Some linear equations have a special relationship. Consider the equations $x = 2$ and $x + 1 = 3$:

$$x = 2 \qquad \text{The solution is 2.} \qquad\qquad x + 1 = 3 \qquad \text{By inspection, we see the solution is 2.}$$
$$2 + 1 \stackrel{?}{=} 3$$
$$3 = 3 \qquad \text{True}$$

The equations $x = 2$ and $x + 1 = 3$ have the same solution and so are *equivalent*.

DEFINITION

Equivalent equations are equations that have the same solution.

To solve an equation means to find the number or constant that, when substituted for the variable, makes the equation a true statement. This solution is found by changing the equation to an equivalent equation of the following form:

$$x = \boxed{}$$

The variable is isolated (alone on one side with coefficient 1).

The number or constant is isolated on the other side.

Using the Addition Property to Solve Linear Equations

One of the properties that we use in solving equations involves adding. Consider the equation $\frac{6}{3} = 2$. Adding 4 to each side of the equation gives us $\frac{6}{3} + 4 = 2 + 4$. Using mental arithmetic, we see that $\frac{6}{3} + 4 = 2 + 4$ is also a true statement. In general, adding the same number to each side of a true equation results in another true equation. So adding -3 to both sides of the equation $x + 3 = 7$ gives us the equivalent equation $x + 3 + (-3) = 7 + (-3)$, or $x = 4$. This example suggests the following property.

> **Addition Property of Equality**
>
> For any real numbers a, b, and c, $a = b$ and $a + c = b + c$ are equivalent.

This property states that when adding any real number to each side of an equation, the result is an equivalent equation. Now, let's apply this property to solving equations.

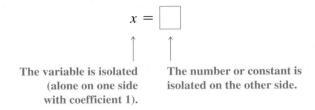

EXAMPLE 3

Solve and check: $x - 5 = -11$

Solution   To solve this equation, we isolate the variable by adding 5, which is the additive inverse of -5, to each side of the equation.

$$x - 5 = -11$$
$$x - 5 + 5 = -11 + 5 \qquad \text{Add 5 to each side of the equation, getting an equivalent equation.}$$
$$x + 0 = -6$$
$$x = -6 \qquad \text{Recall that } x + 0 = x.$$

Check $x - 5 = -11$
$$-6 - 5 \overset{?}{=} -11 \qquad \text{Substitute } -6 \text{ for } x \text{ in the original equation.}$$
$$-11 = -11 \qquad \text{True}$$

So the solution is -6.

PRACTICE 3

Solve and check: $y - 12 = -7$

Can you explain why checking a solution is important?

EXAMPLE 4

Solve and check: $-9 = a + 6$

Solution

$$-9 = a + 6$$
$$-9 + (-6) = a + 6 + (-6) \qquad \text{Add } -6 \text{ to each side of the equation.}$$
$$-15 = a + 0$$
$$-15 = a$$
or $\qquad a = -15$

Check $\quad -9 = a + 6$

$$-9 \overset{?}{=} (-15) + 6 \qquad \text{Substitute } -15 \text{ for } a.$$
$$-9 = -9 \qquad\qquad \text{True}$$

So the solution is -15.

PRACTICE 4

Solve and check: $-2 = n + 15$

Are the solutions to the equations $-9 = a + 6$ and $a + 6 = -9$ the same? Explain.

Because subtracting a number is the same as adding its opposite, the addition property allows us to subtract the same value from each side of an equation. For instance, suppose $a = b$. Subtracting c from each side gives us $a - c = b - c$. So an alternative approach to solving Example 4 is to subtract the same number, namely 6, from each side of the equation, as follows:

$$-9 = a + 6$$
$$-9 - 6 = a + 6 - 6$$
$$-15 = a, \text{ or } a = -15$$

Note that this approach gives us the same solution, namely -15, that we got using the approach in Example 4.

EXAMPLE 5

Solve and check: $r + \dfrac{1}{5} = -\dfrac{3}{5}$

Solution

$$r + \frac{1}{5} = -\frac{3}{5}$$
$$r + \frac{1}{5} - \frac{1}{5} = -\frac{3}{5} - \frac{1}{5} \qquad \text{Subtract } \frac{1}{5} \text{ from each side of the equation.}$$
$$r + 0 = -\frac{4}{5}$$
$$r = -\frac{4}{5}$$

Check $\quad r + \dfrac{1}{5} = -\dfrac{3}{5}$

$$-\frac{4}{5} + \frac{1}{5} \overset{?}{=} -\frac{3}{5} \qquad \text{Substitute } -\frac{4}{5} \text{ for } r.$$
$$-\frac{3}{5} = -\frac{3}{5} \qquad \text{True}$$

PRACTICE 5

Solve and check: $-\dfrac{3}{8} = s + \dfrac{1}{8}$

EXAMPLE 6

Solve and check: $m - (-26.1) = 32$

Solution

$$m - (-26.1) = 32$$
$$m + 26.1 = 32$$
$$m + 26.1 - 26.1 = 32 - 26.1 \qquad \text{Subtract 26.1 from each side}$$
$$\text{of the equation.}$$
$$m = 5.9$$

Check $m - (-26.1) = 32$
$$5.9 - (-26.1) \overset{?}{=} 32 \qquad \text{Substitute 5.9 for } m.$$
$$5.9 + 26.1 \overset{?}{=} 32$$
$$32 = 32 \qquad \text{True}$$

So the solution is 5.9.

PRACTICE 6

Solve and check: $5 = 4.9 - (-x)$

Equations are often useful *mathematical models* that represent real-world situations. Although there is no magic formula for solving applied problems in algebra, it is a good idea to keep the following problem-solving steps in mind.

> ### To Solve an Applied Problem Using an Equation
>
> • Read the problem carefully.
>
> • Translate the problem to an equation.
>
> • Solve the equation.
>
> • Check the solution in the original equation.
>
> • State the conclusion.

We have discussed how to translate word phrases to algebraic expressions. Now, let's look at some examples of translating word sentences to equations, namely those involving addition or subtraction.

Word sentence: A number increased by 1.1 equals 8.6.

Equation: $n \quad + \quad 1.1 \quad = \quad 8.6$

Word sentence: A number minus one-half equals five.

Equation: $y \quad - \quad \dfrac{1}{2} \quad = \quad 5$

Note that in both examples we used a variable to represent the unknown number.

In solving applied problems, first we translate the given word sentences to equations, and then we solve the equations.

EXAMPLE 7

The mean distance between the planet Venus and the Sun is 31.2 million mi more than the mean distance between the planet Mercury and the Sun. (*Source: The New York Times Almanac 2010*)

a. Using the following diagram, write an equation to find Mercury's mean distance from the Sun.

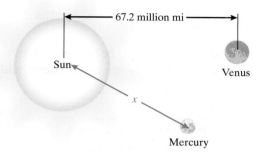

b. Solve this equation.

Solution

a. Let x represent Mercury's mean (or average) distance from the Sun. From the diagram, we see that the mean distance between Venus and the Sun is 67.2 million mi. This distance is 31.2 million mi more than Mercury's mean distance to the Sun. So we translate this sentence to an equation:

Word sentence:	67.2	is	31.2	plus	x
	↓	↓	↓	↓	↓
Equation:	67.2	=	31.2	+	x

b. Solve the equation:

$$67.2 = x + 31.2$$
$$67.2 - 31.2 = x + 31.2 - 31.2 \qquad \text{Subtract 31.2 from each side of the equation.}$$
$$36.0 = x$$

Check
$$67.2 = x + 31.2$$
$$67.2 \overset{?}{=} 36.0 + 31.2 \qquad \text{Substitute 36.0 for } x.$$
$$67.2 = 67.2 \qquad \text{True}$$

So we conclude that Mercury's mean distance from the Sun is 36.0 million mi.

PRACTICE 7

As a result of a chemical reaction, the temperature of a substance rose. The original temperature had been 23.7°C, and the final temperature was 36.0°C. Find the change in temperature.

Mathematically Speaking

Fill in each blank with the most appropriate term or phrase from the given list.

equal equations	each	addition property of equality
one	equation	
isolate	linear equations in one variable	equivalent equations
additive identity property		solve

1. A(n) _____ is a mathematical statement that two expressions are equal.

2. Equations that can be written in the form $ax + b = c$, where a, b, and c are real numbers and $a \neq 0$, are called _____.

3. Equations that have the same solution are called _____.

4. To _____ an equation means to find the value that, when substituted for the variable, makes the equation true.

5. The _____ states that for any real numbers a, b, and c, $a = b$ and $a + c = b + c$ are equivalent.

6. Adding any real number to _____ side of an equation results in an equivalent equation.

Ⓐ *Indicate whether it is true or false that the value of x shown is a solution to the given equation.*

7.

Value of x	Equation	True or False
a. -8	$3x + 13 = -11$	
b. 7	$28 - x = 7 - 4x$	
c. 9	$2(x - 3) = 12$	
d. $\dfrac{2}{3}$	$12x - 2 = 6x + 2$	

8.

Value of x	Equation	True or False
a. -2	$3x - 2 = 10$	
b. 4	$2x + 3 = 5x - 9$	
c. 1	$3(5 - x) = 18$	
d. $-\dfrac{1}{2}$	$6 - 5x = 5x + 11$	

Ⓑ *Indicate what must be done to each side of the equation in order to isolate the variable.*

9. $x + 4 = -6$ 10. $x + 3 = -3$ 11. $z - (-3.5) = 5$ 12. $x - (-15) = 10$

13. $-2 = -1 + x$ 14. $-6 = -5 + x$ 15. $9 = x - 2\frac{1}{5}$ 16. $7 = x - 3\frac{1}{4}$

Solve and check.

17. $y + 9 = -14$ 18. $x + 2 = -10$ 19. $t - 4 = -4$ 20. $m - 6 = -1$

◉ 21. $9 + a = -3$ 22. $10 + x = -4$ ◉ 23. $z - 4 = -10$ 24. $n - 4 = -1$

25. $-30 = x + 12$ 26. $-25 = y + 21$ ◉ 27. $-6 = t - 12$ 28. $-19 = r - 19$

29. $-4 = -10 + r$ 30. $-7 = -2 + m$ 31. $-15 + n = -2$ 32. $-14 + c = -6$

33. $x + \dfrac{2}{3} = -\dfrac{1}{3}$

34. $z + \dfrac{1}{8} = -\dfrac{3}{8}$

35. $8 + y = 4\frac{1}{2}$

36. $9 + z = 6\frac{1}{4}$

37. $m + 2.4 = 5.3$

38. $n + 3.2 = 8.4$

39. $-2.3 + t = -5.9$

40. $-3.4 + r = -9.5$

41. $a - (-35) = 30$

42. $x - (-25) = 24$

43. $m - \left(-\dfrac{1}{4}\right) = -\dfrac{1}{4}$

44. $a - \left(-\dfrac{1}{5}\right) = -\dfrac{1}{5}$

Solve, and round to the nearest hundredth.

45. $y + 2.932 = 4.811$

46. $x + 3.0245 = 9$

Translate each sentence to an equation. Then, solve and check.

47. Two more than a number is 12.

48. 3.2 more than a number is 20.

49. If 4 is subtracted from a number, the result is 21.

50. A number minus $2\frac{1}{7}$ is the same as $1\frac{1}{2}$.

51. If -3 is added to a number, the result is -1.

52. If -5.2 is added to a number, the result is 12.

53. A number plus 7 equals 11.

54. A number plus 1 equals 9.

Choose the equation that best describes the situation.

55. After dieting and exercising, a featherweight boxer lost 6 lb. If the boxer now weighs 127 lb, what was his original weight?

    a. $x - 127 = -6$ **b.** $x + 6 = 127$

    c. $x + 127 = 6$ **d.** $x - 6 = 127$

56. After paying a bill of $5.25, a customer has $2.75 left. How much money did she have prior to paying the bill?

    a. $d - 5.25 = 2.75$ **b.** $d + 2.75 = 5.25$

    c. $d + 5.25 = 2.75$ **d.** $d - 2.75 = -5.25$

57. A digital picture that takes up 4.7 megabytes (Mb) of memory is saved on a computer. If 250 Mb of memory are free after the picture is saved, how many megabytes of free memory did the computer have before the picture was saved?

    a. $x - 4.7 = 250$ **b.** $x - 250 = -4.7$

    c. $x + 4.7 = 250$ **d.** $x + 250 = 4.7$

58. At a college, tuition costs students $3000 a semester. If a student received $2250 in financial aid, how much more money does the student need to pay the balance of the semester's tuition?

    a. $m - 2250 = 3000$ **b.** $m + 2250 = 3000$

    c. $m + 3000 = 2250$ **d.** $m - 3000 = 2250$

Mixed Practice

Indicate what must be done to each side of the equation in order to isolate the variable.

59. $\dfrac{2}{3} + b = 3$

60. $-5 = a - 3$

Solve.

61. Indicate whether the value of x shown is a solution to the given equation by answering true or false.

Value of x	Equation	True or False
a. -4	$6x + 15 = -11$	
b. 9	$4x - 21 = 2x - 3$	
c. 7	$-2(x - 5) = 4$	
d. $\dfrac{3}{4}$	$5 - 8x = -4x + 2$	

Translate each sentence to an equation. Then, solve and check.

62. The difference between 0 and a number is 32.

63. If 2.5 is subtracted from a number, the result is -3.8.

Solve and check.

64. $x - 12 = -7$

65. $-19 = m + 4$

66. $-3.8 + t = -6.6$

67. $7 + n = 2\frac{2}{3}$

68. $x + 7.815 = 3.298$
(Round to the nearest hundredth.)

Applications

C *Write an equation. Solve and check.*

69. If the speed of a car is increased by 10 mph, the speed will be 44 mph. What is the speed of the car now?

70. The difference between the boiling point of gold and the boiling point of aluminum is 337°C. If aluminum boils at 2519°C, the lower temperature of the two metals, what is the boiling point of gold? (*Source: The New York Times Almanac 2010*)

71. A student uses about 370 calories after 1 hr of hiking. This is 190 calories more than is used after 1 hr of stretching. How many calories does the student use in an hour of stretching? (*Source:* U.S. Department of Agriculture)

72. A patient's temperature dropped by 2.4°F to 98.6°F. What had the patient's temperature been?

73. After descending 170 m, a traffic helicopter is 215 m above the ground, as illustrated in the following diagram. What was the original height of the helicopter?

74. In the following diagram, $\angle ABC$ and $\angle CBD$ are complementary angles, that is, the sum of their measures is 90°. Find the number of degrees in $\angle CBD$.

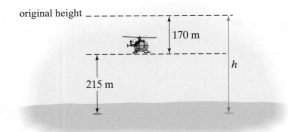

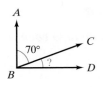

75. In the diagram shown, angles x and y are supplementary angles, that is, the sum of their measures is $180°$. Find the measure of $\angle x$ if the measure of $\angle y$ is $118.5°$.

76. At the close of business today, the price per share of a stock was $21.63. This was down $0.31 from the closing price per share yesterday. What was the closing price per share of the stock yesterday?

• Check your answers on page A-4.

MINDStretchers

Mathematical Reasoning

1. Suppose that $x - a = b$.

a. What happens to x if a decreases and b remains the same?

b. What happens to x if a remains the same and b decreases?

Critical Thinking

2. In the magic square to the right, the sum of each row, column, and diagonal is the same. Calculate that sum. Then, write and solve equations to find v, w, x, y, and z.

v	6	11
w	10	y
x	14	z

Research

3. In your college library or on the Web, investigate the origin of the word *algebra*. Write a few sentences to show what you have learned.

Cultural Note

Just as we solve an equation to identify an unknown number, we use a balance scale to determine an unknown weight. Egyptians 3400 years ago used balance scales to weigh such objects as gold rings.

The balance scale is an ancient measuring device. These scales were used by Sumerians for weighing precious metals and gems at least 9000 years ago.

Source: O. A. W. Dilke, *Mathematics and Measurement* (Berkeley: University of California Press/British Museum, 1987).

2.2   Solving Linear Equations: The Multiplication Property

OBJECTIVES

Ⓐ To solve linear equations using the multiplication property

Ⓑ To solve applied problems using the multiplication property

In the previous section, we used the addition property of equality to solve equations. Another property that is useful in solving equations involves multiplication. Consider the equation $\frac{6}{3} = 2$. If we were to multiply each side of this equation by 9, we would get $\frac{6}{3} \cdot 9 = 2 \cdot 9$. Note that each side of the equation equals 18. In general, multiplying each side of a true equation by the same number results in another true equation. So multiplying both sides of the equation $\frac{x}{2} = 5$ by 2 gives us the equivalent equation $2 \cdot \frac{x}{2} = 2 \cdot 5$, or $x = 10$. This example suggests the following property:

> **Multiplication Property of Equality**
>
> For any real numbers a, b, and c, $c \neq 0$, $a = b$ and $a \cdot c = b \cdot c$ are equivalent.

    This property states that when multiplying each side of an equation by any nonzero real number, the result is an equivalent equation. Let's apply the multiplication property to solving equations.

EXAMPLE 1

Solve and check: $\dfrac{x}{4} = 11$

Solution   In this equation, note that $\frac{x}{4}$ is the same as $\frac{1}{4} \cdot x$. To solve this equation, we isolate the variable by multiplying each side of the equation by 4, the reciprocal of $\frac{1}{4}$.

$$\frac{x}{4} = 11$$

$$4 \cdot \frac{x}{4} = 4 \cdot 11 \qquad \text{Mutiply each side of the equation by 4, getting an equivalent equation.}$$

$$1x = 44 \qquad 4 \cdot \frac{x}{4} = 4 \cdot \frac{1}{4}x = 1x$$

$$x = 44 \qquad 1x = x$$

Check $\dfrac{x}{4} = 11$

$$\frac{44}{4} \overset{?}{=} 11 \qquad \text{Substitute 44 for } x.$$

$$11 = 11 \qquad \text{True}$$

So the solution is 44.

PRACTICE 1

Solve and check: $\dfrac{y}{3} = 21$

EXAMPLE 2

Solve and check: $9x = -72$

Solution

$$9x = -72$$

$$\left(\frac{1}{9}\right)9x = \left(\frac{1}{9}\right)(-72)$$ Multiply each side of the equation by $\frac{1}{9}$.

$$1x = -8 \qquad \frac{1}{9} \cdot 9 = 1$$

$$x = -8$$

Check $9x = -72$

$$9(-8) \stackrel{?}{=} -72$$ Substitute -8 for x.

$$-72 = -72$$ True

So the solution is -8.

PRACTICE 2

Solve and check: $7y = 63$

Because dividing by a number is the same as multiplying by its reciprocal, the multiplication property allows us to divide each side of an equation by a non-zero number. For instance, suppose $a = b$. Multiplying each side by $\frac{1}{c}$ gives us $a \cdot \frac{1}{c} = b \cdot \frac{1}{c}$, for $c \neq 0$. This equation is equivalent to $\frac{a}{c} = \frac{b}{c}$. So an alternative approach to solving Example 2 is to divide each side of the equation by the same number, namely 9, as shown:

$$9x = -72$$

$$\frac{9x}{9} = \frac{-72}{9}$$

$$x = -8$$

Note that this approach gives us the same solution, -8, that we got using the approach in Example 2.

EXAMPLE 3

Solve and check: $-y = -15$

Solution

$$-y = -15$$

$$-1y = -15$$ The coefficient of $-y$ is -1.

$$\frac{-1y}{-1} = \frac{-15}{-1}$$ Divide each side of the equation by -1.

$$y = 15$$

Check $-y = -15$

$$-1(15) \stackrel{?}{=} -15$$ Substitute 15 for y.

$$-15 = -15$$ True

So the solution is 15.

PRACTICE 3

Solve and check: $-x = 10$

EXAMPLE 4

Solve and check: $46 = -4.6n$

Solution

$$46 = -4.6n$$

$$\frac{46}{-4.6} = \frac{-4.6n}{-4.6}$$ **Divide each side of the equation by -4.6.**

$$-10 = n \quad \text{or}$$

$$n = -10$$

Check $46 = -4.6n$

$$46 \overset{?}{=} -4.6(-10) \quad \text{Substitute } -10 \text{ for } n.$$

$$46 = 46 \qquad \text{True}$$

So the solution is -10.

Solve and check: $-11.7 = -0.9z$

EXAMPLE 5

Solve and check: $\dfrac{2w}{3} = 8$

Solution

$$\frac{2w}{3} = 8$$

$$\frac{2}{3}w = 8$$

$$\frac{3}{2} \cdot \frac{2}{3}w = \frac{3}{2} \cdot 8 \quad \text{Multiply each side of the equation by } \frac{3}{2}.$$

$$w = 12 \qquad \frac{3}{2} \cdot \frac{2}{3} = 1 \text{ and } 1w = w.$$

Check $\dfrac{2}{3}w = 8$

$$\frac{2}{3}(12) \overset{?}{=} 8 \quad \text{Substitute 12 for } w.$$

$$2 \cdot 4 \overset{?}{=} 8 \quad \text{Simplify.}$$

$$8 = 8 \qquad \text{True}$$

So the solution is 12.

Solve and check: $\dfrac{6y}{7} = -12$

Can you show another way to solve Example 5? Explain.

Let's now consider applied problems involving the multiplication property. To solve these problems, we translate word sentences to equations involving multiplication or division as follows:

Word sentence: Three times a number x equals -12.

$$3 \cdot x = -12$$

Equation: $3x = -12$

Word sentence: A number d divided by -2 equals 8.

Equation: $d \div -2 = 8$

$$\frac{d}{-2} = 8$$

EXAMPLE 6

A student applies for a job that pays an hourly overtime wage of $15.90. The overtime wage is 1.5 times the regular hourly wage. What is the regular hourly wage for the job?

Solution Letting w represent the regular hourly wage, we write the word sentence, and then translate it to an equation.

Word sentence:
  The overtime wage $15.90 is 1.5 times the regular hourly wage.

Equation: $15.90 = 1.5 \cdot w$

$$15.90 = 1.5w$$

Next, we solve the equation for w.

$$15.9 = 1.5w$$

$$\frac{15.9}{1.5} = \frac{1.5}{1.5}w$$

$$10.6 = w, \text{ or } w = 10.6$$

Check $15.90 = 1.5w$

$15.90 \overset{?}{=} 1.5(10.6)$ Substitute 10.6 for w.

$15.90 = 15.90$ True

So the regular wage is $10.60.

PRACTICE 6

A mechanic billed a customer $189.50 for labor to repair his car. If one-fourth of the bill was for labor, how much was the total bill?

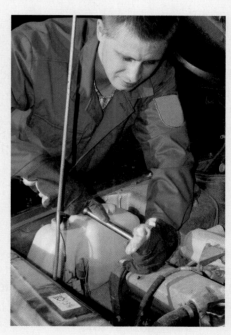

Let's now turn to a particular kind of applied problem—a problem involving *motion*. To solve such problems, we need to use the equation $d = rt$, where d is distance, r is the average rate or speed, and t is time.

EXAMPLE 7

One of the fastest pitchers in the history of Japanese baseball was Yoshinori Sato, whose pitches were clocked at about 147 ft/sec. If the distance from the pitcher's mound to home plate is 60.5 ft, approximate the time it took Sato's pitches to reach home plate, to the nearest hundredth of a second. (*Source:* japantoday.com)

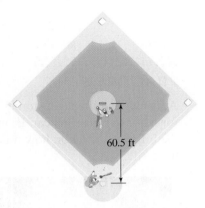

60.5 ft

Solution   The distance between the pitcher's mound and home plate is 60.5 ft and the rate of the ball thrown is 147 ft/sec. Using the equation $d = rt$, we can find the time.

$$d = rt$$
$$60.5 = 147t \quad \text{Substitute 60.5 for } d \text{ and 147 for } r.$$
$$\frac{60.5}{147} = \frac{147t}{147}$$
$$0.41 \approx t, \quad \text{or}$$
$$t \approx 0.41$$

Check   Since the solution of the original equation is a rounded value, the check will not result in an exact equality. To verify the solution, check that the expressions on each side of the equation are approximately equal to one another.

$$60.5 = 147t$$
$$60.5 \stackrel{?}{\approx} 147 \cdot 0.41 \quad \text{Substitute 0.41 for } t.$$
$$60.5 \approx 60.3 \quad \text{True}$$

So Sato's pitches took approximately 0.41 sec to reach home plate.

PRACTICE 7

A driver makes a trip from Washington, D.C., the U.S. capital, to Philadelphia, Pennsylvania, the home of Independence Hall. If she averages 60 mph, how long, to the nearest tenth of an hour, will it take her to get to Philadelphia? (*Source:* mapquest.com)

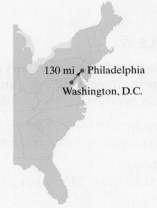

130 mi ▸ Philadelphia
Washington, D.C.

A *Indicate what must be done to each side of the equation in order to isolate the variable.*

1. $\dfrac{x}{3} = -4$
2. $\dfrac{a}{-6} = 1$
3. $-5x = 20$
4. $-4x = 30$
5. $-2.2n = 4$

6. $1.5x = -6$
7. $\dfrac{3}{4}x = 12$
8. $\dfrac{2}{3}x = -6$
9. $-\dfrac{5y}{2} = 15$
10. $-\dfrac{8n}{5} = 4$

Solve and check.

11. $6x = -30$
12. $-8y = 8$
13. $\dfrac{n}{2} = 9$
14. $\dfrac{w}{10} = -21$

15. $\dfrac{a}{4} = 1.2$
16. $\dfrac{n}{7} = -1.3$
17. $-5x = 2.5$
18. $-2y = 0.08$

19. $42 = -6c$
20. $50 = -2x$
21. $11 = -\dfrac{r}{2}$
22. $4 = \dfrac{-m}{3}$

23. $\dfrac{5}{6}x = 10$
24. $\dfrac{3}{4}d = -3$
25. $-\dfrac{2}{5}y = 1$
26. $\dfrac{2}{3}r = -8$

27. $\dfrac{3n}{4} = 6$
28. $\dfrac{5a}{6} = 5$
29. $\dfrac{4c}{3} = -4$
30. $-\dfrac{2z}{7} = 8$

31. $-\dfrac{x}{2.4} = -1.2$
32. $-\dfrac{n}{0.5} = -1.3$
33. $-2.5a = 5$
34. $-2.25 = -1.5t$

35. $\dfrac{2}{3}y = \dfrac{4}{9}$
36. $\dfrac{5}{6}c = \dfrac{2}{3}$

Solve. Round to the nearest hundredth.

37. $\dfrac{x}{-1.515} = 1.515$
38. $\dfrac{n}{-2.968} = -3.85$
39. $-3.14x = 21.4148$
40. $2.54z = 6.4516$

Translate each sentence to an equation. Then, solve and check.

41. The product of -4 and a number is 56.

42. The product of -8 and a number is 72.

43. A number divided by 0.2 is 1.1.

44. A number divided by 0.6 is 1.8.

45. The quotient of a number and 3.5 is 30.

46. The quotient of a number and 2.5 is 40.

47. $\dfrac{1}{6}$ of a number is $2\frac{4}{5}$.

48. $\dfrac{5}{8}$ of a number is 20.

Choose the equation that best describes the situation.

49. A shopper used half his money to buy a backpack. If the backpack cost \$20, how much money did he have prior to this purchase?

   a. $20 = 2m$
   b. $20m = \dfrac{1}{2}$
   c. $20 = \dfrac{m}{2}$
   d. $\dfrac{m}{20} = \dfrac{1}{2}$

50. A child's infant brother weighs 12 lb. If this is $\dfrac{1}{4}$ of her weight, how much does she weigh?

   a. $4w = 12$
   b. $\dfrac{w}{4} = 12$
   c. $12w = \dfrac{1}{4}$
   d. $\dfrac{w}{12} = \dfrac{1}{4}$

51. A student plans to buy a DVD player 6 weeks from now. If the DVD player costs $150, how much money must she save per week in order to buy it?

a. $6p = 150$ **b.** $150p = 6$

c. $\dfrac{p}{6} = 150$ **d.** $\dfrac{p}{150} = 6$

52. The student government at a college sold tickets to a play. From ticket sales, it collected $800, which was twice the cost of the play. How much did the play cost?

a. $\dfrac{c}{800} = 2$ **b.** $800c = 2$

c. $\dfrac{c}{2} = 800$ **d.** $2c = 800$

Mixed Practice

Solve and check.

53. $5x = -20$

54. $\dfrac{2n}{7} = 4$

55. $8 = -\dfrac{a}{3}$

56. $-\dfrac{y}{3.8} = -0.3$

Translate each sentence to an equation. Then, solve and check.

57. The quotient of a number and 5 is equal to 2.

58. $\dfrac{3}{8}$ of a number is 12.

Indicate what must be done to each side of the equation in order to isolate the variable.

59. $-5.2m = 4$

60. $\dfrac{b}{7} = 3$

Applications

B *Write an equation that best describes the situation. Then, solve and check.*

61. According to a geologist, sediment at the bottom of a local lake accumulated at the rate of 0.02 centimeter per year. How long did it take to create a layer of sediment 10.5 cm thick?

62. Because of evaporation, the water level in an aquarium drops at a rate of $\dfrac{1}{10}$ inch per hour. In how many hours will the level drop 2 in.?

63. The bus trip from Miami to San Francisco takes 70 hr. What is the average speed of the bus rounded to the nearest mile per hour? (*Source:* Greyhound)

64. The coat in the ad shown is selling at $\dfrac{1}{4}$ the regular price. What was the regular price?

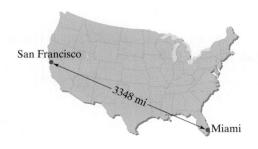

65. A customer has only $20 to spend at a local print shop that charges $0.05 per copy. At this rate, how many copies can she afford to make?

66. Consider the two parcels of land shown—one a rectangle and the other a square. For which value of x do the two parcels have the same area?

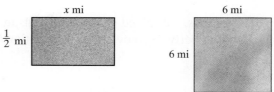

67. A city offers to pay a disposal company $40 per ton to bury 20,000 tons of toxic waste. If this deal represents $\frac{2}{3}$ of the disposal company's projected income, what is that income?

68. The diameter of a tree trunk increases as the tree ages and adds rings. Suppose a trunk's diameter increases by 0.2 inch per year. How many years will it take the tree to increase in diameter from 4 in. to 12 in.?

69. The maximum depth of the Caspian Sea is approximately 1000 m. If this depth is about $\frac{1}{5}$ the maximum depth of the Mediterranean Sea, what is the maximum depth of the Mediterranean Sea? (*Source:* wikipedia. org)

70. A top-secret plane flew 3000 mi in $1\frac{1}{2}$ hr. What was the average speed of this plane?

71. A student takes a job in the college's student center so that she can buy books that cost $187.50. How many hours must she work to make this amount if she earns $7.50/hr?

72. A double-trailer truck is driven at an average speed of 54 mph from Atlanta to Cincinnati, a driving distance of 457 mi. To the nearest tenth of an hour, how long did the trip take? (*Source:* mapquest.com)

73. Last year, a young couple paid a total of $10,020 in rent for their apartment. How much money did they pay per month in rent?

74. An equilateral triangle is a triangle that has three sides equal in length. If the perimeter of an equilateral triangle is $10\frac{1}{2}$ ft, how long is each side?

• Check your answers on page A-4.

MIND*Stretchers*

Mathematical Reasoning

1. In the course of solving a linear equation, you reach the step, $5x = 3x$
   If you then divide both sides by x, you get $5 = 3$, which is impossible. Did you make an error? Explain.

Critical Thinking

2. If you multiply each side of the equation $0.24r = -12.48$ by 100, the result is an equivalent equation. Explain why it is helpful to carry out this multiplication in solving the equation.

Writing

3. Write two different word problems that are applications of each equation.

   a. $6x = 18$ **b.** $\dfrac{x}{3} = 10$

2.3   Solving Linear Equations by Combining Properties

In the previous sections, we solved simple equations involving either the addition property or the multiplication property. We now turn our attention to solving equations that involve both properties.

A To solve a linear equation using both the addition and multiplication properties

Solving Equations Using Both the Addition and Multiplication Properties

B To solve a linear equation involving parentheses and combining like terms

We need to use both the addition property and the multiplication property to solve equations such as:

C To solve applied problems using the addition and multiplication properties, or involving parentheses and combining like terms

$$3x + 4 = 7 \quad \text{and} \quad \frac{r}{2} - 5 = 9$$

To solve these equations, we first use the addition property to get the variable term alone on one side. Then, we use the multiplication property to isolate the variable.

EXAMPLE 1

Solve and check: $3x + 4 = 7$

Solution

$$3x + 4 = 7$$
$$3x + 4 - 4 = 7 - 4 \qquad \text{Subtract 4 from each side of the equation.}$$
$$3x = 3$$
$$\frac{3x}{3} = \frac{3}{3} \qquad \text{Divide each side of the equation by 3.}$$
$$x = 1$$

Check $3x + 4 = 7$

$$3(1) + 4 \stackrel{?}{=} 7 \qquad \text{Substitute 1 for } x.$$
$$3 + 4 \stackrel{?}{=} 7$$
$$7 = 7 \qquad \text{True}$$

So the solution is 1.

PRACTICE 1

Solve and check: $2y + 1 = 9$

In solving Example 1, would we get the same solution if we divided before subtracting? Explain.

EXAMPLE 2

Solve and check: $\dfrac{r}{2} - 5 = 9$

Solution

$$\dfrac{r}{2} - 5 = 9$$

$$\dfrac{r}{2} - 5 + 5 = 9 + 5 \qquad \text{Add 5 to each side of the equation.}$$

$$\dfrac{r}{2} = 14$$

$$2 \cdot \dfrac{r}{2} = 2 \cdot 14 \qquad \text{Multiply each side of the equation by 2.}$$

$$r = 28$$

Check $\dfrac{r}{2} - 5 = 9$

$$\dfrac{28}{2} - 5 \overset{?}{=} 9 \qquad \text{Substitute 28 for } r.$$

$$14 - 5 \overset{?}{=} 9$$

$$9 = 9 \qquad \text{True}$$

So the solution is 28.

PRACTICE 2

Solve and check: $\dfrac{c}{5} - 1 = 8$

EXAMPLE 3

Solve: $-4s + 7 = 3$

Solution

$$-4s + 7 = 3$$

$$-4s + 7 - 7 = 3 - 7 \qquad \text{Subtract 7 from each side of the equation.}$$

$$-4s = -4$$

$$\dfrac{-4s}{-4} = \dfrac{-4}{-4} \qquad \text{Divide each side by } -4.$$

$$s = 1$$

So the solution is 1.

PRACTICE 3

Solve: $-6b - 5 = 13$

Solving Equations by Combining Like Terms

Now, let's consider an equation that has like terms on the same side of the equation. In order to solve this kind of equation, we combine all like terms before using the addition and multiplication properties.

EXAMPLE 4

Solve and check: $2x - 5x = 12$

Solution

$$2x - 5x = 12$$

$-3x = 12$ Combine like terms.

$\dfrac{-3x}{-3} = \dfrac{12}{-3}$ Divide each side of the equation by -3.

$x = -4$

Check

$2x - 5x = 12$

$2(-4) - 5(-4) \overset{?}{=} 12$ Substitute -4 for x.

$-8 + 20 \overset{?}{=} 12$

$12 = 12$ True

So the solution is -4.

PRACTICE 4

Solve and check: $8n + 10n = 24$

EXAMPLE 5

Solve and check: $7x - 3x - 6 = 6$

Solution

$$7x - 3x - 6 = 6$$

$4x - 6 = 6$ Combine like terms.

$4x - 6 + 6 = 6 + 6$ Add 6 to each side of the equation.

$4x = 12$

$\dfrac{4x}{4} = \dfrac{12}{4}$ Divide each side of the equation by 4.

$x = 3$

Check

$7x - 3x - 6 = 6$

$7(3) - 3(3) - 6 \overset{?}{=} 6$ Substitute 3 for x.

$21 - 9 - 6 \overset{?}{=} 6$

$6 = 6$ True

So the solution is 3.

PRACTICE 5

Solve and check: $5 - t - t = -1$

Suppose an equation has like terms that are on opposite sides of the equation. To solve, we use the addition property to get the like terms together on the same side so that they can be combined.

EXAMPLE 6

Solve: $13z + 5 = -z + 12$

Solution

$$13z + 5 = -z + 12$$

$z + 13z + 5 = z + (-z) + 12$ Add z to each side of the equation.

$14z + 5 = 12$ Combine like terms.

$14z + 5 - 5 = 12 - 5$ Subtract 5 from each side of the equation.

$$14z = 7$$

$$\frac{14z}{14} = \frac{7}{14}$$ Divide each side of the equation by 14.

$$z = \frac{1}{2}$$

So the solution is $\frac{1}{2}$.

PRACTICE 6

Solve: $3f - 12 = -f - 15$

Some equations have fractional terms. The key to solving a fractional equation is to *clear the equation of the fractional terms.* We do this by first determining their *least common denominator* (LCD), and then by multiplying both sides of the equation by the LCD.

EXAMPLE 7

Solve and check: $\dfrac{1}{2}x + \dfrac{1}{5}x - \dfrac{7}{10}$

Solution   The fractional terms in this equation are $\dfrac{1}{2}x$, $\dfrac{1}{5}x$ and $\dfrac{7}{10}$.

The denominators of these terms are 2, 5, and 10, so the LCD is 10. To clear the equation of the fractional terms, we multiply each side of the equation by 10.

$$\frac{1}{2}x + \frac{1}{5}x = \frac{7}{10}$$

$10 \cdot \left(\dfrac{1}{2}x + \dfrac{1}{5}x \right) = 10 \cdot \dfrac{7}{10}$ Multiply each side of the equation by the LCD.

$10 \cdot \dfrac{1}{2}x + 10 \cdot \dfrac{1}{5}x = 10 \cdot \dfrac{7}{10}$ Use the distributive property.

$5x + 2x = 7$ Simplify.

$7x = 7$ Combine like terms.

$x = 1$ Divide each side of the equation by 7.

PRACTICE 7

Solve and check: $\dfrac{1}{3}w - \dfrac{1}{6}w = \dfrac{2}{3}$

EXAMPLE 7 (continued)

Check

$$\frac{1}{2}x + \frac{1}{5}x = \frac{7}{10}$$

$$\frac{1}{2}(1) + \frac{1}{5}(1) \stackrel{?}{=} \frac{7}{10} \qquad \text{Substitute 1 for } x.$$

$$\frac{1}{2} + \frac{1}{5} \stackrel{?}{=} \frac{7}{10}$$

$$\frac{5}{10} + \frac{2}{10} \stackrel{?}{=} \frac{7}{10}$$

$$\frac{7}{10} = \frac{7}{10} \qquad \text{True}$$

So the solution is 1.

Solving Equations Containing Parentheses

Some equations contain parentheses. To solve these equations, we first remove the parentheses using the distributive property. Then, we proceed as in previous examples.

EXAMPLE 8

Solve and check: $2c = -3(c - 5)$

Solution

$$2c = -3(c - 5)$$

$$2c = -3c + 15 \qquad \text{Use the distributive property.}$$

$$3c + 2c = 3c - 3c + 15 \qquad \text{Add } 3c \text{ to each side of the equation.}$$

$$5c = 15 \qquad \text{Combine like terms.}$$

$$\frac{5c}{5} = \frac{15}{5} \qquad \text{Divide each side of the equation by 5.}$$

$$c = 3$$

Check

$$2c = -3(c - 5)$$

$$2(3) \stackrel{?}{=} -3(3 - 5) \qquad \text{Substitute 3 for } c.$$

$$6 \stackrel{?}{=} -3(-2)$$

$$6 = 6 \qquad \text{True}$$

So the solution is 3.

PRACTICE 8

Solve and check: $-5(z + 6) = z$

As Example 8 suggests, we use the following procedure to solve linear equations:

To Solve a Linear Equation

- Use the distributive property to clear the equation of parentheses, if necessary.
- Combine like terms where appropriate.
- Use the addition property to isolate the variable term.
- Use the multiplication property to isolate the variable.
- Check by substituting the solution in the original equation.

Let's apply this procedure in the next example. Recall from Section 1.8 that when removing parentheses preceded by a minus sign, all the terms in the parentheses change to the opposite sign.

EXAMPLE 9

Solve: $2(x + 5) - (x - 2) = 4x + 6$

Solution

$$2(x + 5) - (x - 2) = 4x + 6$$

$2x + 10 - x + 2 = 4x + 6$	Use the distributive property.
$x + 12 = 4x + 6$	Combine like terms.
$x - 4x + 12 = 4x - 4x + 6$	Subtract $4x$ from each side of the equation.
$-3x + 12 = 6$	Combine like terms.
$-3x + 12 - 12 = 6 - 12$	Subtract 12 from each side of the equation.
$-3x = -6$	
$\dfrac{-3x}{-3} = \dfrac{-6}{-3}$	Divide each side of the equation by -3.
$x = 2$	

So the solution is 2.

PRACTICE 9

Solve: $2(t - 3) - 3(t - 2) = t + 8$

EXAMPLE 10

Solve: $13 - [4 + 2(x - 1)] = 3(x + 2)$

Solution

$$13 - [4 + 2(x - 1)] = 3(x + 2)$$
$$13 - [4 + 2x - 2] = 3(x + 2)$$
$$13 - [2 + 2x] = 3(x + 2)$$
$$13 - 2 - 2x = 3x + 6$$
$$11 - 2x = 3x + 6$$
$$11 - 2x - 3x = 3x - 3x + 6$$
$$11 - 5x = 6$$
$$11 - 11 - 5x = 6 - 11$$
$$-5x = -5$$
$$\frac{-5x}{-5} = \frac{-5}{-5}$$
$$x = 1$$

So the solution is 1.

PRACTICE 10

Solve: $4[5y - (y - 1)] = 7(y - 2)$

Now, let's consider applications that lead to equations like those that we have discussed in this section.

EXAMPLE 11

An insurance company settles a claim by multiplying the claim by a certain factor and then subtracting the deductible. A payment of $3500 is made on a claim filed for $5000. If the company has a $500 deductible, what factor was used to settle the claim?

Solution First, let x represent the factor used to settle a claim. Then, write the equation:

The claim times the factor less the deductible is the payment.

$$5000 \cdot x - 500 = 3500$$

Next, we solve for x.

$$5000x - 500 = 3500$$
$$5000x - 500 + 500 = 3500 + 500 \qquad \text{Add 500 to each side of the equation.}$$
$$5000x = 4000$$
$$\frac{5000x}{5000} = \frac{4000}{5000} \qquad \text{Divide each side of the equation by 5000.}$$
$$x = \frac{4}{5}, \quad \text{or } 0.8$$

So the factor used to compute the $5000 claim is $\frac{4}{5}$, or 0.8.

PRACTICE 11

A car is purchased for $12,000. The car's value depreciates $1100 in value per year for each of the first 6 yr of ownership. At what point will the car have a value of $6500?

Many motion problems lead to equations of the type that we have discussed in this section, in particular, to equations containing parentheses. Some of these problems involve one or more objects traveling the same distance, at different rates and for different lengths of time. As in other motion problems, we use the formula $d = rt$.

EXAMPLE 12

A bus leaves St. Petersburg traveling at 45 mph. An hour later, a second bus leaves the same city traveling at 55 mph in the same direction. In how many hours will the second bus overtake the first bus?

Solution Let t represent the number of hours traveled by the second bus. Since the first bus left an hour earlier than the second, $t + 1$ represents the number of hours traveled by the first bus.

Recall from Section 2.2 that the distance that an object travels is the product of its average rate of travel (speed) and the time that it has traveled. Putting these quantities in a table clarifies their relationship.

	Rate	· Time	= Distance
First bus	45	$t + 1$	$45(t + 1)$
Second bus	55	t	$55t$

PRACTICE 12

Two friends plan to meet in Boston. A half an hour after one friend took a local train, the other friend takes an Amtrak express train on a parallel track. If both friends leave from the same station, how long will it take the express train to catch up with the local train if their speeds are 60 mph and 50 mph, respectively?

Since the two buses travel the same distance to the point where they meet, we can write the following equation and then solve:

$$45(t + 1) = 55t$$
$$45t + 45 = 55t$$
$$45t - 45t + 45 = 55t - 45t$$
$$45 = 10t$$
$$\frac{45}{10} = \frac{10}{10}t$$
$$4.5 = t, \quad \text{or}$$
$$t = 4.5$$

So the second bus will overtake the first bus in 4.5 hr, or $4\frac{1}{2}$ hr.

EXAMPLE 13

A shopper walks from home to the market at a rate of 3 mph. After shopping, he returns home following the same route walking at 2 mph. If the walk back from the market takes 10 min more than the walk to the market, how far away is the market?

Solution Let t represent the time of the walk to the market. Since the rates are given in miles per hour, we express the time in hours.

The return trip takes 10 min more time. We change 10 min to $\frac{10}{60}$ hr, or $\frac{1}{6}$ hr, getting $t + \frac{1}{6}$ for the time of the return trip.

	Rate ·	Time	= Distance
Going	3	t	$3t$
Returning	2	$t + \frac{1}{6}$	$2\left(t + \frac{1}{6}\right)$

Since the walk to and from the market followed the same route, the distances each way are equal.

$$3t = 2\left(t + \frac{1}{6}\right)$$
$$3t = 2t + \frac{1}{3}$$
$$3t - 2t = 2t - 2t + \frac{1}{3}$$
$$t = \frac{1}{3}$$

It takes $\frac{1}{3}$ hr to walk to the market. So the market is $3 \cdot \frac{1}{3}$, or 1 mi away from home.

PRACTICE 13

On a round trip over the same roads, a car averaged 25 mph going and 30 mph returning. If the entire trip took 5 hr 30 min, what is the distance each way?

In some motion problems, an object travels at different speeds for different parts of the trip. The total distance traveled is the sum of the partial distances.

EXAMPLE 14

A car is driven for 3 hr in a rainstorm. After the weather clears, the car is driven 10 mph faster for 2 more hours, completing a 250-mile trip. How fast was the car driven during the storm?

Solution   Let x represent the speed of the car during the storm. The speed of the car after the storm passes can be represented by $x + 10$. Drawing a diagram helps us to visualize the problem and to see that the total distance driven is the sum of the two partial distances.

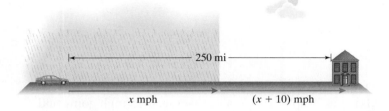

Next, let's complete a table in order to organize the relevant information. Recall that distance is the product of rate and time.

	Rate $\cdot$	Time =	Distance
Storm	x	3	$3x$
Clear	$x + 10$	2	$2(x + 10)$

We are told that the sum of the two partial distances is 250 mi, giving us an equation to solve.

$$3x + 2(x + 10) = 250$$
$$3x + 2x + 20 = 250$$
$$5x + 20 = 250$$
$$5x = 230$$
$$x = \frac{230}{5} = 46$$

So the car was driven at a speed of 46 mph during the storm.

PRACTICE 14

A cyclist pedals uphill for 2 hr. She then continues downhill 10 mph faster for another hour. If the entire trip was 40 mi in length, what was her downhill speed?

A *Solve and check.*

1. $3x - 1 = 8$

2. $7r - 8 = 13$

3. $9t + 17 = -1$

4. $2y + 1 = 9$

5. $20 - 5m = 45$

6. $25 - 3c = 34$

7. $\dfrac{n}{2} - 1 = 5$

8. $\dfrac{s}{3} - 2 = -4$

9. $\dfrac{x}{5} + 15 = 0$

10. $\dfrac{y}{3} + 3 = 42$

11. $3 - t = 1$

12. $2 - x = 2$

13. $-8 - b = 11$

14. $5 - y = 8$

15. $\dfrac{2}{3}x - 9 = 17$

16. $\dfrac{4}{5}d - 3 = 13$

17. $\dfrac{4}{5}r + 20 = -20$

18. $\dfrac{3y}{8} + 14 = -10$

19. $3y + y = -8$

20. $4a + 3a = -21$

21. $7z - 2z = -30$

22. $4x - x = 18$

23. $28 - a + 4a = 7$

24. $5 - 8x - 2x = -25$

25. $1 = 1 - 6t - 4t$

26. $-1 = 5 - z - z$

27. $3y + 2 = -y - 2$

28. $3n + 6 = -n - 6$

29. $5r - 4 = 2r + 6$

30. $7 - m = 5 + 3m$

B *Solve and check.*

31. $4(x + 7) = 7 + x$

32. $3t - 2 = 4(t - 2)$

33. $5(y - 1) = 2y + 1$

34. $5a - 4 = 7(a + 2)$

35. $3a - 2(a - 9) = 4 + 2a$

36. $5 - 2(3x - 4) = 3 - x$

37. $5(2 - t) - (1 - 3t) = 6$

38. $\dfrac{3}{5}(15y + 10) - 3(4y + 3) = 0$

39. $2y - 3(y + 1) = -(5y + 3) + y$

40. $9n + 5(n + 3) = -(n + 13) - 2$

41. $2[3z - 5(2z - 3)] = 3z - 4$

42. $5[2 - (2n - 4)] = 2(5 - 3n)$

43. $-8m - [2(11 - 2m) + 4] = 9m$

44. $7 - [4 + 2(a - 3)] = 11(a + 2)$

Solve. Round to the nearest hundredth.

45. $\dfrac{y}{0.87} + 2.51 = 4.03$

46. $7.02x - 3.64 = 8.29$

47. $7.37n + 4.06 = -1.98n + 6.55$

48. $10.13p = 3.14(p - 7.82)$

Choose the equation that best describes the situation.

49. A car leaves Seattle traveling at a rate of 45 mph. One hour later, a second car leaves from the same place, along the same road, at 54 mph. If the first car travels for t hr, in how many hours will the second car overtake the first car?

a. $54(t - 1) = 45t$ **b.** $45(t + 1) = 54t$

c. $54(t + 1) = 45t$ **d.** $45(t - 1) = 54t$

50. A company budgets $600,000 for an advertising campaign. It must pay $4000 for each television commercial and $1000 per radio commercial. If the company plans to air 50 fewer radio commercials than television commercials, find the number of television commercials t that will be in the advertising campaign.

a. $4000t + 1000(t + 50) = 600,000$ **b.** $4000t + 1000(t - 50) = 600,000$

c. $4000t + 1000t - 50 = 600,000$ **d.** $1000t + 4000(t - 50) = 600,000$

51. A taxi fare is $3.00 for the first mile and $1.25 for each additional mile. If a passenger's total cost was $5.50, how far did she travel in the taxi?

a. $3.00x + 1.25 = 5.50$ **b.** $3.00 + 1.25x = 5.50$

c. $3.00 + 1.25(x + 1) = 5.50$ **d.** $3.00 + 1.25(x - 1) = 5.50$

52. A family's budget allows $\frac{1}{3}$ of the family's monthly income for housing and $\frac{1}{4}$ of its monthly income for food. If a total of $1050 a month is budgeted for housing and food, what is the family's monthly income?

a. $\frac{1}{3}x = 1050 + \frac{1}{4}x$ **b.** $\frac{1}{3}x - \frac{1}{4}x = 1050$

c. $\frac{1}{3}x + \frac{1}{4}x = 1050$ **d.** $\frac{1}{4}x - \frac{1}{3}x = 1050$

Mixed Practice

Solve and check.

53. $5 - 5x + x = 21$

54. $16 - 3t = 31$

55. $4z + 3(5 - z) = -(z - 3) + 8$

56. $\frac{r}{7} + 3 = 11$

57. $-2.31y + 0.14 = -9.23$
(Round to the nearest hundredth.)

58. $7y - 6 = 3(y - 6)$

Applications

C *Write an equation and solve.*

59. A part-time student at a college pays a student fee of $45 plus $135 per credit. How many credits is a part-time student carrying who pays $1260 in all?

60. A health club charges members $10 per month plus $5 per hour to use the facilities. If a member was charged $55 this month, how many hours did she use the facilities?

61. In a local election, a newspaper reported that one candidate received twice as many votes as the other. Altogether, they received a total of 3690 votes. How many votes did each candidate receive?

62. Calcium carbonate (chalk) consists of 10 parts of calcium for each 3 parts of carbon and 12 parts of oxygen by weight. Find the amount of calcium in 75 lb of chalk.

63. A parking garage charges $3 for the first hour and $2 for each additional hour or fraction thereof. If $9 was paid for parking, how many hours was the car parked in the garage?

64. A machinist earns $11.50 an hour for the first 35 hr and $15.30 for each hour over 35 per week. How many hours did he work this week if he earned $555.50?

65. The office manager of an election campaign office needs to print 5000 postcards. It costs 2 cents to print a large postcard and 1 cent to print a small postcard. If $85 is allocated for printing postcards, how many of each type of postcard can be printed?

66. The owner of a small factory has 8 employees. Some of the employees make $10 per hour, whereas the others make $15 per hour. If the total payroll is $105 per hour, how many employees make the higher rate of pay?

67. Twenty minutes after a father left for work on the bus, he noticed that he had left his briefcase at home. His son left home, driving at 36 mph, to catch the bus that was traveling at 24 mph. How long did it take the son to catch the bus?

68. In the rush hour, a commuter drives to work at 30 mph. Returning home off-peak, she takes $\frac{1}{4}$ hr less time driving at 40 mph. What is the distance between the commuter's work and her home?

69. The two snails shown below crawl toward each other at rates that differ by 2 cm/min. If it takes the snails 27 min to meet, how fast is each snail crawling?

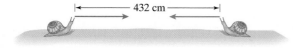

432 cm

70. A signal is sent from a station on the ground to a satellite. The signal bounces off the satellite and then is received at a second ground station. If the signal traveled 2400 mi in all, at what speed was it traveling?

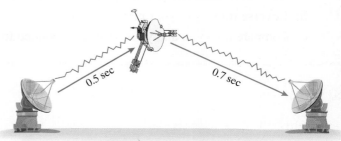

0.5 sec   0.7 sec

71. Two trucks leave a depot at the same time, traveling in opposite directions. One truck goes 4 mph faster than the other. After 2 hr, the trucks are 212 mi apart. What is the speed of the slower truck?

72. If a student drives from college to home at 40 mph, then he is 15 min late. However, if he makes the same trip at 50 mph, he is 12 min early. What is the distance between his college and his home?

• Check your answers on page A-4.

MINDStretchers

Mathematical Reasoning

1. Give an example of an equation that involves combining like terms and that has:

   a. no solution.

   b. an infinite number of solutions.

Patterns

2. Tables can be useful in solving equations.

   a. Complete the following table. After examining your results, identify the solution to the following equation: $3(x - 2) = 2(x + 1)$.

x	0	2	4	6	8	10	12
$3(x - 2)$							
$2(x + 1)$							

   b. Try a similar approach to solving the equation $7x = 5x + 11$. What conclusion can you draw about the solution?

x	0	2	4	6	8	10	12
$7x$							
$5x + 11$							

Groupwork

3. Working with a partner, choose a month on a calendar.

   a. Ask your partner to select four days of the month that form a 2×2 square, but only to tell you the sum of the four days. Determine the four days.

   b. Reverse roles with your partner and repeat part (a).

   c. Compare how you and your partner responded to part (a).

2.4 Solving Literal Equations and Formulas

In many situations, equations describe the relationship between two or more variables. Such equations are called *literal* equations and can be used to describe situations such as how the amount you pay depends on what you buy, how the distance you walk determines how long the walk takes, and how the dosage of a medicine you need to take relates to your weight.

DEFINITION

A **literal equation** is an equation involving two or more variables.

Consider, for instance, the following literal equation:

$$x = y + z$$

Since there is more than one variable, we can solve for one of the variables in terms of the others. For instance, in this equation we can solve for y in terms of x and z.

EXAMPLE 1

Solve $x = y + z$ for y in terms of x and z.

Solution

$$x = y + z$$
$$x - z = y + z - z \quad \text{Subtract } z \text{ from each side of the equation.}$$
$$x - z = y, \text{ or } y = x - z$$

PRACTICE 1

Solve $q = 1 - p$ for p in terms of q.

We see from Example 1 that the solution to a literal equation is not a number, but an algebraic expression.

EXAMPLE 2

Solve $2a + b = c$ for a in terms of b and c.

Solution

$$2a + b = c$$
$$2a + b - b = c - b \quad \text{Subtract } b \text{ from each side of the equation.}$$
$$2a = c - b \quad \text{Simplify.}$$
$$\frac{2a}{2} = \frac{c - b}{2} \quad \text{Divide each side of the equation by 2.}$$
$$a = \frac{c - b}{2}$$

So $a = \dfrac{c - b}{2}$.

PRACTICE 2

Solve $3r - s = t$ for r in terms of s and t.

Note that we can solve literal equations by using the addition and multiplication properties that we have already used to solve other equations. Can you explain how to check your solution to a literal equation?

EXAMPLE 3

Solve $\dfrac{2K}{3m} = n$ for K.

Solution

$$\dfrac{2K}{3m} = n$$

$$3m \cdot \dfrac{2K}{3m} = 3m \cdot n \qquad \text{Multiply each side of the equation by } 3m.$$

$$2K = 3mn$$

$$\dfrac{2K}{2} = \dfrac{3mn}{2} \qquad \text{Divide each side of the equation by 2.}$$

$$K = \dfrac{3mn}{2}$$

So $K = \dfrac{3mn}{2}$.

PRACTICE 3

Solve $\dfrac{4x}{5a} = c$ for x.

EXAMPLE 4

Consider the equation $Ax + By = C$. Solve for y in terms of A, B, C, and x.

Solution

$$Ax + By = C$$

$$Ax - Ax + By = C - Ax \qquad \text{Subtract } Ax \text{ from each side of the equation.}$$

$$By = C - Ax \qquad \text{Simplify.}$$

$$\dfrac{By}{B} = \dfrac{C - Ax}{B} \qquad \text{Divide each side of the equation by } B.$$

$$y = \dfrac{C - Ax}{B}$$

So $y = \dfrac{C - Ax}{B}$.

PRACTICE 4

Consider the equation $y = mx + b$. Solve for x in terms of y, m, and b.

Recall that in Chapter 1, we discussed formulas—a special type of literal equation. Here we focus on solving a formula for one variable in terms of the other variables. When we need to use a formula repeatedly to find the value of a particular variable, the computation can be simplified by first solving for that variable in the formula.

EXAMPLE 5

$P = 2(l + w)$ is the formula for the perimeter P of a rectangle in terms of its length l and width w.

a. Solve for l in terms of P and w.

b. Using the formula found in part (a), find the length of a rectangle with a perimeter of 20 cm and a width of 6 cm.

Solution

a.

$$P = 2(l + w)$$

$$P = 2l + 2w \qquad \text{Use the distributive property.}$$

$$P - 2w = 2l + 2w - 2w \qquad \text{Subtract } 2w \text{ from each side of the equation.}$$

$$P - 2w = 2l$$

$$\frac{P - 2w}{2} = \frac{2l}{2} \qquad \text{Divide each side of the equation by 2.}$$

$$\frac{P - 2w}{2} = l, \text{ or}$$

$$l = \frac{P - 2w}{2}$$

So $l = \dfrac{P - 2w}{2}$.

b. $P = 20$ cm and $w = 6$ cm. We substitute in the formula

$l = \dfrac{P - 2w}{2}$ to find the value of l.

$$l = \frac{P - 2w}{2} = \frac{20 - 2 \cdot 6}{2} = \frac{20 - 12}{2} = \frac{8}{2} = 4$$

So the length of the rectangle is 4 cm.

EXAMPLE 6

$V = lwh$ is the formula for finding the volume of a rectangular solid, where l represents the length of the solid, w the width, and h the height.

a. Solve the formula for h.

b. Using the equation found in part (a), find the value of h for $V = 48$ cu ft, $l = 8$ ft, and $w = 2$ ft.

Solution

a. Solving $V = lwh$ for h, we get:

$$V = lwh$$

$$\frac{V}{lw} = \frac{lwh}{lw}$$

$$\frac{V}{lw} = h$$

So $h = \dfrac{V}{lw}$.

PRACTICE 5

$A = P(1 + rt)$ is the formula for computing the amount in an account earning simple interest. In the formula, A stands for the amount, P for the original principal, r for the annual rate of interest, and t for time.

a. Find a formula for r in terms of A, P, and t.

b. Using the equation found in part (a), evaluate r if $A = \$2100$, $P = \$2000$, and $t = 2$ years.

PRACTICE 6

$A = \dfrac{1}{2}bh$ is the formula for finding the area of a triangle, where b is the base and h is the height.

a. Express h in terms of A and b.

b. Using the formula found in part (a), find the value of h for $A = 63$ in^2 and $b = 9$ in.

EXAMPLE 6 (continued)

b. To find the value of h, we substitute 48 for V, 8 for l, and 2 for w in the formula $h = \dfrac{V}{lw}$.

$$h = \frac{48}{8(2)} = 3$$

The height is 3 ft.

EXAMPLE 7

To convert a temperature expressed in Fahrenheit degrees F to Celsius degrees C, a meteorologist multiplies $\dfrac{5}{9}$ by the difference between the Fahrenheit temperature and 32.

a. Write a formula for this relationship.

b. Solve the formula for F.

c. What Fahrenheit temperature corresponds to a Celsius temperature of 30°?

Solution

a. Stating the rule in words, we get:

Celsius temperature, C, is equal to $\dfrac{5}{9}$ times the quantity Fahrenheit temperature, F, minus 32.

Then, we translate this relationship to a formula.

$$C = \frac{5}{9}(F - 32)$$

b. Now, we solve the formula found in part (a) for F.

$$C = \frac{5}{9}(F - 32)$$

$$\frac{9}{5}C = \frac{9}{5} \cdot \frac{5}{9}(F - 32)$$

$$\frac{9}{5}C = F - 32$$

$$\frac{9}{5}C + 32 = F$$

So $F = \dfrac{9}{5}C + 32$.

c. To find the Fahrenheit temperature that corresponds to a Celsius temperature of 30°, we substitute 30 for C in the formula.

$$F = \frac{9}{5}C + 32 = \frac{9}{5}(30) + 32 = 54 + 32 = 86$$

The corresponding Fahrenheit temperature is, therefore, 86°.

PRACTICE 7

To find the area A of a trapezoid, multiply $\dfrac{1}{2}$ its height h by the sum of the trapezoid's upper base b and lower base B.

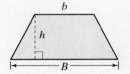

a. Translate this relationship to a formula.

b. Solve the formula for b.

c. What is the upper base of a trapezoid whose area is 32 cm², height is 4 cm, and lower base is 11 cm?

Mathematically Speaking

Fill in each blank with the most appropriate term or phrase from the given list.

number	expression	variable
linear equation	literal equation	formula
constant	algebraic expression	

1. A(n) _____ is an equation involving two or more variables.

2. Literal equations can be solved for one _____ in terms of the others.

3. The solution to a literal equation is a(n) _____.

4. A(n) _____ is a special type of literal equation.

Ⓐ *Solve each equation for the indicated variable.*

5. $y + 10 = x$ for y

6. $b + 13 = a$ for b

7. $d - c = 4$ for d

8. $x - z = -5$ for x

9. $-3y = da$ for d

10. $ax = 5b$ for x

11. $\frac{1}{2}n = 2p$ for n

12. $\frac{3}{2}m = -4l$ for m

◉ 13. $a = \frac{1}{2}xyz$ for z

14. $w = \frac{2}{3}rst$ for r

15. $3x + y = 7$ for x

16. $x + 2y = 5$ for y

◉ 17. $3x + 4y = 12$ for y

18. $5a + 2b = 10$ for a

19. $y - 4t = 0$ for y

20. $6 = p - 4z$ for p

21. $-5b + p = r$ for b

22. $-7a + 3b = c$ for a

23. $h = 2(m - 2l)$ for l

24. $3(a - 2b) = c$ for b

Ⓑ *Solve each formula for the indicated variable.*

25. Uniform motion: $d = rt$ for r

26. Electrical power: $P = iV$ for i

27. Perimeter of a triangle: $P = a + b + c$ for b

28. Perimeter of a rectangle: $P = 2l + 2w$ for w

29. Circumference of a circle: $C = \pi d$ for d

30. Aspect ratio of a hang glider: $R = \frac{s^2}{a}$ for a

31. Power: $P = I^2R$ for R

32. Centripetal force: $F = \frac{mv^2}{r}$ for m

◉ 33. Average of three numbers: $A = \frac{a + b + c}{3}$ for a

34. Distance of a free-falling object: $S = \frac{1}{2}gt^2$ for g

◉ 35. Arithmetic progression: $S = a + (n - 1)d$ for a

36. Simple interest: $A = P(1 + rt)$ for t

Solve each formula for the indicated variable. Then, find the value of this variable using the given information.

37. Simple interest
  a. $I = Prt$ for r
  b. $I = \$12$, $P = \$200$, and $t = 2$ yr

38. Perimeter of a square
  a. $P = 4s$ for s
  b. $P = 60$ ft

39. Area of a parallelogram
  a. $A = bh$ for b
  b. $A = 30 \text{ m}^2$ and $h = 6$ m

40. Average of two numbers
  a. $A = \dfrac{a + b}{2}$ for b
  b. $A = 11.5$ and $a = 10$

Mixed Practice

Solve each formula for the indicated variable.

41. Volume of a cylinder: $V = \pi r^2 h$ for h

42. Area of a rectangle: $A = lw$ for w

Solve each equation for the indicated variable.

43. $m = \dfrac{2}{5}abc$ for b

44. $-10x = yz$ for x

45. $4w + 9z = 3$ for z

46. $a + b = -3$ for a

Solve for the variable shown in color. Then, find the value of this variable for the given values of the other variables.

47. Profit: $P = R - C$ when $P = \$500$ and $C = \$2000$

48. Per capita income: $C = \dfrac{T}{P}$ when $C = \$40,000$ and $P = 300,000,000$

Applications

C *Solve.*

49. In physics, Charles's Law describes the relationship between the volume of a gas and its temperature. The law states that the volume V divided by the temperature T is equal to a constant K.
  a. Write an equation to express this relationship
  b. Solve the relationship for V.

50. In geometry, the volume V of a cylinder is the product of the area B of the circular base and the height h.
  a. Write a formula for this relationship.
  b. Solve the formula for h.

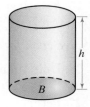

51. In nursing, Clark's Rule for medication expresses a relationship between the recommended dosages for a child and for an adult. The rule states that a child's dosage C equals the product of the weight W of the child in pounds divided by 150 and the adult's dosage A.
  a. Write a formula for this relationship.
  b. Solve the formula for A.

52. The Scholastic Aptitude Test (SAT) is a well-known test taken by high school seniors applying for college admission. Scores on such tests are often converted to standardized scores, or *z-scores*. In 2010, the formula for determining a *z*-score on the mathematics portion of the SAT was $z = \dfrac{x - 516}{116}$, where x is an individual score on the test. The value 516 is the national average score on the test whereas 116 is a measure of how spread out the scores were. (*Source:* professionals.collegeboard.com)
  a. Solve for x in terms of z.
  b. If your *z*-score is 2.2, what was your score on the test, rounded to the nearest whole number?

53. During a storm, the number m of miles away a bolt of lightning strikes can be estimated by first counting the number t of seconds between the bolt of lightning and the associated clap of thunder and then dividing by 5.

   a. Translate this relationship to a formula.

   b. Solve for t in terms of m.

   c. If lightning strikes 2.5 mi away, how many seconds will elapse before the thunder is heard?

54. To estimate a man's shoe size S, triple his foot length l (expressed in inches) and subtract 22.

   a. Express this relationship as a formula.

   b. Estimate the shoe size of a man with a 12-in. foot.

   c. Estimate the foot length of a man with shoe size 12.

55. The circumference C of a circle can be found by doubling the product of the constant π and the circle's radius r.

   a. Express this relationship as a formula.

   b. Solve this formula for r in terms of C and π.

   c. Find the value of r rounded to the nearest tenth if C is 5 ft and π is approximately 3.14.

56. According to Newton's second law of motion, the force F (in newtons) applied to an object equals the product of the object's mass m (in kilograms) and its acceleration a (in m/sec^2).

   a. Translate this relationship to a formula.

   b. Solve for a in terms of m and F.

   c. Find a (in m/sec^2) if m is 3 kg and F is 10 newtons.

• Check your answers on page A-4.

MIND*Stretchers*

Research

1. By examining books in your college library or websites, find several examples of literal equations that relate two or more variables. Write the equation and explain what the variables represent.

Groupwork

2. Working with a partner, give an example of a situation that the following formula might describe.

$$y = mx + b$$

Explain what each variable represents in your example.

Patterns

3. Consider the following table.

x	0	1	2	3	4	$\cdots$	10
y	1	3	5	7	9	$\cdots$	21

Write an equation for y in terms of x.

2.5 Solving Equations Involving Percent

A percent problem typically involves three numbers—the *percent*, the *base*, and the *amount*.

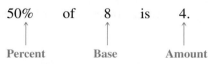

	50%	of	8	is	4.
	↑		↑		↑
	Percent		Base		Amount

We can recognize each number by noting the following:

- The percent typically contains the % sign.
- The base—the number that we are taking the percent of—always follows the word "of" in the statement of the problem.
- The remaining number is called the *amount*.

If any of these numbers is unknown, then we have a percent problem. We can translate the problem into an equation and then solve the equation.

Let's first consider a problem in which we are looking for the base.

EXAMPLE 1

20% of what number is 5?

Solution   Let x represent *what number* (the base). We translate the problem into an equation as shown.

20%	of	what number	is	5?
↓	↓	↓	↓	↓
0.2	·	x	=	5

Then, we solve the equation for x.

$$0.2x = 5$$
$$\frac{0.2x}{0.2} = \frac{5}{0.2} \qquad \text{Divide each side of the equation by 0.2.}$$
$$x = 25$$

So 20% of 25 is 5.

PRACTICE 1

8 is 40% of what number?

EXAMPLE 2

In one year, Steve Nash, an NBA point guard, earned $12.25 million, or 57% of the earnings of another NBA point guard, Jason Kidd. To the nearest million dollars, what were Jason Kidd's earnings? (*Source:* charlotteobserver.com)

PRACTICE 2

In a recent year, Microsoft's profits were about $18 billion, or about 40% of that of ExxonMobil. What were ExxonMobil's profits that year? (*Source: The New York Times Almanac 2010*)

Solution   The question is:

$12.25 million is 57% of what amount?

$$12.25 = 0.57 \cdot x \quad \text{Translate the sentence to an equation.}$$
$$0.57x = 12.25 \quad \text{Rewrite the equation.}$$
$$\frac{0.57x}{0.57} = \frac{12.25}{0.57}$$
$$x \approx 21$$

So Jason Kidd's earnings were about $21,000,000.

Next, let's consider a problem in which we are looking for the amount.

EXAMPLE 3

What number is 3.5% of $200?

Solution   Let x represent *what number* (the amount). Then, we translate the problem into an equation.

What number   is   3.5%   of $200?

$$x \quad = \quad 0.035 \quad \cdot \quad 200$$

Now, we solve for x.

$$x = 0.035 \cdot 200 = 7$$

So $7 is 3.5% of $200.

PRACTICE 3

What is 23% of 45 m?

EXAMPLE 4

The following graph shows the breakdown of the projected U.S. population by gender in the year 2020. If the population is expected to be approximately 340 million people, how many more women than men will there be in 2020? (*Source:* www.census.gov)

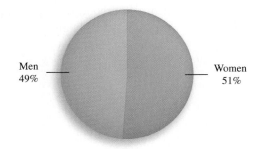

Men 49%   Women 51%

PRACTICE 4

The following graph shows the nations with the biggest share of global arms spending. If the total spending on global arms for all nations is $1.53 trillion, how much more does the U.S. spend than China, to the nearest billion dollars? (*Source: USA Today*)

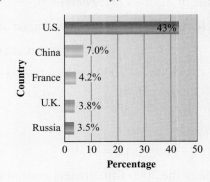

Solution   First, we calculate the number of women in 2020.

What amount   is   51%   of   340,000,000?

$$x \quad = \quad 0.51 \quad \cdot \quad 340,000,000$$
$$x = 0.51 \cdot 340,000,000 = 173,400,000$$

EXAMPLE 4 (continued)

Similarly, the number of men in 2020 is:

$$x = 0.49 \cdot 340{,}000{,}000 = 166{,}600{,}000$$

So there will be $173{,}400{,}000 - 166{,}600{,}000$, or $6{,}800{,}000$ more women than men.

Finally, we consider a problem in which we are looking for the percent.

EXAMPLE 5

What percent of 9 is 6?

Solution   Let x represent *what percent*. Then, we translate the problem into an equation as shown.

$$\underbrace{\text{What percent}}_{x} \quad \text{of} \quad 9 \quad \text{is} \quad 6?$$
$$x \quad \cdot \quad 9 \quad = \quad 6$$

Now, we solve the equation for x.

$$x \cdot 9 = 6$$
$$9x = 6 \qquad \text{$x \cdot 9 = 6$ is equivalent to $9x = 6$}$$
$$\frac{9x}{9} = \frac{6}{9} \qquad \text{Divide each side of the equation by 9.}$$
$$x = \frac{2}{3} = 66\tfrac{2}{3}\% \qquad \text{Change $\tfrac{2}{3}$ to a percent.}$$

So $66\tfrac{2}{3}\%$ of 9 is 6.

PRACTICE 5

14 is what percent of 16?

EXAMPLE 6

Three thousand students in the fifth grade were asked the question, "What job would you most like to have when you grow up?" Of these students, 540 said that they wanted to be teachers. What percent of the fifth graders wanted to be teachers? (*Source: The Second Kids' World Almanac of Records and Facts*)

Solution   The question that we must answer is:

$$540 \text{ is what percent of } 3000?$$

$$540 = x \cdot 3000 \qquad \text{Translate the sentence to an equation.}$$
$$3000x = 540 \qquad \text{Rewrite the equation and solve.}$$
$$\frac{3000}{3000}x = \frac{540}{3000}$$
$$x = 0.18 = 18\%$$

So 18% of the 3000 fifth graders wanted to be teachers.

PRACTICE 6

Of the first 44 U.S. presidents, 14 had also served as vice president. To the nearest percent, what percent of the presidents had been vice president? (*Source:* senate.gov)

Now, let's consider a different type of "what percent" problem that deals with a *changing quantity*. If the quantity is increasing, we have a *percent increase*; if it is decreasing, we have a *percent decrease*.

To Find a Percent Increase or Decrease

- Compute the difference between the two given values.

- Compute what percent this difference is of the *original value*.

EXAMPLE 7

In one year, the total national expenditure for health care in the U.S. was about $75 billion. Ten years later, the amount was approximately $250 billion. What percent increase was this, to the nearest ten percent? (*Source: The New York Times Almanac 2010*)

Solution   First, we compute the difference between values.

$$250 - 75 = 175 \leftarrow \text{Change in value}$$

Then, we answer the question:

What percent of 75 is 175?

$$x \cdot 75 = 175 \qquad \text{Translate the sentence to an equation.}$$
$$75x = 175 \qquad \text{Rewrite the equation and solve.}$$
$$\frac{75x}{75} = \frac{175}{75}$$
$$x = \frac{175}{75}$$
$$= 2.333 \ldots , \text{ or } 233.3 \ldots \%$$

So the total national expenditures for healthcare increased by 230%, to the nearest ten percent.

PRACTICE 7

In 5 yr, the number of nursing homes in Louisiana decreased from 292 to 285. To the nearest tenth of a percent, what was the percent decrease in the number of nursing homes? (*Source:* statehealthfacts.org)

EXAMPLE 8

A college will close if its student enrollment drops by more than 60%. If the enrollment falls from 4000 to 1800, will the college close?

Solution   The student enrollment fell from 4000 to 1800; that is, it fell by 2200. The question is how the percent decrease of the student enrollment compares with 60%. We compute the percent decrease as follows:

What percent of 4000 is 2200?

$$x \cdot 4000 = 2200$$
$$4000x = 2200$$
$$\frac{4000}{4000}x = \frac{2200}{4000}$$
$$x = \frac{22}{40} = \frac{11}{20} = 0.55 = 55\%$$

Since the student enrollment fell by 55%, which is less than 60%, the college will not close.

PRACTICE 8

Financial crashes took place on both Friday, October 29, 1929, and Monday, October 19, 1987. On the earlier date, the stock index dropped from 300 to 230. On the later date, it dropped from 2250 to 1750. As a percent, did the stock index drop more in 1929 or in 1987? (*Source: The Washington Post,* "Sell-Offs Rock Wall Street, World Markets," October 28, 1987)

Some percent problems involve *interest.* When you lend, deposit, or invest money, you *earn* interest. When you borrow, you *pay* interest.

The amount of interest earned (or paid) *I* depends on the *principal P* (the amount of money invested or borrowed), the annual rate of interest *r* (usually expressed as a percent), and the length of time *t* the money is invested or borrowed (generally expressed in years). We compute the amount of interest by multiplying the principal by the rate of interest and the number of years. This kind of interest is called *simple interest.*

EXAMPLE 9

How much simple interest was earned in 2 yr on a principal of $900 in an account with an annual interest rate of 5.25%?

Solution To compute the interest, we use the formula $I = Prt$

$$I = P \cdot r \cdot t.$$
$$= (900)(0.0525)(2)$$
$$= 94.5$$

So the interest earned was $94.50.

PRACTICE 9

In 1 yr, a bank paid $130 in simple interest on an initial balance of $2000. What annual rate of interest was paid?

Some applications of percent involve *multiple investments,* as the following example illustrates.

EXAMPLE 10

A book editor invested part of her $5000 bonus in a fund that paid 4% simple interest and the rest in a CD that paid 6% simple interest. Find the amount invested at each rate if the overall interest earned in 1 yr was $260.

Solution Let's set up a table to organize the given information. Note that the interest earned for each investment is the product of each principal, the rate of interest, and the time.

	Principal ·	Rate of Interest ·	Time =	Interest Earned
Fund	x	0.04	1	$0.04x$
CD	$5000 - x$	0.06	1	$0.06(5000-x)$
Total Interest				260

The interest earned on the total investment is the sum of the interest earned on the fund and the interest earned on the CD.

$$0.04x + 0.06(5000 - x) = 260$$
$$0.04x + 300 - 0.06x = 260$$
$$-0.02x = -40$$
$$\frac{-0.02x}{-0.02} = \frac{-40}{-0.02}$$
$$x = 2000$$

So $2000 was invested in the fund and $5000 − $2000, or $3000, was invested in the CD.

PRACTICE 10

The amount of money a broker invested in bonds was double what she invested in a mutual fund. After 1 yr, the investment in bonds gained 10% and the investment in the mutual fund lost 10%. If the net gain was $700, how much was each of her investments?

Another type of problem involving percent is a *mixture* or *solution problem*. In this kind of problem, the amount of a particular ingredient in the solution or mixture is often expressed as a percent of the total. Chemists and pharmacists commonly need to solve mixture and solution problems.

EXAMPLE 11

A chemist added 2 liters (L) of water to 8 L of a 48% alcohol solution. What is the alcohol concentration of the new solution, rounded to the nearest percent?

Solution By adding 2 L of water, we are diluting 8 L of 48% alcohol solution to a solution with unknown alcohol concentration. Let x represent the alcohol concentration of the new solution expressed as a decimal. It is helpful to organize the given information into a table, as follows:

Action	Substance	Amount (in liters)	Amount of Pure Alcohol (in liters)
Start with	48% alcohol solution	8	$0.48(8)$
Add	water	2	$0(2)$
Finish with	new alcohol solution	8 + 2, or 10	$x(10)$

The amount of pure alcohol in the final solution is equal to the sum of the amount of pure alcohol in the initial solution and the amount of pure alcohol in the water added.

$$0.48(8) + 0(2) = 10x$$

Solving the equation for x, we get:

$$0.48(8) = 10x$$
$$3.84 = 10x$$
$$\frac{3.84}{10} = \frac{10x}{10}$$
$$0.384 = x$$

So the alcohol concentration of the new solution is 38.4%, or, rounded to the nearest percent, 38%.

PRACTICE 11

How many grams of salt must be added to 30 g of a solution that is 20% salt to make a solution that is 25% salt?

How would you explain each entry in the table in Example 11?

Exercises

Mathematically Speaking

Fill in each blank with the most appropriate term or phrase from the given list.

plus	sum of	times
amount	base	difference between

1. In solving the percent problem "30% of what number is 7?", "what number" represents the _____.

2. In solving the percent problem "What is 10% of 4?", "What" represents the _____.

3. In translating the problem "What percent of 36 is 9?", "of" becomes _____.

4. In finding a percent increase or decrease, first compute the _____ the two given values.

A *Find the amount.*

5. What is 75% of 8?

6. What is 50% of 48 ft?

7. Compute 100% of 23.

8. Compute 200% of 6.

9. Find 41% of 7 kg.

10. Find 6% of 9.

11. What is 8% of $500?

12. What is 6% of $200?

13. What is 12.5% of 32?

14. What is 37.5% of 40?

Find the base.

15. 25% of what area is 8 in^2?

16. 30% of what weight is 120 kg?

17. $12 is 10% of how much money?

18. $20 is 10% of how much money?

19. 1% of what salary is $195?

20. 2% of what amount of money is $5?

21. 3.5 is 200% of what number?

22. 8.1 is 150% of what number?

23. 0.5% of what length is 23 m?

24. 0.75% of what number is 24?

Find the percent.

25. 50 cents is what percent of 80 cents?

26. 13 is what percent of 20?

27. 5 is what percent of 15?

28. 10 is what percent of 12?

29. 10 hours is what percent of 8 hours?

30. $30 is what percent of $20?

31. What percent of 5 is $\frac{1}{2}$?

32. What percent of 6 is $\frac{2}{3}$?

33. 2.5 g is what percent of 4 g?

34. 0.1 is what percent of 8?

Mixed Practice

Solve.

35. What is 35% of $400?

36. What is 19% of $10,000?

37. What percent of 50 mi is 20 mi?

38. What percent of $20 is $15?

39. 70% of what amount is 14?

40. 80% of what amount is 28?

41. 0.1% of 35 is what number?

42. 0.5% of 20 is what number?

43. $8 is what percent of $240?

44. 7 is what percent of 21?

45. 3 oz is 20% of what weight?

46. 8 is 50% of what number?

47. $14 is what percent of $8?

48. What is 32% of $9?

49. 80% of what amount is 96?

50. 75 is what percent of 3000?

51. 8% of 120 is what amount?

52. 350% of how much money is $77?

53. 21 is 20% of what number?

54. $\frac{3}{4}$ is what percent of 6?

Applications

B *Solve.*

55. The following graph shows the breakdown of the 213 million eligible voters during a recent United States presidential election.

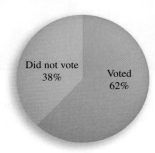

How many more eligible voters voted than did not vote, rounded to the nearest ten million? (*Source:* elections. gmu.edu)

56. In the tax year 2009, a single taxpayer would have paid federal taxes at the rates shown in the following table.

Taxable Income	Tax Rate
0–$8350	10%
$8351–$33,950	15%
$33,951–$82,250	25%
$82,251–$171,550	28%
$171,551–$372,950	33%
$372,951–	35%

For instance, on a taxable income of $10,000, 10% would be owed on the first $8350 and 15% would be owed on the remaining $10,000 − $8350, or $1650. Find the total federal taxes owed on this income. (*Source:* wikipedia.org)

57. A college graduate enlisted in the army for 4 yr. So far, he has served 9 mo. What percent of his enlistment has passed?

58. A questionnaire was mailed to 5000 people. Of these recipients, 3000 responded. What was the response rate, expressed as a percent?

59. Yesterday, 8 employees in an office were absent. If these absentees constitute 25% of the employees, what is the total number of employees?

60. Payroll deductions comprise 40% of a worker's weekly gross income. If these deductions total $240, what is her gross income?

61. A house is purchased for $250,000. What percent of the purchase price is a down payment of $50,000?

62. Last year, an office worker's income was $25,000. If she donated $900 to charity, what percent of her income went to charity?

63. According to a report on a country's economic conditions, 1.5 million people, or 8% of the workforce, were unemployed. How large was the country's workforce?

64. In a heart transplant operation, the donor heart cost $90,000, which was about 12% of the cost of the operation. How much, to the nearest ten thousand dollars, did the heart transplant operation cost? (*Source:* transplantliving.org)

65. In a restaurant, 60% of the tables are in the no-smoking section. If the restaurant has 90 tables, how many tables are there in this section?

66. A shopper lives in a town where the sales tax is 5%. Across the river, the tax is 4%. If it costs her $6 to make the round-trip across the river, on which side of the river should she buy a $250 television set?

67. Suppose that an animal species is considered to be endangered if its population drops by more than 60%. If a species' population fell from 4000 to 1800, should we consider the species endangered?

68. The first commercial telephone exchange was set up in New Haven, Connecticut, in 1878. Between 1880 and 1890, the number of telephone systems in the United States increased from 50 to 200, in round numbers. What percent increase was this? (*Source: The World Almanac 2010*)

69. A student borrowed $2000 for 1 yr at 5% simple interest to buy a computer and printer. How much interest did she pay?

70. How much simple interest is earned on $600 in a bank account at an 8% annual interest rate for 6 mo?

71. How much simple interest is earned on a deposit of $5000 in an online savings account after 1 yr if the interest rate is 5%?

72. An aunt lent her nephew $2000. The nephew agreed to pay his aunt 4% simple interest. If he promised to repay her the entire amount owed at the end of 3 yr, how much money must he pay her?

73. Part of $34,000 is invested at an interest rate of 8% and the rest at an interest rate of 10%. If the total interest earned in 1 yr was $3000, how much money was invested at each rate?

74. The amount of money invested at 5% simple interest was half the amount invested at 7% simple interest. If the total yearly interest earned was $380, how much money was invested at each rate?

75. An amount of $20,000 was invested in a fund with a return of 8%. How much money was invested in a fund with a 5% return if the total return on both investments was $2100?

76. A beneficiary split her $100,000 inheritance between two investments. One of the investments gained 12% and the other lost 8%. If she broke even on the two investments, how much money did she invest in each?

77. A basic lemon vinaigrette salad dressing can be made by mixing 1 cup of olive oil with $\frac{1}{3}$ cup of lemon juice along with pinches of salt and white pepper. How much more olive oil should be added to make a dressing that is 20% lemon juice? (The salt and pepper contribution is negligible and can be left out of the computation.) (*Source: The Sauce Bible*)

78. A brand of antifreeze states that a radiator containing a solution that is 50% antifreeze and 50% water provides protection for a temperature as low as −34°F, whereas a solution that is 70% antifreeze provides protection down to −84°F. A car's radiator has 4 qt of a 50% solution. If the capacity of the radiator is 6 qt and the rest of the radiator is filled with pure antifreeze, what percent of the resulting solution is antifreeze? (*Source: Prestone II Antifreeze*)

79. How many ounces of a 40% acetic acid solution should a photographer add to 30 oz of a 4% acetic acid solution to obtain a 10% solution?

80. A pharmacist has 10 L of a 5% drug solution. How many liters of 2% solution should be added to produce a solution that is 3%?

• Check your answers on page A-4.

MIND*Stretchers*

Writing

1. Write a word problem whose answer is 60%, in which you give the base and the amount.

Mathematical Reasoning

2. What is a% of b% of c% of 1,000,000,000?

Groupwork

3. Some properties of percent are counterintuitive. For example, consider the following two *false* statements. Working with a partner, discuss why they are both false.

  a. If the price of an item is increased by 10% and then decreased by 10%, the final price is the same as the original price.

  b. If the price of an item is increased by 5% and then increased by another 3%, altogether it is increased by 8%.

2.6   Solving Inequalities

In Section 1.1, we used the symbols $<$, $\leq$, $>$, $\geq$, $=$, and $\neq$ to compare two real numbers. For example, with the real numbers -5 and 4, we can write the following statements:

$$-5 < 4 \qquad 4 > -5 \qquad -5 \neq 4$$

A To determine if a number is a solution of an inequality

B To graph the solutions of linear inequalities on the number line

C To solve a linear inequality using the addition and multiplication properties of inequalities

D To solve applied problems involving linear inequalities

DEFINITION

An **inequality** is any mathematical statement containing $<$, $\leq$, $>$, $\geq$, or $\neq$.

Solutions of Inequalities

Now, consider an inequality that involves a variable, say $x < 2$. Let's look at the values of x that make this inequality true.

	Values for x	$x < 2$	True or False?
Values for x that are less than 2	1	$1 < 2$	True
	$\dfrac{1}{2}$	$\dfrac{1}{2} < 2$	True
	0	$0 < 2$	True
	-1	$-1 < 2$	True
Values for x that are not less than 2	2	$2 < 2$	False
	3	$3 < 2$	False
	$3\frac{1}{2}$	$3\frac{1}{2} < 2$	False
	4	$4 < 2$	False

Note that there are many values of x that make $x < 2$ true. Can you name them all? Explain.

DEFINITION

A **solution of an inequality** is any value of the variable that makes the inequality true. To **solve an inequality** is to find all of its solutions.

EXAMPLE 1

Determine whether -3 is a solution of the inequality $2x + 5 \geq -3$.

Solution   To determine if -3 is a solution of the inequality, we substitute -3 for x and simplify.

$$2x + 5 \geq -3$$
$$2(-3) + 5 \overset{?}{\geq} -3 \qquad \text{Substitute } -3 \text{ for } x.$$
$$-6 + 5 \overset{?}{\geq} -3 \qquad \text{Multiply.}$$
$$-1 \geq -3 \qquad \text{True}$$

Because $-1 \geq -3$ is a true statement, -3 is a solution of the inequality.

PRACTICE 1

Is 4 a solution of the inequality
$$\frac{1}{2}x - 2 < -1?$$

For any inequality, we can draw on the number line a picture of its solutions—the *graph* of the inequality. Graphing the solutions of an inequality can be clearer than describing the solutions in symbols or words.

EXAMPLE 2

Draw the graph of $x < 2$.

Solution   The graph of $x < 2$ includes all points on the number line to the left of 2. The open circle on the graph shows that 2 is *not* a solution.

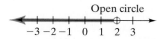

PRACTICE 2

Draw the graph of $x > 1$.

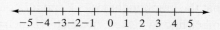

Note that solving an inequality generally results in a range of numbers rather than in a single number.

EXAMPLE 3

Draw the graph of $x \geq -\dfrac{1}{2}$.

Solution   The graph of $x \geq -\dfrac{1}{2}$ includes all points on the number line to the right of $-\dfrac{1}{2}$ and also $-\dfrac{1}{2}$. The *closed circle* shows that $-\dfrac{1}{2}$ is a solution of $x \geq -\dfrac{1}{2}$.

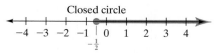

PRACTICE 3

Draw the graph of $x \leq -1\frac{1}{2}$.

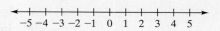

EXAMPLE 4

Draw the graph of $-1 \leq x < 2$.

Solution   This inequality is read either "-1 is less than or equal to x *and* x is less than 2" or "x is greater than or equal to -1 *and* x is less than 2." The solutions of this inequality are all values of x that satisfy both $-1 \leq x$ and $x < 2$, and its graph is the overlap of the graphs of $-1 \leq x$ and $x < 2$.

Graph of $-1 \leq x$:

Graph of $x < 2$:

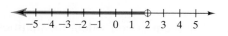

PRACTICE 4

Draw the graph of $-3 < x < 4$.

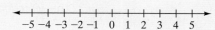

EXAMPLE 4 (continued)

Graph of $-1 \leq x < 2$:

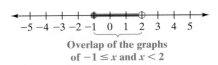

Overlap of the graphs
of $-1 \leq x$ and $x < 2$

Note that the graph includes -1 and all points *between* -1 and 2 on the number line. A closed circle at -1 means that -1 is a solution of the inequality, whereas an open circle at 2 means that 2 is not a solution of the inequality.

Solving Inequalities Using the Addition Property

Now, let's consider what happens when we perform the same operation on each side of an inequality. In the following inequality, we add 4 to each side of the inequality $5 < 12$.

$$5 < 12 \qquad \text{True}$$
$$5 + 4 \overset{?}{<} 12 + 4 \qquad \text{Add 4 to each side of the inequality.}$$
$$9 < 16 \qquad \text{True}$$

Note that $5 < 12$ and $5 + 4 < 12 + 4$ are both true. In the same way, adding -3 to both sides of $x + 3 < 7$ gives us the *equivalent* inequality $x + 3 + (-3) < 7 + (-3)$, or $x < 4$. Alternatively, we could subtract 3 from both sides of the original inequality, getting the same inequality as before. Inequalities are said to be equivalent if they have the same solutions. For instance, 2 is a solution of $x + 3 < 7$ just as it is a solution of $x < 4$.

This example suggests the addition property of inequalities.

Addition Property of Inequalities

For any real numbers a, b, and c,

- $a < b$ and $a + c < b + c$ are equivalent.
- $a > b$ and $a + c > b + c$ are equivalent.

Similar statements hold for $\leq$ and $\geq$.

TIP   When the same number is added to or subtracted from each side of an inequality, the *direction* of the inequality is unchanged.

The way we solve inequalities is similar to the way we solve equations.

Equation	Inequality
$x + 3 = 5$	$x + 3 < 5$
$x + 3 - 3 = 5 - 3$	$x + 3 - 3 < 5 - 3$
$x = 2$ Solution	$x < 2$ Solution
$x + 3 = 5$ is equivalent to $x = 2$.	$x + 3 < 5$ is equivalent to $x < 2$.

We solve inequalities by expressing them as equivalent inequalities in which the variable term is isolated on one side.

EXAMPLE 5

Solve and graph: $y + 5 > 9$

Solution $y + 5 > 9$

$y + 5 - 5 > 9 - 5$ Subtract 5 from each side of the inequality.

$y > 4$

The graph of $y > 4$ is:

$$\xleftarrow{\hspace{0.5cm}} \overset{}{\underset{-1\;\;0\;\;1\;\;2\;\;3\;\;4\;\;5\;\;6\;\;7}{+\!\!+\!\!+\!\!+\!\!+\!\!+\circ\!\!+\!\!+\!\!+}} \xrightarrow{\hspace{0.5cm}}$$

Note that an open circle is drawn at 4 to show that 4 is not a solution.

    Because an inequality has many solutions, we cannot check all of the solutions as we did with an equation. However, we can do a partial check of the solutions of an inequality by substituting points on the graph in the original inequality. For instance, to check that all values for y greater than 4 are the solutions of $y + 5 > 9$, we replace y in the original inequality with some points on the graph and some points not on the graph.

Values for y	$y + 5 > 9$	True or False
6	$6 + 5 > 9$	True
5	$5 + 5 > 9$	True
4	$4 + 5 > 9$	False
3	$3 + 5 > 9$	False

The table confirms that the solution of $y + 5 > 9$ is $y > 4$. That is, any number greater than but not equal to 4 is a solution.

EXAMPLE 6

Solve and graph: $z - 2 \le -3\frac{1}{2}$

Solution $z - 2 \le -3\frac{1}{2}$

$z - 2 + 2 \le -3\frac{1}{2} + 2$ Add 2 to each side of the inequality.

$z \le -1\frac{1}{2}$

So all numbers less than or equal to $-1\frac{1}{2}$ are solutions. The graph of $z \le -1\frac{1}{2}$ is:

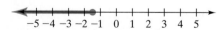

$$\xleftarrow{\hspace{0.5cm}} \underset{-5\;-4\;-3\;-2\;-1\;\;0\;\;1\;\;2\;\;3\;\;4\;\;5}{+\!\!+\!\!+\!\!\bullet\!\!+\!\!+\!\!+\!\!+\!\!+\!\!+} \xrightarrow{\hspace{0.5cm}}$$

PRACTICE 5

Solve and graph: $n + 5 > 4$

PRACTICE 6

Solve and graph: $x - 4 \le 1\frac{1}{2}$

EXAMPLE 7

Solve and graph: $6y - 3 < 5y + 4$

Solution $6y - 3 < 5y + 4$

$6y - 5y - 3 < 5y - 5y + 4$ Subtract 5y from each
side of the inequality.

$y - 3 < 4$

$y - 3 + 3 < 4 + 3$ Add 3 to each side of the
inequality.

$y < 7$

So all numbers less than or equal to 7 are solutions of
$6y - 3 < 5y + 4$. The graph of $y < 7$ is:

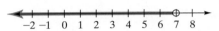

PRACTICE 7

Solve and graph: $4x + 5 > 3x - 2$

Solving Inequalities Using the Multiplication Property

Consider the inequality $12 < 15$. Let's look at what happens when we multiply this inequality by a *positive* number:

$$12 < 15$$
$$12 \cdot 3 \overset{?}{<} 15 \cdot 3 \qquad \text{Multiply each side of the inequality by 3.}$$
$$36 < 45 \qquad \text{True}$$

Note that the direction of the last inequality is the same as that of the first inequality. Now, we consider multiplying each side of the original inequality by a *negative* number:

$$12 < 15$$
$$12(-3) \overset{?}{<} 15(-3) \qquad \text{Multiply each side of the inequality by } -3.$$
$$-36 \overset{?}{<} -45 \qquad \text{False, unless the direction of the inequality sign is reversed}$$
$$-36 > -45 \qquad \text{True}$$

Observe that the direction of the last inequality is the reverse of that of the first inequality. These examples suggest the *multiplication property of inequalities*.

Multiplication Property of Inequalities

For any real numbers a, b, and c,

- If c is positive, then $a < b$ and $ac < bc$ are equivalent.

- If c is negative, then $a < b$ and $ac > bc$ are equivalent.

Similar statements hold for $>$, $\leq$, and $\geq$.

We can demonstrate a similar property for division.

$$12 < 15$$
$$\frac{12}{3} \overset{?}{<} \frac{15}{3} \qquad \text{Divide each side of the}$$
$$\text{inequality by 3.}$$
$$4 < 5 \qquad \text{True}$$

$$12 < 15$$
$$\frac{12}{-3} \overset{?}{<} \frac{15}{-3} \qquad \text{Divide each side of the inequality by } -3.$$
$$-4 \overset{?}{<} -5 \qquad \text{False, so we need to reverse the}$$
$$\text{direction of the inequality sign.}$$
$$-4 > -5 \qquad \text{True}$$

Note that when we multiply or divide each side of an inequality by a positive number, the direction of the inequality remains the same. But when we multiply or divide each side of an inequality by a negative number, the direction of the inequality is *reversed*.

EXAMPLE 8

Solve and graph: $\dfrac{x}{2} < 3$

Solution $\quad \dfrac{x}{2} < 3$

$\qquad 2 \cdot \dfrac{x}{2} < 2 \cdot 3 \qquad$ Multiply each side of the inequality by 2.

$\qquad x < 6$

So any number less than 6 is a solution. The graph of $x < 6$ is:

EXAMPLE 9

Solve and graph: $\quad -4z \le 12$

Solution $\quad -4z \le 12$

$\qquad \dfrac{-4z}{-4} \ge \dfrac{12}{-4} \qquad$ Divide each side of the inequality by -4 and reverse the direction of the inequality.

$\qquad z \ge -3$

The solution of $-4z \le 12$ is $z \ge -3$, so all numbers greater than or equal to -3 are solutions. The graph of $z \ge -3$ is:

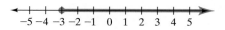

EXAMPLE 10

Solve and graph: $\quad -21 < 6y - 9y$

Solution $\quad -21 < 6y - 9y$

$\qquad -21 < -3y \qquad$ Combine like terms.

$\qquad \dfrac{-21}{-3} > \dfrac{-3y}{-3} \qquad$ Divide each side of the inequality by -3 and reverse the direction of the inequality.

$\qquad 7 > y, \quad$ or $y < 7$

So all numbers less than 7 are solutions. The graph is:

PRACTICE 8

Solve and graph: $\dfrac{x}{3} \le 1$

PRACTICE 9

Solve and graph: $\quad -3x > 15$

PRACTICE 10

Solve and graph: $\quad 10 > 5x - 7x$

As in solving equations, we may need to use more than one property of inequalities to solve some inequalities.

EXAMPLE 11

Solve: $5y - 4 - 6y \leq -8$

Solution

$$5y - 4 - 6y \leq -8$$

$-y - 4 \leq -8$ — Combine like terms.

$-y - 4 + 4 \leq -8 + 4$ — Add 4 to each side of the inequality.

$-y \leq -4$ — Simplify.

$\dfrac{-y}{-1} \geq \dfrac{-4}{-1}$ — Divide each side of the inequality by -1, and reverse the direction of the inequality.

$y \geq 4$

So all numbers greater than or equal to 4 are solutions.

PRACTICE 11

Solve: $-6 \geq 3z + 4 - z$

EXAMPLE 12

Solve: $3n - 2(n + 3) < 14$

Solution

$$3n - 2(n + 3) < 14$$

$3n - 2n - 6 < 14$ — Use the distributive property.

$n - 6 < 14$ — Combine like terms.

$n - 6 + 6 < 14 + 6$ — Add 6 to each side of the inequality.

$n < 20$

So all numbers less than 20 are solutions.

PRACTICE 12

Solve: $7x - (9x + 1) > -5$

Some common word phrases used in applied problems involving inequalities and their translations are shown in the following table:

Word Phrase	Translation
x is less than a	$x < a$
x is less than or equal to a	$x \leq a$
x is greater than a	$x > a$
x is greater than or equal to a	$x \geq a$
x is at most a	$x \leq a$
x is no more than a	$x \leq a$
x is at least a	$x \geq a$
x is no less than a	$x \geq a$

EXAMPLE 13

In geometry, the triangle inequality states that the sum of the lengths of any two sides of a triangle is greater than the length of the third side. In the isosceles triangle shown, write and solve an inequality to find the possible side lengths *a*. Graph the inequality.

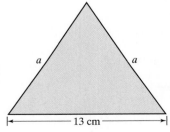

PRACTICE 13

In the triangle shown, for which values of x will the perimeter be greater than or equal to 14 in.?

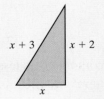

Solution   The sum of the lengths of the two equal sides is greater than the length of the third side.

$$a + a > 13$$
$$2a > 13$$

Solving the inequality, we get:

$$\frac{2a}{2} > \frac{13}{2} \qquad \textbf{Divide each side by 2.}$$

$$a > 6.5$$

The graph is:

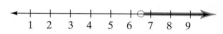

We conclude that the length of each side a is any number greater than 6.5 cm.

EXAMPLE 14

A factory's quality-control department randomly selects a sample of 5 lightbulbs to test. In order to meet quality-control standards, the lightbulbs in the sample must last an average of at least 950 hr. Four of the selected lightbulbs lasted 925 hr, 1000 hr, 950 hr, and 900 hr. How many hours must the fifth lightbulb last if the sample is to meet quality-control standards?

Solution   The average number of hours that the 5 lightbulbs must last is the sum of the hours each lightbulb lasts divided by the number of lightbulbs in the sample, which is 5.

$$\text{Average number of hours} = \frac{925 + 1000 + 950 + 900 + x}{5}$$

where x represents the number of hours the fifth bulb lasts. In order for the average to be *at least* 950, it must be greater than or equal to 950. So we write and solve the inequality.

$$\frac{925 + 1000 + 950 + 900 + x}{5} \geq 950$$

$$5\left(\frac{925 + 1000 + 950 + 900 + x}{5}\right) \geq 950 \cdot 5$$

$$925 + 1000 + 950 + 900 + x \geq 4750$$

$$3775 + x \geq 4750$$

$$3775 - 3775 + x \geq 4750 - 3775$$

$$x \geq 975$$

So the fifth lightbulb in the sample must last at least 975 hr for the sample to meet quality control standards.

PRACTICE 14

A student has two part-time jobs. On the first job, she works 15 hr a week at $8.50 an hour. The second job pays only $7.50 an hour, but she can work as many hours as she wants. To make at least $300, how many hours should she work on the second job?

Mathematically Speaking

Fill in each blank with the most appropriate term or phrase from the given list.

equation	reversed	unchanged
graph	inequality	positive
negative	solution	open
closed		

1. A(n) _____ is any mathematical statement containing $<$, $\leq$, $>$, $\geq$, or $\neq$.

2. A(n) _____ of an inequality is any value of the variable that makes the inequality true.

3. In the graph of the inequality $x > -7$, the circle is _____.

4. In the graph of the inequality $x \geq -7$, the circle is _____.

5. When the same number is added to or subtracted from each side of an inequality, the direction of the inequality is _____.

6. According to the multiplication property of inequalities, if c is _____, then $a \geq b$ and $ac \geq bc$ are equivalent.

7. According to the multiplication property of inequalities, if c is _____, then $a \geq b$ and $ac \leq bc$ are equivalent.

8. When each side of an inequality is divided by a negative number, the direction of the inequality is _____.

A *Indicate whether the value of x shown is a solution of the given inequality by answering true or false.*

9.
Value of x	Inequality	True or False
a. 1	$8 - 3x > 5$	
b. 4	$4x - 7 \leq 2x + 1$	
c. -7	$6(x + 6) < -9$	
d. $-\dfrac{3}{4}$	$8x + 10 \geq 12x + 15$	

10.
Value of x	Inequality	True or False
a. 3	$2x + 1 < -10$	
b. -5	$1 - 3x > 5x + 12$	
c. -2	$-(4x + 8) \geq 0$	
d. $\dfrac{1}{3}$	$9x - 7 \leq 6x - 11$	

B *Graph on the number line.*

11. $x > 3$

12. $x > -1$

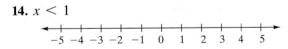

13. $x < -5$

14. $x < 1$

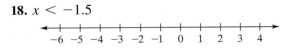

15. $x \geq 2$

16. $x \geq 1$

17. $x < -2.5$

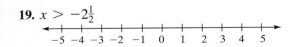

18. $x < -1.5$

19. $x > -2\frac{1}{2}$

20. $x \geq 4\frac{1}{2}$

21. $x \geq -\dfrac{1}{3}$

<-+----+----+----+----+----+----+----+----+----+----+->
　-5　-4　-3　-2　-1　0　1　2　3　4　5

22. $x < \dfrac{3}{4}$

<-+----+----+----+----+----+----+----+----+----+----+->
　-5　-4　-3　-2　-1　0　1　2　3　4　5

23. $-3 < x < 1$

<-+----+----+----+----+----+----+----+----+----+----+->
　-5　-4　-3　-2　-1　0　1　2　3　4　5

24. $0 < x \leq 4$

<-+----+----+----+----+----+----+----+----+----+----+->
　-5　-4　-3　-2　-1　0　1　2　3　4　5

25. $-\dfrac{1}{2} \leq x < 2$

<-+----+----+----+----+----+----+----+----+----+----+->
　-5　-4　-3　-2　-1　0　1　2　3　4　5

26. $-4 \leq x \leq 1\frac{1}{2}$

<-+----+----+----+----+----+----+----+----+----+----+->
　-5　-4　-3　-2　-1　0　1　2　3　4　5

Solve and graph.

27. $v + 2 < -5$

<-+----+----+----+----+----+----+----+----+----+----+->

28. $s + 1 \leq 1$

<-+----+----+----+----+----+----+----+----+----+----+->

29. $y - 5 > -5$

<-+----+----+----+----+----+----+----+----+----+----+->

30. $x - 5 > -1$

<-+----+----+----+----+----+----+----+----+----+----+->

31. $y + 2 \leq 5.5$

<-+----+----+----+----+----+----+----+----+----+----+->

32. $t + 3 \leq 6.5$

<-+----+----+----+----+----+----+----+----+----+----+->

33. $v - 17 \leq -15$

<-+----+----+----+----+----+----+----+----+----+----+->

34. $n - 25 > -30$

<-+----+----+----+----+----+----+----+----+----+----+->

35. $-2 \geq x - 4$

<-+----+----+----+----+----+----+----+----+----+----+->

36. $4 < a - 3$

<-+----+----+----+----+----+----+----+----+----+----+->

37. $\dfrac{1}{3}a < -1$

<-+----+----+----+----+----+----+----+----+----+----+->

38. $\dfrac{1}{2}x \leq 3$

<-+----+----+----+----+----+----+----+----+----+----+->

39. $-5y > 10$

<-+----+----+----+----+----+----+----+----+----+----+->

40. $-7t \leq -21$

<-+----+----+----+----+----+----+----+----+----+----+->

41. $2x \geq 0$

<-+----+----+----+----+----+----+----+----+----+----+->

42. $-3.5m < 0$

<-+----+----+----+----+----+----+----+----+----+----+->

43. $-\dfrac{3}{4}a \geq 3$

<-+----+----+----+----+----+----+----+----+----+----+->

44. $-\dfrac{x}{3} \geq 2$

<-+----+----+----+----+----+----+----+----+----+----+->
　-7　-6　-5　-4　-3　-2　-1　0　1　2　3

45. $6 \leq -\dfrac{2}{3}n$

<-+----+----+----+----+----+----+----+----+----+----+->

46. $4 \leq -\dfrac{y}{2}$

<-+----+----+----+----+----+----+----+----+----+----+->

Solve.

47. $\dfrac{n}{3} + 2 > 3$

48. $\dfrac{1}{2}y + 4 \geq -1$

49. $3x - 12 \leq 6$

50. $2v - 9 > -7$

51. $-21 - 3y > 0$

52. $-36 - 4y > 0$

53. $5n - 11 \geq 2n + 28$

54. $24 - 9s < -13s + 8$

55. $-4m + 8 \leq -3m + 1$

56. $-6y + 13 > y + 6$

57. $-7x + 4x + 23 < 2$

58. $5t - 7 + 2t \leq -14$

59. $-3(z + 5) > -15$

◉ **60.** $2(8 + w) < 22$

61. $0.5(2x + 1) \geq 3x$

62. $-0.2(10d - 5) > 9$

63. $2(x - 2) - 3x \geq -1$

64. $2(4z - 3) < 3(z + 4)$

65. $7y - (9y + 1) < 5$

66. $-(6m - 2) \leq 0$

67. $0.4(5x + 1) \geq 3x$

68. $0.2(y - 3) > 8$

69. $5x + 1 < 3x - 2(4x - 3)$

70. $-2(0.5 - 4t) > -3(4 - 3.5t)$

71. $3 + 5n \leq 6(n - 1) + n$

72. $3(4 - 2m) \geq 2(3m - 6)$

73. $-\dfrac{4}{3}x - 16 > x + \dfrac{1}{3}x$

74. $\dfrac{2}{3}z - 4 < z + \dfrac{1}{3} + \dfrac{1}{3}z$

75. $0.2y > 1500 + 2.6y$

76. $x + 1.6x \leq 52$

Choose the inequality that best describes the situation.

77. To vote in the United States, a citizen must be at least 18 years old.
  a. $a < 18$ **b.** $a > 18$
  c. $a \leq 18$ **d.** $a \geq 18$

78. The number of people seated in a theater is at most 650.
  a. $n < 650$ **b.** $n \leq 650$
  c. $n > 650$ **d.** $n \geq 650$

79. It is generally accepted that a person has a fever if the person's temperature is above 98.6°F. To convert a Celsius temperature to its Fahrenheit equivalent, we use the formula $F = \dfrac{9}{5}C + 32$. For which Celsius temperatures C does a person have a fever?
  a. $\dfrac{9}{5}C + 32 \geq 98.6$ **b.** $\dfrac{9}{5}C + 32 \leq 98.6$
  c. $\dfrac{9}{5}C + 32 < 98.6$ **d.** $\dfrac{9}{5}C + 32 > 98.6$

80. A teenager has a $20 gift certificate at the iTunes store. He decides to purchase three episodes of his favorite television show at $2.99 each, and is also browsing the music store where songs can be downloaded at $1.29 apiece. What is the maximum number of songs he can purchase without exceeding the amount of the gift certificate?
  a. $8.97 + 1.29n > 20$ **b.** $8.97 + 1.29n < 20$
  c. $8.97 + 1.29n \geq 20$ **d.** $8.97 + 1.29n \leq 20$

Mixed Practice

Graph on the number line.

81. $-1 < x \leq 3\frac{1}{2}$

82. $x > -0.5$

Solve.

83. Indicate whether the value of x shown is a solution of the given inequality by answering true or false.

Value of x	Inequality	True or False
a. 2	$4 - 2x < -1$	
b. 6	$3x - 5 \geq 21 - 2x$	
c. -8	$-(2x + 4) \leq 12$	
d. $-\dfrac{1}{2}$	$8x - 5 > 4x - 8$	

Solve and graph.

84. $-3 \leq x - 7$

85. $-\dfrac{2}{3}m \geq 4$

86. $-8t > -24$

Solve.

87. $3(a + 4) \geq 2(3a - 2)$

88. $-5m + 6 \geq -4m + 4$

89. $8x + 7 - 3x < 32$

90. $0.3m - 1200 < 2.8m$

Applications

Write an inequality, and solve.

91. On the last three chemistry exams, a student scored 81, 85, and 91. What score must the student earn on the next exam to have an average above 85?

92. A rectangular deck that is 14 ft long is to be built onto the back of a house. What would be the area of the deck if it is at least 12 ft wide?

93. A novelty store, open Monday through Saturday, must sell an average of $200 worth of merchandise per day to break even on expenses. The following table shows sales for one week.

Day	Monday	Tuesday	Wednesday	Thursday	Friday
Amount of Sales	$250	$250	$150	$130	$180

How much must the store make in sales on Saturday to at least break even for the week?

94. An American tourist about to travel abroad wants to buy euros from one of two currency exchange booths. The first booth charges $1.38 per euro plus a fixed surcharge of $3. The second charges $1.50 per euro with no surcharge. For what kind of transaction does the second booth offer a better deal?

95. A telemarketer claims that every call made to a customer costs at least $2. If a call costs $0.50 plus $0.10 for each minute, how long does each call last?

96. One side of a triangular garden is 2 ft longer than another side. The third side is 4 ft long. What are the maximum lengths of the other two sides if the perimeter of the garden can be no longer than 12 ft?

97. A real estate agent gets $2\frac{1}{2}\%$ on every house she sells plus a $1,000 bonus. Her supervisor offers an alternative deal of 3% on the sale of every house and no bonus. Should she accept the deal? Under what circumstances?

98. A parking garage offers two payment options: a $20 flat fee for the whole day, or $5 plus $2 per hour for each hour or part thereof that a customer parks. Which is the better option for the customer? Explain why.

99. A person weighing 200 lb volunteers for a clinical trial of a new diet pill. If he loses 2.5 lb per month using the diet pill combined with regular exercise, when will he weigh less than 180 lb?

100. A diver scored 9.2, 9.6, 9.7, 9.4, and 9.3 in the first 5 dives in a diving competition, where all scores are given in tenths. To win first place, she must beat a total score of 56.6. What is the lowest score she can get on her last dive to win the competition?

101. A Gallup poll found that between 3 and 9 percent of the U.S. public believed that the Apollo moon landing of July 1969 was faked. Express in terms of inequalities how many millions of Americans p among 200 million members of the public were of this opinion. (*Source:* gallup.com)

102. At least 10,000 hours of study and practice are needed for someone to become an "expert" in a particular field, according to one estimate. If a person has studied a subject for 10 hr/wk, 36 wk per school year, through 12 yr of schooling, how many more hours are needed for him to be an expert? (*Source:* Malcolm Gladwell, *Outliers*)

• Check your answers on page A-4.

MINDStretchers

Mathematical Reasoning

1. Explain for which values of x the following inequality holds: $x + 5 > x + 2$.

Groupwork

2. Working with a partner, explore whether each of the following statements is always, sometimes, or never true. Give examples.
  a. If $a < b$ and $c < d$, then $ac < bd$.

  b. $a - b \leq a + b$

Writing

3. Clearly identifying the variables, give examples of inequalities that
  a. you wish were true.
  b. you wish were false.

Key Concepts and Skills

CONCEPT   SKILL

Concept/Skill	Description	Example
[2.1] **Equation**	A mathematical statement that two expressions are equal.	$y - 6 = 9$
[2.1] **Solution of an equation**	A value of the variable that makes the equation a true statement.	2 is a solution of the equation $3x + 1 = 11 - 2x$.
[2.1] **Linear equation in one variable**	An equation that can be written in the form $ax + b = c$, where a, b, and c are real numbers and $a \neq 0$.	$4x + 1 = 3$
[2.1] **Equivalent equations**	Equations that have the same solution.	$x = 2$ and $x + 1 = 3$
[2.1] **Addition property of equality**	For any real numbers a, b, and c, $a = b$ and $a + c = b + c$ are equivalent.	If $x - 3 = 5$, then $(x - 3) + 3 = 5 + 3$
[2.1] **To solve an applied problem using an equation**	• Read the problem carefully. • Translate the problem into an equation. • Solve the equation. • Check the solution in the original equation. • State the conclusion.	
[2.2] **Multiplication property of equality**	For any real numbers a, b, and c, $c \neq 0$, $a = b$ and $a \cdot c = b \cdot c$ are equivalent.	If $\dfrac{n}{2} = 8$, then $\dfrac{n}{2} \cdot 2 = 8 \cdot 2$
[2.3] **To solve linear equations**	• Use the distributive property to clear the equation of parentheses, if necessary. • Combine like terms where appropriate. • Use the addition property to isolate the variable term. • Use the multiplication property to isolate the variable. • Check by substituting the solution in the original equation.	$2(x + 5) = 6$ $2x + 10 = 6$ $2x + 10 - 10 = 6 - 10$ $2x = -4$ $x = -2$ **Check** $2(x + 5) = 6$ $2(-2 + 5) \stackrel{?}{=} 6$ $6 = 6$ True
[2.4] **Literal equation**	An equation involving two or more variables.	$2t + b = c$
[2.5] **To find a percent increase or decrease**	• Compute the difference between the two given values. • Compute what percent this difference is of the *original value*.	If a quantity changes from 10 to 12, the difference is 2. Since 2 is 20% of 10, the percent increase is 20%.
[2.6] **Inequality**	Any mathematical statement containing $<$, $\leq$, $>$, $\geq$, or $\neq$.	$x \geq -4$
[2.6] **Solution of an inequality**	Any value of the variable that makes the inequality true.	0 is a solution of $x < 7$.

continued

175

Concept/Skill	Description	Example
[2.6] **Addition property of inequalities**	For any real numbers a, b, and c: • $a < b$ and $a + c < b + c$ are equivalent. • $a > b$ and $a + c > b + c$ are equivalent. Similar statements hold for $\leq$ and $\geq$.	If $x < 1$, then $x + 5 < 1 + 5$.

[2.6] **Multiplication property of inequalities**	For any real numbers a, b, and c:
	• If c is positive, then $a < b$ and $ac < bc$ are equivalent.
	• If c is negative, then $a < b$ and $ac > bc$ are equivalent.
	
	Similar statements hold for $>$, $\leq$, and $\geq$.
		If $\dfrac{x}{2} > 4$, then
		
		$2 \cdot \dfrac{x}{2} > 2 \cdot 4$.
		
		If $-2x < 4$, then
		
		$\dfrac{-2x}{-2} > \dfrac{4}{-2}$.

Say Why
Fill in each blank.

1. The number -7 _____ a solution to the equation
   is/is not
   $5x + 1 = 36$ because _____
   _____ .

2. The equation $2 = -\dfrac{3}{4}x$ _____ linear because
   is/is not
   _____
   _____ .

3. When 5 is subtracted from each side of the equation
   $2x + 5 = 10$, the result _____ an equivalent
   is/is not
   equation because _____
   _____ .

4. The equations $5m + 2 = \dfrac{1}{2}$ and $10m + 4 = 1$
   _____ equivalent because _____
   are/are not
   _____
   _____ .

5. The number 8 _____ a solution of
   is/is not
   $x - 6.5 > 1.5$ because _____
   _____ .

6. If $-15x < 3$, then $\dfrac{-15x}{-15}$ _____ less than $\dfrac{3}{-15}$
   is/is not
   because _____
   _____
   _____ .

[2.1]

7. Is 2 a solution of the equation $5x + 3 = 7 - 4x$?

8. Determine whether 0 is a solution of the equation
   $4x - 15 = 5(x - 3)$.

Solve and check.

9. $x - 3 = -12$

10. $t + 10 = 8$

11. $-9 = a + 5$

12. $4 = n - 7$

13. $y - (-3.1) = 11$

14. $r + 4.8 = 20$

[2.2] *Solve and check.*

15. $\dfrac{x}{3} = -2$

16. $\dfrac{z}{2} = -5$

17. $2x = -20$

18. $-5d = 15$

19. $-y = -4$

20. $-x = 3$

21. $20.5 = 0.5n$

22. $30 = -0.2r$

23. $\dfrac{2t}{3} = -6$

24. $\dfrac{5y}{6} = -10$

[2.3] *Solve and check.*

25. $2x + 1 = 7$

26. $-t - 4 = 5$

27. $\dfrac{a}{2} - 3 = -10$

28. $\dfrac{r}{3} - 6 = 12$

29. $-y + 7 = -2$

30. $-2t + 3 = 1$

31. $4x - 2x - 5 = 7$

32. $3y - y + 12 = 6$

33. $z + 1 = -2z + 10$

34. $n - 3 = -n + 7$

35. $c = -2(c + 1)$

36. $p = -(p - 5)$

37. $2(x + 1) - (x - 8) = -x$

38. $-(x + 2) - (x - 4) = -5x$

39. $3[2n - 4(n + 1)] = 6n - 12$

40. $-4(2x - 6) = 7[x - (3x - 1)]$

41. $10 - [3 + (2x - 1)] = 3x$

42. $x - [5 + (3x - 4)] = -x$

[2.4]

43. Solve $a - 5b = 2c$ for a in terms of b and c.

44. Solve $\dfrac{2a}{b} = n$ for a in terms of b and n.

45. An isosceles triangle is a triangle with two sides equal in length. $P = 2a + b$ is a formula for the perimeter P of an isosceles triangle in terms of its equal sides a and the third side b. Solve for a in terms of P and b.

46. $V = \dfrac{1}{3}Bh$ is a formula for the volume V of a cone in terms of the area of its base B and its height h. Solve for h in terms of B and V.

[2.5] *Solve.*

47. 30% of what number is 12?

48. 125% of what number is 5?

49. What percent of 5 is 8?

50. What percent of 8 is 5?

51. What number is 8.5% of \$300?

52. What is 3.5% of \$2000?

[2.6] *Graph each inequality.*

53. $x < 2$

54. $x \geq -4.5$

55. $-2 < x \leq 2$

56. $-0.5 < x < 5$

Solve. Then, graph.

57. $-n \leq 2$

58. $y + 1 > 6$

59. $-\dfrac{1}{2}t + 3 \leq 3$

60. $8y - 2 \leq 6y + 2$

61. $\dfrac{1}{2}(8 - 12x) \leq x - 10$

62. $0.5n - 0.3 < 0.2(2n + 1)$

Mixed Applications

Solve.

63. An air conditioner's energy efficiency ratio (EER) is the quotient of its British thermal unit (Btu) rating and its wattage. What is the Btu rating of a 2000-watt air conditioner if its EER is 8?

64. The sum of the measures of the angles in any triangle is 180°. If the three angles in an equilateral triangle are equal, what is the measure of each of these angles?

65. For a wedding, the reception costs will be $5000 plus $50 per guest. The bride and groom have budgeted $12,000 for the reception. How many guests can the bride and groom invite to the wedding?

66. A polygon is a closed geometric figure with straight sides. In any polygon with n sides, the sum of the measures of its angles is $180(n - 2)$ degrees. If the measures of the angles of a polygon add up to $540°$, how many sides does the polygon have?

67. A newspaper reported that a candidate received 15,360 more votes than her opponent, and that 39,210 votes were cast in the election. How many votes were cast for each candidate?

68. To send a telegram, it costs $2 for the first 10 words in the telegram and y cents for each additional word.
 a. Write an equation to find the cost C of a telegram 26 words long.
 b. Solve this equation for y.

69. The road connecting two factories is 380 mi long. A truck leaves one of the factories traveling toward the other factory at 45 mph, while at the same time a second truck leaves the other factory heading at 50 mph toward the first. How long after the departure will the trucks meet?

70. The rarest blood type is AB^-, which occurs in only 0.6% of people in the United States. Of the 4000 students in your college, how many would you expect to have AB^- blood type? (*Source:* wikipedia.org)

71. A plane flying between two cities at 400 mph arrives half an hour behind schedule. If the plane had flown at a speed of 500 mph, it would have been on time. Find the distance between the cities.

72. Tom Seaver received 425 out of 430 votes electing him to the Baseball Hall of Fame, surpassing the previous voting record for Ty Cobb. To the nearest whole percent, what percent of votes cast did Seaver receive? (*Source:* United Press International)

73. In the decade beginning with 1901, about 8.8 million immigrants entered the United States. In the following decade, this number dropped to approximately 5.7 million people. Find the percent decrease, to the nearest whole percent. (*Source: Statistical Yearbook of the Immigration and Naturalization Service*)

74. Two friends leave a party at 10 P.M. driving in opposite directions. One drives at a speed of 40 mph, whereas the other drives at a speed of 32 mph. At what time are the two friends 18 mi apart?

75. How much interest is earned in 1 year on a principal of $500 at an annual interest rate of 6%?

76. In the presidential election of 1836, Martin Van Buren earned 170 electoral votes. Four years later, Van Buren earned 60 electoral votes. To the nearest whole percent, by what percent did Van Buren's electoral vote count drop? (*Source: Time Almanac 2010*)

77. How much pure alcohol must be mixed with 6 L of a 60% alcohol solution to make a 70% alcohol solution?

78. How many pints of a 1% solution of disinfectant must be combined with 4 pt of a 10% solution to make a 5% solution?

79. To convert a weight expressed in kilograms k to the equivalent number of pounds p, multiply the number of kilograms by 2.2.
 a. Write a formula for this relationship.
 b. Solve this formula for k.

80. To print b books, it costs a publisher $(100,000 + 25b)$ dollars. The publisher receives $50b$ dollars in revenue for selling this number of books. How many books must be sold for the revenue to equal the cost?

• Check your answers on page A-5.

CHAPTER 2   Posttest

FOR EXTRA HELP

CHAPTER **Test Prep** VIDEOS

The Chapter Test Prep Videos with test solutions are available on DVD, in MyMathLab, and on YouTube (search "AkstIntroductory Alg" and click on "Channels").

To see if you have mastered the topics in this chapter, take this test.

1. Determine whether -2 is a solution of the equation $3x - 4 = 6(x + 2)$.

2. What must be done to the equation $x - 1 = -10$ in order to isolate the variable?

Solve and check.

3. $\dfrac{n}{2} = -3$

4. $-y = -11$

5. $\dfrac{3y}{4} = 6$

6. $2x + 5 = 11$

7. $-s + 4 = 2$

8. $10x + 1 = -x + 23$

9. $16a = -4(a - 5)$

10. $2(x + 5) - (x + 4) = 7x + 1$

11. Solve $5n + p = t$ for p in terms of n and t.

12. 40% of what number is 8?

13. What percent of 5 is 10?

14. Draw the graph of $-1 \le x < 3$.

$$\xleftarrow{\hspace{0.3cm}}\overset{\displaystyle +\ +\ +\ +\ +\ +\ +}{\underset{\displaystyle -3\ -2\ -1\ \ 0\ \ 1\ \ 2\ \ 3}{\hspace{3cm}}}\xrightarrow{\hspace{0.3cm}}$$

15. Solve $-2z \le 6$. Then graph.

$$\xleftarrow{\hspace{0.3cm}}\overset{\displaystyle +\ +\ +\ +\ +\ +\ +}{\underset{\displaystyle -3\ -2\ -1\ \ 0\ \ 1\ \ 2\ \ 3}{\hspace{3cm}}}\xrightarrow{\hspace{0.3cm}}$$

16. A taxi charges $4.00 for the first mile plus $1.25 for each additional mile. On a fare of $16.50, how long was the ride?

17. A woman's shoe size S is given by the formula $S = 3L - 21$, where L is her foot length L in inches. Solve for L in terms of S.

18. In a recent year, males experienced 18,400,000, or 40%, of the common operations and procedures performed in the United States. To the nearest million, how many were performed in all? (*Source: The New York Times Almanac 2010*)

19. Two friends live 33 mi apart. They cycled from their homes, riding toward one another and meeting $1\frac{1}{2}$ hr later. If one friend cycled 2 mph faster than the other, what were their two rates?

20. A cell phone service offers two calling plans. Plan A costs $39.99 per month plus $0.79 per minute (or part thereof) for calls outside the network. Plan B costs $54.99 per month plus $0.59 per minute (or part thereof) for calls outside the network. Under what conditions will the monthly cost of Plan A exceed the monthly cost of Plan B?

• Check your answers on page A-5.

Cumulative Review Exercises

To help you review, solve the following:

1. Express as a signed number: a loss of 6 yd in a football play

2. Graph the number -3 on the number line shown.

3. Compute $|-2|$

4. True or false, $1 > -2$

5. Simplify: $2 + (-1) + (-4) + 4$

6. Find the product. $(-0.5)(24)(-0.2)$

7. Divide: $(-10) \div (-2)$

8. The sum S of the first n whole numbers is given by the formula $S = \dfrac{n(n + 1)}{2}$. What is the sum of the first 100 whole numbers?

9. Simplify: $3x + 1 - 7(2x - 5)$

10. Solve and check: $a - (-32) = 12$

11. Solve and check: $\dfrac{5x}{6} = 30$

12. Solve and check: $4m - 2 = 7(m - 2)$

13. Solve the equation for m: $-p = \dfrac{2}{5}mn$

14. 24 is 75% of what number?

15. A dieter gained 3 lb and then lost 5 lb. Express her overall change in weight as a real number.

16. An initial investment of $1000 triples in value every 15 yr. The value of this investment after 60 yr is how many times the initial investment? Write in exponential form.

17. For the first 2 hr of a trip, a driver averaged 55 mph. For the remainder of the trip, he drove at an average of 60 mph. If the entire trip covered 140 mi, how many hours did the trip last?

18. A telephone technician charges $50 for the first hour of work and $25 for each additional hour of work.
  a. Write a formula that expresses the amount of money A that the technician charges in terms of the time t that the technician works.
  b. Solve this equation for t.
  c. If the repairman sent you a bill for $125, how many hours did he claim to work?

19. During what is called the "Group Stage" of soccer's World Cup tournament, games are scored as follows: a win is worth 3 points, a draw 1 point, and a loss 0 points. Each team plays 3 games in the Group Stage. (For example in 2010, the U.S. had 1 win, 2 draws, and 0 losses, giving them a total of 5 points.) If a team wins w games, gets a draw in d games, and loses l games, what is its score? (*Source:* fifa.com)

20. The fuel bill for a condo building was less than 8% of the building's operating expenses. If the fuel bill was $74,000, find the building's operating expenses.

• Check your answers on page A-5.

Graphing Linear Equations and Inequalities

Get Rich with Graphing?

If financial analysts could predict trends on Wall Street, they would know when to buy and when to sell stock in order to maximize profits. To arrive at educated predictions of future trends, analysts use the statistical tool of linear regression.

When using linear regression, an analyst plots points that show a particular stock's recent selling prices. Then, the analyst sketches the straight line that is closest to passing through these points. This **regression line** provides an estimate of how much the stock is likely to increase or decrease in price, and at what pace. If the actual selling prices differ considerably from the predicted prices for a lengthy period of time, the analyst suspects a new trend and recomputes the regression line.

(*Sources:* "Perhaps a Little Lift," *Financial Mail*, December 1, 2000, p. 118; Mario Triola, *Elementary Statistics*, Addison-Wesley, 2012)

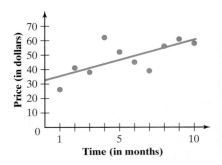

1. On the coordinate plane below, plot the points $A(1, 4)$ and $B(-3, -6)$.

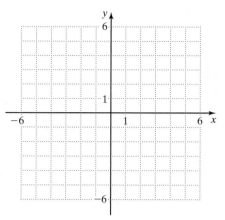

2. In which quadrant is the point $(3, -5)$ located?

3. Given two points $C(7, 5)$ and $D(1, 2)$, compute the slope of the line that passes through the points.

4. For the points $A(2, 1)$, $B(0, 5)$, $C(1, 7)$, and $D(4, 1)$, indicate whether $\overleftrightarrow{AB}$ is parallel to $\overleftrightarrow{CD}$. Explain.

5. For the points $P(-3, 3)$, $Q(1, -1)$, $R(-2, -2)$, and $S(4, 4)$, indicate whether $\overleftrightarrow{PQ}$ is perpendicular to $\overleftrightarrow{RS}$. Explain.

6. In the graph shown, find the x-intercept and the y-intercept.

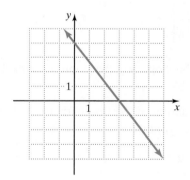

7. For the equation $y = 2x - 5$, complete the following table:

x	4	7		
y			0	-1

Graph each equation.

8. $y = -3$

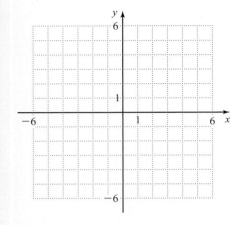

9. $x = 2$

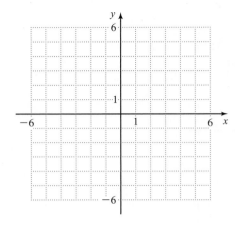

10. $y = -3x + 2$

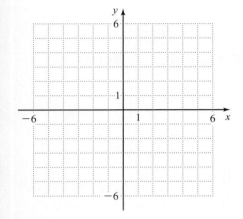

11. $2x - 3y = 6$

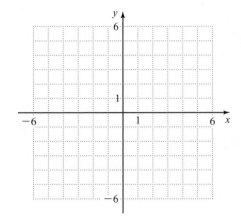

12. What are the slope and the *y*-intercept of the graph of $y = 2x - 5$?

13. Write the equation $5x - y = 8$ in slope-intercept form.

14. Find an equation of the line with slope 2 that passes through the point $(0, 8)$.

15. Find an equation of the line that passes through the points $(4, 1)$ and $(2, -1)$.

16. Graph: $y > 3x + 1$

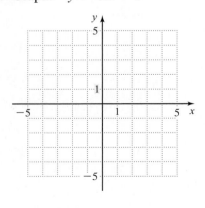

17. In the graph shown, the *y*-axis stands for the number of congressional representatives from a state and the *x*-axis stands for the state's population according to a recent U.S. census.

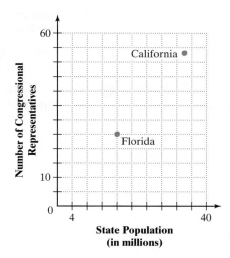

Draw the line that passes through the points. Is this line's slope positive, negative, or zero? How does the state population relate to the number of representatives from that state?

18. Farmers want to have developed varieties of wheat that grow at a faster rate. The graph shown displays the growth pattern of two new varieties of wheat. Which variety grows more quickly?

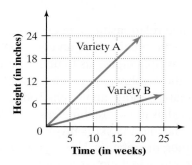

19. A local video store charges a daily rental fee of $2.50 per movie.

a. Express the daily cost *c* of renting *x* movies.

b. Choose an appropriate scale for the axes and graph this relationship.

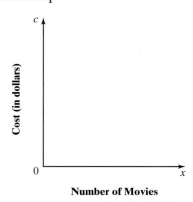

c. What is the slope of the line in part (b)? Explain its significance in this context.

20. A textbook sales representative is driving to a college 200 mi away at a speed of 50 mph.

a. Express the distance *d* (in miles) the sales representative travels in terms of the time *t* (in hours) he has been driving.

b. Choose an appropriate scale for the axes and graph this relationship.

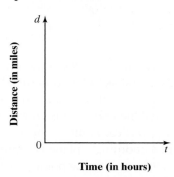

c. What is the slope of the graph? Explain its significance in terms of the trip.

• Check your answers on page A-5.

3.1 Introduction to Graphing

What Graphing Is and Why It Is Important

Many mathematical relationships can be expressed as equations, inequalities, or their graphs. Although lacking the precision of an equation, a graph can clarify at a glance patterns and trends in a relationship, helping us to understand that relationship better.

In the past, the graphing approach to problem solving was usually more time-consuming than the traditional algebraic approach. Today, the use of graphing calculators and computer software packages has made graphing easier. But to utilize these graphing tools, you must first understand the concepts and skills involved in graphing, which are discussed in this chapter.

In this chapter, the relationships that we graph are relatively simple. In Chapters 4 and 9, we will apply graphing techniques to more complex relationships.

A To identify and plot points on a coordinate plane

B To identify the quadrants of a coordinate plane

C To solve applied problems involving graphs

Plotting Points

If you were to enter a theater or sports arena with a ticket for row 5, seat 3, you would know exactly where to sit that is, at $(5, 3)$. Such a system of **coordinates** in which we associate a pair of numbers in a given order with a corresponding location is commonplace.

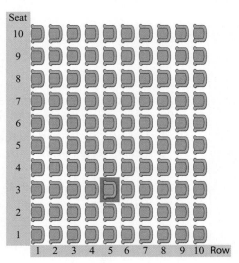

The flat surface on which we draw graphs is called a **coordinate plane**. To create a coordinate plane, we first sketch two perpendicular number lines—one horizontal, the other vertical—that intersect at their zeros. Each number line is called an **axis**. The point where the axes intersect is called the **origin**. It is common practice to refer to the horizontal number line as the **x-axis** and the vertical number line as the **y-axis**.

Each point in a coordinate plane is represented by a pair of numbers called an **ordered pair.** For example, the origin is the point $(0, 0)$. The first number in an ordered pair represents a horizontal distance and is called the **x-coordinate.** The second number represents a vertical distance and is called the **y-coordinate.**

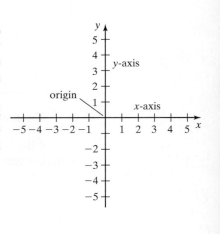

To **plot** a point in the coordinate plane, we find its location represented by its ordered pair. For example, to plot the point $(3, 1)$, we start at the origin and go 3 units *to the right,* then go *up* 1 unit. For this point, we say that $x = 3$ and $y = 1$.

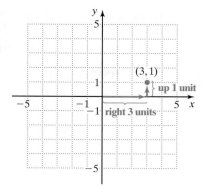

Notice that the two numbers in an ordered pair are written in parentheses, separated by a comma. Do the ordered pairs $(3, 1)$ and $(1, 3)$ correspond to different points? Why?

When an ordered pair has a negative x-coordinate, the corresponding point is to the left of the y-axis, as shown in the following coordinate plane. Similarly, when an ordered pair's y-coordinate is negative, the point is below the x-axis.

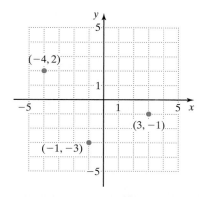

Any ordered pair whose y-coordinate is 0 corresponds to a point that is on the x-axis. For instance, the points $(-4, 0)$, $(0, 0)$, and $(3, 0)$ are on the x-axis, as shown in the graph below on the left. Similarly, any ordered pair whose x-coordinate is 0 corresponds to a point that is on the y-axis. For instance, the points $(0, 2)$, $(0, -1)$, and $(0, -4)$ are on the y-axis, as shown in the middle graph below.

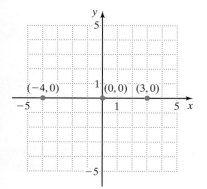

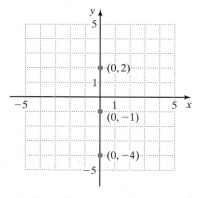

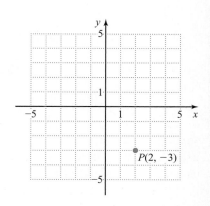

Sometimes we name points with letters. We can refer to a point as P or A or any other letter that we choose. Generally, a capital letter is used. If we want to emphasize that point P has coordinates $(2, -3)$, we can write it as $P(2, -3)$, as shown in the graph above on the right.

If there is a point whose coordinates we do not know, we can refer to it as (x, y) or $P(x, y)$, where x and y are the unknown coordinates.

Now, let's look at some examples of plotting points.

EXAMPLE 1

Plot the following points on a coordinate plane.

a. $(5, 2)$ **b.** $(3, -4)$ **c.** $(-1, 1)$

d. $(0, 3)$ **e.** $(-4, -2)$ **f.** $(0, 0)$

Solution   The points are plotted on the coordinate plane as shown.

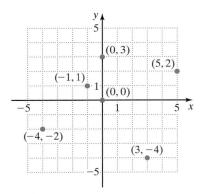

PRACTICE 1

On a coordinate plane, plot the points corresponding to each ordered pair.

a. $(0, -5)$ **b.** $(4, 4)$

c. $(-2, 2)$ **d.** $(5, 0)$

e. $(-3, -2)$ **f.** $(2, -4)$

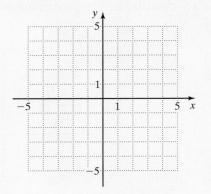

The x- and y-axes are boundaries that separate a coordinate plane into four regions called *quadrants*. These quadrants are named in counterclockwise order starting with Quadrant I, as shown below:

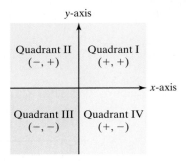

The quadrant in which a point is located tells us something about its coordinates. For instance, any point in Quadrant I is to the right of the y-axis and above the x-axis, so both its coordinates must be positive. Points in this quadrant are of particular interest in applied problems in which all quantities are positive.

Any point in Quadrant II lies to the left of the y-axis and above the x-axis, so its x-coordinate must be negative and its y-coordinate positive. Points in Quadrant III are to the left of the y-axis and below the x-axis, so both coordinates are negative. And finally, any point in Quadrant IV is to the right of the y-axis and below the x-axis, so its x-coordinate must be positive and its y-coordinate negative. Points that lie on an axis are not in any quadrant.

EXAMPLE 2

Determine the quadrant in which each point is located.

a. $(-5, 5)$

b. $(7, 20)$

c. $(1.3, -4)$

d. $(-4, -5)$

Solution

a. $(-5, 5)$ is in Quadrant II.

b. $(7, 20)$ is in Quadrant I.

c. $(1.3, -4)$ is in Quadrant IV.

d. $(-4, -5)$ is in Quadrant III.

PRACTICE 2

In which quadrant is each point located?

a. $\left(-\dfrac{1}{2}, 3\right)$

b. $(6, -7)$

c. $(-1, -4)$

d. $(2, 9)$

Often the points that we are to plot affect how we draw the axes on a coordinate plane. For instance, in the following coordinate planes we choose for each axis an appropriate **scale**—the length between adjacent tick marks—to conveniently plot all points in question.

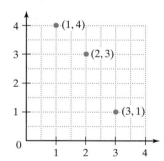
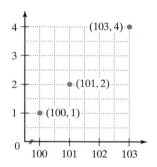

TIP Depending on the location of the points to be plotted, we can choose to show only part of a coordinate plane.

EXAMPLE 3

A young entrepreneur started a dot-com company that made a profit of $10,000 in its first year of business. In the second year, the company's profit grew to $15,000. In the third year, however, the company lost $5000. Plot points on a coordinate plane to display this information.

Solution We let x represent the year of business and y represent the company's profit in dollars that year. The three points to be plotted are:

$(1, 10,000)$, $(2, 15,000)$, and $(3, -5000)$.

PRACTICE 3

The following table shows the average monthly temperatures for the first four months of the year (where month 1 represents January) for Chicago, Illinois. (*Source:* U.S. National Climatic Data Center)

Month m	1	2	3	4
Temperature t (°F)	22	27	37	49

EXAMPLE 3 (continued)

Notice that in the third year, the company's loss is represented by a negative profit. Because the *y*-coordinates are large, we use 5000 as the scale on the *y*-axis. Then, we plot the points on the following coordinate plane.

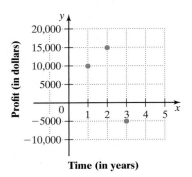

Time (in years)

On the coordinate plane shown, graph the information displayed in the table.

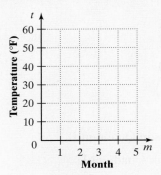

Month

We end this discussion of plotting points with a final comment about variables. On a coordinate plane, the first coordinate of each point is a value of one quantity (or variable), and the second coordinate is a value of another quantity. For instance, in Example 3 we considered the profit that a company makes at various times. One variable represents time and the other variable the profit. Notice that the profit made by the company depends on the time rather than the other way around. So we refer to time as the *independent* variable and the profit as the *dependent* variable. It is customary when plotting points to assign the independent variable to the horizontal axis and the dependent variable to the vertical axis. However, as shown in Practice 3, letters other than *x* and *y* can be used to represent quantities.

Interpreting Graphs

Points plotted on a coordinate plane are merely dots on a piece of paper. However, their significance comes to life when we understand the information that they convey.

Describing the trend on a coordinate plane tells the story of that trend. When key points are missing, as is frequently the case, the story is incomplete. In such cases, we may want to make a prediction, that is, to extend the observed pattern so as to estimate the missing data. Such predictions, while not certain, at least allow us to make decisions based on the best available evidence. We may also want to speculate about the conditions that underlie an observed pattern of plotted points.

The trend among plotted points on a coordinate plane shows a relationship between the two variables. To highlight the relationship, it is common practice either to draw a line that passes through the plotted points or to connect adjacent points with short line segments.

Consider the following examples that involve interpreting trends on a coordinate plane:

EXAMPLE 4

The graph shows the cost C of parking a car at a lot for time t hr. Describe the trend that you observe in terms of both the coordinate plane and the cost of parking.

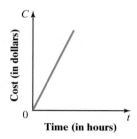

Solution On the graph, the larger C-values correspond to the larger t-values. In terms of parking, we see that the longer a car is parked in the lot, the more it costs to park the car.

PRACTICE 4

The value V of a new car after t years is displayed on the following graph. Describe the line graph in terms of the changing value of the car.

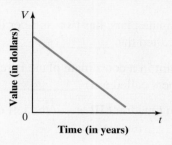

EXAMPLE 5

The following graph shows the cost that a shipping company charges to send a package, depending on the package's weight.

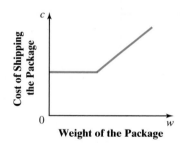

Write a brief story describing the displayed relationship. What business practice does the horizontal line segment reflect?

Solution From the horizontal line segment, we see that the cost of shipping is constant (that is, a flat rate) for lighter packages up to a certain weight. The horizontal line segment indicates that the company established a minimum cost for sending lightweight packages. Since the slanted line segment goes upward to the right, the cost of shipping increases with the weight of heavier packages.

PRACTICE 5

The following graph shows the number of times per minute that a runner's heart beats. Describe in a sentence or two the pattern you observe. Use this pattern to write a scenario as to what the runner might be doing.

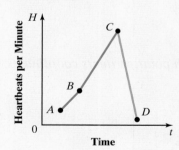

Mathematically Speaking

Fill in each blank with the most appropriate term or phrase from the given list.

above	horizontal	below
dependent	*x*-axis	independent
origin	coordinate	center
ordered pair	vertical	*y*-axis

1. A coordinate plane has two number lines that intersect at a point called the _____.

2. The horizontal number line on a coordinate plane is usually referred to as the _____.

3. Each point in a coordinate plane is represented by a pair of numbers called a(n) _____.

4. The *y*-coordinate of a point on a coordinate plane represents a(n) _____ distance.

5. Points in Quadrant III are to the left of the *y*-axis and _____ the *x*-axis.

6. The _____ variable is usually assigned to the horizontal axis.

Ⓐ *On the coordinate plane below, plot the points with the given coordinates.*

7. $A(0, 5)$ $B(-1, -5)$ $C(1, 4)$
   $D(3, -3)$ $E(-4, 2)$ $F(5, 0)$

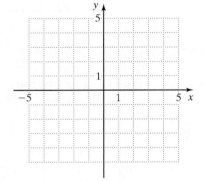

8. $A(-2, 4)$ $B(0, 0)$ $C(2, -1)$
   $D(-3, 0)$ $E(3, 4)$ $F(-4, -2)$

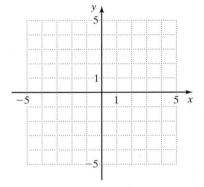

Next to each point, write its coordinates.

9.

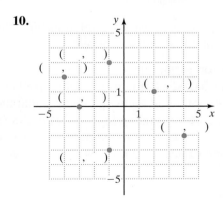

10.

B *Identify the quadrant in which each point is located.*

11. $(-2, -3)$ **12.** $(-13, -24)$ **13.** $(-9, 5)$ ● **14.** $(-5.1, 4)$

15. $\left(3, -\dfrac{1}{2}\right)$ **16.** $\left(3\dfrac{1}{2}, -8\right)$ **17.** $(65, 11)$ **18.** $(8, 6.2)$

Mixed Practice

Solve.

19. Plot the points with the given coordinates on the coordinate plane.

$A(3, 2)$ $\quad B(4, 0)$ $\quad C(-4, -1)$ $\quad D(-2, 3)$ $\quad E(0, -3)$

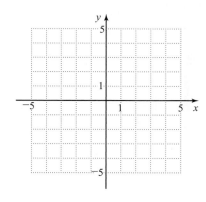

20. Write the coordinates next to each point.

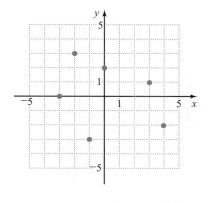

Identify the quadrant in which each point is located.

21. $(-3, 7)$ **22.** $\left(6, -\dfrac{3}{4}\right)$ **23.** $(27, 39)$ **24.** $(-4.2, -3.8)$

Applications

C *Solve.*

25. College students coded A, B, C, and D took placement tests in mathematics and in English. The following coordinate plane displays their scores.

   a. Estimate the coordinates of the plotted points.

   b. Which students scored higher in English than in mathematics?

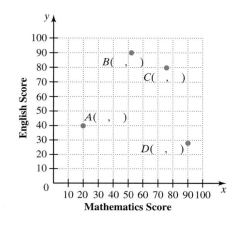

● **26.** Suppose that a financier owns shares of stock in three companies—Dearborn, Inc. (D), Ellsworth Products (E), and Fairfield Publications (F). On the following coordinate plane, the x-value of a point represents the change in value of a share of the indicated stock from the previous day. The y-value stands for the number of shares of that stock owned by the financier.

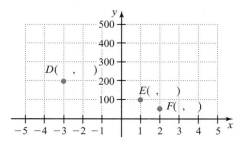

   a. Name the coordinates of the plotted points.

   b. For each point, explain the significance of the product of the point's coordinates.

27. The following table gives the percent of the U.S. adult population that smoked in various years. (*Source:* cdc.gov)

Year	1980	1990	2000	2010
Percent of the Population That Smoked	33%	26%	23%	20%

Plot this information on the coordinate plane below:

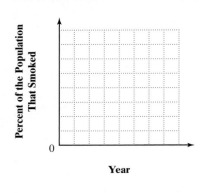

28. The number of electoral votes cast for the winning candidate in recent presidential elections is displayed in the following table:

Year	1996	2000	2004	2008
Electoral Votes	379	271	286	365

(*Source: The New York Times*)

Plot this information in the coordinate plane shown:

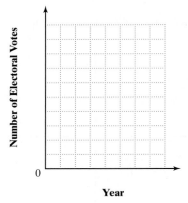

29. A chemist conducts an experiment to measure the melting and boiling points of four substances, as indicated in the following table:

Substance	Symbol	Melting Point (°C)	Boiling Point (°C)
Chlorine	Cl	−101	−35
Oxygen	O	−218	−183
Bromine	Br	−7	59
Phosphorous	P	44	280

On the coordinate plane shown below, an *x*-value represents a substance's melting point and a *y*-value stands for its boiling point, both in degrees Celsius.

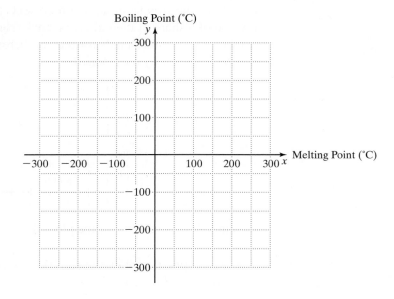

a. Plot points for the four substances. Label each point with the appropriate substance symbol.

b. For each point, which of its coordinates is larger—the *x*-value or the *y*-value? In a sentence, explain this pattern.

30. Meteorologists use the windchill index to determine the windchill temperature (how cold it feels outside) relative to the actual temperature when the wind speed is considered. The following table shows the actual temperatures in degrees Fahrenheit and the related windchill temperatures when the wind speed is 5 mph.

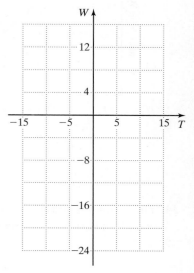

Actual Temperature *T*	−10	−5	0	5	10	15
Windchill Temperature *W*	−22	−16	−11	−5	1	7

a. Plot points (*T*, *W*) on the given coordinate plane.

b. For the plotted points, describe the pattern that you observe.

31. On the following coordinate plane, a *y*-coordinate stands for the number of senators from a state. The corresponding *x*-coordinate represents that state's population according to a recent U.S. census. Describe the pattern that you observe. (*Source:* www.census.gov)

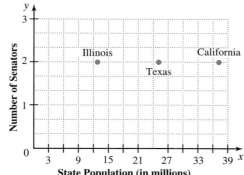

32. Last year's daily closing values (in dollars) of a share of a technology stock are plotted on the graph below. What story is this graph telling?

33. A child walks away from a wall, stands still, and then approaches the wall. In a couple
of sentences, explain which of the graphs below could describe this motion.

a.

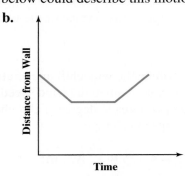

b.

c.

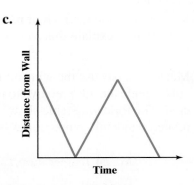

34. The following graph shows the temperature of a patient on a particular day. Describe the
overall pattern you observe in the patient's temperature over the time period.

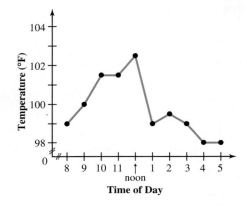

• Check your answers on page A-6.

MINDStretchers

Writing

1. Many situations involve using a coordinate system to identify positions. Two such situations are given below:

   - a chessboard
   - a map

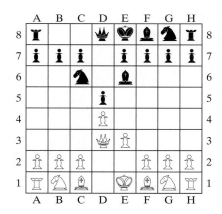
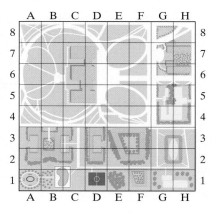

   a. Explain to what extent a chessboard and an atlas map are coordinate systems.

   b. Identify some other examples of coordinate systems in everyday life.

Critical Thinking

2. The map shows a square section of a city. You want to walk along the horizontal and vertical streets from point $(-2, -2)$ to point $(2, 2)$. One possible route is
$(-2, -2) \rightarrow (-1, -2) \rightarrow (0, -2) \rightarrow (0, -1) \rightarrow (0, 0) \rightarrow (0, 1) \rightarrow (0, 2) \rightarrow (1, 2) \rightarrow (2, 2)$
as pictured below.

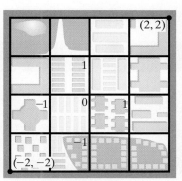

(continued)

This route is 8 blocks long. List four other 8-block routes from $(-2, -2)$ to $(2, 2)$.

$$(-2, -2) \rightarrow (\underline{\quad}, \underline{\quad}) \rightarrow (\underline{\quad}, \underline{\quad}) \rightarrow (\underline{\quad}, \underline{\quad}) \rightarrow (\underline{\quad}, \underline{\quad})$$
$$\rightarrow (\underline{\quad}, \underline{\quad}) \rightarrow (\underline{\quad}, \underline{\quad}) \rightarrow (\underline{\quad}, \underline{\quad}) \rightarrow (2, 2)$$

$$(-2, -2) \rightarrow (\underline{\quad}, \underline{\quad}) \rightarrow (\underline{\quad}, \underline{\quad}) \rightarrow (\underline{\quad}, \underline{\quad}) \rightarrow (\underline{\quad}, \underline{\quad})$$
$$\rightarrow (\underline{\quad}, \underline{\quad}) \rightarrow (\underline{\quad}, \underline{\quad}) \rightarrow (\underline{\quad}, \underline{\quad}) \rightarrow (2, 2)$$

$$(-2, -2) \rightarrow (\underline{\quad}, \underline{\quad}) \rightarrow (\underline{\quad}, \underline{\quad}) \rightarrow (\underline{\quad}, \underline{\quad}) \rightarrow (\underline{\quad}, \underline{\quad})$$
$$\rightarrow (\underline{\quad}, \underline{\quad}) \rightarrow (\underline{\quad}, \underline{\quad}) \rightarrow (\underline{\quad}, \underline{\quad}) \rightarrow (2, 2)$$

$$(-2, -2) \rightarrow (\underline{\quad}, \underline{\quad}) \rightarrow (\underline{\quad}, \underline{\quad}) \rightarrow (\underline{\quad}, \underline{\quad}) \rightarrow (\underline{\quad}, \underline{\quad})$$
$$\rightarrow (\underline{\quad}, \underline{\quad}) \rightarrow (\underline{\quad}, \underline{\quad}) \rightarrow (\underline{\quad}, \underline{\quad}) \rightarrow (2, 2)$$

Groupwork

3. Two points are plotted on a coordinate plane. Discuss with a partner what is special about a third point whose x-coordinate is the average of the first two x-coordinates, and whose y-coordinate is the average of the first two y-coordinates.

Cultural Note

It was the seventeenth-century French mathematician and philosopher René Descartes (pronounced day-KART) who developed the concepts that underlie graphing. The story goes that one morning Descartes, who liked to stay in bed and meditate, began to eye a fly crawling on his bedroom ceiling. In a flash of insight, he realized that it was possible to express mathematically the fly's position in terms of its distance to the two adjacent walls.

3.2 Slope

In the previous section, we discussed points on a coordinate plane. Now, let's look at (straight) lines that pass through points. A key characteristic of a line is its slope. In this section, we focus on the slope of a line and its relationship to the corresponding equation.

Slope

On an airplane, would you rather glide downward gradually or drop like a stone? Would you rather ski down a run that drops precipitously or ski across a gently inclined snowfield? These questions relate to *slope*, the extent to which a line is slanted. In other words, slope is a measure of a line's steepness.

Slope, also called **rate of change**, is an important concept in the study of graphing. Examining the slope of a line can tell us if the quantity being graphed increases or decreases, as well as how fast the quantity is changing. For example, in one application, the slope of a line can represent the speed of a moving object. In another application, the slope can stand for the rate at which a share of stock is changing in value.

To understand exactly what slope means, let's suppose that a straight line on a coordinate plane passes through two arbitrary points. We can call the coordinates of the first point (x_1, y_1), read "x sub 1" and "y sub 1," and the coordinates of the second point (x_2, y_2). These coordinates are written with *subscripts* in order to distinguish them from one another. We can plot these two points on a coordinate plane, and then graph the line passing through them.

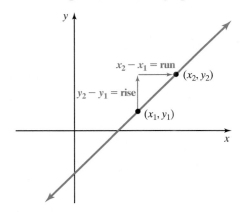

We usually represent the slope of a line by the letter m and define slope to be the ratio of the change in the y-values to the change in the x-values. Using the coordinates of the points (x_1, y_1) and (x_2, y_2) shown in the previous graph gives us the following formula:

$$m = \frac{\text{change in } y\text{-values}}{\text{change in } x\text{-values}} = \frac{y_2 - y_1}{x_2 - x_1}, \quad \text{where } x_1 \neq x_2$$

In this formula, the numerator of the fraction is the vertical change called the *rise* and the denominator is the horizontal change called the *run*. So another way of writing the formula for slope is $m = \dfrac{\text{rise}}{\text{run}}$.

DEFINITION

The **slope** m of a line passing through the points (x_1, y_1) and (x_2, y_2) is defined to be

$$m = \frac{y_2 - y_1}{x_2 - x_1}, \quad \text{where } x_1 \neq x_2.$$

Can you explain why in the definition of slope, x_1 and x_2 must not be equal?

Note that when using the formula for slope, it does not matter which point is chosen for (x_1, y_1) and which point for (x_2, y_2) as long as the order of subtraction of the coordinates is the same in both the numerator and denominator.

EXAMPLE 1

Find the slope of the line that passes through the points $(2, 1)$ and $(4, 2)$. Plot the points, and then sketch the line.

Solution   Let $(2, 1)$ stand for (x_1, y_1) and $(4, 2)$ for (x_2, y_2).

$$\begin{array}{cc} (2, 1) & (4, 2) \\ \uparrow\ \uparrow & \uparrow\ \uparrow \\ x_1\ y_1 & x_2\ y_2 \end{array}$$

Substituting into the formula for slope, we get:

$$m = \frac{y_2 - y_1}{x_2 - x_1} = \frac{2 - 1}{4 - 2} = \frac{1}{2}$$

Now, let's plot $(2, 1)$ and $(4, 2)$, and then sketch the line passing through them.

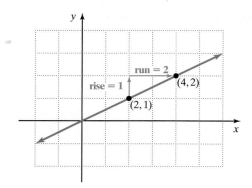

We can also find the slope of a line using its graph. From $(2, 1)$ and $(4, 2)$, we see that the change in y-values (the rise) is $2 - 1$, or 1.

The change in x-values (the run) is $4 - 2$, or 2. So $m = \dfrac{\text{rise}}{\text{run}} = \dfrac{1}{2}$.

Therefore, we get the same answer whether we use the formula

$$m = \frac{y_2 - y_1}{x_2 - x_1} \quad \text{or} \quad m = \frac{\text{rise}}{\text{run}}.$$

PRACTICE 1

Find the slope of a line that contains the points, $(1, 2)$ and $(4, 3)$. Plot the points, and then sketch the line.

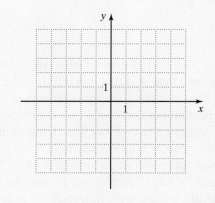

A line rising to the right as shown in Example 1 has a *positive* slope. We say that such a line is *increasing* because as the x-values gets larger, the corresponding y-values also get larger.

EXAMPLE 2

Sketch the line passing through the points $(-3, 1)$ and $(2, -2)$. Find the slope.

Solution First, we plot the points $(-3, 1)$ and $(2, -2)$. Then, we draw a line passing through them.

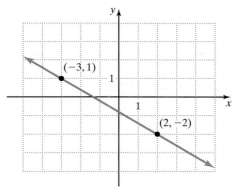

Next, find the slope.

$$(-3, 1) \quad (2, -2)$$
$$\uparrow \uparrow \quad \uparrow \quad \uparrow$$
$$x_1 \, y_1 \quad x_2 \quad y_2$$

$$m = \frac{y_2 - y_1}{x_2 - x_1} = \frac{-2 - 1}{2 - (-3)} = \frac{-3}{5} = -\frac{3}{5}$$

PRACTICE 2

Sketch the line that contains the points $(-2, 1)$ and $(3, -5)$. Find the slope.

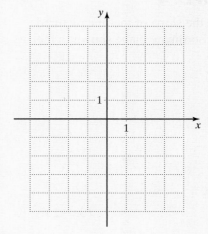

A line falling to the right as shown in Example 2 has a *negative* slope. We say that such a line is *decreasing* because as the x-values get larger, the corresponding y-values get smaller.

EXAMPLE 3

Find the slope of a line that passes through the points $(7, 5)$ and $(-1, 5)$. Plot the points, and then sketch the line.

Solution First, find the slope of the line.

$$(7, 5) \quad (-1, 5)$$
$$\uparrow \uparrow \quad \uparrow \uparrow$$
$$x_1 \, y_1 \quad x_2 \, y_2$$

$$m = \frac{y_2 - y_1}{x_2 - x_1} = \frac{5 - 5}{-1 - 7} = \frac{0}{-8} = 0$$

So the slope of this line is 0.

Next, we plot the points, and then sketch the line passing through them.

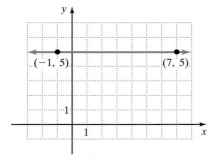

PRACTICE 3

On the following coordinate plane, plot the points $(2, -1)$ and $(6, -1)$. Sketch the line, and then compute its slope.

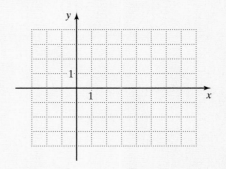

When the slope of a line is 0, its graph is a **horizontal line** as shown in Example 3. All points on a horizontal line have the same *y*-coordinate, that is, the *y*-values are constant for all *x*-values.

EXAMPLE 4

What is the slope of the line pictured on the coordinate plane shown to the right?

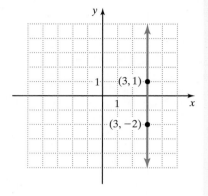

Solution

$$(3, -2) \qquad (3, 1)$$
$$\begin{array}{cc} \uparrow \ \uparrow & \uparrow \ \uparrow \\ x_1 \ \ y_1 & x_2 \ y_2 \end{array}$$

$$m = \frac{y_2 - y_1}{x_2 - x_1} = \frac{1 - (-2)}{3 - 3} = \frac{3}{0}$$

Since division by 0 is undefined, the slope of this line is undefined.

PRACTICE 4

Find the slope of the line that passes through the points $(-2, 7)$ and $(-2, 0)$.

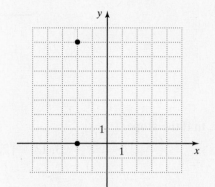

When the slope of a line is undefined, its graph is a **vertical line** as shown in Example 4. All points on a vertical line have the same *x*-coordinate, that is, the *x*-values are constant for all *y*-values.

As we have seen in Examples 1 through 4, the sign of the slope of a line tells us a lot about the line. As we continue graphing lines, it will be helpful to keep in mind the following graphs:

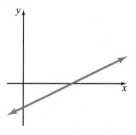

Positive *m*
The line slants upward from left to right.

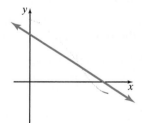

Negative *m*
The line slants downward from left to right.

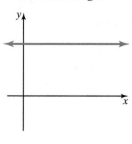

Zero *m*
The line is horizontal.

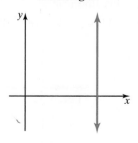

Undefined *m*
The line is vertical.

In the next example, we graph two lines on a coordinate plane.

EXAMPLE 5

Calculate the slopes for the lines shown.

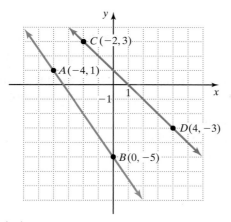

Solution   Line AB, written $\overleftrightarrow{AB}$, passes through $A(-4, 1)$ and $B(0, -5)$. Its slope is:

$$m = \frac{y_2 - y_1}{x_2 - x_1} = \frac{1 - (-5)}{(-4) - 0} = \frac{1 + 5}{-4} = \frac{6}{-4} = -\frac{3}{2}$$

For $\overleftrightarrow{CD}$ passing through $C(-2, 3)$ and $D(4, -3)$, the slope is:

$$m = \frac{y_2 - y_1}{x_2 - x_1} = \frac{3 - (-3)}{(-2) - 4} = \frac{3 + 3}{(-2) - 4} = \frac{6}{-6} = -1$$

Note that both lines have negative slopes and slant downward from left to right.

PRACTICE 5

Compute the slopes for the lines shown.

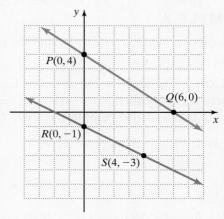

As shown in the next example, how a line slants often helps us to interpret the information given in the graph.

EXAMPLE 6

The following graph shows the amount of money that your dental insurance reimburses you, depending on the amount of your dental bill. Is the slope of the graphed line positive or negative? Explain how you know. What does this mean in terms of insurance reimbursement?

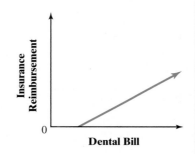

Solution   Since the graphed line slants upward from left to right, its slope is positive. According to this graph, larger x-values correspond to larger y-values. So your dental insurance reimburses you more for larger dental bills.

PRACTICE 6

A doctor is trying to help eliminate an epidemic. Explain, in terms of slope, which scenario would be the most desirable.

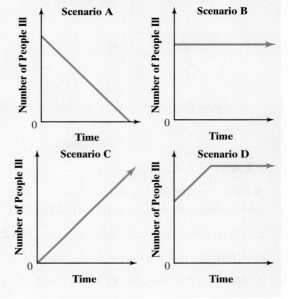

We have already graphed a line by plotting two points and drawing the line passing through them. Now, let's look at graphing a line when given the slope of the line and a point on the line.

EXAMPLE 7

The slope of a line that passes through the point $(2, 5)$ is 3. Graph the line.

Solution   The line in question passes through the point $(2, 5)$. But there are many such lines—which is the right one? We use the slope 3 to find a second point through which the line also passes.

Since 3 can be written as $\dfrac{3}{1}$, we have:

$$\text{slope} = \frac{\text{rise}}{\text{run}} = \frac{3}{1}$$

We first plot the point $(2, 5)$. Starting at $(2, 5)$, we move 3 units up (for a rise of 3) and then 1 unit to the right (for a run of 1). Arriving at the point $(3, 8)$, we sketch the line passing through the points $(2, 5)$ and $(3, 8)$, as shown in the graph on the left below.

Since $\dfrac{3}{1} = \dfrac{-3}{-1}$, we could have started at $(2, 5)$ and moved down 3 units (for a rise of -3) and then 1 unit to the left (for a run of -1). In this case, we would arrive at $(1, 2)$, which is another point on the same line, as shown in the graph on the right below.

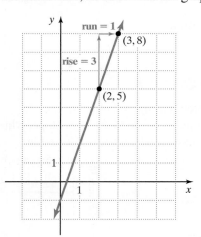
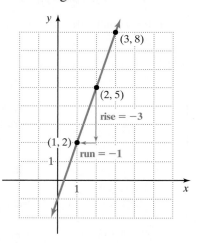

Can you find other points on this line? Explain.

PRACTICE 7

Graph the line with slope 4 that passes through the point $(1, -2)$.

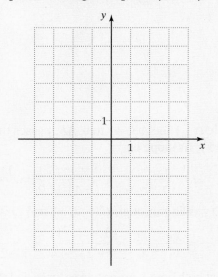

EXAMPLE 8

A family on vacation is driving out of town at a constant speed. At 2 o'clock, they have traveled 110 mi. By 6 o'clock, they have traveled 330 mi.

a. On a coordinate plane, label the axes, and then plot the appropriate points.

b. Compute the slope of the line passing through the points.

c. Interpret the meaning of the slope in this situation.

PRACTICE 8

In 1985, the streetcar running along Boston's Arborway Corridor had a daily ridership of 28,000. The streetcar was then replaced by a bus. Twenty-five years later, daily ridership had dropped to 14,000. (*Source:* arborway.org)

Solution

a. Label the axes on the coordinate plane. Then, plot the points $(2, 110)$ and $(6, 330)$.

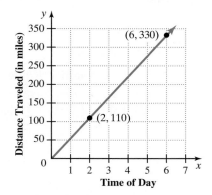

b. The slope of the line through the two points is:

$$m = \frac{y_2 - y_1}{x_2 - x_1}$$

$$= \frac{330 - 110}{6 - 2}$$

$$= \frac{220}{4}$$

$$= 55$$

c. Here the slope is the change in distance divided by the change in time. In other words, the slope is the average speed the family traveled, which is 55 mph.

a. On a coordinate plane, label the axes, and then plot the appropriate points for daily ridership r in y years after 1985.

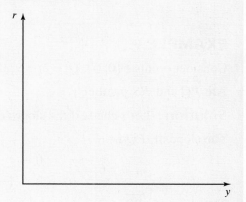

b. Compute the slope of the line that passes through the points.

c. Interpret the meaning of the slope in this situation.

Parallel and Perpendicular Lines

By examining the slopes of straight lines on a coordinate plane, we can solve problems that require us to determine if:

- two given lines are parallel or
- two given lines are perpendicular.

Let's consider parallel lines first.

Since the slope of a line measures its slant, lines in a coordinate plane with equal slopes are parallel, that is, do not intersect. So on the coordinate plane shown below, if we knew the coordinates of points $P, Q, R,$ and S, we could verify that $\overleftrightarrow{PQ}$ and $\overleftrightarrow{RS}$ are parallel by computing their slopes and then checking that these slopes are equal.

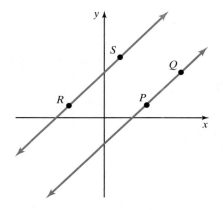

> **Parallel Lines**
>
> Two nonvertical lines are **parallel** if and only if their slopes are equal. That is, if the slopes are m_1 and m_2, then $m_1 = m_2$.

EXAMPLE 9

Consider points $P(0, 0)$, $Q(-2, -5)$, $R(0, 5)$, and $S(-2, 0)$.
Are $\overleftrightarrow{PQ}$ and $\overleftrightarrow{RS}$ parallel?

Solution   Let's check if the slopes of $\overleftrightarrow{PQ}$ and $\overleftrightarrow{RS}$ are equal.
The slope of $\overleftrightarrow{PQ}$ is:

$$m = \frac{y_2 - y_1}{x_2 - x_1} = \frac{0 - (-5)}{0 - (-2)} = \frac{0 + 5}{0 + 2} = \frac{5}{2}$$

The slope of $\overleftrightarrow{RS}$ is:

$$m = \frac{y_2 - y_1}{x_2 - x_1} = \frac{5 - 0}{0 - (-2)} = \frac{5}{0 + 2} = \frac{5}{2}$$

Since the slopes of $\overleftrightarrow{PQ}$ and $\overleftrightarrow{RS}$ are equal, $\overleftrightarrow{PQ}$ and $\overleftrightarrow{RS}$ are parallel.

PRACTICE 9

Decide whether $\overleftrightarrow{EF}$ and $\overleftrightarrow{GH}$ are parallel, given points $E(0, 4)$, $F(4, -1)$, $G(0, 8)$, and $H(8, -2)$.

EXAMPLE 10

A pediatric nurse kept track of the weights of two children who are twins. Use the graph to answer the following questions:

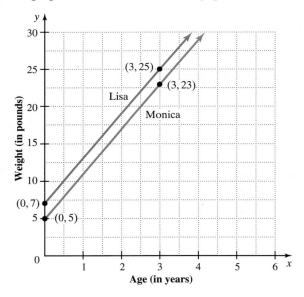

Age (in years)

a. Which twin was heavier at birth?

b. Are the two lines parallel?

c. Which twin grew at a faster rate?

d. Assuming that the rates of growth of the twins remain constant, could this graph be used to project the weight of the twins at age 8? Explain.

PRACTICE 10

The graph shown gives the income of a computer lab technician and a multimedia designer as related to the number of years that they have been employed.

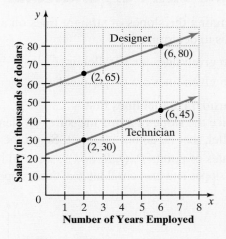

Use this graph to answer the following questions.

Solution

a. At birth, Lisa weighed 7 lb and Monica weighed 5 lb. So Lisa was heavier.

b. To determine if the two lines are parallel, we begin by computing the slope of the line passing through the points $(0, 7)$ and $(3, 25)$.

$$m = \frac{y_2 - y_1}{x_2 - x_1} = \frac{7 - 25}{0 - 3} = \frac{-18}{-3} = 6$$

Now, we compute the slope for the line passing through the points $(0, 5)$ and $(3, 23)$.

$$m = \frac{y_2 - y_1}{x_2 - y_1} = \frac{5 - 23}{0 - 3} = \frac{-18}{-3} = 6$$

Since the slopes of the two lines are equal, the graphed lines are parallel.

c. Since the two lines are parallel, the twins grew at the same rate.

d. If we extend the x- and y-axes, we could project the weight of each child at age 8 by reading the corresponding y-coordinate for x equal to 8.

a. Are the two lines parallel?

b. Does your answer to part (a) agree with your observation of the graph? Explain.

c. Which employee's salary increased at a faster rate?

d. From the graph, estimate the starting salary of the multimedia designer.

Now, let's consider the problem of determining whether two given lines are perpendicular that is, intersect at a right, or 90°, angle.

On a coordinate plane, two lines are perpendicular to one another when the product of their slopes is -1. For instance, the slope of $\overleftrightarrow{PQ}$ in the following graph is:

$$m = \frac{7 - 2}{3 - 1} = \frac{5}{2}$$

The slope of $\overleftrightarrow{QR}$ is:

$$m = \frac{2 - 0}{1 - 6} = \frac{2}{-5} = -\frac{2}{5}$$

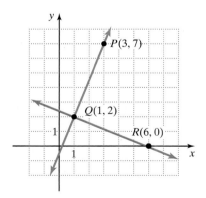

The product of these slopes is:

$$m = \left(\frac{5}{2}\right)\left(-\frac{2}{5}\right) = -1$$

Since the slopes are *negative reciprocals* of each other, the two lines must be perpendicular to one another.

Perpendicular Lines

Two nonvertical lines are **perpendicular** if and only if the product of their slopes is -1. That is, if the slopes are m_1 and m_2, then $m_1 \cdot m_2 = -1$.

EXAMPLE 11

Determine if $\overleftrightarrow{AB}$ and $\overleftrightarrow{BC}$ shown in the graph are perpendicular to one another.

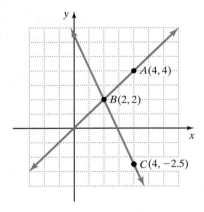

Solution   First, let's compute the slopes of the two lines in question. The slope of $\overleftrightarrow{AB}$ is:

$$m = \frac{y_2 - y_1}{x_2 - x_1} = \frac{4 - 2}{4 - 2} = \frac{2}{2} = 1$$

The slope of $\overleftrightarrow{BC}$ is:

$$m = \frac{y_2 - y_1}{x_2 - x_1} = \frac{2 - (-2.5)}{2 - 4} = \frac{2 + 2.5}{2 - 4} = \frac{4.5}{-2} = -2.25$$

To check if $\overleftrightarrow{AB}$ and $\overleftrightarrow{BC}$ are perpendicular, we find the product of their slopes:

$$(1)(-2.25) = -2.25$$

The lines are not perpendicular to one another since this product is not equal to -1.

PRACTICE 11

Consider points $A(1, 3), B(2, 5),$ and $C(-1, 2)$. Decide if $\overleftrightarrow{AB}$ is perpendicular to $\overleftrightarrow{AC}$.

EXAMPLE 12

On the coordinate plane shown, *x*-values represent streets and *y*-values represent avenues. A road is to be constructed running straight from 4th Street and 3rd Avenue to 9th Street and 9th Avenue. A second road will run from 2nd Street and 10th Avenue to 8th Street and 5th Avenue. Will the roads meet at right angles?

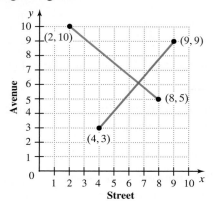

Solution   Let's find the slopes of the two roads. The slope of the road from 9th Street and 9th Avenue to 4th Street and 3rd Avenue is:

$$m = \frac{y_2 - y_1}{x_2 - x_1} = \frac{9 - 3}{9 - 4} = \frac{6}{5}$$

The slope of the road from 2nd Street and 10th Avenue to 8th Street and 5th Avenue is:

$$m = \frac{y_2 - y_1}{x_2 - x_1} = \frac{10 - 5}{2 - 8} = \frac{5}{-6} = -\frac{5}{6}$$

The product of these slopes is:

$$\left(\frac{6}{5}\right)\left(-\frac{5}{6}\right) = -1$$

Therefore, the roads will be perpendicular to one another.

PRACTICE 12

Consider the square 6 units on each side shown in the following diagram. By examining their slopes, determine whether the diagonals of the square are perpendicular to one another.

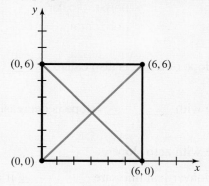

Mathematically Speaking

Fill in each blank with the most appropriate term or phrase from the given list.

y-coordinate	parallel	negative
positive	vertical	x-coordinate
perpendicular	rate of change	run
horizontal	rise	

1. The slope of a line is also called its _____.

2. In the slope formula, the vertical change is called the _____.

3. A line with _____ slope is decreasing.

4. If all points on a line have the same _____, then the line is vertical.

5. A line with zero slope is _____.

6. A line with undefined slope is _____.

7. Two nonvertical lines are _____ if and only if their slopes are equal.

8. Two nonvertical lines are _____ if and only if the product of their slopes is −1.

A *Compute the slope m of the line that passes through the given points. Plot these points on the coordinate plane, and sketch the line that passes through them.*

9. $(2, 3)$ and $(-2, 0)$, $m =$

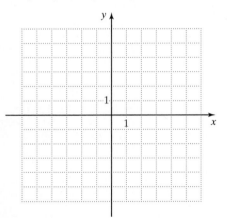

10. $(-1, 4)$ and $(0, 5)$, $m =$

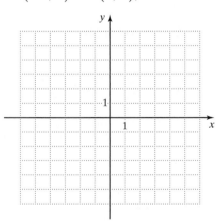

11. $(6, -4)$ and $(6, 1)$, $m =$

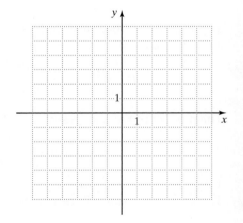

12. $(1, 1)$ and $(1, -3)$, $m =$

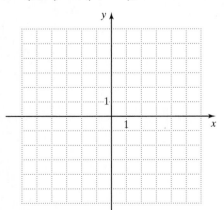

13. $(-2, 1)$ and $(3, -1)$, $m =$

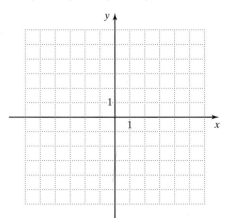

14. $(0, 0)$ and $(-2, 5)$, $m =$

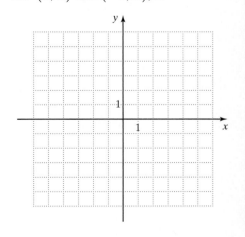

15. $(-1, -4)$ and $(3, -4)$, $m =$

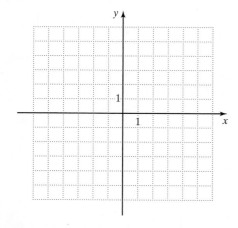

16. $(3, 0)$ and $(5, 0)$, $m =$

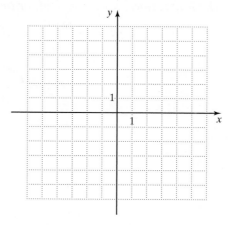

17. $(0.5, 0)$ and $(0, 3.5)$, $m =$

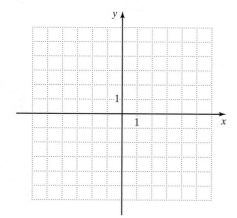

18. $(4, 4.5)$ and $(1, 2.5)$, $m =$

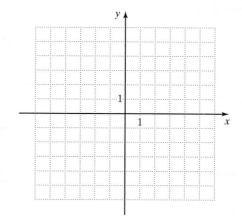

On each graph, calculate the slopes for the lines shown.

19.

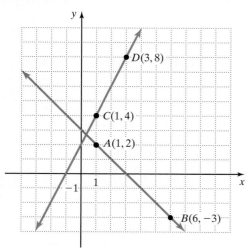

20.

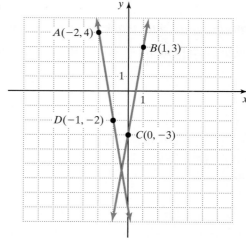

B *Indicate whether the slope of each graph is positive, negative, zero, or undefined. Then, state whether the line is horizontal, vertical, or neither.*

21.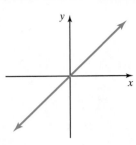

22.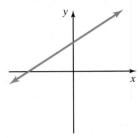

23.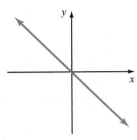

24.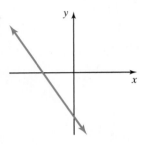

25.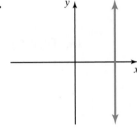

26.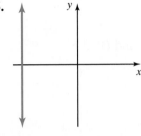

27.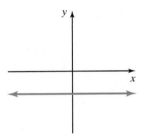

28.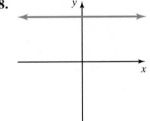

C *Graph the line on the coordinate plane using the given information.*

29. Passes through $(2, 5)$ and $m = 4$ **30.** Passes through $(-1, 1)$ and $m = \frac{1}{2}$ **31.** Passes through $(2, 5)$ and $m = -\frac{4}{3}$

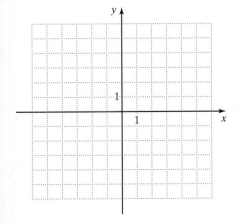

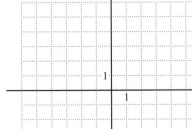

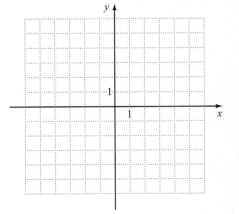

32. Passes through $(1, 5)$ and $m = -3$ **33.** Passes through $(0, -6)$ and $m = 0$ **34.** Passes through $(0, -2)$ and $m = 0$

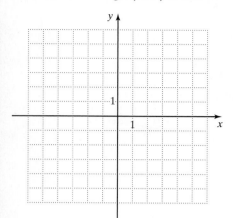   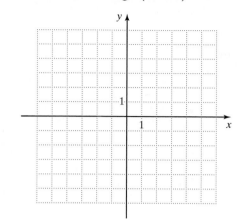   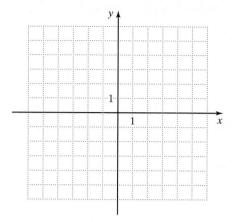

35. Passes through $(-4, 0)$ and the slope is undefined   **36.** Passes through $(-6, 2)$ and the slope is undefined

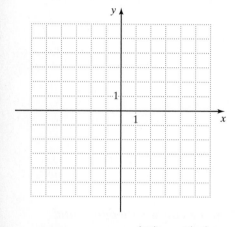   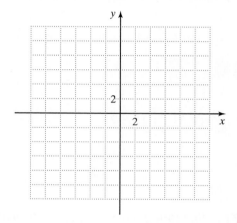

Determine whether $\overleftrightarrow{PQ}$ and $\overleftrightarrow{RS}$ are parallel or perpendicular.

37.

	P	Q	R	S
a.	$(0, -1)$	$(1, 3)$	$(5, 0)$	$(7, 8)$
b.	$(9, 1)$	$(7, 4)$	$(0, 0)$	$(6, 4)$

38.

	P	Q	R	S
a.	$(3, 3)$	$(7, 7)$	$(-5, 5)$	$(2, -2)$
b.	$(8, 0)$	$(0, 4)$	$(0, -4)$	$(-12, 2)$

Mixed Practice

Indicate whether the slope of each graph is positive, negative, zero, or undefined.
Then, state whether the line is horizontal, vertical, or neither.

39.

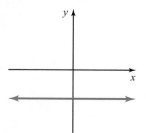

40.

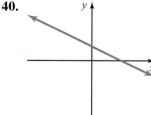

41. Graph the line on the coordinate plane if the line passes through $(-2, 3)$ and $m = 4$.

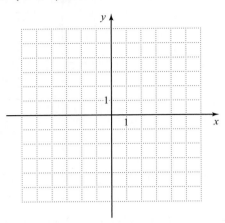

42. Calculate the slopes for the lines shown in the graph.

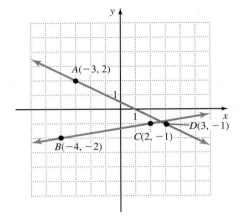

43. Determine whether $\overleftrightarrow{AB}$ and $\overleftrightarrow{CD}$ are parallel or perpendicular.

	A	B	C	D
a.	$(-3, 2)$	$(5, -2)$	$(1, -4)$	$(5, 4)$
b.	$(-1, 5)$	$(5, -3)$	$(2, 2)$	$(5, -2)$

Compute the slope m of the line that passes through the given points. Plot these points on the coordinate plane, and sketch the line that passes through them.

44. $(-4, -5)$ and $(2, -1)$, $m =$

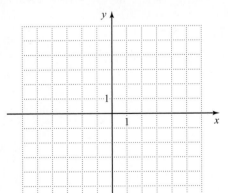

45. $(-3, -2)$ and $(-3, 4)$, $m =$

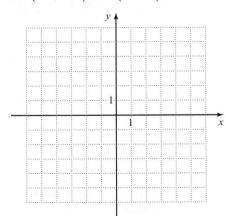

46. $(-4, 4.5)$ and $(-2, -1.5)$, $m =$

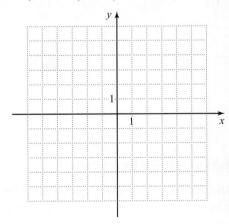

Applications

Solve.

47. A chemist conducts an experiment on the gas contained in a sealed tube. The experiment is to heat the gas and then to measure the resulting pressure in the tube. In the lab manual, points are plotted and the line is sketched to show the gas pressure for various temperatures.

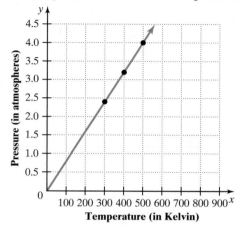

a. Is the slope of this line positive, negative, zero, or undefined?

b. In a sentence, explain the significance of the answer to part (a) in terms of temperature and pressure.

48. To reduce their taxes, many businesses use *the straight-line method of depreciation* to estimate the change in the value over time of equipment that they own. The graph shows the value of equipment owned from the time of purchase to 7 yr later.

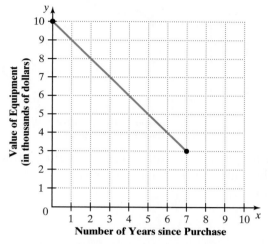

a. Is the slope of this line positive, negative, zero, or undefined?

b. In a sentence or two, explain the significance of the answer to part (a) in terms of the value of the equipment over time.

49. Two motorcyclists leave at the same time, racing down a road. Consider the lines in the graph that show the distance traveled by each motorcycle at various times.

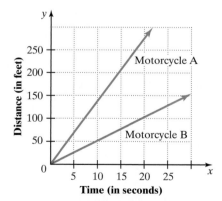

a. Which motorcycle first travels 100 ft?

b. Which motorcycle is traveling more slowly?

c. Explain what the slopes mean in this situation.

50. Most day-care centers charge parents additional fees for arriving late to pick up their children. The following graph shows the late fee for two day-care centers:

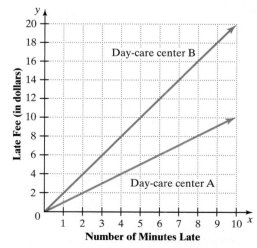

a. Which day-care center charges a higher late fee?

b. Explain what the slopes represent in this situation.

51. The following graph records the amount of garbage deposited in landfills A and B after they are opened.

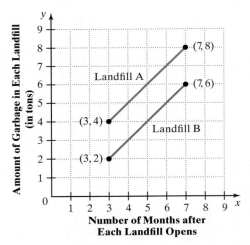

Are the garbage deposits at the two landfills growing at the same rate? Explain in a sentence or two how you know.

52. The weights of a brother and sister from age 3 yr to 7 yr are recorded in the graph. Did their weights increase at the same rate? Explain.

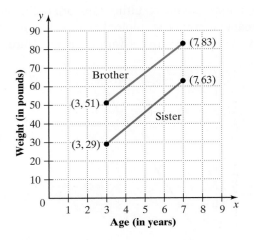

53. After a stock split, the per-share value of the stock increased, as shown in the following table:

Number of Days after the Split	Per-share Stock Value (in dollars)	Point
2	27	$P\,(2, 27)$
4	51	$Q\,(4, 51)$
8	79	$R\,(8, 79)$

a. Choose appropriate scales and label each axis. Plot the points, and then sketch $\overleftrightarrow{PQ}$ and $\overleftrightarrow{QR}$.

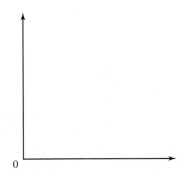

b. Determine whether the rate of increase changed over time. Explain.

54. The position of a dropped object for various times after the object is released is given in the table below:

Time After Release (in seconds)	Position (in feet)	Point
1	−16	$A\,(1, -16)$
2	−64	$B\,(2, -64)$
3	−144	$C\,(3, -144)$

a. Choose appropriate scales and label each axis. Plot the points, and then sketch $\overleftrightarrow{AB}$ and $\overleftrightarrow{BC}$.

b. Compute the slopes of $\overleftrightarrow{AB}$ and $\overleftrightarrow{BC}$.

c. Was the rate of fall for the dropped object constant throughout the experiment? Explain.

55. A hiker is at point A as shown in the graph and wants to take the shortest route through a field to reach a nearby road represented by $\overleftrightarrow{BC}$. The shortest route will be to walk perpendicular to the road. Is $\overleftrightarrow{AD}$ the shortest route? Explain.

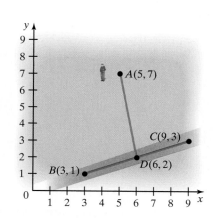

56. The coordinates of the vertices of triangle ABC are shown. Is triangle ABC a right triangle? Explain.

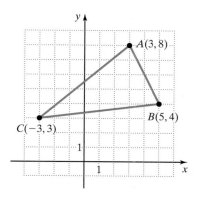

57. On a highway, a driver sets a car's cruise control for a constant speed of 55 mph.

    a. Of the following, which graph shows the distance the car travels? Using the slope of the line, explain.

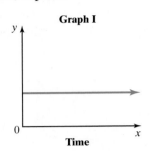

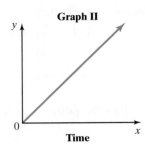

    b. Which of the above graphs shows the speed of the car? Using the slope of the line, explain.

58. City leaders consider imposing an income tax on residents whose income is above a certain amount, as pictured below.

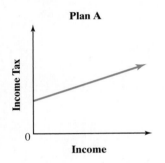

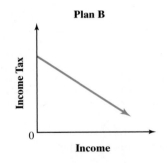

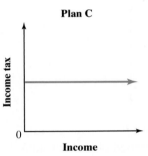

    a. Using the slopes of the lines, describe each plan.

    b. Which plan do you think is the fairest? Explain.

• Check your answers on page A-7.

MINDStretchers

Groupwork

1. A geoboard is a square flat surface with pegs forming a grid pattern. You can stretch rubber bands around the pegs to explore geometric questions such as, "What is the slope of $\overline{AB}$?" Pictured is a geoboard with rubber bands forming a series of steps.

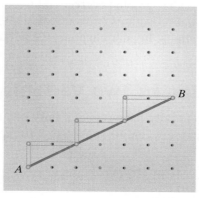

Using a geoboard or graph paper, determine whether the slope of $\overleftrightarrow{AB}$ increases or decreases if:

a. the rise of each step increases by one peg.

b. the run of each step increases by one peg.

Writing

2. Describe a real-world situation that each of the following graphs might illustrate. Explain the significance of slope in the situation that you have described.

a.
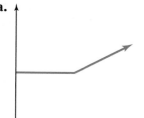

b.

c.
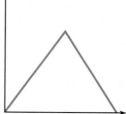

Mathematical Reasoning

3. If the slope of a line is positive, explain why any line perpendicular to it must have a negative slope.

3.3   Linear Equations and Their Graphs

OBJECTIVES

Ⓐ To identify the coordinates of points that satisfy a given linear equation in two variables

Ⓑ To graph a given linear equation in two variables

Ⓒ To solve applied problems involving graphs of linear equations

In the first section of this chapter, we developed the idea of coordinates—an ordered pair of numbers associated with a point on the plane. In the second section, we shifted our attention from points to lines and their slopes. In this section, we focus on Descartes' most startling idea—on a coordinate plane, lines (and indeed other graphs) correspond to equations. Line graphs and their associated equations, called *linear* equations, have important applications in the real world, allowing us to model many situations. In later chapters, we will discuss other types of graphs.

Solutions of a Linear Equation in Two Variables

Recall that linear equations in one variable generally have one and only one solution. For instance, the solution of $2x + 5 = 11$ is 3 because when we substitute 3 for x, the equation is true.

$$2x + 5 = 11$$
$$2 \cdot 3 + 5 \stackrel{?}{=} 11$$
$$11 = 11 \qquad \text{True}$$

We now consider linear equations in *two* variables, for instance $y = 2x + 5$. We can also express $y = 2x + 5$ as $-2x + y = 5$.

$$y = 2x + 5$$
$$-2x + y = 5 \qquad \text{Subtract } 2x \text{ from each side of the equation.}$$

DEFINITION

A **linear equation in two variables**, x and y, is an equation that can be written in the *general form* $Ax + By = C$, where A, B, and C are real numbers and A and B are not both 0.

Note that in the general form of a linear equation, the two variable terms are on one side of the equation and the constant term is on the other. The equation $-2x + y = 5$, for instance, is in general form, with $A = -2$, $B = 1$, and $C = 5$.

Here are some other examples of linear equations in general form:

Equation	A	B	C
$5x + 3y = 10$	5	3	10
$x - 2y = 6 \rightarrow 1x + (-2)y = 6$	1	-2	6
$x = -4 \rightarrow 1x + 0y = -4$	1	0	-4
$y = -4 \rightarrow 0x + 1y = -4$	0	1	-4

Now, let's look at what we mean by a solution of a linear equation in two variables.

DEFINITION

A **solution** of an equation in two variables is an ordered pair of numbers that when substituted for the variables makes the equation true.

Applying this definition to the equation $-2x + y = 5$, we observe that $x = 3$ and $y = 11$ is a solution of the equation because substituting 3 for x and 11 for y makes the equation true.

$$-2x + y = 5$$
$$-2 \cdot 3 + 11 \overset{?}{=} 5$$
$$5 = 5 \qquad \text{True}$$

There are many other solutions of this equation as well. For instance, $x = 0$ and $y = 5$ is another solution.

$$-2x + y = 5$$
$$-2 \cdot 0 + 5 \overset{?}{=} 5$$
$$5 = 5 \qquad \text{True}$$

Unlike a linear equation in one variable that has at most one solution, linear equations in two variables have an infinite number of solutions. Can you explain why this is the case?

Now, how do we *find* solutions of a linear equation in two variables? Typically, we start with the value of one of the variables, and then compute the corresponding value of the other, as the following example illustrates.

EXAMPLE 1

For the equation $4x + y = -5$, find five solutions by completing the following table:

x	-2	3	0		
y				0	1

Solution   In the first column of this table, we substitute -2 for x in the given equation, and then solve for y.

$$4x + y = -5$$
$$4(-2) + y = -5$$
$$-8 + y = -5$$
$$y = 3$$

To find the corresponding y in the second column, we substitute 3 for x:

$$4x + y = -5$$
$$4(3) + y = -5$$
$$12 + y = -5$$
$$y = -17$$

And in the third column, we substitute 0 for x:

$$4x + y = -5$$
$$4(0) + y = -5$$
$$0 + y = -5$$
$$y = -5$$

PRACTICE 1

For the equation $-2x + y = 1$, find the missing values in the following table:

x	0	5	-3		
y				0	-3

EXAMPLE 1 (continued)

The next two columns are different from the earlier ones: they give us y-values and require us to solve for x. First, we see that in the fourth column y is 0.

$$4x + y = -5$$
$$4x + 0 = -5 \qquad \text{Substitute 0 for } y.$$
$$4x = -5$$
$$x = -\frac{5}{4}$$

In the final column of the table, $y = 1$.

$$4x + y = -5$$
$$4x + 1 = -5 \qquad \text{Substitute 1 for } y.$$
$$4x = -6$$
$$x = -\frac{6}{4} = -\frac{3}{2}$$

We have found the missing values, so the table reads:

x	-2	3	0	$-\frac{5}{4}$	$-\frac{3}{2}$
y	3	-17	-5	0	1

The Graph of a Linear Equation in Two Variables

The graph of an equation, more precisely the graph of the *solutions* of that equation, is a kind of picture of the equation.

> **DEFINITION**
> The **graph** of a linear equation in two variables consists of all points whose coordinates make the equation true.

Given a linear equation, how do we find its graph? A general strategy is to first isolate one of the variables, unless it is already done. Then, we identify several solutions of the equation, keeping track of the x- and y-values in a table. We plot the points and then sketch the line passing through them. That line is the graph of the given equation, as the following example illustrates.

Let's graph the equation $y = 3x + 1$. The variable y is already isolated, so we find y-values by substituting arbitrary values of x. For instance, let x equal 0. To find y, we substitute 0 for x.

$$y = 3x + 1 = 3 \cdot 0 + 1 = 0 + 1 = 1$$

So $x = 0$ and $y = 1$ is a solution of this equation. We say that the ordered pair $(0, 1)$ is a solution of $y = 3x + 1$.

Let's choose three other values of x, say -1, 1, and 2. Substituting -1 for x in the equation, we get:

$$y = 3x + 1 = 3(-1) + 1 = -2$$

Substituting 1 for x, we get:

$$y = 3x + 1 = 3 \cdot 1 + 1 = 4$$

Substituting 2 for x gives us:

$$y = 3x + 1 = 3 \cdot 2 + 1 = 7$$

Next, we enter these results in a table.

x	-1	0	1	2
y	-2	1	4	7

Then, we plot on a coordinate plane the four points $A(-1, -2)$, $B(0, 1)$, $C(1, 4)$, and $D(2, 7)$. If we have not made a mistake, the points will all lie on the same line. We know that the line segments $\overline{AB}$, $\overline{BC}$, and $\overline{CD}$ are part of the same line because they all have equal slopes. The graph of the equation $y = 3x + 1$ is the line passing through these points. So any point on this line satisfies the equation $y = 3x + 1$.

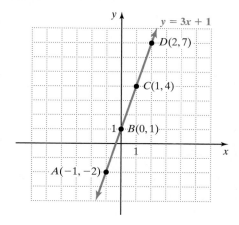

Do you think that if we had chosen three other x-values we would have gotten the same graph? Check to see that this is the case.

This example suggests the following procedure for graphing a linear equation:

To Graph a Linear Equation in Two Variables

- Isolate one of the variables—usually y—if it is not already done.

- Choose three x-values, entering them in a table.

- Complete the table by calculating the corresponding y-values.

- Plot the three points—two to draw the line and the third to serve as a *checkpoint*.

- Check that the points seem to lie on the same line.

- Draw the line passing through the points.

A couple of observations about the preceding example are worth making:

- The slope of the line graphed is 3. We can see this by taking any pair of points on the line, for example $(0, 1)$ and $(1, 4)$, and computing the slope of the line between them.

$$m = \frac{y_2 - y_1}{x_2 - x_1} = \frac{1 - 4}{0 - 1} = \frac{-3}{-1} = 3$$

- This slope is identical to the coefficient of x in the equation $y = 3x + 1$.

We also see that the point where the graph crosses the y-axis is $(0, 1)$. This point is called the *y-intercept*. Note that the constant term in the equation $y = 3x + 1$ is also 1.

These relationships are more than a coincidence, and we will say more about them in Section 3.4.

EXAMPLE 2

Graph the equation $y = -\frac{3}{2}x$ by choosing three points whose coordinates satisfy the equation.

Solution   We begin by choosing x-values. In this case, we choose multiples of 2 for the x-values. Then, we find the corresponding y-values.

x	$y = -\frac{3}{2}x$	(x, y)
0	$y = -\frac{3}{2}(0) = 0$	$(0, 0)$
2	$y = -\frac{3}{2}(2) = -3$	$(2, -3)$
4	$y = -\frac{3}{2}(4) = -6$	$(4, -6)$

We plot the points, and then draw a line passing through them to get the desired line.

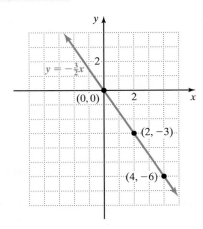

PRACTICE 2

Graph the equation $y = -\frac{3}{5}x$.

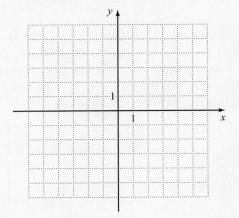

Note that in Example 2, we chose multiples of 2 for the x-values. Can you explain why?

EXAMPLE 3

Consider the equation $2x + y = 3$.

a. Graph the equation.
b. Find the slope of the line.

Solution

a. We begin by solving the equation for y.

$$2x + y = 3$$
$$y = -2x + 3 \quad \text{Subtract } 2x \text{ from each side.}$$

Next, we choose three values for x, for instance, -1, 0, and 2. Then, we enter them into a table and find their corresponding y-values as follows:

x	$y = -2x + 3$	(x, y)
-1	$y = -2(-1) + 3 = 2 + 3 = 5$	$(-1, 5)$
0	$y = -2(0) + 3 = 0 + 3 = 3$	$(0, 3)$
2	$y = -2(2) + 3 = -4 + 3 = -1$	$(2, -1)$

PRACTICE 3

Consider the equation $-2x + y = -5$.

a. Graph the equation.

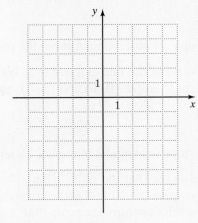

b. Find the slope of the line.

Now, we plot the points on the coordinate plane. Since the points seem to lie on the same line, we draw a line passing through the points.

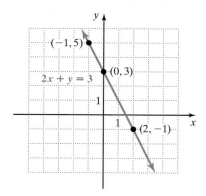

b. To find the slope of the line, we can consider the points $(0, 3)$ and $(2, -1)$.

$$m = \frac{3 - (-1)}{0 - 2} = \frac{4}{-2} = -2$$

Note that the slope is the coefficient of x in the equation $y = -2x + 3$.

We have graphed equations in general form by first isolating y and then computing y-values for arbitrary x-values. Now, we graph equations in general form with a different approach using x- and y-intercepts. Note that intercepts stand out on a graph and are easy to plot. Since an x-intercept lies on the x-axis, its y-value must be 0. Similarly, since a y-intercept lies on the y-axis, the x-value of a y-intercept must be 0.

DEFINITION

The **x-intercept** of a line is the point where the graph crosses the x-axis. The **y-intercept** is the point where the graph crosses the y-axis.

The following graph shows a line passing through two points, $(0, 4)$ and $(3, 0)$, which are both intercepts.

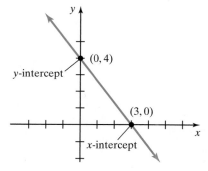

Using the x- and y-intercepts to graph equations can save work, especially when the coefficients of the two variables are factors of the constant term, as shown in the next example.

EXAMPLE 4

Consider the equation $2x + 3y = 6$. Find the x- and y-intercepts. Then, graph.

Solution   Since the y-intercept has x-value 0, we let $x = 0$ and then solve for y.

For $x = 0$:

$$2x + 3y = 6$$
$$2 \cdot 0 + 3y = 6 \qquad \text{Substitute 0 for } x.$$
$$3y = 6$$
$$\frac{3y}{3} = \frac{6}{3}$$
$$y = 2$$

So the y-intercept is $(0, 2)$.

Similarly, the x-intercept has y-value 0. So we let $y = 0$ and then solve for x.

For $y = 0$:

$$2x + 3y = 6$$
$$2x + 3 \cdot 0 = 6 \qquad \text{Substitute 0 for } y.$$
$$2x = 6$$
$$\frac{2x}{2} = \frac{6}{2}$$
$$x = 3$$

So the x-intercept is $(3, 0)$.

Before graphing, we choose a third point to be used as a checkpoint.

For $2x + 3y = 6$, let $x = 6$:

$$2x + 3y = 6$$
$$2 \cdot 6 + 3y = 6 \qquad \text{Substitute 6 for } x.$$
$$12 + 3y = 6$$
$$3y = -6$$
$$y = -2$$

So the checkpoint is $(6, -2)$.

Plotting the points $(0, 2)$, $(3, 0)$, and $(6, -2)$ on a coordinate plane, we confirm that they seem to lie on the same line. Finally, we draw a line through the points, getting the desired graph.

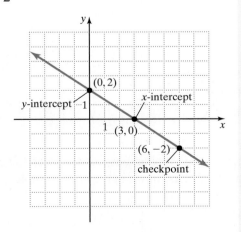

PRACTICE 4

Consider the equation $x - 2y = 4$. Find the x- and y-intercepts. Then, graph.

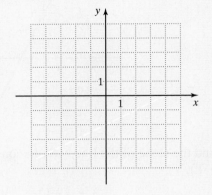

Example 4 suggests the following procedure:

To Graph a Linear Equation in Two Variables Using the x- and y-intercepts

- Let $x = 0$, and find the y-intercept.
- Let $y = 0$, and find the x-intercept.
- Find a checkpoint.
- Plot the three points.
- Check that the points seem to lie on the same line.
- Draw the line passing through the points.

We know that the general form of a linear equation in two variables is $Ax + By = C$. Sometimes in a linear equation one of the two variables is missing, that is, the coefficient of one of the two variables is zero. Consider the following equations:

$$y = 9 \quad \rightarrow \quad 0x + y = 9$$
$$x = -5.8 \quad \rightarrow \quad x + 0y = -5.8$$

Let's look at the graphs of these equations.

EXAMPLE 5

Graph:

a. $y = 9$ **b.** $x = -5.8$

Solution

a. For the line $y = 9$, the coefficient of the x-term is 0. The x-value can be any real number and the y-value is always 9. So the graph is as follows:

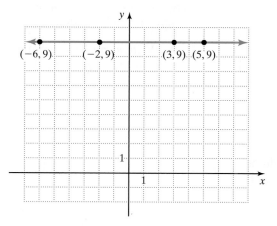

The graph of this equation is a horizontal line. Recall that the slope of a horizontal line is 0.

PRACTICE 5

On the given coordinate plane, graph:

a. $y = -1$ **b.** $x = 2.5$

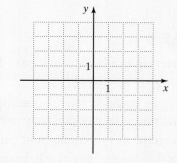

EXAMPLE 5 (continued)

b. For the line $x = -5.8$, the coefficient of the y-term is 0. The y-value can be any real number and the x-value is always -5.8. So the graph is as follows:

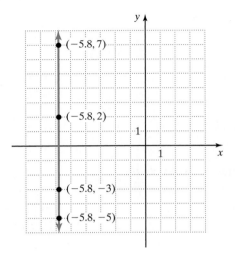

The graph of this equation is a vertical line. The slope of this line is undefined.

In general, the graph of the equation $x = a$ is a vertical line passing through the point $(a, 0)$, and the graph of the equation $y = b$ is a horizontal line passing through the point $(0, b)$. What are the equations of the x- and y-axes?

Now, let's use our knowledge of graphing linear equations to solve applied problems.

EXAMPLE 6

For cable television, a homeowner pays $30 per month plus $5 for each pay-per-view movie ordered.

a. Express as an equation the relationship between the monthly bill B and the number n of pay-per-view movies.

b. Draw the graph of this equation in Quadrant I of a coordinate plane.

c. Compute the slope of this graph. In terms of the cable TV bill, explain the significance of the slope.

d. In terms of the cable TV bill, explain the significance of the B-intercept of the graph.

e. From the graph in part (b), estimate what the cable bill would be if the homeowner had ordered 15 pay-per-view movies that month.

Solution

a. The monthly bill (in dollars) amounts to the sum of 30 and 5 times the number of pay-per-view movies which the homeowner ordered, so

$$B = 5n + 30.$$

PRACTICE 6

A stockbroker charges as her commission on stock transactions $40 plus 3% of the value of the sale.

a. Write as an equation the commission C in terms of the sales s.

b. To draw the graph of this equation in Quadrant I, we enter several nonnegative n-values, say 0, 10, and 20, in a table and then compute the corresponding B-values.

n	$B = 5n + 30$	(n, B)
0	$B = 5(0) + 30 = 30$	$(0, 30)$
10	$B = 5(10) + 30 = 80$	$(10, 80)$
20	$B = 5(20) + 30 = 130$	$(20, 130)$

Since the monthly bill B depends on the number n of pay-per-view movies ordered each month, we label the horizontal axis with the independent variable n and the vertical axis with the dependent variable B. Now, we choose an appropriate scale for each axis and plot $(0, 30)$, $(10, 80)$, and $(20, 130)$. Then, we draw the line passing through these points.

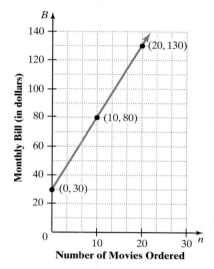

We restricted the graph to Quadrant I because the number of movies ordered n and the corresponding bill B are always nonnegative.

c. Substituting n and B for x and y, respectively, in the slope formula, we get the following slope:

$$m = \frac{B_2 - B_1}{n_2 - n_1} = \frac{80 - (30)}{10 - (0)} = \frac{50}{10} = 5.$$

Note that we could have predicted this answer since we know that the bill increases by \$5 for every additional pay-per-view movie ordered.

d. Since the B-intercept is the point $(0, 30)$, \$30 would be the amount of the bill if the homeowner had watched no pay-per-view movies at all during the month.

e. From the graph, it appears that if $n = 15$, then $B = 105$, that is, the cable bill would be \$105.

b. Draw a graph showing this relationship on sales up to \$1000. Be sure to choose an appropriate scale for each axis.

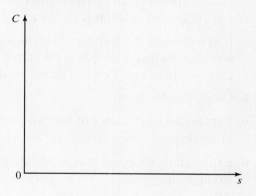

c. Compute the slope of this graph. In terms of the stock broker's commission, explain the significance of the slope.

d. If the value of a sale is \$500, estimate from the graph in part (b) the broker's commission.

EXAMPLE 7

A dietician uses milk and cottage cheese as sources of calcium in her diet. One serving of milk contains 300 mg of calcium, and one serving of cottage cheese contains 100 mg of calcium. The recommended daily amount (RDA) of calcium is 1000 mg.

a. If m represents the number of servings of milk and c the number of servings of cottage cheese in a diet that contains the RDA of calcium, write an equation that relates m and c.

b. Graph this equation.

c. Explain the significance of the two intercepts in terms of the number of servings.

d. Explain how we could have predicted that the slope of this graph would be negative.

Solution

a. The amount of calcium in the milk is $300m$, and the amount of calcium in the cottage cheese is $100c$. Since the RDA of calcium is 1000 mg, the following equation holds:

$$300m + 100c = 1000, \text{ or } 3m + c = 10$$

b. To graph, we identify the m- and c-intercepts, as well as a third point, say with $m = 1$.

m	0	$3\frac{1}{3}$	1
c	10	0	7

Next, we choose an appropriate scale for each axis and plot the three points, checking that the points all lie on the same line. Then, we draw the line passing through them.

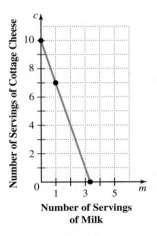

c. The m-intercept represents the number of servings she would need to meet the daily minimum requirement if she uses only milk as her source of calcium. The c-intercept represents the number of servings she would need to meet the RDA if she uses only cottage cheese as her source of calcium.

d. Even without drawing this line, we know that its slope has to be negative for the following reason: The RDA of calcium is a fixed amount (1000 mg). So larger values of m must correspond to smaller values of c. The line will therefore have to be decreasing, falling to the right and with a negative slope.

PRACTICE 7

An athlete has just signed a contract with a total value of $10 million. According to the terms of the contract, she earns $2 million in some years and $1 million in other years.

a. Let x stand for the number of years in which she earns $2 million and y the number of $1 million years. Write an equation that relates x and y.

b. Choose an appropriate scale for each axis on the coordinate plane. Graph the equation found in part (a).

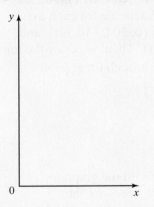

c. Describe in terms of the contract the significance of the slope of the graph.

d. Describe in terms of the contract the significance of the x- and y-intercepts of the graph.

Mathematically Speaking

Fill in each blank with the most appropriate term or phrase from the given list.

three *x*-values	vertical	*x*-intercept
graph	solution	horizontal
y-intercept	three points	

1. A(n) _____ of an equation in two variables is an ordered pair of numbers that when substituted for the variables makes the equation true.

2. The _____ of a linear equation in two variables consists of all points whose coordinates satisfy the equation.

3. One way of graphing a linear equation in two variables is to plot _____.

4. The _____ of a line is the point where the graph crosses the *x*-axis.

5. One way to graph a linear equation using intercepts is to first let $x = 0$ and then find the _____.

6. The graph of the equation $y = c$ is a(n) _____ line passing through the point $(0, c)$.

A *Complete each table so that the ordered pairs are solutions of the given equation.*

7. $y = 3x - 8$

x	4	7	
x			0

8. $y = 2x - 5$

x	0		
y		15	17

9. $y = -5x$

x	3.5	6		
y			$\frac{1}{2}$	-8

10. $y = -10x$

x	$\frac{1}{5}$	2.9		
y			-6	-1

11. $3x + 4y = 12$

x	0	-4		
y			-3	0

12. $4x + y = 8$

x	5	0		
y			0	16

13. $y = \frac{1}{3}x - 1$

x	3	6	-3	
y				-1

14. $y = -\frac{3}{2}x + 2$

x	$\frac{4}{3}$	6	-2	
y				2

B *Graph each equation by finding three points whose coordinates satisfy the equation.*

15. $y = x$

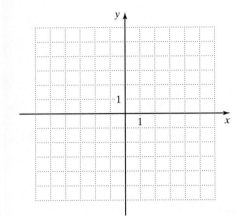

16. $y = 3x$

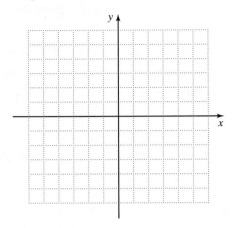

17. $y = \dfrac{1}{2}x$

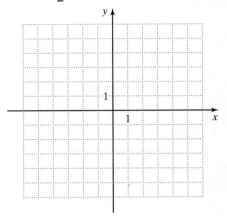

18. $y = \dfrac{1}{4}x$

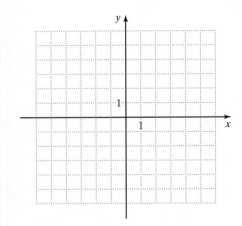

19. $y = -\dfrac{5}{4}x$

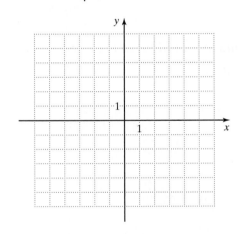

20. $y = -\dfrac{3}{2}x$

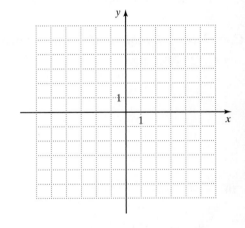

21. $y = 2x + 1$

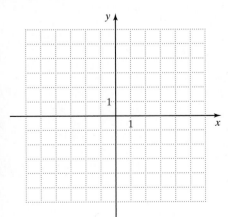

22. $y = 3x + 1$

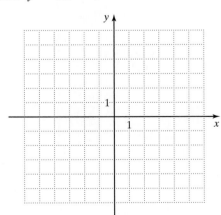

23. $y = -\dfrac{1}{3}x + 1$

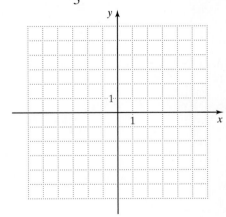

24. $y = -\dfrac{3}{4}x + 2$

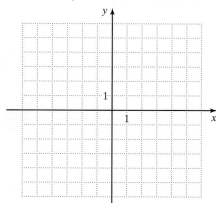

25. $y - 2x = -3$

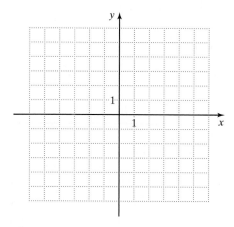

26. $y - 3x = 2$

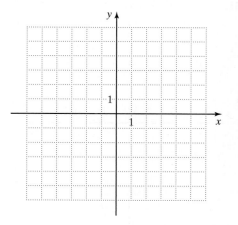

27. $x + y = 6$

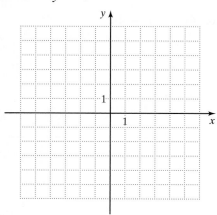

28. $3x + y = 4$

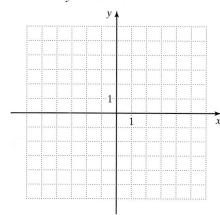

29. $x - 2y = 4$

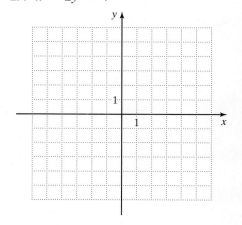

30. $x - 3y = 15$

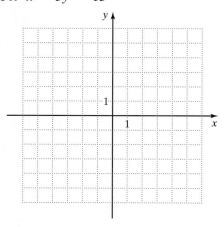

For each equation, find the x- and y-intercepts. Then, use the intercepts to graph the equation.

31. $5x + 3y = 15$

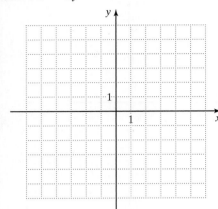

32. $4x + 5y = 20$

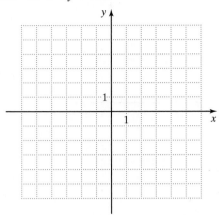

33. $3x - 6y = 18$

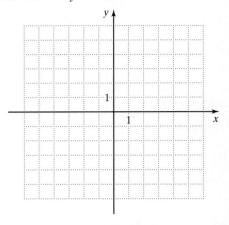

34. $7x - 2y = -7$

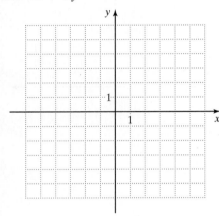

35. $3y - 2x = -6$

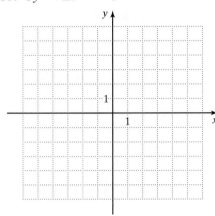

36. $4y - 5x = 10$

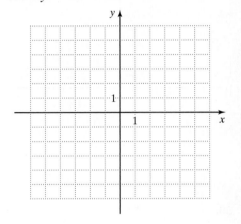

37. $9y + 6x = -9$

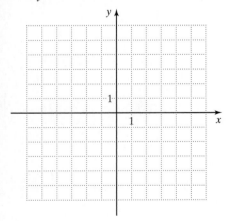

38. $4y + 8x = -4$

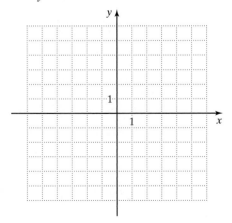

39. $y = \dfrac{1}{2}x + 2$

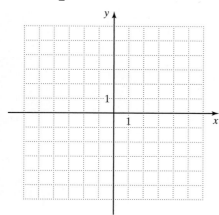

40. $y = \dfrac{5}{4}x - 5$

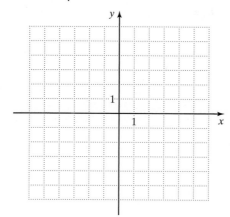

Graph.

41. $y = -2$

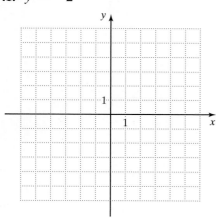

42. $y = 0$

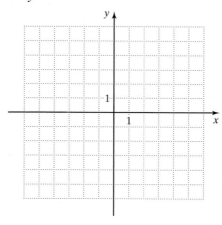

43. $x = 3$

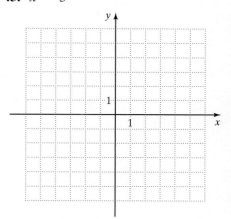

44. $x = 0$

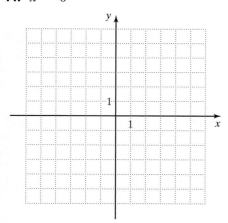

45. $x = -5.5$

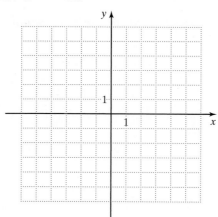

46. $y = -0.5$

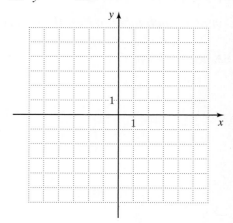

47. $y = \dfrac{1}{2}x + 3$

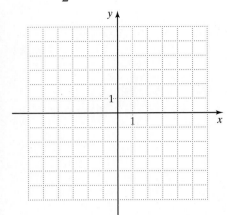

48. $y = \dfrac{1}{2}x + 6$

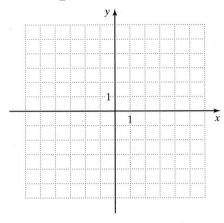

49. $3x + 5y = -15$

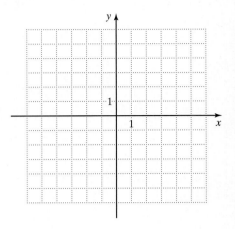

50. $3y - 5x = 15$

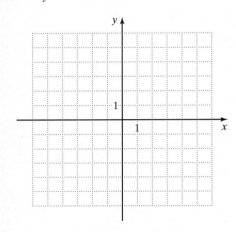

51. $y = -\dfrac{3}{5}x + \dfrac{2}{5}$

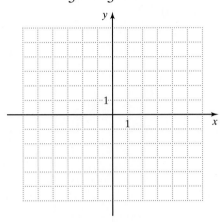

52. $y = -\dfrac{1}{4}x + \dfrac{3}{4}$

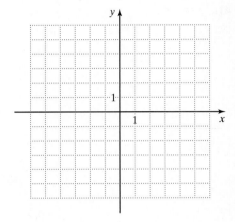

Mixed Practice

Complete each table so that the ordered pairs are solutions of the given equation.

53. $y = -2x + 6$

x	−3	$\dfrac{5}{2}$		
y			−10	4

54. $3x - 4y = 6$

x	0		−6	
y		0		3

For each equation, find the x- and y-intercepts, and another point whose coordinates satisfy the equation. Then, use these points to graph the equation.

55. $y = -\dfrac{1}{2}x + 2$

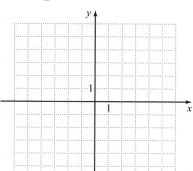

56. $2x - y = 4$

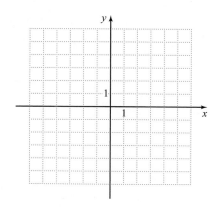

57. $6x - 2y = -12$

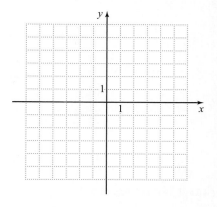

58. $y = \dfrac{3}{2}x - 3$

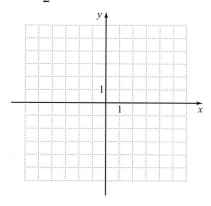

59. $y = -4x$

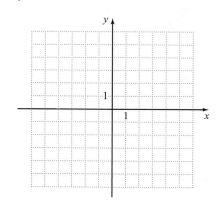

60. $y = \dfrac{3}{4}x$

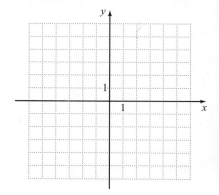

61. $3y + 6x = -3$

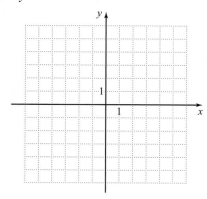

Graph.

62. $x = -3\dfrac{1}{2}$

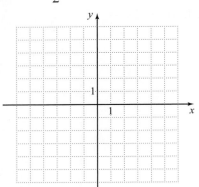

63. $y = -0.5x + 3$

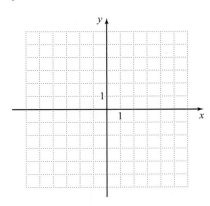

64. $2y - 4x = 8$

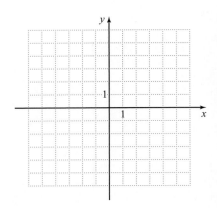

Applications

C *Solve.*

65. In a physics lab, students study the mathematics of motion. They learn that if an object is tossed straight upward with an initial velocity of 10 ft/sec, then after t sec the object will be traveling at a velocity of v ft/sec, where

$$v = 10 - 32t.$$

a. Complete the table at the right.

t	0		1	1.5	2
v		-6			

Explain what a positive value of v means. What does a negative value of v mean?

b. Choose an appropriate scale for each axis, and then graph this equation.

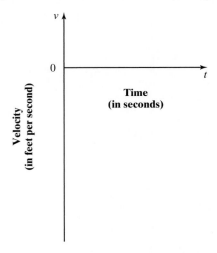

c. In terms of the object's motion, explain the significance of the v-intercept.

d. In terms of the object's motion, explain the significance of the t-intercept.

66. Each day, a local grocer varies the price p of an item in dollars, and then keeps track of the number s of items sold. According to his records, the following equation describes the relationship between s and p:

$$s = -2p + 12$$

a. Complete the table shown at the right.

p	1	3	5
s			0

b. Choose appropriate scales for the axes, and then graph the equation.

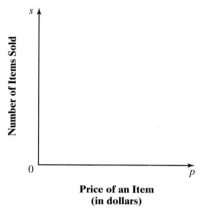

**Price of an Item
(in dollars)**

c. Explain why it makes sense to consider the graph only in Quadrant I.

d. From the graph, estimate the price required to sell 4 items.

67. A young couple buys furniture for $2000, agreeing to pay $500 down and $100 at the end of each month until the entire debt is paid off.

a. Express the amount P paid off in terms of the number m of monthly payments.

b. Complete the table shown at the right.

m	1	2	3
P			

c. Choose an appropriate scale for the axes, and then graph this equation.

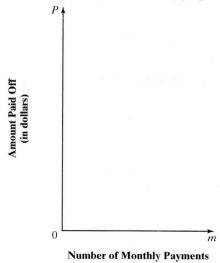

Number of Monthly Payments

68. Students studying forensic science know that when the femur bone of an adult female is unearthed, a good estimate of her height h is 29 more than double the length l of the femur bone, where all measurements are in inches. (*Source:* nsbri.org)

   a. Express this relationship as a formula.

   b. Complete the table shown at the right.

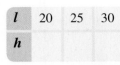

l	20	25	30
h			

   c. Choose an appropriate scale for the axes, and then graph this relationship.

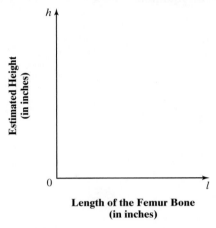

Length of the Femur Bone
(in inches)

   d. Use the graph to estimate the height of a woman whose femur bone was 14 in. in length.

69. The coins in a cash register, with a total value of $2, consist of n nickels and d dimes.

   a. Represent this relationship as an equation.

   b. Graph the equation found in part (a).

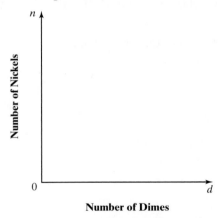

Number of Dimes

   c. Explain in a sentence or two why not every point on this graph in Quadrant I is a reasonable solution to the problem.

70. On the first leg of a trip, a truck driver drove for x hr at a constant speed of 50 mph. On the second leg of the trip, he drove for y hr consistently at 40 mph. In all, he drove 1000 mi.

a. Translate this information into an equation.

b. Choose appropriate scales for the axes, and then graph the equation found in part (a).

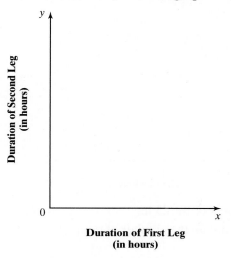

c. What are the x- and y-intercepts of this graph? Explain their significance in terms of the trip.

d. Find the slope of the line. Explain whether you would have expected the slope to be positive or negative, and why.

71. At a computer rental company, the fee F for renting a laptop is $40 plus $5 for each of the d days that the laptop is rented.

a. Express this relationship as an equation.

b. Choose appropriate scales for the axes, and then graph the equation expressed in part (a).

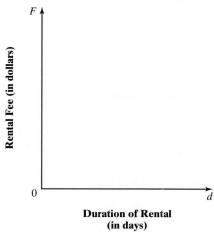

c. Explain the significance of the F-intercept in this context.

72. At a local community center, the annual cost c to use the swimming pool includes an annual membership fee of $75 plus $5 per hour for h hr of pool time.

   a. Write an equation for the annual cost of swimming at the community center in terms of the number of hours of pool time.

   b. Choose appropriate scales for the axes, and then graph the equation for up to and including 150 hr.

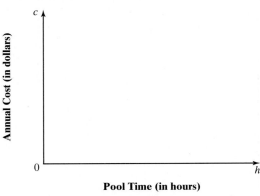

   c. Use the graph to estimate the annual cost of using the pool for 25 hr.

   d. Suppose the annual cost for swimming was $500. Estimate the number of hours of pool time.

• Check your answers on page A-9.

MIND*Stretchers*

Groupwork

1. Not all graphs are linear. For example, the graph of the equation $y = x^2 - 4$ is nonlinear, as the graph at the right illustrates:

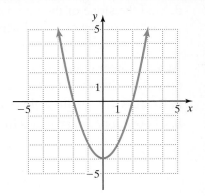

   a. Identify the x- and y-intercepts for the graph shown.

   b. Show that the x- and y-intercepts found in part (a) satisfy the equation $y = x^2 - 4$.

Writing

2. Give some advantages and disadvantages of graphing a linear equation by finding three arbitrary points versus using the intercepts.

Mathematical Reasoning

3. Recall that for a linear equation to be in general form, it must be written as $Ax + By = C$, where A, B, and C are real numbers and A and B are not both 0. What would the graph of this equation look like if A and B were both 0?

3.4 More on Linear Equations and Their Graphs

In the previous section of this chapter, we considered linear equations written in the general form and how to graph them. Now, we will discuss two other forms of linear equations: the *slope-intercept form* and the *point-slope form*. Using these two forms, we will continue to show how line graphs and their corresponding equations have real-world applications.

Slope-Intercept Form

Recall from Section 3.3 that one approach to graphing an equation written in general form is to isolate y.

$$-5x + y = 2 \qquad \text{An equation in general form}$$
$$y = 5x + 2 \qquad \text{Solve for } y.$$

The linear equation $y = 5x + 2$ is said to be in slope-intercept form. The graph of this equation has slope 5, which is equal to the coefficient of x, and y-intercept $(0, 2)$, where 2 is the constant term of the equation.

DEFINITION

A linear equation is in **slope-intercept form** if it is written as

$$y = mx + b,$$

where m and b are constants. In this form, m is the slope and $(0, b)$ is the y-intercept of the graph of the equation.

This form is used for identifying the slope and y-intercept of the graph of a linear equation without drawing the graph of the equation. The coefficient of x is the slope m, and the constant term b is the y-coordinate of the y-intercept $(0, b)$.

The following table gives additional examples of equations written in slope-intercept form:

Equation	Slope m	y-intercept $(0, b)$
$y = \frac{1}{2}x + 1$	$\frac{1}{2}$	$(0, 1)$
$y = 2x - 1$	2	$(0, -1)$
$y = -7x \rightarrow y = -7x + 0$	-7	$(0, 0)$
$y = 5 \rightarrow y = 0x + 5$	0	$(0, 5)$

EXAMPLE 1

Find the slope and y-intercept of the equation $y = 3x - 5$.

Solution The equation $y = 3x - 5$ or $y = 3x + (-5)$ is already in slope-intercept form, $y = mx + b$. The slope m is 3, and the y-intercept is $(0, -5)$ since the equation has constant term -5.

PRACTICE 1

Find the slope and y-intercept of the equation $y = -2x + 3$.

EXAMPLE 2

For the graph of $y = 3x$, find the slope and y-intercept.

Solution   We can rewrite $y = 3x$ as $y = 3x + 0$. Now, the equation $y = 3x + 0$ is in slope-intercept form, with slope $m = 3$ and $b = 0$. Since $b = 0$, the y-intercept is $(0, 0)$. That is, the graph passes through the origin.

PRACTICE 2

Find the slope and y-intercept of the graph of $y = -x$.

EXAMPLE 3

Express $3x + 5y = 6$ in slope-intercept form.

Solution   Since the slope-intercept form of an equation is $y = mx + b$, we need to solve the given equation for y.

$$3x + 5y = 6$$
$$5y = -3x + 6$$
$$\frac{5y}{5} = \frac{-3}{5}x + \frac{6}{5}$$
$$y = -\frac{3}{5}x + \frac{6}{5}$$

So $y = -\frac{3}{5}x + \frac{6}{5}$ is the equation written in slope-intercept form, where m is $-\frac{3}{5}$ and b is $\frac{6}{5}$.

PRACTICE 3

Express $3x - 2y = 4$ in slope-intercept form.

EXAMPLE 4

Write $y - 1 = 5(x - 1)$ in slope-intercept form.

Solution   To get the equation in the form $y = mx + b$, we must solve for y.

$$y - 1 = 5(x - 1)$$
$$y - 1 = 5x - 5$$
$$y = 5x - 5 + 1$$
$$y = 5x - 4$$

So $y = 5x - 4$ is the equation written in slope-intercept form, where m is 5 and b is -4.

PRACTICE 4

Change the equation $y - 2 = 4(x + 1)$ to slope-intercept form.

We can use the slope-intercept form to write the equation of a line when given its slope and y-intercept.

EXAMPLE 5

Write the equation of the line with slope $-\dfrac{4}{5}$ and y-intercept $(0, -3)$.

Solution We are given that the slope is $-\dfrac{4}{5}$ and the y-intercept is $(0, -3)$. So $m = -\dfrac{4}{5}$ and $b = -3$. We substitute these values in the slope-intercept form:

$$y = mx + b$$
$$y = -\frac{4}{5}x + (-3), \quad \text{or } y = -\frac{4}{5}x - 3$$

PRACTICE 5

A line on a coordinate plane has slope 1 and intersects the y-axis 2 units above the origin. Write its equation in slope-intercept form.

Recall our discussion in Section 3.2 of parallel and perpendicular lines. The next two examples deal with the equations of lines that are parallel or perpendicular.

EXAMPLE 6

Find an equation of the line that is parallel to the graph of $y = 4x + 1$ and has y-intercept $(0, 3)$.

Solution The line $y = 4x + 1$ is in slope-intercept form. The slope of this line is 4. Since parallel lines have the same slope, the line we want will also have slope $m = 4$. Since its y-intercept is $(0, 3)$, $b = 3$. Therefore, the desired equation is $y = 4x + 3$.

PRACTICE 6

What is the equation of the line parallel to the graph of $y = -2x + 3$ with y-intercept $(0, -1)$?

EXAMPLE 7

What is the equation of the line that is perpendicular to the graph of $y = 3x - 1$ and has y-intercept $(0, 1)$?

Solution The line $y = 3x - 1$ is written in slope-intercept form. So its slope must be 3. We know that the slopes of two perpendicular lines are negative reciprocals of each other. Therefore, the slope of the line we want has slope $m = -\dfrac{1}{3}$. Since its y-intercept is $(0, 1)$, $b = 1$.

So the desired equation is $y = -\dfrac{1}{3}x + 1$.

PRACTICE 7

Find the equation of the line that is perpendicular to the graph of $y = 2x + 5$ and has y-intercept $(0, -2)$.

EXAMPLE 8

At her college, a student pays $75 per credit-hour plus a flat student fee of $100. Find an equation in slope-intercept form that relates the amount A that she pays to the number h of credit-hours in her program.

Solution The amount A in dollars that the student pays is the sum of 75 times the number h of credit-hours in her program and 100. So we have:

$$A = 75h + 100$$

This equation is written in slope-intercept form.

PRACTICE 8

A bathtub, which has a capacity of 45 gal, is filled to the top with water. The tub starts to drain at a rate of 3 gal/min. Write an equation in slope-intercept form that expresses the amount of water w left in the tub (in gallons) in terms of the time t that the tub has been draining (in minutes).

We have already graphed equations of the form $y = mx + b$ by finding three points whose coordinates satisfy the equation. Now, let's focus on graphing such equations using the slope and the y-intercept.

Consider the equation $y = 3x - 1$, which has slope 3 and y-intercept $(0, -1)$. Since 3 is $\frac{3}{1}$, we know from the definition of slope that the *rise* is 3 and the *run* is 1. Starting at $(0, -1)$, we move up 3 units and then 1 unit to the right to find a second point $(1, 2)$ on the line. Then, we draw the line through the two points $(0, -1)$ and $(1, 2)$.

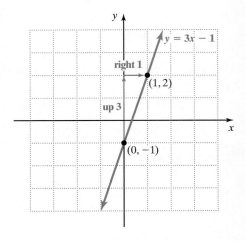

This example leads us to the following rule:

To Graph a Linear Equation in Two Variables Using the Slope and the y-intercept

- First, locate the y-intercept.
- Then, use the slope to find a second point on the line.
- Finally, draw the line through the two points.

EXAMPLE 9

Graph $y = -\frac{3}{2}x + 2$ using the slope and y-intercept.

Solution   Since $y = -\frac{3}{2}x + 2$ is in slope-intercept form, the slope is $-\frac{3}{2}$ and the y-intercept is $(0, 2)$. First, we locate the y-intercept $(0, 2)$. Since the slope $-\frac{3}{2}$ equals $\frac{-3}{2}$, from the point $(0, 2)$ we move *down* 3 units and then 2 units to the *right* to find the second point $(2, -1)$. Then, we draw the line through the points $(0, 2)$ and $(2, -1)$ as shown in the graph.

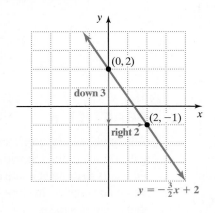

PRACTICE 9

Use the slope and y-intercept to graph $y = -\frac{1}{3}x - 4$.

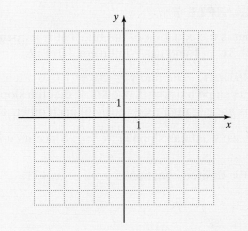

Note that since $\frac{-3}{2} = \frac{3}{-2}$, from the point $(0, 2)$ we could have moved *up* 3 units, and then 2 units to the *left* to find the second point $(-2, 5)$. Then we could have drawn the line through $(0, 2)$ and $(-2, 5)$ to obtain the same graph of the equation, as shown on the coordinate plane to the right.

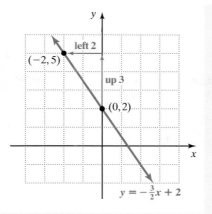

Recall that in Section 3.3 we graphed $2x + 3y = 6$ using the intercepts. We can also graph this equation using the slope and y-intercept. Which method do you prefer? Explain why.

Point-Slope Form

The last form of a linear equation that we will discuss is called the point-slope form.

DEFINITION

The **point-slope form** of a linear equation is written as

$$y - y_1 = m(x - x_1),$$

where x_1, y_1, and m are constants. In this form, m is the slope and (x_1, y_1) is a point that lies on the graph of the equation.

This form is useful for finding the equation of a line in two particular situations:

- when we know the slope of the line and a point on it, or
- when we know two points on the line.

EXAMPLE 10

A line with slope -2 passes through the point $(-1, 5)$. Find the equation of this line written in point-slope form.

Solution   Since we know a point on the line and the slope of the line, we can substitute directly into the point-slope form, where $x_1 = -1$, $y_1 = 5$, and $m = -2$.

$$y - y_1 = m(x - x_1)$$
$$y - 5 = -2[x - (-1)]$$
$$y - 5 = -2(x + 1) \qquad \text{Point-slope form}$$

We can leave this equation in point-slope form, or we can simplify the equation and write it in either general form

$$y - 5 = -2x - 2$$
$$2x + y = 3 \qquad \text{General form}$$

or in slope-intercept form

$$y = -2x + 3 \qquad \text{Slope-intercept form}$$

PRACTICE 10

A line passing through the point $(7, 0)$ has slope 2. Find its equation in point-slope form.

EXAMPLE 11

What is the equation of the line passing through the points $(3, 5)$ and $(2, 1)$?

Solution   Since we know the coordinates of two points on the line, we can find its slope.

$$m = \frac{y_2 - y_1}{x_2 - x_1} = \frac{5 - 1}{3 - 2} = \frac{4}{1} = 4$$

The line with slope $m = 4$ passing through the point $(3, 5)$ is:

$$y - y_1 = m(x - x_1)$$
$$y - 5 = 4(x - 3)$$

PRACTICE 11

Find the equation in point-slope form of the line passing through $(7, 7)$ and the origin.

The equation found in Example 11 is $y = 4x - 7$ in slope-intercept form. Had we substituted the point $(2, 1)$ rather than the point $(3, 5)$ into the point-slope form, would the resulting equation be the same?

EXAMPLE 12

An accountant's computer decreases in value by $400 a year. The computer was worth $1600 one year after he bought it. Write an equation that gives the value V of the computer in terms of the number of years t since the purchase.

Solution   Each year that passes, t increases by 1 and V decreases by 400. Therefore, the graph of the equation we are seeking has slope -400. Because the computer is worth $1600 one year after purchase, the graph passes through the point $(1, 1600)$. Since we know a point on the line as well as its slope, we can find the point-slope form of the equation.

$$V - V_1 = m(t - t_1)$$
$$V - 1600 = -400(t - 1)$$

If we like, we can simplify the equation and write it in slope-intercept form:

$$V - 1600 = -400t + 400$$
$$V = -400t + 2000$$

PRACTICE 12

The total weight of a box used for shipping baseballs increases by 5 oz for each baseball that is packed in the box. A box with 4 balls weighs 27 oz. Write an equation that expresses the total weight w of the box in terms of the number of baseballs b packed in the box.

Can you explain why the slope of the line in Example 12 is negative?

Using a Calculator or Computer to Graph Linear Equations

Calculators with graphing capabilities and computers with graphing software allow us to graph equations at the push of a key, even those with complicated coefficients. Although they vary somewhat in terms of features and commands, these machines all graph the equation of your choice on a coordinate plane.

To graph an equation, begin by making certain that the equation is in slope-intercept form. On many graphers, pressing the $\boxed{Y=}$ *key results in a screen being displayed on which you enter the equation. For instance, if you wanted to graph* $2x - y = 1$, *you would first solve for y, resulting in* $y = 2x - 1$, *and then enter* $2x - 1$ *to the right of* \Y1 = *on the screen. Pressing the* $\boxed{GRAPH}$ *key displays a coordinate plane in which the graph of* $y = 2x - 1$ *is sketched, as we see on the screen to the right.*

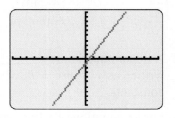

Many graphers have a **TRACE** *feature that highlights a point on the graph and displays its coordinates. As you hold down an arrow key, you can see how the coordinates change as the highlighted point moves along the graph.*

> **TIP** The viewing window allows you to set the range and scales for the axes. Before you graph an equation, be sure to set the viewing window in which you would like to display the graph.

EXAMPLE 13

Graph $y - x = 2$, and then use the **TRACE** feature to identify the y-intercept.

Solution   First, solve for y: $y = x + 2$. Next, press $\boxed{Y=}$ and enter $x + 2$ to the right of \Y1 =. Then, set the viewing window in which you want to display the graph. Finally, press $\boxed{GRAPH}$ to display the graph of the equation. If the graph of $y = x + 2$ does not appear, check your grapher's instruction manual.

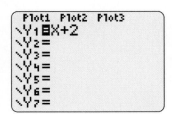

With the **TRACE** feature, run the cursor along the graph until it appears to be on the y-axis. The displayed coordinates of this y-intercept are approximately $x = 0$ and $y = 2$.

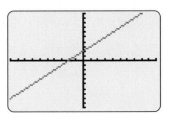

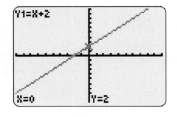

PRACTICE 13

Graph $2y + x = 3$, and then find the y-intercept with the **TRACE** feature.

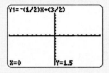

Mathematically Speaking

Fill in each blank with the most appropriate term or phrase from the given list.

standard	point-slope	slope
x-intercept	y-intercept	slope-intercept

1. A linear equation is in _____ form if it is written as $y = mx + b$, where m and b are constants.

2. For an equation of a line written in slope-intercept form, m is the _____ of the line.

3. For an equation of a line written in slope-intercept form, $(0, b)$ is the _____ of the line.

4. The _____ form of a linear equation is written as $y - y_1 = m(x - x_1)$, where x_1, y_1, and m are constants.

Ⓐ *Complete each table.*

5.

Equation	Slope m	y-intercept $(0, b)$	Which Graph Type Best Describes the Line? $\diagup \diagdown - \vert$	x-intercept
$y = 3x - 5$				
$y = -2x$				
$y = 0.7x + 3.5$				
$y = \frac{3}{4}x - \frac{1}{2}$				
$6x + 3y = 12$				
$y = -5$				
$x = -2$				

6.

Equation	Slope m	y-intercept $(0, b)$	Which Graph Type Best Describes the Line? $\diagup \diagdown - \vert$	x-intercept
$y = -3x + 5$				
$y = 2x$				
$y = 1.5x + 6$				
$y = \frac{2}{3}x + \frac{1}{2}$				
$4x + 6y = 24$				
$y = 0.3$				
$x = 2$				

Find the slope and y-intercept of each equation.

7. $y = -x + 2$

8. $y = -\frac{1}{2}x + 3$

9. $y = 3x - 4$

10. $y = 4x - 2$

B *Write the following equations in slope-intercept form.*

11. $x - y = 10$

12. $3x - y = 15$

13. $x + 10y = 10$

14. $x + y = 7$

15. $6x + 4y = 1$

16. $3x + 5y = 15$

17. $2x - 5y = 10$

18. $4x - 8y = 12$

19. $y + 1 = 3(x + 5)$

20. $y - 1 = 3(x - 5)$

C *Match the equation to its graph.*

21. $4x - 2y = 6$

22. $-2x + 4y = 8$

23. $2y - x = 8$

24. $6x + 3y = -9$

a.

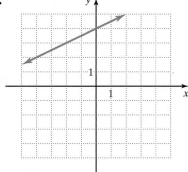

b.

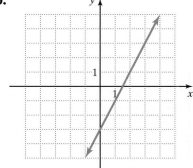

c.

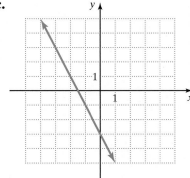

d.

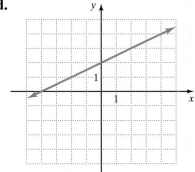

Graph the following equations using the slope and y-intercept.

25. $y = 2x + 1$

26. $y = 3x + 1$

27. $y = -\dfrac{2}{3}x + 6$

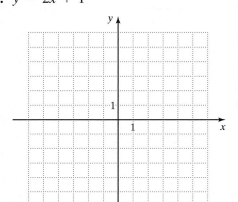

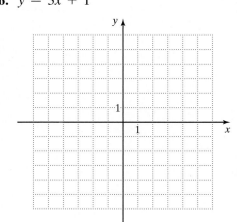

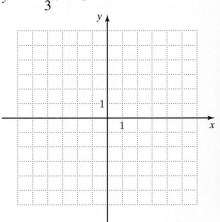

28. $y = -\dfrac{3}{2}x - 6$

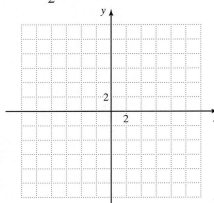

29. $x + y = 1$

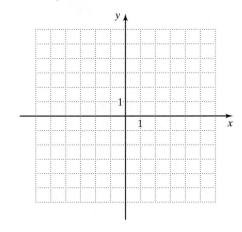

30. $x + y = -4$

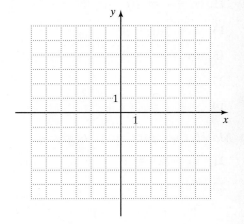

31. $y = -\dfrac{3}{4}x$

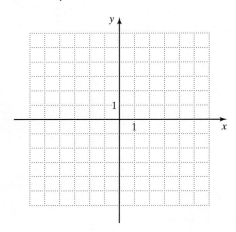

32. $y = -\dfrac{1}{2}x$

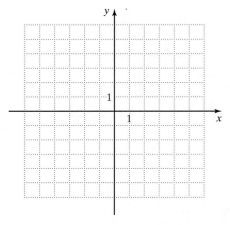

33. $x + 2y = 4$

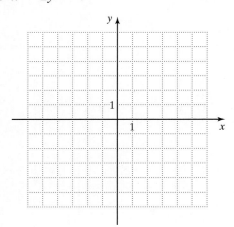

34. $2x + 3y = 12$

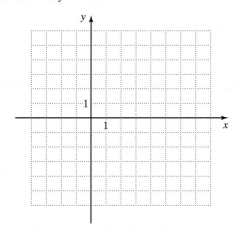

35. $y = 3.735x + 1.056$

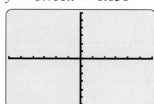

36. $y = -0.875x + 2.035$

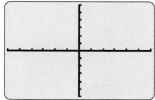

Solve.

37. Find the equation of the line with slope 3 that passes through the point $(0, 7)$.

38. What is the equation of the line that has slope -1 with y-intercept $(0, -2)$?

39. Find the equation of the line that is parallel to the graph of $y = 5x - 1$ and has y-intercept $(0, -20)$.

40. What is the equation of the line that is parallel to the graph $y = \frac{1}{3}x - 1$ and has y-intercept $(0, 4)$?

41. What is the equation of the line that is perpendicular to the graph of $y = 2x$ and that passes through $(-2, 5)$?

42. Find the equation of the line that is perpendicular to the graph $y = -x$ and that passes through the point $(1, -3)$.

43. What is the equation of the line passing through the points $(2, 1)$ and $(1, 2)$?

44. The points $(5, 1)$ and $(2, -3)$ lie on a line. Find its equation.

45. Find the equation of the line passing through points $(-1, -5)$ and $(-7, -6)$.

46. What is the equation of the line passing through the origin and the point $(3, 5)$?

47. Write the equation of the vertical line that passes through the point $(-3, 5)$.

48. What is the equation of the vertical line passing through the point $(1, -8)$?

49. What is the equation of the horizontal line passing through the point $(2, -6)$?

50. What is the equation of the horizontal line passing through the point $(-4, 7)$?

Find the equation of each graph.

51.

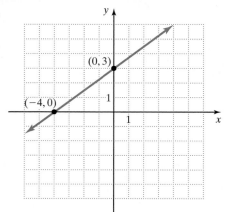

52.

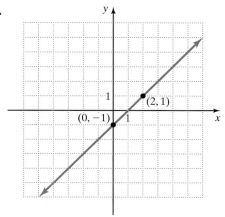

53.

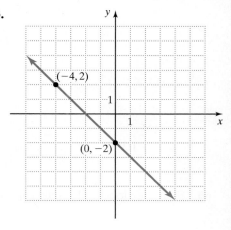

54.

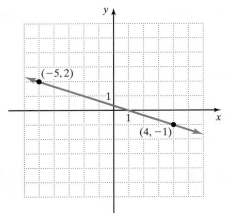

55.

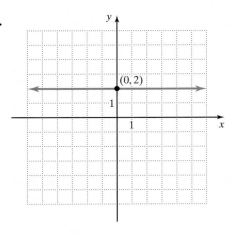

56.

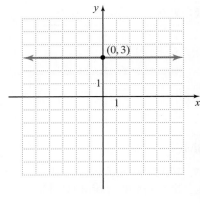

57.

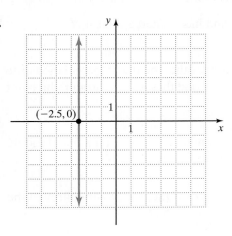

58.

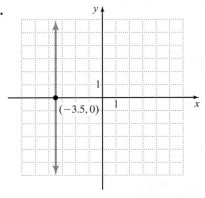

Mixed Practice

59. Complete the table.

Equation	Slope m	y-intercept $(0, b)$	Which Graph Type Best Describes the Line? $\diagup \diagdown - \mid$	x-intercept
$y = -7x + 2$				
$y = 4x$				
$y = 2.5x + 10$				
$y = \dfrac{2}{3}x - \dfrac{1}{4}$				
$5x + 4y = 20$				
$x = 9$				
$y = -3.2$				

60. Find the slope and y-intercept of $y = -\dfrac{2}{5}x + 3$.

Write the following equations in slope-intercept form.

61. $4x - y = 5$

62. $3x - 6y = 8$

63. Which equation describes the graph?

a. $-4x + y = 5$

b. $4x + y = -5$

c. $-5x + y = -4$

d. $5x + y = -4$

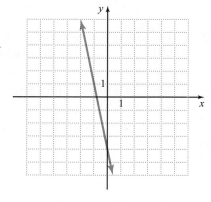

64. What is the equation of the line that is parallel to the graph $y = \dfrac{1}{2}x + 2$ and has x-intercept $(3, 0)$?

65. What is the equation of the line that is perpendicular to the graph of $y = -2x + 1$ that passes through the point $(4, 1)$?

66. What is the equation of the line that passes through the points $(2, -2)$ and $(-2, 1)$?

Graph the following equations using the slope and y-intercept.

67. $y = \dfrac{3}{5}x - 2$

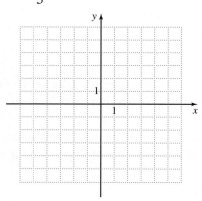

68. $3x + 2y = 4$

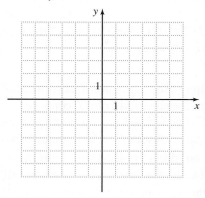

Find the equation of each graph.

69.

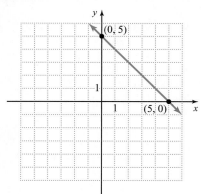

70.

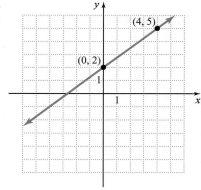

Applications

E *Solve.*

71. The following graph describes the relationship between Fahrenheit temperature F and Celsius temperature C.

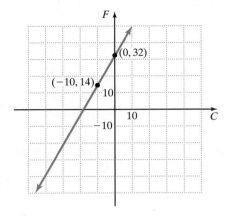

  a. Find the slope of the line.

  b. Find the equation of the line in slope-intercept form.

  c. Water boils at 212°F. Use part (b) to find the Celsius temperature at which water boils.

72. The owner of a shop buys a piece of machinery for $1500. The value V of the machinery declines by $150 per year.

  a. Write an equation for V after t years in slope-intercept form.

  b. Graph the equation found in part (a).

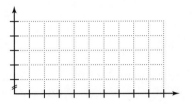

  c. Explain the significance of the two intercepts in this context.

73. Each month, a utility company charges its residential customers a flat fee for electricity plus 6 cents per kilowatt-hour (kWh) consumed. Last month, a customer used 500 kWh of electricity, and his bill amounted to $45.

  a. Express as an equation in point-slope form the relationship between the customer's monthly bill y in cents and the number x of kilowatt-hours of electricity consumed.

  b. Express the equation found in part (a) in slope-intercept form.

  c. What does the y-intercept represent in this situation?

74. A condo unit has been appreciating in value at $5000 per year. Three years after it was purchased, it was worth $65,000.

  a. Find an equation that expresses the value y of the condo in terms of the number x of years since it was purchased.

  b. Graph the equation found in part (a).

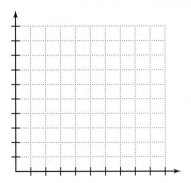

  c. What is the significance of the y-intercept in this situation?

75. A salesperson earns a salary of $1500 per month plus a commission of 3% of the total monthly sales.

   a. Write a linear equation giving the salesperson's total monthly income I in terms of sales S.

   b. Graph the equation found in part (a).

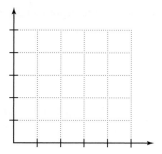

   c. Find the salesperson's income on monthly sales of $6200.

76. When the brakes on a train are applied, the speed of the train decreases by the same amount every second. Two seconds after applying the brakes, the train's speed is 88 mph. After 4 sec, its speed is 60 mph.

   a. Write an equation expressing the speed s of the train in terms of the time t seconds after applying the brakes.

   b. Graph the equation found in part (a).

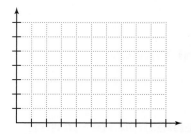

   c. What was the speed of the train when the brakes were first applied?

77. Pressure under water increases with greater depth. The pressure P on an object and the depth d below sea level are related by a linear equation. The pressure at sea level is 1 atmosphere (atm), whereas 33 ft below sea level the pressure is 2 atm. Find the equation expressing P in terms of d.

78. The length of a heated object and the temperature of the object are related by a linear equation. A rod at 0° Celsius is 10 m long, and at 25° Celsius it is 10.1 m long. Write an equation for length in terms of temperature.

79. When a force is applied to a spring, its length changes. The length L and the force F are related by a linear equation. The spring shown here was initially 10 in. long when no force was applied. What is the equation that expresses length in terms of force?

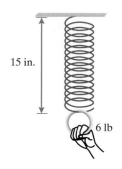

15 in.

6 lb

80. A company purchased a computer workstation for $8000. After 3 yr, the estimated value of the workstation was $4400. If the value V in dollars and the age a of the workstation are related by a linear equation, find an equation that expresses V in terms of a.

• Check your answers on page A-11.

MIND*Stretchers*

Technology

1. Consider $2x - 7 = 0$, which is a linear equation in x.
   a. Solve the equation.
   b. On a graphing calculator or computer with graphing software, graph $y = 2x - 7$. Then, find the x-intercept of the line. Explain in a sentence or two how you can use this approach to solve the equation $2x - 7 = 0$.

Critical Thinking

2. Consider the equation $y = mx + b$. Explain under what circumstances its graph lies completely in Quadrants I and II.

Mathematical Reasoning

3. What kind of line corresponds to an equation that can be written in *general form* but in neither slope-intercept form nor point-slope form?

3.5 Linear Inequalities and Their Graphs

In Section 2.6, we showed how to graph inequalities in one variable on a number line. In such inequalities, the solutions are real numbers. For instance, the graph of $x \leq 2$ is shown below.

Now, we consider graphing inequalities in two variables on a coordinate plane. In this case, the solutions are ordered pairs of real numbers. For instance, if we want to graph the solutions to $x \leq 2$ in the coordinate plane, we would shade the region to the left of the vertical line $x = 2$ because every ordered pair in this region has an x-coordinate that is less than 2. Every point on the line is also a solution to $x \leq 2$ because each pair on this line has an x-coordinate equal to 2.

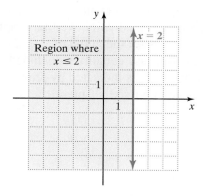

In this section, we focus our attention on graphing linear inequalities in *two* variables, such as $2x + 3y > 1$.

DEFINITION

A **linear inequality in two variables** is an inequality that can be written in the form $Ax + By < C$, where A, B, and C are real numbers and A and B are not both 0. The inequality symbol can be $<$, $>$, $\leq$, or $\geq$.

As we will see, a linear inequality has as its graph a half-plane, which is a region of the coordinate plane bounded on one side by a straight line.

Consider some situations that give rise to such inequalities:

The number of men m and women w invited to a party is at most 20. $\longrightarrow m + w \leq 20$

Your income i exceeds your expenses e by more than \$1000. $\longrightarrow i > e + 1000$

Let's look at what we mean by a *solution* of such inequalities.

DEFINITION

A **solution of an inequality in two variables** is an ordered pair of numbers that when substituted for the variables makes the inequality a true statement.

EXAMPLE 1

Is the ordered pair $(2, 5)$ a solution to the inequality $y \geq x + 1$?

Solution   When we substitute 2 for x and 5 for y in the given inequality, it becomes:

$$y \geq x + 1$$
$$5 \overset{?}{\geq} 2 + 1$$
$$5 \geq 3 \quad \text{True}$$

Since $5 \geq 3$ is true, the ordered pair $(2, 5)$ is a solution to $y \geq x + 1$. Is the ordered pair $(6, 5)$ a solution to this inequality?

PRACTICE 1

Is $(1, 3)$ a solution to the inequality $y < x - 1$?

By the *graph* of an inequality in two variables, we mean the set of all points on the plane whose coordinates satisfy the inequality. To explore what such a graph looks like, let's consider $y \geq 2x$. To find the graph of this inequality, we first graph the corresponding equation $y = 2x$. Since the inequality symbol is $\geq$, the graph of $y \geq 2x$ includes points for which y is either greater than or equal to $2x$. Because equality is included, points on the line $y = 2x$ are part of the graph of the inequality, which is indicated by drawing the *boundary line $y = 2x$* as a solid line. This line cuts the plane into two half-planes.

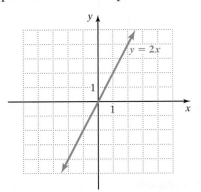

Next, we take an arbitrary point on either side of the boundary line—a *test point*. If the coordinates of the test point satisfy the inequality, then the desired graph contains the half-plane in which the test point lies. Otherwise, the desired graph contains the other half-plane. Suppose that we take $(4, 0)$ as our test point.

$$y \geq 2x$$
$$0 \overset{?}{\geq} 2(4)$$
$$0 \geq 8 \quad \text{False}$$

Since the inequality does not hold for the point $(4, 0)$, the half-plane that we want is the region above the graph of $y = 2x$, so we shade this region. Therefore, the graph of $y \geq 2x$ is the boundary line and the shaded region.

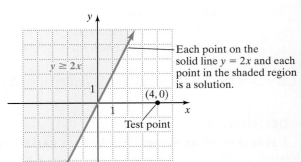

Each point on the solid line $y = 2x$ and each point in the shaded region is a solution.

If the inequality had been $y > 2x$ instead of $y \geq 2x$, the boundary line would not have been part of the graph. We would have indicated the exclusion of the boundary with a broken line, as in the following diagram:

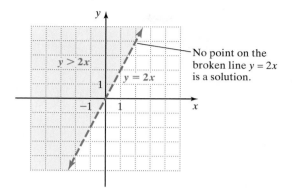

To Graph a Linear Inequality in Two Variables

- Graph the corresponding linear equation. For an inequality with the symbol $\leq$ or $\geq$, draw a solid line; for an inequality with the symbol $<$ or $>$, draw a broken line. This line is the boundary between two half-planes.

- Choose a test point in either half-plane and substitute the coordinates of this point in the inequality. If the resulting inequality is true, then the graph of the inequality is the half-plane containing the test point. If it is not true, then the other half-plane is the graph. A solid line is part of the graph, and a broken line is not.

EXAMPLE 2

Find the graph of $y - 2x < 4$.

Solution First, we graph the equation $y - 2x = 4$. Solving for y gives $y = 2x + 4$. We draw a broken line since the original inequality symbol is $<$. Then, we choose in either half-plane a test point, say $(0, 0)$.

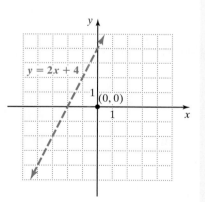

We substitute $x = 0$ and $y = 0$ into the inequality:

$$y - 2x < 4$$
$$0 - 2 \cdot 0 \overset{?}{<} 4$$
$$0 < 4 \qquad \text{True}$$

Since the inequality is true, the graph of the inequality is the half-plane containing the test point. So the half-plane below the line is

PRACTICE 2

Graph the inequality $y + 3x \geq 6$.

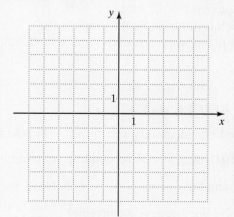

EXAMPLE 2 (continued)

our graph. Note that the graph does not include the boundary line since the original inequality symbol is $<$.

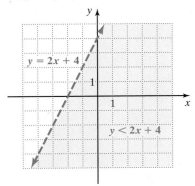

We can test any point on the graph. Why is $(0, 0)$ a good choice?

EXAMPLE 3

Graph $x < 0$.

Solution The boundary line $x = 0$ is the y-axis. It is drawn as a broken line, since the original inequality symbol is $<$. We need to select a test point on either side of the boundary line. Let's take the point $(1, 0)$ as the test point, which is in the half-plane to the right of the line $x = 0$. Substituting into the inequality $x < 0$, we get $1 < 0$, which is not true. So the graph is the half-plane to the left of the line $x = 0$, but not including this line.

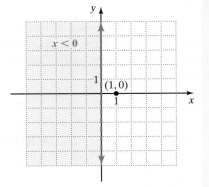

EXAMPLE 4

Graph the inequality $x - 3y \leq -9$ in the first quadrant.

Solution The corresponding equation is $x - 3y = -9$.

Solving for y gives $y = \frac{1}{3}x + 3$. Next, we graph this line, drawing a solid line because the original inequality symbol is $\leq$.

Taking $(0, 0)$ as the test point, we check whether $0 - 3(0) \leq -9$. Since this inequality does not hold, the test point is not part of the graph. So the graph in Quadrant I is the region in the quadrant above the line $y = \frac{1}{3}x + 3$, and including it.

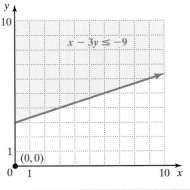

PRACTICE 3

Find the graph of $y \geq -5$.

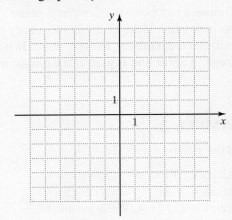

PRACTICE 4

Find the graph of $x - 2y > -6$ in Quadrant I.

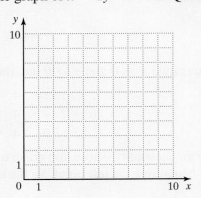

EXAMPLE 5

An online perfumery makes $50 on each bottle of perfume and $20 on each bottle of cologne. To earn a profit, the total amount of money made daily must exceed the perfumery's daily overhead of $1000.

a. Express as an inequality: The amount of money made on selling p bottles of perfume and c bottles of cologne is greater than the overhead.

b. Graph this inequality.

c. Use this graph to decide whether selling 10 bottles of perfume and 12 bottles of cologne results in a profit.

Solution

a. The factory makes $50 on each bottle of perfume sold. Therefore, the factory makes $50p$ dollars by selling p bottles of perfume. Similarly, the factory makes $20c$ dollars by selling c bottles of cologne. Since the total amount of money made must be greater than the overhead of $1000, the inequality is $50p + 20c > 1000$. If we divide both sides of this inequality by the common factor 10, we get $5p + 2c > 100$.

b. To graph the inequality $5p + 2c > 100$, we first graph the corresponding equation, $5p + 2c = 100$. We restrict our attention to the portion of the graph in Quadrant I, since the variables can only assume nonnegative values. The boundary line is not included since the symbol in the linear inequality is $>$. Using $(0, 0)$ as a test point, we see that the inequality $5 \cdot 0 + 2 \cdot 0 > 100$ is false. So the graph of $5p + 2c > 100$ is the region in Quadrant I above the boundary line, but not including the line.

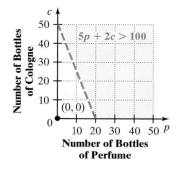

c. To decide whether there was a profit when selling 10 bottles of perfume and 12 bottles of cologne, we plot the point $(10, 12)$. Since the point lies *outside* the graph of our inequality, there was no profit.

PRACTICE 5

Each day, a refinery can produce both diesel fuel and gasoline, with a total maximum output of 3000 gal.

a. Express this relationship as an inequality, where d represents the amount of diesel fuel produced and g the amount of gasoline produced, both in gallons.

b. Graph this inequality.

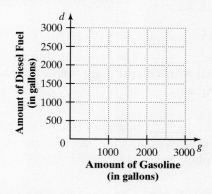

c. Explain the significance of the intercepts of this graph.

Mathematically Speaking

Fill in each blank with the most appropriate term or phrase from the given list.

broken	solution	solid
is	graph	is not
half-plane	ray	

1. The graph of a linear inequality in two variables is a(n) _____.

2. A(n) _____ of an inequality in two variables is an ordered pair of numbers that when substituted for the variables makes the inequality a true statement.

3. The _____ of an inequality in two variables is the set of all points on the plane whose coordinates satisfy the inequality.

4. A(n) _____ boundary line is drawn when graphing a linear inequality that involves the symbol $\leq$ or the symbol $\geq$.

5. A(n) _____ boundary line is drawn when graphing a linear inequality that involves the symbol $<$ or the symbol $>$.

6. If the boundary line is broken, it _____ part of the graph.

A *Decide if the given ordered pair is a solution to the inequality.*

7. $y < 3x$ $(0, 0)$

8. $y > -5x$ $(-1, 4)$

9. $y \geq 2x - 1$ $(-\frac{1}{2}, -2)$

10. $y \leq -\frac{2}{3}x + 5$ $(6, 1)$

11. $2x - 3y > 10$ $(10, 8)$

12. $5x + 3y \geq 12$ $(0, -2)$

B *Each solid or broken line is the graph of $y = x$. Shade in the graph of the given inequality.*

13. $y > x$

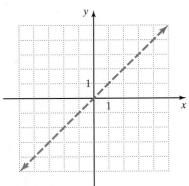

14. $y \geq x$

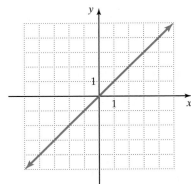

15. $x \leq y$

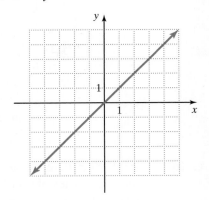

16. $x < y$

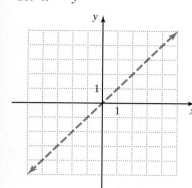

17. $y < x$

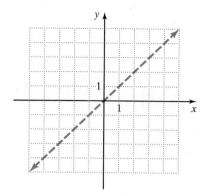

18. $y \leq x$

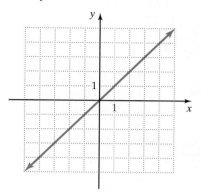

Match each inequality to its graph.

19. $y < \dfrac{1}{4}x - 1$

20. $y > -2x + 3$

21. $x - 4y \leq 4$

22. $2x + y \geq 3$

a.

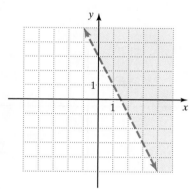

b.

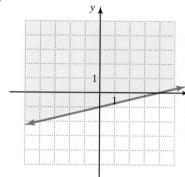

c.

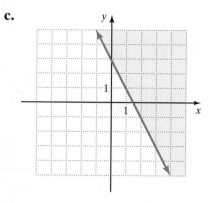

d.

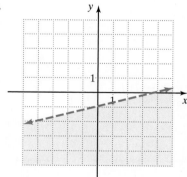

Graph the linear inequality.

23. $x > -5$

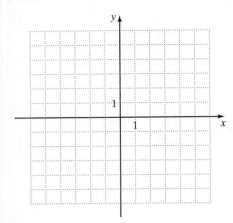

24. $x > 3$

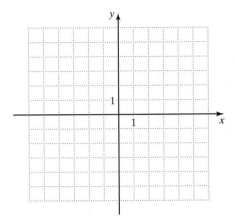

25. $y < 0$

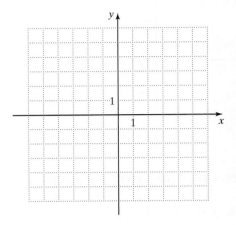

26. $y < 4$

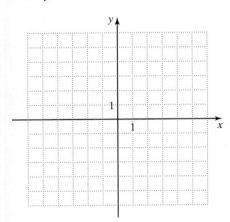

27. $y \le 3x$

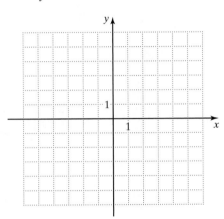

28. $y \le -x$

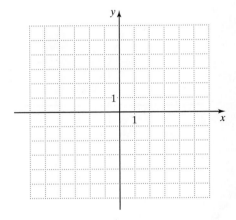

29. $y \ge -2x$

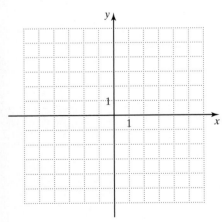

30. $y \ge 4x$

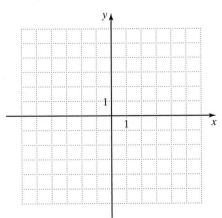

31. $y \le \dfrac{1}{2}x$

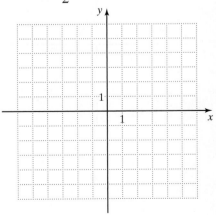

32. $y > -\dfrac{2}{3}x$

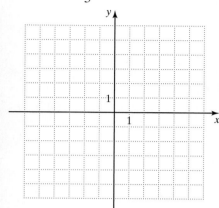

33. $y > 3x + 5$

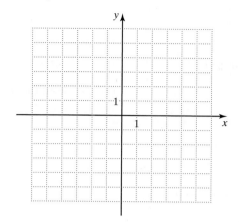

34. $y \geq -x - 1$

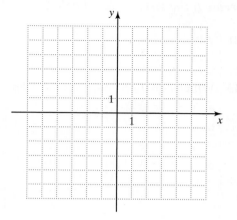

35. $5y - x > 10$

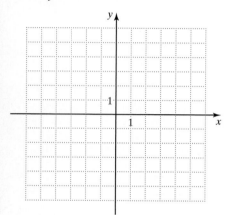

36. $4y + x < -12$

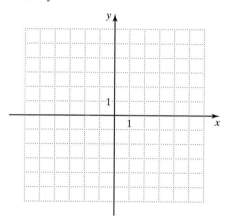

37. $2x - 3y \geq 3$

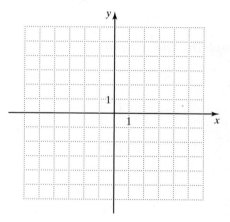

38. $3x - 2y \leq 4$

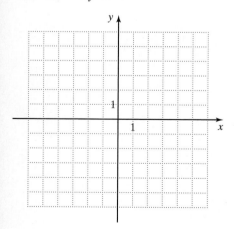

39. $4x - 5y \leq 10$

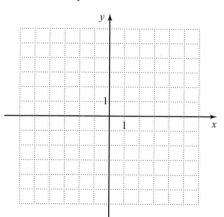

40. $-3x + 2y < 1$

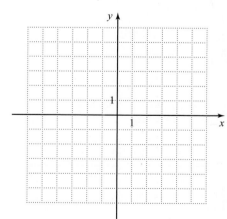

Mixed Practice

Decide if the given ordered pair is a solution to the inequality.

41. $y > -\dfrac{1}{2}x + 2$ $(4, 0)$

42. $x - 2y \leq -2$ $(8, 6)$

43. Which equation describes the graph?

    a. $y + 4x \geq 2$

    b. $y - 4x \geq 2$

    c. $y > 4x + 2$

    d. $y < 4x + 2$

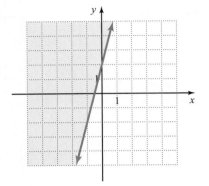

44. Shade in the graph of $x > y$.

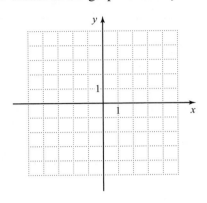

Graph the linear inequality.

45. $y < -3$

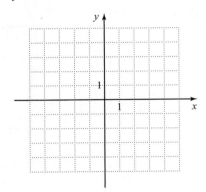

46. $y \leq \dfrac{2}{3}x$

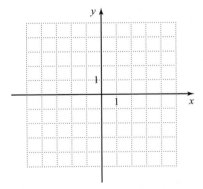

47. $y > -2x - 2$

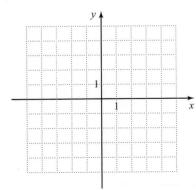

48. $2x - 3y < 6$

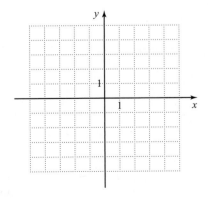

Applications

Solve.

49. According to a guideline, a family's housing expenses h should be less than $\frac{1}{4}$ of the family's combined income i.

  a. Express this guideline as an inequality.

  b. Graph this inequality.

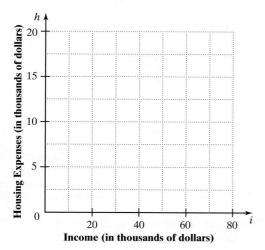

  c. Choose a point on the graph. For this point, explain why the guideline holds.

50. To purchase an apartment in a particular building, a buyer is allowed to take out a mortgage m that is at most 75% of the apartment's selling price s.

  a. Express this relationship as an inequality.

  b. Graph this inequality.

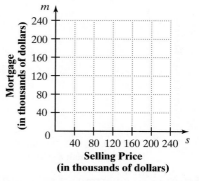

  c. Plot the point $(100, 90)$. For this point, explain in terms of the mortgage policy of the building whether the buyer will be able to buy the apartment.

51. A printing company ships x copies of a college's student handbook to the uptown campus and y copies to the downtown campus. The company must ship a total of at least 200 copies to these two locations.

  a. Express this relationship as an inequality.

  b. Graph this inequality.

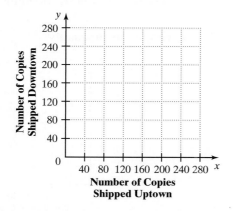

  c. Give the coordinates of a point that satisfies the inequality. Check that the coordinates satisfy the company's shipping requirement.

52. In September 2010, an investor had a maximum of $1500 to purchase stocks. She wanted to buy x shares of General Electric (GE) and y shares of Ford Motor Company (F). General Electric was selling at approximately $16 per share and Ford at approximately $12 per share. (*Source:* NYSE.com)

  a. Write as an inequality the possible amount of each stock that she could have bought.

  b. Solve this problem graphically.

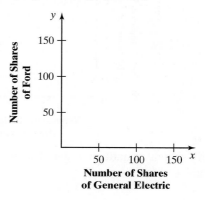

53. A local gourmet coffee shop sells small and large gift baskets. A small gift basket sells for $30 and a large gift basket sells for $75. The coffee shop would like a revenue of at least $1500 per month on the sale of gift baskets.

   a. Write an inequality, where x is the number of small gift baskets sold in a month and y is the number of large gift baskets sold, to represent this situation.

   b. Graph this inequality.

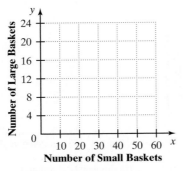
Number of Small Baskets

   c. Use the graph to determine if selling 20 small gift baskets and 20 large gift baskets will generate the desired revenue.

54. In moving into a new apartment, a young couple needs to borrow money for both furniture and a car. The loan for furniture has a 10% annual interest rate, whereas the car loan has a 5% annual interest rate. The couple can afford at most $2000 in interest payments for the year.

   a. Express the given information as an inequality, representing the car loan amount by c and the furniture loan amount by f.

   b. Graph this inequality.

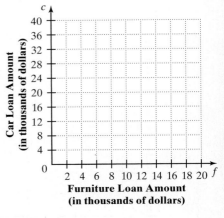
Furniture Loan Amount (in thousands of dollars)

   c. What is the maximum car loan amount that the couple can afford?

   d. If the couple borrows $20,000 for the car loan, what is the most that they can afford to borrow for furniture?

55. A plane is carrying bottled water and cases of medicine to victims of a flood. Each bottle of water weighs 10 lb, and each case of medicine weighs 15 lb. The plane can carry a maximum of 50,000 lb of cargo.

   a. Express this weight limitation of the cargo as an inequality in terms of the number of bottles of water w and the number of cases of medicine m in the plane.

   b. Graph this inequality.

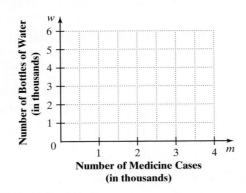
Number of Medicine Cases (in thousands)

   c. Identify several quantities of water and medicine that the plane can carry.

56. An elevator has a maximum capacity of 1600 lb. Suppose that the average weight of an adult is 160 lb and the average weight of a child is 40 lb.

   a. Write an inequality that relates the number of adults a and the number of children c who can ride an elevator without overloading it.

   b. Graph this inequality.

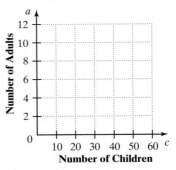
Number of Children

   c. What is an example of a group of people who will overload the elevator?

57. A student has two part-time jobs. One pays $8 per hour, and the other $10 per hour. Between the two jobs, the student needs to earn at least $200 per week.

a. Write an inequality that shows the number of hours that the student can work at each job.

b. Graph this inequality.

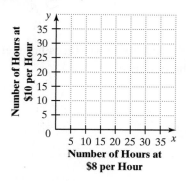

c. Give some examples of the number of hours that the student can work at each job.

58. Scientists who study weather have developed linear models that relate a region's weather conditions to the kind of vegetation that grows in the region. One such model predicts desert conditions if $3t - 35p > 140$, where t represents the average annual temperature (in degrees Fahrenheit) and p the annual precipitation (in inches).

a. Graph this relationship on the coordinate plane.

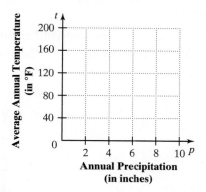

b. Give some examples of weather conditions that this model predicts will lead to desert conditions.

59. While on a diet, a model wants to snack on fresh apples and bananas. An apple contains 60 calories and a banana contains 100 calories. If she wants to consume fewer than 300 calories, find the number of apples a and the number of bananas b that she can eat. Solve this problem graphically.

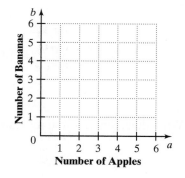

60. A student on spring break wants to drive no more than 300 mi in one day. The trip is along two highways. On the first highway, the student drives at an average speed of 60 mph for x hr and on the second highway at a speed of 50 mph for y hr. What are some possible times that the student can drive? Solve this problem graphically.

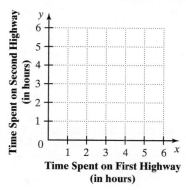

• Check your answers on page A-12.

MINDStretchers

Mathematical Reasoning

1. Consider the three graphs: $y < b$, $y = b$, and $y > b$, where b is a positive number. If you were to graph $y < b$, $y = b$, and $y > b$ on the same coordinate plane, what would you get? Would you get the same answer if b were negative?

Groupwork

2. In playing a carnival game, you roll a pair of dice—a red die and a blue die—each with six faces numbered 1, 2, 3, 4, 5, and 6. The grid below shows all the possible outcomes when you roll the two dice.

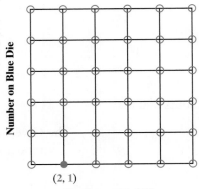

(2, 1)

Number on Red Die

For each point on the grid, the first coordinate represents the roll on the red die and the second coordinate represents the roll on the blue die. For instance, the point $(2, 1)$ corresponds to rolling a 2 on the red and a 1 on the blue.

   a. How many points in all are there on the grid?

   b. If the number on the blue die is greater than the number on the red die, you will win a prize. Fill in the points on the grid that correspond to winning a prize. How many points did you fill in?

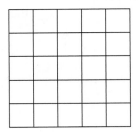

   c. What fraction of the total number of points on the grid did you fill in? What does this fraction represent?

Writing

3. Compare solving a linear *equation* in two variables by graphing with solving a linear *inequality* in two variables by graphing. What are the similarities? What are the differences?

Key Concepts and Skills

Concept/Skill	Description	Example
[3.1] **Coordinate plane**	A flat surface on which we draw graphs. The **coordinate plane** is formed by two perpendicular lines called **axes** which intersect at the **origin**.	y-axis, x-axis, origin
[3.1] **Coordinates**	An **ordered pair** of numbers that represents a point in the coordinate plane.	$(-4, 2)$, $(3, -1)$, $(-1, -3)$
[3.1] **Quadrant**	One of four regions of a coordinate plane separated by axes.	Quadrant II $(-,+)$ Quadrant I $(+,+)$ Quadrant III $(-,-)$ Quadrant IV $(+,-)$
[3.1] **Scale**	The length between adjacent tick marks on an axis in a coordinate plane.	Scale
[3.2] **Slope**	The **slope** m of a line passing through the points (x_1, y_1) and (x_2, y_2): $$m = \frac{y_2 - y_1}{x_2 - x_1}, \quad \text{where } x_1 \neq x_2$$	For $(1, 5)$ and $(-2, 6)$, $$m = \frac{6 - 5}{-2 - 1} = \frac{1}{-3} = -\frac{1}{3}$$

continued

273

Concept/Skill	Description	Example
[3.2] **Horizontal line**	A line whose slope is 0.	Horizontal line
[3.2] **Vertical line**	A line whose slope is undefined.	Vertical line
[3.2] **Parallel lines**	Two nonvertical lines are **parallel** if and only if their slopes are equal. That is, if the slopes are m_1 and m_2, then $m_1 = m_2$.	The line passing through $(0, 1)$ and $(2, 5)$ and the line passing through $(3, 6)$ and $(1, 2)$ are parallel since both lines have slope 2.
[3.2] **Perpendicular lines**	Two lines are **perpendicular** if and only if the product of their slopes is -1. That is, if the slopes are m_1 and m_2, then $m_1 \cdot m_2 = -1$.	The line passing through $(0, 3)$ and $(1, 4)$ and the line passing through $(2, 8)$ and $(3, 7)$ are perpendicular since the product of their slopes, 1 and -1, is -1.
[3.3] **Linear equation in two variables, x and y**	An equation that can be written in the *general form* $Ax + By = C$, where A, B, and C are real numbers and A and B are not both 0.	$3x + 5y = 7$
[3.3] **Solution of an equation in two variables**	An ordered pair of numbers that when substituted for the variables makes the equation true.	$(1, 5)$ is a solution of the equation $y = x + 4$: $$5 \stackrel{?}{=} 1 + 4$$ $$5 = 5 \quad \text{True}$$
[3.3] **Graph of a linear equation in two variables**	All points whose coordinates satisfy the equation.	

Concept/Skill	Description	Example
[3.3] To graph a linear equation in two variables	• Isolate one of the variables—usually *y*—if it is not already done. • Choose three *x*-values, entering them in a table. • Complete the table by calculating the corresponding *y*-values. • Plot the three points—two to draw the line and the third to serve as a *checkpoint*. • Check that the points seem to lie on the same line. • Draw the line passing through the points.	To graph $y - 3x = 1$: $y = 3x + 1$ | x	$y = 3x + 1$	(x, y)
0	$y = 3(0) + 1$	$(0, 1)$
1	$y = 3(1) + 1$	$(1, 4)$
2	$y = 3(2) + 1$	$(2, 7)$ | 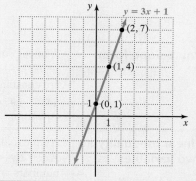
[3.3] Intercepts of a line	The *x*-intercept: the point where the graph crosses the *x*-axis. The *y*-intercept: the point where the graph crosses the *y*-axis.	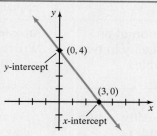
[3.3] To graph a linear equation in two variables using the *x*- and *y*-intercepts	• Let $x = 0$, and find the *y*-intercept. • Let $y = 0$, and find the *x*-intercept. • Find a checkpoint. • Plot the three points. • Check that the points seem to lie on the same line. • Draw the line passing through the points.	To graph $2x + 3y = 6$: |	$2x + 3y = 6$	(x, y)
$x = 0$	$2 \cdot 0 + 3y = 6; y = 2$	$(0, 2)$
$y = 0$	$2x + 3 \cdot 0 = 6; x = 3$	$(3, 0)$
$x = 6$	$2 \cdot 6 + 3y = 6; y = -2$	$(6, -2)$ | 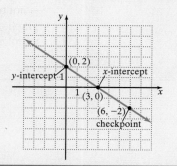
[3.4] Slope-intercept form	A linear equation written as $y = mx + b$, where m and b are constants. In this form, m is the slope and $(0, b)$ is the *y*-intercept of the graph of the equation.	The line with slope 5 and *y*-intercept $(0, 2)$: $y = mx + b$ $y = 5x + 2$

continued

Concept/Skill	Description	Example
[3.4] **To graph a linear equation in two variables using the slope and y-intercept**	• First, locate the y-intercept. • Then, use the slope to find a second point on the line. • Finally, draw the line through the two points.	To graph $y = 3x - 1$: The y-intercept is $(0, -1)$ and the slope is 3.
[3.4] **Point-slope form**	A linear equation written as $y - y_1 = m(x - x_1)$, where x_1, y_1, and m are constants. In this form, m is the slope and (x_1, y_1) is a point that lies on the graph of the equation.	The line with slope 5 passing through the point $(2, 1)$: $$y - y_1 = m(x - x_1)$$ $$y - 1 = 5(x - 2)$$
[3.5] **Linear inequality in two variables**	An inequality that can be written in the form $Ax + By < C$, where A, B, and C are real numbers and A and B are not both 0. The inequality symbol can be $<$, $>$, $\geq$, or $\leq$.	$$5x + 3y < 1$$
[3.5] **Solution of an inequality in two variables**	An ordered pair of numbers that when substituted for the variables makes the inequality a true statement.	$(3, 1)$ is a solution to the inequality $x < 5y$: $$3 < 5(1)$$ $$3 < 5 \qquad \text{True}$$
[3.5] **To graph a linear inequality in two variables**	• Graph the corresponding linear equation. For an inequality with the symbol $\leq$ or $\geq$, draw a solid line; for an inequality with the symbol $<$ or $>$, draw a broken line. This line is the boundary between two half-planes. • Choose a test point in either half-plane and substitute the coordinates of this point in the inequality. If the resulting inequality is true, then the graph of the inequality is the half-plane containing the test point. If it is not true, then the other half-plane is the graph. A solid line is part of the graph, and a broken line is not.	To graph $y > x + 1$, first graph the line $y = x + 1$. The inequality does not hold for the test point $(0, 0)$. The graph of $y > x + 1$ is the half-plane above and excluding the line $y = x + 1$.

Say Why
Fill in each blank.

1. The points $(2, -5)$ and $(3, 5)$ _____ in the same $\frac{}{\text{are/are not}}$ quadrant because _____ _____.

2. The point $(3, -1)$ _____ lie on the graph of $\frac{}{\text{does/does not}}$ the equation $2x - 5y = 1$ because _____ _____.

3. The graph of the equation $y = -2x + 7$ _____ $\frac{}{\text{is/is not}}$ increasing because _____ _____.

4. The graph of the equation $y = 5$ _____ horizontal $\frac{}{\text{is/is not}}$ because _____.

5. The x-intercept of $y = 2x + 6$ _____ -3 because $\frac{}{\text{is/is not}}$ _____.

6. The slope of a vertical line _____ undefined $\frac{}{\text{is/is not}}$ because _____ _____.

7. A line with slope 5 _____ perpendicular to a line $\frac{}{\text{is/is not}}$ with slope $\frac{1}{5}$ because _____ _____.

8. $(0, 0)$ _____ a solution to the linear inequality $\frac{}{\text{is/is not}}$ $2x - 5y \geq 0$ because _____ _____.

[3.1]

9. Plot the points with the given coordinates.
$A(0, 5)$ $B(-1, -6)$ $C(3, -4)$ $D(-2, 2)$

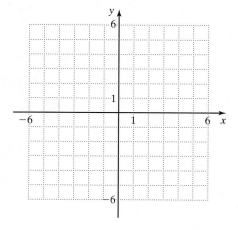

10. Fill in the coordinates of each point.

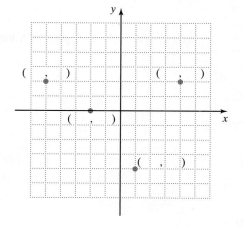

Identify the quadrant in which each point is located:

11. $(5, -1)$

12. $(-7, -2)$

[3.2] *Compute the slope m of the line that passes through the given points. Plot these points on the coordinate plane, and draw the line.*

13. $(2, 0)$ and $(3, 5)$, $m =$

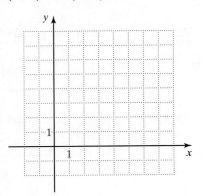

14. $(5, 7)$ and $(2, 7)$, $m =$

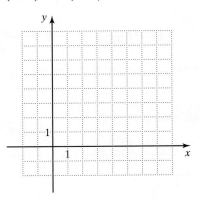

Draw the line on the coordinate plane based on the given information.

15. Passes through $(3, -1)$ and $m = 4$

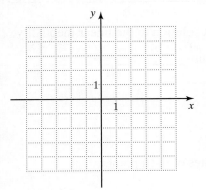

16. Passes through $(0, 0)$ and $m = -\dfrac{1}{2}$

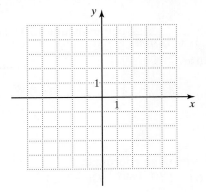

Determine whether the slope of each graph is positive, negative, zero, or undefined.

17.

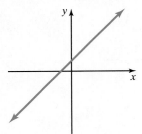

18.

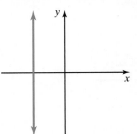

19.

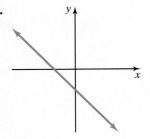

20.

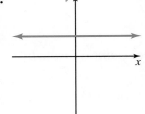

Determine whether $\overleftrightarrow{AB}$ and $\overleftrightarrow{CD}$ are parallel or perpendicular.

21. $A(5, 0)$ $B(3, 0)$ $C(-3, -2)$ $D(1, -2)$

22. $A(4, 8)$ $B(5, 9)$ $C(2, -3)$ $D(0, -1)$

For the following graphed line, find:

23. the *x*-intercept.

24. the *y*-intercept.

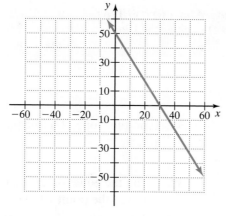

[3.3] *Complete each table of values for the given equation.*

25. $y = 2x - 5$

x	0	1		
y			0	1

26. $y = -x + 3$

x	2	5		
y			7	-5

Graph.

27. $4x - 3y = -12$

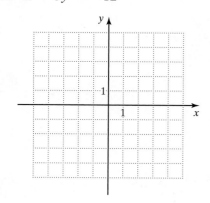

28. $x + 2y = -6$

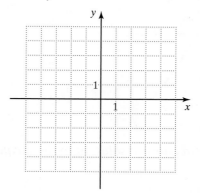

29. $y = \dfrac{1}{2}x$

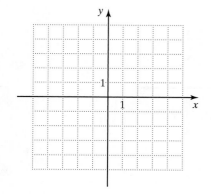

30. $y = -x + 2$

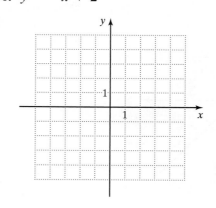

[3.4] *Write each equation in slope-intercept form.*

31. $x - y = 10$

32. $x + 2y = -1$

Complete the following table:

	Equation	Slope m	y-intercept $(0, b)$	Indicate Which Graph Type Best Describes the Line. $\diagup \diagdown - \mid$	x-intercept
33.	$y = 4x - 16$				
34.	$y = -\frac{1}{3}x$				

35. Find the slope of a line perpendicular to the line $x - 2y = 4$.

36. Find the slope of a line parallel to the line $3x - y = 1$.

37. Find the equation of the line with slope -1 that passes through the point $(3, 5)$.

38. Write the equation of the horizontal line that passes through the point $(3, 0)$.

39. The points $(2, 0)$ and $(1, 5)$ lie on a line. Find its equation.

40. What is an equation of the line passing through the points $(3, 1)$ and $(-2, 0)$?

Find the equation of each graph.

41.

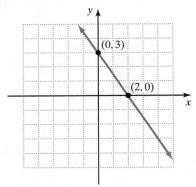

42.

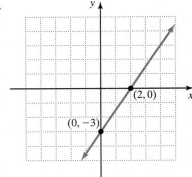

[3.5] *Decide if the ordered pair is a solution to the inequality.*

43. $(-2, 7), x + y < 1$

44. $(1, -4), 2x - 3y \geq 14$

Each line is the graph of $y = -x$. Shade in the graph of the given inequality.

45. $y > -x$

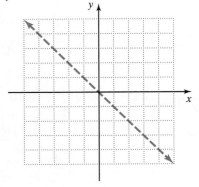

46. $y < -x$

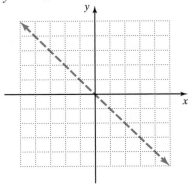

Graph the linear inequality.

47. $y \leq 2x$

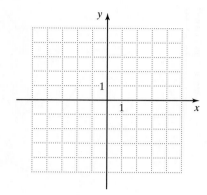

48. $y - x > -1$

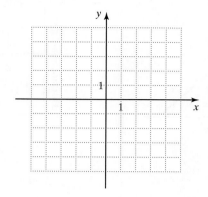

Mixed Applications

Solve.

49. The following table shows the amount A (in dollars) that a bed and breakfast charges when renting a room for s days.

Length of Stay s	Rental Amount, A
2	180
5	450

   a. Graph the points given in the table. Draw a line passing through the points.

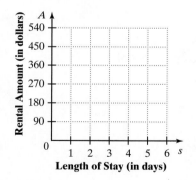

   b. What is the A-intercept of this line? Explain its significance in terms of renting a room.

50. The following table shows the cost C (in cents) of duplicating q flyers at a print shop.

Quantity q	Cost C
1	4
10	40

   a. Plot the points given in the table and draw the line passing through them.

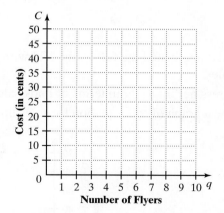

   b. Calculate the slope of this line. Explain its significance in terms of the price structure at the print shop.

51. A man drives toward a town, stops, and then again drives toward the town. Which of the following graphs could describe this motion? Explain your answer.

a.

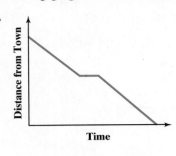

b.

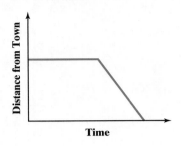

c.

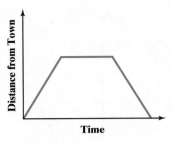

52. The graph below shows the altitude of an airplane during a flight. Write a brief story describing the altitude of the plane relative to the duration of the flight.

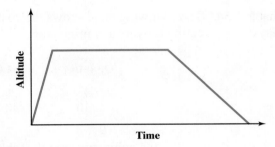

53. A novelist negotiated the following deal with her publisher: a $20,000 bonus plus 9% of book sales.

  a. Express her income i in terms of book sales s.

  b. Draw a graph of this equation for sales up to $500,000.

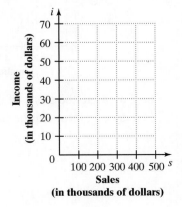

54. A bank account with an initial balance of $100 earns simple interest at an annual rate of 4%. The amount A in the account after t years is given by:

$$A = 100 + 4t$$

a. Graph this equation.

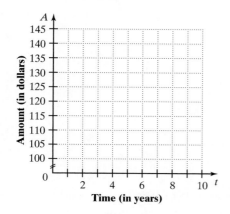

b. What is the A-intercept of this graph? Explain its significance in terms of the bank account.

55. The width of a jewel case is $\frac{1}{4}$ in. for a single compact disc and $\frac{1}{2}$ in. for double compact discs. If there are s single jewel cases and d double jewel cases on a shelf 30 in. long, then

$$\frac{1}{4}s + \frac{1}{2}d < 30.$$

a. Graph this inequality.

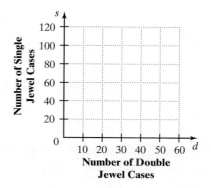

b. From this graph, identify one possible combination of single and double jewel cases that will fit on the shelf.

56. To be able to catch up and pass a friend driving away at a speed of 50 mph, it is necessary to cover a distance of d mi in t hr, where

$$d > 50t.$$

a. Graph this inequality.

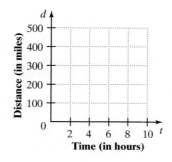

b. From this graph, choose a point in the shaded region. Explain what its coordinates mean in terms of catching up and passing the friend.

• Check your answers on page A-13.

CHAPTER 3   Posttest

FOR
EXTRA
HELP

CHAPTER
Test Prep
VIDEOS

The Chapter Test Prep Videos with test solutions are available on DVD, in MyMathLab, and on You Tube (search "AkstIntroductory Alg" and click on "Channels").

To see whether you have mastered the topics in this chapter, take this test.

1. On the coordinate plane shown, plot the points $A(-2, 0)$ and $B(5, 3)$.

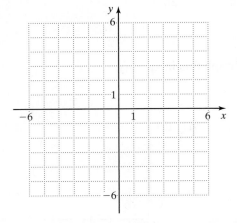

2. In which quadrant is the point $(-5, 3)$ located?

3. Given two points $C(8, 1)$ and $D(3, -4)$, compute the slope of the line that passes through the points.

4. Are the graphs of $y = 3x + 1$ and $y = 3x - 2$ parallel? Explain how you know.

5. For the points $A(0, 1)$, $B(2, 8)$, $C(0, 6)$, and $D(7, 4)$, indicate whether $\overleftrightarrow{AB}$ is perpendicular to $\overleftrightarrow{CD}$. Explain.

6. In the following graph, find the x-intercept and the y-intercept.

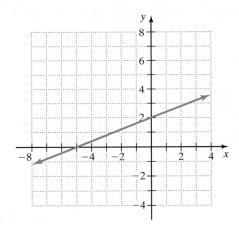

7. The graph on the following coordinate plane shows how the rental cost at a local car rental establishment relates to the number of miles that the car has been driven.

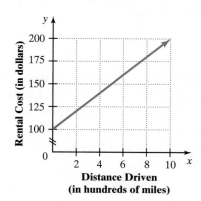

Is the slope of the graphed line positive, negative, zero, or undefined? Describe in a sentence or two the relationship between the rental cost and the number of miles driven.

8. Do the points $(0, 0)$, $(-2, -4)$, and $(1, 2)$ lie on the same line? Explain.

9. For the equation $y = -3x + 1$, complete the following table.

x	−3	5		
y			0	−2

Graph the equation.

10. $y = 2$

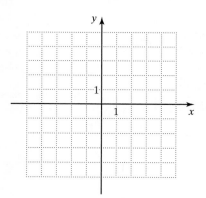

11. $x = -4$

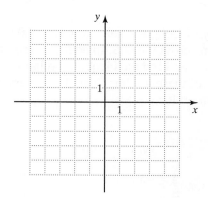

12. $y = -x - 3$

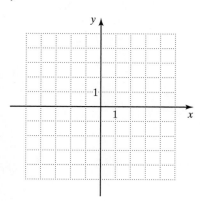

13. $3x - 2y = 6$

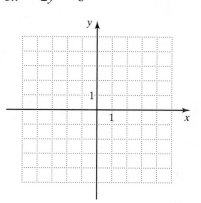

14. What are the slope and the y-intercept of the graph of $y = 3x + 1$?

15. Write the equation $2x - y = 5$ in slope-intercept form.

16. Find the equation of the line with slope -1 that passes through the point $(0, -3)$.

17. The points $(3, 5)$ and $(-4, 2)$ lie on a line. Find its equation.

18. Graph $y \leq -\dfrac{1}{2}x + 1$.

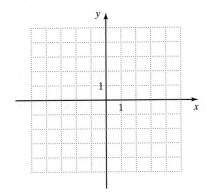

19. An entrepreneur is establishing a small business to manufacture leather bags. Her initial investment is $1000, and her unit cost to manufacture each bag is $30. Write an equation that gives the total cost C of manufacturing b bags. Plot the total cost of manufacturing 100, 200, and 300 bags.

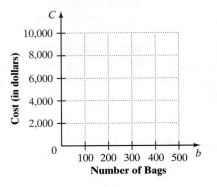

20. In 2010, the Library of Congress held over 124,000 telephone books and microfilmed city directories from across the U.S. In addition, the library acquires over 8,000 more of these holdings each year. Find the inequality relating the number of years x passed since 2010 to the number of these holdings y in the Library of Congress' collection, and graph this inequality. (*Source:* loc.gov)

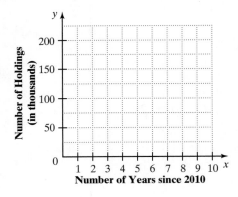

• Check your answers on page A-14.

Cumulative Review Exercises

To help you review, solve the following:

1. Replace ▢ with $>, <,$ or $=$ to make a true statement:
   $$-2.3 \; ▢ \; -2\frac{1}{3}$$

2. Find the difference: $3.9 - 8.2$

3. Calculate: $(7 - 9) \div (-2 - 6)$

4. Evaluate: $2 \cdot 6 - 6^2$

5. Find the value of $x^2 - 4x + 1$ if $x = -3$.

6. Simplify: $x - 3 - 3(x - 2) + 2(5x - 2)$

7. Solve and check: $-1.69 = 1.3x$

8. Solve for x: $2x - 1 = 5x + 11$

9. Solve for x: $2x - 3(5 - x) = 4x + 2$

10. Solve for x and graph: $3x + 1 > 7$

11. Identify the quadrant in which each point is located:
    a. $(-8, 6)$
    b. $(8, -6)$

12. Graph: $2x + 5y = -12$

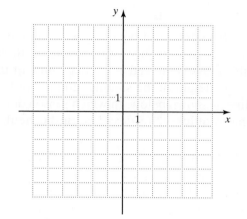

13. What is the equation of the line passing through the points $(-1, 7)$ and $(3, 4)$?

14. Graph the following: $x - 3y > -3$

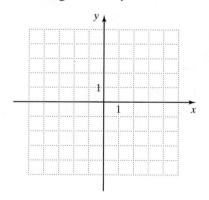

15. An air conditioner can reduce the temperature in a room by 8°F every 5 min. The temperature in the room was 62°F after the air conditioner had been running for 10 min. Write a linear equation that expresses the temperature in the room in terms of the time that the air conditioner has been running.

16. The speed of sound S (in meters per second) in air at temperature T can be approximated by the following formula:
    $$S = 0.6T + 331.$$
    Solve for T in terms of S. (*Source:* Peter J. Nolan, *Fundamentals of College Physics*)

17. In a recent year, the combined cost of Medicare and Medicaid amounted to $462 billion, accounting for 13% of all federal budget outlays. Find the total of all federal budget outlays, rounded to the nearest trillion dollars. (*Source:* cnn.com)

18. The following graph shows the average weight for American males, age 20 and over. Describe the overall pattern in this population's weight over a lifespan. (*Source:* cdc.gov)

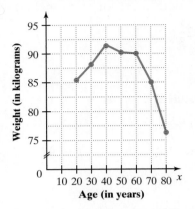

19. St. Augustine grass is a type of grass found in the southern United States. Its growth rate varies with the season. After this grass is mowed at the beginning of the season, its average height in the summer and the fall is graphed over time as shown below. In which season does the grass grow more quickly? (*Source:* aggie-horticulture.tamu.edu)

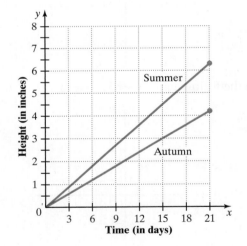

20. A drama club washes cars as a fund-raising activity. The club charges $6 to wash each car.

a. Write an equation that expresses the club's income y in terms of the number x of cars they wash.

b. Graph the equation.

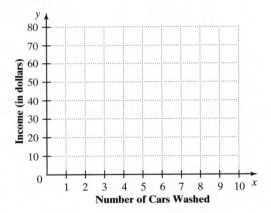

c. What are the x- and y-intercepts of the graph?

• Check your answers on page A-15.

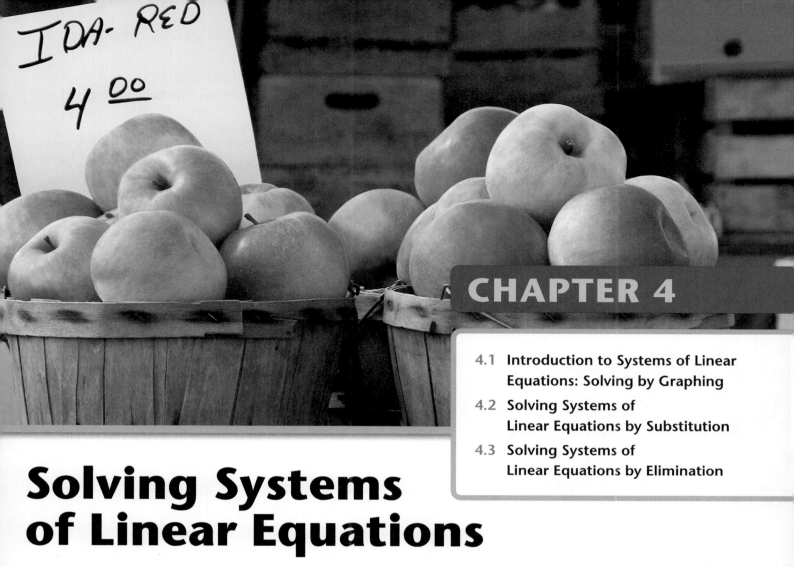

CHAPTER 4

Solving Systems of Linear Equations

Economics and Linear Curves

In a market, sellers can set the price at which their goods are offered for sale. How does this price affect the number of goods that buyers are willing to purchase? How does it affect the number of goods that producers are willing to supply the sellers?

Generally, as the *price* of an item increases, the *quantity* of items sold declines. This trend is captured in a **demand curve**—commonly approximated by a straight line with a negative slope. The coordinates of each point on this line correspond to the price at which retailers sell items and the quantity of items they sell.

By contrast, the **supply curve** has a positive slope, meaning that as selling prices increase, wholesalers are inclined to make more goods available to retailers. The coordinates of a point on this line represent the price at which retailers sell items and the quantity of items that wholesalers supply to the retailers.

Graphing both the supply curve and the demand curve on the same coordinate plane shows the price at which the market is **at equilibrium**. At this point of equilibrium, the quantity supplied is equal to the quantity demanded.

(*Source:* Michael Parkin, *Economics*, Pearson Addison-Wesley, 2010)

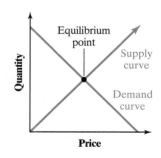

1. Determine which ordered pair is a solution of the following system:

$$x + 2y = 5$$
$$5x - y = -8$$

a. $(5, 0)$

b. $(-1, 3)$

c. $(1, -3)$

2. For the system graphed, indicate the number of solutions.

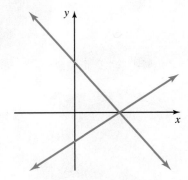

Solve each system by graphing.

3. $x + y = -2$
  $y = x + 4$

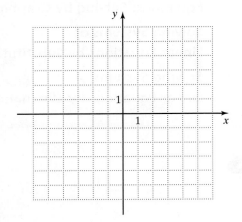

4. $x - 2y = 1$
  $y = 2$

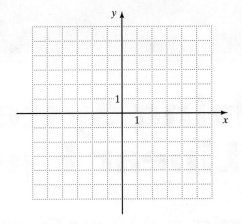

5. $x - 2y = -4$
  $4y = 2x + 8$

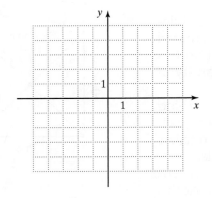

6. $3y = 6 - x$
  $x + 3y = 3$

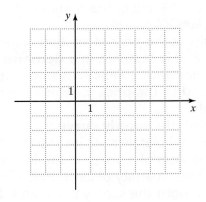

Solve each system by substitution.

7. $x - 2y = 7$
   $y = -11 - x$

8. $7x - 4y = 10$
   $x - 2y = 0$

9. $a + 3b = -2$
   $a = 2b - 7$

Solve each system by elimination.

10. $2x + 5y = -13$
    $-2x + 6y = -20$

11. $6x - 8y = 36$
    $1.5x - 2y = 9$

12. $3x - 7y = -19$
    $2x + 3y = -5$

Solve each system.

13. $4x + y = 0$
    $5y = 12 - 8x$

14. $-3n + 5m = 10$
    $-4m = -2(n + 1)$

15. $x + 9 = 2y$
    $8y - 13 = 4x$

16. $6x + 10y - 12 = 0$
    $3x + 2.5y - 6 = 0$

Solve.

17. At a college commencement, four times as many bachelor's degrees as associate's degrees were awarded to graduating students. If 3095 students graduated, how many bachelor's degrees and associate's degrees were awarded?

18. For a student club fund-raiser, the number of $2 raffle tickets printed was three times the number of $5 tickets. If all of the tickets are sold, receipts from the $5 tickets will be $50 less than those from the $2 tickets. How many $5 tickets were printed?

19. A lottery winner invested $200,000 of her winnings in two funds earning 5% and 6% simple interest, respectively. If after one year she earned $11,200 in interest, how much did she invest in each fund?

20. On a boating trip, it took 2 hr to travel 13 mi with the current. It took the same amount of time to travel 11 mi against the current on the return trip. Find the speed of the boat and the speed of the current.

• Check your answers on page A-15.

OBJECTIVES

A To decide whether an ordered pair is a solution of a system of linear equations in two variables

B To determine the number of solutions of a system of linear equations

C To solve a system of linear equations by graphing

D To solve applied problems involving systems of linear equations

What Systems of Linear Equations Are and Why They Are Important

Recall from previous chapters that some situations are described by a linear equation in one variable, say, $2x + 1 = 10$, whereas others are described by a single linear equation in *two* variables, for instance, $y = 3x - 5$.

Now, let's consider situations in which the relationship between two variables is described by a *pair* of linear equations, for instance:

$$x + y = 7$$
$$x - y = 3$$

Groups of equations, called *systems*, serve as a model for a wide variety of applications in fields such as business and science. A system can represent the conditions that must be satisfied in a particular situation. For instance, the system might describe how the cost of products relate to one another or the motion of an airplane in various wind conditions.

In this chapter, we deal with three approaches to solving systems of linear equations, namely by *graphing*, *substitution*, and *elimination*.

Introduction to Systems of Linear Equations

We begin by focusing on the meaning of a system of equations.

DEFINITION

A **system of equations** is a group of two or more equations solved simultaneously.

Systems of equations are sometimes written with large braces:

$$\begin{cases} x + y = 7 \\ x - y = 3 \end{cases} \quad \text{or} \quad \begin{cases} x + y = 7 \\ x - y = 3 \end{cases}$$

Braces are used to emphasize that any solution of a system must satisfy *all* the equations in the system. For instance, $x = 5$ and $y = 2$ is a solution of the system above, because when we substitute 5 for x and 2 for y into the equations, *both* equations are true:

$$x + y = 7 \quad \rightarrow \quad 5 + 2 \overset{?}{=} 7 \quad \text{True}$$
$$x - y = 3 \quad \rightarrow \quad 5 - 2 \overset{?}{=} 3 \quad \text{True}$$

A solution of a system of two equations is commonly represented as an ordered pair of numbers. For instance, the solution of the system just mentioned can be written as $(5, 2)$. Can you explain why $(2, 5)$ is not a solution of this system?

DEFINITION

A **solution** of a system of two linear equations in two variables is an ordered pair of numbers that makes both equations in the system true.

EXAMPLE 1

Consider the system:

$$x - 2y = 6$$
$$2x + 5y = 3$$

a. Is $(4, -1)$ a solution of the system?

b. Is $(2, -2)$ a solution of the system?

Solution

a. To decide if the ordered pair $(4, -1)$ is a solution of this system, we substitute the x-coordinate 4 for x and the y-coordinate -1 for y in the equations and check if both equations are true.

$x - 2y = 6 \quad \rightarrow \quad 4 - 2(-1) \overset{?}{=} 6 \quad \xrightarrow{\text{Simplifies to}} \quad 6 = 6 \qquad$ True

$2x + 5y = 3 \quad \rightarrow \quad 2(4) + 5(-1) \overset{?}{=} 3 \quad \xrightarrow{\text{Simplifies to}} \quad 3 = 3 \qquad$ True

The ordered pair $(4, -1)$ satisfies both equations and so is a solution of the system.

b. To see if $(2, -2)$ is a solution, we substitute 2 for x and -2 for y in the equations and check if they are both true.

$x - 2y = 6 \quad \rightarrow \quad 2 - 2(-2) \overset{?}{=} 6 \quad \xrightarrow{\text{Simplifies to}} \quad 6 = 6 \qquad$ True

$2x + 5y = 3 \quad \rightarrow \quad 2(2) + 5(-2) \overset{?}{=} 3 \quad \xrightarrow{\text{Simplifies to}} \quad -6 = 3 \qquad$ False

The ordered pair $(2, -2)$ is not a solution of the system because it does not satisfy both equations.

PRACTICE 1

Consider the following system:

$$3x + 2y = 5$$
$$4x - 2y = -5$$

Determine whether the following ordered pairs are solutions of the system.

a. $(0, 2.5)$

b. $(1, -1)$

Number of Solutions of a System

In solving a system of linear equations, a question that immediately comes to mind is how many solutions the system has. Let's consider this question graphically. Since each equation is linear, both graphs are lines. Now, suppose that we graph the two equations on the same coordinate plane. Any point at which the two graphs of the system intersect is a solution of the system, because that point must satisfy both equations.

For instance, let's reconsider the system discussed on the previous page

$$x + y = 7$$
$$x - y = 3$$

and solve it by graphing both equations.

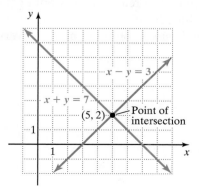

Note that the lines intersect at $(5, 2)$—precisely the ordered pair that we have previously determined to be a solution of this system. Since the two lines meet at a single point, the system has *exactly one* solution.

Not all systems have one solution. For instance, consider the following system in which there are *no* solutions:

$$3x - y = 2$$
$$3x - y = 4$$

Graphing this system, we get:

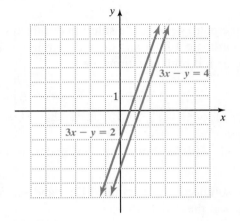

We observe that the lines are parallel and, therefore, do not intersect. So in this case, the system has no solutions.

Finally, let's examine a system that has more than one solution.

$$2x - 2y = 6$$
$$y = x - 3$$

The graph of this system is as follows:

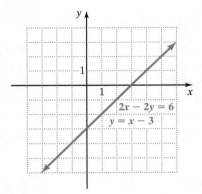

Note that only one line is shown. The reason is that both equations in the system have the same graph. Any point on this line, for instance $(3, 0)$, is a solution of both equations. A system such as this one has *infinitely many* solutions, namely every point on the line.

Every system of linear equations has one solution, no solution, or infinitely many solutions. The following table summarizes the main features of the three types of systems:

Number of Solutions	Description of the System's Graph	Possible Graph
One solution	The lines intersect at exactly one point.	
No solution	The lines are parallel.	
Infinitely many solutions	The lines coincide, that is, they are the same line.	

EXAMPLE 2

For each system graphed, determine the number of solutions.

a.

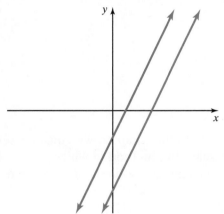

PRACTICE 2

Determine the number of solutions of each of the following systems:

a.

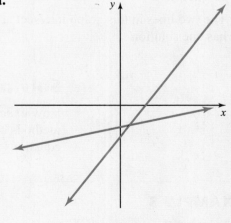

EXAMPLE 2 (continued)

b.

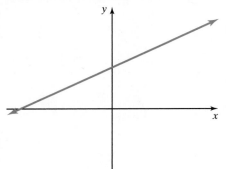

c.

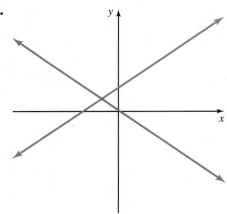

b.

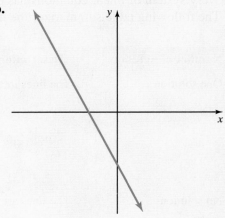

c.

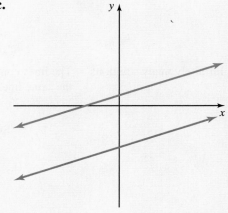

Solution

a. The graph of the system consists of two lines that appear to be parallel. Since the lines do not intersect, the system has no solution.

b. This graph is a single line. Therefore, the system has infinitely many solutions.

c. The two lines in this graph intersect at a single point. So the system has one solution.

Solving Systems by Graphing

How exactly are systems of linear equations solved? In this chapter, we consider several methods of solving systems. Let's first discuss the **graphing method** in which we graph the equations that make up the system. Any point of intersection is a solution of the system.

EXAMPLE 3

Solve the following system by graphing:

$$x + y = 6$$
$$x - y = -4$$

Solution Let's graph each linear equation by using the x- and y-intercept method and then sketching the line that passes through these points.

PRACTICE 3

On the given coordinate plane, solve the following system by graphing:

$$x + y = 2$$
$$x - y = 4$$

$$x + y = 6 \qquad\qquad x - y = -4$$

x	y
0	6
6	0
3	3

x	y
0	4
−4	0
2	6

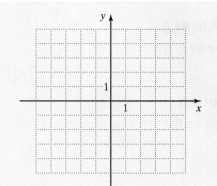

On the same coordinate plane, we plot the points, and then graph both equations.

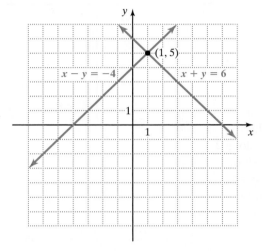

The corresponding lines appear to intersect at the point $(1, 5)$, giving $x = 1$ and $y = 5$.

Check Since solving a system of equations by graphing may result in approximate solutions, we confirm that $(1, 5)$ is the solution by substituting these values into the original equations:

$x + y = 6$ $\xrightarrow{\text{Substitute 1 for } x \text{ and 5 for } y.}$ $1 + 5 \stackrel{?}{=} 6$

$6 = 6$ True

$x - y = -4$ $\xrightarrow{\text{Substitute 1 for } x \text{ and 5 for } y.}$ $1 - 5 \stackrel{?}{=} -4$

$-4 = -4$ True

So $(1, 5)$ is the solution of the system.

To Solve a System of Linear Equations by Graphing

* Graph both equations on the same coordinate plane.

* There are three possibilities:

    a. If the lines intersect, then the solution is the ordered pair of coordinates for the point of intersection. Check that these coordinates satisfy both equations.

    b. If the lines are parallel, then there is no solution of the system.

    c. If the lines coincide, then there are infinitely many solutions, namely all the ordered pairs of coordinates that represent points on the line.

EXAMPLE 4

Solve by graphing.

$$y = 2x + 5$$
$$2x - y = 2$$

Solution

Graphing the two equations, we get:

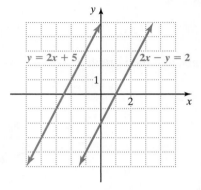

The lines appear to be parallel, which suggests that the system has no solution. To confirm that the lines are parallel, we can check that their slopes are equal. The graph of the first equation, $y = 2x + 5$, has slope 2. To find the slope of the second equation, we write $2x - y = 2$ in slope-intercept form, getting $y = 2x - 2$. The graph of this equation also has slope 2. Therefore, the lines are parallel and the system has no solution.

PRACTICE 4

Solve for x and y by graphing.

$$y = x - 6$$
$$x - y = 4$$

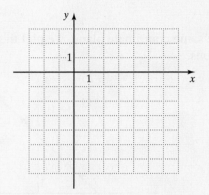

EXAMPLE 5

Solve by graphing.

$$2y = -8x + 2$$
$$-4x - y = -1$$

Solution

When we graph the two equations in this system, we get the same graph.

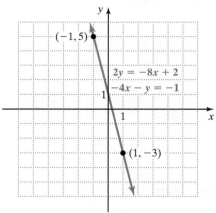

Note that when the two equations in the system are changed to slope-intercept form, the equations are identical.

$$2y = -8x + 2 \xrightarrow{\text{Isolate } y.} y = -4x + 1$$

$$-4x - y = -1 \xrightarrow{\text{Isolate } y.} y = -4x + 1$$

We conclude that the system has infinitely many solutions. All points on the line, some of which are indicated, are solutions. Can you identify another point on the graph and confirm that it is a solution to the system?

PRACTICE 5

Solve by graphing.

$$6x = 15 - 3y$$
$$y = 5 - 2x$$

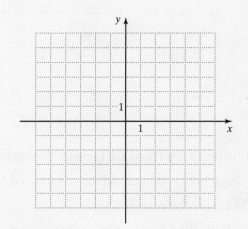

EXAMPLE 6

The U.S. House of Representatives has 435 members. On a certain bill, all the representatives voted, and 15 more representatives voted for the bill than against the bill. (There were no abstentions.)

a. If s represents the number of representatives who *supported* the bill, and n the number of representatives who did *not* support the bill, express the given information as a system of equations.

b. On a coordinate plane, graph the system found in part (a).

c. Find the coordinates of the point of intersection.

d. In this problem, what is the significance of the coordinates of the point of intersection?

Solution

a. The given information can be expressed algebraically as:

$$s + n = 435$$
$$s = n + 15$$

b. First, let's graph s along the vertical axis and n along the horizontal axis. Since the number of representatives voting for or against a bill is between 0 and 435, we then choose an appropriate scale and label the two axes accordingly. Next, we graph the two equations.

c. The lines intersect at the point $(210, 225)$, that is, $n = 210$ and $s = 225$.

d. We conclude that 210 representatives voted against the bill and 225 voted for the bill.

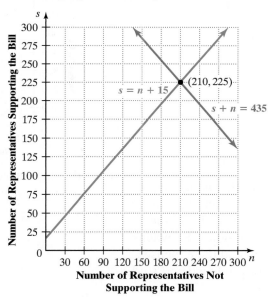

PRACTICE 6

A liberal arts student transferring to a four-year institution took a test of verbal skills and a test of mathematical skills. Her total score was 1150, and the verbal score v was 100 less than the math score m.

a. Express the given information as a system of equations.

b. On a coordinate plane, graph the system found in part (a).

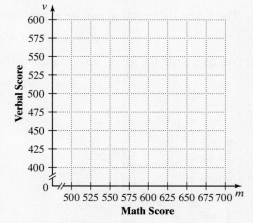

c. Identify the coordinates of the point of intersection.

d. In this situation, what is the significance of the coordinates of the point of intersection?

When running a business, it is important to determine both the income that the business makes and the expenses that it takes to run the business. The business' income and its expenses depend on the number of items produced and sold. These quantities can be graphed on a coordinate plane, as shown in the next example. The point at which the income for a business equals its expenses is called the *break-even point*.

EXAMPLE 7

For a start-up business, an entrepreneur determined that to produce computer-generated, silk-screen T-shirts it will cost $3.25 a shirt plus $450 in fixed overhead. Each shirt produced is sold at $5.50.

a. If x represents the number of T-shirts sold and y the amount it costs to produce the T-shirts, write an equation that expresses y in terms of x.

b. If x represents the number of T-shirts produced and y the amount of income from selling the T-shirts, write an equation that relates x and y.

c. On a coordinate plane, graph the lines found in parts (a) and (b).

d. Find the break-even point for producing the T-shirts. Explain its significance in terms of the x- and y-coordinates.

Solution

a. The given information can be expressed as:
$$y = 3.25x + 450$$

b. We can write the given information as:
$$y = 5.50x$$

c.

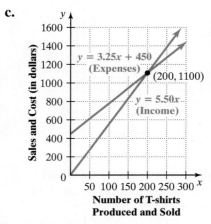

Number of T-shirts
Produced and Sold

d. Since the break-even point is the point where the income equals the expenses, we must find the point of intersection of the lines $y = 3.25x + 450$ and $y = 5.50x$. The intersection of the lines is the point $(200, 1100)$, which is the break-even point. So when 200 shirts are produced and sold, the income and the expenses will be the same, $1100. After 200 shirts are produced and sold, the business will start making a profit.

PRACTICE 7

To print a newsletter costs $450 fixed overhead plus $1.50 a copy. The newsletters sell for $3 each.

a. If x represents the number of copies printed and y the amount of money it costs to print the newsletter, write an equation that expresses y in terms of x.

b. If x represents the number of copies printed and y the amount of income from newsletter sales, write an equation that relates x and y.

c. On the coordinate plane below, graph the lines found in parts (a) and (b).

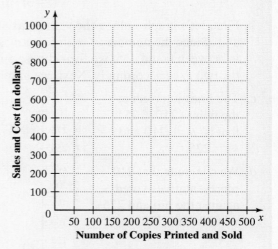

Number of Copies Printed and Sold

d. Find the break-even point for printing the newsletter.

In this section, we have examined the graphing method of solving systems of linear equations. A major advantage of this method over alternative methods is that it helps us to visualize the problem and its solution. However, a disadvantage of this approach is that our reading of a graph may be inaccurate, particularly when a coordinate of the point of intersection is not an integer or is very large. So a solution found by the graphing method may not be exact.

Solving Systems of Linear Equations on a Grapher

Both graphing calculators and computers with graphing software can help facilitate the process of solving systems of linear equations. As with the paper-and-pencil approach, a grapher displays the graphs of the equations that make up a system on the same coordinate plane. We can then use one of the special features of the grapher to read the coordinates of the point at which the graphed lines intersect, that is, the solution of the system.

The most common features of a grapher that help us to read the coordinates of the point of intersection are TRACE, ZOOM, and INTERSECT.

- *With the TRACE feature, the cursor runs along either of the graphed lines until it is positioned on or near the point of intersection; the coordinates of that point are then displayed.*

- *The ZOOM feature lets us position the cursor as close as we want to the point of intersection.*

- *The INTERSECT feature automatically calculates the point of intersection.*

Note that each of the features may give only an approximation for the point of intersection. However, the most accurate approximation of the point of intersection is given by the INTERSECT feature.

EXAMPLE 8

Use either a graphing calculator or graphing software to solve.
$$4x - y = 11$$
$$x = y + 6$$

Solution   Begin by solving each equation for y.

$$4x - y = 11 \quad \xrightarrow{\text{Isolate } y.} \quad y = 4x - 11$$

$$x = y + 6 \quad \xrightarrow{\text{Isolate } y.} \quad y = x - 6$$

Then, press the $\boxed{Y=}$ key, and enter $4x - 11$ to the right of **Y1** = and $x - 6$ to the right of **Y2** =. Set the viewing window. Then press the $\boxed{\text{GRAPH}}$ key to display the coordinate plane on which the two equations are graphed. The **TRACE** feature can be used to move a cursor along one of the lines toward the intersection of the graphs by holding down an arrow key. Note that as the cursor is moved, the changing coordinates of its position will be displayed on the screen. Once the cursor reaches the point of intersection, we can read the coordinates on the screen.

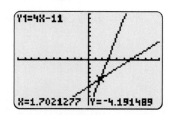

Using the TRACE feature

To get a better approximation of the solution, we can either activate the **ZOOM** feature to zoom in on the intersection point or activate the **INTERSECT** feature.

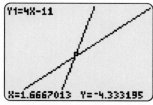

Using the ZOOM feature

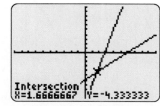

Using the INTERSECT feature

So the approximate solution is $(1.667, -4.333)$.

PRACTICE 8

Use a grapher to solve the following system of equations:
$$8x - y = 1$$
$$y = x + 5$$

Mathematically Speaking

Fill in each blank with the most appropriate term or phrase from the given list.

are parallel	solution	coincide
graph	set of equations	system of equations

1. A(n) _____ is a group of two or more equations solved simultaneously.

2. A(n) _____ of a system of two linear equations in two variables is an ordered pair of numbers that makes both equations in the system true.

3. If a system of linear equations has no solution, its graph consists of lines that _____.

4. If a system has infinitely many solutions, its graph consists of lines that _____.

A *Indicate whether each ordered pair is or is not a solution of the given system.*

5. $x + y = 3$
$2x - y = 6$
 a. $(0, 3)$ _____
 b. $(3, 3)$ _____
 c. $(3, 0)$ _____

6. $x - 6y = 3$
$x - y = -7$
 a. $(-2, -9)$ _____
 b. $(-9, -2)$ _____
 c. $(9, -2)$ _____

7. $4x + 5y = 0$
$7x - y = 0$
 a. $(1, 7)$ _____
 b. $(-5, 4)$ _____
 c. $(0, 0)$ _____

8. $2x - 2y = 30$
$8x + 2y = -10$
 a. $(1, -9)$ _____
 b. $(16, 1)$ _____
 c. $(2, -11)$ _____

B *Match each system with the appropriate graph.*

9. **a.** A system with solution $(1, 3)$
 c. A system with infinitely many solutions

 b. A system with solution $(-1, 3)$
 d. A system with no solution

I

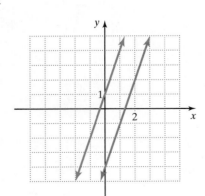

II

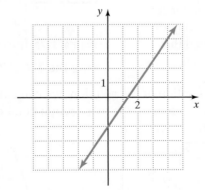

III

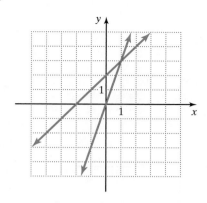

IV

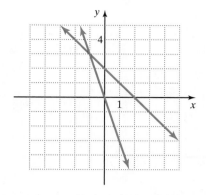

10. a. A system with solution $\left(-\frac{1}{4}, -1\frac{1}{4}\right)$

c. A system with infinitely many solutions

b. A system with solution $(-1, 3)$

d. A system with no solution

I

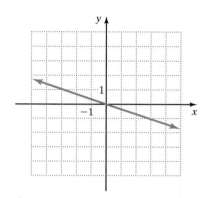

II

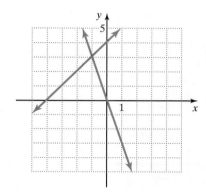

III

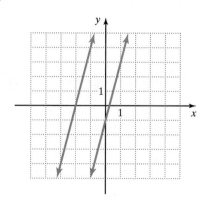

IV

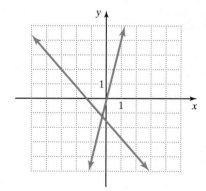

C *Solve by graphing.*

11. $x - y = 2$
$\quad\ \ x + y = 4$

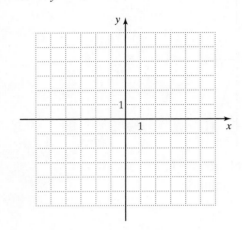

12. $x + 2y = 3$
$\quad\ \ x +\ \ y = 2$

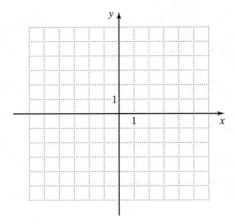

13. $\quad\quad y = x + 4$
$\quad\ \ x + y = 4$

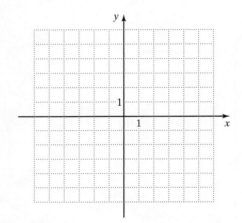

14. $-5 = 2x + y$
$\quad\quad\ \ y = -x$

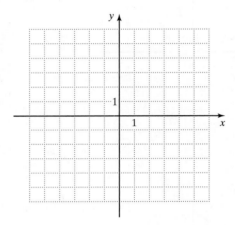

15. $y = -x + 6$
$\quad\ \ y = -3x + 8$

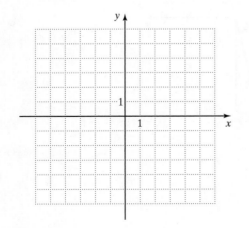

16. $y = x + 1$
$\quad\ \ y = -x - 3$

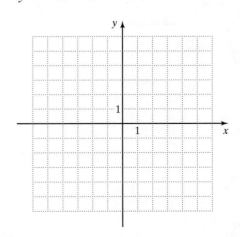

17. $y = -\frac{1}{2}x + 1$
$y = 2x + 1$

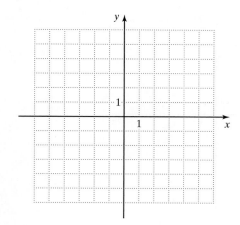

18. $y = 2x - 6$
$y = 3 - \frac{1}{4}x$

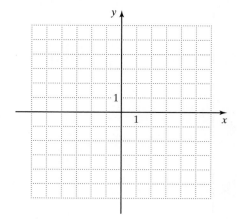

19. $y = 5 + 3x$
$x + y = -3$

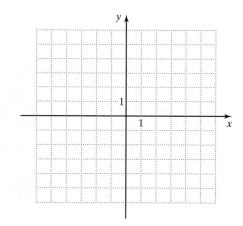

20. $2x + y = -3$
$y = -(x + 4)$

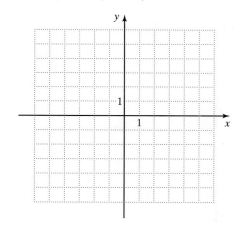

21. $2y = 6x + 2$
$3y - 9x = 3$

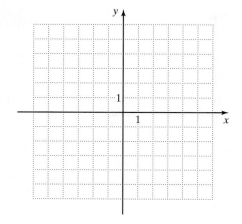

22. $4x = 8y + 4$
$5x - 10y = 5$

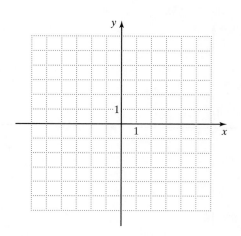

23. $2x + y = -4$
$\quad\quad y = -2x + 3$

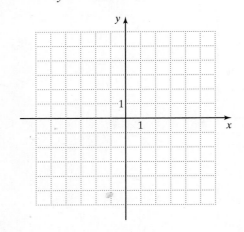

24. $3x - y = 1$
$\quad\quad 6x + 4 = 2y$

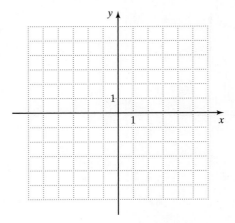

25. $\quad\quad\quad x = 5 + y$
$\quad -2x + 2y = -10$

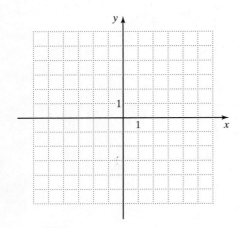

26. $x + \ y = 6$
$\quad 3y - 18 = -3x$

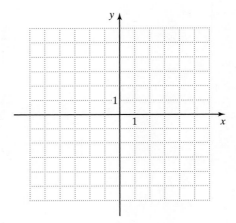

27. $x - y = -1$
$\quad x - y = 4$

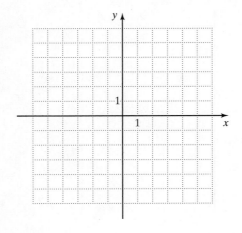

28. $x + 5y = 6$
$\quad x + 5y = 0$

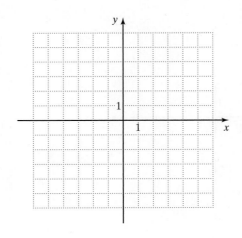

29. $3x + 2y = -10$
$5x - y = -8$

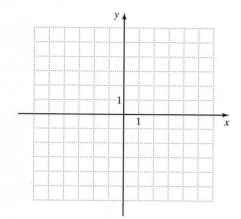

30. $-x - 2y = 8$
$-6x + 4y = 0$

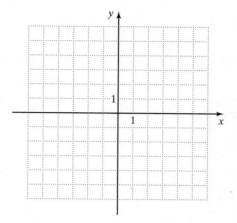

31. $-3x + y - 14 = 0$
$3x - y - 11 = 0$

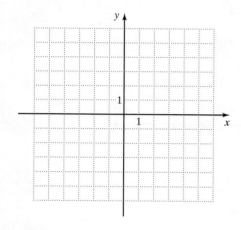

32. $7x - y - 2 = 0$
$14x - 2y - 12 = 0$

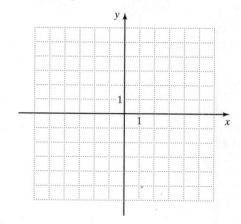

33. $10x + 2y = -6$
$y = 2$

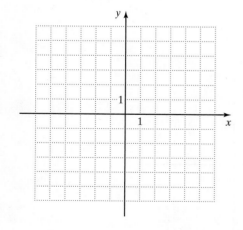

34. $x + 5y = -15$
$x = 5$

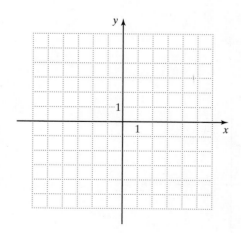

$$x = 0$$
$$x - 2y = 4$$

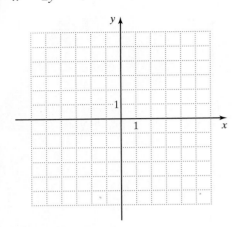

36. $y = -3$
$y = 2x + 3$

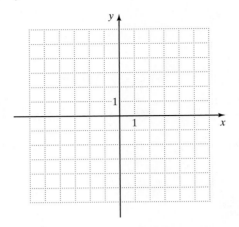

37. $4x - 4y = -8$

$$y = \frac{2}{3}x$$

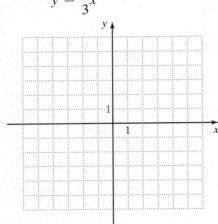

38. $3x + 2y = 6$

$$y = -\frac{3}{4}x$$

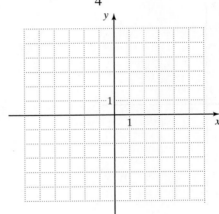

Mixed Practice

Solve by graphing.

39. $2y - 4x = -4$
$y = -2 + 2x$

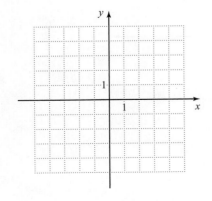

40. $4x + 2 = 2y$
$2x - y = 3$

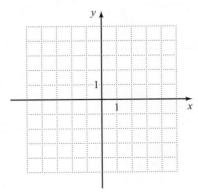

41. $y = -3 + \dfrac{1}{3}x$

$y = -2x + 4$

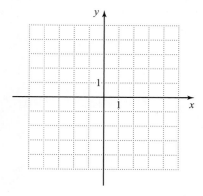

42. $-4 = 2x - y$

$x = 2y + 1$

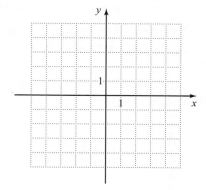

43. $3x - 2y = 2$

$4x + y = 10$

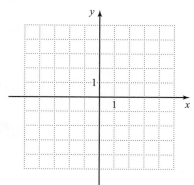

44. $x - 3y = -6$

$x = -3$

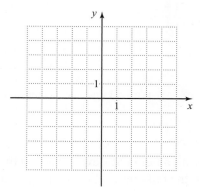

Solve.

45. Which system matches the given graph?

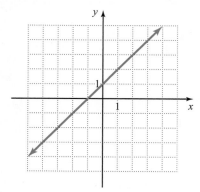

a. A system with no solution

b. A system with infinitely many solutions

c. A system with exactly one solution

46. Indicate whether each ordered pair is or is not a solution to the system.

$$x - 2y = -2$$
$$x - y = 4$$

a. $(-2, -6)$

b. $(12, 8)$

c. $(10, 6)$

Applications

D   *Solve.*

47. A young married couple had a combined annual income of $57,000.

     a. If the wife made $3000 more than the husband, write these relationships as a system of equations. Let *x* represent the husband's income and *y* the wife's income.

     b. Graph the equations.

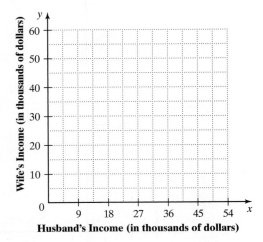

     c. Find the two incomes.

48. An airplane flying with a tailwind flew at a speed of 450 mph, relative to the ground. When flying against the tailwind, it flew at a speed of 350 mph.

     a. Express these relationships as equations, where *x* represents the speed of the airplane in calm air and *y* the speed of the wind.

     b. Graph these equations.

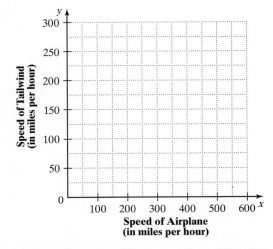

     c. Find the speed of the airplane in calm air and the speed of the wind.

49. Mike the plumber charges $75 for a house call and then $40 per hour for labor. Sally the plumber charges $100 for a house call and then $30 per hour for labor.

a. Write a cost equation for each plumber, where y is the total cost of plumbing repairs and x is the number of hours of labor.

b. Graph the two equations.

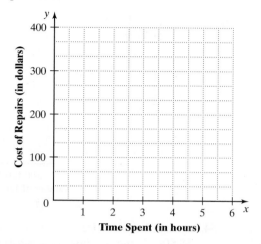

c. Determine the number of hours of plumbing repairs that would be required for the two plumbers to charge the same amount.

d. Determine from the graph which plumber charges less if the estimated amount of time to carry out the plumbing repairs is 5 hr.

50. To connect to the Web, customers must choose between two Internet service providers (ISPs). Flat ISP charges its customers a monthly flat fee of $10 regardless of how many hours they connect to the Web. A competing company, Variable ISP, charges $2.50 per month plus $0.25 for each hour of connection time.

a. Express each company's price structure p in terms of hours connected h.

b. Draw a graph that shows how each company's price structure relates to connection time.

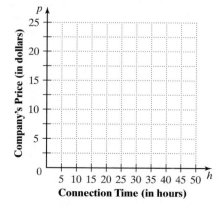

c. For which connection time do the two companies charge the same?

51. A small company duplicates DVDs. The cost of duplicating is $30 fixed overhead plus $0.25 per DVD duplicated. The company generates revenues of $1.50 per DVD. Use a graph to determine the break-even point for duplicating DVDs.

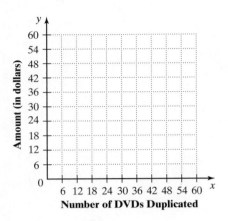

Number of DVDs Duplicated

52. A clothing company sells jackets for $140 per jacket. The company's fixed costs are $9000 and the variable costs are $50 per jacket. Use a graph to determine the break-even point for production.

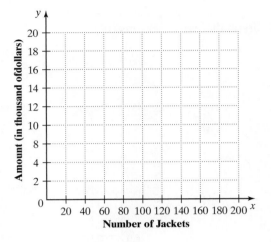

Number of Jackets

53. A movie fan rented 6 films at a local video store for one day. The daily rental charge was $2 on some films and $4 on others. If the total rental charge was $22, use a graph to determine how many $4 films were rented.

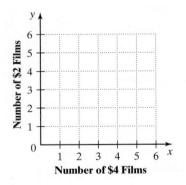

Number of $4 Films

54. An appliance store sells washer-dryer combos for $1500. If the washer costs $200 more than the dryer, use a graph to find the cost of each appliance.

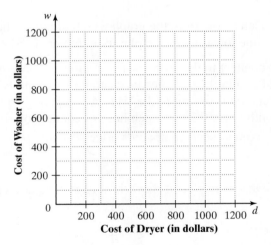

Cost of Dryer (in dollars)

55. Silver's Gym charges a $300 initiation fee plus $30 per month. DeLuxe Fitness Center has an initial charge of $400 but only charges $25 per month. Use a graph to determine for what number of months both health clubs will charge the same amount.

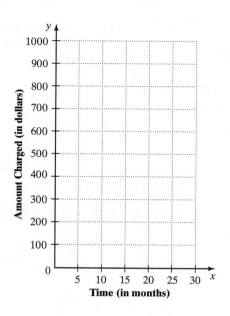

56. A plant nursery is selling a 7-foot specimen of a tree that grows about 1.5 ft/yr and a 6-foot specimen of a tree that grows 2 ft/yr. Use a graph to determine in how many years the two trees will be the same height.

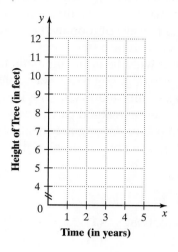

• Check your answers on page A-16.

MIND*Stretchers*

Critical Thinking

1. Write a system of linear equations that has $(2, 5)$ as its only solution.

Writing

2. Is it possible for a system of two linear equations to have exactly two solutions? If not, explain why.

Mathematical Reasoning

3. Not every system of equations is linear. For example the system

$$y = 2x$$
$$y = x^2$$

has the graph shown to the right. How many solutions does this system have? Explain how you know.

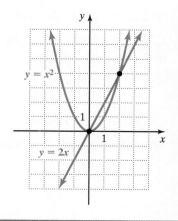

4.2 Solving Systems of Linear Equations by Substitution

OBJECTIVES

This section deals with solving a system of equations by the **substitution method**. As in the previous section, we restrict the discussion to systems of two linear equations in two variables. In applying the substitution method, we solve for one variable in terms of the other in one linear equation, getting a much simpler problem.

A To solve a system of linear equations by substitution

B To solve applied problems involving systems of linear equations

As compared with the graphing method, the substitution method has the advantage of being faster and also of giving us an exact solution. This consideration is especially important if the coordinates of the solution are not integral, or in an applied problem where precision counts. The substitution method particularly lends itself to solving systems in which a variable is isolated in one of the original equations.

To see how solving by substitution works, let's look at some examples.

EXAMPLE 1

Solve by substitution:

(1) $x + y = 10$
(2) $y = 2x + 4$

Solution Notice that Equation (2) is solved for y in terms of x. So we can substitute the expression $2x + 4$ from Equation (2) for y in Equation (1).

(1) $x + y = 10$

$x + (2x + 4) = 10$ Substitute $2x + 4$ for y.

We now have an equation that contains only one variable, namely x, and so is easy to solve.

$$x + (2x + 4) = 10$$
$$3x + 4 = 10$$
$$3x = 6$$
$$x = 2$$

Now that we have found the value of x, let's substitute it in either of the original equations in order to determine the corresponding value of y. Substituting 2 for x in Equation (2), we get:

(2) $y = 2x + 4$
$\quad = 2(2) + 4$
$\quad = 8$

So the solution to the system is $(2, 8)$, that is, $x = 2$ and $y = 8$.

Check We can check the solution by substituting 2 for x and 8 for y in the original equations.

(1) $x + y = 10$ (2) $y = 2x + 4$
$\quad 2 + 8 \overset{?}{=} 10$ $8 \overset{?}{=} 2 \cdot 2 + 4$
$\quad\quad 10 = 10$ True $8 = 8$ True

The check confirms the solution $(2, 8)$.

PRACTICE 1

Use the substitution method to solve.

(1) $x - y = 7$
(2) $x = -y + 1$

In Example 1, would we have gotten the same solution if we had solved for y in the *first* equation and then replaced y in the second equation?

The preceding example suggests the following procedure for solving systems of linear equations in two variables.

To Solve a System of Linear Equations by Substitution

- In one of the equations, solve for either variable in terms of the other variable.

- In the other equation, substitute the expression equal to the variable found in the previous step. Then, solve the resulting equation for the remaining variable.

- Substitute the value found in the previous step in either of the original equations and solve for the other variable.

- Check by substituting the values in both equations of the original system.

EXAMPLE 2

Use the substitution method to solve the following system:

$$\textbf{(1)} \qquad d = 3q - 8$$
$$\textbf{(2)} \quad 3d - 4q = 10$$

Solution Since d is isolated in Equation (1), we can substitute the expression $3q - 8$ for d in Equation (2).

$$\textbf{(2)} \qquad 3d - 4q = 10$$
$$3(3q - 8) - 4q = 10 \qquad \text{Substitute } 3q - 8 \text{ for } d.$$
$$9q - 24 - 4q = 10$$
$$5q - 24 = 10$$
$$5q = 34$$
$$q = \frac{34}{5}, \text{ or } 6.8$$

Now, let's solve for d by substituting 6.8 for q in Equation (1).

$$\textbf{(1)} \quad d = 3q - 8$$
$$= 3(6.8) - 8$$
$$= 20.4 - 8$$
$$= 12.4$$

The solution is $q = 6.8$ and $d = 12.4$. Note that neither value is an integer. So if we solve this system by the graphing method, we have to estimate the coordinates of a point.

Check We can confirm the solution by substituting these values in both of the original equations.

$$\textbf{(1)} \quad d = 3q - 8 \qquad\qquad \textbf{(2)} \qquad 3d - 4q = 10$$
$$12.4 \overset{?}{=} 3(6.8) - 8 \qquad\qquad 3(12.4) - 4(6.8) \overset{?}{=} 10$$
$$12.4 \overset{?}{=} 20.4 - 8 \qquad\qquad 37.2 - 27.2 \overset{?}{=} 10$$
$$12.4 = 12.4 \qquad \text{True} \qquad\qquad 10 = 10 \qquad \text{True}$$

So the solution is confirmed.

PRACTICE 2

Solve for m and n by substitution.

$$\textbf{(1)} \qquad m = -5n + 1$$
$$\textbf{(2)} \quad 2m + 3n = 7$$

In Example 2, would we have gotten the same solution to the system if the variables had been called x and y instead of q and d? Explain.

EXAMPLE 3

Solve the system by substitution.

$$(1) \quad 5x - 3y = 5$$
$$(2) \quad 2x - y = 1$$

Solution First, we need to solve for x or y in either of the equations. Let's solve for y in Equation (2) where the coefficient of y is -1.

$$(2) \quad 2x - y = 1$$
$$-y = -2x + 1$$
$$y = 2x - 1 \qquad \text{Divide each side by } -1.$$

Next, we substitute the expression $2x - 1$ for y in Equation (1) and solve for x.

$$(1) \qquad 5x - 3y = 5$$
$$5x - 3(2x - 1) = 5 \qquad \text{Substitute } 2x - 1 \text{ for } y.$$
$$5x - 6x + 3 = 5$$
$$-x + 3 = 5$$
$$-x = 2$$
$$x = -2$$

Finally, we solve for y by substituting -2 for x in Equation (2).

$$(2) \qquad 2x - y = 1$$
$$2(-2) - y = 1$$
$$-4 - y = 1$$
$$-y = 5$$
$$y = -5$$

So the solution is $(-2, -5)$. Check this in the original system.

EXAMPLE 4

Solve by substitution.
$$(1) \qquad x - 4y = 15$$
$$(2) \quad -2x + 8y = 5$$

Solution We begin by solving for either x or y in one of the equations. Let's solve for x in Equation (1) since the coefficient of x in this equation is 1.

$$(1) \quad x - 4y = 15$$
$$x = 4y + 15$$

Next, we substitute the expression $4y + 15$ for x in equation (2).

$$(2) \qquad -2x + 8y = 5$$
$$-2(4y + 15) + 8y = 5 \qquad \text{Substitute } 4y + 15 \text{ for } x.$$
$$-8y - 30 + 8y = 5$$
$$-30 = 5 \qquad \text{False}$$

Getting a false statement means that no value of y makes this last equation true. So the original system has no solution.

PRACTICE 3

Solve by substitution.

$$(1) \quad 2x - 7y = 7$$
$$(2) \quad 6x - y = 1$$

PRACTICE 4

Use the substitution method to solve the following system:

$$(1) \qquad 3x + y = 10$$
$$(2) \quad -6x - 2y = 1$$

What do you think the graph of the system in Example 4 looks like?

EXAMPLE 5

Solve by substitution.

$$(1) \quad 6x + 2y = 4$$
$$(2) \quad -y = 3x - 2$$

Solution Since the coefficient of y in Equation (2) is -1, let's solve this equation for y.

$$(2) \quad -y = 3x - 2$$
$$y = -3x + 2 \quad \text{Divide each side by } -1.$$

Now, we substitute $-3x + 2$ for y in Equation (1).

$$(1) \quad 6x + 2y = 4$$
$$6x + 2(-3x + 2) = 4 \quad \text{Substitute } -3x + 2 \text{ for } y.$$
$$6x - 6x + 4 = 4$$
$$4 = 4 \quad \text{True}$$

The original system has infinitely many solutions, since every x-value makes the last equation true. For each x-value, the corresponding y-value can be found by substituting it in either of the original equations and solving for y. Some solutions are $(0, 2)$ and $(1, -1)$.

PRACTICE 5

Solve for x and y.

$$(1) \quad y = -2x + 4$$
$$(2) \quad 10x + 5y = 20$$

In Example 5, what do you think the graph of the system looks like? Can you identify a particular solution to the system?

TIP When solving a system of linear equations in two variables by substitution:
- if we get a false statement, then the system has no solution.
- if we get a true statement, then the system has infinitely many solutions.

Now, let's use the substitution method to solve some applications.

EXAMPLE 6

A car rental agency has two plans:

- In the Ambassador Plan, renting a car for one day costs $35 plus $0.25 per mile driven.
- In the Diplomat Plan, a one-day car rental costs $50 plus $0.10 per mile driven.

a. For each plan, write a linear equation that relates a day's price p for renting a car to the number of miles driven n. Express the given information as a system of equations.

b. Use the substitution method to solve the system of linear equations.

c. In the context of this problem, what is the significance of the solution?

PRACTICE 6

To watch movies on premium channels, a couple decides to choose between two television cable deals:

- the TV Deal that costs $20 installation and $35 per month, and
- the Movie Deal that costs $30 installation and $25 per month.

a. Write an equation for each deal, expressing the cost of a deal c in terms of the number of months n for which the couple signs up.

EXAMPLE 6 (continued)

Solution

a. The Ambassador Plan can be expressed as $p = 0.25n + 35$; the Diplomat Plan becomes $p = 0.10n + 50$. The system representing both plans is therefore:

$$\textbf{(1)} \quad p = 0.25n + 35$$
$$\textbf{(2)} \quad p = 0.10n + 50$$

b. To solve the system, we can set the two expressions for p equal to each other.

$$0.25n + 35 = 0.10n + 50$$

Solving for n gives us:

$$0.25n + 35 = 0.10n + 50$$
$$0.15n = 15$$
$$n = \frac{15}{0.15}$$
$$n = 100$$

To solve for p, we can substitute 100 for n in Equation (1).

$$\textbf{(1)} \quad p = 0.25n + 35$$
$$= 0.25(\textbf{100}) + 35$$
$$= 25 + 35$$
$$= 60$$

So the solution to the system is $n = 100$ and $p = 60$.

c. With the appropriate units, the solution is $n = 100$ mi and $p = \$60$. This means that the cost of a one-day rental on the two plans is the same amount of money, namely \$60, only when the car is driven 100 mi. For other distances driven, the plans charge different amounts.

b. Solve the system of linear equations by substitution.

c. In the context of this problem, what is the significance of the solution?

Recall that we discussed mixture problems involving a single equation in Section 2.5. The following example shows how we can apply our knowledge of solving systems of linear equations to these problems.

EXAMPLE 7

How much 30% alcohol solution and 50% alcohol solution must be mixed to get 10 gal of 42% solution?

Solution   We solve this problem as we did earlier mixture problems, namely by organizing the given information in a table. Let's represent the amount of 30% solution by x and the amount of 50% solution by y.

Action	Percent of Alcohol	Amount of Solution (in gallons)	Amount of Alcohol (in gallons)
Start with	30%	x	$0.3x$
Add	50%	y	$0.5y$
Finish with	42%	10	0.42(10), or 4.2

PRACTICE 7

A chemist wishes to combine an alloy that is 20% copper with one that is 50% copper to obtain 15 oz of an alloy that is 25% copper. Find the quantities of the alloys required.

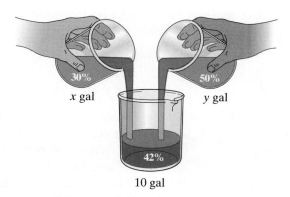

The amount of alcohol in the 30% solution is 30% of x, or $0.3x$. The amount of alcohol in the 50% solution is 50% of y, or $0.5y$. The amount of alcohol in the 42% solution is 42% of 10, or 4.2. Since the total amount of the combined solutions is 10 gal and the total amount of alcohol is 4.2 gal, we get the following system:

$$
\begin{aligned}
\textbf{(1)} \quad & x + y = 10 \\
\textbf{(2)} \quad & 0.3x + 0.5y = 4.2
\end{aligned}
$$

In applying the substitution method, we begin by solving for y in Equation (1).

$$
\begin{aligned}
\textbf{(1)} \quad x + y &= 10 \\
y &= -x + 10
\end{aligned}
$$

We then substitute $-x + 10$ for y in Equation (2).

$$
\begin{aligned}
\textbf{(2)} \qquad 0.3x + 0.5y &= 4.2 \\
0.3x + 0.5(-x + \mathbf{10}) &= 4.2 \\
0.3x - 0.5x + 5 &= 4.2 \\
-0.2x &= -0.8 \\
x &= \frac{-0.8}{-0.2} \\
x &= 4
\end{aligned}
$$

After replacing x by 4 in Equation (1), we solve for y.

$$
\begin{aligned}
\textbf{(1)} \quad x + y &= 10 \\
4 + y &= 10 \\
y &= 6
\end{aligned}
$$

So the solution to the system is $(4, 6)$. In other words, 4 gal of 30% solution and 6 gal of 50% solution are needed to produce 10 gal of the 42% solution.

A system of linear equations can also serve as a model for investment problems, as the following example illustrates:

EXAMPLE 8

A stockbroker had $10,000 to invest for her client. The broker invested part of this amount at a low-risk, low-yield 5% rate of return per year and the rest at a high-risk, high-yield 7% rate. If the client earned a return of $550 in one year, how much money did the broker invest at each rate?

Solution   The following table reflects the given information. Here, x stands for the amount of the investment at a 5% return, and y the investment at a 7% return.

Rate of Return	Amount of Investment ($)	Amount of Return ($)
5%	x	$0.05x$
7%	y	$0.07y$
TOTAL	10,000	550

The amount of return on each investment is the product of the rate of return and the amount of the investment. We add the amount of the individual investments to find the total investment, and the amount of returns on each investment to find the total amount of return. Since the total investment is $10,000 and the total return is $550, we get the following system:

$$\text{(1)} \qquad x + y = 10,000$$
$$\text{(2)} \quad 0.05x + 0.07y = 550$$

Now, let's solve for y in Equation (1).

$$\text{(1)} \quad x + y = 10,000$$
$$y = 10,000 - x$$

We then substitute $10,000 - x$ for y in Equation (2).

$$\text{(2)} \qquad 0.05x + 0.07y = 550$$
$$0.05x + 0.07(10,000 - x) = 550$$
$$0.05x + 700 - 0.07x = 550$$
$$-0.02x + 700 = 550$$
$$-0.02x = -150$$
$$x = \frac{-150}{-0.02}$$
$$x = 7500$$

After substituting 7500 for x in Equation (1), we solve for y.

$$\text{(1)} \qquad x + y = 10,000$$
$$7500 + y = 10,000$$
$$y = 2500$$

Therefore, the solution to the system is (7500, 2500). In other words, $7500 was invested at 5% and $2500 at 7%.

PRACTICE 8

During the recession of the early 1990s, the stock prices of fast-food companies suffered. Between January 5, 1990 and January 4, 1991, McDonald's stock fell approximately 15% and Wendy's fell about 77%. If a financier invested $120,000 in these two companies in January 1990, and ended up with stock worth only $77,200 in January 1991, how much did the financier invest in each company? (*Source:* google.com/finance)

A *Solve by substitution and check.*

1. $x + y = 10$
$y = 2x + 1$

2. $x - y = 7$
$x = 5y + 3$

3. $y = -3x - 15$
$y = -x - 7$

4. $y = -2x - 21$
$y = 5x$

5. $-x - y = 8$
$x = -3y$

6. $x - y = 15$
$y = -4x$

7. $4x + 2y = 10$
$x = 2$

8. $2y + x = 10$
$y = -5$

9. $-x + 20y = 0$
$x - \quad y = 0$

10. $5x + 3y = 0$
$x + \quad y = 0$

11. $6x + 4y = 2$
$2x + \quad y = 0$

12. $x - 3y = 0$
$2x - 3y = 6$

13. $3x + 5y = -12$
$x + 2y = -6$

14. $3x + 5y = -1$
$3x + \quad y = -5$

15. $m = 20 - 2n$
$2m + 4n = -22$

16. $x + 2y = 4$
$3x + 6y = 3$

17. $7x - 3y = 26$
$3x - \quad y = 11$

18. $x + 3y = 1$
$-3x - 5y = -2$

19. $8x + 2y = -1$
$y = -4x + 1$

20. $4x + 2y = 4$
$y = 5 - 2x$

21. $6x - 2y = 2$
$y = 3x - 1$

22. $2x - 6y = -12$
$x = 3y - 6$

23. $2x + 6y = 12$
$x + 3y = 6$

24. $3u + 6v = 60$
$-2u = 4v - 40$

25. $p + 2q = 13$
$q + 7 = 4p$

26. $a - b = -1$
$6b = 5a$

27. $s - 3t + 5 = 0$
$-4s + t - 9 = 0$

28. $2l - 3w + 6 = 0$
$l - w - 10 = 0$

Mixed Practice

Solve by substitution and check.

29. $y = 2x + 12$
$y = -3x - 3$

30. $2y - 8x = -2$
$y = 4x - 1$

31. $6x + 2y = 2$
$y = -3x + 2$

32. $-x - y = 12$
$x = -5y$

33. $5x + 3y = -6$
$-3x + y = 5$

34. $2y - 3x = 6$
$y - 3x = 0$

Applications

B *Solve.*

35. Two taxi companies compete in the same neighborhood. One of these companies charges $3 for the taxi drop plus $1.25 for each mile driven, while the other charges $2 for the taxi drop, plus $1.50 for each mile driven.

   a. Express these relationships as an algebraic system, where m represents miles driven and c represents cost.

   b. Solve the system and interpret the results.

36. Two electricians make house calls. One charges $75 for a visit plus $50 per hour of work. The other charges $95 per visit plus $40 per hour of work.

   a. Letting c represent cost and h represent hours worked, write a linear equation for each electrician that expresses the charge for a house call in terms of the length of a visit.

   b. For how many hours of work do the two electricians charge the same?

37. On a particular airline route, a full-price coach ticket costs $310 and a discounted coach ticket costs $210. On one of these flights, there were 172 passengers in coach, which resulted in a total ticket income of $44,120. How many full-price tickets were sold?

38. In 2010, the combined cost of two military aircraft—a Charger and a Fighter—was $260 million. A year later, the cost of a Charger increased by 10% and the cost of a Fighter decreased by 10%. If the combined cost in 2011 was $246 million, find the cost of each type of aircraft in 2010.

39. A laboratory technician needs to make a 10-liter batch of antiseptic that is 60% alcohol. How can she combine a batch of antiseptic that is 30% alcohol with another that is 70% to get the desired concentration?

40. A bottle of fruit juice contains 20% water. How much water must be added to this bottle to produce 8 L of fruit juice that is 50% water?

41. A corporation merged two departments into one. In one department, 5% of the employees were women, whereas in the other department, 80% were women. When the departments were merged, 50% of the 150 employees were women. How many women were in each department before the merger?

42. A hospital needs 30 L of a 10% solution of disinfectant. How many liters of a 20% solution and a 4% solution should be mixed to obtain this 10% solution?

43. A student took out two loans totaling $5000. She borrowed the maximum amount she could at 6% and the remainder at 7% interest per year. At the end of the first year, she owed $310 in interest. How much was loaned at each rate?

44. A man invested three times as much money in a bond fund that earned 8% in a year as he did in a mutual fund that returned 4% in the year. How much money did he invest in each fund if the total earnings for the year were $112?

45. A $40,000 investment was split so that part was invested at a 7% annual rate of interest and the rest at 9%. If the total annual earnings were $3140, how much money was invested at each rate?

46. A financial adviser counseled a client to invest $15,000, split between two stocks. At the end of one year, the investment in one stock increased in value by 4%, and the investment in the second stock increased in value by 8%. If the total increase in value of the investment was $1120, how much money was invested in each stock?

• Check your answers on page A-18.

MIND Stretchers

Groupwork

1. Cramer's Rule is a formula that can be used to solve a system of linear equations for x and y.
   Consider the following system:

   $$ax + by = c$$
   $$dx + ey = f$$

   The formula states that $x = \dfrac{ce - bf}{ae - bd}$ and $y = \dfrac{af - cd}{ae - bd}$. Note that this mechanical approach allows machines to solve systems of equations.

   a. Working with a partner, make up your own values for a, b, c, d, e, and f, and substitute these values in the system.

   $$\underline{\qquad}x + \underline{\qquad}y = \underline{\qquad}$$
   $$\underline{\qquad}x + \underline{\qquad}y = \underline{\qquad}$$

   b. Use Cramer's Rule to calculate x and y.

   $$x = \frac{ce - bf}{ae - bd} = \underline{\qquad}$$

   $$y = \frac{af - cd}{ae - bd} = \underline{\qquad}$$

   c. By substitution, check whether (x, y) is in fact a solution to the system.

Mathematical Reasoning

2. On the coordinate plane, consider the quadrilateral $ABCD$ shown.
   At what point do the diagonals $\overline{AC}$ and $\overline{BD}$ intersect? Explain how to find the answer exactly.

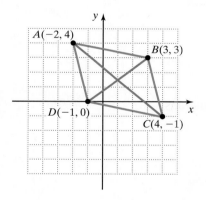

Critical Thinking

3. For what value of k will the system shown have infinitely many solutions?

   $$kx - 2y = 10$$
   $$4x - y = 5$$

4.3 Solving Systems of Linear Equations by Elimination

The final method of solving a system of linear equations that we consider is called the **elimination** (or **addition**) **method**. As in the substitution method discussed in the previous section, the elimination method involves changing a given system of two equations in two variables to one equation in one variable. A system in which the coefficients of a variable are opposites particularly lends itself to the elimination method.

Recall that in solving linear equations in one variable, we frequently used the addition property of equality: the equations $a = b$ and $a + c = b + c$ are equivalent. That is, if we add the same number to both sides of an equation, we get an equivalent equation.

The elimination method for solving *systems* is based on a closely related property of equality: If $a = b$ and $c = d$, then $a + c = b + d$. This property allows us to "add equations."

Let's look at examples of applying this property in the elimination method.

EXAMPLE 1

Solve the following system by the elimination method.

$$\textbf{(1)} \quad x + y = 4$$
$$\textbf{(2)} \quad x - y = 2$$

Solution First, we decide which variable to eliminate. Since the coefficients of the y-terms in the two equations are opposites (namely $+1$ and -1), the y-terms are eliminated if we add the equations.

$$\textbf{(1)} \quad x + y = 4$$
$$\textbf{(2)} \quad \underline{x - y = 2}$$
$$2x + 0y = 6$$
$$2x = 6$$
$$x = 3$$

Next, to find y, we substitute 3 for x in either of the original equations. Substituting in Equation (1) we get:

$$\textbf{(1)} \quad x + y = 4$$
$$3 + y = 4$$
$$y = 4 - 3$$
$$y = 1$$

So $x = 3$ and $y = 1$. That is, the solution is $(3, 1)$.

Check We check the solution by substituting these values for x and y in both of the original equations.

$$\textbf{(1)} \quad x + y = 4 \qquad\qquad \textbf{(2)} \quad x - y = 2$$
$$3 + 1 \overset{?}{=} 4 \qquad\qquad\qquad 3 - 1 \overset{?}{=} 2$$
$$4 = 4 \quad \text{True} \qquad\qquad\qquad 2 = 2 \quad \text{True}$$

So our solution $(3, 1)$ is confirmed.

PRACTICE 1

Solve for x and y.

$$\textbf{(1)} \quad x + y = 6$$
$$\textbf{(2)} \quad x - y = -10$$

EXAMPLE 2

Use the elimination method to solve for x and y.

$$(1) \quad 3x + 2y = 14$$
$$(2) \quad 5x + 2y = -8$$

Solution In this system, adding the two given equations does not eliminate either variable. However, note that the two y-terms have the same coefficient, namely 2. So if we multiply Equation (1) by -1 and then add the equations, the y-terms will cancel out.

$(1) \quad 3x + 2y = 14 \xrightarrow{\text{Multiply by } -1.} -3x - 2y = -14$

$(2) \quad 5x + 2y = -8 \qquad\qquad\qquad \underline{5x + 2y = -8} \qquad \text{Add the equations.}$

$$2x \qquad\quad = -22$$
$$x = -11$$

Next, we substitute -11 for x in either of the original equations. Let's choose Equation (1), and then solve for y.

$$(1) \qquad\quad 3x + 2y = 14$$
$$3(-11) + 2y = 14$$
$$-33 + 2y = 14$$
$$2y = 14 + 33$$
$$2y = 47$$
$$y = \frac{47}{2} = 23.5$$

So the solution is $(-11, 23.5)$.

PRACTICE 2

Solve the following system using the elimination method:

$$(1) \quad 4x + 3y = -7$$
$$(2) \quad 5x + 3y = -5$$

EXAMPLE 3

Solve.

$$(1) \quad 3x - y = -2$$
$$(2) \quad x + 5y = 10$$

Solution Note that the coefficient of x in Equation (2) is $+1$. By multiplying this equation by -3 and then adding the two equations, we eliminate the x-terms.

$(1) \quad 3x - y = -2 \qquad\qquad\qquad 3x - y = -2$

$(2) \quad x + 5y = 10 \xrightarrow{\text{Multiply by } -3.} \underline{-3x - 15y = -30}$

$$-16y = -32 \qquad \text{Add the equations.}$$
$$y = 2$$

Next, we substitute 2 for y in Equation (2) and then solve for x.

$$(2) \qquad x + 5y = 10$$
$$x + 5(2) = 10$$
$$x + 10 = 10$$
$$x = 0$$

So the solution is $(0, 2)$.

PRACTICE 3

Solve for x and y.

$$(1) \quad x - 3y = -18$$
$$(2) \quad 5x + 2y = 12$$

How could we have solved the system in Example 3 another way?

EXAMPLE 4

Use the elimination method to solve the following system of linear equations:

$$(1) \quad 4x + 3y = -19$$
$$(2) \quad 3x - 2y = -10$$

Solution This system is more complicated to solve than the previous examples because there is no single integer that we can multiply either equation by that will eliminate a variable when we add the equations. Instead, we must multiply *both* equations by integers that lead to the elimination of a variable. There are a number of possible strategies to accomplish this. We can, for instance, multiply Equation (1) by 2 and Equation (2) by 3 to eliminate the y-terms when the equations are added.

$$(1) \quad 4x + 3y = -19 \xrightarrow{\text{Multiply by 2.}} \quad 8x + 6y = -38$$
$$(2) \quad 3x - 2y = -10 \xrightarrow{\text{Multiply by 3.}} \quad \underline{9x - 6y = -30}$$
$$\phantom{(2) \quad 3x - 2y = -10 \xrightarrow{\text{Multiply by 3.}} \quad} 17x = -68 \quad \text{Add the equations.}$$
$$\phantom{(2) \quad 3x - 2y = -10 \xrightarrow{\text{Multiply by 3.}} \quad 17x +} x = -4$$

Now, let's substitute -4 for x in Equation (1), and then solve for y.

$$(1) \qquad 4x + 3y = -19$$
$$4(-4) + 3y = -19$$
$$-16 + 3y = -19$$
$$3y = 16 + (-19)$$
$$3y = -3$$
$$y = -1$$

So the solution is $(-4, -1)$.

PRACTICE 4

Solve by elimination.

$$(1) \quad 5x - 7y = 24$$
$$(2) \quad 3x - 5y = 16$$

How could we have solved the system in Example 4 by eliminating x instead of y?

To Solve a System of Linear Equations by Elimination

- Write both equations in the general form $Ax + By = C$.
- Choose the variable that you want to eliminate.
- If necessary, multiply one or both equations by appropriate numbers so that the coefficients of the variable to be eliminated are opposites.
- Add the equations. Then, solve the resulting equation for the remaining variable.
- Substitute the value found in the previous step in either of the original equations and solve for the other variable.
- Check by substituting the values in both equations of the original system.

EXAMPLE 5

Solve by elimination.

$$(1) \qquad 5x = 3y$$
$$(2) \quad -3x + 2y = 9$$

Solution Equation (1) is not in the form $Ax + By = C$, so let's begin by rewriting it in general form.

$$(1) \qquad 5x = 3y \xrightarrow{\text{Write in general form.}} 5x - 3y = 0$$
$$(2) \quad -3x + 2y = 9 \qquad\qquad\qquad -3x + 2y = 9$$

Now, suppose we choose to eliminate the x-terms. To do this, we can multiply Equation (1) by 3, Equation (2) by 5, and then add the equations.

$$(1) \quad 5x - 3y = 0 \xrightarrow{\text{Multiply by 3.}} 15x - 9y = 0$$
$$(2) \quad -3x + 2y = 9 \xrightarrow{\text{Multiply by 5.}} \underline{-15x + 10y = 45} \quad \text{Add the equations.}$$
$$y = 45$$

To solve for x, let's substitute 45 for y in Equation (1).

$$(1) \quad 5x = 3y$$
$$5x = 3(45)$$
$$5x = 135$$
$$x = 27$$

So the solution is $(27, 45)$.

EXAMPLE 6

Solve by elimination.

$$(1) \quad 4x - 6y + 12 = 0$$
$$(2) \quad 2x - 3y = -4$$

Solution We begin by writing Equation (1) in general form.

$$(1) \quad 4x - 6y + 12 = 0 \xrightarrow{\text{Write in general form.}} 4x - 6y = -12$$
$$(2) \quad 2x - 3y = -4 \qquad\qquad\qquad 2x - 3y = -4$$

Now, let's eliminate the y-terms. To do this, we can multiply Equation (2) by -2, and then add the equations.

$$(1) \quad 4x - 6y = -12 \qquad\qquad 4x - 6y = -12$$
$$(2) \quad 2x - 3y = -4 \xrightarrow{\text{Multiply by } -2.} \underline{-4x + 6y = 8}$$
$$0 = -4 \quad \text{False}$$

Since adding the equations yields a false statement, the original system has no solution.

Let's use our knowledge of the elimination method to solve some applied problems, beginning with a motion problem on the following page.

PRACTICE 5

Use the elimination method to solve the following system:

$$(1) \quad -2x + 5y = 20$$
$$(2) \qquad\quad 3x = 7y - 26$$

PRACTICE 6

Solve:

$$3x = 4 + y$$
$$9x - 3y = 12$$

EXAMPLE 7

It takes a plane 3 hr to fly between two air-ports, traveling with a tailwind at a ground speed of 500 mph. The plane then takes 4 hr to make the return trip against the same wind. What is the speed of the plane in still air? What is the speed of the wind?

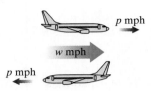

Solution   Let p represent the speed of the plane in still air and w represent the speed of the wind. On the initial flight, the wind is with the plane so its speed relative to the ground is $p + w$. When returning, the wind is against the plane so its ground speed is $p - w$. We can organize the given information in the following table:

	Ground speed	· Time =	Distance
Going	$p + w$	3	$3(p + w)$
Returning	$p - w$	4	$4(p - w)$

Note that since the distance the plane travels is the product of its ground speed and the time it travels, we can compute each entry in the distance column of the table by multiplying the corresponding entries in the ground speed and the time columns.

     Since we are told that the speed going is 500 mph, we have:

$$p + w = 500$$

But the distance going and the distance returning are equal, so

$$3(p + w) = 4(p - w)$$

Now, we have a system of two equations, which we must solve.

$$
\begin{aligned}
\textbf{(1)} \quad & p + w = 500 \\
\textbf{(2)} \quad & 3(p + w) = 4(p - w)
\end{aligned}
$$

We can write Equation (2) in general form, by simplifying.

$$
\begin{aligned}
3(p + w) &= 4(p - w) \\
3p + 3w &= 4p - 4w \\
3p - 4p + 3w + 4w &= 0 \\
-p + 7w &= 0
\end{aligned}
$$

The system then becomes:

$$
\begin{aligned}
\textbf{(1)} \quad & p + w = 500 \\
\textbf{(2)} \quad & -p + 7w = 0
\end{aligned}
$$

Adding the equations eliminates the p-terms.

$$
\begin{aligned}
8w &= 500 \\
w &= 62.5
\end{aligned}
$$

Finally, let's substitute 62.5 for w in Equation (1) and solve for p.

$$
\begin{aligned}
\textbf{(1)} \quad p + 62.5 &= 500 \\
p &= 437.5
\end{aligned}
$$

So the speed of the plane in still air is 437.5 mph, and the wind speed is 62.5 mph.

PRACTICE 7

A whale swimming with the current traveled 80 mi in 2 hr. Swimming against the current, the whale traveled only 40 mi in the same amount of time. Find the whale's speed in calm water and the speed of the current.

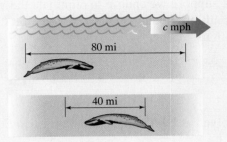

EXAMPLE 8

A student had two part-time jobs in a restaurant. One week she earned a total of $306, working 12 hr as a cashier and 10 hr as a cook. The next week, she worked 14 hr as a cashier and 22 hr cooking, earning $512. What is her hourly wage as a cashier? As a cook?

Solution   Let x represent the student's hourly wage as a cashier and y represent the student's hourly wage as a cook. The first week, the student earned $306, and the second week, $512. So we must solve the following system:

$$\textbf{(1)} \quad 12x + 10y = 306$$
$$\textbf{(2)} \quad 14x + 22y = 512$$

We can divide each equation by 2 to simplify.

$$\textbf{(1)} \quad 6x + 5y = 153$$
$$\textbf{(2)} \quad 7x + 11y = 256$$

Let's eliminate the x-terms by multiplying Equation (1) by 7 and Equation (2) by -6. Then, we add the equations.

$$\textbf{(1)} \quad 6x + 5y = 153 \xrightarrow{\text{Multiply by 7.}} 42x + 35y = 1071$$
$$\textbf{(2)} \quad 7x + 11y = 256 \xrightarrow{\text{Multiply by } -6.} -42x - 66y = -1536$$

Adding the equations eliminates the x-terms.

$$-31y = -465$$
$$y = \frac{-465}{-31}$$
$$y = 15$$

Finally, we substitute 15 for y in the original Equation (1) and solve for x.

$$\textbf{(1)} \quad 12x + 10y = 306$$
$$12x + 10(\mathbf{15}) = 306$$
$$12x + 150 = 306$$
$$12x = 156$$
$$x = \frac{156}{12}$$
$$x = 13$$

So the student earned $13 per hour as a cashier and $15 per hour as a cook.

PRACTICE 8

Admission prices at a high school football game were $10 for adults and $6 for students. The total value of the 175 tickets sold was $1450. How many adults and how many students attended the game?

In this chapter, we have discussed three methods of solving a system of linear equations—the graphing method, the substitution method, and the elimination (or addition) method. To help in deciding which method to apply in a given problem, listed in the table below are some advantages and disadvantages of each method.

Method	Advantages	Disadvantages
Graphing Method	• Provides a picture that makes relationships understandable.	• Approximates solutions, particularly when they are not integers or are large. • Can be time consuming if not using a grapher.
Substitution Method	• Gives exact solutions. • Is easy to use when a variable in one of the original equations is isolated.	• No picture. • Can result in complicated equations with parentheses or fractions.
Elimination Method	• Gives exact solutions. • Is easy to use when the two coefficients of a variable are opposites.	• No picture.

A *Solve.*

1. $x + y = 3$
$x - y = 7$

2. $x - y = 10$
$x + y = -8$

3. $x + y = -4$
$-x + 3y = -6$

4. $5x - y = 8$
$2x + y = -1$

5. $10p - q = -14$
$-4p + q = -4$

6. $a + b = -4$
$-a + 2b = -8$

7. $3x + y = -3$
$4x + y = -4$

8. $x + 4y = -3$
$x - 7y = 19$

9. $3x + 5y = 10$
$3x + 5y = -5$

10. $8x + 2y = 3$
$4x + y = -9$

11. $9x + 6y = -15$
$-3x - 2y = 5$

12. $4x + y = -3$
$8x + 2y = -6$

13. $5x + 2y = -9$
$-5x + 2y = 11$

14. $7x + 4y = -6$
$-x + 4y = 10$

15. $2s + d = -2$
$5s + 3d = -6$

16. $-5x + 8y = -7$
$-6x + 9y = -9$

17. $3x - 5y = 1$
$7x - 8y = 17$

18. $3x + 2y = 9$
$-2x + 3y = -19$

19. $5x + 2y = -1$
$4x - 5y = -14$

20. $10x - 3y = 9$
$3x - 2y = -5$

21. $7p + 3q = 15$
$-5p - 7q = 16$

22. $8a + 2b = 18$
$4a - 3b = -15$

23. $6x + 5y = -8.5$
$8x + 10y = -3$

24. $6x - 6y = -3.6$
$-4x + 8y = -16$

25. $3.5x + 5y = -3$
$2x = -2y$

26. $3x - 3y = 0$
$1.5y = -6x + 30$

27. $2x - 4 = -y$
$x + 2y = 0$

28. $y = -3x + 7$
$4x + 2y = 11$

29. $8x + 10y = 1$
$-4x - 5y + 6 = 0$

30. $x - y = 6$
$3x = 3y + 10$

Mixed Practice

Solve

31. $4a - b = -10$
$-3a + b = 7$

32. $9x - 6y = 3$
$3x - 2y = 6$

33. $3x - 5y = 4$
$-6x + 10y = -8$

34. $7p + 4q = -12$
$4p + q = -3$

35. $5x + 3y = -3$
$-7x - 5y = 4$

36. $4x + 3y = -5.5$
$5x + 6y = -3.5$

Applications

B *Solve.*

37. A quarterback throws a pass that travels 40 yd with the wind in 2.5 sec. If he had thrown the same pass against the wind, the football would have traveled 20 yd in 2 sec. Find the speed of a pass that the quarterback would throw if there were no wind.

38. A crew team rows in a river with a current. When the team rows with the current, the boat travels 14 mi in 2 hr. Against the current, the team rows 6 mi in the same amount of time. At what speed does the team row in still water?

39. To enter a zoo, adult visitors must pay $5, whereas children and seniors pay only half price. On one day, the zoo collected a total of $765. If the zoo had 223 visitors that day, how many half-price admissions and how many full-price admissions did the zoo collect?

40. The molecular weight of a compound such as water or hydrogen peroxide is found by adding the atomic weights of all the atoms in the compound. A water molecule consists of two hydrogen atoms and one oxygen atom, with a molecular weight of approximately 18. A hydrogen peroxide molecule consists of two hydrogen atoms and two oxygen atoms, with a molecular weight of about 34. What is the atomic weight of hydrogen? Of oxygen? (*Source:* John T. Moore, *Chemistry Made Simple*)

41. The British Parliament consists of two houses—the House of Lords (consisting of peers who are appointed) and the House of Commons (consisting of Members of Parliament, or MPs, who are elected). In November 2010, there were a total of 1388 peers and MPs, with 88 more peers than MPs. How many MPs and how many peers were there? (*Source:* www.parliament.uk)

42. Compact discs are stored in single jewel cases, which are 0.375 in. thick, and in multiple-CD jewel cases, which have a thickness of 0.875 in. If 86 of these jewel cases exactly fit on the storage shelf shown, how many of the jewel cases are single?

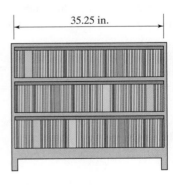

43. A particular computer takes 43 nanoseconds to carry out 5 sums and 7 products, and 42 nanoseconds to perform 2 sums and 9 products. How long does the computer take to carry out one sum? To carry out one product?

44. A wholesale novelty shop sells some embroidered scarves for $12 each and others for $15 each. A customer pays $234 for seventeen scarves. How many scarves at each price did she buy?

45. One issue of a journal has 3 full-page ads and 5 half-page ads, generating $6075 in advertising revenue. The next issue has 4 full-page ads and 4 half-page ads, resulting in advertising revenue of $6380. Determine the advertising rates in this journal for full-page and half-page ads.

46. According to a law of physics, the lever shown will balance when the products of each weight and the length of its force arm are equal.

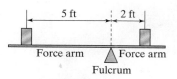

The weights shown above balance. If 10 lb are added to the left weight, which is then moved 1 ft closer to the fulcrum, the lever will again balance. Find the two original weights.

• Check your answers on page A-18.

MIND*Stretchers*

Groupwork

1. Your friend performs the following magic trick: She asks you to think of two numbers but not to tell her what they are. Instead, you tell her the sum and the difference of the two numbers. She promptly tells you what the two original numbers were. Explain how your friend does the trick. Working with partners, try to perform this trick.

Writing

2. Consider the following system of equations:

$$\textbf{(1)} \qquad 5x - 8y = 4$$
$$\textbf{(2)} \qquad 12x + 24y = 11$$

Explain how the concept of LCM relates to solving this system by elimination.

Mathematical Reasoning

3. The elimination method can be extended to three linear equations in three variables. Solve the following system:

$$4x - y - 3z = 30$$
$$3x - 2y - 6z = -5$$
$$x - z = 5$$

Cultural Note

arl Friedrich Gauss (1777–1855) is generally considered to be one of the greatest mathematicians in history. Among his many mathematical contributions was the *Gaussian elimination method* of solving systems of linear equations, discussed in this chapter. Gauss (rhymes with "house") is credited with being the first to prove the major mathematical result known as the Fundamental Theorem of Algebra. He was also an important scientist. In astronomy, he laid the theoretical foundation for predicting a planet's orbit. To honor this German's groundbreaking work in physics, his name is given to the unit (*gauss*) used today to express the strength of a magnetic field. Also, Gauss' portrait appears on the German ten-mark bill, as shown to the left.

(*Sources:* Roger Cooke, *The History of Mathematics*, John Wiley, 1997; D. E. Smith, *History of Mathematics*, Dover Publications, 1923)

Key Concepts and Skills

Concept/Skill	Description	Example
[4.1] System of equations	Two or more equations considered simultaneously, that is, together.	$x + y = 7$ $x - y = 1$
[4.1] Solution of a system of linear equations in two variables	An ordered pair of numbers that makes both equations in the system true.	Is $(4, 3)$ a solution of the following system? $x + y = 7 \rightarrow 4 + 3 \stackrel{?}{=} 7$ True $x - y = 1 \rightarrow 4 - 3 \stackrel{?}{=} 1$ True Yes, $(4, 3)$ is a solution of the system.
[4.1] To solve a system of linear equations by graphing	• Graph both equations on the same coordinate plane. • There are three possibilities: **a.** If the lines intersect, then the solution is the ordered pair of coordinates for the point of intersection. Check that these coordinates satisfy both equations. **b.** If the lines are parallel, then there is no solution of the system. **c.** If the lines coincide, then there are infinitely many solutions, namely all the ordered pairs of coordinates that represent points on the line.	$x + y = 7$ $x - y = 3$ $3x - y = 2$ $3x - y = 4$ $x - y = 3$ $2x - 2y = 6$
[4.2] To solve a system of linear equations by substitution	• In one of the equations, solve for either variable in terms of the other variable.	**(1)** $2x + 3y = 2$ **(2)** $\quad x + 4y = 6$ **(2)** $\quad x + 4y = 6$ $\qquad\qquad x = -4y + 6$

continued

Concept/Skill	**Description**	**Example**
	• In the other equation, substitute the expression equal to the variable found in the previous step. Then solve the resulting equation for the remaining variable. • Substitute the value found in the previous step in either of the original equations and solve for the other variable. • Check by substituting the values in both equations of the original system.	**(1)** $\quad\quad 2x + 3y = 2$ $\quad 2(-4y + 6) + 3y = 2$ $\quad\quad -8y + 12 + 3y = 2$ $\quad\quad\quad -5y + 12 = 2$ $\quad\quad\quad\quad -5y = -10$ $\quad\quad\quad\quad\quad y = 2$ **(2)** $\quad\quad x + 4y = 6$ $\quad\quad x + 4(2) = 6$ $\quad\quad x + 8 = 6$ $\quad\quad x = -2$ **Check** **(1)** $\quad\quad\quad 2x + 3y = 2$ $\quad 2(-2) + 3(2) \overset{?}{=} 2$ $\quad\quad\quad -4 + 6 \overset{?}{=} 2$ $\quad\quad\quad\quad 2 = 2 \quad$ True **(2)** $\quad\quad\quad x + 4y = 6$ $\quad\quad -2 + 4(2) \overset{?}{=} 6$ $\quad\quad -2 + 8 \overset{?}{=} 6$ $\quad\quad\quad 6 = 6 \quad$ True The solution is $(-2, 2)$.
[4.3] To solve a system of linear equations by elimination	• Write both equations in the general form $Ax + By = C$. • Choose the variable that you want to eliminate. • If necessary, multiply one or both equations by appropriate numbers so that the coefficients of the variable to be eliminated are opposites. • Add the equations. Then, solve the resulting equation for the remaining variable. • Substitute the value found in the previous step in either of the original equations and solve for the other variable. • Check by substituting the values in both equations of the original system.	**(1)** $3x - 2y = -5$ **(2)** $2x - 4y = 2$ Multiply Equation (1) by 2. $\quad 6x - 4y = -10$ Multiply Equation (2) by -1. $\quad \underline{-2x + 4y = -2}$ $\quad\quad\quad\quad\quad\quad 4x \quad\quad = -12$ $\quad\quad\quad\quad\quad\quad\quad\quad x = -3$ **(2)** $\quad\quad\quad 2x - 4y = 2$ $\quad\quad 2(-3) - 4y = 2$ $\quad\quad\quad -6 - 4y = 2$ $\quad\quad\quad\quad -4y = 8$ $\quad\quad\quad\quad\quad y = -2$ **Check** **(1)** $\quad\quad\quad 3x - 2y = -5$ $\quad 3(-3) - 2(-2) \overset{?}{=} -5$ $\quad\quad\quad -9 + 4 \overset{?}{=} -5$ $\quad\quad\quad\quad -5 = -5 \quad$ True **(2)** $\quad\quad\quad 2x - 4y = 2$ $\quad 2(-3) - 4(-2) \overset{?}{=} 2$ $\quad\quad -6 + 8 \overset{?}{=} 2$ $\quad\quad\quad 2 = 2 \quad$ True The solution is $(-3, -2)$.

Say Why

Fill in each blank.

1. The ordered pair $(-1, 2)$ _____ a solution of the
   is/is not

   system of equations $\begin{array}{l} 3x + 4y = 5 \\ -x + 2y = 3 \end{array}$ because _____

   _____

   _____.

2. If the graph of a system of equations consists of parallel
   lines, then the system _____ any solutions
   has/does not have

   because _____

   _____.

3. If the graph of a system of equations consists of exactly
   one line, then the system _____ infinitely
   has/does not have

   many solutions because _____

   _____.

4. In solving the system of equations $\begin{array}{l} 2x - y = 7 \\ -2x + 3y = 10 \end{array}$

   adding the equations _____ eliminate the
   does/does not

   x-terms because _____

   _____.

[4.1]

5. Consider the following system:

$$x + 2y = -4$$
$$3x - y = 3$$

   Is $(2, -3)$ a solution of the system?

6. For each of the systems graphed, determine the number of solutions.

   a.

   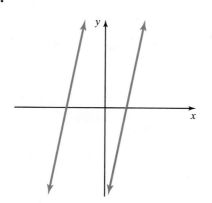

   b.

   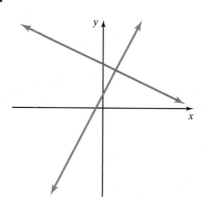

   c.

   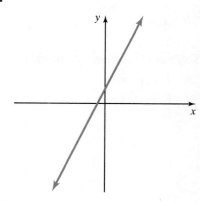

7. Match each system with the appropriate graph.

   a. A system with solution $(2, 1)$

   b. A system with solution $(1, 2)$

   c. A system with infinitely many solutions

   d. A system with no solution

I

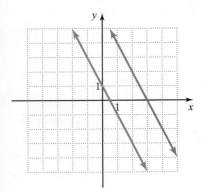

II

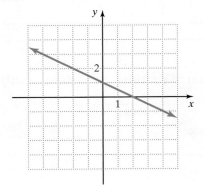

III

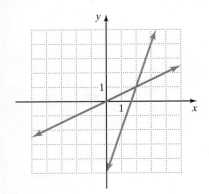

IV

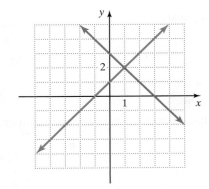

Solve each system by graphing.

8. $x + y = 6$

    $x - y = -4$

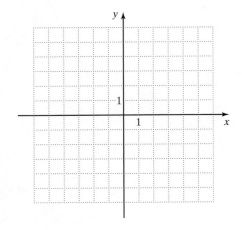

9. $y = 2x$

    $6x - 3y = 3$

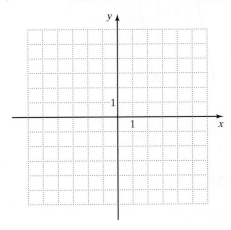

10.
$$2y = -8x + 2$$
$$-4x - y = -1$$

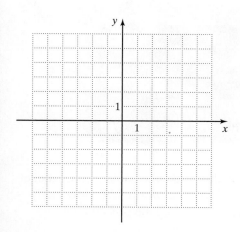

11. $x - 3y = -15$
$$y - x = 5$$

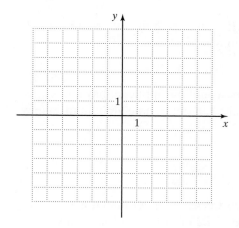

[4.2] *Solve each system by substitution.*

12. $x + y = 3$
$$y = 2x + 6$$

13. $a = 3b - 4$
$$a + 4b = 10$$

14. $x - 3y = 1$
$$-2x + 6y = 7$$

15. $10x + 2y = 14$
$$-y = 5x - 7$$

[4.3] *Solve each system by elimination.*

16. $x + y = 1$
$$x - y = 7$$

17. $2x + 3y = 8$
$$4x + 6y = 16$$

18. $4x = 9 - 5y$
$$2x + 3y = 3$$

19. $3x + 2y = -4$
$$4x - 3y = 23$$

Mixed Applications

Solve.

20. A student starts a typing service. He buys a computer and a printer for $1750, and then charges $5.50 per page for typing. Expenses for ink, paper, and electricity amount to $0.50 per page.

   a. Write the given information as a system of equations.

   b. How many pages must the student type to break even?

21. A job applicant must choose between two sales positions. One position pays $10/hr, and the other $8/hr plus a base pay of $50/wk.

    a. Express the weekly salaries of the two positions as a system of equations.

    b. How many hours would the applicant have to work per week in order to earn the same amount of money at each position?

22. The movie screen shown has a perimeter of 332 ft. If the length is 26 ft more than the width, find the area of the screen.

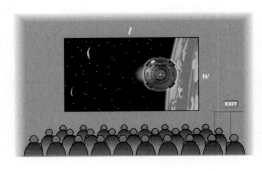

23. A doubles tennis court is 42 ft longer than it is wide. If the court's perimeter is 228 ft, find its dimensions.

24. The coin box of a vending machine contains only nickels and dimes. There are 350 coins worth $25. How many nickels and how many dimes did the box contain?

25. Two trains start 500 mi apart and speed toward each other. The difference between the average speeds of the trains is 5 mph. If they pass one another after 4 hr, find the rate of each train.

26. During a season, a college basketball team scored 2437 points on a combination of three-point and two-point baskets. If the team made 1085 baskets, how many two-point baskets and how many three-point baskets did the team make?

27. A pharmacist has 10% and 30% alcohol solutions in stock. To prepare 200 mL of a 25% solution, how much of each should the pharmacist mix?

28. A farmer is preparing an insecticide by mixing a 50% solution with water. How much of this solution and how much water are needed to fill a 2000-liter tank with a 35% solution?

29. Last year, a financial adviser recommended that her client invest part of his $50,000 in secure municipal bonds that paid 6% and the rest in corporate stocks that paid 8%. How much money did the client put into each type of investment if the total annual return was $3200?

30. An investor split $10,000 between a high-risk mutual fund and a low-risk mutual fund. Last year, the high-risk fund paid 12% and the low-risk fund paid 2%, for a total of $900. How much money was invested in each fund?

31. Two airplanes leave an airport at the same time, one flying 100 mph faster than the other. The planes travel in opposite directions, and after 2 hr they are 1800 mi apart. Determine the speed of the slower plane.

32. Flying with the wind, a bird flew 13 mi in half an hour. On the return trip against the wind, it was able to travel only 8 mi in the same amount of time. Find the speed of the bird in still air and the speed of the wind.

33. The U.S. Senate has 100 members. After debating the merits of a treaty, all the senators voted, and 14 more voted for the treaty than against. None of the senators abstained.

   a. How many senators voted *for* the treaty?

   b. How many senators voted *against* the treaty?

34. In a chemistry lab, a piece of copper starting at 2°C is heated at the rate of 3° per minute. At the same time, a piece of iron starting at 86°C is being cooled at the rate of 4° per minute.

   a. After how much time will the two metals be at the same temperature?

   b. After how much time will the iron be 14° colder than the copper?

• Check your answers on page A-18.

FOR
EXTRA
HELP

CHAPTER
Test Prep
VIDEOS

The Chapter Test Prep Videos with test solutions are available on DVD, in MyMathLab, and on You Tube (search "AkstIntroductory Alg" and click on "Channels").

To see if you have mastered the topics in this chapter, take this test.

1. Indicate which ordered pair is a solution of the system:

$$x + y = -1$$
$$3x - y = 1$$

a. $(0, -1)$

b. $(-1, 0)$

c. $(2, -3)$

2. How many solutions does the graphed system appear to have?

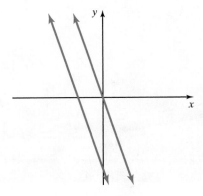

Solve each system by graphing.

3. $x - y = 3$
   $x + y = 3$

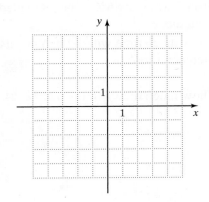

4. $5 = 2x - y$
   $y = 2x$

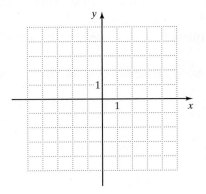

5. $2x + 3y = 4$
   $3x - y = -5$

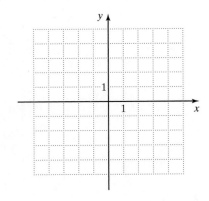

6. $6y = 4 - 2x$
   $x + 3y = 2$

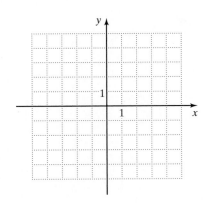

Solve each system by substitution.

7. $x = 3y - 7$
$\quad y = x + 5$

8. $3x - 5y = -12$
$\quad x + 2y = 7$

9. $u - 3v = -12$
$\quad 5u + v = 8$

Solve each system by elimination.

10. $4x + y = 3$
$\quad 7x - y = 19$

11. $\quad x - y = 5$
$\quad 2x - 2y = 5$

12. $-5p + 2q = 1$
$\quad 4p + 3q = 1.5$

Solve each system.

13. $7x \quad\quad = 2y$
$\quad -4x + y = 1$

14. $\quad\quad 4l = -(m + 3)$
$\quad 8l + 2m = -6$

15. $5x + 2y = -1$
$\quad x - 1 = y$

16. $\quad 5x + 3y - 9 = 0$
$\quad 2x - 7y - 20 = 0$

Solve.

17. In a local election, the ratio of votes for the winning candidate to the losing candidate was 2 to 1. If 6306 votes were cast in the election, how many votes did the winning candidate get?

18. Nutritional information for 3-oz servings of turkey and of salmon is given in the following table:

	Turkey, light meat	Salmon
Amount of Fat (in grams)	· 3	3
Number of Calories	135	99

How many servings of turkey and of salmon would it take to get 9 g of fat and 333 cal?

19. A 20% iodine solution is mixed with a 60% iodine solution to produce 4 gal of a 50% iodine solution. How many gallons of each solution are needed?

20. A small airplane traveled 170 mph with a tailwind and 130 mph with a headwind. Find the speed of the wind and the speed of the airplane in still air.

• Check your answers on page A-19.

Cumulative Review Exercises

To help you review, solve the following.

1. True or false: $-6 < -5$

2. Calculate: $-4 \div 2 + 3\,(-1)(8)$

3. Evaluate $\dfrac{2}{3}(x + y)^2$ if $x = 4$ and $y = -7$.

4. Simplify: $2(4x - 1) - 3(y + x)$

5. Is 5 a solution of the equation $3p + 1 = 9 - p$?

6. Solve: $5(x + 1) - (x - 2) = x - 2$

7. Solve and graph: $3x - 7 < 4(x - 2)$

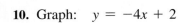

8. Find the slope of the following line:

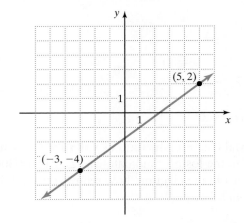

9. Find the slope and *y*-intercept of the graph whose equation is $3x + 6y = 12$.

10. Graph: $y = -4x + 2$

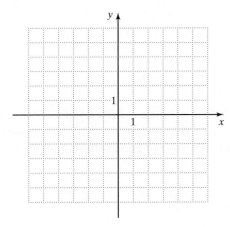

11. Graph the following linear inequality: $4y - x \leq 8$

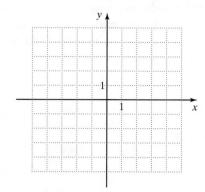

12. Solve by graphing.

$$y = 2x + 3$$
$$y = -\frac{1}{2}x - 2$$

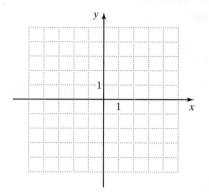

13. Solve by substitution and check.

$$3a - 4b = 1$$
$$a - 2b = -1$$

14. Solve by elimination.

$$3x + 2y = -10$$
$$-3x + 2y = 14$$

15. In 2009, the amount of Social Security tax withheld from a paycheck was 6.2%. The amount of Medicare tax withheld was 1.45%. For a 2009 gross income of g, how much was withheld for both Social Security and Medicare? (*Source:* irs.gov)

16. In renting a car for a day, you must choose between two local car agencies. One agency charges $35 per day plus $0.20 per mile and the other charges $50 per day plus $0.15 per mile. Under what circumstances do the two agencies charge the same amount for a one-day rental?

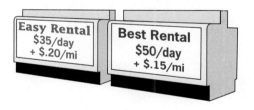

17. The maximum speed of a supersonic airplane S in miles per hour is commonly represented by its Mach number M where

$$M = \frac{S}{740}.$$

What is the maximum speed of an airplane flying at Mach 2.1?

18. An average adult male has about 5 L of blood, consisting of plasma and cells. About 90% of the plasma is water, which accounts for half of the total amount of blood. How much blood, to the nearest liter, is plasma, and how much is cells? (*Source:* americasblood.org)

19. The following table displays the approximate number of medical doctors in the United States for various years.

Year	1980	1990	2000	2007
Number of Doctors (in hundreds of thousands)	5	6	8	9

(*Source:* American Medical Association, *Physician Characteristics and Distribution in the U.S., 2010*)

Plot this information in the coordinate plane below.

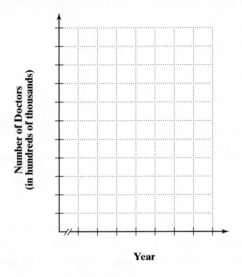

20. The state of Colorado is shaped like a rectangle, with borders that are latitude and longitude lines. The north-south border is about 100 mi shorter than the east-west border, and the perimeter is approximately 1320 mi. Estimate the area of the state. (*Source:* www.census.gov)

• Check your answers on page A-19.

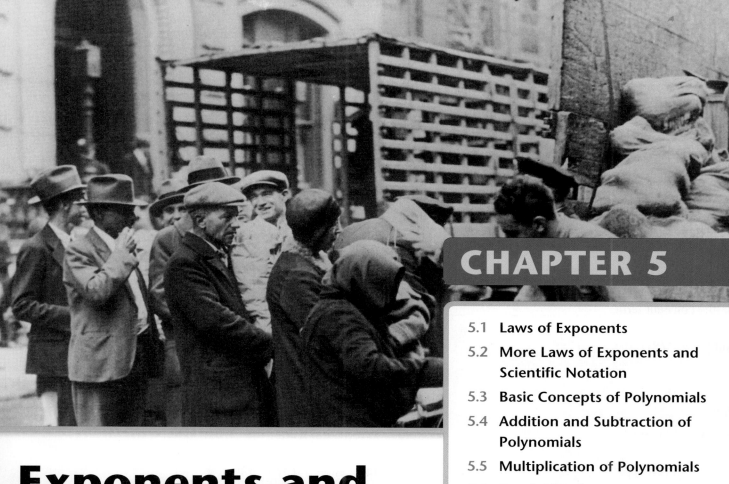

CHAPTER 5

Exponents and Polynomials

Layoffs and Polynomials

Layoffs are common whenever the economy contracts. For instance, in 1933, the worst year of the Great Depression, so many American workers were laid off that one-fourth of the U.S. labor force was unemployed.

Suppose that 100 employees work in a factory. If the fraction f of these employees are laid off, the factory will still employ $100(1 - f)$, or $100 - 100f$, workers. If the factory again lays off the same fraction of employees, the remaining number of workers can be represented by $100(1 - f)(1 - f)$, an expression that can also be written as the *polynomial* $100 - 200f + 100f^2$.

(*Source:* Michael Parkin, *Economics*, Addison-Wesley, 1999)

To see if you have already mastered the topics in this chapter, take this test.

Simplify.

1. $x^5 \cdot x^4$

2. $y^7 \div y^3$

3. $-3a^0$

4. $(4x^4y^3)^2$

5. $\left(\dfrac{a}{b^5}\right)^3$

6. $(5x^{-1}y^4)^{-2}$

7. For the polynomial $6x^4 + 5x^3 + x^2 - 7x + 8$, identify:

   a. the terms _____

   b. the coefficients _____

   c. the degree _____

   d. the constant term _____

8. Find the sum: $(2n^2 + 7n - 10) + (n^2 - 6n + 12)$

9. Subtract: $(8x^2 - 9) - (7x^2 - x - 5)$

10. Combine: $(6a^2b + ab - a^2) + (2a^2 + 3a^2b - 5b^2) - (7a^2 - 2ab - b^2)$

Multiply.

11. $3x^2(x^2 - 4x + 9)$

12. $(n + 3)(2n^2 + n - 6)$

13. $(4x + 9)(x - 3)$

14. $(3y - 7)(3y + 7)$

15. $(5 - 2n)^2$

Divide.

16. $\dfrac{9t^4 - 18t^3 - 45t^2}{9t^2}$

17. $(4x^2 - 3x - 10) \div (x - 2)$

Solve.

18. In chemistry, a mole (mol) of any material contains 6×10^{23} molecules. Express in scientific notation the number of molecules that 200 mol of hydrogen will contain. (*Source:* Karen Timberlake, *Chemistry: An Introduction to General Organic and Biological Chemistry*)

19. A student deposits $100 in an account at interest rate r (in decimal form) compounded annually. The amount in the account after 2 years is given by:

$$A = 100(1 + r)^2$$

Write A as a polynomial in r without parentheses.

20. The polynomial $0.08x + 6.07$ approximates the world population (in billions) for a particular year, where x represents the number of years after 2000. According to this model, find the world population in 2030 to the nearest billion. (*Source:* census.gov)

• Check your answers on page A-19.

5.1 Laws of Exponents

What Exponents Are and Why They Are Important

In Chapter 1, we considered algebraic expressions. In this chapter, we discuss a particular kind of expression called a *polynomial* and operations involving polynomials. Central to our discussion of these operations is a thorough understanding of the laws of exponents.

Exponents play an important role in arithmetic and algebra. Powers of 10 are key to the decimal place-value system that underlies the reading, writing, and computation of real numbers. There are many other applications of exponents in the sciences and in business, such as compound interest, population growth, and scientific notation as used in astronomy and physics.

Ⓐ To simplify an expression with exponents, including the exponent zero or one

Ⓑ To simplify an expression by using the product or quotient rule of exponents

Ⓒ To simplify an expression with negative exponents

Ⓓ To solve applied problems involving exponents

Exponents

Recall the definition of an exponent from Section 1.6: An exponent (or power) is a number that indicates how many times another number (called the *base*) is used as a factor. For instance,

$$\overset{\text{Exponent}}{\underset{\text{Base}\quad\text{3 factors of 5}}{5^3 = \underbrace{5 \cdot 5 \cdot 5}}}$$

In general, if a is a positive integer, the expression x^a means

$$x^a = \underbrace{x \cdot x \cdot x \cdot \ \cdots \ \cdot x}_{a \text{ factors}},$$

where the exponent a indicates that there are a factors of x.

It follows from the definition of an exponent that raising a base to the power 1 has 1 factor of the base.

Exponent 1

For any real number x, $x^1 = x$.

In words, any number raised to the power 1 is the number itself. For example, $7^1 = 7$, $(-5)^1 = -5$, and $x^1 = x$.

EXAMPLE 1

Multiply.

a. 2^5 **b.** $\left(-\dfrac{1}{3}\right)^4$ **c.** $(-x)^1$ **d.** $(-x)^2$

Solution

a. $2^5 = 2 \cdot 2 \cdot 2 \cdot 2 \cdot 2 = 32$

b. $\left(-\dfrac{1}{3}\right)^4 = \left(-\dfrac{1}{3}\right)\left(-\dfrac{1}{3}\right)\left(-\dfrac{1}{3}\right)\left(-\dfrac{1}{3}\right) = \dfrac{1}{81}$

c. $(-x)^1 = -x$

d. $(-x)^2 = (-x)(-x) = x^2$

PRACTICE 1

Multiply.

a. 10^4

b. $\left(-\dfrac{1}{2}\right)^5$

c. $(-y)^6$

d. $(-y)^1$

The expression 10^0 involves raising a number to the 0 power. To understand the value of this expression, consider the following pattern:

$$10^5 = 10 \cdot 10 \cdot 10 \cdot 10 \cdot 10 = 100,000$$
$$10^4 = 10 \cdot 10 \cdot 10 \cdot 10 = 10,000$$
$$10^3 = 10 \cdot 10 \cdot 10 = 1000$$
$$10^2 = 10 \cdot 10 = 100$$
$$10^1 = 10$$
$$10^0 = ?$$

Note that to go from 10^5 to 10^4 we divide the first number by 10. The pattern continues, and going from 10^2 to 10^1, we divide 10^2 by 10. So to go from 10^1 to 10^0, it seems reasonable for us to divide 10^1 by 10, which is 1. Thus, we take 10^0 to be equal to 1. Would we have gotten the same result had we considered powers of 2 instead of powers of 10 in order to determine the value of 2^0?

> **Exponent 0**
>
> For any nonzero real number x, $x^0 = 1$.

In words, any nonzero number raised to the power 0 is equal to 1. For example, $8^0 = 1$ and $(-100)^0 = 1$. Note that the expression 0^0 is undefined. Throughout the remainder of this text, we will assume that any variable raised to the 0 power represents a nonzero number.

EXAMPLE 2

Simplify.

a. 25^0 **b.** $(-3.5)^0$ **c.** a^0

d. $-a^0$ **e.** $4x^0$

Solution

a. $25^0 = 1$ **b.** $(-3.5)^0 = 1$ **c.** $a^0 = 1$

d. $-a^0 = -1 \cdot a^0 = -1 \cdot 1 = -1$

e. $4x^0 = 4 \cdot 1 = 4$

PRACTICE 2

Simplify.

a. 8^0

b. $\left(-\dfrac{2}{3}\right)^0$

c. y^0

d. $-y^0$

e. $(4x)^0$

Laws of Exponents

The laws of exponents are rules that apply to exponents. These rules are not arbitrary but follow logically from the definition of exponent.

Let's first discuss the *product rule*. This rule applies when we multiply two powers with the same base. Consider the expression $x^4 \cdot x^6$. Using the definition of an exponent we get:

$$x^4 \cdot x^6 = \underbrace{\overbrace{(x \cdot x \cdot x \cdot x)}^{4 \text{ factors}} \overbrace{(x \cdot x \cdot x \cdot x \cdot x \cdot x)}^{6 \text{ factors}}}_{10 \text{ factors}} = x^{10}$$

Since 4 factors of x and 6 additional factors of x make 10 factors of x, it follows that

$$x^4 \cdot x^6 = x^{4+6} = x^{10}.$$

This result can be generalized as follows:

Product Rule of Exponents

For any nonzero real number x and for any nonnegative integers a and b,

$$x^a \cdot x^b = x^{a+b}.$$

In words, to multiply powers of the same base, add the exponents and keep the base the same.

EXAMPLE 3

Simplify using the product rule, if possible.

a. $2^3 \cdot 2^5$ **b.** $(-5)^2 \cdot (-5)^8$ **c.** $x^8 \cdot x^{10}$

d. $m^4 \cdot m$ **e.** $x^2 \cdot y^3$

Solution When we multiply powers of the same base, the product rule tells us to add the exponents but *not to change the base.*

a. $2^3 \cdot 2^5 = 2^{3+5} = 2^8$

b. $(-5)^2 \cdot (-5)^8 = (-5)^{2+8} = (-5)^{10}$

c. $x^8 \cdot x^{10} = x^{8+10} = x^{18}$

d. $m^4 \cdot m = m^4 \cdot m^1 = m^{4+1} = m^5$

e. In the expression $x^2 \cdot y^3$, we cannot apply the product rule because the bases are not the same.

PRACTICE 3

Simplify using the product rule, if possible.

a. $10^8 \cdot 10^4$

b. $(-4)^3 \cdot (-4)^3$

c. $n^3 \cdot n^7$

d. $y^5 \cdot y^0$

e. $a \cdot b^4$

Now, we discuss another law of exponents—the *quotient rule.* This rule applies when we divide two powers of the same base. Consider the expression $\dfrac{x^6}{x^2}$. Using the definition of exponent, we get:

$$\frac{x^6}{x^2} = \frac{\overbrace{x \cdot x \cdot x \cdot x \cdot \overset{1}{\cancel{x}} \cdot \overset{1}{\cancel{x}}}^{6 \text{ factors}}}{\underset{2 \text{ factors}}{\underset{1 \quad\ 1}{\cancel{x} \cdot \cancel{x}}}} = \underset{4 \text{ factors}}{\underline{x \cdot x \cdot x \cdot x}} = x^4$$

So

$$\frac{x^6}{x^2} = x^{6-2} = x^4,$$

which suggests the following rule:

Quotient Rule of Exponents

For any nonzero real number x and for any positive integers a and b,

$$\frac{x^a}{x^b} = x^{a-b}.$$

In words, to divide powers of the same base, subtract the exponent in the denominator from the exponent in the numerator, and keep the base the same.

EXAMPLE 4

Simplify using the quotient rule, if possible.

a. $\dfrac{10^5}{10^2}$ **b.** $(-2)^5 \div (-2)^5$ **c.** $\dfrac{p^{12}}{p^7}$

d. $\dfrac{y^{13}}{y}$ **e.** $\dfrac{x^4}{y^2}$

Solution   When we divide powers of the same base, the quotient rule tells us to subtract the exponents but *not to change the base*.

a. $\dfrac{10^5}{10^2} = 10^{5-2} = 10^3$

b. $(-2)^5 \div (-2)^5 = \dfrac{(-2)^5}{(-2)^5} = (-2)^{5-5} = (-2)^0 = 1$

c. $\dfrac{p^{12}}{p^7} = p^{12-7} = p^5$

d. $\dfrac{y^{13}}{y} = \dfrac{y^{13}}{y^1} = y^{13-1} = y^{12}$

e. In the expression $\dfrac{x^4}{y^2}$, we cannot apply the quotient rule because the bases are not the same.

PRACTICE 4

Simplify using the quotient rule, if possible.

a. $\dfrac{7^7}{7^2}$

b. $(-9)^6 \div (-9)^5$

c. $\dfrac{s^{10}}{s^{10}}$

d. $\dfrac{r^8}{r}$

e. $\dfrac{a^5}{b^3}$

Note in Example 4(b) that the quotient rule confirms the fact that any nonzero real number raised to the 0 power is 1. That is,

$$\frac{(-2)^5}{(-2)^5} = \frac{-32}{-32} = 1 \quad \text{and} \quad \frac{(-2)^5}{(-2)^5} = (-2)^{5-5} = (-2)^0 = 1$$

EXAMPLE 5

Simplify.

a. $x^3 \cdot x \cdot x^5$ **b.** $(a^2 b)(ab^4)$ **c.** $\dfrac{t^3 \cdot t^5}{t^2}$

Solution

a. $x^3 \cdot x \cdot x^5 = x^9$ Use the product rule.

b. $(a^2 b)(ab^4) = a^2 \cdot a^1 \cdot b^1 \cdot b^4$ Rearrange the factors.

              $= a^3 b^5$ Use the product rule.

c. $\dfrac{t^3 \cdot t^5}{t^2} = \dfrac{t^8}{t^2} = t^6$ Use the product rule in the numerator. Then, use the quotient rule.

PRACTICE 5

Simplify.

a. $y^2 \cdot y^3 \cdot y^4$

b. $(x^3 y^3)(x^2 y^3)$

c. $\dfrac{a^7}{a \cdot a^4}$

Negative Exponents

Until now, we have considered only exponents that were either positive integers or 0. What meaning should we give to *negative exponents*?

The quotient rule is the key to answering this question. Consider, for instance, the quotient $\dfrac{6^4}{6^7}$. On the one hand, we can simplify this fraction by using the definition of exponent and canceling the common factors, getting:

$$\frac{6^4}{6^7} = \frac{\overset{1}{\cancel{6}} \cdot \overset{1}{\cancel{6}} \cdot \overset{1}{\cancel{6}} \cdot \overset{1}{\cancel{6}}}{\underset{1}{\cancel{6}} \cdot \underset{1}{\cancel{6}} \cdot \underset{1}{\cancel{6}} \cdot \underset{1}{\cancel{6}} \cdot 6 \cdot 6 \cdot 6} = \frac{1}{6^3}$$

On the other hand, we can simplify by using the quotient rule, getting:

$$\frac{6^4}{6^7} = 6^{4-7} = 6^{-3}$$

Since

$$\frac{6^4}{6^7} = 6^{-3} \qquad \text{and} \qquad \frac{6^4}{6^7} = \frac{1}{6^3},$$

we conclude that $6^{-3} = \dfrac{1}{6^3}$, which suggests the following rule:

Negative Exponent

For any nonzero real number x and for any integer a,

$$x^{-a} = \frac{1}{x^a}.$$

In words, to evaluate x^{-a}, take the reciprocal of x^{-a} and change the sign of the exponent.

In general, an expression with exponents is considered *simplified* when it is written with only positive exponents.

EXAMPLE 6

Simplify.

a. 5^{-2} **b.** p^{-8} **c.** $-(8x)^{-1}$ **d.** $(-4)^{-2}$

Solution

a. $5^{-2} = \dfrac{1}{5^2} = \dfrac{1}{25}$ **b.** $p^{-8} = \dfrac{1}{p^8}$

c. $-(8x)^{-1} = -1(8x)^{-1} = \dfrac{-1}{(8x)^1} = -\dfrac{1}{8x}$

d. $(-4)^{-2} = \dfrac{1}{(-4)^2} = \dfrac{1}{16}$

PRACTICE 6

Simplify.

a. 9^{-2}

b. n^{-5}

c. $-(3y)^{-1}$

d. $(5)^{-3}$

TIP A negative exponent indicates a reciprocal. For example, $5^{-2} = \dfrac{1}{25}$.

The product rule of exponents and the quotient rule of exponents, which were defined for positive-integer exponents, also hold for negative-integer exponents.

EXAMPLE 7

Simplify by writing each expression using only positive exponents.

a. $2^{-1}q$ **b.** $5x^{-2}$ **c.** $\dfrac{y^6}{y^{10}}$

d. $4^{-1} \cdot x^{-5} \cdot x^2$ **e.** $\dfrac{1}{x^{-2}}$

Solution

a. $2^{-1}q = \dfrac{1}{2} \cdot q = \dfrac{q}{2}$

b. $5x^{-2} = 5 \cdot \dfrac{1}{x^2} = \dfrac{5}{x^2}$

c. $\dfrac{y^6}{y^{10}} = y^{6-10} = y^{-4} = \dfrac{1}{y^4}$

d. $4^{-1} \cdot x^{-5} \cdot x^2 = 4^{-1} \cdot x^{-5+2} = \dfrac{1}{4}x^{-3} = \dfrac{1}{4} \cdot \dfrac{1}{x^3} = \dfrac{1}{4x^3}$

e. $\dfrac{1}{x^{-2}} = 1 \div x^{-2} = 1 \div \dfrac{1}{x^2} = 1 \cdot \dfrac{x^2}{1} = x^2$

PRACTICE 7

Write as expressions using only positive exponents.

a. $8^{-1}s$

b. $3x^{-1}$

c. $\dfrac{r^3}{r^9}$

d. $3^2 \cdot g^{-1} \cdot g^{-4}$

e. $\dfrac{1}{x^{-3}}$

The definition of a negative exponent and Example 7(e), in which we saw that $\dfrac{1}{x^{-2}} = x^2$, suggest the following:

> ### Reciprocal of x^{-a}
> For any nonzero real number x and any integer a,
> $$\dfrac{1}{x^{-a}} = x^a.$$

In words, the reciprocal of x^{-a} is x^a.

EXAMPLE 8

Write as an expression using positive exponents.

a. $\dfrac{5}{x^{-4}}$ **b.** $\dfrac{1}{3y^{-1}}$ **c.** $\dfrac{a^2}{b^{-3}}$

Solution

a. $\dfrac{5}{x^{-4}} = 5 \cdot \dfrac{1}{x^{-4}} = 5 \cdot x^4 = 5x^4$

b. $\dfrac{1}{3y^{-1}} = \dfrac{1}{3} \cdot \dfrac{1}{y^{-1}} = \dfrac{1}{3} \cdot y^1 = \dfrac{y}{3}$

c. $\dfrac{a^2}{b^{-3}} = a^2 \cdot \dfrac{1}{b^{-3}} = a^2 \cdot b^3 = a^2b^3$

PRACTICE 8

Write as an expression using positive exponents.

a. $\dfrac{1}{a^{-3}}$

b. $\dfrac{2}{5x^{-2}}$

c. $\dfrac{r^3}{2s^{-1}}$

EXAMPLE 9

Physicists study different kinds of electromagnetic waves, including radio waves, X-rays, and gamma rays. The diagram below, called the *electromagnetic spectrum*, shows the relationship among the wavelengths of these waves.

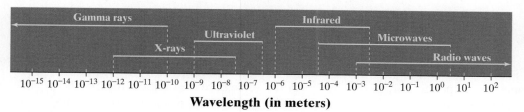

Wavelength (in meters)

Consider a particular X-ray whose wavelength is 10^{-10} m and a gamma ray whose wavelength is 10^{-15} m. (*Source:* Arthur Beiser, *The Mainstream of Physics*)

a. Which of these rays has a greater length?

b. What is the ratio of the longer to the shorter wavelength?

Solution

a. The wavelengths are 10^{-10} m for the X-ray and 10^{-15} m for the gamma ray. Converting these expressions with negative exponents to equivalent expressions with positive exponents, the wavelengths can be written as $\frac{1}{10^{10}}$ m and $\frac{1}{10^{15}}$ m, respectively. Since $\frac{1}{10^{10}}$ has the smaller denominator, it is the larger number.

Therefore, the X-ray has the greater wavelength.

b. We use the quotient rule to compute the ratio of the greater wavelength, 10^{-10} m, to the shorter wavelength, 10^{-15} m:

$$\frac{10^{-10}}{10^{-15}} = \frac{10^{15}}{10^{10}} = 10^{15-10} = 10^5$$

So the ratio is $\frac{10^5}{1}$, or 10^5 to 1.

PRACTICE 9

A computer's memory is often measured in *bits*, *bytes*, and *megabytes*. A bit (short for "binary digit") is the smallest unit of data in the memory of the computer. A byte is equal to 2^3 bits, whereas a megabyte (MB) is equal to 2^{20} bytes.

a. How many bits are in a megabyte? Write the answer as a power of 2.

b. In most computers, the hard drive's capacity is expressed in *giga-bytes*. A gigabyte (GB) is 2^{10} megabytes. How many bytes are there in a gigabyte? Express the result as a power of 2.

Mathematically Speaking

Fill in each blank with the most appropriate term or phrase from the given list.

the opposite of	added	power of 1
multiplied	0 power	subtracted
1 divided by	divided	

1. The product rule of exponents states that when powers of the same base are multiplied, the exponents are _____ and the base is left the same.

2. The quotient rule of exponents states that when powers of the same base are divided, the exponents are _____ and the base is left the same.

3. Any nonzero real number raised to the _____ is 1.

4. For any nonzero real number x and for any integer a, x^{-a} can be written as _____ x^a.

A *Multiply.*

5. 5^3

6. 1^8

7. $\left(\dfrac{3}{4}\right)^2$

8. $\left(\dfrac{7}{8}\right)^2$

9. $(0.4)^2$

10. $(0.2)^2$

11. $-(0.5)^2$

12. $-(0.3)^2$

13. $(-2)^3$

14. $(-4)^3$

15. $(-3)^4$

16. $(-6)^4$

17. $\left(-\dfrac{1}{2}\right)^3$

18. $\left(-\dfrac{4}{5}\right)^3$

19. $(-x)^4$

20. $(-y)^3$

Simplify.

21. $(pq)^1$

22. $(xy)^1$

23. $(-3)^0$

24. $(-4)^0$

25. $-a^0$

26. $-b^0$

B *Simplify using the product rule, if possible.*

27. $10^9 \cdot 10^2$

28. $5^2 \cdot 5^3$

29. $4^3 \cdot 4^5$

30. $6^2 \cdot 6^8$

31. $a^4 \cdot a^2$

32. $x \cdot x^8$

33. $x^3 \cdot y^5$

34. $a^4 \cdot b^5$

35. $n^6 \cdot n$

36. $y \cdot y^0$

37. $x^2 y$

38. $a^4 b^3$

Simplify using the quotient rule, if possible.

39. $\dfrac{8^5}{8^3}$

40. $\dfrac{3^8}{3^2}$

41. $\dfrac{5^4}{5}$

42. $\dfrac{2^9}{2^3}$

43. $\dfrac{y^6}{y^5}$

44. $\dfrac{x^{12}}{x^{12}}$

45. $\dfrac{a^{10}}{a^4}$

46. $\dfrac{x^7}{x^5}$

47. $\dfrac{y^8}{x^4}$

48. $\dfrac{a^5}{b}$

49. $\dfrac{x^6}{x^6}$

50. $r^4 \div r^0$

Simplify.

51. $y^2 \cdot y^3 \cdot y$

52. $t \cdot t \cdot t^2$

53. $(p^2 q^3)(p^5 q^2)$

54. $(xy^2)(x^2 y^6)$

55. $(yx^2)(xz^2)(yz)$

56. $a(a^4 b^2)(bc^3)$

57. $\dfrac{a^2 \cdot a^3}{a^4}$

58. $\dfrac{t^4}{t^3 \cdot t}$

59. $\dfrac{x^2 \cdot x^4}{x^3 \cdot x}$

60. $\dfrac{y^5 \cdot y^5}{y^2 \cdot y^3}$

Write as an expression using only positive exponents.

61. 5^{-1}

62. 7^{-1}

63. x^{-1}

64. a^{-1}

65. $(-3a)^{-1}$

66. $-(5y)^{-1}$

67. 2^{-4}

68. 7^{-3}

69. -3^{-4}

70. -5^{-2}

71. $8n^{-3}$

72. $2y^{-3}$

73. $(-x)^{-2}$

74. $(-a)^{-4}$

75. $-3^{-2}x$

76. $-4^{-1}y$

77. $x^{-2} y^3$

78. xy^{-3}

79. qr^{-1}

80. rs^{-1}

81. $4x^{-1} y^2$

82. $-5a^2 b^{-4}$

83. $p^{-2} \cdot p^{-3}$

84. $t^{-3} \cdot t^{-3}$

85. $p^{-1} \cdot p^4$

86. $s^4 \cdot s^{-2}$

87. $\dfrac{a^3}{a^4}$

88. $\dfrac{n}{n^5}$

89. $\dfrac{2}{n^{-4}}$

90. $\dfrac{3}{n^{-1}}$

91. $\dfrac{p^4}{q^{-1}}$

92. $\dfrac{a}{b^{-6}}$

93. $\dfrac{t^{-2}}{t^3}$

94. $\dfrac{x^{-2}}{x^5}$

95. $\dfrac{x^5}{x^{-2}}$

96. $\dfrac{n^7}{n^{-1}}$

97. $\dfrac{a^{-4}}{a^{-5}}$

98. $\dfrac{y^{-1}}{y^{-6}}$

99. $\dfrac{a^{-3}}{b^{-3}}$

100. $\dfrac{x^{-4}}{y^{-2}}$

Mixed Practice

Simplify.

101. $(s^2 t^4)(st^2)$

102. $\dfrac{a^6 \cdot a}{a^2 \cdot a^3}$

Solve.

103. Multiply: $\left(-\dfrac{2}{3}\right)^3$

104. Simplify $\dfrac{a^5}{a^4}$ using the quotient rule, if possible.

105. Simplify $x^2 y^3$ using the product rule, if possible.

106. Simplify: -5^0

Write as an expression using only positive exponents.

107. $(-y)^{-6}$

108. $\dfrac{3}{4k^{-2}}$

109. $q^{-2} \cdot q^{-3}$

110. $a^{-3} b^2$

111. $\dfrac{x^4}{y^{-3}}$

112. $\dfrac{m^{-5}}{n^{-3}}$

Applications

D *Solve.*

113. The first day of an epidemic, 35 people got sick. Each day thereafter, the number of people who got ill doubled.

  a. How many people were ill on the sixth day of the epidemic? On the tenth day?

  b. How many times as great was the number of people ill on the tenth day as compared to the number ill on the sixth day?

114. The value of a new car t years after it is purchased is given by the expression $28{,}000(1.25)^{-t}$.

  a. What is the value of the car one year after it was purchased?

  b. Evaluate the expression for $t = 0$. Explain the significance of this value.

115. The concentration of a pollutant in a pond is 60 parts per million (ppm). The pollution level drops by 5% each month, so the amount of pollutant each month is 95% of the amount in the previous month, as shown in the following table:

Month	Pollution Level (ppm)
1	60
2	60×0.95
3	$60 \times (0.95)^2$
4	$60 \times (0.95)^3$
5	$60 \times (0.95)^4$
6	$60 \times (0.95)^5$
7	$60 \times (0.95)^6$

What will the pollution level be in the twelfth month?

116. The population of the United States in 1820 was approximately 10^7. One hundred years later, it was approximately 10^8. By what factor did the U.S. population grow during this century? (*Source:* U.S. Bureau of the Census)

117. A small, cube-shaped box is packed within a larger, cube-shaped box and is surrounded by Styrofoam peanuts. Since the side length of the larger box is $\dfrac{5}{2}$ that of the smaller box, we can represent the two side lengths as $5x$ and $2x$.

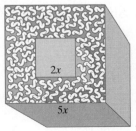

  a. Write an expression for the volume of each box.

  b. How many times the volume of the small box is the volume of the large box?

118. The *multiplication principle* states that the number of possible ways to do multiple things is the product of the number of ways to do each individual thing. For instance, if one die has 6 possible results, then two dice have 6×6 or 36 possible results. While playing a board game, a player rolls 3 red dice and 2 white dice. Each of the 6 faces of a die has a 1, 2, 3, 4, 5, or 6.

   a. How many different possible rolls are there for the 5 dice?

   b. How many different possible rolls are there where all the red dice come up 6, and both of the white dice come up as something other than 6?

• Check your answers on page A-19.

MINDStretchers

Writing

1. Identify the errors that were made resulting in the following false statements:

   a. $4^{-3} = -\dfrac{1}{64}$

   b. $x^{-2} \cdot x^{-3} = x^6$

   c. $4^3 \cdot 4^{-5} = 16^{-2}$

   d. $5^4 \div 5 = 1^4$

Groupwork

2. Which is larger: x^2 or x^{-2}? Explain your answer, and give some examples.

Mathematical Reasoning

3. Explain why a thousand million is the same as a billion.

5.2 More Laws of Exponents and Scientific Notation

In this section, we consider several additional laws of exponents, as well as an important application of exponents known as scientific notation.

OBJECTIVES

A To simplify an expression by using the power rule

B To simplify an expression by raising a product to a power

C To simplify an expression by raising a quotient to a power

D To write a number in scientific notation or standard notation

E To solve applied problems involving laws of exponents or scientific notation

Additional Laws of Exponents

In the previous section, we discussed the product rule for the product of powers and the quotient rule for the quotient of powers. We now consider a third rule known as the *power rule*. The power rule deals with expressions in which a power is raised to a power.

Let's consider, for instance, the expression $(x^2)^3$. Using the definition of an exponent gives us:

$$(x^2)^3 = \underbrace{x^2 \cdot x^2 \cdot x^2}_{3 \text{ factors of } x^2} = x^{2+2+2} = x^6$$

So $(x^2)^3 = x^6$. We can generalize this result as follows:

Power Rule of Exponents

For any nonzero real number x and any integers a and b,

$$(x^a)^b = x^{ab}.$$

In words, to raise a power to a power, *multiply* the exponents and *keep the base the same*.

EXAMPLE 1

Simplify using the power rule of exponents.

a. $(5^2)^2$ **b.** $(2^{-3})^2$ **c.** $-(p^4)^5$ **d.** $(q^2)^{-1}$

Solution We apply the power rule and then simplify.

a. $(5^2)^2 = 5^{2 \cdot 2} = 5^4 = 625$

b. $(2^{-3})^2 = 2^{-3 \cdot 2}$

$= 2^{-6}$

$= \dfrac{1}{2^6}$ Take the reciprocal of 2^{-6}, and change the exponent from -6 to 6.

$= \dfrac{1}{64}$

c. $-(p^4)^5 = -(p^{4 \cdot 5}) = -p^{20}$

d. $(q^2)^{-1} = q^{2 \cdot (-1)} = q^{-2} = \dfrac{1}{q^2}$

PRACTICE 1

Simplify using the power rule of exponents.

a. $(2^3)^2$

b. $(7^3)^{-1}$

c. $(q^2)^4$

d. $-(p^3)^{-5}$

TIP   Be sure to distinguish between the *product* rule and the *power* rule.

Product rule: $x^a \cdot x^b = x^{a+b}$ **Power rule:** $(x^a)^b = x^{ab}$

Add the exponents. ⤴   **Multiply the exponents.** ⤴

Another law of exponents has to do with *raising a product to a power*. For instance, consider the expression $(5x)^3$.

Rearrange the factors.

$$(5x)^3 = (5x)(5x)(5x) = (5 \cdot 5 \cdot 5)(x \cdot x \cdot x) = 5^3 \cdot x^3 = 125x^3$$

So we see that $(5x)^3$ is the same as 5^3 times x^3. We can generalize this result as follows:

Raising a Product to a Power

For any nonzero real numbers x and y and any integer a,

$$(xy)^a = x^a \cdot y^a.$$

In words, to raise a product to a power, raise each factor to that power.

EXAMPLE 2

Simplify using the rule for raising a product to a power.

a. $(2y)^4$ **b.** $(-3a)^2$ **c.** $-(3a)^2$

Solution   We apply the rule for raising a product to a power, and then simplify.

a. $(2y)^4 = 2^4 \cdot y^4 = 16y^4$

b. $(-3a)^2 = (-3)^2(a)^2 = 9a^2$

c. $-(3a)^2 = -3^2 \cdot a^2 = -9a^2$

PRACTICE 2

Simplify.

a. $(7a)^2$

b. $(-4x)^3$

c. $-(4x)^3$

EXAMPLE 3

Simplify.

a. $(2x^4)^5$ **b.** $(p^3q^5)^3$ **c.** $-7(m^5n^{10})^2$ **d.** $(5a^{-2}c^4)^{-2}$

Solution

a. $(2x^4)^5 = 2^5(x^4)^5$ Use the rule for raising a product to a power.

        $= 32x^{20}$ Use the power rule.

b. $(p^3q^5)^3 = (p^3)^3(q^5)^3 = p^9q^{15}$

c. $-7(m^5n^{10})^2 = -7(m^5)^2(n^{10})^2 = -7m^{10}n^{20}$

d. $(5a^{-2}c^4)^{-2} = 5^{-2}(a^{-2})^{-2}(c^4)^{-2}$ Use the rule for raising a product to a power.

        $= 5^{-2}(a^4)(c^{-8})$ Use the power rule.

        $= \dfrac{1}{25} \cdot a^4 \cdot \dfrac{1}{c^8}$ Use the definition of a negative exponent.

        $= \dfrac{a^4}{25c^8}$

PRACTICE 3

Simplify.

a. $(-6a^9)^2$

b. $(q^8r^{10})^2$

c. $-2(ab^7)^3$

d. $(7a^{-1}c^{-5})^2$

The final law of exponents that we discuss is *raising a quotient to a power*. For instance, consider the expression $\left(\dfrac{a}{b}\right)^4$, where a divided by b is raised to the fourth power. By the definition of exponent, we get:

$$\left(\frac{a}{b}\right)^4 = \frac{a}{b}\cdot\frac{a}{b}\cdot\frac{a}{b}\cdot\frac{a}{b} = \frac{a\cdot a\cdot a\cdot a}{b\cdot b\cdot b\cdot b} = \frac{a^4}{b^4}$$

So $\left(\dfrac{a}{b}\right)^4 = \dfrac{a^4}{b^4}$. We generalize this result as follows:

Raising a Quotient to a Power

For any nonzero real numbers x and y and any integer a,

$$\left(\frac{x}{y}\right)^a = \frac{x^a}{y^a}.$$

In words, to raise a quotient to a power, raise both the numerator and the denominator to that power.

EXAMPLE 4

Simplify by using the rule for raising a quotient to a power.

a. $\left(\dfrac{x}{5}\right)^3$ **b.** $\left(\dfrac{-a}{b}\right)^4$ **c.** $\left(\dfrac{5}{x}\right)^{-3}$ **d.** $\left(\dfrac{-3r^2}{st^4}\right)^3$ **e.** $\left(\dfrac{9u}{v^{-1}}\right)^2$

Solution Here we use the rule for raising a quotient to a power, and then simplify.

a. $\left(\dfrac{x}{5}\right)^3 = \dfrac{x^3}{5^3} = \dfrac{x^3}{125}$ **b.** $\left(\dfrac{-a}{b}\right)^4 = \dfrac{(-a)^4}{b^4} = \dfrac{a^4}{b^4}$

c. $\left(\dfrac{5}{x}\right)^{-3} = \dfrac{5^{-3}}{x^{-3}}$

$= 5^{-3}\cdot\dfrac{1}{x^{-3}}$

$= \dfrac{1}{5^3}\cdot x^3$

$= \dfrac{x^3}{125}$

d. $\left(\dfrac{-3r^2}{st^4}\right)^3 = \dfrac{(-3r^2)^3}{(st^4)^3} = \dfrac{(-3)^3(r^2)^3}{s^3(t^4)^3} = \dfrac{-27r^6}{s^3t^{12}}$

e. $\left(\dfrac{9u}{v^{-1}}\right)^2 = \dfrac{(9u)^2}{(v^{-1})^2}$

$= \dfrac{9^2u^2}{v^{-2}}$

$= 81u^2\cdot\dfrac{1}{v^{-2}}$

$= 81u^2v^2$

PRACTICE 4

Simplify.

a. $\left(\dfrac{y}{3}\right)^2$ **b.** $\left(\dfrac{-u}{v}\right)^{10}$

c. $\left(\dfrac{3}{y}\right)^{-2}$ **d.** $\left(\dfrac{-10a^5}{3b^2c}\right)^2$

e. $\left(\dfrac{5x}{y^{-2}}\right)^3$

Note that the simplified form of the expression in Example 4(a) is the same as the simplified form of the expression in Example 4(c). Since $\left(\dfrac{5}{x}\right)^{-3} = \dfrac{x^3}{125}$ and $\left(\dfrac{x}{5}\right)^3 = \dfrac{x^3}{125}$, we conclude that $\left(\dfrac{5}{x}\right)^{-3} = \left(\dfrac{x}{5}\right)^3$. This conclusion leads us to another law of exponents— *raising a quotient to a negative power.*

Raising a Quotient to a Negative Power

For any nonzero real numbers x and y and any integer a,

$$\left(\frac{x}{y}\right)^{-a} = \left(\frac{y}{x}\right)^{a}.$$

In words, to raise a quotient to a negative power, take the reciprocal of the quotient and change the sign of the exponent.

EXAMPLE 5

Simplify.

a. $\left(\dfrac{2}{x}\right)^{-3}$ b. $\left(\dfrac{3r}{10s}\right)^{-1}$ c. $\left(\dfrac{x^4}{y^2}\right)^{-2}$

Solution Use the rule for raising a quotient to a negative power.

a. $\left(\dfrac{2}{x}\right)^{-3} = \left(\dfrac{x}{2}\right)^3 = \dfrac{x^3}{2^3} = \dfrac{x^3}{8}$

b. $\left(\dfrac{3r}{10s}\right)^{-1} = \left(\dfrac{10s}{3r}\right)^1 = \dfrac{10s}{3r}$

c. $\left(\dfrac{x^4}{y^2}\right)^{-2} = \left(\dfrac{y^2}{x^4}\right)^2 = \dfrac{(y^2)^2}{(x^4)^2} = \dfrac{y^4}{x^8}$

PRACTICE 5

Simplify.

a. $\left(\dfrac{5}{a}\right)^{-2}$

b. $\left(\dfrac{4u}{v}\right)^{-1}$

c. $\left(\dfrac{a^5}{b^3}\right)^{-2}$

Scientific Notation

Scientific notation is an important application of exponents—whether they are positive, negative, or zero. Scientists use this notation to abbreviate very large or very small numbers. Note that scientific notation is based on powers of 10.

Example	Standard Notation	Scientific Notation
The speed of light	983,000,000 ft/sec	9.83×10^8 ft/sec
The length of a virus	0.000000000001 m	1×10^{-12} m

Scientific notation has several advantages over standard notation. For example, when a number contains a long string of 0's, writing it in scientific notation can take fewer digits. Also, numbers written in scientific notation can be relatively easy to multiply or divide.

DEFINITION

A number is in **scientific notation** if it is written in the form

$$a \times 10^n$$

where n is an integer and a is greater than or equal to 1 but less than 10 ($1 \le a < 10$).

Note that any value of a that satisfies the inequality $1 \leq a < 10$ must have *one nonzero digit* to the left of the decimal point. For instance, 7.3×10^5 is written in scientific notation. Do you see why the numbers 0.83×10^2, 5×3^7, and 13.8×10^{-4} are *not* written in scientific notation?

> **TIP** When written in scientific notation, large numbers have positive powers of 10, whereas small numbers have negative powers of 10. For instance, 3×10^{23} is large, whereas 3×10^{-23} is small.

Let's now consider how to change a number from scientific notation to standard notation.

EXAMPLE 6

Change the number 2.41×10^5 from scientific notation to standard notation.

Solution To express this number in standard notation, we need to multiply 2.41 by 10^5. Since $10^5 = 100,000$, multiplying 2.41 by $100,000$ gives:

$$2.41 \times 10^5 = 2.41 \times 100,000 = 241,000.00 = 241,000$$

The number $241,000$ is written in standard notation.

Note that the power of 10 here is *positive* and that the decimal point is moved five places *to the right*. So a shortcut for expressing 2.41×10^5 in standard notation is to move the decimal point in 2.41 five places to the right.

$$2.41 \times 10^5 = 2\underset{\smile}{41000.} = 241,000$$

PRACTICE 6

Express 2.539×10^2 in standard notation.

EXAMPLE 7

Convert 3×10^{-5} to standard notation.

Solution Using the definition of a negative exponent, we get:

$$3 \times 10^{-5} = 3 \times \frac{1}{10^5}, \text{ or } \frac{3}{10^5}$$

Since $10^5 = 100,000$, dividing 3 by $100,000$ gives us:

$$\frac{3}{10^5} = \frac{3}{100,000} = 0.00003$$

Here we note that the power of 10 is *negative* and that the decimal point, which is understood to be at the right end of a whole number, is moved five places *to the left*. So a shortcut for expressing 3×10^{-5} in standard notation is to move the decimal point in $3.$ five places to the left.

$$3 \times 10^{-5} = 3. \times 10^{-5} = \underset{\smile}{.00003} = .00003, \text{ or } 0.00003$$

PRACTICE 7

Change 4.3×10^{-9} to standard notation.

> **TIP** When converting a number from scientific notation to standard notation, move the decimal point to the *right* if the power of 10 is *positive* and to the *left* if the power of 10 is *negative*.

Now, let's consider the reverse situation, namely changing a number in standard notation to scientific notation.

EXAMPLE 8

Express 37,000,000,000 in scientific notation.

Solution For a number to be written in scientific notation, it must be of the form

$$a \times 10^n$$

where n is an integer and $1 \le a < 10$. We know that 37,000,000,000 and 37,000,000,000. are the same. We move the decimal point *to the left* so that there is one nonzero digit to the left of the decimal point. The power of 10 by which we multiply is the same as the number of places moved.

$$37{,}000{,}000{,}000 = 3.7\,0\,0\,0\,0\,0\,0\,0\,0\,0 \times 10^{10}$$

Move 10 places to the *left*.

$$= 3.7 \times 10^{10}$$

Since 3.7 and 3.7000000000 are equivalent, we can drop the trailing zeros. So 37,000,000,000 expressed in scientific notation is 3.7×10^{10}.

PRACTICE 8

Write 8,000,000,000,000 in scientific notation.

EXAMPLE 9

Convert 0.00000000000000002 to scientific notation.

Solution We must write the number 0.00000000000000002 in the form

$$a \times 10^n$$

where n is an integer and $1 \le a < 10$. We move the decimal point *to the right* so that there is one nonzero digit to the left of the decimal point. The power of 10 by which we multiply is the number of places moved, preceded by a *negative* sign.

$$0.00000000000000002 = 0\,0\,0\,0\,0\,0\,0\,0\,0\,0\,0\,0\,0\,0\,0\,0\,2. \times 10^{-17}$$

Move 17 places to the *right*.

$$= 2. \times 10^{-17} = 2 \times 10^{-17}$$

PRACTICE 9

Express 0.000000000071 in scientific notation.

Next, let's consider calculations involving numbers written in scientific notation. We focus on the operations of multiplication and division.

EXAMPLE 10

Calculate, writing the result in scientific notation.
a. $(4 \times 10^{-1})(2.1 \times 10^6)$
b. $(1.2 \times 10^5) \div (2 \times 10^{-4})$

PRACTICE 10

Calculate, writing the result in scientific notation.

a. $(7 \times 10^{-2})(3.52 \times 10^3)$

b. $(2.4 \times 10^3) \div (6 \times 10^{-9})$

EXAMPLE 10 (continued)

Solution

a. $(4 \times 10^{-1})(2.1 \times 10^6)$

$\quad = (4 \times 2.1)(10^{-1} \times 10^6)$ Regroup the factors.

$\quad = 8.4 \times 10^{-1+6}$ Use the product rule.

$\quad = 8.4 \times 10^5$

b. $(1.2 \times 10^5) \div (2 \times 10^{-4})$

$\quad = \dfrac{1.2 \times 10^5}{2 \times 10^{-4}}$

$\quad = \dfrac{1.2}{2} \times \dfrac{10^5}{10^{-4}}$ Rewrite the quotient as a product of quotients.

$\quad = 0.6 \times 10^{5-(-4)}$

$\quad = 0.6 \times 10^9$ Use the quotient rule.

Note that 0.6×10^9 is not in scientific notation because 0.6 is not between 1 and 10, that is, it does not have one nonzero digit to the left of the decimal point. To write 0.6×10^9 in scientific notation, we convert 0.6 to scientific notation and then simplify the product.

$\quad \mathbf{0.6 \times 10^9} = (6 \times 10^{-1}) \times 10^9$ Convert 0.6 to scientific notation.

$\quad\quad\quad\quad\quad\quad = 6 \times (10^{-1} \times 10^9)$

$\quad\quad\quad\quad\quad\quad = 6 \times 10^8$ Use the product rule.

So the answer is 6×10^8.

EXAMPLE 11

There are about 5×10^6 red blood cells per cubic millimeter of blood. Each of these red blood cells contains about 2×10^8 hemoglobin molecules. Calculate the approximate number of hemoglobin molecules per cubic millimeter of blood, writing the result in scientific notation. (*Source:* Sylvia Mader, *Inquiry into Life*)

Solution   We need to find the product of the number of red blood cells per cubic millimeter of blood and the number of hemoglobin molecules per red blood cell:

$$(5 \times 10^6)(2 \times 10^8) = \underbrace{(5 \times 2)}\underbrace{(10^6 \times 10^8)}$$

$$= \quad 10 \quad \times \quad 10^{6+8}$$

$$= \quad 10 \quad \times \quad 10^{14}$$

To write this number in scientific notation, we convert 10 to scientific notation and then simplify the product.

$$10 \times 10^{14} = (1.0 \times 10^1) \times 10^{14}$$

$$= 1.0 \times (10^1 \times 10^{14})$$

$$= 1.0 \times 10^{15}$$

So there are approximately 1×10^{15} hemoglobin molecules per cubic millimeter of blood.

PRACTICE 11

The number of hairs on the average human head is estimated to be about 1.5×10^5. If there are approximately 6×10^9 people in the world, estimate the number of human head hairs in the world. (*Source: Time Almanac 2011*)

EXAMPLE 12

A certain DVD holds 9.4×10^9 bytes of information. How many files, each containing 9.4×10^6 bytes, can the DVD hold?

Solution   To determine the number of files that will fit on the DVD, we divide:

$$
\begin{aligned}
(9.4 \times 10^9) \div (9.4 \times 10^6) &= \frac{9.4 \times 10^9}{9.4 \times 10^6} \\
&= \frac{9.4}{9.4} \times \frac{10^9}{10^6} \\
&= 1 \times 10^{9-6} \\
&= 1 \times 10^3
\end{aligned}
$$

So the DVD can hold 1×10^3, or 1000 files.

PRACTICE 12

At the very best, a light microscope can distinguish points 2×10^{-7} m apart, whereas an electronic microscope can distinguish points that are 2×10^{-10} m apart. The second number is how many times the first number? (*Source:* Sylvia Mader, *Inquiry into Life*)

Calculators and Scientific Notation

Calculators vary as to how numbers are displayed or entered in scientific notation.

Display

In order to avoid an overflow error, many calculator models change to scientific notation an answer that is either too small or too large to fit into the calculator's display. Calculators generally use the base 10 without displaying it. Some calculators display scientific notation with either an E or e, and others show a space. For example, 3.1E–4 or 3.1   – 4 can represent 3.1×10^{-4}. What other differences do you see between written scientific notation and displayed scientific notation?

EXAMPLE 13

Multiply 1,000,000,000 by 2,000,000,000.

Solution

Press

1000000000 $\boxed{\times}$ 2000000000 $\boxed{\text{ENTER}}$

Display

```
1000000000 * 2000000000
                     2 E 18
```

Does your calculator display the product in scientific notation, that is, as 2E18, 2e18, or 2.   18?

PRACTICE 13

Square 0.000000005. How is the answer displayed?

Enter

Some calculators give the wrong answer to a computation if very large or very small numbers are entered in standard form rather than in scientific notation. To enter a number in scientific notation, many calculators have a key labeled $\boxed{\text{EE}}$*,* $\boxed{\text{EXP}}$*, or* $\boxed{\text{EEX}}$*. For a negative exponent, a key labeled* $\boxed{+/-}$ *or* $\boxed{(-)}$ *must be pressed either before or after the exponent key, depending on the calculator.*

EXAMPLE 14

Enter the number 5,000,000,000,000 in scientific notation.

Solution

Press	Display
5 $\boxed{\text{EE}}$ 12 $\boxed{\text{ENTER}}$	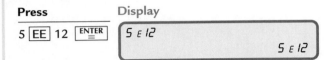

PRACTICE 14

In your calculator, enter in scientific notation the number 0.00000000073.

EXAMPLE 15

Multiply 3.5×10^4 by 2.1×10^7 on a calculator.

Solution

Press	Display
3.5 $\boxed{\text{EE}}$ 4 $\boxed{\times}$ 2.1 $\boxed{\text{EE}}$ 7 $\boxed{\text{ENTER}}$	

So the answer is 7.35×10^{11}. If your calculator has enough places in the display, it may give the answer to this problem in standard form: 735,000,000,000.

PRACTICE 15

Use a calculator to divide 9.2×10^{12} by 2×10^4.

The following table summarizes the laws of exponents considered in Sections 5.1 and 5.2. For any nonzero numbers x and y, and any integers a and b:

Exponent 1	$x^1 = x$ (x can be any real number.)
Exponent 0	$x^0 = 1$
Product rule of exponents	$x^a \cdot x^b = x^{a+b}$
Quotient rule of exponents	$\dfrac{x^a}{x^b} = x^{a-b}$
Negative exponents	$x^{-a} = \dfrac{1}{x^a}$
Reciprocal of x^{-a}	$\dfrac{1}{x^{-a}} = x^a$
Power rule of exponents	$(x^a)^b = x^{ab}$
Raising a product to a power	$(xy)^a = x^a \cdot y^a$
Raising a quotient to a power	$\left(\dfrac{x}{y}\right)^a = \dfrac{x^a}{y^a}$
Raising a quotient to a negative power	$\left(\dfrac{x}{y}\right)^{-a} = \left(\dfrac{y}{x}\right)^a$

Mathematically Speaking

Fill in each blank with the most appropriate term or phrase from the given list.

add the factors	raise both the numerator and the denominator	left
right		factors
terms	raise each factor to that power	multiply the exponents
power form		raise the reciprocal of the quotient
scientific notation	add the powers	

1. The expression $(x^3)^2$ contains two
   _____ of x^3.

2. The power rule of exponents states that to raise a power to a power, _____ and leave the base the same.

3. To raise a product to a power, _____.

4. To raise a quotient to a power, _____ _____ to that power.

5. To raise a quotient to a negative power, _____ _____ to the opposite of the given power.

6. A number is written in _____ if it is in the form $a \times 10^n$, where n is an integer and a is greater than or equal to 1 but less than 10.

7. To convert a number from scientific to standard notation, move the decimal point to the _____ if the power of 10 is negative.

8. To convert a number from standard to scientific notation, move the decimal point to the _____ if the number is less than 1.

A *Simplify.*

9. $(2^2)^4$

10. $(3^3)^2$

11. $(5^2)^2$

12. $(2^3)^3$

13. $(10^5)^2$

14. $(0^5)^3$

15. $(4^{-2})^2$

16. $(2^{-3})^4$

17. $(x^4)^6$

18. $(p^2)^{10}$

19. $(y^4)^2$

20. $(n^3)^3$

21. $(x^{-2})^3$

22. $(y^{-5})^6$

23. $(n^{-2})^{-2}$

24. $(a^{-5})^{-4}$

B *Express in simplest form.*

25. $(4x)^3$

26. $(2y)^5$

27. $(-8y)^2$

28. $(-7a)^2$

29. $-(4n^5)^3$

30. $-(5x^3)^3$

31. $4(-2y^2)^4$

32. $2(-3t)^3$

33. $(3a)^{-2}$

34. $-(5t)^{-3}$

35. $(pq)^{-7}$

36. $(mn)^{-6}$

37. $(r^2t)^6$

38. $(a^3b^5)^4$

39. $(-2p^5q)^2$

40. $(-3a^2b^3)^4$

41. $-2(m^4n^8)^3$

42. $4(x^2y^3)^2$

43. $(-4m^5n^{-10})^3$

44. $(3a^{-3}c^8)^2$

45. $(a^3b^2)^{-4}$

46. $(p^3q^4)^{-2}$

47. $(4x^{-2}y^3)^2$

48. $(2x^{-2}y^3)^2$

C *Simplify.*

49. $\left(\dfrac{5}{b}\right)^3$

50. $\left(\dfrac{x}{4}\right)^2$

51. $\left(\dfrac{c}{b}\right)^2$

52. $\left(\dfrac{t}{s}\right)^5$

53. $-\left(\dfrac{a}{b}\right)^7$

54. $-\left(\dfrac{x}{y}\right)^3$

55. $\left(\dfrac{a^2}{3}\right)^3$

56. $\left(\dfrac{y^6}{4}\right)^3$

57. $\left(-\dfrac{p^3}{q^2}\right)^5$ **58.** $\left(-\dfrac{x^2}{y^3}\right)^4$ **59.** $\left(\dfrac{a}{4}\right)^{-1}$ **60.** $\left(\dfrac{b}{3}\right)^{-1}$

61. $\left(\dfrac{2x^5}{y^2}\right)^3$ **62.** $\left(\dfrac{n^2}{3w^5}\right)^2$ **63.** $\left(\dfrac{pq}{p^2q^2}\right)^5$ **64.** $\left(\dfrac{s^2t^3}{st^2}\right)^2$

65. $\left(\dfrac{3x}{y^{-3}}\right)^4$ **66.** $\left(\dfrac{p^{-1}}{5q^5}\right)^2$ **67.** $\left(\dfrac{-u^2v^3}{4vu^4}\right)^2$ **68.** $\left(\dfrac{2xy^3}{xy^2}\right)^4$

69. $\left(-\dfrac{x^{-2}y}{2z^{-4}}\right)^4$ **70.** $-\left(\dfrac{4a^{-4}}{bc^{-2}}\right)^2$ **71.** $\left(\dfrac{r^5}{t^6}\right)^{-2}$ **72.** $\left(\dfrac{y^3}{x^3}\right)^{-2}$

73. $\left(\dfrac{2a^4}{b^2}\right)^{-3}$ **74.** $\left(\dfrac{q^5}{5p^4}\right)^{-2}$

D *Express in standard notation.*

75. 3.17×10^8 **76.** 9.1×10^5 **77.** 1×10^{-6} **78.** 8.33×10^{-4}

79. 6.2×10^6 **80.** 7.55×10^{10} **81.** 4.025×10^{-5} **82.** 2.1×10^{-3}

Express in scientific notation.

83. 420,000,000   **84.** 100,000,000   **85.** 0.0000035   **86.** 0.00017

87. 217,000,000,000   **88.** 154,800,000,000   **89.** 0.00000000731   **90.** 0.00000005672

Complete each table.

91.

Standard Notation	Scientific Notation (written)	Scientific Notation (displayed on a calculator)
975,000,000		
	4.87×10^8	
		1.652E−10
0.000000067		
	1×10^{-13}	
		3.281E9

92.

Standard Notation	Scientific Notation (written)	Scientific Notation (displayed on a calculator)
975,000,000,000		
	5×10^8	
		4.988E−7
0.0000048		
	9.34×10^{-9}	
		9.772E6

🖩 *Calculate, writing the result in scientific notation.*

93. $(3 \times 10^2)(3 \times 10^5)$

94. $(5 \times 10^4)(7.1 \times 10^3)$

95. $(2.5 \times 10^{-2})(8.3 \times 10^{-3})$

96. $(9.1 \times 10^{-13})(6.3 \times 10^{-10})$

97. $(2.1 \times 10^4)(8 \times 10^{-4})$

98. $(8.6 \times 10^9)(4.4 \times 10^{-12})$

99. $(2.5 \times 10^8) \div (2 \times 10^{-2})$

100. $(3.0 \times 10^4) \div (1 \times 10^3)$

101. $(6 \times 10^5) \div (2 \times 10^3)$

102. $(4.8 \times 10^{-3}) \div (8 \times 10^2)$

103. $(9.6 \times 10^{20}) \div (3.2 \times 10^{12})$

104. $(8.4 \times 10^6) \div (4.2 \times 10^7)$

Mixed Practice

Simplify.

105. Express 3.067×10^{-4} in standard notation.

106. Express 895,600,000 in scientific notation.

🖩 **107.** Complete the table.

Standard Notation	Scientific Notation (written)	Scientific Notation (displayed on a calculator)
428,000,000,000		
	3.24×10^6	
		5.224E−6
0.000000057		
	6.82×10^{-7}	
		4.836E7

Simplify, using only positive exponents.

108. $(a^{-3})^5$

109. $\left(\dfrac{a^2}{3b^5}\right)^{-2}$

110. $(2r^7 s^{-3})^3$

111. $-(4y)^{-3}$

112. $\left(-\dfrac{m^2}{n^3}\right)^5$

113. $-\left(\dfrac{2x^{-2}}{y^{-3}z}\right)^4$

114. $2(-3x^4)^3$

🖩 *Calculate, writing the result in scientific notation.*

115. $(6.3 \times 10^{-4}) \div (9 \times 10^3)$

116. $(4.1 \times 10^{-3})(2.7 \times 10^{-2})$

Applications

E *Solve.*

117. Consider the two boxes shown. How many times the volume of the smaller box is the volume of the larger box?

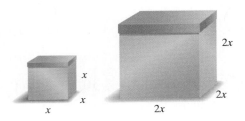

118. The infectious part of a virus is typically between 2.5×10^{-8} m and 2×10^{-7} m in size. Express these quantities in standard notation. (*Source:* Sylvia Mader, *Inquiry into Life*)

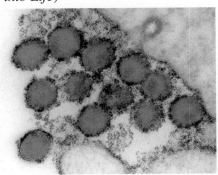

119. A DVD holds between 4×10^9 and 1.7×10^{10} bytes of data. Express these quantities in standard notation.

120. After a flood, the radius of a circular pond doubles. How does the area of the pond change?

121. The wavelength of red light is 0.0000007 m. Write this length in scientific notation.

122. In a recent year, the U.S. federal budget had a deficit of $1,400,000,000,000. Express this amount in scientific notation. (*Source: The 2011 Statistical Abstract of the United States*)

123. The cell is considered the basic unit of life. Each day, the body destroys and replaces more than 200 billion cells. Write this quantity in scientific notation.

124. The diameter of an atom is about 1.1×10^{-10} m. What is this quantity in standard notation? (*Source:* Peter J. Nolan, *Fundamentals of College Physics*)

125. The mass of a proton is about 1.7×10^{-24} g. Rewrite this quantity in standard notation. (*Source:* Karen Timberlake, *Chemistry*)

126. To measure vast distances, astronomers use a unit called a *parsec*, which is equal to about 3.086×10^{18} cm. Express this quantity in standard form. (*Source:* Derek McNally, *Positional Astronomy*)

127. The world population is projected to be 7.6×10^9 in 2020. What is this population expressed in standard notation? (*Source:* census.gov)

128. The diameter of a water molecule is about 2.8 angstroms, where one angstrom is 0.00000001 cm. What is the radius of a water molecule in centimeters, expressed in scientific notation? (*Source:* answers.com)

129. For each pound of body weight, a human body contains about 3.2×10^4 microliters (μL) of blood. In turn, a microliter of blood contains about 5×10^6 red blood cells. A person weighing 100 lb has approximately how many red blood cells?

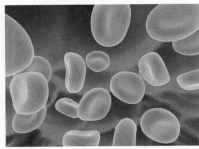

130. On the television series *Star Trek: The Next Generation*, the android Data could carry out 60 trillion operations per second. Express this rate in scientific notation.

131. Light travels through a vacuum at a speed of 186,000 mi/sec.

   a. Express this speed in scientific notation.

   b. How long will it take for light to travel to Earth from the star Vega, which is 1.58×10^{14} mi from Earth? (*Source: The New York Times Almanac 2011*)

132. There are 26,890,000,000,000,000,000 molecules of a gas in a cubic meter.

   a. Rewrite this quantity in scientific notation.

   b. What volume is required for 3.4×10^{20} molecules of the gas?

• Check your answers on page A-20.

MIND*Stretchers*

Investigation

1. On a scientific calculator, enter the number 2. Double that number. Then, keep doubling the result. After how many doublings does your calculator display the number in scientific notation? Explain how you could have predicted that result.

Critical Thinking

2. What is the mathematical relationship between $(a^m)^n$ and $(a^m)^{-n}$? Justify your answer.

Research

3. In your college library or on the Web, determine the annual national debt for the United States for 5 consecutive years. Would you use scientific notation or standard notation to express these amounts? Explain why.

5.3   Basic Concepts of Polynomials

What Polynomials Are and Why They Are Important

In Chapter 1, we discussed algebraic expressions in general. We now consider a particular kind of algebraic expression called a *polynomial*.

    Just as whole numbers are fundamental to arithmetic, polynomials play a similarly key role in algebra.

$$\text{Whole number:} \quad 3 \cdot 10^2 + 7 \cdot 10^1 + 8 \cdot 10^0 = 378$$
$$\text{Polynomial:} \quad 3 \cdot x^2 + 7 \cdot x^1 + 8 \cdot x^0 = 3x^2 + 7x + 8$$

In fact, a good deal of algebra is devoted to studying properties of polynomials and operations on polynomials.

    Many phenomena in the sciences and business can be described by polynomial expressions. Even when the description is only approximate, the polynomial approximation is often a good one.

Monomials

We begin by considering algebraic expressions called *monomials*.

> **DEFINITION**
> A **monomial** is an expression that is the product of a real number and variables raised to nonnegative integer powers.

Some examples of monomials are:

$$5x \quad -7t^4 \quad \frac{4}{5}x^2 \quad -2p^2q^5$$

Recall that in the expression $5x$, 5 is called the *coefficient*.

    Note that a constant such as -12 can be thought of as $-12x^0$. So any constant is also considered a monomial.

    The expression $5x + 3$ is not a monomial since it is the sum of $5x$ and 3, which are called *terms*. A monomial consists only of factors. Can you explain whether $\dfrac{2}{x}$ is a monomial?

    Monomials serve as building blocks (or terms) for the larger set of polynomials. Polynomials are formed by adding and subtracting monomial terms.

> **DEFINITION**
> A **polynomial** is an algebraic expression with one or more monomials added or subtracted.

Here are some examples of polynomials:	Here are some examples of algebraic expressions that are *not* polynomials:
$3x^2 - 5x + 7$	$2x^{-1}$
$4t^2 - 3$	$\dfrac{t^2}{3} + \dfrac{4}{t}$
$-8x^2$	$\dfrac{n + 1}{n}$
$17x^4 + 5x^3 - 8x^2 + x - 1$	$\dfrac{5x^2}{2x + 9}$
$20pq - p^2 - 7q^2 + 6$	$2\sqrt{x} + 1$

EXAMPLE 1

Consider the polynomial $3x^5 + 2x^3 - 8$.

a. Identify the terms of the polynomial.

b. For each term, identify its coefficient.

Solution

a. The terms are $3x^5$, $2x^3$, and -8.

b. The coefficients of the terms are 3, 2, and -8, respectively.

PRACTICE 1

For the polynomial $-10x^2 + 4x + 20$, find (a) the terms and (b) their coefficients.

Classification of Polynomials

There are several ways to classify polynomials. One way is according to the number of variables in the polynomial, and another is according to the number of terms in the polynomial. Finally, a third kind of classification is according to the degree of the polynomial. We consider each of these classifications in turn.

Number of Variables

A polynomial such as $3x^2 - 6x + 9$ is said to be *in one variable,* namely in x. The polynomial $t^5 + 11t^4 - 7t^3 - t^2 + 10t - 50$ is also in one variable, namely in t. On the other hand, the polynomial $3x^4y - 5x^2y^3 + 9$ is in *two* variables, x and y. Throughout this text, we focus on polynomials in one variable.

Number of Terms

As we have seen, a polynomial with just one term is called a *monomial*. A polynomial with *two* terms is called a **binomial**. A **trinomial** is a polynomial with *three* terms. Polynomials with four or more terms are simply called *polynomials*.

$$10x \longleftarrow \text{Monomial}$$
$$3x - 2 \longleftarrow \text{Binomial}$$
$$x^2 + 9x + 6 \longleftarrow \text{Trinomial}$$
$$-x^3 + 8x^2 + x - 19 \longleftarrow \text{Polynomial}$$

EXAMPLE 2

Classify each polynomial according to the number of terms.

Polynomial	Monomial	Binomial	Trinomial
$3x - 5$			
$7x^2 - 3x + 10$			
$10a^2$			

Solution

Polynomial	Monomial	Binomial	Trinomial
$3x - 5$		✓	
$7x^2 - 3x + 10$			✓
$10a^2$	✓		

PRACTICE 2

Classify each polynomial according to the number of terms.

Polynomial	Monomial	Binomial	Trinomial
$5x^2 + 2x + 9$			
$-4x^2$			
$12p - 1$			

Degree and Order of Terms

Let's consider a monomial in one variable. The **degree** of the monomial is the power of the variable in the monomial.

$$3x^4 \longleftarrow \text{Of the fourth degree or of degree 4}$$
$$-7y^2 \longleftarrow \text{Of the second degree or of degree 2}$$

Recall that a constant, such as 3, can be thought of as $3x^0$ and so is considered to be of degree 0.

8 is a monomial of degree **0**.

-12 is a monomial of degree **0**.

Polynomials are also classified by their degree. The degree of a polynomial is the highest degree of any of its terms. For instance, the degree of $8x^3 + 9x^2 - 7x - 1$ is **3** since 3 is the highest degree of any term.

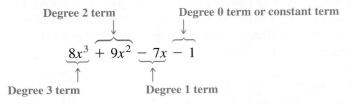

EXAMPLE 3

Identify the degree of each polynomial.

a. $-20x^3$ **b.** $5 + x$ **c.** $-7x^2 + 3x + 10$ **d.** -36

Solution

a. $-20x^3$ is of degree **3**.

b. $5 + x = 5 + x^1$, which is of degree **1**.

c. $-7x^2 + 3x + 10$ is of degree **2**.

d. $-36 = -36x^0$, which is of degree **0**.

PRACTICE 3

Indicate the degree of each polynomial.

a. $7n$

b. $x - x^2$

c. $2 + 4x^2 - 10x^3$

d. -8

The **leading term** of a polynomial is the term in the polynomial with the highest degree, and the coefficient of that term is called the **leading coefficient**. The term of degree 0 is called the **constant term**. So in the polynomial $3t^2 - 8t + 1$, the leading term is $3t^2$, the leading coefficient is 3, and the constant term is 1.

EXAMPLE 4

Complete the table.

Polynomial	Constant Term	Leading Term	Leading Coefficient
$2x^{10}$			
$4x + 25$			
$3 - 10x + 8x^2$			
$7x^3 - x - 8$			

PRACTICE 4

Complete the table.

Polynomial	Constant Term	Leading Term	Leading Coefficient
$-3x^7 + 9$			
x^5			
$x^4 - 7x - 1$			
$3x + 5x^3 + 20$			

Solution

Polynomial	Constant Term	Leading Term	Leading Coefficient
$2x^{10}$	0	$2x^{10}$	2
$4x + 25$	25	$4x$	4
$3 - 10x + 8x^2$	3	$8x^2$	8
$7x^3 - x - 8$	-8	$7x^3$	7

The terms of a polynomial are usually arranged in *descending order of degree*. That is, we write the leading term on the left, then the term of the next highest degree, and so forth.

$x^3 + 5x^2 - 3x - 10$ Descending order of degree—the exponents get smaller from left to right.

Occasionally, however, the terms of a polynomial are written in *ascending order of degree*.

$-10 - 3x + 5x^2 + x^3$ Ascending order of degree—the exponents get larger from left to right.

EXAMPLE 5

Rearrange each polynomial in descending order.

a. $3x^4 + 9x^2 - 7x^3 + x + 10$

b. $-7x + 6x^3 + 10$

Solution

a. We rewrite the polynomial so that the term with the highest exponent of x is on the left, the next highest exponent comes second, and so on.

$$3x^4 + 9x^2 - 7x^3 + x + 10 = 3x^4 - 7x^3 + 9x^2 + x^1 + 10x^0$$
$$= 3x^4 - 7x^3 + 9x^2 + x + 10$$

b. We rewrite the polynomial so that the exponents get smaller from left to right.

$$-7x + 6x^3 + 10 = 6x^3 + \boxed{} - 7x + 10$$

(Degree 3, Degree 2, Degree 1, Degree 0)

Note that a term with coefficient 0 is usually not written. The unwritten term is said to be a *missing term*. For instance, in the polynomial

$$6x^3 - 7x + 10$$

$0x^2$ is the missing term. So we can write this polynomial as $6x^3 + 0x^2 - 7x + 10$.

PRACTICE 5

Write each polynomial in descending order.

a. $-8x + 9x^5 - 7x^4 + 9x^2 - 6$

b. $x^3 + 7x^5 - 3x^2 + 8$

The concept of missing terms is important in the division of polynomials, as we will see later in this chapter.

Simplifying and Evaluating Polynomials

Recall that in Section 1.8, we discussed how to simplify algebraic expressions by combining like terms.

EXAMPLE 6

Simplify, and then put in descending order.

$$8 + 3x^3 + 9x + 1 - 8x + 7x^2 - 3x^2$$

Solution

$$
\begin{aligned}
&8 + 3x^3 + 9x + 1 - 8x + 7x^2 - 3x^2 \\
&= 8 + 3x^3 + 9x + 1 - 8x + 7x^2 - 3x^2 \\
&= 9 + 3x^3 + x + 4x^2 \qquad \text{Combine like terms.} \\
&= 3x^3 + 4x^2 + x + 9
\end{aligned}
$$

PRACTICE 6

Combine like terms, and then write in descending order.

$$2x^2 + 3x - x^2 + 5x^3 + 3x - 5x^3 + 20$$

Recall, also, our discussion of evaluating algebraic expressions in Section 1.8. Polynomials, like other algebraic expressions, are evaluated by replacing each variable with the given number and then carrying out the computation.

EXAMPLE 7

Find the value of $2x^2 - 8x - 5$ when:

a. $x = 3$ **b.** $x = -3$

Solution

a. $2x^2 - 8x - 5 = 2(3)^2 - 8(3) - 5$ Substitute 3 for x.

$$
\begin{aligned}
&= 2(9) - 8(3) - 5 \\
&= 18 - 24 - 5 = -11
\end{aligned}
$$

b. $2x^2 - 8x - 5 = 2(-3)^2 - 8(-3) - 5$ Substitute -3 for x.

$$
\begin{aligned}
&= 2(9) - 8(-3) - 5 \\
&= 18 + 24 - 5 \\
&= 37
\end{aligned}
$$

PRACTICE 7

Find the value of $x^2 - 5x + 5$ when:

a. $x = 2$

b. $x = -2$

EXAMPLE 8

If $1000 is deposited in a savings account that pays compound interest at a rate r compounded annually, then after 2 years the balance in the account will be represented by the polynomial $(1000r^2 + 2000r + 1000)$ dollars. Find the balance if $r = 0.05$.

Solution   We need to replace r by 0.05 in the polynomial.

$$
\begin{aligned}
1000r^2 + 2000r + 1000 &= 1000(0.05)^2 + 2000(0.05) + 1000 \\
&= 1000(0.0025) + 2000(0.05) + 1000 \\
&= 2.5 + 100 + 1000 \\
&= 1102.5
\end{aligned}
$$

So the account balance is $1102.50.

PRACTICE 8

If an object is dropped from a height of 500 ft above the ground, its height (in feet above the ground) after t sec is given by the expression $500 - 16t^2$. How high above the ground is the object after 3 sec?

EXAMPLE 9

The polynomial $-322x^2 + 3027x + 15{,}530$ models the total number of adoptions for a particular year to the United States from other countries, where x represents the number of years after 1999. According to this model, how many of these intercountry adoptions were there, to the nearest thousand, in 2009? (*Source:* adoption.state.gov)

Solution   Since the year 2009 is 10 years after 1999, we substitute 10 for x in the polynomial.

$$-322x^2 + 3027x + 15{,}530 = -322(\mathbf{10})^2 + 3027(\mathbf{10}) + 15{,}530$$
$$= -32{,}200 + 30{,}270 + 15{,}530$$
$$= 13{,}600 \approx 14{,}000$$

So there were, to the nearest thousand, 14,000 adoptions in 2009 to the United States from other countries.

PRACTICE 9

The polynomial $0.02x^3 - 0.75x^2 + 6.89x + 45.17$ can be used to model the temperature (in degrees Fahrenheit) at a particular time in Anaheim, California on November 26, 2010, where x represents the number of hours past 8:00 A.M. According to this model, find the temperature in Anaheim at 11:00 A.M. to the nearest degree. (*Source:* forecast.weather.gov)

Mathematically Speaking

Fill in each blank with the most appropriate term or phrase from the given list.

ascending	constant term	descending
power	coefficient	leading term
degree	leading coefficient	
polynomial	monomial	

1. A(n) _____ is an expression that is the product of a real number and variables raised to non-negative integer powers.

2. The _____ of the expression $-\dfrac{3}{4}x^3$ is $-\dfrac{3}{4}$.

3. The degree of a monomial in one variable is the _____ of the monomial's variable.

4. The _____ of a polynomial is the highest degree of any of its terms.

5. The _____ of a polynomial is the term in the polynomial with the highest degree.

6. The coefficient of the leading term is called the _____.

7. The term of degree 0 in a polynomial is called the _____.

8. Polynomial terms are usually written in _____ order of degree.

A *Indicate whether each of the following is a polynomial.*

9. $7x^2$

10. $2x^2$

11. $x - 7\sqrt{x} + 1$

12. $\dfrac{2}{x+3}$

13. $10p + q$

14. $4a - 3a^2$

15. $2x^{-1}y - x^2$

16. $3y^{-1} + y$

For each polynomial, find (a) the terms and (b) their coefficients.

17. $5x^4 - 2x^3 + 1$

18. $-6y^3 + 4y - 3$

Classify each polynomial according to the number of terms.

19.

Polynomial	Monomial	Binomial	Trinomial
$5x - 1$			
$-5a^2$			
$-6a + 3$			
$x^3 + 4x^2 + 2$			

20.

Polynomial	Monomial	Binomial	Trinomial
$5x + x^2$			
$3x$			
$12p - 1$			
$2x^5 - x^3 + x$			

Rearrange each polynomial in descending order, and then identify its degree.

◉ **21.** $3x^2 - 2x + 8 - 4x^3$

22. $5x^3 + 7x + 1 - 7x^2$

23. $2 - 3y$

24. $7 - 5x$

◉ **25.** $4p^2 - p^4 + 3p^3 + 10 - p$

26. $25 - y + 2y^2 + y^3 - 3y^4$

◉ **27.** $-y^3 - 2y + 2 - 4y^5$

28. $5x^3 + 3x^5 + 8x + 3$

29. $5a^2 - a$

30. $-8x^2 + 6x - 2$

31. $3p^3$

32. $5x^2$

Complete the following table.

33.

Polynomial	Constant Term	Leading Term	Leading Coefficient
$-x^7 + 2$			
$2x - 30$			
$-5x + 1 + x^2$			
$7x^3 - 2x - 3$			

34.

Polynomial	Constant Term	Leading Term	Leading Coefficient
$5x^3 + 8$			
$-x + 10$			
$2x^2 - 3x + 4$			
$-5x + x^4 - 9$			

B *Simplify.*

35. $9x^3 - 7x^2 + 1 + x^3 + 10x + 5$

36. $2y^3 - 7y^2 + 1 + 2y^2 + 3y + 8$

37. $r^3 + 2r^2 + 15 + r^2 - 8r - 1$

38. $n^4 - n^3 - 7n^2 + n^2 + 10n + 3$

Combine like terms. Then, write the polynomial in descending order of powers.

39. $4x^2 - 2x - x^2 - 10 - 3x + 4$

40. $x^3 + 5x - 7x^2 - 1 - x^3 + x$

41. $6n^3 + 20n - n^2 + 2 - 4n^3 + 15n^2 + 8$

42. $8y^2 + y^3 - 8y + 20 + 3y + 9y^2$

Identify the missing terms of each polynomial.

43. $x^3 - 7x - 2$

44. $n^2 + 7n$

45. $6x^3 + 8x^2 + 1$

46. $x^4 - 3x$

C *Find the value of each polynomial for the given values of the variable.*

47. $7x - 3$, for $x = 2$ and $x = -2$

48. $5a + 11$, for $a = 0$ and $a = 2$

◉ **49.** $n^2 - 3n + 9$, for $n = 7$ and $n = -7$

50. $3y^2 + 2y + 1$, for $y = 2$ and $y = -1$

▦ **51.** $2.1x^2 + 3.9x - 7.3$, for $x = 2.37$ and $x = -2.37$

▦ **52.** $0.1x^3 + 4.1x - 9.1$, for $x = 3.14$ and $x = -3.14$

Mixed Practice

Rearrange each polynomial in descending order, and then identify its degree.

53. $-a^3 + 2a - 4 + 5a^4$

54. $8x + 6x^3$

55. Complete the following table:

Polynomial	Constant Term	Leading Term	Leading Coefficient
$4x - 20$			
$-7x^6 + 9$			
$3x + 2 + x^2$			
$6x - x^5 + 11$			

56. Classify each polynomial according to the number of terms.

Expression	Monomial	Binomial	Trinomial
$2p - 3p^2$			
$-8x^3$			
$a^2 - 4a + 5$			

57. Is $x^2 + 4x^{-2} - 2$ a polynomial?

Find the value of each polynomial for the given values of the variable.

58. $5y^2 - 8y - 9$, for $y = 3$ and $y = -2$

59. $5.3x^2 - 2.7x - 6.8$, for $x = 1.25$ and $x = -3.87$

60. Simplify: $4n^3 - 8n^4 + 3 + 6n^2 + 5n^4 - 4n^2$

61. Identify the missing terms of the polynomial $9x^4 + x^2 - 3x$.

62. Combine like terms; then, write the polynomial in descending order of powers:
$3m^2 - m^3 + 9m - 11m^2 - 15 + 7m$

Applications

D *Solve.*

63. The polynomial $1 + x^2 + x^{15} + x^{16}$ is used by computer scientists to detect errors in computer data. Classify this polynomial in terms of its variables and its degree.

64. The owner of a factory estimates that her profit (in dollars) is
$$0.003x^3 - 1.4x^2 + 300x - 1000$$
where x is the number of items that the factory produces. Describe this polynomial in terms of its variables and its degree.

65. The polynomial $x + \dfrac{x^2}{20}$ is the *stopping distance* of a car in feet after the brakes are applied, where the variable x is the speed of the car in miles per hour before braking. Find the stopping distance for a car that had been traveling at 40 mph.

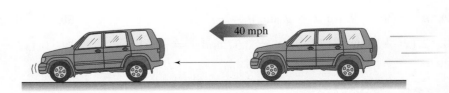

66. There are *n* teams that compete in a sports league, where each team plays every other team once. The following polynomial gives the total number of games that must be played:

$$0.5n^2 - 0.5n$$

If the league has 20 teams, how many games are played?

67. The average monthly cell telephone bill (in dollars) in the U.S. for a particular year is approximated by the polynomial $-0.18x^2 + 1.91x + 45.45$, where *x* represents the number of years after 2000. Find the average monthly cell telephone bill in 2003 to the nearest dollar. (*Source:* ctia.org)

68. The payroll (in millions of dollars) of the Philadelphia Flyers ice hockey team for a particular year is modeled by the polynomial $0.58x^2 + 4.96x + 41.48$, where *x* represents the number of years after 2007. According to this model, find the team payroll in 2015 to the nearest million dollars. (*Source:* usatoday.com)

69. The number of U.S. radio stations with a Spanish format for a particular year is approximated by the polynomial $-0.82x^3 + 6.09x^2 + 22.07x + 631.59$, where *x* represents the number of years after 2003. According to this model, find to the nearest hundred the number of these stations in 2005. (*Source:* mstreet.net)

70. The average basic cable television subscribers (in thousands) in the U.S. for a particular year can be approximated by the polynomial $-88.167x^3 + 184.500x^2 - 114.333x + 65,337$, where *x* represents the number of years after 2005. According to this model, how many basic cable television subscribers to the nearest million were there in 2006? (*Source:* census.gov)

• Check your answers on page A-20.

MINDStretchers

Research

1. The following prefixes are used with polynomials. Use a dictionary to fill in the table.

Prefix	Meaning of This Prefix	Three Words Beginning with This Prefix
Mono-		
Bi-		
Tri-		
Poly-		

Groupwork

2. There are some polynomials whose value is a prime number for many values of the variable. For instance, consider the second-degree polynomial $n^2 + n + 41$.

  a. Check that for *n* equal to a whole number between 0 and 39, the value of this polynomial is a prime number.

  b. Is the value of this polynomial a prime number for $n = 40$? Explain your answer.

(continued)

Patterns

3. The degree of a monomial in more than one variable is the sum of the powers of the variables in that term. Recall that the degree of a polynomial is the highest degree of any of its terms. The following tables show polynomials that represent the area or the volume of various common geometric figures.

Area

Geometric Figure	Polynomial	Degree of the Polynomial
Square	s^2	
Triangle	$0.5bh$	
Trapezoid	$0.5hb + 0.5hB$	
Circle	$3.14r^2$	
Rectangle	lw	

Volume

Geometric Figure	Polynomial	Degree of the Polynomial
Cube	e^3	
Rectangular solid	lwh	
Sphere	$\frac{4}{3}\pi r^3$	
Cylinder	$\pi r^2 h$	

a. Complete the tables by finding the degree of each polynomial.

b. Describe the pattern you observe in the table for area. Explain your observation.

c. Describe the pattern you observe in the table for volume. Explain your observation.

Cultural Note

Muhammad ibn Musa al-Khwarizmi, a ninth-century mathematician, wrote *al-Kitab al-mukhtasar fi hisab al-jabr wa'l-muqabala* (*The Compendious Book on Calculation by Completion and Balancing*)—one of the earliest treatises on algebra and the source of the word *algebra*. This work dealt with solving equations as well as with practical applications of algebra. Al-Khwarizmi, from whose name the word *algorithm* derives, also wrote influential works on astronomy and on the Hindu numeration system.

(*Sources:* Jan Gullberg, *Mathematics From the Birth of Numbers*, W. W. Norton, 1997 and Morris Kline, *Mathematics, A Cultural Approach,* Addison-Wesley, 1962).

5.4 Addition and Subtraction of Polynomials

In this section, we consider the addition and subtraction of polynomials and their applications to real-world situations. As our discussion proceeds, note the similarity between adding and subtracting whole numbers in arithmetic and these operations on polynomials in algebra.

A To add polynomials

B To subtract polynomials

C To solve applied problems involving the addition or subtraction of polynomials

Adding Polynomials

The key to adding polynomials is identifying like terms and then adding them.

> **To Add Polynomials**
>
> • Add the like terms.

As with whole numbers, we can add polynomials using either a horizontal or vertical format. To add polynomials horizontally, we simply remove the parentheses and combine like terms.

EXAMPLE 1

Find the sum: $(8x^2 + 3x + 4) + (-12x^2 + 7)$

Solution   Recall that when a plus sign precedes terms in parentheses, we remove the parentheses and keep the sign of each term.

$$\overbrace{(8x^2 + 3x + 4)}^{\substack{\text{First}\\\text{polynomial}}} + \overbrace{(-12x^2 + 7)}^{\substack{\text{Second}\\\text{polynomial}}}$$

$$= 8x^2 + 3x + 4 - 12x^2 + 7 \quad \text{Remove parentheses.}$$

$$= -4x^2 + 3x + 11 \quad \text{Combine like terms.}$$

PRACTICE 1

Add: $6x - 3$ and $9x^2 - 3x - 40$

EXAMPLE 2

Combine: $(3st^2 - 4st + t^2) + (8s^2t - 3t^2) + (10t^2 - 5st + 7s^2)$

Solution

$(3st^2 - 4st + t^2) + (8s^2t - 3t^2) + (10t^2 - 5st + 7s^2)$

$$= 3st^2 - 4st + t^2 + 8s^2t - 3t^2 + 10t^2 - 5st + 7s^2$$

$$= 3st^2 + 8s^2t - 9st + 7s^2 + 8t^2$$

PRACTICE 2

Combine: $(9p^2 + 4pq + 2q^2) + (-p^2 - 5q^2) + (2p^2 - 3pq - 7q^2)$

Now, let's look at how to add polynomials in a vertical format. Recall that in adding whole numbers, for instance 329 and 50, we position the addends so that digits with the same place value are in the same column.

$$\begin{array}{r}\text{Hundreds}\underset{\xrightarrow{}}{} \overset{\text{Tens}}{\underset{\downarrow\downarrow}{}} \xleftarrow{}\text{Ones}\\ 3\,2\,9 \\ +\ \ 5\,0 \\ \hline \end{array}$$

Similarly, when we add polynomials vertically, we position the polynomials so that like terms are in the same column. Suppose, for example, we want to add $7x^5 + x^3 - 3x^2 + 8$

and $2x^5 - 7x^4 + 9x - 9$. In general, a polynomial is considered to be in simplest form when each term is simplified and like terms are combined. Usually the terms are then rearranged in descending order. So, first we make sure that both polynomials are in descending order. Then, we write the polynomials as follows:

$$7x^5 + 0x^4 + \ x^3 - 3x^2 + 0x + 8$$
$$2x^5 - 7x^4 + 0x^3 + 0x^2 + 9x - 9$$

Note that we could have just left a space for each missing term of the polynomials. Do you see that each column contains like terms? We then add the terms in each column, as shown.

$$7x^5 \qquad\quad + x^3 - 3x^2 \qquad + 8$$
$$\underline{2x^5 - 7x^4 \qquad\qquad\qquad + 9x - 9}$$
$$9x^5 - 7x^4 + x^3 - 3x^2 + 9x - 1$$

So the sum is $9x^5 - 7x^4 + x^3 - 3x^2 + 9x - 1$.

Let's consider some more examples.

EXAMPLE 3

Add vertically: $3x^2 - 5x - 6$ and $10x + 20$

Solution   First, we check that both polynomials are in descending powers of x. Next, we rewrite the polynomials vertically, with like terms positioned in the same column:

First-degree terms

Second-degree term ⟶   ↓   ⟵ Zero-degree (constant) terms

$$3x^2 - 5x - \ 6$$
$$10x + 20$$

We add within columns.

$$3x^2 - \ 5x - \ 6$$
$$\underline{\qquad\ 10x + 20}$$
$$3x^2 + \ 5x + 14$$

The sum is $3x^2 + 5x + 14$, which is in simplest form.

PRACTICE 3

Find the sum of $8n^2 + 2n - 1$ and $3n^2 - 2$ using a vertical format.

EXAMPLE 4

Find the sum of $7x^3 - 10x^2y + 8xy^2 + 13y^3$, $14xy^2 - 1$ and $3x^2y - 5xy^2 - y^3 + 2$ using a vertical format.

Solution

$$7x^3 - 10x^2y + \ 8xy^2 + 13y^3$$
$$14xy^2 \qquad\qquad - 1$$
$$\underline{\quad\ 3x^2y - \ 5xy^2 - \ y^3 + 2}$$
$$7x^3 - \ 7x^2y + 17xy^2 + 12y^3 + 1$$

PRACTICE 4

Add vertically:

$7p^3 - 8p^2q - 3pq^2 + 20$,
$10p^2q + pq^2 - q^3 + 5$, and $p^3 - q^3$

Is the sum in Example 4 in simplified form? Explain.

Subtracting Polynomials

Recall that when a minus sign precedes terms in parentheses, we remove the parentheses and change the sign of each term.

EXAMPLE 5

Remove the parentheses, and simplify.

a. $2x - (3x + 4y)$

b. $(5n + 2m) - (n + m) + (3n - 4m)$

Solution

a. In the expression $2x - (3x + 4y)$, note that $(3x + 4y)$ is preceded by a minus sign. So we remove the parentheses, and change the sign of each term in parentheses. Then, we combine like terms.

$$2x - (3x + 4y) = 2x - 3x - 4y$$
$$= -x - 4y$$

b. $(5n + 2m) \underbrace{- (n + m)}_{\substack{\text{For a polyno-}\\\text{mial preceded}\\\text{by a minus}\\\text{sign, } change\\\text{signs of terms.}}} \underbrace{+ (3n - 4m)}_{\substack{\text{For a polyno-}\\\text{mial preceded}\\\text{by a plus sign,}\\keep \text{ signs of}\\\text{terms.}}} = 5n + 2m - n - m$$
$$+ 3n - 4m = 7n - 3m$$

PRACTICE 5

Remove the parentheses, and simplify.

a. $-(4r - 3s) + 7r$

b. $(2p + 5q) + (p - 6q) - (3p + 2q)$

To subtract real numbers, we change the number being subtracted to its opposite, and then add. Subtraction of polynomials works very much in the same way.

To Subtract Polynomials

- Change the sign of each term of the polynomial being subtracted.

- Add the like terms.

For instance, suppose that we want to subtract the polynomial $2x^4 - 5x^3 + 4x^2 + x + 1$ from $3x^4 + x^3 - 4x^2 + 8x - 9$.

$$\overbrace{(3x^4 + x^3 - 4x^2 + 8x - 9)}^{\substack{\text{The polynomial from}\\\text{which we are subtracting}}} - \overbrace{(2x^4 - 5x^3 + 4x^2 + x + 1)}^{\substack{\text{The polynomial being}\\\text{subtracted}}}$$

$$= (3x^4 + x^3 - 4x^2 + 8x - 9)$$
$$+ (-2x^4 + 5x^3 - 4x^2 - x - 1)$$

Change the sign of each term of the polynomial being subtracted, and then add.

$$= 3x^4 + x^3 - 4x^2 + 8x - 9 - 2x^4 + 5x^3 - 4x^2 - x - 1$$

Remove parentheses.

$$= x^4 + 6x^3 - 8x^2 + 7x - 10$$

Combine like terms.

EXAMPLE 6

Subtract: $(5x^2 - 3x + 7) - (-2x^2 + 8x + 9)$

Solution

$$(5x^2 - 3x + 7) - (-2x^2 + 8x + 9) = (5x^2 - 3x + 7) + (2x^2 - 8x - 9)$$
$$= 5x^2 - 3x + 7 + 2x^2 - 8x - 9$$
$$= 7x^2 - 11x - 2$$

PRACTICE 6

Find the difference:
$$(2x - 1) - (3x^2 + 15x - 1)$$

Thus far, we have subtracted polynomials using a horizontal format. However, we can also subtract polynomials vertically, a skill that comes up when dividing polynomials. As in the case of the vertical addition of polynomials, the key in vertical subtraction is to position the polynomials so that like terms are in the same column. Then, we change the sign of each term of the polynomial being subtracted, and add.

EXAMPLE 7

Subtract using a vertical format: $(7x^2 - 3x + 7) - (x^2 + 8x - 9)$

Solution

$$
\begin{array}{r}
7x^2 - 3x + 7 \\
-(x^2 + 8x - 9) \\
\end{array}
$$
   Position like terms in the same columns.

$$
\begin{array}{r}
7x^2 - 3x + 7 \\
-x^2 - 8x + 9 \\
\end{array}
$$
   Change the sign of each term of the polynomial being subtracted.

$$
\begin{array}{r}
7x^2 - 3x + 7 \\
-x^2 - 8x + 9 \\
\hline
6x^2 - 11x + 16 \\
\end{array}
$$
   Add.

PRACTICE 7

Subtract vertically:
$$(20x - 13) - (5x^2 - 12x + 13)$$

EXAMPLE 8

Subtract $10x^2 + 8xy + y^2$ from $4y^2 - 9x^2$, using a vertical format.

Solution

$$
\begin{array}{r}
-9x^2 \quad\quad + 4y^2 \\
-(10x^2 + 8xy + y^2) \\
\end{array}
$$
   Position like terms in the same column.

$$
\begin{array}{r}
-9x^2 \quad\quad + 4y^2 \\
-10x^2 - 8xy - y^2 \\
\end{array}
$$
   Change the sign of each term in the polynomial being subtracted.

$$
\begin{array}{r}
-9x^2 \quad\quad + 4y^2 \\
-10x^2 - 8xy - y^2 \\
\hline
-19x^2 - 8xy + 3y^2 \\
\end{array}
$$
   Add.

PRACTICE 8

Find the difference using a vertical format:
$$(2p^2 - 7pq + 5q^2) - (3p^2 + 4pq - 12q^2)$$

EXAMPLE 9

The polynomial $0.96x + 33.05$ approximates the number of residential natural gas consumers (in thousands) for a particular year in the state of Vermont x years after 2005. The corresponding polynomial for the state of Alaska is $0.35x + 10.85$. Find the polynomial that approximates the total number of these consumers for the two states. (*Source:* eia.gov)

Solution   To determine the total number of residential natural gas consumers from either Vermont or Alaska, we add the number from Vermont to the number from Alaska. These numbers are approximated by the given polynomials.

$$
\begin{array}{r}
0.96x + 33.05 \\
\underline{0.35x + 10.85} \\
1.31x + 43.9
\end{array}
$$

So $1.31x + 43.9$ represents the total number (in thousands) of these consumers for Vermont or Alaska.

PRACTICE 9

The polynomial $0.01x + 4.03$ approximates the yearly stadium attendance (in millions) at Yankee Stadium for a particular year x years after 2004. The corresponding polynomial for Fenway Park is $0.05x + 2.81$. Find the polynomial that approximates how much greater the attendance (in millions) is at Yankee Stadium than at Fenway Park. (*Source:* baseball-almanac.com)

A *Add horizontally.*

1. $3x^2 + 6x - 5$ and $-x^2 + 2x + 7$

2. $10x^2 + 3x + 9$ and $-x^2 - 5x + 1$

3. $2n^3 + n$ and $3n^3 + 8n$

4. $9y + 2y^2$ and $-y - 3y^2$

5. $10p + 3 + p^2$ and $p^2 - 7p - 4$

6. $x^2 + 3x - 8$ and $10 - 3x + 4x^2$

7. $8x^2 + 7xy - y^2$ and $3x^2 - 10xy + 3y^2$

8. $20p^2 + 15q^4 + 30pq$ and $-4pq + 10q^4 - p^2$

9. $2p^3 - p^2q - 5pq^2 + 1, 3p^2q + 2pq^2 - 4q^3 + 4,$ and $p^3 + q^3$

10. $2x^3 - 4x^2y + xy^2 + y^3, 3xy^2 - 6,$ and $2x^2y - xy^2 - 2y^3 + 3$

Add vertically.

11. $10x^2 - 3x - 8$ and $20x + 3$

12. $t^2 + 4t + 5$ and $-t + 10$

13. $5x^3 + 7x - 1$ and $x^2 + 2x + 3$

14. $2r^3 + r + 2$ and $-r^3 - 8r^2 + 5r - 6$

15. $5ab^2 - 3a^2 + a^3$ and $2ab^2 + 9a^2 - 4a^3$

16. $p^2q^3 - p^2 - q^3$ and $5p^2q^3 + p^2 + q^3$

B *Subtract horizontally.*

17. $2x^2 + 3x - 7$ from $x^2 + x + 4$

18. $2x^3 + 7x^2 + 3x$ from $8x^3 - 10x^2 + x$

19. $3x^3 + x^2 + 5x - 8$ from $x^3 + 10x^2 - 8x + 3$

20. $5t^2 - 7t - 1$ from $8t^2 - 3t + 2$

21. $5x + 9$ from $x^2 + 3x$

22. $3x - 7$ from $x^2 - x + 4$

23. $4y^2 - 6xy - 3$ from $1 - 6xy + 5x^2 - y^2$

24. $p^4 - 7p^2q^2 + q^4$ from $8p^4 - 3q^4$

Subtract vertically.

25. $7p^2 - 10p - 1$ from $2p^2 - 3p + 5$

26. $x^2 - 5x + 2$ from $3x^2 + 10x - 2$

27. $8t^3 - 5$ from $9t^3 - 12t^2 + 3$

28. $10x + 7$ from $x^2 + 2x - 6$

29. $r^3 - 3r^2s - 5$ from $4r^3 - 20r^2s - 7$

30. $-5x^3 - 2y^2$ from $13xy^3 + 7x^3 - 10y^2$

Remove the parentheses and simplify, if possible.

31. $7x - (8x + r)$

32. $8x - (9x + y)$

33. $2p - (3q + r)$

34. $t - (4r - s)$

35. $(4y - 1) + (3y^2 - y + 5)$

36. $(2x + 6) + (x^2 - 4x + 3)$

37. $(m^3 - 6m + 7) - (-9 + 6m)$

38. $(p^2 + 3p - 5) - (p^3 + 6 - p^2)$

39. $(2x^3 - 7x + 8) - (5x^2 + 3x - 1)$

40. $(n^3 - 4n + 2) - (n^2 - 8n + 1)$

41. $(8x^2 + 3x) + (x - 2) + (x^2 + 9)$

42. $(5y^2 + y) + (9y - 1) + (y^2 + 2)$

43. $(3x - 7) + (2x + 9) - (7x - 10)$

44. $(5n + 1) - (3n + 1) + (2n + 5)$

45. $(7x^2y^2 - 10xy + 4) - (2xy + 8) + (x^2y^2 - 10)$

46. $(2m^2n^2 - mn + 7) - (mn - 3) + (m^2n^2 + 7)$

Mixed Practice

Subtract horizontally.

47. $3m^2 - 4m + 2$ from $5m^2 - 9m - 7$

48. $x^2 - 5x$ from $3x - 4$

Remove the parentheses and simplify, if possible.

49. $(5x^3 - x - 8) - (-3x^2 - 3x + 4)$

50. $3a - (4b - c)$

51. $(3m - 1) - (4m + 5) + (6m - 2)$

52. $(6t - 3) + (2t^2 - 3t + 4)$

Add horizontally.

53. $3x^2 + 5x - 4$ and $-5x^2 + x + 7$

54. $6a^2 - 8b^3 - 7ab$ and $-5ab + 3b^3 - 11a^2$

Solve.

55. Subtract $p^2 - 4p + 3$ from $6p^2 + 7p - 4$ vertically.

56. Add $t^3 - 3t^2 + t + 9$ and $-8t^3 + t^2 - 2$ vertically.

Applications

C *Solve.*

57. The surface area of the cylindrical can shown is approximated by the polynomial expression $3.14r^2 + 3.14r^2 + 6.28rh$.

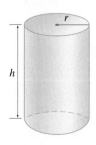

a. Simplify this polynomial expression.

b. Find a polynomial expression for the surface area of a can where the radius r and height h are equal.

58. The room shown is in the shape of a cube.

a. Write a simplified expression for the surface area of the four walls.

b. Find the surface area of the four walls when e is 9 ft, x is 3 ft, and y is 7 ft.

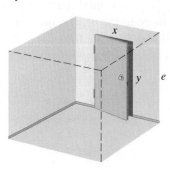

59. The total U.S. imports of petroleum (in thousands of barrels per day) in a given year are approximated by the polynomial $-37x^3 + 23x^2 - 13x + 13{,}718$, where x represents the number of years after 2005. The corresponding total of exports is modeled by the polynomial $-9x^3 + 84x^2 + 28x + 1175$. Write a polynomial that represents how many more thousands of barrels per day were imported than exported in a given year. (*Source:* eia.doe. gov)

60. The polynomial $327x^3 - 1651x^2 + 3x + 178{,}279$ models the number (in thousands) of male commissioned officers in the U.S. Department of Defense in a given year, where x represents the number of years after 2004. The corresponding polynomial for female commissioned officers is $149x^3 - 745x^2 + 310x + 34{,}022$. Find a polynomial that represents how many more thousands of male commissioned officers than female commissioned officers there were for a given year. (*Source:* prhome. defense.gov)

61. The polynomial $-2.4x^3 + 7.8x^2 - 6.0x + 68.3$ models the amount (in millions of dollars) spent on the Texas Rangers baseball team for a particular year, where x represents the number of years after 2007. The corresponding polynomial for the San Francisco Giants is $-1.6x^3 + 14.6x^2 - 26.6x + 90.2$. (*Source:* usatoday.com)

   a. Find the polynomial that represents how much more the payroll is for the Giants than the Rangers.

   b. Using the polynomial found in part (a), find how much more to the nearest million dollars the Giants payroll was than the Rangers in 2010.

62. The amount spent (in billions of dollars) on home health care in the U.S. for a particular year is modeled by the polynomial $0.1x^2 - 5.1x + 42.7$, where x represents the number of years after 2004. The corresponding polynomial for nursing home care is $0.3x^2 + 4.7x + 115.3$. Find the polynomial that represents the total expenditures for home health care and nursing home care during a given year. (*Source:* cms.gov)

• Check your answers on page A-20.

Patterns

1. A Fibonacci sequence is a list of numbers with the following property: After the first two numbers, every other number on the list is the sum of the two previous numbers. The following, for example, is a Fibonacci sequence:

$$7 + 11$$
$$\downarrow$$
$$7, 11, 18, 29, 47, 76, 123, \ldots$$
$$\uparrow$$
$$11 + 18$$

   a. What is the next number in this sequence?

   b. If the first two numbers in a Fibonacci sequence are a and b, find the third and fourth numbers. Check that the tenth number in the sequence is given by the polynomial $21a + 34b$.

Mathematical Reasoning

2. Is it possible to add two polynomials, each of degree 4, and have the sum be a polynomial of degree 2? If so, give an example. If not, explain why not.

Critical Thinking

3. Is the subtraction of polynomials a commutative operation? Give an example to support your answer.

OBJECTIVES

A To multiply monomials

B To multiply a monomial by a polynomial

C To multiply binomials

D To multiply polynomials

E To solve applied problems involving the multiplication of polynomials

In this section, we discuss how to multiply polynomials. We start by finding the product of two monomials.

Multiplying Monomials

When multiplying monomials such as $3x^2$ and $2x^5$, the product rule of exponents helps us to find the product.

$$
\begin{aligned}
(3x^2)(2x^5) &= (3 \cdot 2) \cdot (x^2 \cdot x^5) && \text{The commutative and associative properties of multiplication} \\
&= (3 \cdot 2) \cdot (x^{2+5}) && \text{The product rule of exponents} \\
&= 6x^7
\end{aligned}
$$

To Multiply Monomials

- Multiply the coefficients.
- Multiply the variables, using the product rule of exponents.

EXAMPLE 1

Multiply: $-2x \cdot 8x$

Solution
$$
\begin{aligned}
-2x \cdot 8x &= (-2 \cdot 8) \cdot (x \cdot x) \\
&= (-2 \cdot 8)(x^{1+1}) \\
&= -16x^2
\end{aligned}
$$

PRACTICE 1

Find the product of $(-10x^2)$ and $(-4x^3)$.

EXAMPLE 2

Multiply: $(-2x^3y)(-4x^2y^4)(10xy)$

Solution

$$
\begin{aligned}
(-2x^3y)(-4x^2y^4)(10xy) &= (-2 \cdot -4 \cdot 10) \cdot (x^3 \cdot x^2 \cdot x) \cdot (y \cdot y^4 \cdot y) \\
&= 80(x^{3+2+1})(y^{1+4+1}) \\
&= 80x^6y^6
\end{aligned}
$$

Note that the variables in a product are generally written in alphabetical order.

PRACTICE 2

Find the product:
$(7ab^2)(10a^2b^3)(-5a)$

EXAMPLE 3

Simplify: $(-3p^2r)^3$

Solution

$$
\begin{aligned}
(-3p^2r)^3 &= (-3)^3(p^2)^3(r)^3 && \text{Use the rule for raising a product to a power.} \\
&= -27p^6r^3
\end{aligned}
$$

PRACTICE 3

Find the square of $-5xy^2$.

Multiplying a Monomial by a Polynomial

Now, let's use our knowledge of multiplying monomials to find the product of a monomial and a polynomial.

Consider, for instance, the product $(7x)(9x^2 + 5)$. We use the distributive property to find this product.

$$(7x)(9x^2 + 5) = (7x)(9x^2) + (7x)(5) = 63x^3 + 35x$$

Let's look at some more examples.

EXAMPLE 4

Multiply: $(-8x + 9)(-3x^2)$

Solution
$$\begin{aligned}(-8x + 9)(-3x^2) &= (-3x^2)(-8x + 9)\\ &= (-3x^2)(-8x) + (-3x^2)(9)\\ &= 24x^3 - 27x^2\end{aligned}$$

PRACTICE 4

Find the product: $(10s^2 - 3)(7s)$

EXAMPLE 5

Multiply: $3p^2q(5p^3 - 2pq + q^3)$

Solution

$$\begin{aligned}3p^2q(5p^3 - 2pq + q^3) &= 3p^2q(5p^3) + 3p^2q(-2pq) + 3p^2q(q^3)\\ &= 15p^5q - 6p^3q^2 + 3p^2q^4\end{aligned}$$

PRACTICE 5

Simplify:
$$-2m^3n^2(-6m^3n^5 + 2mn^2 + n)$$

EXAMPLE 6

Simplify: $8x^2(3x + 1) + x^2(5x^2 - 6x + 5)$

Solution

$$\begin{aligned}8x^2(3x + 1) &+ x^2(5x^2 - 6x + 5)\\ &= 8x^2(3x) + 8x^2(1) + x^2(5x^2) + x^2(-6x) + x^2(5)\\ &= 24x^3 + 8x^2 + 5x^4 - 6x^3 + 5x^2\\ &= 5x^4 + 18x^3 + 13x^2 \quad \text{Combine like terms.}\end{aligned}$$

PRACTICE 6

Simplify:
$$7s^3(-2s^2 + 5s + 4) - s^2(s^2 + 6s - 1)$$

Multiplying Two Binomials

Now, we extend the discussion to the multiplication of binomials. As in the case of multiplying monomials and binomials, we use the distributive property.

Consider, for example, the product $(x + 4)(7x + 2)$. To apply the distributive property, we can think of the first factor $(x + 4)$ as a single number multiplied by the binomial $(7x + 2)$.

$$(a) \cdot (b + c) = (a) \cdot (b) + (a) \cdot (c)$$

$$\begin{aligned}(x + 4)(7x + 2) &= (x + 4)(7x) + (x + 4)(2) &&\text{Use the distributive property.}\\ &= 7x(x + 4) + 2(x + 4) &&\text{Use the commutative property.}\\ &= 7x \cdot x + 7x \cdot 4 + 2 \cdot x + 2 \cdot 4 &&\text{Use the distributive property.}\\ &= 7x^2 + 28x + 2x + 8\\ &= 7x^2 + 30x + 8 &&\text{Combine like terms.}\end{aligned}$$

Would we get the same answer if we multiplied $(7x + 2)$ by $(x + 4)$?

Let's consider some other examples.

EXAMPLE 7

Find the product: $(3x + 1)(5x - 2)$

Solution

$$
\begin{aligned}
(3x + 1)(5x - 2) &= (3x + 1)(5x) + (3x + 1)(-2) \\
&= 5x(3x + 1) + (-2)(3x + 1) \\
&= 5x \cdot 3x + 5x \cdot 1 + (-2) \cdot 3x + (-2) \cdot 1 \\
&= 15x^2 + 5x - 6x - 2 \\
&= 15x^2 - x - 2
\end{aligned}
$$

PRACTICE 7

Multiply: $(a - 1)(2a + 3)$

Another way to multiply two binomials is called the **FOIL method,** which is derived from the distributive property. With this method, we can memorize a formula that makes multiplying binomials quick and easy.

FOIL stands for **F**irst, **O**uter, **I**nner, and **L**ast. Let's see how this method works, applying it to the product $(x + 4)(7x + 2)$ that we discussed above.

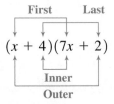

The two **first** terms in the binomials are x and $7x$, and their product is $7x^2$.

$$(x + 4)(7x + 2) \qquad x \cdot 7x = 7x^2$$

The two **outer** terms are x and 2, and their product is $2x$.

$$(x + 4)(7x + 2) \qquad x \cdot 2 = 2x$$

The two **inner** terms are 4 and $7x$, and their product is $28x$.

$$(x + 4)(7x + 2) \qquad 4 \cdot 7x = 28x$$

Next, the two **last** terms are 4 and 2, and their product is 8.

$$(x + 4)(7x + 2) \qquad 4 \cdot 2 = 8$$

Finally, we add the four products, combining any like terms.

$$7x^2 + 2x + 28x + 8 = 7x^2 + 30x + 8$$

Note that, as expected, this answer is the same as the one found using the distributive property.

To Multiply Two Binomials Using the FOIL Method

Consider $(a + b)(c + d)$.

- Multiply the two *first* terms in the binomials.

$$(a + b)(c + d) \qquad \text{Product is } ac.$$
$$\text{F}$$

- Multiply the two *outer* terms.

$$(a + b)(c + d) \qquad \text{Product is } ad.$$
$$\text{O}$$

- Multiply the two *inner* terms.

$$(a + b)(c + d) \qquad \text{Product is } bc.$$
$$\text{I}$$

- Multiply the two *last* terms.

$$(a + b)(c + d) \qquad \text{Product is } bd.$$
$$\text{L}$$

- Find the sum of these four products.

$$(a + b)(c + d) = ac + ad + bc + bd$$

With some practice, the FOIL method can be done mentally.

EXAMPLE 8

Multiply: $(8x - 3)(2x - 1)$

Solution   Using the FOIL method, we get:

$(8x - 3)(2x - 1)$

F: $(8x)(2x) = 16x^2$
O: $(8x)(-1) = -8x$
I: $(-3)(2x) = -6x$
L: $(-3)(-1) = 3$

So $(8x - 3)(2x - 1) = 16x^2 - 8x - 6x + 3 = 16x^2 - 14x + 3$.
Note that the middle term, $-14x$, is the sum of the outer and inner products, which are like terms.

PRACTICE 8

Find the product of $8x + 3$ and $2x - 1$.

EXAMPLE 9

Multiply: $(3a + b)(2a - b)$

Solution

$$
\begin{aligned}
(3a + b)(2a - b) &= \overset{\text{F}}{(3a)(2a)} + \overset{\text{O}}{(3a)(-b)} + \overset{\text{I}}{(b)(2a)} + \overset{\text{L}}{(b)(-b)} \\
&= 6a^2 - 3ab + 2ab - b^2 \\
&= 6a^2 - ab - b^2
\end{aligned}
$$

PRACTICE 9

Find the product: $(7m - n)(2m + n)$

Multiplying Polynomials

Finally, let's consider multiplying polynomials in general. Here we extend the previous discussion of multiplying binomials to multiplying polynomials that can have any number of terms. Let's consider the following example in which we multiply two polynomials written in a horizontal format.

EXAMPLE 10

Multiply: $(x^2 + 3x - 1)(x + 2)$

Solution

$(x^2 + 3x - 1)(x + 2)$

$= (x^2 + 3x - 1)(x) + (x^2 + 3x - 1)(2)$ Use the distributive property.

$= x^3 + 3x^2 - x + 2x^2 + 6x - 2$ Use the distributive property.

$= x^3 + 5x^2 + 5x - 2$ Combine like terms.

PRACTICE 10

Multiply $(n^2 + n - 2)(n - 4)$ using a horizontal format.

Instead of multiplying horizontally, we can multiply two polynomials in a vertical format similar to that used for multiplying whole numbers.

$$
\begin{array}{r}
x^2 + 3x - 1 \\
x + 2 \\
\hline
2x^2 + 6x - 2 \\
x^3 + 3x^2 - x \\
\hline
x^3 + 5x^2 + 5x - 2
\end{array}
$$

Multiply $x^2 + 3x - 1$ by 2.

Multiply $x^2 + 3x - 1$ by x.

Add like terms.

Note that we positioned the terms of the two "partial products" in the shaded area so that each column contains like terms. Finally, to find the product of the original polynomials, we add down each column in the shaded area.

EXAMPLE 11

Find the product of $3y^2 + y + 5$ and $4y - 1$.

Solution
We begin by rewriting the problem in a vertical format.

$$
\begin{array}{r}
3y^2 + y + 5 \\
4y - 1 \\
\hline
-3y^2 - y - 5 \\
12y^3 + 4y^2 + 20y \\
\hline
12y^3 + y^2 + 19y - 5
\end{array}
$$

So we conclude:

$$(3y^2 + y + 5)(4y - 1) = 12y^3 + y^2 + 19y - 5$$

PRACTICE 11

Multiply: $(8n^2 - n + 3)(n + 2)$

EXAMPLE 12

Multiply: $(4x^3 - 2x + 1)(x + 5)$

Solution   Let's multiply vertically.

Use $+ 0x^2$ for the missing x^2 term.

$$
\begin{array}{r}
4x^3 + 0x^2 - 2x + 1 \\
x + 5 \\
\hline
20x^3 + 0x^2 - 10x + 5 \\
4x^4 + 0x^3 - 2x^2 + x \\
\hline
4x^4 + 20x^3 - 2x^2 - 9x + 5
\end{array}
$$

Note that we wrote $0x^2$ for the missing second-degree term in the top polynomial. (Alternatively, we could have left a blank space there.) Similarly, we write a term with a 0 coefficient for each missing term in the partial products.

PRACTICE 12

Find the product of $8x^3 + 9x - 1$ and $3x + 7$.

EXAMPLE 13

Find the product of $a^2 + 2ab + b^2$ and $a + b$.

Solution

$$
\begin{array}{r}
a^2 + 2ab + b^2 \\
a + b \\
\hline
a^2b + 2ab^2 + b^3 \\
a^3 + 2a^2b + ab^2 \\
\hline
a^3 + 3a^2b + 3ab^2 + b^3
\end{array}
$$

PRACTICE 13

Multiply $p^2 - 2pq + q^2$ by $p - q$.

EXAMPLE 14

A box factory makes an open-top box from a piece of cardboard by cutting out squares that are x ft by x ft from each corner, as shown below. The area of the base of the box is given by $(4 - 2x)(3 - 2x)$ ft^2. Rewrite the area of the base without parentheses.

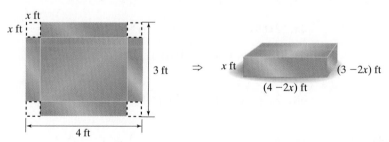

Solution   To remove parentheses, we multiply the binomials using the FOIL method.

$$
\begin{array}{cccc}
\text{F} & \text{O} & \text{I} & \text{L}
\end{array}
$$
$$
(4 - 2x)(3 - 2x) = (4)(3) + (4)(-2x) + (-2x)(3) + (-2x)(-2x)
$$
$$
= 12 - 8x - 6x + 4x^2
$$
$$
= 12 - 14x + 4x^2, \text{ or } 4x^2 - 14x + 12
$$

So the area of the base is $(4x^2 - 14x + 12)$ ft^2.

PRACTICE 14

At the end of 2 years, the amount of money in a savings account is given by

$$(P + Pr) + (P + Pr)r$$

where P is the initial balance and r is the annual rate of interest compounded annually. Rewrite this expression by removing parentheses and simplifying.

Multiply.

1. $(6x)(-4x)$

2. $(-3x)(2x)$

3. $(9t^2)(-t^3)$

4. $(-6y^2)(7y^5)$

5. $(-5x^2)(-4x^4)$

6. $(-2h^2)(-3h^3)$

7. $(10x^3)(-7x^5)$

8. $(20a)(-5a^{99})$

9. $(4pq^2)(4p^2qr^2)$

10. $(8st^2)(6s^4t^7)$

11. $(-8x)^2$

12. $(-10p)^2$

13. $\left(\frac{1}{2}t^4\right)^3$

14. $\left(\frac{1}{3}n^3\right)^3$

15. $(7a)(10a^2)(-5a)$

16. $(-8n^3)(2n^3)(-n)$

17. $(2ab^2)(-3abc)(4a^2)$

18. $(4mn^3)(-m)(7mn^2)$

Find the product.

19. $(7x - 5)x$

20. $(3y - 7)y$

21. $(9t + t^2)(5t)$

22. $(5x - x^2)(4x)$

23. $6a^3(4a^2 - 7a)$

24. $2y(8y - y^3)$

25. $4x^2(3x - 2)$

26. $-5p^7(9p - 3)$

27. $x^3(x^2 - 2x + 4)$

28. $t^2(t^2 + 8t + 1)$

29. $5x(3x^2 + 5x + 6)$

30. $4x(3x^2 - x + 2)$

31. $(5x^2 - 3x - 7)(-9x)$

32. $(10x^2 + x - 1)(2x)$

33. $6x^2(x^3 + 4x^2 - x - 1)$

34. $(x^3 + 6x^2 - 9x + 10)(-3x^4)$

35. $4p(7q - p^2)$

36. $(v + 3w^2)(-7v)$

37. $-pq^2(3p^3 - 9q)$

38. $(m^2 + n^2)\,3mn^5$

39. $2a^2b^3(3a^4b^2 + 10ab^5)$

40. $-4x^2y^2(7x^2y^3 - 4x^3y^4)$

Simplify.

41. $10x + 2x(-3x + 8)$

42. $-7x + 3x(5x - 9)$

43. $-x + 8x(x^2 - 2x + 1)$

44. $2x(8x^2 + 7x - 2) - 6x^2$

45. $9x(x^2 + 3x - 5) + 8x(-4x^2 + x)$

46. $x(x^2 + 11x) - 10x(2x^2 + 3)$

47. $-4xy(2x^2 + 4xy) + x^2y(7x^2 - 2y)$

48. $2s^3t(5s - t^2) - 7s^2t^2(9s^2 - 10t)$

49. $5a^2b^2(3ab^4 - a^3b^2) + 4a^2b^2(9ab^4 - 10a^3b^2)$

50. $(3p - 8pq)(5p^2) - (4p^3 + 1)(7q)$

C *Multiply.*

51. $(y + 2)(y + 3)$

52. $(x + 1)(x + 4)$

53. $(x - 3)(x - 5)$

54. $(n - 4)(n - 2)$

◉ **55.** $(a - 2)(a + 2)$

56. $(x + 3)(x - 3)$

57. $(w + 3)(2w - 7)$

58. $(8x + 5)(x + 4)$

59. $(3 - 2y)(5y - 1)$

60. $(4u - 1)(3 - 2u)$

61. $(10p - 4)(2p - 1)$

62. $(7x + 1)(7x - 3)$

63. $(u + v)(u - v)$

64. $(x + y)(x - y)$

65. $(2p - q)(q - p)$

66. $(x + 4y)(x - y)$

67. $(3a - b)(a - 2b)$

68. $(5x + 4y)(x - y)$

◉ **69.** $(p - 8)(4q + 3)$

70. $(x + 7)(6y - 1)$

D *Find the product.*

71. $(x - 3)(x^2 - 3x + 1)$

72. $(a + 2)(a^2 - 4a + 4)$

◉ **73.** $(2x - 1)(x^2 + 3x - 5)$

74. $(8n + 3)(2n^2 - 9n - 1)$

75. $(a - b)(a^2 + ab + b^2)$

76. $(x^2 + xy + y^2)(y - x)$

77. $(3x)(x + 5)(x - 7)$

78. $(y^2)(8 - 3y)(8 + y)$

79. $(3n)(2n - 1)(2n + 1)$

80. $(-a)(a + 2b)(a - 3b)$

Mixed Practice

Simplify.

81. $m^2n(8m^2 - 6n) - 2mn(3m^2 - 7mn)$

82. $8t^2 + 5t(3t^2 - 2t + 1)$

Multiply.

83. $(-8s^3)(-5s^4)$

84. $3p(8p - p^2)$

85. $(9y^2 + 7y - 8)(-4y)$

86. $(u - 5)(u - 7)$

87. $(3x^4)(-8x^5)(x^2)$

88. $(2w - 3)(4 - w)$

89. $-a^2b(4a - 2b^3)$

90. $(3h^2k^7)(-9hk^5l)$

91. $(x - 1)(x^2 - 3x + 2)$

92. $(a + 6)(8a - 3)$

Applications

Solve.

93. Backgammon is one of the world's oldest board games. The length and width of the distinctive board (shown below) differ by 30 mm. Find the area of the board in terms of x without using parentheses.

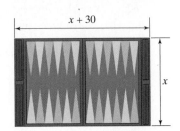

$x + 30$

x

94. A raindrop with diameter d has a cross-sectional area A that can be modeled using the formula $6A - 25 = 2d(5d - 12)$. Solve this formula for A. (*Source:* Brian Lim, "Derivation of the Shape of a Raindrop," cs.cmu.edu)

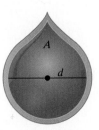

A

d

95. A factory has been selling 1000 color laser printers per year for $1500 each. The company's market research indicates that for each $100 that the price is raised, sales will fall by 30 units. The expression $(1500 + 100x)(1000 - 30x)$ gives the estimated revenue that the company will take in if it adjusts the price of a printer, where x represents the number of $100 increases in the price. Rewrite this expression, multiplying out the factors.

96. A company's total revenue R is given by the equation

$$R = px$$

where p is the price of each item and x is the number of items sold. Write a polynomial for the revenue if

$$x = -\frac{1}{4}p + 100.$$

97. Investment brokers use the formula $A = P(1 + r)^t$ for the amount of money A in a client's account that earns compound interest. In this formula, P is the principal (the original amount of money that the client invested), r is the rate of return per time period (in decimal form), and t is the number of time periods that the money has been invested.

   a. If the client invested $5000 for 3 periods, write this formula as a polynomial in r without parentheses.

   b. If a client invested $5000, how much greater is the amount of money in the account after 3 periods as compared with 2 periods? Write your answer without parentheses.

   c. If the client's rate of return on the investment is 10%, how much money is represented by the expression in part (b)?

98. The expression for the volume of a sphere is $\frac{4}{3}\pi r^3$, where r is the radius of the sphere.

   Assume that the shape of the Earth is approximately a sphere of radius r miles.

   a. Find and simplify the expression for the volume of the sphere formed by everywhere rising 5 mi above the surface of the Earth.

   b. Find and simplify the expression for the volume of the sphere formed by everywhere descending 5 mi below the surface of the Earth.

   c. According to some scientists, all the life discovered so far in the universe is found in the layer around the Earth that extends 5 mi above the Earth's surface to 5 mi below. What is the volume of this layer?

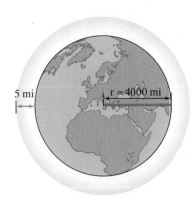

5 mi

$r = 4000$ mi

• Check your answers on page A-21.

MINDStretchers

Mathematical Reasoning

1. We can draw rectangles to visualize the product of two binomials. Consider, for example, the product $(x + 9)(x + 2)$. Using the diagram at the right, explain how the product can be expressed as the sum of four areas.

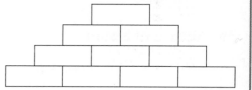

Compare your answer with the result of using the FOIL method to multiply the two binomials.

Patterns

2. Simplify the following polynomials:
$$(x + y)^0 =$$
$$(x + y)^1 =$$
$$(x + y)^2 =$$
$$(x + y)^3 =$$

In the following table, enter the coefficients from these polynomials. What pattern (known as Pascal's Triangle) do you observe in this table?

Historical

3. *Lattice multiplication* is a procedure that originated in India in the twelfth century. In this procedure, whole numbers are multiplied as if they were binomials. Each digit of each factor is multiplied separately. The products are recorded in little cells within a lattice, and then added along the diagonals. For instance, to find $29 \cdot 47$, the four products are placed in the small, diagonally split squares. The product of 2 and 4, shown in red, is 8.

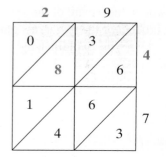

In the diagram to the right, the square in the upper left shows $^0/_8$, which represents 8. Since the product of 9 and 4 is 36, the square in the upper right contains $^3/_6$. The lower products are $^1/_4$ or 14 and $^6/_3$ or 63. These products are all added along the diagonal. For instance, in the diagonal shaded green, the sum $6 + 6 + 4$ is 16; the 6 is written below and the 1 is regrouped into the diagonal above and added into that diagonal: $1 + (3 + 8 + 1)$. The product, 1363, appears down the left side of the lattice and across the bottom.

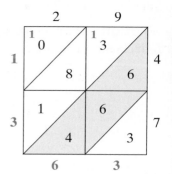

Use lattice multiplication to find the following products:

a. $53 \cdot 89$

b. $61 \cdot 94$

OBJECTIVES

A   To use the formulas for squaring a binomial

B   To use the formula for multiplying the sum and the difference of two terms

C   To solve applied problems involving formulas for special products

Recall that in the previous section, we discussed two ways of multiplying binomials—first by directly applying the distributive property, and second by using the FOIL method. In this section, we focus on yet a third approach—using formulas. These formulas provide a shortcut for finding special products, that is, special cases of multiplying binomials.

The Square of a Binomial

To square a binomial means to multiply that binomial by itself. Consider, for instance, $(x + 5)^2$, in which we square the *sum* of two terms. We can rewrite this expression as a binomial multiplied by itself and then use the FOIL method to find the product.

$$
\begin{aligned}
(x + 5)^2 &= (x + 5)(x + 5) \\
&= x^2 + 5x + 5x + 25 \\
&= x^2 + 10x + 25
\end{aligned}
$$

Note that the expression $x^2 + 10x + 25$ is equal to $(x)^2 + 2(x)(5) + (5)^2$. So the square of the sum of x and 5 equals the square of x plus twice the product of x and 5 plus the square of 5. This observation suggests the following formula for squaring the sum of two terms.

> ### The Square of a Sum
>
> $$(a + b)^2 = a^2 + 2ab + b^2$$

In words, the square of the *sum* of two terms is equal to the square of the first term plus twice the product of the two terms plus the square of the second term.

    Can you explain why this formula follows from the FOIL method of multiplying binomials?

EXAMPLE 1

Simplify: $(x + 7)^2$

Solution   Since we are squaring the sum of x and 7, we can use the formula for the square of a sum.

First term	Second term	The square of the first term		The square of the second term
↓	↓	↓		↓
$(x$	$+ \quad 7)^2$	$= \quad x^2$	$+ \quad \underbrace{2(x)(7)}_{\text{Twice the product of the two terms}}$	$+ \quad (7)^2$

$$= x^2 + 14x + 49$$

PRACTICE 1

Simplify: $(p + 10)^2$

Note in Example 1 that $(x + 7)^2 = x^2 + 14x + 49$, whereas $x^2 + 7^2 = x^2 + 49$. So we see that $(x + 7)^2 \neq x^2 + (7)^2$.

EXAMPLE 2

Simplify: $(p + q)^2$

Solution In $(p + q)^2$, the first term is p and the second term is q.

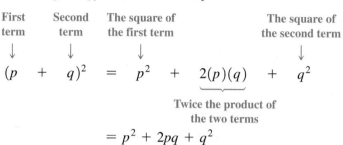

$$= p^2 + 2pq + q^2$$

PRACTICE 2

Simplify: $(s + t)^2$

EXAMPLE 3

Simplify: $(2x + 3y)^2$

Solution

$$(2x + 3y)^2 = (2x)^2 + 2(2x)(3y) + (3y)^2$$
$$= 4x^2 + 12xy + 9y^2$$

PRACTICE 3

Simplify: $(4p + 5q)^2$

Now, let's examine a second and related formula, which involves the square of the *difference* of two terms rather than their sum. For instance, consider $(x - 5)^2$. Again, we rewrite the square of the binomial and then apply the FOIL method:

$$(x - 5)^2 = (x - 5)(x - 5)$$
$$= x^2 - 5x - 5x + 25$$
$$= x^2 - 10x + 25$$

Note that the expression $x^2 - 10x + 25$ is equal to $x^2 - 2(x)(5) + 25$. So the square of the difference of x and 5 equals the square of x minus twice the product of x and 5 plus the square of 5. This example leads to the following formula for squaring the difference of two terms:

The Square of a Difference

$$(a - b)^2 = a^2 - 2ab + b^2$$

In words, the square of the difference of two terms is equal to the square of the first term minus twice the product of the two terms plus the square of the second term.

If we compare the formula for the square of a sum with the formula for the square of a difference, we see that the signs of the middle term of the resulting trinomial differ. That is, for $(a + b)^2$, the middle term is positive, whereas for $(a - b)^2$, the middle term is negative.

EXAMPLE 4

Simplify: $(8a - 1)^2$

Solution Here, we are squaring the difference of two terms, so we use the formula for the square of a difference.

First term	Second term	The square of the first term		The square of the second term
↓	↓	↓		↓

$$(8a - 1)^2 = (8a)^2 - \underbrace{2(8a)(1)}_{\substack{\text{Twice the product of} \\ \text{the two terms}}} + 1^2$$

$$= 64a^2 - 16a + 1$$

PRACTICE 4

Simplify: $(5x - 2)^2$

EXAMPLE 5

Simplify: $(p - q)^2$

Solution

$$(p - q)^2 = p^2 - 2(p)(q) + q^2$$
$$= p^2 - 2pq + q^2$$

PRACTICE 5

Simplify: $(u - v)^2$

EXAMPLE 6

Simplify: $(3a - 4b)^2$

Solution

$$(3a - 4b)^2 = (3a)^2 - 2(3a)(4b) + (4b)^2$$
$$= 9a^2 - 24ab + 16b^2$$

PRACTICE 6

Simplify: $(2x - 9y)^2$

The Product of the Sum and Difference of Two Terms

The third special binomial formula relates to multiplying *the sum of two terms by the difference of the same two terms*. Explain why neither of the two previous formulas applies in this situation. For example, consider the product $(x + 5)(x - 5)$. Using the FOIL method we get:

$$(x + 5)(x - 5) = x \cdot x + x \cdot (-5) + 5 \cdot x - 5 \cdot 5$$
$$= x^2 - \underline{5x + 5x} - 25$$

The middle terms cancel each other out.

$$= x^2 - 25$$

The Product of the Sum and Difference of Two Terms

$$(a + b)(a - b) = a^2 - b^2$$

In words, the product of the sum and difference of the *same* two terms is equal to the square of the first term minus the square of the second term.

EXAMPLE 7

Multiply: $(x + 11)(x - 11)$

Solution

First term	Second term		The square of the first term		The square of the second term
↓	↓		↓		↓

$$(x + 11)(x - 11) = x^2 - (11)^2$$
$$= x^2 - 121$$

PRACTICE 7

Find the product of $(t + 10)$ and $(t - 10)$.

EXAMPLE 8

Multiply.

a. $(p - q)(p + q)$ **b.** $(3m + 2n)(3m - 2n)$

Solution

a. Since $(p - q)(p + q) = (p + q)(p - q)$, the formula for finding the product of the sum and difference of two terms applies.

$$(p - q)(p + q) = p^2 - q^2$$

b. $(3m + 2n)(3m - 2n) = (3m)^2 - (2n)^2 = 9m^2 - 4n^2$

PRACTICE 8

Find the product.

a. $(r - s)(r + s)$

b. $(8s - 3t)(8s + 3t)$

EXAMPLE 9

Find the product: $(3a^2 - 5)(3a^2 + 5)$

Solution $(3a^2 - 5)(3a^2 + 5) = (3a^2)^2 - (5)^2 = 9a^4 - 25$

PRACTICE 9

Multiply: $(10 - 7k^2)(10 + 7k^2)$

EXAMPLE 10

The nineteenth-century French physician Jean Louis Poiseuille investigated the flow of blood in the smaller blood vessels of the body. He discovered that the speed of blood varies from 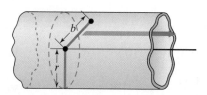 point to point within a blood vessel. In a blood vessel of radius r at a point b units from the center of the blood vessel, the blood flow speed is given by the expression $k(r + b)(r - b)$, where k is a constant. Write this expression without parentheses.

Solution   We need to multiply out $k(r + b)(r - b)$.

$$k(r + b)(r - b) = k(r^2 - b^2)$$
$$= kr^2 - kb^2$$

So the expression is $kr^2 - kb^2$.

PRACTICE 10

The area of the square wooden frame shown can be represented by the polynomial $(S + s)(S - s)$, where s is the side length of the smaller square and S is the side length of the larger square. Write this expression without parentheses.

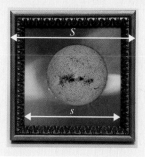

Memorize the formulas for the square of a sum, the square of a difference, and the product of the sum and difference of two terms. With sufficient practice, you will be able to find these special products mentally.

Mathematically Speaking

Fill in each blank with the most appropriate term or phrase from the given list.

| minus | plus | positive |
| negative | times | divided by |

1. The square of the sum of two terms is equal to the square of the first term _____ twice the product of the two terms plus the square of the second term.

2. In the formula for the square of the sum of two terms, the middle term of the trinomial is _____.

3. In the formula for the square of a difference of two terms, the middle term of the trinomial is _____.

4. The product of the sum and difference of the same two terms is equal to the square of the first term _____ the square of the second term.

Simplify.

5. $(y + 2)^2$

6. $(a + 3)^2$

7. $(x + 4)^2$

8. $(n + 8)^2$

9. $(x - 11)^2$

10. $(b - 10)^2$

11. $(6 - n)^2$

12. $(9 - y)^2$

13. $(x + y)^2$

14. $(s + t)^2$

15. $(3x + 1)^2$

16. $(5x + 3)^2$

17. $(4n - 5)^2$

18. $(2x - 1)^2$

19. $(9x + 2)^2$

20. $(11m + 3)^2$

21. $\left(a + \dfrac{1}{2}\right)^2$

22. $\left(b + \dfrac{1}{5}\right)^2$

23. $(8b + c)^2$

24. $(3s + t)^2$

25. $(5x - 2y)^2$

26. $(3m - 4n)^2$

27. $(-x + 3y)^2$

28. $(-p + 2q)^2$

29. $(4x^3 + y^4)^2$

30. $(5a^2 - c^2)^2$

Multiply.

31. $(a + 1)(a - 1)$

32. $(r + 8)(r - 8)$

33. $(4x - 3)(4x + 3)$

34. $(7y - 2)(7y + 2)$

35. $(10 + 3y)(3y - 10)$

36. $(-1 + 9x)(9x + 1)$

37. $\left(m - \dfrac{1}{2}\right)\left(m + \dfrac{1}{2}\right)$

38. $\left(m - \dfrac{1}{4}\right)\left(m + \dfrac{1}{4}\right)$

39. $(n + 0.3)(n - 0.3)$

40. $(y + 0.5)(y - 0.5)$

41. $(4a + b)(4a - b)$

42. $(p - 3q)(p + 3q)$

43. $(3x - 2y)(3x + 2y)$

44. $(10t + 3s)(10t - 3s)$

45. $(1 - 5n)(5n + 1)$

46. $(2s + 3)(3 - 2s)$

47. $x(x + 5)(x - 5)$

48. $y(y - 7)(y + 7)$

49. $5n^2(n + 7)^2$

50. $-8y^3(2y - 1)^2$

51. $(n^2 - m^4)(n^2 + m^4)$

52. $(x^3 + y^5)(x^3 - y^5)$

53. $(a - b)(a + b)(a^2 + b^2)$

54. $(x + y)(x - y)(x^2 + y^2)$

Mixed Practice

Simplify each expression.

55. $(6n + 4)^2$

56. $(2h - 7k)(2h + 7k)$

57. $(4p - 9)(4p + 9)$

58. $(w + 7)^2$

59. $(8 - a)^2$

60. $(3a + 8)(8 - 3a)$

61. $-2x^2(4x - 3)^2$

62. $(3x - 5y)^2$

Applications

Solve.

63. The mayor of a city plans for a square-shaped park, as shown on the grid to the right. The area of the park can be modeled by the expression $(x - 5)^2 + (y - 1)^2$. Rewrite this expression by removing parentheses and simplifying.

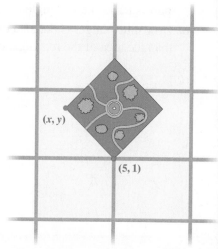

64. A Rubik's Cube is a three-dimensional mechanical puzzle considered to be the world's best-selling toy, with more than 300 million sold worldwide. On each of the 6 faces of the cube is an *n*-by-*n* grid of colored squares. (For instance, a cube with 3-by-3 grids is shown at the right.) Each face of the cube turns independently. The puzzle is solved when on each face all the colored squares are the same color, although different faces will have different colors. (*Source:* wikipedia.org)

  a. How many more colored squares does a Rubik's Cube with an $(n + 1)$-by-$(n + 1)$ grid have than one with an *n*-by-*n* grid?

  b. How many more colored squares in all on its faces does a Rubik's Cube with 4×4 faces have than one with 3×3 faces?

65. An investment of *A* dollars increases in value by *P*% for each of two years. The value of the investment at the end of the two years can be represented by

$$A\left(1 + \frac{P}{100}\right)\left(1 + \frac{P}{100}\right)$$

Multiply out this expression.

66. A carpenter measures the side of a wooden square to be *x* with the margin of error *e*.

  a. What are the longest possible true dimensions of the wooden square? The shortest possible true dimensions?

  b. What is the difference in area between the wooden square with the longest possible dimensions and the wooden square with the shortest possible dimensions in terms of *e* and *x*?

67. To measure the spread of data, statisticians compute the *sample variance* of the data. For a sample of size 3, they use the following formula:

$$\text{Sample variance} = \frac{(a - m)^2 + (b - m)^2 + (c - m)^2}{2}$$

Rewrite this formula without parentheses, combining like terms.

68. As the temperature of a lightbulb's filament changes from T_1 to T_2, the energy that the filament radiates changes by the quantity

$$a(T_1 - T_2)(T_1 + T_2)(T_1^2 + T_2^2)$$

Multiply to find this change.

69. Consider an $(x + y)$ by $(x - y)$ rectangle.
   a. What is the perimeter of this rectangle?
   b. Does a square with side length x have the same perimeter as the rectangle in part (a)?
   c. How much larger is the area of the square in part (b) than the area of the rectangle in part (a)?

70. When a radioactive substance decays, the remaining amount A of the substance after t min can be modeled by the formula $A = m(1 - r)^t$, where m is the initial mass of the substance, t is the elapsed time in minutes, and the rate of decay is r per minute. (*Source:* Atomic Mass Data Center)
   a. Write this formula without parentheses for the amount of 1000 micrograms of a radioactive substance that remains after 2 minutes.
   b. A substance decays at the rate of 3%. After 2 minutes, how much of 1000 micrograms of this substance is left, rounded to the nearest microgram? (*Hint:* Before substituting into the formula, change the rate from a percent to a decimal.)

- Check your answers on page A-21.

MIND*Stretchers*

Patterns

1. Mentally compute each product. (*Hint:* Think of these computations as "special products.")

   a. $9999 \times 10{,}001$

   b. $30\frac{1}{10} \times 29\frac{9}{10}$

Mathematical Reasoning

2. Suppose that you square two consecutive whole numbers and subtract the smaller square from the larger. Is it possible that the difference is an even number? Explain your answer.

Historical

3. By the year 2000 B.C., the astronomers of Mesopotamia knew the relationship $(a - b)(a + b) = a^2 - b^2$. They could demonstrate this relationship by a geometric model. Find the area of the remaining region if the yellow square is removed from the figure shown. Show that this model verifies the relationship $(a - b)(a + b) = a^2 - b^2$. (*Hint:* Find the area in two ways.)

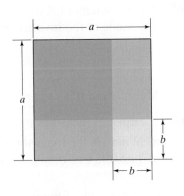

OBJECTIVES

A To divide monomials

B To divide a polynomial by a monomial

C To divide a polynomial by a binomial

D To solve applied problems involving the division of polynomials

The final section of this chapter deals with dividing polynomials. We progress from dividing a monomial by another monomial to dividing a polynomial by a binomial.

Dividing Monomials

The ability to divide monomials—the simplest of polynomial divisions—depends on our knowledge of both fractions and exponents, as the following example illustrates:

EXAMPLE 1

Simplify: $8x^6 \div 4x^2$

Solution   We begin by rewriting the problem in fractional form.

$$8x^6 \div 4x^2 = \frac{8x^6}{4x^2}$$

$$\frac{8x^6}{4x^2} = \frac{8}{4} \cdot \frac{x^6}{x^2}$$

$$= 2x^{6-2} \qquad \text{Divide the coefficients and use the quotient rule of exponents.}$$

$$= 2x^4$$

PRACTICE 1

Find the quotient: $-12n^6 \div 3n$

EXAMPLE 2

Divide: $\dfrac{-15a^2b^3c}{10ab^2}$

Solution $\dfrac{-15a^2b^3c}{10ab^2} = \dfrac{-15}{10} \cdot \dfrac{a^2b^3c}{ab^2}$

$$= \frac{-3}{2} \cdot a^{2-1}b^{3-2}c$$

$$= \frac{-3}{2}abc$$

$$= -\frac{3abc}{2}$$

PRACTICE 2

Find the quotient:
$(20p^3q^2r^4) \div (-5p^2q^2r)$

Dividing a Polynomial by a Monomial

Now, we extend our discussion of division to finding the quotient of a polynomial and a monomial. We consider an example, written in fractional form.

EXAMPLE 3

Simplify the quotient: $(10x^2 + 15x) \div (5x)$

Solution   Let's write the given expression in fractional form.

$$(10x^2 + 15x) \div (5x) = \frac{\overbrace{10x^2 + 15x}^{\text{Polynomial}}}{\underbrace{5x}_{\text{Monomial}}}$$

Since we are dividing a polynomial by a monomial, we can rewrite the given fraction as the sum of two fractions with the same denominator. Then, we simplify each fraction.

$$\frac{10x^2 + 15x}{5x} = \frac{10x^2}{5x} + \frac{15x}{5x}$$
$$= 2x + 3$$

PRACTICE 3

Divide: $(21x^3 - 14x^2) \div 7x$

Example 3 suggests that when dividing a polynomial by a monomial, we can divide each term in the polynomial by the monomial, and then add. Let's apply this shortcut in the following examples.

EXAMPLE 4

Divide: $\dfrac{9x^4 - 6x^3 + 12x^2}{-3x^2}$

Solution   Divide each term in the numerator by the denominator. Then, add the quotients:

$$\frac{9x^4 - 6x^3 + 12x^2}{-3x^2} = \frac{9x^4}{-3x^2} - \frac{6x^3}{-3x^2} + \frac{12x^2}{-3x^2}$$
$$= -3x^2 + 2x - 4$$

PRACTICE 4

Simplify: $\dfrac{14x^8 + 10x^5 - 8x^3}{-2x^3}$

EXAMPLE 5

Divide: $\dfrac{10p^4q^5 + 12p^2q^4 - pq^3}{2pq^2}$

Solution

$$\frac{10p^4q^5 + 12p^2q^4 - pq^3}{2pq^2} = \frac{10p^4q^5}{2pq^2} + \frac{12p^2q^4}{2pq^2} - \frac{pq^3}{2pq^2}$$
$$= 5p^3q^3 + 6pq^2 - \frac{q}{2}$$

PRACTICE 5

Find the quotient:

$$\frac{-5a^7b^6 + a^2b^4 - 15ab^3}{5ab^3}$$

Dividing a Polynomial by a Binomial

Now, let's take a look at the general case—how to divide one polynomial by another polynomial. We will restrict our attention to dividing a polynomial by a *binomial*, but these remarks apply to dividing by polynomials with three or more terms as well.

     The procedure for dividing a polynomial by a binomial is similar to that of dividing whole numbers, which is commonly called *long division*.

$$
\begin{array}{r}
12 \longleftarrow \text{Quotient} \\
\text{Divisor} \longrightarrow 13\overline{)\ 158} \longleftarrow \text{Dividend} \\
-13 \quad\quad \\
\overline{28} \quad\quad \\
-26 \quad\quad \\
\overline{2} \longleftarrow \text{Remainder}
\end{array}
$$

We stop dividing because the remainder 2 is smaller than the divisor 13. So $158 \div 13 = 12$ with remainder 2.

     Now, let's consider the following problem in dividing polynomials:

$$(x^2 - 5x - 24) \div (x + 3)$$

We set up the problem just as if we were dividing whole numbers, being careful to correctly distinguish between the dividend and the divisor. In this case, $x^2 - 5x - 24$ is the dividend and $x + 3$ is the divisor.

$$
\begin{array}{r}
x \\
x + 3\overline{)x^2 - 5x - 24} \\
\underline{x^2 + 3x}
\end{array}
$$

Divide the first term in the dividend by the first term in the divisor: Think $x^2 \div x = x$. Place x in the quotient above the x term in the dividend. Then, *multiply* x in the quotient by the divisor: $x(x + 3) = x^2 + 3x$.

$$
\begin{array}{r}
x \\
x + 3\overline{)x^2 - 5x - 24} \\
\underline{x^2 + 3x} \\
-8x - 24
\end{array}
$$

Subtract $(x^2 + 3x)$ from $(x^2 - 5x)$ by changing the signs of each term in $(x^2 + 3x)$ and then adding to get $-8x$. Bring down the next term, -24. We have to divide again because the degree of $-8x - 24$ is equal to the degree of $x + 3$. Both have degree 1.

$$
\begin{array}{r}
x - 8 \\
x + 3\overline{)x^2 - 5x - 24} \\
\underline{x^2 + 3x} \\
-8x - 24 \\
\underline{-8x - 24} \\
0
\end{array}
$$

The remainder is 0.

Divide x into $-8x$: Think $-8x \div x = -8$. Place -8 in the quotient above the constant term in the dividend. Then, *multiply* -8 in the quotient by the divisor: $-8(x + 3) = -8x - 24$. Finally, *subtract* $(-8x - 24)$ from $(-8x - 24)$ by changing the signs of each term in $(-8x - 24)$ and then adding to get 0.

Note that the degree of the remainder is less than the degree of the divisor because the remainder 0 is a constant, which has degree 0, and the divisor $x + 3$ has degree 1.

     So $(x^2 - 5x - 24) \div (x + 3) = x - 8$. Can you explain how to check this quotient?

     This example suggests the following general method for dividing a polynomial by a polynomial.

To Divide a Polynomial by a Polynomial

- Arrange each term of the dividend and divisor in descending order.
- Divide the first term of the dividend by the first term of the divisor. The result is the first term of the quotient.
- Multiply the first term of the quotient by the divisor and place the product under the dividend.
- Subtract the product, found in the previous step, from the dividend.
- Bring down the next term to form a new dividend.
- Repeat the process until the degree of the remainder is less than the degree of the divisor.

EXAMPLE 6

$$2x - 1 \overline{)6x^2 + 9x - 6}$$

Solution The dividend and divisor are already in descending order.

$$
\begin{array}{r}
3x \\
2x - 1 \overline{)6x^2 + 9x - 6} \\
\underline{6x^2 - 3x} \\
12x - 6
\end{array}
$$

Divide $6x^2$ by $2x$, getting $3x$.
Multiply $3x$ by $(2x - 1)$, getting $(6x^2 - 3x)$.
Subtract $(6x^2 - 3x)$ from $(6x^2 + 9x)$ and bring down -6.

$$
\begin{array}{r}
3x + 6 \\
2x - 1 \overline{)6x^2 + 9x - 6} \\
\underline{6x^2 - 3x} \\
12x - 6 \\
\underline{12x - 6} \\
0
\end{array}
$$

Divide $12x$ by $2x$, getting 6.
Multiply 6 by $(2x - 1)$, getting $(12x - 6)$.
Subtract $(12x - 6)$ from $(12x - 6)$.

Since the degree of 0 is less than the degree of $(2x - 1)$, the process stops.
So $(6x^2 + 9x - 6) \div (2x - 1) = 3x + 6$.

PRACTICE 6

Find the quotient:
$(10x^2 + 17x + 3) \div (5x + 1)$

Now, we focus on problems in dividing polynomials that have *remainders*.

Dividing whole numbers:

$$
\begin{array}{r}
21 \\
45 \overline{)956} \\
\underline{90} \\
56 \\
\underline{45} \\
11 \longleftarrow \text{Remainder}
\end{array}
$$

Check

$$45 \cdot 21 + 11 \overset{?}{=} 956$$
$$945 + 11 = 956 \qquad \text{True}$$

So $956 \div 45 = 21\dfrac{11}{45}$.

Dividing polynomials:

$$
\begin{array}{r}
x + 3 \\
x + 5 \overline{)x^2 + 8x + 16} \\
\underline{x^2 + 5x} \\
3x + 16 \\
\underline{3x + 15} \\
1 \longleftarrow \text{Remainder}
\end{array}
$$

Check

$$(x + 5)(x + 3) + 1 \overset{?}{=} x^2 + 8x + 16$$
$$x^2 + 8x + 15 + 1 = x^2 + 8x + 16 \qquad \text{True}$$

So $(x^2 + 8x + 16) \div (x + 5) = x + 3 + \dfrac{1}{x + 5}$.

Note that we can check a problem involving division of polynomials in the same way that we check division of whole numbers:

$$\text{Divisor} \cdot \text{quotient} + \text{remainder} = \text{Dividend}$$

EXAMPLE 7

Find the quotient: $(x^3 + 3x^2 - 8x + 2) \div (x + 5)$

Solution

$$
\begin{array}{r}
x^2 - 2x + 2 \\
x + 5\overline{\smash{)}x^3 + 3x^2 - 8x + 2} \\
\underline{x^3 + 5x^2} \\
-2x^2 - 8x \\
\underline{-2x^2 - 10x} \\
2x + 2 \\
\underline{2x + 10} \\
-8
\end{array}
$$

Since the degree of the remainder is less than the degree of the divisor, we stop.

Check

$$
\begin{aligned}
(x + 5)(x^2 - 2x + 2) + (-8) &\overset{?}{=} x^3 + 3x^2 - 8x + 2 \\
x^3 + 3x^2 - 8x + 2 &= x^3 + 3x^2 - 8x + 2 \quad \text{True}
\end{aligned}
$$

So we write the answer as:

$$x^2 - 2x + 2 + \frac{-8}{x + 5}$$

PRACTICE 7

Divide
$(3x^3 + 7x^2 + 11x + 5)$ by
$(3x + 1)$.

Some problems in dividing polynomials involve terms that are not in *descending order*.

EXAMPLE 8

Divide $(6 + 8x^2 - 14x)$ by $(2x - 3)$.

Solution   Before dividing, we place the terms in both the divisor and the dividend in descending order. Here we need to rearrange the terms in the dividend:

$$
\begin{array}{r}
4x - 1 \\
2x - 3\overline{\smash{)}8x^2 - 14x + 6} \\
\underline{8x^2 - 12x} \\
-2x + 6 \\
\underline{-2x + 3} \\
3 \quad \text{The remainder is 3.}
\end{array}
$$

So $(8x^2 - 14x + 6) \div (2x - 3) = 4x - 1 + \dfrac{3}{2x - 3}$.

How would you check this solution?

PRACTICE 8

Divide:
$(-21s + 10 + 9s^2) \div (3s - 2)$

In dividing polynomials, we may have *missing terms* in the dividend, as shown in the following example.

EXAMPLE 9

$x + 3\overline{)2x^3 + 7x^2 - 9}$

Solution   Since there is no x-term in the dividend, we can insert $0x$ as a placeholder for the missing term.

$$
\begin{array}{r}
2x^2 + x - 3 \\
x + 3\overline{)2x^3 + 7x^2 + 0x - 9} \\
\underline{2x^3 + 6x^2} \\
x^2 + 0x \\
\underline{x^2 + 3x} \\
-3x - 9 \\
\underline{-3x - 9} \\
0
\end{array}
$$

So $\dfrac{2x^3 + 7x^2 - 9}{x + 3} = 2x^2 + x - 3.$

PRACTICE 9

Divide: $\dfrac{4n^3 - 19n^2 - 4}{4n - 3}$

EXAMPLE 10

A homeowner wishes to increase the length of a flower bed by twice as much as the increase in the width. The area of the new flower bed is given by $(2x^2 + 20x + 48)$ ft^2.

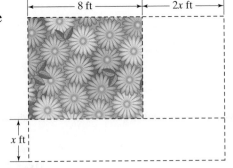

8 ft   2x ft

x ft

a. Use long division to find the width of the new flower bed in terms of x.

b. If the homeowner increases the width by 1 ft, then what are the dimensions of the new flower bed?

Solution

a. To find the width of the new rectangular flower bed, we can use the formula $A = lw$, where the area A and the length l are given by $(2x^2 + 20x + 48)$ ft^2 and $(2x + 8)$ ft respectively. Since

$A = lw, w = \dfrac{A}{l}.$

Using $w = \dfrac{A}{l}$, we conclude that $w = \dfrac{2x^2 + 20x + 48}{2x + 8}$. Dividing the numerator by the denominator, we find that $w = x + 6$. So the width is $(x + 6)$ ft.

b. So if the owner increases the width of the flower bed by 1 foot, the dimensions of the new flower bed are:

$(1 + 6)$ ft by $(2 \cdot 1 + 8)$ ft, or 7 ft by 10 ft.

PRACTICE 10

If $10 is invested at an interest rate of r per year and compounded annually, the future value S in dollars at the end of the nth year is given by:

$$S = 10(1 + r)^n$$

a. What is the future value of the investment after 1 yr? After 2 yr?

b. Write the answers to part (a) without parentheses and in descending order.

c. Using your answer in part (b), determine how many times as great the future value of the investment is after 2 yr as compared to the future value after 1 yr.

Mathematically Speaking

Fill in each blank with the most appropriate term or phrase from the given list.

powers	polynomial by the monomial	quotient
dividend	coefficients	divisor
remainder	monomial by the polynomial	

1. To divide a monomial by a monomial, divide the _____ and use the quotient rule of exponents.

2. To divide a polynomial by a monomial, divide each term in the _____ and then add.

3. In dividing one polynomial by another, repeat the process until the degree of the _____ is less than the degree of the divisor.

4. To check if the quotient of two polynomials is correct, see if the _____ is the sum of the remainder and the product of the divisor and the quotient.

A *Simplify.*

5. $\dfrac{10x^4}{5x^2}$

6. $\dfrac{6x^3}{2x^2}$

7. $\dfrac{16a^8}{-4a}$

8. $\dfrac{-35y^2}{7y}$

9. $\dfrac{-8x^5}{-6x^4}$

10. $\dfrac{-9p^5}{-12p^2}$

11. $\dfrac{12p^2q^3}{3p^2q}$

12. $\dfrac{9a^4b}{3a^2b}$

13. $\dfrac{-24u^6v^4}{-8u^4v^2}$

14. $\dfrac{-4x^5y^4}{-2x^2y}$

⊙ 15. $\dfrac{-15a^2b^5}{7ab^3}$

16. $\dfrac{21x^3y^5}{-10xy^2}$

17. $\dfrac{-6u^5v^3w^3}{4u^2vw^3}$

18. $\dfrac{-10x^3yz^2}{8x^2yz}$

B *Divide.*

19. $\dfrac{6n^2 + 10n}{2n}$

20. $\dfrac{12m^4 + 15m^3}{3m}$

21. $\dfrac{20b^4 - 10b}{10b}$

22. $\dfrac{2x^2 - 8x}{2x}$

23. $\dfrac{18a^2 + 12a}{-3a}$

24. $\dfrac{16x^2 + 10x}{-2x}$

25. $\dfrac{9x^5 - 6x^7}{3x^5}$

26. $\dfrac{6a^3 - 4a^2}{2a^2}$

27. $\dfrac{12a^4 - 18a^3 + 30a^2}{6a^2}$

28. $\dfrac{8x^5 + 4x^4 - 16x^3}{4x^2}$

29. $\dfrac{n^5 - 10n^4 - 5n^3}{-5n^3}$

30. $\dfrac{9y^6 - 3y^5 - 2y^4}{-3y^4}$

31. $\dfrac{20a^2b + 4ab^3}{8ab}$

32. $\dfrac{14xy^2 - 21x^3y^4}{7xy^2}$

⊙ 33. $\dfrac{12x^2y^3 - 9xy - 3xy^2}{-3xy}$

34. $\dfrac{10ab^8 - 4ab^6 + 6ab^4}{-2ab^4}$

35. $\dfrac{8p^2q^3 - 4p^3q^3 + 6p^4q}{4p^2q}$

36. $\dfrac{6x^2y^3 - 18x^3y^4 + 9x^4y^5}{6x^2y^3}$

Find the quotient.

37. $(x^2 - 4x - 21) \div (x + 3)$

38. $(x^2 + 6x - 40) \div (x + 10)$

39. $(56x^2 - 23x + 2) \div (8x - 1)$

40. $(30x^2 + 13x - 3) \div (6x - 1)$

41. $(6x^2 + 13x - 5) \div (2x + 5)$

42. $(10x^2 - x - 2) \div (5x + 2)$

43. $(-2x + 5x^2 - 3) \div (x - 1)$

44. $(19x + 2x^2 + 35) \div (x + 7)$

45. $(4 + 20x + 21x^2) \div (2 + 3x)$

46. $(-3 + x + 2x^2) \div (3 + 2x)$

Divide.

47. $\dfrac{x^2 + 2x + 5}{x + 2}$

48. $\dfrac{x^2 + 2x + 7}{x - 2}$

49. $\dfrac{-3 - 5x + 2x^2}{x - 3}$

50. $\dfrac{-2x + 5x^2 - 3}{x - 1}$

51. $\dfrac{8x^2 - 6x - 11}{4x + 3}$

52. $\dfrac{3x^2 - x - 8}{3x - 1}$

53. $\dfrac{-x + x^3 - 5x^2 + 5}{x + 1}$

54. $\dfrac{-5 + 11x - 7x^2 + x^3}{x - 5}$

55. $\dfrac{6x^3 - 11x^2 - 5x + 19}{3x - 4}$

56. $\dfrac{2x^3 + x^2 - 4x - 8}{2x + 1}$

57. $\dfrac{5x^2 - 2}{x - 4}$

58. $\dfrac{10x^2 - 2x}{x + 3}$

59. $\dfrac{4x^3 - x + 3}{2x - 3}$

60. $\dfrac{3x^3 + x^2 - 4}{x + 1}$

61. $\dfrac{x^3 + 27}{x + 3}$

62. $\dfrac{x^3 - 1}{x - 1}$

Mixed Practice

Simplify.

63. $\dfrac{56r^2}{-8r}$

64. $\dfrac{42x^5y^2}{7xy^2}$

65. $\dfrac{-18a^2b^3c}{27ab^2c}$

Divide.

66. $\dfrac{15a^8b - 21a^4b - 24a^3b}{3a^3b}$

67. $\dfrac{18n^6 - 48n^4 - 2n^3}{-6n^3}$

68. $(36x^2 + x - 2) \div (4x + 1)$

69. $(5 + 13x + 6x^2) \div (5 + 3x)$

70. $\dfrac{10t^5 - 6t^3}{2t}$

71. $\dfrac{4x^3 + x^2 - 4x + 2}{4x + 1}$

72. $\dfrac{4x^2 - 5x}{x - 3}$

73. $\dfrac{32m^3 - 72m}{-8m}$

74. $\dfrac{2x^2 - 13x - 24}{x - 8}$

Applications

Ⓓ *Solve.*

75. The number of typed words that fit on a page depends on the font used as well as the font's point size. For the popular font Times New Roman, the average number of words that fit on a (single-spaced) page can be modeled by the polynomial $9x^2 - 340x + 3600$, where x is the point size of the font. (*Source:* writersservices.com)

  a. Use this formula to approximate the number of words on a page for 10-point Times New Roman.

  b. What is the average number of words on a page *per point* of font size?

  c. What is the average number of words on a page per point for 10-point Times New Roman?

76. A prepaid phone card company charges a $0.49 connection fee for each call plus $0.01 for each minute.

  a. Write an expression for the cost of a call in terms of x, the number of minutes the call lasts.

  b. Write an expression for the cost per minute of a call.

  c. Use division to rewrite the expression in part (b).

77. The formula $d = rt$ can be used to find the time t when given the distance d and the rate r.

  a. Solve the equation $d = rt$ for t.

  b. Use the answer from part (a) to find an expression for the time it takes to travel a distance of $(t^3 - 6t^2 + 7t + 14)$ mi at a rate of $(t + 1)$ mph.

78. A *geometric series* is a sum of terms where each term is formed by multiplying the previous term by a constant. For example, the series $5 + 5r + 5r^2$ has three terms where the first term is 5 and each of the other terms is r times the previous term. Use long division to show that the sum of the first three terms can be calculated from the formula $\dfrac{-5r^3 + 5}{-r + 1}$.

79. The polynomial $3x^3 + 348x^2 + 5250x + 45,000$ models the total expenditures on home health care (in millions of dollars) for a particular year in the United States, where x represents the number of years after 2004. The polynomial $3x + 300$ approximates the U.S. population (in millions) during the same years. Write a polynomial that approximates the total expenditures on home heath care per person. (*Source:* cms.gov)

80. The polynomial $-9x^3 - 750x^2 + 16,800x + 180,000$ models the total expenditures on prescription drugs (in millions of dollars) in the United States for a particular year, where x represents the number of years after 2004. The polynomial $3x + 300$ approximates the U.S. population (in millions) during the same years. Write a polynomial that approximates the total expenditures on prescription drugs per person. (*Source:* cms.gov)

• Check your answers on page A-21.

Mathematical Reasoning

1. When you divide a polynomial by a trinomial, explain how you know if the trinomial is a factor of the polynomial.

Patterns

2. Consider the following table:

Divisor	Dividend	Quotient
$x + 1$	$x^2 - x + 1$	
$x + 1$	$x^3 - x^2 + x - 1$	
$x + 1$	$x^4 - x^3 + x^2 - x + 1$	

   a. Complete the table by dividing each dividend by the divisor.
   b. Predict the result of dividing $x^5 - x^4 + x^3 - x^2 + x - 1$ by $x + 1$.
      Verify your prediction.

Critical Thinking

3. Divide $(2y^2 + 5y + 3)$ by $(y + 1)$. For which values of y is the quotient larger in value than the divisor?

Concept/Skill	Description	Example
[5.1] **Exponent (or Power)**	A number that indicates how many times another number (called the *base*) is used as a factor.	$4^3 = \underbrace{4 \cdot 4 \cdot 4}_{\text{3 factors}}$ with Exponent labeling the 3 and Base labeling the 4
[5.1] **Exponent 1**	For any real number x, $x^1 = x$.	$2^1 = 2$ $(ab)^1 = ab$
[5.1] **Exponent 0**	For any nonzero real number x, $x^0 = 1$.	$5^0 = 1$ $(ab)^0 = 1$
[5.1] **Product rule of exponents**	For any nonzero real number x and for any integers a and b, $$x^a \cdot x^b = x^{a+b}$$	$3^2 \cdot 3^3 = 3^{2+3} = 3^5 = 243$ $x \cdot x = x^{1+1} = x^2$
[5.1] **Quotient rule of exponents**	For any nonzero real number x and for any integers a and b, $$\frac{x^a}{x^b} = x^{a-b}$$	$2^5 \div 2^3 = \frac{2^5}{2^3} = 2^{5-3} = 2^2 = 4$ $\frac{x^6}{x} = x^{6-1} = x^5$
[5.1] **Negative exponents**	For any nonzero real number x and for any integer a, $$x^{-a} = \frac{1}{x^a}$$	$6^{-2} = \frac{1}{6^2} = \frac{1}{36}$ $(2x)^{-2} = \frac{1}{(2x)^2} = \frac{1}{4x^2}$
[5.1] **Reciprocal of x^{-a}**	For any nonzero real number x and for any integer a, $$\frac{1}{x^{-a}} = x^a$$	$\frac{1}{5^{-3}} = 5^3 = 125$ $\frac{x^3}{y^{-2}} = x^3 y^2$
[5.2] **Power rule of exponents**	For any nonzero real number x and for any integers a and b, $$(x^a)^b = x^{ab}$$	$(2^3)^2 = 2^{3 \cdot 2} = 2^6 = 64$ $(p^7)^3 = p^{7 \cdot 3} = p^{21}$
[5.2] **Raising a product to a power**	For any nonzero real numbers x and y and any integer a, $$(xy)^a = x^a \cdot y^a$$	$(4y)^3 = 4^3 y^3 = 64y^3$ $(c^4 d^3)^5 = c^{4 \cdot 5} d^{3 \cdot 5}$ $= c^{20} d^{15}$
[5.2] **Raising a quotient to a power**	For any nonzero real numbers x and y and any integer a, $$\left(\frac{x}{y}\right)^a = \frac{x^a}{y^a}$$	$\left(\frac{2}{3}\right)^3 = \frac{2^3}{3^3} = \frac{8}{27}$ $\left(\frac{-5}{b}\right)^4 = \frac{(-5)^4}{b^4} = \frac{625}{b^4}$

continued

Concept/Skill	Description	Example
[5.2] **Raising a quotient to a negative power**	For any nonzero real numbers x and y and any integer a, $$\left(\frac{x}{y}\right)^{-a} = \left(\frac{y}{x}\right)^{a}$$	$$\left(\frac{2}{3}\right)^{-1} = \left(\frac{3}{2}\right)^{1} = \frac{3}{2}$$ $$\left(\frac{p}{q}\right)^{-3} = \left(\frac{q}{p}\right)^{3} = \frac{q^3}{p^3}$$
[5.2] **Scientific notation**	A number is in scientific notation if it is written in the form $$a \times 10^n$$ where n is an integer and a is greater than or equal to 1 but less than 10 ($1 \le a < 10$).	5.3×10^9 and 2.41×10^{-5} are in scientific notation.
[5.3] **Monomial**	An expression that is the product of a real number and variables raised to nonnegative integer powers.	$3x^3$ $-4a^2b$
[5.3] **Polynomial**	An algebraic expression with one or more monomials added or subtracted.	$-5x$ ⟵ Monomial $2x + 1$ ⟵ Binomial $x^2 - 9x + 2$ ⟵ Trinomial $-x^3 + 7x^2 - x + 19$ Polynomial
[5.3] **Degree of a monomial**	The power of the variable in a monomial.	$3x^5$ is of degree 5.
[5.3] **Degree of a polynomial**	The highest degree of any of its terms.	$4x^2 - x + 3$ is of degree 2.
[5.3] **Leading term of a polynomial**	The term in a polynomial with the highest degree. Its coefficient is called the **leading coefficient** of the polynomial. The term of degree 0 is called the **constant term**.	In $-8x^2 - 7x + 3$, $-8x^2$ is the leading term, -8 is the leading coefficient, and 3 is the constant term.
[5.4] **To add polynomials**	• Add the like terms.	Find the sum horizontally: $(5x^2 + 6x - 9) + (-10x^2 + 7)$ $= 5x^2 + 6x - 9 - 10x^2 + 7$ $= -5x^2 + 6x - 2$ Add vertically: $(5x^2 + 6x - 9) + (-10x^2 + 7)$ $5x^2 + 6x - 9$ $\underline{-10x^2 + 7}$ $-5x^2 + 6x - 2$

Concept/Skill	Description	Example
[5.4] To subtract polynomials	• Change the sign of each term of the polynomial being subtracted. • Add the like terms.	Subtract horizontally: $$(2x^2 - 6x + 1) - (-x^2 + 4x + 5)$$ $$= 2x^2 - 6x + 1 + x^2 - 4x - 5$$ $$= 3x^2 - 10x - 4$$ Subtract vertically: $$(2x^2 - 6x + 1) - (-x^2 + 4x + 5)$$ $$\begin{array}{rr} 2x^2 - 6x + 1 & 2x^2 - 6x + 1 \\ -(-x^2 + 4x + 5) & \underline{x^2 - 4x - 5} \\ & 3x^2 - 10x - 4 \end{array}$$
[5.5] To multiply monomials	• Multiply the coefficients. • Multiply the variables, using the product rule of exponents.	Multiply: $(-3x^2y)(5x^3y^2)$ $= (-3 \cdot 5)(x^2 \cdot x^3)(y \cdot y^2) = -15x^5y^3$

[5.5] To multiply two binomials using the FOIL method	Consider $(a + b)(c + d)$.
	• Multiply the two first terms in the binomials.
	
	    $(a + b)(c + d)$ Product is ac.
	      F
	
	• Multiply the two outer terms.
	
	    $(a + b)(c + d)$ Product is ad.
	      O
	
	• Multiply the two inner terms.
	
	    $(a + b)(c + d)$ Product is bc.
	      I
	
	• Multiply the two last terms.
	
	    $(a + b)(c + d)$ Product is bd.
	      L
	
	The product of the two binomials is the sum of these four products.
	$$(a + b)(c + d) = ac + ad + bc + bd$$
		Last
		First
		$$(3x - 1)(2x + 5)$$
		Inner
		Outer
		
		F $(3x)(2x) = 6x^2$
		O $(3x)(5) = 15x$
		I $(-1)(2x) = -2x$
		L $(-1)(5) = -5$
		
		The product $(3x - 1)(2x + 5)$ is the sum of these four products:
		$$6x^2 + 15x - 2x - 5$$
		$$= 6x^2 + 13x - 5$$

[5.6] The square of a sum	$$(a + b)^2 = a^2 + 2ab + b^2$$	$$(x + 3)^2 = x^2 + 2(x)(3) + (3)^2$$ $$= x^2 + 6x + 9$$ $$(7y + 5)^2 = (7y)^2 + 2(7y)(5) + (5)^2$$ $$= 49y^2 + 70y + 25$$
[5.6] The square of a difference	$$(a - b)^2 = a^2 - 2ab + b^2$$	$$(y - 6)^2 = y^2 - 2(y)(6) + (6)^2$$ $$= y^2 - 12y + 36$$ $$(4x - 1)^2 = (4x)^2 - 2(4x)(1) + (1)^2$$ $$= 16x^2 - 8x + 1$$

continued

Concept/Skill	Description	Example
[5.6] **The product of the sum and difference of two terms**	$(a + b)(a - b) = a^2 - b^2$	$(2t + s)(2t - s) = (2t)^2 - (s)^2$ $$= 4t^2 - s^2$$
[5.7] **To divide a polynomial by a polynomial**	• Arrange each term of the dividend and divisor in descending order. • Divide the first term of the dividend by the first term of the divisor. The result is the first term of the quotient. • Multiply the first term of the quotient by the divisor and place the product under the dividend. • Subtract the product, found in the previous step, from the dividend. • Bring down the next term to form a new dividend. • Repeat the process until the degree of the remainder is less than the degree of the divisor.	$(3y^2 - 4y - 7) \div (y - 2)$ Quotient $3y + 2$ $y - 2 \overline{)3y^2 - 4y - 7}$ ⟵ Dividend Divisor $\dfrac{3y^2 - 6y}{2y - 7}$ $\dfrac{2y - 4}{}$ Remainder ⟶ -3

Say Why

Fill in each blank.

1. $3^2 \cdot 3^4$ _____ equal to 3^8 because _____
   is/is not

   _____

   _____.

2. $(-5)^7 \div (-5)^4$ _____ equal to $(-5)^3$ because
   is/is not

   _____

   _____.

3. The expression $\dfrac{1}{a^3}$ _____ be simplified as a^{-3}
   can/cannot

   because _____

   _____.

4. The number 0.24×10^3 _____ written in scientific
   is/is not

   notation because _____.

5. $-\dfrac{2}{3}x^{-3}$ _____ a monomial because _____
   is/is not

   _____.

6. In the polynomial $x^2 + 5$, the constant term

   _____ have degree 0 because
   does/does not

   _____.

[5.1] *Simplify.*

7. $(-x)^3$

8. -31^0

9. $n^4 \cdot n^7$

10. $x^6 \cdot x$

11. $\dfrac{n^8}{n^5}$

12. $p^{10} \div p^7$

13. $y^4 \cdot y^2 \cdot y$

14. $(a^2 b)(ab^2)$

15. $x^0 y$

16. $\dfrac{n^4 \cdot n^7}{n^9}$

Write as an expression using only positive exponents.

17. $(5x)^{-1}$

18. $-3n^{-2}$

19. $8^{-2} v^4$

20. $\dfrac{1}{y^{-4}}$

21. $x^{-8} \cdot x^7$

22. $5^{-1} \cdot y^6 \cdot y^{-3}$

23. $\dfrac{a^5}{a^{-5}}$

24. $\dfrac{t^{-2}}{t^4}$

25. $\dfrac{x^{-2}}{y}$

26. $\dfrac{x^2}{y^{-1}}$

[5.2] *Simplify.*

27. $(10^2)^4$

28. $-(x^3)^3$

29. $(2x^3)^2$

30. $(-4m^5 n)^3$

31. $3(x^{-2})^6$

32. $(a^3 b^{-4})^{-2}$

33. $\left(\dfrac{x}{3}\right)^4$

34. $\left(\dfrac{-a}{b^3}\right)^2$

35. $\left(\dfrac{x}{y}\right)^{-6}$

36. $\left(\dfrac{x^2}{y^{-1}}\right)^5$

37. $\left(\dfrac{4a^3}{b^4 c}\right)^2$

38. $\left(\dfrac{-u^{-5} v^2}{7w}\right)^2$

Express in standard notation.

39. 3.7×10^{10}

40. 1.63×10^9

41. 5.022×10^{-5}

42. 6×10^{-11}

Express in scientific notation.

43. $1,200,000,000,000$ **44.** $427,000,000$ **45.** 0.00000000000004 **46.** 0.00000056

Perform the indicated operation. Then, write the result in scientific notation.

47. $(1.4 \times 10^6)(4.2 \times 10^3)$ **48.** $(3 \times 10^{-2})(2.1 \times 10^5)$

49. $(1.8 \times 10^4) \div (3 \times 10^{-3})$ **50.** $(9.6 \times 10^{-4}) \div (1.6 \times 10^6)$

[5.3]   *Indicate whether the expression is a polynomial.*

51. $3x^4 - 5x^3 + \dfrac{x^2}{4} - 8$ **52.** $-2x^2 - \dfrac{7}{x} + 1$

Classify each polynomial according to the number of terms.

53. $2x^5 + 7x^2 - 5$ **54.** $16 - 4t^2$

Write the polynomial in descending order. Then, identify the degree, leading term, and leading coefficient of the polynomial.

55. $8y - 3y^3 + y^2 - 1$ **56.** $n^4 - 6n^2 - 7n^3 + n$

Simplify. Then, write the polynomial in descending order of powers.

57. $10x - 8x^2 - 8x + 9x^2 - x^3 + 13$ **58.** $4n^3 - 7n + 9 - 3n^2 - n^3 + 7n^2 - 5 + n$

Evaluate the polynomial for the given values of the variable.

59. $2n^2 - 7n + 3$ for $n = -1$ and $n = 3$ **60.** $x^3 - 8$ for $x = 2$ and $x = -2$

[5.4]   *Perform the indicated operations.*

61. $(4x^2 - x + 4) + (-3x^2 + 9)$ **62.** $(5y^4 - 2y^3 + 7y - 11) + (6 - 8y - y^2 - 5y^4)$

63. $(a^2 + 5ab + 6b^2) + (3a^2 - 9b^2) + (-7ab - 3a^2)$ **64.** $(5s^3t - 2st + t^2) + (s^2t - 5t^2) + (t^2 - 4st + 9s^2)$

65. $(x^2 - 5x + 2) - (-x^2 + 3x + 10)$ **66.** $(10n^3 + n^2 - 4n + 1) - (11n^3 - 2n^2 - 5n + 1)$

67. $\begin{aligned} 5y^4 - 4y^3 \qquad + y - 6 \\ -(\qquad y^3 - 2y^2 + 7y - 3) \end{aligned}$ **68.** $\begin{aligned} -9x^3 + 8x^2 - 11x - 12 \\ +(11x^3 \qquad - x + 15) \end{aligned}$

Simplify.

69. $14t^2 - (10t^2 - 4t)$ **70.** $-(5x - 6y) + (3x - 7y)$

71. $(3y^2 - 1) - (y^2 + 3y + 2) + (-2y + 5)$ **72.** $(1 - 4x - 6x^2) - (7x - 8) - (-11x - x^2)$

[5.5]   *Multiply.*

73. $-3x^4 \cdot 2x$ **74.** $(3ab)(8a^2b^3)(-6b)$ **75.** $2xy^2(4x - 5y)$

76. $(x^2 - 3x + 1)(-5x^2)$ **77.** $(n + 3)(n + 7)$ **78.** $(3x - 9)(x + 6)$

79. $(2x - 1)(4x - 1)$

80. $(3a - b)(3a + 2b)$

81. $(2x^3 - 5x + 2)(x + 3)$

82. $(y - 2)(y^2 - 7y + 1)$

Simplify.

83. $-y + 2y(-3y + 7)$

84. $4x^2(2x - 6) - 3x(3x^2 - 10x + 2)$

[5.6] *Simplify.*

85. $(a - 1)^2$

86. $(s + 4)^2$

87. $(2x + 5)^2$

88. $(3 - 4t)^2$

89. $(5a - 2b)^2$

90. $(u^2 + v^2)^2$

Multiply.

91. $(m + 4)(m - 4)$

92. $(6 - n)(6 + n)$

93. $(7n - 1)(7n + 1)$

94. $(2x + y)(2x - y)$

95. $(4a - 3b)(4a + 3b)$

96. $x(x + 10)(x - 10)$

97. $-3t^2(4t - 5)^2$

98. $(p^2 - q^2)(p + q)(p - q)$

[5.7] *Divide.*

99. $12x^4 \div 4x^2$

100. $\dfrac{-20a^3b^5c}{10ab^2}$

101. $(18x^3 - 6x) \div (3x)$

102. $\dfrac{10x^5 + 6x^4 - 4x^3 - 2x^2}{2x^2}$

103. $(3x^2 + 8x - 35) \div (x + 5)$

104. $\dfrac{13 - 5x^2 + 2x^3}{2x - 1}$

Mixed Applications

Solve.

105. The half-life of the element thorium-232 is 13,900,000,000 yr. Express this length of time in scientific notation. (*Source:* Peter J. Nolan, *Fundamentals of College Physics*)

106. Physicists use both the joule (J) and the electron volt (eV) as units of work, where 1 J is equal to 6.24×10^{18} eV. Rewrite this quantity in standard notation. (*Source:* Peter J. Nolan, *Fundamentals of College Physics*)

107. A grain of bee pollen is about 0.00003 m in diameter. Express this length in scientific notation.

108. The area of the United States is approximately 3.7×10^6 mi². Express this area in standard notation. (*Source: The New York Times Almanac,* 2011)

109. At a party there are n people present. If everyone shakes hands with everyone else, then the polynomial

$$\frac{n^2}{2} - \frac{n}{2}$$

gives the total number of handshakes. If 9 people are at the party, how many handshakes will there be?

110. An object falling from an altitude of 500 m will be $(-4.9t^2 + 500)$ m above the ground after t sec. What is the altitude of the object 2 sec into the fall?

111. The polynomial $0.02x^3 - 0.71x^2 + 5.95x + 65.08$ can be used to model the temperature (in degrees Fahrenheit) at a particular time in Orlando, Florida on November 26, 2010, where x represents the number of hours past 8:00 A.M. According to this model, find the temperature in Orlando at 11:00 A.M. to the nearest degree. (*Source:* forecast.weather.gov)

112. The polynomial $3.4x^3 - 20.1x^2 + 27.7x + 1683$ approximates the number of two-year colleges in the United States for a particular year, where x represents the number of years after 2005. Use this polynomial to approximate, to the nearest hundred, the number of two-year colleges in 2007. (*Source:* nces.ed.gov)

113. The area of a triangle can be represented by the expression $\frac{1}{2}bh$, where b is the length of the base and h, the height.

The surface area of a pyramid is the sum of the area of its square base and the areas of the four (identical) triangular faces. In the pyramid located by the entrance to Paris' Louvre Museum, the triangular faces have a height 8 m less than the side length of its square base. (*Source:* Pei Cobb Fried & Partners)

a. Write a polynomial for the surface area of the Louvre Pyramid in terms of b, the side length of its square base.

b. The side length of the pyramid's base is 35 m. Find the surface area of the pyramid.

114. The polynomial $39x^2 + 4740x + 84,000$ approximates the total expenditures (in millions of dollars) on dental services in the U.S. for a particular year, where x represents the number of years after 2004. The polynomial $3x + 300$ approximates the U.S. population (in millions) during the same years. (*Source:* cms.gov)

a. Write a polynomial that approximates the total annual expenditures on dental services per person.

b. Using this model, find the amount of expenditures on dental services per person in 2008 to the nearest hundred dollars.

115. The volume of a box is the product of its length, width, and height.

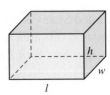

a. Write an expression for the volume (in cubic inches) of a box with length l in., width w in., and height h in.

b. Write an expression for the volume (in cubic inches) of a box with length l ft, width w ft, and height h ft. There are 12 in. in a foot.

116. The cylindrical storage vat pictured has height r and a base with area πr^2. The volume V of the vat is the product of its height and the area of the base. Write a formula for V in terms of r.

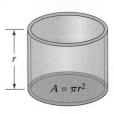

$A = \pi r^2$

• Check your answers on page A-21.

CHAPTER 5 Posttest

FOR
EXTRA
HELP

Test Prep in MyMathLab, and on You Tube* (search "AkstIntroductory
VIDEOS Alg" and click on "Channels").

To see if you have mastered the topics in this chapter, take this test.

Simplify.

1. $x^6 \cdot x$

2. $n^{10} \div n^4$

3. $7a^{-1}b^0$

4. $(-3x^2 y)^3$

5. $\left(\dfrac{x^2}{y^3}\right)^4$

6. $\left(\dfrac{3x^2}{y}\right)^{-3}$

7. For the polynomial $-x^3 + 2x^2 + 9x - 1$, identify:

  a. the terms _____

  b. the coefficients _____

  c. the degree _____

  d. the constant term _____

8. Find the sum: $(y^2 - 1) + (y^2 - y + 6)$

9. Subtract: $(x^2 - 7x - 4) - (2x^2 - 8x + 5)$

10. Combine: $(4x^2 y^2 - 6xy - y^2) - (3x^2 + x^2 y^2 - 2y^2) -$
$(x^2 - 6xy + y^2)$

Multiply.

11. $(2mn^2)(5m^2 n - 10mn + mn^2)$

12. $(y^3 - 2y^2 + 4)(y - 1)$

13. $(3x - 1)(2x + 7)$

14. $(7 - 2n)(7 + 2n)$

15. $(2m - 3)^2$

Divide.

16. $\dfrac{12s^3 + 15s^2 - 27s}{-3s}$

17. $(3t^3 - 5t^2 - t + 6) \div (3t - 2)$

Solve.

18. Medical X-rays, with a wavelength of about 10^{-10} m, can penetrate the flesh (but not the bones) of your body. Ultraviolet rays, which cause sunburn by penetrating only the top layer of skin, have a wavelength about 1000 times as long as X-rays. Find the length of ultraviolet rays. Write the answer in scientific notation. (*Source:* Peter J. Nolan, *Fundamentals of College Physics*)

19. A real estate broker sells two houses. The first house sells for $140,000 and is expected to increase in value by $1500/yr. The second house is purchased for $90,000 and will likely appreciate by $800/yr. Write an expression for the value of each house after x yr.

20. A young couple is saving up to purchase a car. They deposit $1000 in an account that has an annual interest rate of r (in decimal form). At the end of 2 yr, the value of the account will be $1000(1 + r)^2$ dollars. Find the account balance at that time if the interest rate is 3%.

• Check your answers on page A-22.

Cumulative Review Exercises

To help you review, solve the following.

1. Find the product: $-3.2(-2.3)$

2. Evaluate $3a^2 - 5ab + b^2$ for $a = -3$ and $b = 2$.

3. Simplify: $-4(y - 5) + 2(3y - 1)$

4. Solve for x: $2x + 18 - 9x = 5(4 - 3x) + 6x$

5. Solve $y = mx + b$ for m.

6. Solve: $-7x + 2(3x - 2) \geq 1$

7. Compute the slope of the line that passes through $(-3, 4)$ and $(1, 2)$. Plot the points and sketch the line.

8. Find the slope and y-intercept of the line $2x - 3y = 6$.

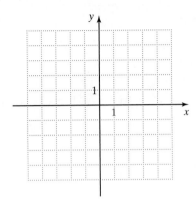

9. Graph the inequality: $y < 2x + 1$

10. Solve by graphing: $\begin{aligned} y &= 1 + 3x \\ x + y &= -3 \end{aligned}$

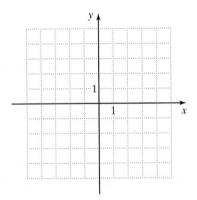

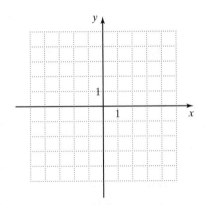

11. Solve by elimination:

$$3x - 2y = 10$$
$$2x + 3y = -2$$

12. Find the difference: $(3m^2 - 8m + 7) - (2m^2 + 8m - 9)$

13. Multiply: $(4a + 7)(7 - 4a)$

14. Find the quotient: $(-15x^2 - 7x + 4) \div (5x + 4)$

15. The Winter Olympics occur every 4 yr.

   a. If x represents the first year that they took place, write expressions for the next three Winter Olympic years.

   b. If the Winter Olympics were held in 1972, were they also held in 1980? Explain.

16. An executive goes out to dinner and leaves a 20% tip for the service.

   a. The bill for the meal without tip is represented by b. Write an expression for the amount of the tip in terms of b.

   b. The total cost c of the meal is the original bill plus the tip. Write an equation describing this situation.

17. On a travel website, the cost C of a vacation package is $300 for airfare plus $125 per day for d days at the hotel.

 a. Represent this relationship as an equation.

 b. Choose appropriate scales for the axes, and then graph the equation found in part (a).

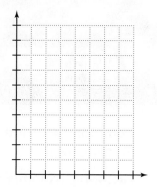

 c. Explain the significance of the C-intercept in the context.

18. The cost of a small one-bedroom apartment is $1500 security deposit and $1000 rent per month. By contrast, large studio apartment costs $1800 security deposit and $900 rent per month.

 a. For each apartment, write a linear equation that expresses the cost of an apartment c (in dollars) in terms of the number of months m the apartment is rented.

 b. Use the substitution method to solve the system of linear equations.

 c. In the context of this problem, what is the significance of the solution?

19. According to Albert Einstein's famous equation $E = mc^2$, all objects, even resting ones, contain energy E. If the mass m of a raisin is 10^{-3} kg, and the speed c of light is about 3×10^8 m/sec, find the amount of energy the raisin contains. Write the answer in scientific notation.

20. The expression for the surface area of a cylinder is $2\pi rh + 2\pi r^2$, where r is the radius of the base and h is the height of the cylinder. If the radius of a cylindrical gift box is $(2x + 3)$ in. and its height is twice the radius, what is the surface area of the box? Write the answer as a polynomial in descending order of powers of x.

• Check your answers on page A-22.

The Enigma machine, used by the German military during World War II for coding and decoding secret messages.

02075 / bac

A

Factoring Polynomials

Factoring and Cryptography

Cryptography, the science of coding and decoding messages, has been important throughout history. Sending secret messages in concealed form is particularly useful during wartime. For instance, the United States entered World War I in part because British intelligence intercepted and deciphered a message sent to the German minister in Mexico; the message called for a German–Mexican alliance against the United States. During World War II, the ability of the Allies to decode Nazi secrets allowed their commanders to eavesdrop on German plans and may have shortened the war by several years.

In peacetime, cryptographic techniques, used in electronic banking and in Web-based credit card purchases, play an increasingly important role in our lives.

Of the numerous cryptographic techniques for coding messages today, one important technique is *prime number encryption*. To crack a message coded in this way depends on finding the prime factors of a given, very large, whole number. Whereas multiplying two prime numbers is easy, reversing the process is difficult and drawn out. For instance, it has been estimated that finding the prime factorization of a five-digit whole number takes some 14 billion mathematical steps.

(*Sources:* Rudolf Kippenhahn, *Code Breaking: A History and Exploration*, The Overlook Press, 1999; F. H. Hinsley and Alan Stripp, Editors, *Codebreakers: The Inside Story of Bletchley Park*, Oxford University Press, 1994)

1. Find the greatest common factor of $18ab$ and $36a^4$.

Factor.

2. $4pq + 16p$

3. $10x^2y - 5x^3y^3 + 5xy^2$

4. $3x^2 + 6x + 2x + 4$

5. $n^2 - 11n + 24$

6. $4a + a^2 - 21$

7. $9y - 12y^2 + 3y^3$

8. $5a^2 + 6ab - 8b^2$

9. $-12n^2 + 38n + 14$

10. $4x^2 - 28x + 49$

11. $25n^2 - 9$

12. $x^2y - 4y^3$

13. $y^6 - 9y^3 + 20$

Solve.

14. $n(n - 6) = 0$

15. $3x^2 + x = 2$

16. $(y + 4)(y - 2) = 7$

17. The lateral surface area of a rectangular solid is given by the formula $A = 2lw + 2lh + 2wh$. Solve this formula for h in terms of A, w, and l.

18. A baseball player hits a pop-up fly ball with an initial velocity of 63 ft/sec from a height of 4 ft above the ground. The height of the ball (in feet) t sec after it is hit is given by the expression $-16t^2 + 63t + 4$. Write this expression in factored form.

19. A homeowner wants to fence off part of her yard to build a square play area for her children. Write an expression, in factored form, for the area of the yard *not* covered by the play area.

20. Find the dimensions of the LCD television screen shown if the length is 8 in. longer than the height.

S ft

15 ft

15 ft

S ft

40 in.

• Check your answers on page A-22.

OBJECTIVES

A To find the greatest common factor (GCF) of two or more integers or terms

B To factor out the greatest common factor from a polynomial

C To factor by grouping

D To solve applied problems involving factoring

What Factoring Is and Why It Is Important

In the previous chapter, we discussed how to *multiply* two factors in order to find their polynomial product.

Multiplying

$$(3x + 4)(2x + 1) = 6x^2 + 11x + 4$$

Factor   Factor   Product

In this chapter, we reverse the process, beginning with a polynomial and expressing it as a product of factors. Rewriting a polynomial as a product is called *factoring* the polynomial.

Factoring

$$6x^2 + 11x + 4 = (3x + 4)(2x + 1)$$

Polynomial   Factor   Factor

Just as factoring integers plays a key role in arithmetic, factoring polynomials is an important skill in algebra. Factoring polynomials helps us to simplify certain algebraic expressions and also to solve various types of equations.

Finding the Greatest Common Factor of Two or More Integers or Terms

We have already shown that every composite number can be written as the product of prime factors, called its *prime factorization*. For instance, the prime factorization of 15 is $3 \cdot 5$, and the prime factorization of 35 is $5 \cdot 7$. Since 5 appears in both factorizations, 5 is said to be a *common factor* of 15 and 35.

DEFINITION

A **common factor** of two or more integers is an integer that is a factor of each integer.

We can use the concept of common factor to find a *greatest common factor*.

DEFINITION

The **greatest common factor (GCF)** of two or more integers is the greatest integer that is a factor of each integer.

Let's consider an example of finding the greatest common factor of three numbers.

EXAMPLE 1

Find the GCF of 45, 63, and 81.

Solution   First, we find the prime factorization of each number.

$$45 = 3 \cdot 3 \cdot 5 = 3^2 \cdot 5$$
$$63 = 3 \cdot 3 \cdot 7 = 3^2 \cdot 7$$
$$81 = 3 \cdot 3 \cdot 3 \cdot 3 = 3^4$$

Then, we look for the greatest common factor. We see that $3 \cdot 3 = 3^2$ is a factor of each number. Since no power of 3 higher than 3^2 and no prime number other than 3 is a factor of *all* three numbers, the GCF of 45, 63, and 81 is 3^2, or 9.

PRACTICE 1

What is the GCF of 24, 72, and 96?

We can extend the concept of greatest common factor to monomials. For instance, to find the greatest common factor of $-21x^2$ and $35xy^2$, we begin by writing the monomials in factored form.

$$-21x^2 = -1 \cdot 3 \cdot 7 \cdot x \cdot x$$
$$35xy^2 = 5 \cdot 7 \cdot x \cdot y \cdot y$$

From the factorizations we see that each monomial has a factor of 7 and a factor of x in common. Note that 7 is the greatest factor of the coefficients and x^1 is the highest power of x that is a factor of each monomial. So the greatest common factor of $-21x^2$ and $35xy^2$ is the product of 7 and x, or $7x$.

EXAMPLE 2

Find the GCF of x^4, x^3, and x^2.

Solution   We write each monomial in factored form.

$$x^4 = x \cdot x \cdot x \cdot x$$
$$x^3 = x \cdot x \cdot x$$
$$x^2 = x \cdot x$$

From the factored forms, we see that the monomials have at most two factors of x in common. So the GCF of x^4, x^3, and x^2 is $x \cdot x$, or x^2.

PRACTICE 2

What is the GCF of a^3, a^2, and a?

Note that we could have written the factored forms of the monomials in Example 2 as:

$$x^4 = x^2 \cdot x^2$$
$$x^3 = x^2 \cdot x^1$$
$$x^2 = x^2$$

From these factored forms we see that x^2 is the highest power of the variable factor common to all three monomials. So the greatest common factor of the monomials is the lowest power of x in any of the monomials.

Now, let's consider the greatest common factor of expressions involving more than one variable.

EXAMPLE 3

Identify the GCF of $3a^3b$ and $-15a^2b^4$.

Solution First, we write each monomial in factored form.

$$3a^3b = 3 \cdot a^2 \cdot a \cdot b$$
$$-15a^2b^4 = -1 \cdot 3 \cdot 5 \cdot a^2 \cdot b \cdot b^3$$

The greatest common factor of the coefficients is 3 and the common variable factors to the lowest powers are a^2 and b. So the GCF of $3a^3b$ and $-15a^2b^4$ is $3a^2b$.

PRACTICE 3

Find the GCF of $-18x^3y^4$ and $12xy^2$.

Factoring Out the Greatest Common Factor from a Polynomial

Recall from the previous chapter that we used the distributive property to multiply a monomial by a polynomial. For example, consider the product of $2x$ and $(x + 7)$.

$$2x(x + 7) = 2x \cdot x + 2x \cdot 7 = 2x^2 + 14x$$

When the terms of a polynomial have common factors, we can factor the polynomial by using the distributive property. So we can write $2x^2 + 14x$ in factored form by dividing out $2x$, the GCF of the terms $2x^2$ and $14x$.

$$2x^2 + 14x = 2x \cdot x + 2x \cdot 7 = 2x(x + 7)$$
$$\uparrow$$
$$\text{GCF}$$

So the factored form of $2x^2 + 14x$ is $2x(x + 7)$.

EXAMPLE 4

Factor: $25x^3 + 10x^2$

Solution The GCF of $25x^3$ and $10x^2$ is $5x^2$.

$25x^3 + 10x^2 = 5x^2(5x) + 5x^2(2)$ Factor out the GCF $5x^2$ from each term.

$\qquad\qquad\quad = 5x^2(5x + 2)$ Use the distributive property.

So the factorization of $25x^3 + 10x^2$ is $5x^2(5x + 2)$.

PRACTICE 4

Factor: $10y^2 + 8y^5$

Now, let's consider some examples of factoring polynomials in more than one variable.

EXAMPLE 5

Factor: $12c^2 - 2cd$

Solution   The GCF of $12c^2$ and $-2cd$ is $2c$.

$12c^2 - 2cd = 2c(6c) - 2c(d)$ **Factor out the GCF $2c$ from each term.**

$\qquad\qquad\ = 2c(6c - d)$ **Use the distributive property.**

PRACTICE 5

Factor: $21a^2b - 14a$

EXAMPLE 6

Factor: $3x^2y^4 + 9xy^2$

Solution

$$3x^2y^4 + 9xy^2 = 3xy^2(xy^2) + 3xy^2(3)$$
$$= 3xy^2(xy^2 + 3)$$

PRACTICE 6

Factor: $8a^2b^2 - 6ab^3$

EXAMPLE 7

Express in factored form: $20y^2 - 5y + 15$

Solution

$$20y^2 - 5y + 15 = 5(4y^2) - 5(y) + 5(3)$$
$$= 5(4y^2 - y + 3)$$

PRACTICE 7

Factor: $24a^2 - 48a + 12$

When solving some literal equations, it may be necessary to factor out common monomial factors.

EXAMPLE 8

Solve $ax + b = cx + d$ for x in terms of a, b, c, and d.

Solution   To solve for x, we bring all the terms involving x to the left side of the equation and the other terms to the right side of the equation.

$ax + b = cx + d$

$ax + b - b = cx + d - b$ **Subtract b from each side of the equation.**

$\qquad\quad ax = cx + d - b$

$ax - cx = cx - cx + d - b$ **Subtract cx from each side of the equation.**

$ax - cx = d - b$

$x(a - c) = d - b$ **Factor out x on the left side of the equation.**

$\dfrac{x(a - c)}{a - c} = \dfrac{d - b}{a - c}$ **Divide each side of the equation by $(a - c)$, the coefficient of x.**

$\qquad\quad x = \dfrac{d - b}{a - c}$

PRACTICE 8

Solve $ab = s^2 - ac$ for a in terms of b, c, and s.

Factoring by Grouping

Recall that the factorization of $2x^2 + 14x$ is $2x(x + 7)$. The factor $(x + 7)$ is called a *binomial* factor. The distributive property can be used to divide out not only a common monomial factor but also a common binomial factor, if there is one. For instance, let's consider the expression $x(x + 5) + 2(x + 5)$.

$$\underbrace{x(x + 5)}_{\text{First term}} + \underbrace{2(x + 5)}_{\text{Second term}}$$

In this polynomial, the binomial factor $(x + 5)$ is common to both terms. We can factor out $(x + 5)$, getting:

$$x(x + 5) + 2(x + 5) = (x + 5)(x + 2)$$

So the factored form of $x(x + 5) + 2(x + 5)$ is $(x + 5)(x + 2)$.

EXAMPLE 9

Factor: $x(x + 4) - 5(x + 4)$

Solution   Factoring out $(x + 4)$, we get:

$$x(x + 4) - 5(x + 4) = (x + 4)(x - 5)$$

PRACTICE 9

Factor: $4(y - 3) + y(y - 3)$

In some algebraic expressions, such as $x(a - 7) + 3(7 - a)$, the binomial factors are opposites. In order to factor out a *common* binomial factor, we must rewrite one of the binomials by factoring out -1, as shown in the next example.

EXAMPLE 10

Factor: $x(a - 7) + 3(7 - a)$

Solution   The binomial factors $(a - 7)$ and $(7 - a)$ are opposites, so we factor out -1 from the binomial $(7 - a)$ and rewrite the original expression. Note that $(7 - a) = -1(a - 7)$.

$$x(a - 7) + 3(7 - a) = x(a - 7) + 3\left[-1(a - 7)\right] \quad \text{Factor out } -1 \text{ from } (7 - a).$$
$$= x(a - 7) - 3(a - 7) \quad \text{Simplify.}$$
$$= (a - 7)(x - 3) \quad \text{Factor out } (a - 7).$$

PRACTICE 10

Factor: $3y(x - 1) + 2(1 - x)$

EXAMPLE 11

Factor: $5n(4n - 1) - (4n - 1)$

Solution

$$5n(4n - 1) - (4n - 1) = 5n(4n - 1) - 1(4n - 1)$$
$$= (4n - 1)(5n - 1)$$

PRACTICE 11

Factor: $(4 - 3x) + 2x(4 - 3x)$

When trying to factor a polynomial that has four terms, it may be possible to group pairs of terms in such a way that a common binomial factor can be found. This method is called **factoring by grouping**.

EXAMPLE 12

Factor by grouping: $xy - 4x + 3y - 12$

Solution

$$xy - 4x + 3y - 12 = (xy - 4x) + (3y - 12)$$ Group the first two terms and the last two terms.

$$= x(y - 4) + 3(y - 4)$$ Factor out the GCF from each group.

$$= (y - 4)(x + 3)$$ Write in factored form.

PRACTICE 12

Factor: $4b - 20 + ab - 5a$

EXAMPLE 13

Express in factored form: $6h - 6k - h^2 + hk$

Solution

$$6h - 6k - h^2 + hk = (6h - 6k) + (-h^2 + hk)$$ Group the first two terms and the last two terms.

$$= (6h - 6k) - (h^2 - hk)$$ Factor out -1 in the second group.

$$= 6(h - k) - h(h - k)$$ Factor out the GCF from each group.

$$= (h - k)(6 - h)$$ Write in factored form.

PRACTICE 13

Factor: $5y - 5z - y^2 + yz$

EXAMPLE 14

Each week, a sales associate receives a salary of d dollars as well as 5% commission on the value of the sales that she makes. Last week, sales amounted to x dollars, and this week, sales rose to y dollars. How much greater was the sales associate's total income this week than last week? Express this amount in factored form.

Solution   Last week, the sales associate made d dollars in salary and $0.05x$ in commission. This week, the associate made d dollars in salary and $0.05y$ in commission. So the difference in total income is:

$$(d + 0.05y) - (d + 0.05x) = d + 0.05y - d - 0.05x$$
$$= 0.05y - 0.05x$$
$$= 0.05(y - x)$$

So the sales associate made $0.05(y - x)$ dollars more this week than last week.

PRACTICE 14

The distance an object under constant acceleration travels in time t is given by the expression $v_0 t + \frac{1}{2}at^2$, where v_0 is the object's initial velocity and a is its acceleration. Factor this expression.

Mathematically Speaking

Fill in each blank with the most appropriate term or phrase from the given list.

sum	greatest common factor	factoring
greatest factor	(GCF)	product
multiplying	common factor	

1. Rewriting a polynomial as a product is called _____ the polynomial.

2. A(n) _____ of two or more integers is an integer that is a factor of each integer.

3. The _____ of two or more integers is the greatest integer that is a factor of each integer.

4. The greatest common factor of two or more monomials is the _____ of the greatest common factor of the coefficients and, for each variable, the variable to the lowest power to which it is raised in any of the monomials.

Find the greatest common factor of each group of terms.

5. 27, 54, and 81

6. 28, 35, 63

7. x^4, x^6, x^3

8. y^2, y, y^5

9. $16b, 8b^3, 12b^2$

10. $3a, 7a^2, 5a^4$

11. $-12x^5y^7, 4y^3$

12. $9m^3, 6m^2n$

13. $18a^5b^4, -6a^4b^3, 9a^2b^2$

14. $24mn, 32mn^2, 16m^2n$

15. $x(3x - 1)$ and $8(3x - 1)$

16. $6(5n + 2)$ and $n(5n + 2)$

17. $4x(x + 7)$ and $9x(x + 7)$

18. $y(y - 4)$ and $6y(y - 4)$

Factor out the greatest common factor.

19. $3x + 6$

20. $10y + 15$

21. $24x^2 + 8$

22. $30y^2 - 6$

23. $27m - 9n$

24. $16r - 8t$

25. $7x^2 - 2x$

26. $3b^2 - 18b$

27. $5b^2 - 6b^3$

28. $4z^5 - 12z^2$

29. $10x^3 - 15x$

30. $12a^2 - 18a$

31. $a^2b^2 + ab$

32. $xy + x^2y^2$

33. $6xy^2 + 7x^2y$

34. $3p^3q^2 + 5p^2q^3$

35. $27pq^2 - 18p^2q$

36. $45c^2d - 15cd^2$

37. $2x^3y + 12x^3y^4$

38. $7a^2b^3 + 9a^4b^3$

39. $3c^3 + 6c^2 + 12$

40. $5y^2 - 20y + 10$

41. $9b^4 - 3b^3 + b^2$

42. $8y^5 - y^4 - 4y^2$

43. $2m^4 + 10m^3 - 6m^2$

44. $3x^3 - 9x^2 - 27x$

45. $5b^5 - 3b^3 + 2b^2$

46. $9c^4 + c^3 + 6c^2$

47. $15x^4 - 10x^3 - 25x$

48. $12m^6 + 9m^5 + 15m^3$

49. $4a^2b + 8a^2b^2 - 12ab$

50. $5m^2n - 15mn^2 + 10mn$

51. $9c^2d^2 + 12c^3d + 3cd^3$

52. $18x^2y^4 - 24xy^3 + 30x^3y$

C *Factor by grouping.*

53. $x(x - 1) + 3(x - 1)$

54. $2(n + 4) + n(n + 4)$

55. $5a(a - 1) - 3(a - 1)$

56. $4x(x + 3) - 7(x + 3)$

57. $r(s + 7) - 2(7 + s)$

58. $a(6 + b) - 7(b + 6)$

59. $a(x - y) - b(x - y)$

60. $y(a - z) - x(a - z)$

61. $3x(y + 2) - (y + 2)$

62. $(n - 1) - 2m(n - 1)$

63. $b(b - 1) + 5(1 - b)$

64. $x(x - 3) + 2(3 - x)$

65. $y(y - 1) - 5(1 - y)$

66. $n(n - 9) - 4(9 - n)$

67. $(t - 3) - t(3 - t)$

68. $w(w - 4) + (4 - w)$

69. $9a(b - 7) + 2(7 - b)$

70. $2y(x - 2) + 3(2 - x)$

71. $rs + 3s + rt + 3t$

72. $mn + 2m + np + 2p$

73. $xy + 6y - 4x - 24$

74. $ab - 5b - 2a + 10$

75. $15xy - 9yz + 20xz - 12z^2$

76. $6ab + 12ac - 5bc - 10c^2$

77. $2xz + 8x + 5yz + 20y$

78. $3ab + 9a + 4bc + 12c$

Solve for the indicated variable.

79. $TM = PC + PL$ for P

80. $S = a + Nd - d$ for d

81. $S = 2lw + 2lh + 2wh$ for l

82. $S = a + ar^n$ for a

Mixed Practice

Factor out the greatest common factor.

83. $16p^3 + 24p$

84. $4u^4 - 28u^2 + 36u$

85. $48rs^2 - 60r^2s$

86. $7m^3 - 4m^2$

87. $42j^2 - 6$

Solve.

88. $A = \frac{1}{2}hb_1 + \frac{1}{2}hb_2$ for h

Factor by grouping.

89. $st - 3t - 7s + 21$

90. $7b(b + 2) - 5(b + 2)$

91. $3x(y - 4) + 5(4 - y)$

92. $2bc + 8ab - 3ac - 12a^2$

Find the greatest common factor of each group of terms.

93. $16a^5b^3, -12a^2b^3, 20a^3b$

94. $14x^4, 21xy^3$

Applications

Solve.

95. When an object with mass m increases in velocity from v_1 to v_2, its momentum increases by $mv_2 - mv_1$. Factor this expression.

96. One item sells for p dollars and another for q dollars. In addition, an 8% sales tax is charged on all items sold. An expression for the total selling price is $1.08p + 1.08q$. Write this expression in factored form.

97. In a meeting of diplomats, all diplomats must shake hands with one another. If there are n diplomats, the expression $0.5n^2 - 0.5n$ represents the total number of handshakes at the meeting. Factor this expression.

98. For an investment earning simple interest, the future value of the investment is represented by the expression $P + Prt$, where P is the present value of the investment, r is the annual interest rate, and t is the time in years. Factor this expression.

99. In a polygon with n sides, the number of diagonals is given by the expression $\frac{1}{2}n^2 - \frac{3}{2}n$. Write this expression in factored form.

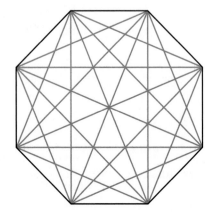

100. In a polygon with n sides, the interior angles (measured in degrees) add up to $180n - 360$. Find an equivalent expression by factoring out the GCF.

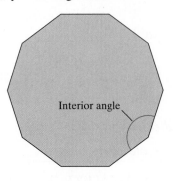

Interior angle

101. Consider the formula $P = nC + nT + D$, where P is the total price of a purchase, n is the number of items purchased, C is the cost per item, T is the tax on each item, and D is the total delivery charge. Solve for n in terms of P, C, T, and D.

102. The *harmonic mean* H of two numbers x and y is a kind of average. Solve for H in the formula $Hx + Hy = 2xy$.

103. For more than 30 yr, *Wheel of Fortune* has been a popular show on syndicated television. In one game, a player earned $700 for each of n copies of the letter N and $900 for each of s copies of the letter S, spending $250 for each of e copies of the vowel E. Find in factored form the amount that the player won. (*Source: Wheel of Fortune*)

104. Many companies pay their hourly employees "time-and-a-half" during overtime hours, that is, during hours worked more than 40 per week. The regular hourly rate for employees is h dollars. During a particular week, x employees work 40 hours and y employees work 50 hours. What is the total cost of the company payroll for that week, expressed in factored form?

• Check your answers on page A-22.

Mathematical Reasoning

1. A four-digit whole number can be represented by the expression

$$1000d + 100c + 10b + a$$

where a is the digit in the units place, b is the digit in the tens place, c is the digit in the hundreds place, and d is the digit in the thousands place.

a. Consider several four-digit whole numbers (for instance, 8351). Then, for each of these numbers, reverse the order of the digits, forming a second number $1000a + 100b + 10c + d$ (here, 1538). Subtract the smaller number from the larger ($8351 - 1538 = 6813$). Then, check whether this difference is divisible by 9.

b. Show that when you reverse the order of the digits of *any* four-digit whole number and subtract the two four-digit numbers, their difference must be divisible by 9.

Critical Thinking

2. Factor the expression $a^{n+2}b^n - a^n b^{n+1}$.

Groupwork

3. Working with a partner, for each polynomial list three numbers or monomials that when placed in the ▢ will make the polynomial factorable.

a. $2xy - 7x + \boxed{} - 14$

b. $xy^2 + \boxed{} + 3y^2 - 48$

n this section, we move on to another kind of factoring—factoring trinomials of the form $ax^2 + bx + c$, where $a = 1$. Recall that the coefficient a of the leading term ax^2 is called the leading coefficient. In other words, we are examining trinomials of the form $x^2 + bx + c$, where the leading coefficient is 1. First, we factor trinomials in which the constant term is positive, such as:

$$x^2 + 5x + 6 \quad \text{and} \quad x^2 - 8x + 16$$

$$\underset{\text{Constant term}}{\underline{\qquad\qquad\qquad}}$$

Next, we factor trinomials in which the constant term is negative, for example:

$$x^2 + 3x - 10 \quad \text{and} \quad x^2 - 5x - 24$$

$$\underset{\text{Constant term}}{\underline{\qquad\qquad\qquad}}$$

In Section 6.5, we will see how factoring trinomials helps us to solve related equations and applied problems.

Factoring $x^2 + bx + c$, for c Positive

Recall from Section 5.5 the FOIL method of multiplying two binomials.

$$\overset{\text{F} \quad \text{O} \quad \text{I} \quad \text{L}}{(x + 2)(x + 3) = x^2 + 3x + 2x + 6}$$
$$= x^2 + 5x + 6$$

Note that when the leading coefficient of both binomials is 1, their product also has leading coefficient 1. This suggests that factoring a trinomial of the form $x^2 + bx + c$ gives the product of two binomials of the form $(x + ?)(x + ?)$, where each question mark represents an integer. To find the binomials, we apply the FOIL method in reverse.

For instance, let's factor $x^2 + 5x + 6$. Since the leading term of the trinomial is x^2, we apply the FOIL method to two binomial factors, placing x as the first term in each of the factors.

$$\overset{\text{F} \quad \overbrace{\text{O} \quad \text{I}} \quad \text{L}}{\underset{\downarrow \qquad \downarrow \qquad \downarrow}{}}$$
$$x^2 + 5x + 6 = (x + ?)(x + ?) = x^2 + 5x + 6$$

We see that the product of the constant terms of the binomial factors must be 6, the constant term of the trinomial. The sum of the outer and inner products must be $5x$. So we need to find two integers whose product is 6 and whose sum is 5.

To find these integers, consider all possible factors of 6, that is, of $+6$.

Factors of 6	Possible Binomial Factors	Sum of Outer and Inner Products
1, 6	$(x + 1)(x + 6)$	$6x + x = 7x$
−1, −6	$(x - 1)(x - 6)$	$-6x - x = -7x$
2, 3	$(x + 2)(x + 3)$	$3x + 2x = 5x$ ← The correct middle term
−2, −3	$(x - 2)(x - 3)$	$-3x - 2x = -5x$

So $x^2 + 5x + 6 = (x + 2)(x + 3)$.

We check that the factors are correct by multiplying.

$$(x + 2)(x + 3) = x^2 + \underbrace{3x + 2x}_{\text{Sum of the outer and inner products}} + 6 = x^2 + \underbrace{5x}_{\text{Middle term}} + 6$$

This way of factoring a trinomial, listing all the possibilities, is sometimes called the *trial-and-error method*. Factoring a trinomial by the trial-and-error method involves recognizing patterns, looking for clues, and, once the factorization is found, multiplying to check. Note that by the commutative property of multiplication, we can also write the product $(x + 2)(x + 3)$ as $(x + 3)(x + 2)$.

EXAMPLE 1

Factor: $y^2 + 7y + 10$

Solution Applying the FOIL method in reverse, we know that the first term of each factor is y. So the factors are of the form $(y + ?)(y + ?)$. Using the trial-and-error method, we need to find two integers whose product is 10 and whose sum is 7. The following table shows how to test pairs of factors of 10, the constant term of the trinomial.

Factors of 10	Sum of Factors
1, 10	11
−1, −10	−11
2, 5	7 ← Factors 2 and 5 have a sum of 7.
−2, −5	−7

So $y^2 + 7y + 10 = (y + 2)(y + 5)$, or $(y + 5)(y + 2)$.
Same sign

PRACTICE 1

Factor: $x^2 + 5x + 4$

Can you explain why in the preceding list of possible factors we need only have tested the positive factors of the constant term?

EXAMPLE 2

Factor: $x^2 - 10x + 16$

Solution   The constant term of the trinomial is positive, and the coefficient of the x-term is negative. So the constant terms of the binomial factors must both be negative.

$$x^2 - 10x + 16 = (x - ?)(x - ?)$$

We need to find two negative integers whose product is 16 and whose sum is -10. In this case we need to consider only negative factors of 16.

Factors of 16	Sum of Factors
$-1, -16$	-17
$-2, -8$	-10
$-4, -4$	-8

$\longleftarrow$ Factors -2 and -8 have a sum of -10.

So $x^2 - 10x + 16 = (x - 2)(x - 8)$, or $(x - 8)(x - 2)$.

PRACTICE 2

Factor: $y^2 - 9y + 20$

TIP   When the constant term c of the trinomial $x^2 + bx + c$ is positive,
- the constant terms of the binomial factors are both positive when b, the coefficient of the x-term in the trinomial, is positive, and
- the constant terms are both negative when b is negative.

Not every trinomial can be expressed as the product of binomial factors with coefficients that are integers. Polynomials that are not factorable are called **prime polynomials**. For instance, the trinomial $x^2 + 5x + 1$ is a prime polynomial. Can you explain why this trinomial is prime?

EXAMPLE 3

Factor: $x^2 + 4x + 6$

Solution   Both the constant term and the coefficient of the x-term are positive. So if the trinomial is factorable, the constant terms of its binomial factors must both be positive as well.

$$x^2 + 4x + 6 = (x + ?)(x + ?)$$

We need to find two positive integers whose product is 6 and whose sum is 4.

Factors of 6	Sum of Factors
1, 6	7
2, 3	5

Since neither sum of the factors yields the correct coefficient of the x-term, we can conclude that the polynomial is not factorable. Therefore, this is a *prime polynomial*.

PRACTICE 3

Factor: $x^2 + 3x + 5$

When the terms of a trinomial are not in descending order, we usually rewrite the trinomial before factoring, as shown in the next example.

EXAMPLE 4

Factor: $12 - 8x + x^2$

Solution   We begin by writing the terms in descending order.

$$12 - 8x + x^2 = x^2 - 8x + 12$$
$$= (x - ?)(x - ?)$$

Since the middle term of the trinomial is negative and the constant term is positive, we are looking for two negative integers. So we need to consider only negative factors of 12.

Factors of 12	Sum of Factors
$-1, -12$	-13
$-2, -6$	-8
$-3, -4$	-7

←—— Factors -2 and -6 have a sum of -8.

So $x^2 - 8x + 12 = (x - 2)(x - 6)$.

PRACTICE 4

Factor: $32 - 12y + y^2$

Note that in Example 4, we could also have factored $12 - 8x + x^2$ as $(2 - x)(6 - x)$, without rearranging the terms of the trinomial. Can you explain why the two solutions $(x - 2)(x - 6)$ and $(2 - x)(6 - x)$ are equal?

Now, let's consider a trinomial of the form $x^2 + bxy + cy^2$. Note that this trinomial contains more than one variable. When factoring these trinomials, we use the trial-and-error method where the binomial factors are of the form $(x + ?y)(x + ?y)$ and the question marks represent factors of c, the coefficient of the y^2 term, whose sum is b, the coefficient of the xy-term.

EXAMPLE 5

Factor: $x^2 + 3xy + 2y^2$

Solution $x^2 + 3xy + 2y^2 = (x + ?y)(x + ?y)$

Since the middle term and the last term are both positive, we look for only positive factors of 2, the coefficient of y^2, whose sum is 3, the coefficient of the xy-term.

Factors of 2	Sum of Factors
1, 2	3

So $x^2 + 3xy + 2y^2 = (x + 1y)(x + 2y) = (x + y)(x + 2y)$, or $(x + 2y)(x + y)$.

PRACTICE 5

Factor: $p^2 - 4pq + 3q^2$

Factoring $x^2 + bx + c$, for c Negative

Now, let's consider how to factor trinomials in which the coefficient of the first term is 1 and the sign of the constant term is negative.

EXAMPLE 6

Factor: $x^2 + 2x - 3$

Solution   The goal is to find factors of -3 whose sum is 2.

Factors of -3	Sum of Factors
$-1, 3$	2	← Factors -1 and 3 have a sum of 2.
$1, -3$	-2

So $x^2 + 2x - 3 = (x - 1)(x + 3)$.

Check   We check by multiplying.

$$(x - 1)(x + 3) = x^2 \underbrace{+ 3x - x}_{\substack{\text{Sum of the} \\ \text{outer and} \\ \text{inner products}}} - 3 = x^2 \underbrace{+ 2x}_{\substack{\text{Middle} \\ \text{term}}} - 3$$

PRACTICE 6

Factor: $x^2 + x - 6$

TIP   When the constant term c of the trinomial $x^2 + bx + c$ is negative, the constant terms of the binomial factors have opposite signs.

EXAMPLE 7

Factor: $y^2 - 3y - 10$

Solution   We must find two integers whose product is -10 and whose sum is -3. Since the constant term is negative, one of its factors must be positive and the other negative.

Factors of -10	Sum of Factors
$1, -10$	-9
$-1, 10$	9
$2, -5$	-3	← Factors 2 and -5 have a sum of -3.
$-2, 5$	3

So $y^2 - 3y - 10 = (y + 2)(y - 5)$.
   Note that since the sum of the two factors of -10 is negative, the negative factor must have a larger absolute value than the positive factor.

PRACTICE 7

Factor: $x^2 - 21x - 46$

EXAMPLE 8

Factor: $x^2 - 12 + x$

Solution

$$x^2 - 12 + x = x^2 + x - 12 \qquad \text{Write the terms in descending order.}$$
$$= (x + ?)(x - ?) \qquad \text{Since the constant term is negative, one of its factors is positive and the other factor is negative.}$$

Now, we must find two factors of -12 whose sum is 1. So the positive factor must have a larger absolute value than the negative factor. Thus, we consider only factors of -12 for which the positive factor has the larger absolute value.

Factors of -12	Sum of Factors
$-1, 12$	11
$-2, 6$	4
$-3, 4$	1 ⟵ Factors -3 and 4 have a sum of 1.

So $x^2 + x - 12 = (x - 3)(x + 4)$.

PRACTICE 8

Factor: $y^2 - 24 + 2y$

Next, let's consider factoring a trinomial in two variables.

EXAMPLE 9

Factor: $x^2 - 5xy - 14y^2$

Solution $x^2 - 5xy - 14y^2 = (x + ?y)(x - ?y)$

We must find two factors of -14 whose sum is -5. Since the product of these factors is a negative number, one of the factors must be positive and the other negative. Since the sum of the factors is negative, the negative factor must have a larger absolute value than the positive factor. Thus, we consider only factors of -14 for which the negative factor has the larger absolute value.

Factors of -14	Sum of Factors
$1, -14$	-13
$2, -7$	-5 ⟵ Factors 2 and -7 have a sum of -5.

So $x^2 - 5xy - 14y^2 = (x + 2y)(x - 7y)$.

PRACTICE 9

Factor: $a^2 - 5ab - 24b^2$

Some trinomials have a common factor. When factoring such a trinomial, we factor out the GCF before trying to factor the trinomial into the product of two binomials, as shown in the next example.

EXAMPLE 10

Factor: $3x^2 + 6x - 24$

Solution Each term of the trinomial has a common factor of 3, so we begin by factoring out this factor.

$$3x^2 + 6x - 24 = 3(x^2 + 2x - 8)$$
$$= 3(x + ?)(x - ?)$$

Since the product of the two missing integers is negative, we need to consider only factors of -8, one positive and the other negative, where the positive factor has a larger absolute value than the negative factor.

Factors of -8	Sum of Factors
$-1, 8$	7
$-2, 4$	2

$\longleftarrow$ Factors -2 and 4 have a sum of 2.

So $3x^2 + 6x - 24 = 3(x^2 + 2x - 8)$
$$= 3(x - 2)(x + 4).$$

Note that after factoring out the GCF 3, neither of the remaining factors of the polynomial has a common factor.

PRACTICE 10

Factor: $y^3 - 9y^2 - 10y$

EXAMPLE 11

Express in factored form: $3a^4 - 21a^3 - 24a^2$

Solution

$3a^4 - 21a^3 - 24a^2 = 3a^2(a^2 - 7a - 8)$ Factor out the GCF $3a^2$ from each term.

$\qquad = 3a^2(a + 1)(a - 8)$ Factor $a^2 - 7a - 8$.

So $3a^4 - 21a^3 - 24a^2 = 3a^2(a + 1)(a - 8)$.

PRACTICE 11

Factor: $8x^3 + 8x^2 - 16x$

As in the previous examples, after factoring out the GCF the remaining trinomial can sometimes still be factored. A polynomial is **factored completely** when it is expressed as the product of a monomial and one or more prime polynomials. Throughout the remainder of this text, *factor* means to factor completely.

EXAMPLE 12

Factor: $-x^2 + 9x - 14$

Solution

$-x^2 + 9x - 14 = -1(x^2 - 9x + 14)$ Factor out -1 so that the leading coefficient is 1.

$\qquad = -1(x - 7)(x - 2)$
$\qquad = -(x - 7)(x - 2)$

PRACTICE 12

Write in factored form:
$-x^2 - 10x + 11$

EXAMPLE 13

Show that for any whole number n, the number represented by $n^2 + 7n + 12$ can be expressed as the product of two consecutive whole numbers.

Solution Since we want to represent the expression $n^2 + 7n + 12$ as a product, we factor it.

$$n^2 + 7n + 12 = (n + 3)(n + 4)$$

Note that the two factors, $n + 3$ and $n + 4$, are whole numbers that differ by 1.

$$(n + 4) - (n + 3) = n + 4 - n - 3 = 1$$

So $n^2 + 7n + 12$ can be expressed as the product of two consecutive whole numbers.

PRACTICE 13

A ball is tossed upward with a velocity of 32 ft/sec from a roof 48 ft above the ground. The expression $-16t^2 + 32t + 48$ approximates the height of the ball above the ground in feet after t sec. Factor this expression.

Mathematically Speaking

Fill in each blank with the most appropriate term or phrase from the given list.

are both negative	not factorable	opposite signs
a factor greater than 0	are both positive	ascending
	descending	the same sign
factorable	a common factor	

1. Polynomials that are _____ are called prime polynomials.

2. Polynomials to be factored are generally written in _____ order.

3. If $c < 0$ in the trinomial $x^2 + bx + c$, then the constant terms of the binomial factors have _____.

4. If each term in a trinomial has _____, factor out the GCF before trying to factor the trinomial into the product of two binomials.

Match each trinomial with its binomial factors.

5. $x^2 - 16x + 28$ **a.** $(x - 1)(x + 28)$

6. $x^2 + 12x - 28$ **b.** $(x - 7)(x + 4)$

7. $x^2 + 29x + 28$ **c.** $(x - 4)(x - 7)$

8. $x^2 + 27x - 28$ **d.** $(x - 2)(x + 14)$

9. $x^2 - 3x - 28$ **e.** $(x + 28)(x + 1)$

10. $x^2 - 11x + 28$ **f.** $(x - 14)(x - 2)$

Find the missing factor.

11. $x^2 - 3x - 10 = (x + 2)(\quad)$

12. $x^2 - 11x + 18 = (x - 9)(\quad)$

13. $x^2 + 5x + 4 = (x + 1)(\quad)$

14. $x^2 - x - 12 = (x - 4)(\quad)$

15. $x^2 + 5x - 6 = (x + 6)(\quad)$

16. $x^2 + 4x - 21 = (x - 3)(\quad)$

Factor, if possible.

17. $x^2 + 6x + 8$ **18.** $x^2 + 9x + 8$ **19.** $x^2 + 5x - 6$ **20.** $x^2 + 2x - 3$

21. $x^2 + x + 2$ **22.** $x^2 + 13x + 7$ **23.** $x^2 + 2x - 8$ **24.** $x^2 + 4x - 45$

25. $x^2 - 4x + 3$ **26.** $x^2 - 6x + 5$ **27.** $y^2 - 12y + 32$ **28.** $m^2 - 5m + 4$

29. $t^2 - 4t - 5$ **30.** $s^2 - 4s - 12$ **31.** $x^2 + 2x - 1$ **32.** $x^2 + x + 7$

33. $y^2 - 9y + 20$ **⊙ 34.** $a^2 - 7a + 10$ **35.** $y^2 - 13y - 48$ **36.** $x^2 - 3x - 18$

37. $b^2 + 11b + 28$ **38.** $p^2 + 13p + 22$ **39.** $x^2 - 14x + 49$ **40.** $x^2 - 12x + 36$

41. $-y^2 + 5y + 50$ **42.** $-x^2 + 5x + 84$ **43.** $x^2 + 64 - 16x$ **44.** $a^2 + 20 - 12a$

45. $16 - 10x + x^2$ **46.** $20 + b^2 - 21b$ **47.** $81 - 30w + w^2$ **48.** $y^2 + 72 - 17y$

49. $p^2 - 8pq + 7q^2$ **50.** $s^2 - 10st + 25t^2$ **51.** $p^2 - 4pq - 5q^2$ **52.** $x^2 - 4xy - 12y^2$

53. $m^2 - 12mn + 35n^2$ **54.** $r^2 - rs - 30s^2$ **55.** $x^2 + 9xy + 8y^2$ **56.** $a^2 + 12ab + 27b^2$

57. $5x^2 - 5x - 30$ **58.** $4y^2 + 12y - 40$ **59.** $2x^2 + 10x - 28$ **60.** $8r^2 - 56r - 64$

61. $12 - 18t + 6t^2$ **62.** $8 - 10x + 2x^2$ **63.** $3x^2 + 24 + 18x$ **64.** $5z^2 - 15 - 10z$

65. $y^3 + 3y^2 - 10y$ **66.** $x^3 - 6x^2 - 7x$ **67.** $a^3 + 8a^2 + 15a$ **68.** $q^3 - q^2 - 42q$

69. $t^4 - 14t^3 + 24t^2$ **70.** $x^4 + 6x^3 - 27x^2$ **71.** $4a^3 - 12a^2 + 8a$ **72.** $3y^3 - 18y^2 + 24y$

73. $2x^3 + 30x + 16x^2$ **74.** $5b^3 + 10b + 15b^2$ ◉ **75.** $4x^3 + 48x - 28x^2$ **76.** $3r^3 - 42r + 15r^2$

77. $-56s + 6s^2 + 2s^3$ **78.** $-20y - 16y^2 + 4y^3$ **79.** $2c^4 + 4c^3 - 70c^2$ **80.** $4t^4 + 24t^3 - 64t^2$

81. $ax^3 - 18ax^2 + 32ax$ **82.** $b^2y^3 - 5b^2y^2 - 36b^2y$

Mixed Practice

Factor, if possible.

83. $x^2 + 6x + 3$

84. $x^2 - xy - 12y^2$

85. $5x^4 - 15x^3 - 50x^2$

86. $3z^3 + 24z^2 + 36z$

87. $-w^2 + 6w + 40$

88. $b^2 - 6b + 5$

89. $6m^2 - 6m - 36$

90. $s^4 - 17s^3 + 72s^2$

91. $t^2 + 60 - 17t$

92. $n^2 - 11n + 18$

Solve.

93. Choose the correct binomial factors of the trinomial $x^2 - 5x - 24$.
   a. $(x + 8)$ and $(x - 3)$ **b.** $(x + 3)$ and $(x - 8)$
   c. $(x - 6)$ and $(x + 4)$ **d.** $(x - 4)$ and $(x + 6)$

94. Find the missing factor.
   $$x^2 + 5x - 36 = (x - 4)(\qquad)$$

Applications

Solve.

95. Scientists who study genetics use the equation $p^2 + 2pq + q^2 = 1$, where p represents a certain dominant gene and q represents a recessive gene. Rewrite the equation so that the left side is factored.

96. Show that for any whole number n, the number represented by $n^2 + 11n + 30$ can be expressed as the product of two consecutive whole numbers.

97. A child throws a stone downward with an initial velocity of 48 ft/sec from a height of 160 ft. One step in figuring out how long it takes for the stone to reach the ground is to factor the expression $16t^2 + 48t - 160$. What is its factorization?

98. A statistician found that the cost in dollars for a company to produce x units of a certain product can be approximated by $C = x^2 - 14x + 45$. Factor the expression on the right side of this equation.

99. According to specifications, a box manufacturer makes a closed box with a length and width that are 3 in. longer than the height. Let x equal the height of the box.

  a. Find the surface area of the box.

  b. Factor this polynomial.

100. For any whole number n, show that the number represented by $4n^2 + 20n + 24$ can be expressed as the product of two consecutive even numbers.

101. The Meteor Crater in Arizona is approximately circular, as shown in the photograph below.

The depth (in feet) of the crater below any point on a diameter can be modeled by the expression $\frac{57}{400,000}(x^2 - 2000x - 3,000,000)$, where x is the distance to the point on the diameter 1000 ft in from the edge. Write this expression in factored form. (*Source:* barringercrater.com)

102. The first field goal longer than 60 yd in National Football League history occurred on November 8, 1970. In one possible path the ball could have traveled, the height of the ball (in feet) is modeled by the polynomial $\left(-\frac{1}{18}\right)(x^2 + 57x - 180)$, where x is the ground distance (in yards) of the ball to the goal post. Factor this expression. (*Source:* nfl.com)

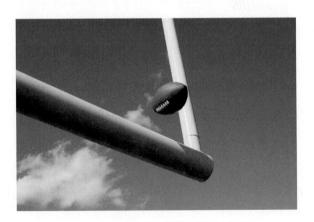

• Check your answers on page A-22.

Groupwork

1. Work with a partner. Next to each trinomial, list at least two integers that when inserted in the box will make the trinomial factorable.

   a. $x^2 + \boxed{}\, x + 60$

   b. $x^2 - x + \boxed{}$

Investigation

2. Algebra tiles give a physical representation of algebraic concepts. Below, the green tile (x by x) represents a square which models x^2, the square-term of the polynomial $x^2 + 5x + 6$. Each blue tile (x by 1) represents $1x$, or x. So the five blue tiles model $5x$, the linear term of the polynomial. Each orange tile (1 by 1) represents 1. Therefore, the six orange squares model the constant term 6. These tiles can be placed and moved according to the rules of algebra. Copy or trace the algebra tiles below onto a piece of paper. Then, cut out each tile separately. Finally, position all the tiles like a jigsaw puzzle so as to form a rectangle. (*Hint:* Factor the polynomial $x^2 + 5x + 6$.)

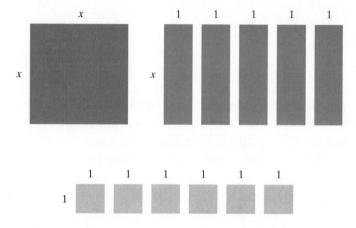

Mathematical Reasoning

3. In arithmetic, we simplify a fraction such as $\dfrac{20}{24}$ by writing the numerator and denominator in factored form and then dividing out common factors. For example, $\dfrac{20}{24} = \dfrac{2 \cdot 2 \cdot 5}{2 \cdot 2 \cdot 2 \cdot 3} = \dfrac{5}{6}$. Assuming $x \neq 1$ and $x \neq 3$, simplify $\dfrac{x^2 - 3x + 2}{x^2 - 4x + 3}$.

In the previous section, we discussed factoring trinomials of the form $ax^2 + bx + c$, where $a = 1$. Here, we consider polynomials whose leading coefficient is not 1, that is, trinomials such as:

$$2x^2 + 5x + 3 \quad \text{and} \quad 5x^2 - 13x - 6$$

$$\uparrow \qquad\qquad\qquad \uparrow$$

The coefficient of the leading term is not 1.

The method of factoring that we use is, again, trial and error. However, we also discuss an alternative procedure that is based on factoring by grouping, a method covered in Section 6.1.

Factoring $ax^2 + bx + c$, for $a \neq 1$

Consider the product of two binomials:

$$\begin{array}{cccc} \text{F} & \text{O} & \text{I} & \text{L} \end{array}$$
$$(2x + 3)(x + 1) = (2x)(x) + (2x)(1) + (3)(x) + (3)(1)$$
$$= 2x^2 \quad + \quad 2x \quad + \quad 3x \quad + \quad 3$$
$$= 2x^2 + 5x + 3$$

Now, to reverse the process, we start with the product $2x^2 + 5x + 3$. To factor this polynomial, first check for common factors. Since there are no common factors other than 1 and -1, we use the FOIL method in reverse. Here, however, the leading coefficient is not 1, so we consider the factors of both 2 and 3. We then list and test all combinations of these factors to see if any will give us the desired middle term $5x$. In other words, we are looking for four integers so that:

$$2x^2 + 5x + 3 = (?x + ?)(?x + ?)$$

Factors of 2	Factors of 3	Possible Binomial Factors	Sum of Outer and Inner Products
2, 1	3, 1	$(2x + 3)(x + 1)$	$2x + 3x = 5x$ ⟵ Correct middle term
		$(2x + 1)(x + 3)$	$6x + x = 7x$
2, 1	$-3, -1$	$(2x - 3)(x - 1)$	$-2x - 3x = -5x$
		$(2x - 1)(x - 3)$	$-6x - x = -7x$

Using this trial-and-error method, we find that $(2x + 3)(x + 1)$ is the correct factorization.

Note that this trial-and-error method for factoring a trinomial such as $2x^2 + 5x + 3$ is similar to the method for factoring trinomials with leading coefficient 1: We list and test all the possible factors of both the leading term and the constant term of the trinomial. We are looking for a combination of factors where the sum of the outer and inner products is the middle term of the trinomial. Practice and experience will shorten the process.

EXAMPLE 1

Factor: $3x^2 + 11x + 10$

Solution The terms of $3x^2 + 11x + 10$ have no common factors. So we proceed to use the trial-and-error method to factor this trinomial. Note that both the middle term $11x$ and the constant term 10 are positive. So we need to consider only combinations of the positive factors of 3 and positive factors of 10 that will give us a middle term with coefficient 11.

$$3x^2 + 11x + 10 = (?x + ?)(?x + ?)$$

Factors of 3	Factors of 10	Possible Binomial Factors	Middle Term
3, 1	2, 5	$(3x + 2)(x + 5)$	$15x + 2x = 17x$
		$(3x + 5)(x + 2)$	$6x + 5x = 11x$ ← Correct middle term
	10, 1	$(3x + 10)(x + 1)$	$3x + 10x = 13x$
		$(3x + 1)(x + 10)$	$30x + x = 31x$

So $3x^2 + 11x + 10 = (3x + 5)(x + 2)$.

PRACTICE 1

Factor: $5x^2 + 14x + 8$

EXAMPLE 2

Express in factored form: $15 - 17x + 4x^2$

Solution First, we rewrite the terms of the trinomial in descending order: $4x^2 - 17x + 15$. These terms have no common factor. To find the factorization of $4x^2 - 17x + 15$, we consider combinations of factors of 4 and factors of 15 that will result in a middle term with the coefficient -17.

Factors of 4	Factors of 15	Possible Binomial Factors	Middle Term
2, 2	$-3, -5$	$(2x - 3)(2x - 5)$	$-10x - 6x = -16x$
	$-15, -1$	$(2x - 15)(2x - 1)$	$-2x - 30x = -32x$
4, 1	$-3, -5$	$(4x - 3)(x - 5)$	$-20x - 3x = -23x$
		$(4x - 5)(x - 3)$	$-12x - 5x = -17x$ ← Correct middle term
	$-15, -1$	$(4x - 15)(x - 1)$	$-4x - 15x = -19x$
		$(4x - 1)(x - 15)$	$-60x - x = -61x$

So $4x^2 - 17x + 15 = (4x - 5)(x - 3)$.

PRACTICE 2

Factor: $21 - 25x + 6x^2$

Note that in Examples 1 and 2 we could have stopped the process of testing possible factorizations after finding the correct middle term.

EXAMPLE 3

Factor: $2y^2 + 19y - 10$

Solution The terms of $2y^2 + 19y - 10$ have no common factors. So we factor the trinomial by considering combinations of the factors of 2 and factors of -10 that will give us the middle term with coefficient 19.

Factors of 2	Factors of −10	Possible Binomial Factors	Middle Term
2, 1	2, −5	$(2y + 2)(y - 5)$	$-10y + 2y = -8y$
		$(2y - 5)(y + 2)$	$4y - 5y = -y$
	−2, 5	$(2y - 2)(y + 5)$	$10y - 2y = 8y$
		$(2y + 5)(y - 2)$	$-4y + 5y = y$
	10, −1	$(2y + 10)(y - 1)$	$-2y + 10y = 8y$
		$(2y - 1)(y + 10)$	$20y - y = 19y$ ← Correct middle term
	−10, 1	$(2y - 10)(y + 1)$	$2y - 10y = -8y$
		$(2y + 1)(y - 10)$	$-20y + y = -19y$

So $2y^2 + 19y - 10 = (2y - 1)(y + 10)$.

PRACTICE 3

Factor: $7y^2 + 47y - 14$

Can you explain why $(2y + 2)$, $(2y - 2)$, $(2y + 10)$, and $(2y - 10)$ can be immediately eliminated as possible factors of $2y^2 + 19y - 10$ in Example 3?

Consider the following possible binomial factors in Example 3.

The signs of the constant terms are reversed.

$$(2y - 1)(y + 10) = 2y^2 + 19y - 10$$
$$(2y + 1)(y - 10) = 2y^2 - 19y - 10$$

The sign of the middle term changes.

Comparing these possible factors suggest the following shortcut:

> **TIP** Reversing the signs of the constant terms in binomial factors has the effect of switching the sign of the middle term in their product.

EXAMPLE 4

Factor: $5x^2 - 13x - 6$

Solution Since the terms of $5x^2 - 13x - 6$ have no common factors, let's look at the combinations of factors of 5 and factors of -6 that will give us the middle term with coefficient -13.

Factors of 5	Factors of -6	Possible Binomial Factors	Middle Term	
5, 1	2, -3	$(5x + 2)(x - 3)$	$-15x + 2x = -13x$	← Correct middle term
		$(5x - 3)(x + 2)$	$10x - 3x = 7x$	
	$-2, 3$	$(5x - 2)(x + 3)$	$15x - 2x = 13x$	
		$(5x + 3)(x - 2)$	$-10x + 3x = -7x$	
	$-6, 1$	$(5x - 6)(x + 1)$	$5x - 6x = -x$	
		$(5x + 1)(x - 6)$	$-30x + x = -29x$	
	6, -1	$(5x + 6)(x - 1)$	$-5x + 6x = x$	
		$(5x - 1)(x + 6)$	$30x - x = 29x$	

So $5x^2 - 13x - 6 = (5x + 2)(x - 3)$. Note that it was unnecessary to examine any factors after the first trial, since we found the correct combination for the middle term $-13x$.

PRACTICE 4

Factor: $2x^2 - x - 10$

EXAMPLE 5

Factor: $12y^3 + 2y^2 - 2y$

Solution Since $2y$ is the GCF of the trinomial, first we factor it out getting:

$$12y^3 + 2y^2 - 2y = 2y(6y^2 + y - 1)$$

Next, we factor $6y^2 + y - 1$, looking for a combination of factors of 6 and factors of -1 that will give us the middle term with coefficient 1.

Factors of 6	Factors of -1	Possible Binomial Factors	Middle Term	
6, 1	1, -1	$(6y + 1)(y - 1)$	$-6y + y = -5y$	
		$(6y - 1)(y + 1)$	$6y - y = 5y$	
3, 2	1, -1	$(3y + 1)(2y - 1)$	$-3y + 2y = -y$	
		$(3y - 1)(2y + 1)$	$3y - 2y = y$	← Correct middle term

So $12y^3 + 2y^2 - 2y = 2y(6y^2 + y - 1) = 2y(3y - 1)(2y + 1)$.

PRACTICE 5

Factor: $18x^3 - 21x^2 - 9x$

Now, we consider factoring a trinomial of the form $ax^2 + bxy + cy^2$. This type of trinomial contains more than one variable, so we need to look for a factorization of the form $(?x + ?y)(?x + ?y)$.

EXAMPLE 6

Factor: $12x^2 + 28xy + 8y^2$

Solution   Since 4 is the GCF of the trinomial, let's first factor it out.

$$12x^2 + 28xy + 8y^2 = 4(3x^2 + 7xy + 2y^2)$$

Next, we factor $3x^2 + 7xy + 2y^2$. We look for the combination of factors of 3 and factors of 2 that will result in the middle term $7xy$.

$$4(3x^2 + 7xy + 2y^2) = 4(?x + ?y)(?x + ?y)$$

Factors of 3	Factors of 2	Possible Binomial Factors	Middle Term	
3, 1	2, 1	$(3x + 2y)(x + y)$	$3xy + 2xy = 5xy$	Correct
		$(3x + y)(x + 2y)$	$6xy + xy = 7xy$	← middle term

So $12x^2 + 28xy + 8y^2 = 4(3x^2 + 7xy + 2y^2) = 4(3x + y)(x + 2y)$.

PRACTICE 6

Factor: $36c^2 - 12cd - 15d^2$

Now, let's consider an alternative procedure for factoring a trinomial $ax^2 + bx + c$ based on *grouping*. This method, which the next example illustrates, is sometimes called the *ac method*.

EXAMPLE 7

Factor: $2x^2 + 5x - 3$

Solution   First, we check that the terms of $2x^2 + 5x - 3$ have no common factors. Next, instead of listing the factors of 2 and the factors of -3 as in the trial-and-error method, we begin by finding their product:

$$ac = (2)(-3) = -6$$

We then look for two factors of the number ac (that is, -6) that add up to b (that is, 5). The numbers 6 and -1 satisfy these conditions, since $(6)(-1) = -6$ and $(6) + (-1) = 5$. We use these numbers to split up the middle term in the original trinomial, and then rewrite the trinomial.

$$
\begin{aligned}
2x^2 + 5x - 3 &= 2x^2 + 6x + (-1)x - 3 && \text{Split up the middle term:}\\
&&& 5x = 6x + (-1)x.\\
&= [2x^2 + 6x] + [(-1)x - 3] && \text{Group the first two terms}\\
&&& \text{and the last two terms.}\\
&= 2x(x + 3) + (-1)(x + 3) && \text{Factor out the GCF from}\\
&&& \text{each group.}\\
&= (x + 3)(2x - 1) && \text{Write in factored form.}
\end{aligned}
$$

So $2x^2 + 5x - 3 = (x + 3)(2x - 1)$. As usual, we can check that this factorization is correct by multiplication.

PRACTICE 7

Factor: $2x^2 - 7x - 4$

Now, let's solve some applied problems involving the factoring of trinomials.

EXAMPLE 8

A ball is thrown upward at 40 ft/sec from the top of a building 24 ft above the ground. The height of the ball above the ground (in feet) t sec after the ball is thrown is given by the expression $-16t^2 + 40t + 24$. Write this expression in factored form.

Solution   We factor the expression $-16t^2 + 40t + 24$:

$$-16t^2 + 40t + 24 = -8(2t^2 - 5t - 3)$$
$$= -8(2t + 1)(t - 3)$$

So the factorization of $-16t^2 + 40t + 24$ is $-8(2t + 1)(t - 3)$, which can also be written

$$8(-2t - 1)(t - 3),$$

or

$$8(2t + 1)(-t + 3).$$

PRACTICE 8

A bin is made from a 7-foot by 5-foot sheet of metal by cutting out squares of equal size from each corner and then turning up the sides. The volume of the resulting bin can be represented by the expression $4x^3 - 24x^2 + 35x$. Rewrite this expression in factored form.

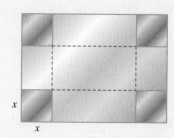

Match each trinomial with its binomial factors.

1. $2x^2 + 3x - 9$ a. $(2x - 9)(x + 1)$

2. $2x^2 - 19x + 9$ b. $(2x - 9)(x - 1)$

3. $2x^2 - 11x + 9$ c. $(x - 9)(2x + 1)$

4. $2x^2 - 3x - 9$ d. $(x - 9)(2x - 1)$

5. $2x^2 - 17x - 9$ e. $(2x - 3)(x + 3)$

6. $2x^2 - 7x - 9$ f. $(x - 3)(2x + 3)$

Find the missing factor.

7. $3x^2 + 16x + 5 = (x + 5)(\quad\quad)$

8. $2x^2 + 11x + 12 = (2x + 3)(\quad\quad)$

9. $5x^2 - 13x - 6 = (5x + 2)(\quad\quad)$

10. $2x^2 - x - 6 = (2x + 3)(\quad\quad)$

11. $3x^2 - 11x + 6 = (3x - 2)(\quad\quad)$

12. $6x^2 - 7x + 2 = (2x - 1)(\quad\quad)$

Factor, if possible.

13. $3x^2 + 8x + 5$

14. $2x^2 + 15x + 7$

15. $2y^2 - 11y + 5$

16. $3y^2 - 10y + 7$

17. $3x^2 + 14x + 8$

18. $2x^2 + 11x + 9$

19. $5x^2 + 9x - 6$

20. $4y^2 - 16y - 7$

21. $6y^2 - y - 5$

22. $5x^2 + 17x - 12$

23. $2y^2 - 11y + 14$

24. $7y^2 - 19y + 10$

25. $9a^2 - 18a - 16$

26. $10m^2 - m - 21$

27. $4x^2 - 13x + 3$

28. $4n^2 - 9n + 2$

29. $6 + 17y + 12y^2$

30. $4 + 16n + 15n^2$

31. $-17m + 21 + 2m^2$

32. $-16x + 5 + 3x^2$

33. $-6a^2 - 7a + 3$

34. $-5b^2 - 14b + 3$

35. $8y^2 + 5y - 22$

36. $6x^2 + 5x - 25$

37. $7y^2 + 36y - 5$

38. $2y^2 + 27y + 14$

39. $8a^2 + 65a + 8$

40. $8n^2 + 33n + 4$

41. $6x^2 + 25x - 9$

42. $10x^2 + 21x - 10$

43. $8y^2 - 26y + 15$

44. $4m^2 - 16m - 9$

45. $14y^2 - 38y + 20$

46. $9y^2 - 24y + 15$

47. $28a^2 + 24a - 4$

48. $6x^2 + 40x - 14$

49. $-6b^2 + 40b + 14$

50. $-25m^2 + 65m + 30$

51. $12y^3 + 50y^2 + 28y$

52. $6x^3 + 45x^2 + 21x$

53. $14a^4 - 38a^3 + 20a^2$

54. $10n^4 - 35n^3 + 15n^2$

55. $2x^3y + 13x^2y + 15xy$

56. $3xy^3 + 10xy^2 + 3xy$

57. $6ab^3 - 44ab^2 + 14ab$

58. $12a^3b - 34a^2b + 24ab$

59. $20c^2 - 9cd + d^2$

60. $12a^2 - 25ab + 12b^2$

61. $2x^2 - 5xy - 3y^2$ **62.** $6s^2 - st - 12t^2$ **63.** $8a^2 - 6ab + b^2$ **64.** $3m^2 - 8mn + 5n^2$

65. $18x^2 + 3xy - 6y^2$ **66.** $4s^2 + 10st - 24t^2$ **67.** $16c^2 - 44cd + 30d^2$ **68.** $16a^2 - 48ab + 36b^2$

69. $27u^2 + 18uv + 3v^2$ **70.** $4a^2 + 26ab - 48b^2$ **71.** $42x^3 + 45x^2y - 27xy^2$ **72.** $60p^3 + 28p^2q - 16pq$

73. $-30x^4y + 35x^3y^2 + 15x^2y^3$ **74.** $-24x^3y^2 - 6x^2y^3 + 18xy^4$

75. $5ax^2 - 28axy - 12ay^2$ **76.** $3cx^2 + 7cxy - 20cy^2$

Mixed Practice

Factor, if possible.

77. $-5m + 2 + 3m^2$ **78.** $3y^2 + 2y + 5$ **79.** $8x^2 - 2x - 3$ **80.** $8x^2 + 36x - 20$

81. $14x^3 + 44x^2 + 6x$ **82.** $30a^3b - 55a^2b + 15ab$ **83.** $8m^2 - 18mn + 9n^2$ **84.** $24a^3 + 18a^2b - 15ab^2$

85. $7r^2 - 9r + 2$ **86.** $10a^2 + 11a - 6$

Solve.

87. Factor: $3x^2 - 22x + 7$.
  a. $(3x + 7)(x + 1)$ **b.** $(x + 7)(3x + 1)$
  c. $(3x - 7)(x - 1)$ **d.** $(x - 7)(3x - 1)$

88. Find the missing factor:
  $5x^2 - 2x - 3 = (5x + 3)($ $)$

Applications

B *Solve.*

89. An object is thrown upward so that its height in meters above the ground at time t sec is represented by the expression $-5t^2 - 21t + 20$. Factor this expression.

90. A box with width w has a volume that can be expressed as $3w^3 - 2w^2 - w$. Rewrite this expression in factored form.

91. A homeowner decides to increase the area of his 8-ft by 10-ft deck by increasing both the length and the width, as shown in the diagram below:

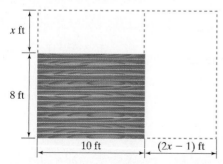

  a. Find the area of the expanded deck by adding the areas of the three rectangles.
  b. Express the area of the expanded deck in factored form.

92. The diagram below shows a circular pad with radius r on the square top of a dining room table.

  a. Find the area of the region of the table top not covered by the circular pad.
  b. Express the area in part (a) in factored form.

93. Show that the expression $4n^2 - 12n + 5$ can be written as the product of two integers that differ by 4, no matter what integer n represents.

94. The squares of the first n whole numbers add up to $\dfrac{2n^3 + 3n^2 + n}{6}$. Write this expression so that the numerator is in factored form.

- Check your answers on page A-23.

MINDStretchers

Groupwork

1. Work with a partner. Next to each polynomial, list at least two integers that, when inserted in the box, will make the polynomial factorable.

   a. $2x^2 + \boxed{}\ x + 5$ 　　　 **b.** $\boxed{}\ x^2 - 4x + 1$ 　　　 **c.** $3x^2 - x + \boxed{}$

Investigation

2. Copy or trace the following algebra tiles onto a piece of paper. Then, cut out each tile separately. Finally, position all the tiles like a jigsaw puzzle so as to form a rectangle. (*Hint:* Factor the polynomial $2x^2 + 7x + 3$.)

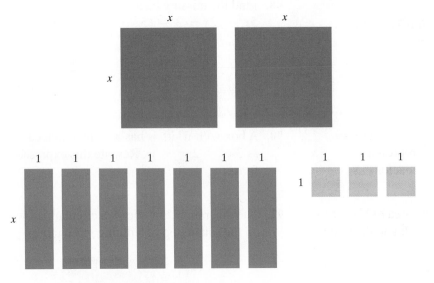

Writing

3. Do you prefer to factor trinomials using the trial-and-error method or the *ac* method? Explain why.

Recall that in Section 5.6, we considered formulas for multiplying binomials in certain special cases: the square of a sum, the square of a difference, and the product of the sum and the difference of the same two terms. In this section, we show that these formulas also allow us to factor special polynomials, called *perfect square trinomials* and the *difference of squares*. Recognizing these special polynomials makes it easier to factor them.

Factoring Perfect Square Trinomials

We have seen that squaring the sum or the difference of two terms gives us:

$$(a + b)^2 = a^2 + 2ab + b^2 \quad \text{and} \quad (a - b)^2 = a^2 - 2ab + b^2$$

Each of these products is called a **perfect square trinomial**. Such trinomials may be factored by reversing the multiplication process.

Factoring a Perfect Square Trinomial

$$a^2 + 2ab + b^2 = (a + b)^2$$
$$a^2 - 2ab + b^2 = (a - b)^2$$

The first formula shows us how to factor a trinomial that happens to be the sum of the squares of two terms *plus* twice their product. In this formula, the terms are a and b, the sum of their squares is $a^2 + b^2$, and $+2ab$ is twice their product. In words, the formula states that the factorization of such a trinomial is the square of the *sum* of the two terms, namely $(a + b)^2$.

The second formula applies when we want to factor a trinomial that is the sum of the squares of two terms *minus* twice their product. In this formula, the terms are again a and b, the sum of their squares is $a^2 + b^2$, and $-2ab$ is minus twice their product. In words, this formula states that the factorization of such a trinomial is the square of the *difference* of the two terms, namely $(a - b)^2$.

Keep in mind that the formulas $a^2 + 2ab + b^2 = (a + b)^2$ and $a^2 - 2ab + b^2 = (a - b)^2$ only apply when a polynomial is a perfect square trinomial. Let's consider an example of recognizing these trinomials.

EXAMPLE 1

Determine whether each polynomial is a perfect square trinomial.

a. $x^2 - 8x + 16$

b. $-x^2 + 2x + 1$

c. $x^2 + 4x + 1$

d. $n^2 - 6n - 9$

e. $9x^2 + 30xy + 25y^2$

PRACTICE 1

Indicate whether each trinomial is a perfect square.

a. $x^2 + 6x + 9$

b. $-4t^2 - 4t + 1$

c. $y^2 - 14y + 49$

d. $x^2 - 2x - 1$

e. $4p^2 - 4pq + q^2$

EXAMPLE 1 (continued)

Solution

a. In the polynomial $x^2 - 8x + 16$, x^2 and 16 are perfect squares and correspond to a^2 and b^2, respectively, in the formula $a^2 - 2ab + b^2$. So a corresponds to x, and b corresponds to 4. Since the middle term, $-8x$, or $-2 \cdot x \cdot 4$, corresponds to $-2ab$, the polynomial is a perfect square trinomial.

$$x^2 - 8x + 16 = x^2 - \underbrace{2 \cdot x \cdot 4}_{} + (4)^2$$
$$\quad\quad\quad\quad\quad\;\; a^2 \quad - \quad 2ab \quad + \quad b^2$$

b. In the polynomial $-x^2 + 2x + 1$, $-x^2$ is not a perfect square since its coefficient is negative. Therefore, $-x^2 + 2x + 1$ is not a perfect square trinomial.

c. For the polynomial $x^2 + 4x + 1$, x^2 and 1 are both perfect squares with a corresponding to x, and b to 1 in the formula $a^2 + 2ab + b^2$. However, the middle term $+4x$ is not twice the product of x and 1. So $x^2 + 4x + 1$ is not a perfect square trinomial.

d. The constant term of a perfect square trinomial, b^2, must be positive. So $n^2 - 6n - 9$ is not a perfect square, since its constant term -9 is negative.

e. In the polynomial $9x^2 + 30xy + 25y^2$, we know that $9x^2$ is equal to $(3x)^2$ and $25y^2$ equals $(5y)^2$. So a corresponds to $3x$ and b corresponds to $5y$ in the formula $a^2 + 2ab + b^2$. The middle term $+30xy$ can be expressed as $+2 \cdot 3x \cdot 5y$. Therefore, the trinomial is a perfect square.

$$9x^2 + 30xy + 25y^2 = (3x)^2 + \underbrace{2(3x)(5y)}_{} + (5y)^2$$
$$\quad\quad\quad\quad\quad\quad\quad\;\; a^2 \quad + \quad 2ab \quad + \quad b^2$$

Now, let's practice *factoring* perfect square trinomials.

EXAMPLE 2

Factor: $x^2 + 12x + 36$

Solution   For the trinomial $x^2 + 12x + 36$, the first term x^2 and the last term 36, or 6^2, are perfect squares. The middle term, $12x$ or $2 \cdot x \cdot 6$, is twice the product of x and 6. It follows that $x^2 + 12x + 36$ is a perfect square trinomial. Since its middle term is positive, we apply the formula for the square of a sum.

$$x^2 + 12x + 36 = x^2 + \underbrace{2 \cdot x \cdot 6}_{} + (6)^2 = (x + 6)^2$$
$$\quad\quad\quad\quad\quad\; a^2 \; + \quad 2ab \quad + \quad b^2 \;\; = \;\; (a + b)^2$$

Check   We can confirm our answer by multiplying out $(x + 6)^2$.

$(x + 6)^2 = (x + 6)(x + 6) = x^2 + 2 \cdot 6x + 36 = x^2 + 12x + 36$

PRACTICE 2

Factor: $n^2 + 20n + 100$

EXAMPLE 3

Write as the square of a binomial: $x^2 + 9 - 6x$

Solution   Let's begin by rewriting the trinomial in descending order: $x^2 - 6x + 9$.

$$x^2 - 6x + 9 = x^2 - \underbrace{2 \cdot x \cdot 3}_{} + (3)^2 = (x - 3)^2$$

$$\underset{\uparrow}{a^2} - \underset{\uparrow}{2ab} + \underset{\uparrow}{b^2} = \underset{\uparrow \quad \uparrow}{(a - b)^2}$$

PRACTICE 3

Factor: $t^2 + 4 - 4t$

EXAMPLE 4

Express $9x^2 - 6xy + y^2$ as the square of a binomial.

Solution   Here the trinomial contains more than one variable.

$$9x^2 - 6xy + y^2 = (3x)^2 - \underbrace{2 \cdot 3x \cdot y}_{} + y^2 = (3x - y)^2$$

$$\underset{\uparrow}{a^2} - \underset{\uparrow}{2ab} + \underset{\uparrow}{b^2} = \underset{\uparrow \quad \uparrow}{(a - b)^2}$$

PRACTICE 4

Write $25c^2 - 40cd + 16d^2$ as the square of a binomial.

EXAMPLE 5

Factor: $y^{10} + 16y^5 + 64$

Solution   Since $y^{10} = (y^5)^2$ and $64 = (8)^2$, we know that y^{10} and 64 are perfect squares. Also, $16y^5 = 2 \cdot 8 \cdot y^5$, that is, $16y^5$ is twice the product of y^5 and 8. So we conclude that $y^{10} + 16y^5 + 64$ is a perfect square trinomial.

$$y^{10} + 16y^5 + 64 = (y^5)^2 + \underbrace{2 \cdot y^5 \cdot 8}_{} + (8)^2 = (y^5 + 8)^2$$

$$\underset{\uparrow}{a^2} + \underset{\uparrow}{2ab} + \underset{\uparrow}{b^2} = \underset{\uparrow \quad \uparrow}{(a + b)^2}$$

PRACTICE 5

Express $x^4 + 8x^2 + 16$ as the square of a binomial.

In factoring a perfect square trinomial, how do we know whether the binomial squared is the sum or the difference of two terms?

Factoring the Difference of Squares

Recall that when finding the product of the sum and the difference of the same terms, we get:

$$(a + b)(a - b) = a^2 - b^2$$

The product is a binomial that is called a **difference of squares**. We can factor such a binomial by reversing the multiplication process.

> **Factoring the Difference of Squares**
>
> $$a^2 - b^2 = (a + b)(a - b)$$

This formula is a shortcut for factoring a binomial equal to the square of one term *minus* the square of another term. The terms are a and b, and so their squares are a^2 and b^2. In words, the formula states that the factorization of a binomial that is the difference of the squares of two terms is the sum of the two terms times the difference of the same two terms, that is, $(a + b)(a - b)$.

EXAMPLE 6

Indicate whether each binomial is a difference of squares.

a. $x^2 - 81$ **b.** $x^2 + y^2$

c. $x^2 - y^3$ **d.** $4p^6 - q^2$

Solution

a. In the binomial $x^2 - 81$, both x^2 and 81 are perfect squares and correspond to a^2 and b^2, respectively, in our formula.

$$x^2 - 81 = x^2 - 9^2$$
$$\underset{a^2}{\uparrow} \quad \underset{b^2}{\uparrow}$$

Here, a corresponds to x, and b corresponds to 9. So $x^2 - 81$ is a difference of squares.

b. In the expression $x^2 + y^2$, both x^2 and y^2 are perfect squares, where x corresponds to a and y to b. However, the binomial is the sum and not the difference of squares.

c. For $x^2 - y^3$, x^2 is a perfect square, but y^3 is not. So $x^2 - y^3$ is not a difference of squares.

d. The binomial $4p^6 - q^2$ can be rewritten as $(2p^3)^2 - q^2$, and so is a difference of squares.

PRACTICE 6

Determine whether each binomial is a difference of squares.

a. $x^2 - 64$

b. $x^2 + 49$

c. $x^3 - 16$

d. $r^4 - 9s^6$

Note, as the preceding example suggests, that even powers of a variable are perfect squares whereas odd powers of a variable are not. Can you explain why?

EXAMPLE 7

Factor: $x^2 - 100$

Solution Since x^2 and 100 are perfect squares, $x^2 - 100$ is a difference of squares.

$$x^2 - 100 = x^2 - 10^2 = \underbrace{(x + 10)}\underbrace{(x - 10)}$$
$$\underset{a^2}{\uparrow} \quad \underset{b^2}{\uparrow} \quad = \quad \underset{(a + b)}{} \quad \underset{(a - b)}{}$$

Check We can verify that $(x + 10)(x - 10)$ is the factorization of $x^2 - 100$ by multiplying.

$$(x + 10)(x - 10) = x^2 - 10x + 10x - 100 = x^2 - 100$$

PRACTICE 7

Factor: $y^2 - 121$

EXAMPLE 8

Write $16x^2 - 49y^2$ in factored form.

Solution Since $16x^2 = (4x)^2$ and $49y^2 = (7y)^2$, we know that $16x^2$ and $49y^2$ are perfect squares. So $16x^2 - 49y^2$ is a difference of squares.

$$16x^2 - 49y^2 = (4x)^2 - (7y)^2 = \underbrace{(4x + 7y)}\underbrace{(4x - 7y)}$$

$$\underset{a^2}{\uparrow} \quad - \quad \underset{b^2}{\uparrow} \quad = \quad \underset{(a + b)}{\uparrow} \quad \underset{(a - b)}{\uparrow}$$

PRACTICE 8

Express $9x^2 - 25y^2$ in factored form.

EXAMPLE 9

Factor: $4x^4 - 9y^6$

Solution Because $4x^4 = (2x^2)^2$ and $9y^6 = (3y^3)^2$, we see that $4x^4$ and $9y^6$ are both perfect squares. So $4x^4 - 9y^6$ is a difference of squares.

$$4x^4 - 9y^6 = (2x^2)^2 - (3y^3)^2 = \underbrace{(2x^2 + 3y^3)}\underbrace{(2x^2 - 3y^3)}$$

$$\underset{a^2}{\uparrow} \quad - \quad \underset{b^2}{\uparrow} \quad = \quad \underset{(a + b)}{\uparrow} \quad \underset{(a - b)}{\uparrow}$$

PRACTICE 9

Factor: $64x^8 - 81y^2$

EXAMPLE 10

Find an expression for the area of the cross section of the pipe pictured. Write this expression in factored form.

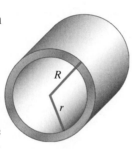

Solution The cross section is a ring-shaped region between two circles with the same center. The radius of the inner circle is r and the radius of the outer circle is R. Using the formula for the area of a circle helps us to find an expression for the area of the cross section:

Area of the cross section

$= $ Area of the large circle $-$ Area of the small circle

$= \qquad \pi R^2 \qquad - \qquad \pi r^2$

$= \pi(R^2 - r^2)$

$= \pi(R + r)(R - r)$

So in factored form, an expression for the area of the cross section of the pipe is $\pi(R + r)(R - r)$.

PRACTICE 10

A stone is dropped from a bridge 256 ft above a river. The height of the stone above the river (in feet) t sec after it is dropped is given by the expression $256 - 16t^2$. Factor this expression.

Mathematically Speaking

Fill in each blank with the most appropriate term or phrase from the given list.

odd	the square	even
perfect square trinomial	half the product	twice the product
square of the sum of two terms	difference of squares	

1. A(n) _____ of the form $a^2 + 2ab + b^2$ can be factored as $(a + b)^2$.

2. The trinomial $a^2 + 5ab + 25b^2$ is not a perfect square, because the middle term is not _____ of a and $5b$.

3. The binomial $a^2 - b^2$, a(n) _____, can be factored as $(a + b)(a - b)$.

4. Powers of a variable that are _____ are not perfect squares.

Determine whether each polynomial is a perfect square trinomial, a difference of squares, or neither.

5. $x^2 + 2x + 1$

6. $y^2 + 8y + 16$

7. $-t^2 - 4t + 1$

8. $-y^2 + 2y + 9$

9. $x^2 - 6x + 9$

10. $n^2 - 2n + 1$

11. $x^2 - 25$

12. $y^2 - 49$

13. $81x^2 - 36y^2$

14. $4x^2 - 100y^2$

15. $25x^2 - 20x + 4$

16. $16x^2 - 24x + 9$

17. $y^2 + 1$

18. $y^2 + 3$

19. $25x^2 + 10xy + y^2$

20. $49x^2 + 14xy + y^2$

21. $x^3 - 1$

22. $x^4 + 9$

Factor, if possible.

23. $x^2 - 12x + 36$

24. $y^2 - 14y + 49$

25. $y^2 + 20y + 100$

26. $x^2 + 2x + 1$

27. $a^2 - 4a + 4$

28. $b^2 - 22b + 121$

29. $x^2 - 6x - 9$

30. $y^2 - 10y - 25$

31. $4a^2 - 36a + 81$

32. $25b^2 - 20b + 4$

33. $49x^2 + 28x + 4$

34. $9y^2 + 24y + 16$

35. $36 - 60x + 25x^2$

36. $49 - 42y + 9y^2$

37. $m^2 + 26mn + 169n^2$

38. $4a^2 + 36ab + 81b^2$

39. $225a^2 - 30ab + b^2$

40. $25s^2 - 40st + 16t^2$

41. $y^4 + 2y^2 + 1$

42. $x^4 + 4x^2 + 4$

43. $6x^2 + 12x + 6$

44. $12y^2 + 24y + 12$

45. $27m^3 - 36m^2 + 12m$

46. $48y^3 - 24y^2 + 3y$

47. $4s^2t^3 + 80s^2t^2 + 400s^2t$

48. $2x^3y^2 - 52x^2y^2 + 338xy^2$

C Factor, if possible.

49. $m^2 - 64$

50. $n^2 - 1$

51. $y^2 - 81$

52. $x^2 - 16$

53. $144 - x^2$

54. $225 - t^2$

55. $100m^2 - 81$

56. $16n^2 - 25$

57. $36x^2 + 121$

58. $64n^2 + 169$

59. $1 - 9x^2$

60. $81 - 4y^2$

61. $x^2 - 4y^2$

62. $49c^2 - d^2$

63. $100x^2 - 9y^2$

64. $36a^2 - 121b^2$

65. $3k^3 - 147k$

66. $5m^3 - 125m$

67. $4y^4 - 36y^2$

68. $3t^5 - 300t^3$

69. $27x^2y - 3x^2y^3$

70. $50xy - 18x^3y$

71. $2a^2b^2 - 98$

72. $9x^4y^2 - 81$

73. $256 - r^4$

74. $625 - t^4$

75. $5x^4 - 80y^4$

76. $64s^4 - 4t^4$

77. $x^2(c - d) - 4(c - d)$

78. $y^2(a - b) - (a - b)$

79. $16(x - y) - a^2(x - y)$

80. $9(y - c) - x^2(y - c)$

Mixed Practice

Factor, if possible.

81. $9c^2 + 48cd + 64d^2$

82. $169m^2 - 49n^2$

83. $a^2 - 225b^4$

84. $8t^3 - 56t^2 + 98t$

85. $54a^4b^2 - 36a^2b + 6$

86. $12xy - 27xy^3$

87. $81 - w^4$

88. $196 - r^2$

89. $36u^2 + 60u + 25$

90. $121s^2 + 36$

Determine whether each polynomial is a perfect square trinominal, a difference of squares, or neither.

91. $j^2 - 169$

92. $81a^2 - 90ab + 25b^2$

Applications

D Solve.

93. If the radius of a balloon decreases from radius r_1 to radius r_2, then the drop in the balloon's surface area is given by the expression $4\pi r_1^2 - 4\pi r_2^2$. Write this expression in factored form.

94. The height (in feet) of a stone dropped from a cliff 100 ft above a river, as shown in the illustration, is given by the expression $100 - 16t^2$, where t is time (in seconds). Factor this expression.

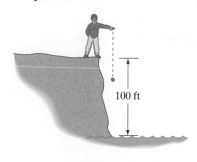

100 ft

95. An open box is made from a 2-foot by 2-foot piece of cardboard by cutting out equal squares from each of the four corners and turning up the sides. The volume of the resulting box can be modeled by the polynomial $4x - 8x^2 + 4x^3$. Factor the expression.

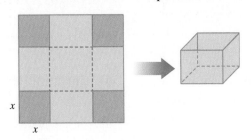

96. Find an expression, in factored form, for the area of the wooden border of the square picture frame shown.

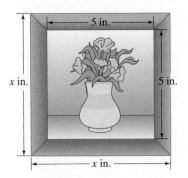

97. When the velocity of a rocket increases from v_1 to v_2, the force caused by air resistance increases by $kv_2^2 - kv_1^2$ for a constant k. Write this polynomial in factored form.

98. A $16,000 investment grew by an average annual rate of return of r. After two years, the value of the investment in dollars was $16{,}000 + 32{,}000r + 16{,}000r^2$. What is the factorization of this expression?

• Check your answers on page A-23.

MIND*Stretchers*

Patterns

1. Find two factors of 3599 both of which are greater than 1. (*Hint:* $3599 = 3600 - 1$.)

Groupwork

2. Try this trick with a partner:

   a. Take your partner's age in years.
   b. Square it.
   c. Subtract 9.
   d. Divide the result by 3 less than your partner's age.
   e. Subtract 53.
   f. Add your partner's age.
   g. Divide by 2.
   h. Add 5^2.
   i. Check that you wind up where you started—with your partner's age.

In the table, record the results for three different ages. Then, in the fourth row, repeat the steps with a variable x representing your partner's age. Explain why this trick works.

(a)	(b)	(c)	(d)	(e)	(f)	(g)	(h)	(i)
x								

Writing

3. In a few sentences, explain how the diagram at the right shows that $(a + b)^2 = a^2 + 2ab + b^2$.

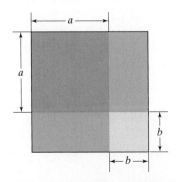

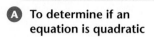

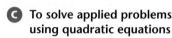
This section deals with a kind of equation that we have not previously considered, namely, a *quadratic equation*. Such equations come up in physics, in finance, and other fields as well.

Consider, for instance, a situation involving the movement of a rocket. If the rocket is shot straight upward from ground level with an initial velocity of 80 ft/sec, the rocket's height above the ground h can be modeled by the expression $80t - 16t^2$, where t is the elapsed time in seconds and h is measured in feet. To find the time at which the rocket falls back and hits the ground (that is, when $h = 0$), we need to be able to solve the quadratic equation $80t - 16t^2 = 0$.

DEFINITION

A **second-degree** or **quadratic equation** is an equation that can be written in the form $ax^2 + bx + c = 0$, where $a, b,$ and c are real numbers and $a \neq 0$.

Some examples of quadratic equations are:

$$x^2 - x + 6 = 0 \qquad 3x^2 - 12 = 0 \qquad (x - 1)^2 = 0$$

Can you explain why these polynomials are of the second degree?

EXAMPLE 1

Determine if the following equations are linear or quadratic.

a. $5x^2 - x = 3$ **b.** $\dfrac{x - 3}{2} = 4$

Solution

a. $5x^2 - x = 3$ is equivalent to $5x^2 - x - 3 = 0$. Since this equation is of the form $ax^2 + bx + c = 0$, where $a = 5$, $b = -1$, and $c = -3$, the equation is quadratic.

b. $\dfrac{x - 3}{2} = 4$ is equivalent to $x - 3 = 8$. This equation is of the form $ax + b = c$, where $a = 1$, $b = -3$, and $c = 8$. So the equation is linear.

PRACTICE 1

Determine whether these equations are quadratic or linear.

a. $3(x + 1) = 6$

b. $2x^2 = x^2 + 7x$

As in the case of a linear equation, a value is said to be a *solution* of a quadratic equation if substituting the value for the variable makes the equation a true statement.

Using Factoring to Solve Quadratic Equations

In this section, we consider those quadratic equations that can be solved by factoring. In Chapter 9, we will consider additional approaches to solving quadratic equations.

The key to solving quadratic equations by factoring is to apply the *zero-product property*.

> ## The Zero-Product Property
> If $ab = 0$, then $a = 0$ or $b = 0$, or both a and $b = 0$.

In words, this property states that if the product of two factors is zero, then either one or both of the factors must be zero.

Consider these examples of the zero-product property.

- If $2x = 0$, then x must be 0 (since $2 \neq 0$).
- If $x(3x - 1) = 0$, then either $x = 0$ or $3x - 1 = 0$.
- If $(x - 3)(x + 2) = 0$, then either $x - 3 = 0$ or $x + 2 = 0$.

Let's see how to use the zero-product property to solve a quadratic equation that is already in factored form.

EXAMPLE 2

Solve: $(2x - 1)(x + 6) = 0$

Solution   Since $(2x - 1)(x + 6) = 0$, the zero-product property tells us that at least one of the factors must equal zero, that is either $2x - 1 = 0$ or $x + 6 = 0$.

$$(2x - 1)(x + 6) = 0$$

$2x - 1 = 0$ or $x + 6 = 0$ Set each factor equal to 0.

$\quad\quad 2x = 1 \quad\quad\quad\quad x = -6$ Solve each equation for x.

$$x = \frac{1}{2}$$

Check   We replace x by the values $\frac{1}{2}$ and -6 in the original equation.

Substitute $\frac{1}{2}$ for x.

$$(2x - 1)(x + 6) = 0$$

$$\left[2 \cdot \left(\frac{1}{2}\right) - 1\right]\left(\frac{1}{2} + 6\right) \overset{?}{=} 0$$

$$(1 - 1)(6\tfrac{1}{2}) \overset{?}{=} 0$$

$$0 \cdot 6\tfrac{1}{2} \overset{?}{=} 0$$

$$0 = 0 \quad \text{True}$$

Substitute -6 for x.

$$(2x - 1)(x + 6) = 0$$

$$[(2)(-6) - 1](-6 + 6) \overset{?}{=} 0$$

$$(-13)(0) \overset{?}{=} 0$$

$$0 = 0 \quad \text{True}$$

So the solutions of the equation $(2x - 1)(x + 6) = 0$ are $\frac{1}{2}$ and -6.

PRACTICE 2

Solve: $(3x - 1)(x + 5) = 0$

In Example 2, note that we found the solutions of a quadratic equation by solving two linear equations: $2x - 1 = 0$ and $x + 6 = 0$. Explain how we know that these equations are linear.

When the quadratic expression in a second-degree equation is not given in factored form, we need to factor it before solving.

EXAMPLE 3

Solve: $y^2 - 5y = 0$

Solution

$$y^2 - 5y = 0$$
$$y(y - 5) = 0 \qquad \text{Factor the left side of the equation.}$$

Next, we set each factor equal to 0. Then, we solve for y.

$$y = 0 \quad \text{or} \quad y - 5 = 0$$
$$y = 5$$

Check   We verify our solutions in the original equation.

Substitute 0 for y.

$$y^2 - 5y = 0$$
$$(0)^2 - 5(0) \overset{?}{=} 0$$
$$0 - 0 \overset{?}{=} 0$$
$$0 = 0 \qquad \text{True}$$

Substitute 5 for y.

$$y^2 - 5y = 0$$
$$(5)^2 - 5(5) \overset{?}{=} 0$$
$$25 - 25 \overset{?}{=} 0$$
$$0 = 0 \qquad \text{True}$$

The solutions are 0 and 5.

PRACTICE 3

Solve: $y^2 + 6y = 0$

In order to apply the zero-product property, the product of the factors of a quadratic must equal zero. This implies that a quadratic equation must be written in *standard form*, $ax^2 + bx + c = 0$, before it can be solved.

EXAMPLE 4

Solve: $2x^2 + x = 1$

Solution

$$2x^2 + x = 1$$
$$2x^2 + x - 1 = 0 \qquad \text{Write in standard form by adding } -1 \text{ to each side.}$$

$$(2x - 1)(x + 1) = 0 \qquad \text{Factor the left side of the equation.}$$
$$2x - 1 = 0 \quad \text{or} \quad x + 1 = 0 \qquad \text{Set each factor to equal to 0.}$$
$$2x = 1 \qquad\qquad x = -1 \qquad \text{Solve for } x.$$
$$x = \frac{1}{2}$$

PRACTICE 4

Solve: $4y^2 - 11y = 3$

EXAMPLE 4 (continued)

Check We verify our solutions in the original equation.

Substitute $\frac{1}{2}$ for x.

$$2x^2 + x = 1$$

$$2\left(\frac{1}{2}\right)^2 + \frac{1}{2} \overset{?}{=} 1$$

$$2\left(\frac{1}{4}\right) + \frac{1}{2} \overset{?}{=} 1$$

$$1 = 1 \quad \text{True}$$

Substitute -1 for x.

$$2x^2 + x = 1$$

$$2(-1)^2 + (-1) \overset{?}{=} 1$$

$$2(1) + (-1) \overset{?}{=} 1$$

$$1 = 1 \quad \text{True}$$

So the solutions are $\frac{1}{2}$ and -1.

EXAMPLE 5

Solve: $2x(x - 3) = 8$

Solution We begin by writing the equation in standard form.

$$2x(x - 3) = 8$$

$$2x^2 - 6x = 8 \quad \text{Multiply.}$$

$$2x^2 - 6x - 8 = 0 \quad \text{Write in standard form by adding } -8 \text{ to each side.}$$

$$2(x^2 - 3x - 4) = 0 \quad \text{Factor out the GCF.}$$

$$2(x + 1)(x - 4) = 0 \quad \text{Write in factored form.}$$

Next, we set factors containing variables equal to 0. Then, we solve for x.

$$x + 1 = 0 \quad \text{or} \quad x - 4 = 0$$

$$x = -1 \quad\quad\quad x = 4$$

Check

Substitute -1 for x.

$$2x(x - 3) = 8$$

$$2(-1)(-1 - 3) \overset{?}{=} 8$$

$$(-2)(-4) \overset{?}{=} 8$$

$$8 = 8 \quad \text{True}$$

Substitute 4 for x.

$$2x(x - 3) = 8$$

$$2(4)(4 - 3) \overset{?}{=} 8$$

$$8(1) \overset{?}{=} 8$$

$$8 = 8 \quad \text{True}$$

So the solutions are -1 and 4.

PRACTICE 5

Solve: $3t(t + 4) = 15$

These examples lead us to the following strategy (or rule) for solving a quadratic equation by factoring:

To Solve a Quadratic Equation by Factoring

- If necessary, rewrite the equation in standard form with 0 on one side.
- Factor the other side.
- Use the zero-product property to get two simple linear equations.
- Solve the linear equations.
- Check by substituting the solutions in the original quadratic equation.

EXAMPLE 6

A homeowner has a square garden that she wants to make longer. If she extends one side by 5 ft and the adjacent side by 2 ft, the resulting garden would be rectangular with an area of 130 ft². How much fencing will she need to enclose the enlarged garden?

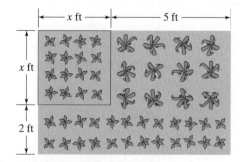

Solution   Let's represent the length of each side of the square by x. The resulting rectangular garden will have dimensions $(x + 5)$ and $(x + 2)$. The area of a rectangle can be computed by multiplying its length and its width, and we are told that this area is 130.

$$(x + 5)(x + 2) = 130$$
$$x^2 + 7x + 10 = 130$$
$$x^2 + 7x - 120 = 0$$
$$(x + 15)(x - 8) = 0$$
$$x + 15 = 0 \quad \text{or} \quad x - 8 = 0$$
$$x = -15 \qquad x = 8$$

Check

Substitute -15 for x.

$$(x + 5)(x + 2) = 130$$
$$(-15 + 5)(-15 + 2) \overset{?}{=} 130$$
$$(-10)(-13) \overset{?}{=} 130$$
$$130 = 130 \quad \text{True}$$

Substitute 8 for x.

$$(x + 5)(x + 2) = 130$$
$$(8 + 5)(8 + 2) \overset{?}{=} 130$$
$$(13)(10) \overset{?}{=} 130$$
$$130 = 130 \quad \text{True}$$

So the two solutions of the equation are -15 and 8. Since x represents a length, we can reject the negative solution -15. We conclude that the length of each side of the square is 8 ft. To compute the perimeter of the rectangle, we can substitute into the following formula:

$$P = 2l + 2w$$
$$= 2(8 + 5) + 2(8 + 2)$$
$$= 2(13) + 2(10)$$
$$= 26 + 20$$
$$= 46$$

Therefore, 46 ft of fencing is needed to enclose the enlarged garden.

PRACTICE 6

A framemaker is planning to frame a rectangular painting that has an area of 80 in².

If she has 3 ft of framing to put around the picture, what should the dimensions of the frame be? (Ignore the frame's thickness.)

We can also apply what we know about solving quadratic equations to problems involving the Pythagorean theorem. This theorem relates to the lengths of the three sides of a right triangle.

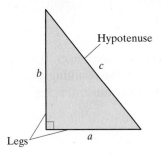

Hypotenuse

c

b

Legs

a

A right triangle has one 90° angle. The side opposite the 90° angle, the longest side, is called the *hypotenuse*. The other sides are called *legs*.

The Pythagorean theorem states that for every right triangle, the sum of the squares of the lengths of the legs equals the square of the length of the hypotenuse: $a^2 + b^2 = c^2$.

EXAMPLE 7

How far from the base of a building should a painter place a 17-ft ladder so that it reaches 15 ft up the building?

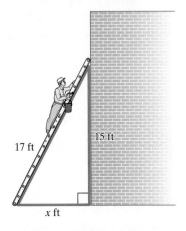

17 ft

15 ft

x ft

Solution   Let x represent the distance from the base of the building to the bottom of the ladder. Note that the ladder, the ground, and the building form a right triangle. Using the Pythagorean theorem, we get:

$$a^2 + b^2 = c^2$$
$$x^2 + 15^2 = 17^2$$
$$x^2 + 225 = 289$$
$$x^2 - 64 = 0$$
$$(x + 8)(x - 8) = 0$$
$$x + 8 = 0 \quad \text{or} \quad x - 8 = 0$$
$$x = -8 \qquad\qquad x = 8$$

Since x represents a distance, we consider only the positive value of x, namely 8.

Check   Substitute 8 for x:

$$x^2 + 15^2 = 17^2$$
$$8^2 + 15^2 \stackrel{?}{=} 17^2$$
$$64 + 225 \stackrel{?}{=} 289$$
$$289 = 289 \quad \text{True}$$

So the painter should place the ladder 8 ft from the base of the building.

PRACTICE 7

Two scooters, traveling at constant rates, leave an intersection at the same time. One scooter travels north while the other travels east. When the scooter traveling east has gone 5 mi, the distance between the two scooters is 13 mi. How far has the scooter going north traveled?

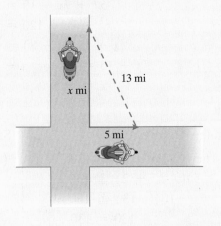

x mi

13 mi

5 mi

Mathematically Speaking

Fill in each blank with the most appropriate term or phrase from the given list.

sum of the squares of the legs	binomial equation	either one or both
in standard form	square of the sum of the legs	a perfect square
both	quadratic equation	

1. A second-degree or _____ is an equation that can be written in the form $ax^2 + bx + c = 0$, where a, b, and c are real numbers and $a \neq 0$.

2. The zero-product property states that if the product of two factors is zero, then _____ of the factors must be zero.

3. In using the zero-product property to solve a quadratic equation, the equation must be _____.

4. The Pythagorean theorem states that for every right triangle, the _____ equals the square of the hypotenuse.

Ⓐ *Indicate whether each equation is linear or quadratic.*

5. $x^2 - 3x + 2 = 0$

6. $x^2 = 6x$

7. $(x + 3)(x - 4) = x^2$

8. $\dfrac{x + 1}{4} = 2$

9. $2x^2 + 12x = -10$

10. $(x + 4)(x - 1) = 14$

Ⓑ *Solve.*

⊙ 11. $(x + 3)(x - 4) = 0$

12. $(x - 2)(x - 1) = 0$

13. $4(x - 1) = 0$

14. $-7(2t + 3) = 0$

15. $y(3y + 5) = 0$

16. $2y(5y - 4) = 0$

⊙ 17. $(2t + 1)(t - 5) = 0$

18. $(t - 3)(3t - 1) = 0$

19. $(2x + 3)(2x - 3) = 0$

20. $(5 - 4x)(5 + 4x) = 0$

21. $t(2 - 3t) = 0$

22. $t(1 + 2t) = 0$

⊙ 23. $y^2 - 2y = 0$

24. $y^2 + 3y = 0$

25. $5x - 25x^2 = 0$

26. $3t + 6t^2 = 0$

⊙ 27. $x^2 + 5x + 6 = 0$

28. $x^2 + x - 2 = 0$

29. $x^2 + x - 56 = 0$

30. $y^2 - y - 90 = 0$

31. $2x^2 - 5x - 3 = 0$

32. $2y^2 + 5y - 12 = 0$

⊙ 33. $6x^2 - x - 2 = 0$

34. $4t^2 - 8t + 3 = 0$

35. $0 = 36x^2 - 12x + 1$

36. $0 = 25y^2 + 10y + 1$

37. $r^2 - 121 = 0$

38. $t^2 - 49 = 0$

39. $0 = (2x - 3)^2$

40. $0 = (4x + 1)^2$

41. $16x^2 - 16x + 4 = 0$

42. $9t^2 + 18t + 9 = 0$

43. $9m^2 + 15m - 6 = 0$

44. $6n^2 - 32n + 10 = 0$

45. $r^2 - r = 6$

46. $r^2 + 3r = 10$

47. $y^2 - 7y = -12$

48. $y^2 - y = 12$

49. $n^2 + 2n = 8$

50. $t^2 - 3t = 18$

51. $3y^2 + 4y = -1$

52. $3k^2 + 17k = -10$

53. $4x^2 + 6x = -2$

54. $12x^2 + 33x = 9$

55. $2n^2 = -10n$

56. $3m^2 = 6m$

57. $4x^2 = 1$

58. $9y^2 = 16$

59. $8y^2 = 2$

60. $12x^2 = 48$

61. $3r^2 + 6r = 2r^2 - 9$

62. $5n^2 + 36 = 12n + 4n^2$

63. $x(x - 1) = 12$

64. $r(r + 3) = 10$

65. $4t(t - 1) = 24$

66. $2n(n + 7) = -24$

67. $(y + 3)(y - 2) = 14$

68. $(m - 6)(m + 1) = -10$

69. $(3n - 2)(n + 5) = -14$

70. $(t - 5)(t + 2) = 18$

71. $(n + 2)(n + 4) = 12n$

72. $(3x + 5)(x - 1) = 16x$

73. $3x(2x - 5) = x^2 - 10$

74. $n^2 + 8 = 3n(n - 2)$

Mixed Practice

Solve.

75. $4r^2 + 11r - 3 = 0$

76. $0 = 9u^2 + 6u + 1$

77. $10x^2 + 25x - 15 = 0$

78. $n^2 + 3n = 28$

79. $5a^2 + 19a = 4$

80. $9b^2 = 27b$

81. $3s^2 + 64 = 16s + 2s^2$

82. $(x - 6)(x + 2) = 33$

83. $6t(3t - 5) = 0$

84. $(7 - 3m)(7 + 3m) = 0$

85. $4j + 8j^2 = 0$

Solve.

86. Is the equation $(x - 3)(x + 1) = 5$ linear or quadratic?

Applications

Solve.

87. In a certain league, the teams play each other twice in a season. It can be shown that if there are n teams in the league, the teams must play $n^2 - n$ games. If the league plays 210 games in a season, how many teams were in the league?

88. The number of ways to pair n students in a physics lab can be represented by the expression $\frac{1}{2}n(n - 1)$. If there are 325 different ways to pair students, how many students are in the lab?

89. Two cars leave an intersection, one traveling west and the other south. After some time, the faster car is 2 mi farther away from the intersection than the slower car. At that time, the two cars are 10 mi apart. How far did each car travel?

90. The sail on a sailboat is a right triangle in which the hypotenuse is called the leech. A 12-foot tall mainsail has a leech length of 13 ft. If sailcloth costs $10 per square foot, what is the cost of a new mainsail?

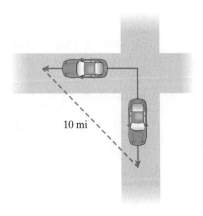

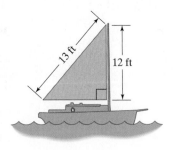

91. The base of 1 World Trade Center in New York City is a cube. The area of each side of the base is 40,000 ft^2. What is the height of the base of the tower? (*Source: worldconstructionnetwork.com*)

92. A diver jumps from a diving board 24 ft above a pool. After t sec, the diver's height h above the water (in feet) is given by the expression $-16t^2 + 8t + 24$. In how many seconds will the diver hit the water?

93. A businessman invested $8000 in a high-risk growth fund that after two years was worth $12,500. His broker used the equation $8000(1 + r)^2 = 12,500$ to find the average annual rate of return. What is this rate?

94. An artist wants to frame an 8-inch by 10-inch painting with a uniform border around the painting. She only has enough materials to cover 40 in.^{2} of border. How wide can the border be based on her amount of materials?

95. The St. Louis Gateway Arch is the tallest national monument in the United States. At x yards (on the ground) from the center, the height of the arch (in yards) can be approximated by the expression $\frac{2}{105}\left[(105)^2 - x^2\right]$. (*Source: gatewayarch.com*)

$\frac{2}{105}\left[(105)^2 - x^2\right]$

x

a. Factor this expression.
b. Find the width of the arch.

96. A standard DVD has a radius of 6 cm. The amount of data, in gigabytes, stored on the DVD from the center to r cm out can be modeled by the expression $0.04677\pi(r^2 - 4)$ except on the center ring on which the DVD spins, where no data can be stored. (*Source: iso.org*)

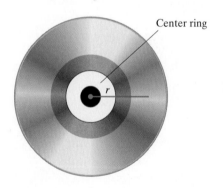

Center ring

r

a. Write this expression in factored form.

b. When r is equal to the radius of the center ring, the amount of data stored is 0. What is the radius of the center ring?

• Check your answers on page A-23.

MINDStretchers

Mathematical Reasoning

1. Give an example of an equation:
   a. whose solutions are 2 and 3.
   b. whose only solution is 5.
   c. whose solutions have opposite signs.
   d. whose solutions are 2, 3, and 4.

Technology

2. Consider the quadratic equation $x^2 - 3x - 10 = 0$.
   a. Solve this equation.
   b. On a calculator or a computer, graph $y = x^2 - 3x - 10$. Use the graph to find the x-intercepts.

   c. Explain how you can use these intercepts to solve the original equation $x^2 - 3x - 10 = 0$.

Writing

3. Explain whether the zero-product property is true for more than two factors.

Concept/Skill	Description	Example
[6.1] Common factor of two or more integers	An integer that is a factor of each integer.	5 is a common factor of 15 and 50.
[6.1] Greatest common factor (GCF) of two or more integers	The greatest integer that is a factor of each integer.	The GCF of 45, 63, and 81 is 9.
[6.1] Greatest common factor (GCF) of two or more monomials	The product of the greatest common factor of the coefficients and, for each variable, the variable to the lowest power to which it is raised in any of the monomials.	The GCF of $6x^4$, $8x^3$, and $12x^2$ is $2x^2$.
[6.1] Factoring by grouping	Group pairs of terms and factor out a GCF in each group, if necessary. Then, factor out the common binomial factor.	$xy + x - 4y - 4$ $= (xy + x) + (-4y - 4)$ $= x(y + 1) - 4(y + 1)$ $= (y + 1)(x - 4)$
[6.2] Factoring a trinomial of the form $ax^2 + bx + c$, where $a = 1$	• List and test the factors of c to find two integers for $(x + ?)(x + ?)$ whose product is c and whose sum is b.	$x^2 - 8x + 12 = (x - 2)(x - 6)$ because Factors of 12 $(-2) \cdot (-6) = 12$ and $(-2) + (-6) = -8$ Sum of factors
[6.2] Prime polynomial	A polynomial that is not factorable.	$x^2 + 5x + 1$
[6.3] Factoring a trinomial $ax^2 + bx + c$, where $a \neq 1$ (Trial-and-Error Method)	• List and test the factors of a and of c to find four integers for $(?x + ?)(?x + ?)$ so that the product of the leading coefficients of the binomial factors is a, the product of the constant terms of the binomial factors is c, and the coefficients of the inner and outer products add up to b.	$2x^2 + 15x + 7 = (2x + 1)(x + 7)$ because $2 \cdot 1 = 2$, $1 \cdot 7 = 7$, and $2 \cdot 7 + 1 \cdot 1 = 15$.
[6.3] Factoring a trinomial $ax^2 + bx + c$, where $a \neq 1$ (ac Method)	• Form the product ac. • Find two factors of ac that add up to b. • Use these factors to split up the middle term in the original trinomial. • Group the first two terms and the last two terms. • From each group, factor out the common factor. • Factor out the common binomial factor.	For $2x^2 + 15x + 7$, $ac = 2 \cdot 7 = 14$ $1 \cdot 14 = 14$ and $1 + 14 = 15$ $2x^2 + 15x + 7$ $= 2x^2 + x + 14x + 7$ $= (2x^2 + x) + (14x + 7)$ $= x(2x + 1) + 7(2x + 1)$ $= (2x + 1)(x + 7)$
[6.4] Factoring a perfect square trinomial	$a^2 + 2ab + b^2 = (a + b)^2$ $a^2 - 2ab + b^2 = (a - b)^2$	$x^2 + 12x + 36 = (x + 6)^2$ $x^2 - 6x + 9 = (x - 3)^2$
[6.4] Factoring the difference of squares	$a^2 - b^2 = (a + b)(a - b)$	$x^2 - 100 = (x + 10)(x - 10)$

continued

Concept/Skill	Description	Example
[6.5] **Second-degree or quadratic equation**	An equation that can be written in the form $ax^2 + bx + c = 0$, where a, b, and c are real numbers and $a \neq 0$.	$x^2 - x + 6 = 0$
[6.5] **The zero-product property**	If $ab = 0$, then $a = 0$ or $b = 0$, or both a and $b = 0$.	If $x(3x - 1) = 0$, then either $x = 0$ or $3x - 1 = 0$.
[6.5] **To solve a quadratic equation by factoring**	• If necessary, rewrite the equation in standard form with 0 on one side. • Factor the other side. • Use the zero-product property to get two simple linear equations. • Solve the linear equations. • Check by substituting the solutions in the original quadratic equation.	$$x(x - 3) = 4$$ $$x^2 - 3x = 4$$ $$x^2 - 3x - 4 = 0$$ $$(x - 4)(x + 1) = 0$$ $$x - 4 = 0 \quad \text{or} \quad x + 1 = 0$$ $$x = 4 \qquad\qquad x = -1$$ **Check** Substitute 4 for x. $$4(4 - 3) \overset{?}{=} 4$$ $$4(1) \overset{?}{=} 4$$ $$4 = 4 \qquad \text{True}$$ Substitute -1 for x. $$(-1)(-1 - 3) \overset{?}{=} 4$$ $$(-1)(-4) \overset{?}{=} 4$$ $$4 = 4 \qquad \text{True}$$

Say Why
Fill in each blank.

1. The trinomial $x^2 + 2x + 2$ _____ a prime
   is/is not
   polynomial because _____
   _____.

2. The polynomial $4x^2 - 2x - 6$ _____ factored
   is/is not
   completely as $2(2x^2 - x - 3)$ because
   _____
   _____.

3. The trinomial $x^2 - 4xy + 4y^2$ _____ a perfect
   is/is not
   square trinomial because _____
   _____.

4. The binomial $9x^2 - 4y^2$ _____ be factored as
   can/cannot
   $(3x + 2y)(3x - 2y)$ because _____
   _____.

[6.1] *Find the greatest common factor.*

5. 48, 36, and 60

6. $9m^3n$, $24m^4$, and $15m^2n^2$

Factor.

7. $3x - 6y$

8. $16p^3q^2 + 18p^2q - 4pq^2$

9. $(n - 1) + n(n - 1)$

10. $xb - 5b - 2x + 10$

Solve for the indicated variable.

11. $d = rt_1 + rt_2$ for r

12. $ax + y = bx + c$ for x

[6.2] *Factor, if possible.*

13. $x^2 + x + 1$

14. $m^2 - m + 3$

15. $y^2 + 42 + 13y$

16. $m^2 - 7mn + 10n^2$

17. $24 - 8x - 2x^2$

18. $-15xy^2 + 3x^3 - 12x^2y$

[6.3] *Factor, if possible.*

19. $3x^2 + 5x - 2$

20. $5n^2 + 13n + 6$

21. $3n^2 - n - 1$

22. $6x^2 - x - 12$

23. $2a^2 + 3ab - 35b^2$

24. $16a - 4a^2 - 15$

25. $9y^3 - 21y + 60y^2$

26. $2p^2q - 3pq^2 - 2q^3$

[6.4] *Factor, if possible.*

27. $b^2 - 6b + 9$

28. $64 - x^2$

29. $25y^2 - 20y + 4$

30. $9a^2 + 24ab + 16b^2$

31. $81p^2 - 100q^2$

32. $4x^8 - 28x^4 + 49$

33. $48x^4 - 3y^4$

34. $x^2(x - 1) - 9(x - 1)$

35. $(x + 2)(x - 1) = 0$ **36.** $t(t - 4) = 0$ **37.** $3x^2 + 18x = 0$ **38.** $4x^2 + 4x + 1 = 0$

39. $y^2 - 10y = -16$ **40.** $3k^2 - k = 2$ **41.** $4n(2n + 3) = 20$ **42.** $(y - 1)(y + 2) = 10$

Mixed Applications

Solve.

43. The length of an object varies with its temperature. The expression $aLt_2 - aLt_1$ represents the change in the length of an object heated to temperature t_2, where L is its length at temperature t_1 and a is the *coefficient of linear expansion*, a constant that depends on the material that the object is made of. Write this expression in factored form.

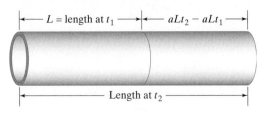

44. If a ball is thrown straight upward at v ft/sec, its height above the point of release is given by the expression $vt - 16t^2$, where t is the number of seconds after release. Write an expression in factored form for the distance between the object's location at t_1 and at t_2.

45. Find the distance between the two intersections shown on the city grid pictured if each block is 500 ft long.

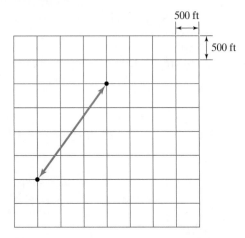

46. A kite maker designs the diamond-shaped kite shown. The diagonals of the kite cross at right angles. The vertical diagonal is 52 in. long. What is the length of the horizontal diagonal of the kite shown?

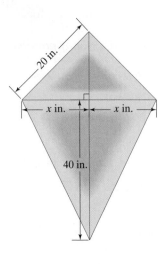

47. After t sec, the height of a rocket launched straight upward from ground level with an initial velocity of 76 ft/sec can be modeled by the polynomial $76t - 16t^2$. After how many seconds will the rocket reach a height of 18 ft above the launch (that is, equal to +18)?

48. The formula for the area of the trapezoid shown is $A = \frac{1}{2}hb + \frac{1}{2}hB$. Solve this formula for h in terms of A, b, and B.

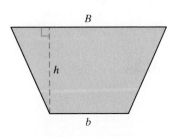

• Check your answers on page A-23.

CHAPTER 8 Posttest

EXTRA
HELP

Test Prep
VIDEOS

Alg" and click on "Channels").

To see whether you have mastered the topics in this chapter, take this test.

1. Find the greatest common factor of $12x^3$ and $15x^2$.

Factor.

2. $2xy - 14y$

3. $6pq^2 + 8p^3 - 16p^2q$

4. $ax - bx + by - ay$

5. $n^2 - 13n - 48$

6. $-8 + x^2 - 2x$

7. $15x^2 - 5x^3 + 20x$

8. $4x^2 + 13xy - 12y^2$

9. $-12x^2 + 36x - 27$

10. $9x^2 + 30xy + 25y^2$

11. $121 - 4x^2$

12. $p^2q^2 - 1$

13. $y^4 - 8y^2 + 16$

Solve.

14. $(n + 8)(n - 1) = 0$

15. $6x^2 + 10x = 4$

16. $(2n + 1)(n - 1) = 5$

17. The energy it takes to lift an object of mass m from level y_1 to level y_2 is $mgy_2 - mgy_1$, where g is a constant. Factor this expression.

18. In the right triangle shown, find the length of the missing side.

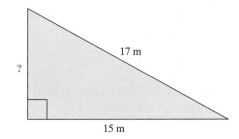

19. A rectangular garden measures 25 ft by 30 ft. The gardener wishes to surround the garden with a border of mulch x ft wide, as shown. Write an expression in factored form for the area of the mulch border in terms of x.

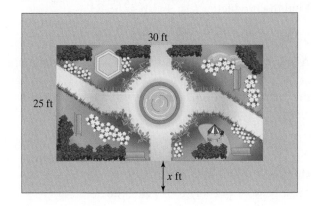

20. In a physics experiment, a weight is dropped from a platform 9 ft above the ground. The time t (in seconds) it takes the weight to reach the ground may be found by solving the equation $9 - 16t^2 = 0$. Solve for t.

• Check your answers on page A-23.

To help you review, solve the following:

1. Evaluate $2x + y - z$ if $x = -4$, $y = 0$, and $z = -5$.

2. Simplify: $5a + 2 - 2b + 3a + 9b$

3. Solve: $z + 4 = 5z - 2(z + 6)$

4. 4 oz is what percent of 2 lb?

5. Graph $2x + 4y = 12$.

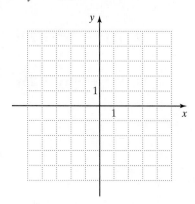

6. Graph the line $4x - 3y = 12$ on a coordinate plane.

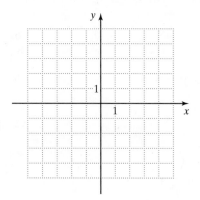

7. Solve by substitution and check:

$$4x + 2y = 1$$
$$y = 3 - 2x$$

8. Solve the system: **(1)** $x + y = 6$

   (2) $y = 2x + 9$

9. Simplify: $\dfrac{n^2 \cdot n^3 \cdot n}{n^5}$

10. Find the value of $5a^3 + 3a^2 + 2a + 1$ if $a = -2$.

11. Find the product: $(3x + 1)(7x - 2)$

12. Factor, if possible: $3x^2 - x - 14$

13. Factor, if possible: $16x^2 - 24x + 9$

14. Solve: $8x^2 + 22x = 6$

15. A share of Proctor & Gamble stock (ticker PG) opened at $61.91. Five days later it closed at $61.16. What was the average daily change in the price of a share? (*Source: New York Times*)

16. A part-time college student pays a student fee of f dollars plus c dollars per credit. What is the charge for a student with an 8-credit schedule?

17. The owner of a shirt factory has fixed expenses of $500 per day. It costs the factory $15 to produce each shirt. If the shirts are sold wholesale at $25 apiece, how many shirts must be sold per day to break even?

18. The following graph shows the average number of hours that American Internet users spent online per week in various years. (*Source: The World Almanac and Book of Facts, 2011*)

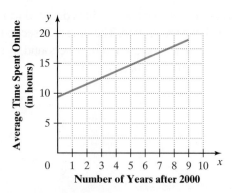

Is the slope of this line positive, negative, or zero?

19. The distance from the Earth to the Sun is approximately 150,000,000 km. Express this distance in scientific notation. (*Source:* Bennett et al., *The Cosmic Perspective*)

20. An expression for the total surface area of the tin can shown below is $2\pi r^2 + 2\pi rh$. Rewrite this expression in factored form.

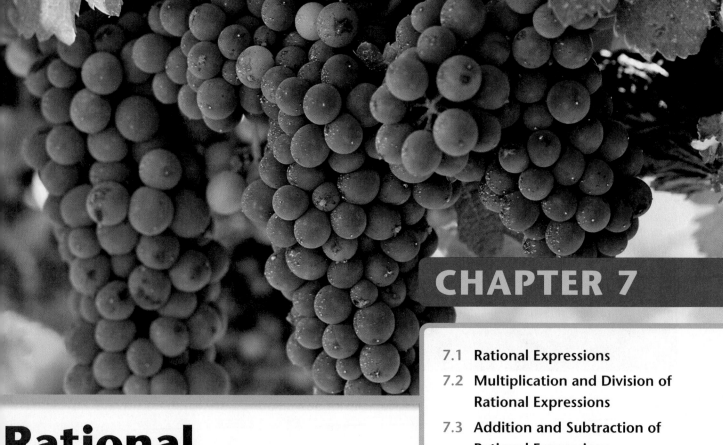

Rational Expressions and Equations

Rational Expressions and Grapes

Science plays a role in viticulture, the art of growing grapes. For instance, scientists have used DNA fingerprinting—the same technique used in paternity suits and criminal trials—to trace the parentage of numerous grape varieties.

Viticulture also makes use of algebra. Since much of the flavor and color of grapes is in their skin, wine growers want to raise grapes with an increased surface-to-volume ratio. This ratio is larger for smaller grapes, which have proportionately more skin than larger grapes.

We can see why this is true by assuming that a grape is approximately a sphere of radius r. The ratio of a sphere's surface area to its volume is represented by the rational expression $\dfrac{4\pi r^2}{\frac{4}{3}\pi r^3}$, which simplifies to $\dfrac{3}{r}$. Substituting smaller values for the radius r in this expression gives larger values of the expression. Thus, the smaller the grape is, the larger the surface-to-volume ratio.

(*Source:* David L. Wheeler, "Scholars Marry the Science and the Art of Winemaking," *Chronicle of Higher Education*, May 30, 1997)

1. Identify the values for which the rational expression

   $\dfrac{5}{x + 6}$ is undefined.

2. Show that the rational expressions $\dfrac{n^2 - 2n}{3n}$ and $\dfrac{n - 2}{3}$

   are equivalent if $n \neq 0$.

Simplify.

3. $\dfrac{24x^2y^3}{6xy^5}$

4. $\dfrac{4a^2 - 8a}{a - 2}$

5. $\dfrac{w^2 - 6w}{36 - w^2}$

6. $\dfrac{\frac{5n}{8}}{\frac{n^3}{16}}$

7. Write $\dfrac{2}{3a}$ and $\dfrac{1}{9a - 18}$ in terms of their LCD.

Perform the indicated operation.

8. $\dfrac{12y}{y + 1} - \dfrac{7y + 2}{y + 1}$

9. $\dfrac{1}{6x^2} + \dfrac{5}{4x}$

10. $\dfrac{3}{c - 3} - \dfrac{1}{c + 3}$

11. $\dfrac{1}{x^2 - 2x + 1} - \dfrac{2}{1 - x^2}$

12. $\dfrac{15a^3b}{20n^4} \cdot \dfrac{16n^2}{9ab}$

13. $\dfrac{y - 4}{5y^2 + 10y} \cdot \dfrac{y^2 - 2y - 8}{y^2 - 16}$

14. $\dfrac{x^2 - x - 2}{x^2 + 5x + 4} \div \dfrac{x^2 - 7x + 10}{x - 5}$

Solve and check.

15. $\dfrac{3x - 8}{x^2 - 4} + \dfrac{2}{x - 2} = \dfrac{7}{x + 2}$

16. $\dfrac{x}{x + 4} - 1 = \dfrac{2}{x - 1}$

17. $\dfrac{9}{2x} = \dfrac{x}{x - 1}$

Solve.

18. An ice cream company has fixed costs of \$1500 and variable costs of \$2 for each gallon of ice cream it produces. The cost per gallon of producing x gallons of ice cream is given by the expression $2 + \dfrac{1500}{x}$. Write this cost as a single rational expression.

19. On the first part of a 360-mi trip, a family drives 195 mi at an average speed of r mph. They drive the remainder of the trip at an average speed that is 10 mph less than their speed during the first part of the trip. If the entire trip took 6 hr, what was the average speed during each part of the trip?

20. It takes a photocopier t min to make 30 copies. If at the same rate it takes 5 min longer to make 90 copies, how long does it take the photocopier to make 30 copies?

• Check your answers on page A-24.

7.1 Rational Expressions

OBJECTIVES

A To identify values for which a rational expression is undefined

B To determine whether rational expressions are equivalent

C To simplify a rational expression

D To solve applied problems involving rational expressions

What Rational Expressions Are and Why They Are Important

In this chapter, we move beyond our previous discussion of polynomials to consider a type of algebraic expression called a *rational expression*. Rational expressions, sometimes called *algebraic fractions*, are useful in many disciplines, including the sciences, the social sciences, medicine, and business.

For instance, the work of some anthropologists who study the history of the human species involves rational expressions. In studying a fossil record, anthropologists examine ancient skulls and compute $\dfrac{W}{L}$, the ratio of each skull's width W to its length L. This expression is an example of a rational expression.

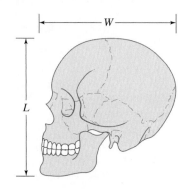

In this chapter, we discuss how to simplify rational expressions, as well as how to carry out operations on rational expressions. We also focus on how to solve equations involving rational expressions.

Introduction to Rational Expressions

Rational expressions in algebra are similar to fractions in arithmetic. In arithmetic, a fraction such as $\dfrac{3}{4}$ is the quotient of two integers, whereas in algebra, a rational expression such as $\dfrac{W}{L}$ is the quotient of two polynomials.

DEFINITION

A *rational expression* $\dfrac{P}{Q}$ is an algebraic expression that can be written as the quotient of two polynomials, P and Q, where $Q \neq 0$.

Other examples of rational expressions are:

$$\frac{5}{x-2} \qquad \frac{n^2 + 2n - 1}{n+1} \qquad \frac{-a^2}{7bc}$$

A rational expression can be written in terms of division. For example,

$$\frac{5}{x-2} \text{ can be written as } 5 \div (x-2), \text{ and}$$

$$\frac{n^2 + 2n - 1}{n + 1} \text{ can be written as } (n^2 + 2n - 1) \div (n + 1).$$

Since we can write a rational expression as division, we must be sure that its denominator does not equal 0. *When a variable is replaced with a value that makes the denominator 0, the rational expression is undefined.*

For instance, consider the rational expression $\frac{5}{x-2}$. For what values of x is this expression undefined? Setting the denominator $x - 2$ equal to 0 gives us:

$$x - 2 = 0$$
$$x = 2$$

When x is replaced with 2 in the rational expression, we get:

$$\frac{5}{x-2} = \frac{5}{2-2} = \frac{5}{0} \quad \leftarrow \text{ Undefined}$$

So $\frac{5}{x-2}$ is undefined when x is equal to 2.

EXAMPLE 1

Identify all numbers for which the following rational expressions are undefined.

a. $\dfrac{3}{x+5}$

b. $\dfrac{8x}{x^2 - 3x + 2}$

Solution   First, we find the values for which the rational expressions are undefined by setting each denominator equal to 0. Then, we solve for x.

a. $\dfrac{3}{x+5}$

$$x + 5 = 0$$
$$x = -5$$

So $\dfrac{3}{x+5}$ is undefined when x is equal to -5.

b. $\dfrac{8x}{x^2 - 3x + 2}$

$$x^2 - 3x + 2 = 0$$
$$(x - 2)(x - 1) = 0 \quad \text{Factor.}$$
$$x - 2 = 0 \quad \text{or} \quad x - 1 = 0 \quad \text{Set each factor equal to 0.}$$
$$x = 2 \qquad\qquad x = 1 \quad \text{Solve for } x.$$

So $\dfrac{8x}{x^2 - 3x + 2}$ is undefined when x is equal to 1 or 2.

PRACTICE 1

Indicate the values of the variable for which each rational expression is undefined.

a. $\dfrac{n+1}{n-4}$

b. $\dfrac{6}{n^2 - 9}$

Note that throughout the following discussion, we assume that all rational expressions are defined.

Equivalent Expressions

Recall from Chapter R that equivalent fractions, such as $\frac{1}{2}$ and $\frac{3}{6}$, are fractions that have the same value even though they are written differently. Similarly, **equivalent rational expressions** are rational expressions that have the same value, no matter what value replaces the variable.

> ### To Find an Equivalent Rational Expression
>
> Multiply the numerator and denominator of $\frac{P}{Q}$ by the same polynomial R.
>
> $$\frac{P}{Q} = \frac{PR}{QR},$$
>
> where $Q \neq 0$ and $R \neq 0$.

In words, this rule states that to find an equivalent rational expression, multiply the numerator and denominator of a rational expression by the same nonzero polynomial, as shown in Example 2 below.

EXAMPLE 2

Indicate whether each pair of rational expressions is equivalent.

a. $\frac{2}{3}$ and $\frac{2x}{3x}$ **b.** $\frac{d-5}{3}$ and $\frac{4d^2 - 20d}{12d}$

Solution

a. Multiplying both the numerator and denominator of $\frac{2}{3}$ by x, we see

that $\frac{2}{3} = \frac{2x}{3x}$. So the expressions $\frac{2}{3}$ and $\frac{2x}{3x}$ are equivalent.

b. Multiplying both the numerator and denominator of $\frac{d-5}{3}$ by $4d$, we get:

$$\frac{(d-5) \cdot 4d}{3 \cdot 4d} = \frac{4d^2 - 20d}{12d}$$

So $\frac{d-5}{3}$ and $\frac{4d^2 - 20d}{12d}$ are equivalent.

PRACTICE 2

Determine whether each pair of rational expressions is equivalent.

a. $\frac{b}{3}$ and $\frac{3b^2}{9b}$

b. $\frac{5}{a+7}$ and $\frac{5a}{a^2 + 7a}$

Note that we can also use the rule to find an equivalent rational expression by dividing the numerator and denominator of the rational expression by the same nonzero polynomial.

$$\frac{x^2 - 4}{x^2 + 5x + 6} = \frac{(x - 2)(x + 2)}{(x + 3)(x + 2)}$$ Factor the numerator and denominator.

$$= \frac{(x - 2)\overset{1}{\cancel{(x + 2)}}}{(x + 3)\underset{1}{\cancel{(x + 2)}}}$$ Divide out the common factor $(x + 2)$.

$$= \frac{x - 2}{x + 3}$$

So $\dfrac{x^2 - 4}{x^2 + 5x + 6}$ is equivalent to $\dfrac{x - 2}{x + 3}$.

> **TIP** When simplifying a rational expression, do not divide out common terms in a sum or difference in the numerator and denominator. For instance, do not divide out the x's in $\dfrac{x - 2}{x + 3}$.

A rational expression is said to be *in simplest form* (or *reduced to lowest terms*) when its numerator and denominator have no common factor other than 1 or −1. Throughout the remainder of this book, we generally simplify any answer that is a rational expression. Can you explain the similarities and differences in simplifying a fraction in arithmetic and simplifying a rational expression in algebra?

To Simplify a Rational Expression

- Factor the numerator and denominator.
- Divide out any common factors in the numerator and denominator.

EXAMPLE 3

Write in simplest form.

a. $\dfrac{10x^2}{-5xy}$ **b.** $\dfrac{-4a^3b^2}{-2ab}$

Solution

a. $\dfrac{10x^2}{-5xy} = \dfrac{5x(2x)}{5x(-y)}$ Factor the numerator and denominator.

$$= \frac{\overset{1}{\cancel{5x}}(2x)}{\underset{1}{\cancel{5x}}(-y)}$$ Divide out the common factor $5x$.

$$= -\frac{2x}{y}$$ Simplify.

b. $\dfrac{-4a^3b^2}{-2ab} = \dfrac{-2ab(2a^2b)}{-2ab}$ Factor the numerator and denominator.

$$= \frac{\overset{1}{\cancel{-2ab}}(2a^2b)}{\underset{1}{\cancel{-2ab}}} = 2a^2b$$ Divide out the common factor $-2ab$, and simplify.

PRACTICE 3

Express in lowest terms.

a. $\dfrac{-12n^3}{-6mn}$

b. $\dfrac{-3x^4y}{2x^2y}$

EXAMPLE 4

Write in simplest form, if possible.

a. $\dfrac{3x - 6}{9x + 12}$ **b.** $\dfrac{n + 3}{2n + 6}$ **c.** $\dfrac{t + 3}{t - 1}$

Solution

a. $\dfrac{3x - 6}{9x + 12} = \dfrac{3(x - 2)}{3(3x + 4)}$ Factor the numerator and the denominator.

$$= \dfrac{\overset{1}{\cancel{3}}(x - 2)}{\underset{1}{\cancel{3}}(3x + 4)}$$ Divide out the common factor 3.

$$= \dfrac{x - 2}{3x + 4}$$

b. $\dfrac{n + 3}{2n + 6} = \dfrac{n + 3}{2(n + 3)}$ Factor the denominator.

$$= \dfrac{\overset{1}{\cancel{n + 3}}}{2(\underset{1}{\cancel{n + 3}})}$$ Divide out the common factor $n + 3$.

$$= \dfrac{1}{2}$$

c. $\dfrac{t + 3}{t - 1}$ The numerator $t + 3$ and the denominator $t - 1$ have no common factor (other than 1). This expression cannot be simplified.

EXAMPLE 5

Simplify.   **a.** $\dfrac{ab - ac}{ax + 2ay}$ **b.** $\dfrac{2t^2 - 2}{t^2 + t - 2}$ **c.** $\dfrac{3x^2 + 2x - 1}{3x^2 - 4x + 1}$

Solution

a. $\dfrac{ab - ac}{ax + 2ay} = \dfrac{a(b - c)}{a(x + 2y)}$

$$= \dfrac{\overset{1}{\cancel{a}}(b - c)}{\underset{1}{\cancel{a}}(x + 2y)}$$

$$= \dfrac{b - c}{x + 2y}$$

b. $\dfrac{2t^2 - 2}{t^2 + t - 2} = \dfrac{2(t^2 - 1)}{(t - 1)(t + 2)}$

$$= \dfrac{2(t - 1)(t + 1)}{(t - 1)(t + 2)}$$

$$= \dfrac{2(\overset{1}{\cancel{t - 1}})(t + 1)}{(\underset{1}{\cancel{t - 1}})(t + 2)}$$

$$= \dfrac{2(t + 1)}{t + 2}$$

c. $\dfrac{3x^2 + 2x - 1}{3x^2 - 4x + 1} = \dfrac{(3x - 1)(x + 1)}{(3x - 1)(x - 1)}$

$$= \dfrac{(\overset{1}{\cancel{3x - 1}})(x + 1)}{(\underset{1}{\cancel{3x - 1}})(x - 1)} = \dfrac{x + 1}{x - 1}$$

PRACTICE 4

Write in lowest terms.

a. $\dfrac{2y - 8}{4y + 6}$

b. $\dfrac{v - 4}{v + 3}$

c. $\dfrac{x + 2}{3x + 6}$

PRACTICE 5

Simplify.

a. $\dfrac{wt - wx}{wz - 3wg}$

b. $\dfrac{4n^2 - 4}{n^2 + 3n + 2}$

c. $\dfrac{2y^2 - y - 1}{2y^2 + 7y + 3}$

The following examples involve factors such as $a - b$ and $b - a$ that are opposites of each other; that is, they differ only in sign. Recall that since $(-1)(a - b) = b - a$,

it follows that $\dfrac{a - b}{b - a} = -1$ because $\dfrac{a - b}{b - a} = \dfrac{\overset{1}{\cancel{(a - b)}}}{-1\underset{1}{\cancel{(a - b)}}} = \dfrac{1}{-1} = -1.$

EXAMPLE 6

Write in lowest terms.

a. $\dfrac{s - 5}{5 - s}$ 　　　　　 **b.** $\dfrac{2p - 10}{-p + 5}$

c. $\dfrac{3x - 6}{4 - x^2}$ 　　　　 **d.** $\dfrac{1 - x^2}{x^2 - 3x + 2}$

Solution

a. $\dfrac{s - 5}{5 - s} = \dfrac{s - 5}{-1(s - 5)}$ 　　Write $5 - s$ as $-1(s - 5)$.

$\qquad = \dfrac{\overset{1}{\cancel{s - 5}}}{-1\underset{1}{\cancel{(s - 5)}}}$ 　　Divide out the common factor $s - 5$.

$\qquad = \dfrac{1}{-1}$ 　　Simplify.

$\qquad = -1$

b. $\dfrac{2p - 10}{-p + 5} = \dfrac{2(p - 5)}{-1(p - 5)} = \dfrac{2\overset{1}{\cancel{(p - 5)}}}{-1\underset{1}{\cancel{(p - 5)}}} = \dfrac{2}{-1} = -2$

c. $\dfrac{3x - 6}{4 - x^2} = \dfrac{3(x - 2)}{(2 - x)(2 + x)}$ 　　Factor the numerator and denominator.

$\qquad = \dfrac{3(x - 2)}{-1(x - 2)(2 + x)}$ 　　Write $(2 - x)$ as $-1(x - 2)$.

$\qquad = \dfrac{3\overset{1}{\cancel{(x - 2)}}}{-\underset{1}{\cancel{(x - 2)}}(2 + x)}$ 　　Divide out the common factor $(x - 2)$.

$\qquad = -\dfrac{3}{x + 2}$ 　　Simplify.

d. $\dfrac{1 - x^2}{x^2 - 3x + 2} = \dfrac{(1 + x)(1 - x)}{(x - 2)(x - 1)} = \dfrac{-(x + 1)(x - 1)}{(x - 2)(x - 1)}$

$\qquad = \dfrac{-(x + 1)\overset{1}{\cancel{(x - 1)}}}{(x - 2)\underset{1}{\cancel{(x - 1)}}} = -\dfrac{x + 1}{x - 2}$

PRACTICE 6

Simplify.

a. $\dfrac{-y - 1}{1 + y}$

b. $\dfrac{3x - 12}{-x + 4}$

c. $\dfrac{3n - 15}{25 - n^2}$

d. $\dfrac{4 - 9s^2}{3s^2 + s - 2}$

EXAMPLE 7

It costs a television manufacturer $\dfrac{95x + 10{,}000}{x}$ dollars per television to produce x televisions.

a. For which value of x is this rational expression undefined?

b. Explain in a sentence or two why you think that it makes sense for the cost per television set to be undefined for the value of x found in part (a).

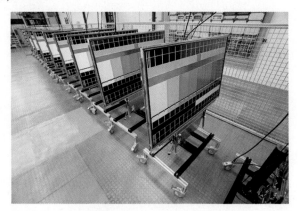

Solution

a. A rational expression is undefined when its denominator is equal to 0. For the expression $\dfrac{95x + 10{,}000}{x}$, the denominator is 0 when $x = 0$.

b. When $x = 0$, the manufacturer is producing *no* television sets. So it makes no sense to speak of the cost per television set.

PRACTICE 7

For a circle of radius r, the ratio of its area to its circumference is given by the expression $\dfrac{\pi r^2}{2\pi r}$.

a. Simplify this expression.

b. For which value of r is the original rational expression undefined?

Mathematically Speaking

Fill in each blank with the most appropriate term or phrase from the given list.

exponential expression	undefined	rational expression
in simplest form	equivalent	equal to 0

1. A(n) _____, $\dfrac{P}{Q}$, is an algebraic expression that can be written as the quotient of two polynomials, P and Q, where $Q \neq 0$.

2. A rational expression is _____ if a variable is replaced with a value that makes its denominator 0.

3. Rational expressions are _____ if they have the same value no matter what value replaces the variable.

4. A rational expression is said to be _____ (or reduced to lowest terms) when its numerator and denominator have no common factor other than 1 or -1.

A *Identify the values for which the given rational expression is undefined.*

5. $\dfrac{7}{x}$

6. $\dfrac{-2}{c}$

7. $\dfrac{8}{y-2}$

8. $\dfrac{y-6}{y-1}$

9. $\dfrac{4}{x+2}$

10. $\dfrac{x-3}{x+5}$

⊙ 11. $\dfrac{n+11}{2n-1}$

12. $\dfrac{x-4}{3x-2}$

13. $\dfrac{x^2+1}{x^2-1}$

14. $\dfrac{n+2}{n^2-16}$

⊙ 15. $\dfrac{x^2+x+1}{x^2-x-20}$

16. $\dfrac{p^2+7}{p^2-4p-21}$

B *Indicate whether each pair of rational expressions is equivalent.*

17. $\dfrac{p}{q}$ and $\dfrac{pr}{qr}$

18. $\dfrac{8}{3y}$ and $\dfrac{16}{6y}$

⊙ 19. $\dfrac{3t+5}{t+1}$ and $\dfrac{3t^2+5t}{t^2+t}$

20. $\dfrac{2x-1}{x-4}$ and $\dfrac{14x-7}{7x-28}$

⊙ 21. $\dfrac{x-1}{x+3}$ and $-\dfrac{1}{3}$

22. $\dfrac{n+4}{2n+4}$ and $\dfrac{1}{2}$

23. $\dfrac{x-2}{2-x}$ and $\dfrac{2-x}{x-2}$

24. $\dfrac{y-x}{y+x}$ and $\dfrac{y+x}{y-x}$

25. $\dfrac{x^2+4}{x+2}$ and $x+2$

26. $\dfrac{(y+5)^2}{y+5}$ and $y+5$

C *Simplify.*

27. $\dfrac{10a^4}{12a}$

28. $\dfrac{4b}{6b^3}$

⊙ 29. $\dfrac{3x^2}{12x^5}$

30. $\dfrac{-2y^6}{2y^4}$

31. $\dfrac{9s^3t^2}{6s^5t}$

32. $\dfrac{10r^3s^4}{5r^3s^4}$

33. $\dfrac{24a^4b^5}{3ab^2}$

34. $\dfrac{14yz^2}{21y^2z^2}$

35. $\dfrac{-2p^2q^3}{-10pq^4}$

36. $\dfrac{-3t^4s^5}{t^2s}$

37. $\dfrac{5x(x+8)}{4x(x+8)}$

38. $\dfrac{7n(n-1)}{2n(n-1)}$

39. $\dfrac{8x^2(5-2x)}{3x(2x-5)}$

40. $\dfrac{(a-b)ab^2}{3a^3(b-a)}$

41. $\dfrac{5x-10}{5}$

42. $\dfrac{3y+12}{3}$

43. $\dfrac{2x^2+2x}{4x^2+6x}$

44. $\dfrac{7y^3-5y^2}{3y^2+y}$

45. $\dfrac{a^2-4a}{ab-4b}$

46. $\dfrac{a^2+2b}{4b+2a^2}$

47. $\dfrac{6x-4y}{9x-6y}$

48. $\dfrac{10m-2n}{n-5m}$

49. $\dfrac{t^2-1}{t+1}$

50. $\dfrac{b^2-1}{b-1}$

51. $\dfrac{p^2-q^2}{q^2-p^2}$

52. $\dfrac{y^2-4x^2}{4x^2-y^2}$

53. $\dfrac{n-1}{2-2n}$

54. $\dfrac{x-3}{3-x}$

55. $\dfrac{(b-4)^2}{b^2-16}$

56. $\dfrac{m^2-1}{(m+1)^2}$

⊙ 57. $\dfrac{2x^2+5x-3}{10x-5}$

58. $\dfrac{6p+12}{p^2-p-6}$

59. $\dfrac{a^3+9a^2+14a}{a^2-10a-24}$

60. $\dfrac{x^2+3x-4}{x^2+2x-3}$

⊙ 61. $\dfrac{t^2-4t-5}{t^2-3t-10}$

62. $\dfrac{y^2+8y+15}{y^2-2y-15}$

63. $\dfrac{9-16d^2}{16d^2-24d+9}$

64. $\dfrac{6n^2+7n-10}{25-36n^2}$

65. $\dfrac{8s-2s^2}{2s^2-11s+12}$

66. $\dfrac{3x^2+2x-1}{12x^2-4x}$

67. $\dfrac{6x^2+5x+1}{6x^2-x-1}$

68. $\dfrac{2n^3+2n^2-4n}{n^3+2n^2-3n}$

69. $\dfrac{6y^2-7y+2}{6y^3+5y^2-6y}$

70. $\dfrac{3x^2+xy}{3x^2+7xy+2y^2}$

71. $\dfrac{2ab^2+4a^2b}{2b^2+5ab+2a^2}$

72. $\dfrac{p^2-4pq-12q^2}{2p^2-15pq+18q^2}$

73. $\dfrac{m^2+3mn-28n^2}{2m^2+4mn-48n^2}$

74. $\dfrac{3a^2+5ab-2b^2}{3a^2+8ab-3b^2}$

Mixed Practice

Simplify.

75. $\dfrac{12x^2y^6}{18x^3y^4}$

76. $\dfrac{-5a^7b^4}{a^3b^3}$

77. $\dfrac{3x^3-9x^2+6x}{x^3+x^2-6x}$

78. $\dfrac{x^2-3x-10}{x^2+4x+4}$

79. $\dfrac{w^2-2y^2}{2y^2-w^2}$

80. $\dfrac{h+3k^2}{12k^2+4h}$

81. $\dfrac{2x^2+x-3}{3x^2-3x}$

82. $\dfrac{r^2s(r-s)}{2rs^3(s-r)}$

Identify the values for which the given rational expression is undefined.

83. $\dfrac{p-5}{2p-3}$

84. $\dfrac{m+3}{m^2-25}$

Indicate whether each pair of rational expressions is equivalent.

85. $\dfrac{6y+7}{y+1}$ and $\dfrac{6y^2+7y}{y^2+y}$

86. $\dfrac{2u-9}{5u-9}$ and $\dfrac{2}{5}$

Applications

D *Solve.*

87. The surface area of a person's skin is important in both clothing design and medicine. A rough estimate of the skin area for someone with height h in. and waist circumference C in. is Ch in.^{2} Use this estimate to approximate the ratio of skin area to height. (*Source:* Weinstein and Adam, *Guesstimation: Solving the World's Problems on the Back of a Cocktail Napkin*)

88. The ratio of a company's stock price P to its earnings E, both per share, is called the company's *price-to-earnings* (or *P/E) ratio*. When is this ratio undefined? (*Source:* investopedia.com)

89. When a mathematics department had more faculty and staff members, each had an office with dimensions x ft by x ft. Now that the department has gotten smaller, its offices are being enlarged. Each faculty and each staff office will be made 2 ft wider, whereas each faculty office will also be made 5 ft longer.

    a. Write an expression for the area (in square feet) of an enlarged faculty office.

    b. Write an expression for the area (in square feet) of an enlarged staff office.

    c. Write an expression for the ratio of the area of an enlarged faculty office to the area of an enlarged staff office. Simplify.

    d. Find the value of the expression in part (c) if the length of each office had been 8 ft.

90. The baking time for bread depends on the ratio of its volume V to its surface area S. For each of the following types of bread, express the ratio $\dfrac{V}{S}$ in simplest form.

    a. A rectangular loaf of bread

    b. A cylindrical bread stick

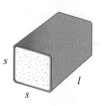

$$V = s^2 l$$
$$S = 2s^2 + 4sl$$

$$V = \pi r^2 h$$
$$S = 2\pi r^2 + 2\pi rh$$

91. An archer shoots an arrow at the target shown.

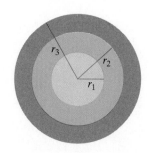

    a. Find and simplify the ratio of the area of the inner circle to the area of the adjacent ring.

    b. Find and simplify the ratio of the area of the smaller ring to the area of the larger ring.

92. A basketball with radius r is packed in a cubic box with side $2r$. The volume of the ball is $\dfrac{4}{3}\pi r^3$ and the volume of the box is $(2r)^3$.

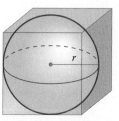

    a. The smaller volume is what fraction of the larger volume? Simplify your answer.

    b. Does the answer to part (a) depend on the value of r?

• Check your answers on page A-24.

MINDStretchers

Technology

1. Using graphing calculator or software, display the graphs of $y = \dfrac{x^2 - 1}{x - 1}$ and $y = x + 1$ on the same screen.

   Compare the two graphs. What conclusion can you draw from this comparison?

Patterns

2. Some rational expressions, when simplified, are equivalent to polynomials. Check the simplifications of the following rational expressions:

   a. $\dfrac{x^2 - 1}{x - 1} = 1 + x$ **b.** $\dfrac{x^3 - 1}{x - 1} = 1 + x + x^2$ **c.** $\dfrac{x^4 - 1}{x - 1} = 1 + x + x^2 + x^3$

   Identify this pattern and extend it.

Writing

3. Sometimes, arithmetic fractions are written with a slanted fraction line, as in 3/4, rather than a horizontal fraction line, as in $\dfrac{3}{4}$. In a sentence or two, explain the disadvantage of writing a rational expression such as $\dfrac{x + 2}{x - 3}$ with a slanted fraction line.

Multiplying Rational Expressions

Operations on rational expressions in algebra are similar to those on fractions in arithmetic. For instance, to multiply arithmetic fractions, we multiply their numerators to get the numerator of the product and multiply their denominators to get the denominator of the product.

$$\frac{2}{3} \cdot \frac{5}{7} = \frac{2 \cdot 5}{3 \cdot 7} = \frac{10}{21}$$

In algebra, rational expressions are multiplied in the same way.

$$\frac{2}{a} \cdot \frac{b}{y} = \frac{2b}{ay}$$

Recall from arithmetic that it is often easier to first divide out any common factors and then multiply the numerators and denominators:

$$\frac{2}{9} \cdot \frac{3}{10} = \frac{2}{\underset{3}{\cancel{9}}} \cdot \frac{\overset{1}{\cancel{3}}}{\underset{5}{\cancel{10}}} = \frac{1}{15}$$

Similarly, in algebra:

$$\frac{5a}{b} \cdot \frac{b}{2a^2} = \frac{5\overset{1}{\cancel{a}}}{\underset{1}{\cancel{b}}} \cdot \frac{\overset{1}{\cancel{b}}}{\underset{a}{2a^2}} = \frac{5}{2a}$$

EXAMPLE 1

Multiply.

a. $\dfrac{3}{x} \cdot \dfrac{5}{x}$ **b.** $\dfrac{4x^2}{7y^3} \cdot \dfrac{3y}{8x}$

Solution

a. There are no common factors in the numerators and denominators.

$$\frac{3}{x} \cdot \frac{5}{x} = \frac{3 \cdot 5}{x \cdot x}$$ Multiply the numerators, and multiply the denominators.

$$= \frac{15}{x^2}$$ Simplify.

b. $\dfrac{4x^2}{7y^3} \cdot \dfrac{3y}{8x} = \dfrac{\overset{1}{\cancel{4x^2}}\,\overset{x}{}}{\underset{y^2}{7y^3}} \cdot \dfrac{\overset{1}{\cancel{3y}}}{\underset{2\,1}{8x}}$ Divide out all common factors in the numerators and denominators.

$$= \frac{x \cdot 3}{7y^2 \cdot 2}$$ Multiply the numerators, and multiply the denominators.

$$= \frac{3x}{14y^2}$$ Simplify.

PRACTICE 1

Multiply.

a. $\dfrac{n}{7} \cdot \dfrac{2}{m}$

b. $\dfrac{p^2}{6q^2} \cdot \dfrac{2q}{5p}$

The following rule describes how we find the product of rational expressions, such as:

$$\frac{2n}{n^2 - 5n} \quad \text{and} \quad \frac{3n - 15}{4n}$$

To Multiply Rational Expressions

- Factor the numerators and denominators.
- Divide the numerators and denominators by all common factors.
- Multiply the remaining factors in the numerators and the remaining factors in the denominators.

EXAMPLE 2

Multiply.

a. $\dfrac{2n}{n^2 - 5n} \cdot \dfrac{3n - 15}{4n}$

b. $\dfrac{10}{9 - x^2} \cdot \dfrac{6 + 2x}{5}$

Solution

a. $\dfrac{2n}{n^2 - 5n} \cdot \dfrac{3n - 15}{4n} = \dfrac{2n}{n(n - 5)} \cdot \dfrac{3(n - 5)}{4n}$ Factor the numerator and the denominator.

$$= \frac{\overset{1}{\overset{1}{\cancel{2n}}}}{\underset{1}{\underset{1}{\cancel{n(n - 5)}}}} \cdot \frac{\overset{1}{3\cancel{(n - 5)}}}{\underset{2}{\cancel{4}n}}$$ Divide out common factors.

$$= \frac{3}{2n}$$ Multiply the factors in the numerators and in the denominators.

b. $\dfrac{10}{9 - x^2} \cdot \dfrac{6 + 2x}{5} = \dfrac{10}{(3 - x)(3 + x)} \cdot \dfrac{2(3 + x)}{5}$

$$= \frac{\overset{2}{\cancel{10}}}{(3 - x)\cancel{(3 + x)}} \cdot \frac{\overset{1}{2\cancel{(3 + x)}}}{\underset{1}{\cancel{5}}} = \frac{4}{3 - x}$$

PRACTICE 2

Multiply.

a. $\dfrac{3t}{t^2 + 5t} \cdot \dfrac{3t + 15}{6t}$

b. $\dfrac{8}{36g^2 - 1} \cdot \dfrac{1 + 6g}{4}$

EXAMPLE 3

Find the product.

a. $\dfrac{x^2 - x - 20}{5x + 5} \cdot \dfrac{15x^2 - 15}{2x + 8}$

b. $\dfrac{16 + 6a - a^2}{a^2 - 10a - 24} \cdot \dfrac{a^2 - 6a - 27}{a^2 - 17a + 72}$

PRACTICE 3

Find the product.

a. $\dfrac{y^2 + 4y - 21}{3y - 9} \cdot \dfrac{6y^2 - 24}{y + 2}$

b. $\dfrac{x^2 - x - 30}{x^2 + 10x + 9} \cdot \dfrac{18 - 7x - x^2}{x^2 - 8x + 12}$

Solution

a. $\dfrac{x^2 - x - 20}{5x + 5} \cdot \dfrac{15x^2 - 15}{2x + 8}$

$$= \frac{(x - 5)(x + 4)}{5(x + 1)} \cdot \frac{15(x^2 - 1)}{2(x + 4)}$$

$$= \frac{(x - 5)(x + 4)}{5(x + 1)} \cdot \frac{15(x - 1)(x + 1)}{2(x + 4)}$$

$$= \frac{(x - 5)\cancel{(x + 4)}}{\cancel{5}\cancel{(x + 1)}} \cdot \frac{\overset{3}{\cancel{15}}(x - 1)\cancel{(x + 1)}}{\cancel{2}\cancel{(x + 4)}}$$

$$= \frac{3(x - 5)(x - 1)}{2}$$

b. $\dfrac{16 + 6a - a^2}{a^2 - 10a - 24} \cdot \dfrac{a^2 - 6a - 27}{a^2 - 17a + 72}$

$$= \frac{(8 - a)(2 + a)}{(a - 12)(a + 2)} \cdot \frac{(a - 9)(a + 3)}{(a - 8)(a - 9)}$$

$$= \frac{(8 - a)\cancel{(2 + a)}}{(a - 12)\cancel{(a + 2)}} \cdot \frac{\cancel{(a - 9)}(a + 3)}{(a - 8)\cancel{(a - 9)}}$$

$$= \frac{-1(a - 8)}{a - 12} \cdot \frac{a + 3}{a - 8}$$

$$= \frac{-\cancel{(a - 8)}}{a - 12} \cdot \frac{a + 3}{\cancel{a - 8}}$$

$$= \frac{-(a + 3)}{a - 12}$$

$$= -\frac{a + 3}{a - 12}$$

Dividing Rational Expressions

In dividing arithmetic fractions, we take the reciprocal of the divisor and change the operation to multiplication. Recall that the reciprocal of a fraction is formed by interchanging its numerator and denominator.

The divisor   The reciprocal of the divisor

$$\frac{2}{3} \div \frac{3}{4} = \frac{2}{3} \cdot \frac{4}{3} = \frac{2 \cdot 4}{3 \cdot 3} = \frac{8}{9}$$

Division   Multiplication

Rational expressions are divided in the same way.

$$\frac{p}{q} \div \underset{\substack{\uparrow \\ \text{Division}}}{\overset{\substack{\text{The} \\ \text{divisor} \\ \downarrow}}{\frac{r}{s}}} = \frac{p}{q} \cdot \underset{\substack{\uparrow \\ \text{Multiplication}}}{\overset{\substack{\text{The reciprocal} \\ \text{of the divisor} \\ \downarrow}}{\frac{s}{r}}} = \frac{p \cdot s}{q \cdot r} = \frac{ps}{qr}$$

In general, to divide rational expressions, we apply the following rule:

To Divide Rational Expressions

- Take the reciprocal of the divisor and change the operation to multiplication.
- Follow the rule for multiplying rational expressions.

EXAMPLE 4

Divide. **a.** $\dfrac{3}{x} \div \dfrac{y}{2}$ **b.** $\dfrac{6a}{b} \div \dfrac{3a^2}{b^3}$

Solution

a. $\dfrac{3}{x} \div \dfrac{y}{2} = \dfrac{3}{x} \cdot \dfrac{2}{y}$ Take the reciprocal of the divisor, and change the operation to multiplication.

$\qquad = \dfrac{6}{xy}$ Multiply.

b. $\dfrac{6a}{b} \div \dfrac{3a^2}{b^3} = \dfrac{6a}{b} \cdot \dfrac{b^3}{3a^2}$

$\qquad = \dfrac{\overset{2}{\cancel{6}}\overset{1}{a}}{\underset{1}{\cancel{b}}} \cdot \dfrac{b^{3\,b^2}}{\underset{1}{\cancel{3}}\underset{a}{a^2}}$

$\qquad = \dfrac{2b^2}{a}$

EXAMPLE 5

Find the quotient.

a. $\dfrac{a + 1}{a - 2} \div \dfrac{a + 2}{a - 1}$ **b.** $\dfrac{6x + 3y}{4x - 12y} \div \dfrac{10x + 5y}{2x - 6y}$

c. $\dfrac{x^2 + 3x + 2}{3x + 12} \div \dfrac{x^2 - 1}{7x}$

Solution

a. $\dfrac{a + 1}{a - 2} \div \dfrac{a + 2}{a - 1} = \dfrac{a + 1}{a - 2} \cdot \dfrac{a - 1}{a + 2}$ Take the reciprocal, and change the operation to multiplication.

$\qquad = \dfrac{(a + 1)(a - 1)}{(a - 2)(a + 2)}$ Multiply.

PRACTICE 4

Find the quotient.

a. $\dfrac{a}{4} \div \dfrac{b}{6}$

b. $\dfrac{8q^3}{p^2} \div \dfrac{2q^4}{p^4}$

PRACTICE 5

Find the quotient.

a. $\dfrac{x + 3}{x - 10} \div \dfrac{x + 1}{x + 3}$

b. $\dfrac{p + 4q}{3p - 6q} \div \dfrac{2p + 8q}{2p - 4q}$

c. $\dfrac{y^2 + 4y + 3}{5y + 10} \div \dfrac{y^2 - 1}{y}$

b. $\dfrac{6x + 3y}{4x - 12y} \div \dfrac{10x + 5y}{2x - 6y}$

$= \dfrac{6x + 3y}{4x - 12y} \cdot \dfrac{2x - 6y}{10x + 5y}$ Take the reciprocal, and change the operation to multiplication.

$= \dfrac{3(2x + y)}{4(x - 3y)} \cdot \dfrac{2(x - 3y)}{5(2x + y)}$ Factor the numerators and denominators.

$= \dfrac{\overset{1}{3(2x + y)}}{\underset{2}{4(x - 3y)}} \cdot \dfrac{\overset{1}{2(x - 3y)}}{\underset{1}{5(2x + y)}}$ Divide out the common factors.

$= \dfrac{3}{10}$ Multiply.

c. $\dfrac{x^2 + 3x + 2}{3x + 12} \div \dfrac{x^2 - 1}{7x} = \dfrac{x^2 + 3x + 2}{3x + 12} \cdot \dfrac{7x}{x^2 - 1}$

$= \dfrac{(x + 1)(x + 2)}{3(x + 4)} \cdot \dfrac{7x}{(x + 1)(x - 1)}$

$= \dfrac{\overset{1}{(x + 1)}(x + 2)}{3(x + 4)} \cdot \dfrac{7x}{\underset{1}{(x + 1)}(x - 1)}$

$= \dfrac{7x(x + 2)}{3(x + 4)(x - 1)}$

EXAMPLE 6

According to the physicist Isaac Newton, the force of gravitation between two objects with mass m and M is given by the expression $G \cdot \dfrac{m}{d} \cdot \dfrac{M}{d}$, where d is the distance between them and G is the gravitational constant. Simplify this expression.

Solution

$$G \cdot \dfrac{m}{d} \cdot \dfrac{M}{d} = \dfrac{G}{1} \cdot \dfrac{m}{d} \cdot \dfrac{M}{d} = \dfrac{GmM}{d^2}$$

PRACTICE 6

When a body moves in a circular path, a force called the *centripetal force* is directed toward the center of the circle. This force has magnitude equal to the product of the object's mass m and its acceleration a. If $m = \dfrac{W}{g}$ and $a = \dfrac{v^2}{r}$, find the magnitude of the centripetal force in terms of W, g, v, and r, which represent the weight of the object, a constant due to gravity, the velocity of the object, and the radius of the circle, respectively.

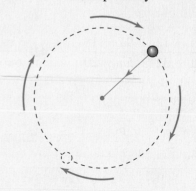

Multiply. Express the product in lowest terms.

1. $\dfrac{1}{t^2} \cdot \dfrac{t}{4}$

2. $-\dfrac{y}{8} \cdot \dfrac{10}{y^3}$

3. $\dfrac{2}{a} \cdot \dfrac{3}{b}$

4. $\dfrac{5x}{2} \cdot \dfrac{2y}{3}$

5. $\dfrac{2x^4}{3x^5} \cdot \dfrac{5}{x^8}$

6. $\dfrac{4c}{d^3} \cdot \dfrac{3d}{8c^4}$

7. $-\dfrac{7x^2y}{3} \cdot \dfrac{6}{x^3y}$

8. $-\dfrac{5p^2}{10pq^2} \cdot \left(-\dfrac{6pq}{5p^3q}\right)$

9. $\dfrac{x}{x-2} \cdot \dfrac{5x-10}{x^4}$

10. $\dfrac{t}{t+3} \cdot \dfrac{4t+12}{t}$

11. $\dfrac{8n-3}{n^2} \cdot n$

12. $\dfrac{s+1}{s} \cdot s^2$

13. $\dfrac{8x-6}{5x+20} \cdot \dfrac{2x+8}{4x-3}$

14. $\dfrac{10y-1}{3y-6} \cdot \dfrac{5y-10}{10y-1}$

15. $\dfrac{5n-1}{6n+4} \cdot \dfrac{3n+2}{1-5n}$

16. $\dfrac{-4x-2}{10-x} \cdot \dfrac{6x+1}{2x+1}$

17. $\dfrac{x^2-4y^2}{x+y} \cdot \dfrac{3x+3y}{4x-8y}$

18. $\dfrac{a^2-4b^2}{6a-6b} \cdot \dfrac{10a-10b}{3a+6b}$

19. $\dfrac{p^4-1}{p^4-16} \cdot \dfrac{p^2+4}{p^2+1}$

20. $\dfrac{x^4-81}{x^2-x-12} \cdot \dfrac{x^2-16}{x^2+9}$

21. $\dfrac{n^2-2n-24}{n^2+6n+8} \cdot \dfrac{n^2+5n+6}{n^2-5n-6}$

22. $\dfrac{t^2+4t-21}{t^2+2t-15} \cdot \dfrac{t^2+t-20}{t^2+3t-28}$

23. $\dfrac{2y^2-y-6}{2y^2+y-3} \cdot \dfrac{2y^2-3y+1}{2y^2-9y+10}$

24. $\dfrac{2m+1}{2m^2+7m+3} \cdot \dfrac{m^2+2m-3}{2m+4m^2}$

25. $\dfrac{2}{x^3} \cdot \dfrac{4x}{5} \cdot \dfrac{10}{x^2}$

26. $\dfrac{a-3}{a^2} \cdot \dfrac{2a}{5} \cdot \dfrac{10a}{3}$

27. $\dfrac{x^2-7x+10}{2x-2} \cdot \dfrac{6x}{x^2-2x-15} \cdot \dfrac{x^2+2x-3}{x-2}$

28. $\dfrac{p^2-q^2}{p^2-pq} \cdot \dfrac{q}{2p^2-pq-q^2} \cdot \dfrac{6p+3q}{p}$

Divide. Express the quotient in lowest terms.

29. $\dfrac{7}{a} \div \dfrac{14}{a}$

30. $\dfrac{-5}{n} \div \dfrac{n}{2}$

31. $\dfrac{p^3}{10} \div \dfrac{p^3}{20}$

32. $\dfrac{-s}{v^2} \div \dfrac{s^2}{v}$

33. $\dfrac{12}{x^3} \div \dfrac{6}{5x^2}$

34. $\dfrac{1}{a^2} \div \dfrac{2}{3a}$

35. $-\dfrac{3}{t} \div t$

36. $\dfrac{5}{s} \div s$

37. $\dfrac{9xy^2}{2x^3} \div \dfrac{3x^2y}{4y}$

38. $\dfrac{10a^3}{7ab^2} \div \dfrac{-6a^2b^2}{14ab}$

39. $\dfrac{c+3}{c-5} \div \dfrac{c+9}{c-7}$

40. $\dfrac{t+4}{t-6} \div \dfrac{2t-8}{2t-4}$

41. $\dfrac{6a-12}{8a+32} \div \dfrac{9a-18}{5a+20}$

42. $\dfrac{3x+6}{6x+18} \div \dfrac{2x+4}{x+3}$

43. $\dfrac{x+1}{10} \div \dfrac{1-x^2}{5}$

44. $\dfrac{4x+8}{8} \div \dfrac{4-x^2}{4}$

45. $\dfrac{p^2-1}{1-p} \div \dfrac{p+1}{p}$

46. $\dfrac{y+6}{3y} \div \dfrac{y^2-36}{6-y}$

47. $\dfrac{x^2y+3xy^2}{x^2-9y^2} \div \dfrac{5x^2y}{x^2-2xy-3y^2}$

48. $\dfrac{2x+1}{2x^2+7x+3} \div \dfrac{2x+4x^2}{x^2+2x-3}$

49. $\dfrac{2t^2 - 3t - 2}{2t + 1} \div (4 - t^2)$

50. $(y - x) \div \dfrac{x^2 - y^2}{x^2 + xy}$

51. $\dfrac{x^2 - 11x + 28}{x^2 - x - 42} \div \dfrac{x^2 - 2x - 8}{x^2 + 7x + 10}$

52. $\dfrac{a^2 - a - 56}{a^2 + 8a + 7} \div \dfrac{a^2 - 13a + 40}{a^2 - 4a - 5}$

53. $\dfrac{3p^2 - 3p - 18}{p^2 + 2p - 15} \div \dfrac{2p^2 + 6p - 20}{2p^2 - 12p + 16}$

54. $\dfrac{3y^2 + 13y + 4}{16 - y^2} \div \dfrac{3y^2 - 5y - 2}{3y - 12}$

Mixed Practice

Divide. Express the quotient in lowest terms.

55. $\dfrac{r^2}{s} \div \left(-\dfrac{r^4}{s} \right)$

56. $\dfrac{x^2 + 5x - 24}{x^2 + 9x + 8} \div \dfrac{x^2 - 10x + 21}{x^2 - 6x - 7}$

57. $\dfrac{3x + 9}{12} \div \dfrac{9 - x^2}{8}$

58. $\dfrac{h^2 + 2h - 8}{h + 4} \div (4 - h^2)$

59. $-\dfrac{6q^2}{15p^2q} \div \dfrac{12p^2q^2}{5p^2q}$

Multiply. Express the product in lowest terms.

60. $\dfrac{9a}{2b^2} \cdot \dfrac{b}{3a^3}$

61. $\dfrac{3y}{4} \cdot \dfrac{y + 3}{2y^2} \cdot \dfrac{6y}{3}$

62. $\dfrac{2c + 2d}{6c - 3d} \cdot \dfrac{4c^2 - d^2}{c + d}$

63. $\dfrac{x^2 + x - 2}{x + 1} \cdot \dfrac{3x + 3x^2}{3x^2 + 7x + 2}$

64. $\dfrac{u^2}{u + 5} \cdot \dfrac{5u + 25}{u}$

Applications

Solve.

65. An investment of p dollars is growing at the annual simple interest rate r. The number of years that it will take the investment to be worth A dollars is given by the expression:

$$\left(\dfrac{A - p}{p} \right) \div r$$

Simplify this expression.

66. A store is having a sale, with each item selling at a discount of $x\%$.

a. With this discount, what percent of the normal price is a customer paying on each item?

b. What is the sale price of 10 items that normally sell for z dollars each?

67. A company has annual expenses totaling B dollars, of which $p\%$ goes toward rents. Of the rental expenses, $q\%$ goes toward the head office. Write as a rational expression the annual cost of the head office rental.

68. Physicists studying momentum may use the expression

$$\dfrac{W}{g} \cdot \dfrac{1}{t} \cdot (v_2 - v_1),$$

where W is the weight of an object, g is a constant, t is time, v_2 is the final velocity of the object, and v_1 is the initial velocity of the object. Write as a single rational expression.

69. Electricians use the following formula when studying the resistance in a heating element:

$$P = \frac{V^2}{R + r} \cdot \frac{r}{R + r}$$

Multiply out the right-hand side of this formula.

70. Suppose that the chance that one event occurs is $\frac{a}{b}$ and the chance that an independent (or unrelated) event occurs is $\frac{c}{d}$.

a. The chance that both events will occur is the product of their individual chances. Write this product as a rational expression.

b. The chance that neither event will occur can be represented by $\left(1 - \frac{a}{b}\right)\left(1 - \frac{c}{d}\right)$. Write this product as a single rational expression.

71. The following diagram shows a cylinder and its inscribed sphere, where the radius of the sphere is r and the height of the cylinder is h.

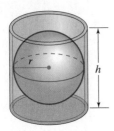

Volume of a sphere $= \frac{4}{3}\pi r^3$

Volume of a cylinder $= \pi r^2 h$

Surface area of a sphere $= 4\pi r^2$

Surface area of a cylinder $= 2\pi rh + 2\pi r^2$

Use these formulas to answer the following questions.

a. Find the ratio of the volume of the sphere to the volume of the cylinder.

b. Find the ratio of the surface area of the sphere to the surface area of the cylinder.

c. Divide the expression in part (a) by the expression in part (b).

72. In 1934, Harold Urey won the Nobel Prize in chemistry for discovering deuterium (also called *heavy hydrogen*). This discovery involved reducing the mass of an electron in an atom of deuterium.

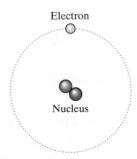

If m stands for the mass of an electron, then the reduced mass r can be represented by

$$r = \frac{mM}{M + m},$$

where M is the mass of the atom's nucleus. Find an expression for the quotient r divided by m in terms of m and M, and simplify. (*Source:* wikipedia.org)

• Check your answers on page A-24.

Critical Thinking

1. Even though the operation of division is not commutative, can you find four different polynomials P, Q, R, and S for which $\dfrac{P}{Q} \div \dfrac{R}{S} = \dfrac{R}{S} \div \dfrac{P}{Q}$? Explain.

Mathematical Reasoning

2. True or false, the expressions $\dfrac{\left(\dfrac{a}{b}\right)}{c}$ and $\dfrac{a}{\left(\dfrac{b}{c}\right)}$ are equivalent. Explain.

Groupwork

3. Working with a partner, find the following:
   a. Two different pairs of rational expressions whose quotient is $\dfrac{x+4}{x+6}$.

   b. Two different pairs of rational expressions whose product is $\dfrac{x^2 - x - 6}{2x^2 + 5x - 3}$.

In the last section, we discussed the multiplication and division of rational expressions. In this section, we consider how to find their sums and differences.

Adding and Subtracting Rational Expressions with the Same Denominator

Recall from arithmetic that to add like fractions, we add the numerators and keep the same denominator.

$$\frac{3}{5} + \frac{1}{5} = \frac{3+1}{5} = \frac{4}{5}$$

To subtract like fractions, we subtract the numerators and keep the same denominator.

$$\frac{3}{5} - \frac{1}{5} = \frac{3-1}{5} = \frac{2}{5}$$

In algebra, the process of adding and subtracting rational expressions with the same denominator is similar:

$$\frac{3}{x} + \frac{1}{x} = \frac{3+1}{x} = \frac{4}{x} \quad \text{and} \quad \frac{3}{x} - \frac{1}{x} = \frac{3-1}{x} = \frac{2}{x}$$

A To add or subtract rational expressions with the same denominator

B To find the least common denominator (LCD) of two or more rational expressions

C To add or subtract rational expressions with different denominators

D To solve applied problems involving the addition or subtraction of rational expressions

To Add (or Subtract) Rational Expressions with the Same Denominator

- Add (or subtract) the numerators and keep the same denominator.

- Simplify, if possible.

EXAMPLE 1

Add.

a. $\dfrac{3}{x-1} + \dfrac{8}{x-1}$ **b.** $\dfrac{3x}{4y} + \dfrac{x}{4y}$

c. $\dfrac{3x}{x+1} + \dfrac{3}{x+1}$ **d.** $\dfrac{2t^2 + 5t - 7}{3t - 1} + \dfrac{7t^2 - 5t + 6}{3t - 1}$

Solution

a. $\dfrac{3}{x-1} + \dfrac{8}{x-1} = \dfrac{3+8}{x-1}$ Add the numerators, and keep the same denominator.

$= \dfrac{11}{x-1}$ Simplify.

b. $\dfrac{3x}{4y} + \dfrac{x}{4y} = \dfrac{4x}{4y}$ Add the numerators, and keep the same denominator.

$= \dfrac{\overset{1}{\cancel{4}}x}{\underset{1}{\cancel{4}}y}$ Divide out the common factor.

$= \dfrac{x}{y}$ Simplify.

PRACTICE 1

Add.

a. $\dfrac{9}{y+2} + \dfrac{1}{y+2}$

b. $\dfrac{10r}{3s} + \dfrac{5r}{3s}$

c. $\dfrac{6t-7}{t-1} + \dfrac{1}{t-1}$

d. $\dfrac{n^2 + 10n + 1}{n+5} + \dfrac{2n^2 + 4n - 6}{n+5}$

c. $\dfrac{3x}{x+1} + \dfrac{3}{x+1} = \dfrac{3x+3}{x+1}$ Add the numerators, and keep the same denominator.

$$= \dfrac{3\cancel{(x+1)}^{1}}{\cancel{x+1}_{1}}$$ Factor the numerator, and divide out the common factor.

$$= 3$$ Simplify.

d. $\dfrac{2t^2 + 5t - 7}{3t - 1} + \dfrac{7t^2 - 5t + 6}{3t - 1} = \dfrac{(2t^2 + 5t - 7) + (7t^2 - 5t + 6)}{3t - 1}$

$$= \dfrac{9t^2 - 1}{3t - 1}$$ Combine like terms in the numerator.

$$= \dfrac{\cancel{(3t-1)}^{1}(3t + 1)}{\cancel{3t-1}_{1}}$$

$$= 3t + 1$$

EXAMPLE 2

Subtract.

a. $\dfrac{6}{z} - \dfrac{4}{z}$ **b.** $\dfrac{5y}{8} - \dfrac{3y}{8}$ **c.** $\dfrac{x}{3y} - \dfrac{2x}{3y}$

Solution

a. $\dfrac{6}{z} - \dfrac{4}{z} = \dfrac{6 - 4}{z}$ Subtract the numerators, and keep the same denominator.

$$= \dfrac{2}{z}$$ Simplify.

b. $\dfrac{5y}{8} - \dfrac{3y}{8} = \dfrac{5y - 3y}{8} = \dfrac{\cancel{2y}^{1}}{\cancel{8}_{4}} = \dfrac{y}{4}$

c. $\dfrac{x}{3y} - \dfrac{2x}{3y} = \dfrac{x - 2x}{3y} = \dfrac{-x}{3y} = -\dfrac{x}{3y}$

EXAMPLE 3

Find the difference.

a. $\dfrac{x + 5y}{2y} - \dfrac{x - 11y}{2y}$ **b.** $\dfrac{3ax + bx}{a + 2b} - \dfrac{2ax - bx}{a + 2b}$

c. $\dfrac{6x + 12}{x^2 - x - 6} - \dfrac{x + 2}{x^2 - x - 6}$

PRACTICE 2

Find the difference.

a. $\dfrac{12}{v} - \dfrac{7}{v}$

b. $\dfrac{7t}{10} - \dfrac{t}{10}$

c. $\dfrac{7p}{3q} - \dfrac{8p}{3q}$

PRACTICE 3

Subtract.

a. $\dfrac{7a - 4b}{3a} - \dfrac{a - 4b}{3a}$

b. $\dfrac{9xy - 5xz}{4y - z} - \dfrac{xy - 3xz}{4y - z}$

EXAMPLE 3 (continued)
Solution

c. $\dfrac{2x + 13}{x^2 - 7x + 10} - \dfrac{5x + 7}{x^2 - 7x + 10}$

a. $\dfrac{x + 5y}{2y} - \dfrac{x - 11y}{2y}$

$= \dfrac{(x + 5y) - (x - 11y)}{2y}$ Subtract the numerators, and keep the same denominator.

$= \dfrac{x + 5y - x + 11y}{2y}$ Remove the parentheses.

$= \dfrac{16y}{2y}$ Combine like terms.

$= \dfrac{\overset{8}{\cancel{16}}\overset{1}{\cancel{y}}}{\underset{1}{\cancel{2}}\underset{1}{\cancel{y}}}$ Divide out the common factors.

$= 8$ Simplify.

b. $\dfrac{3ax + bx}{a + 2b} - \dfrac{2ax - bx}{a + 2b} = \dfrac{(3ax + bx) - (2ax - bx)}{a + 2b}$

$= \dfrac{3ax + bx - 2ax + bx}{a + 2b}$

$= \dfrac{ax + 2bx}{a + 2b}$

$= \dfrac{x(a + 2b)}{a + 2b}$

$= \dfrac{x\overset{1}{\cancel{(a + 2b)}}}{\underset{1}{\cancel{a + 2b}}} = x$

c. $\dfrac{6x + 12}{x^2 - x - 6} - \dfrac{x + 2}{x^2 - x - 6} = \dfrac{(6x + 12) - (x + 2)}{x^2 - x - 6}$

$= \dfrac{6x + 12 - x - 2}{x^2 - x - 6}$

$= \dfrac{5x + 10}{x^2 - x - 6}$

$= \dfrac{5(x + 2)}{(x - 3)(x + 2)}$

$= \dfrac{5\overset{1}{\cancel{(x + 2)}}}{(x - 3)\underset{1}{\cancel{(x + 2)}}} = \dfrac{5}{x - 3}$

The Least Common Denominator of Rational Expressions

Recall that combining unlike arithmetic fractions involves finding their least common denominator (LCD). We write the fractions in terms of their LCD, and then add these like fractions.

$$\frac{1}{2} + \frac{4}{5} = \frac{1}{2} \cdot \frac{5}{5} + \frac{4}{5} \cdot \frac{2}{2} = \frac{5}{10} + \frac{8}{10} = \frac{13}{10}$$

10 is the LCD of $\frac{1}{2}$ and $\frac{4}{5}$.

To find the LCD of rational expressions, we begin by factoring their denominators completely. For instance, consider the rational expressions $\dfrac{2}{9a^2}$ and $\dfrac{1}{12a^4}$, which we can write as:

$$\underbrace{\frac{2}{3 \cdot 3 \cdot a \cdot a}}_{} \quad \text{and} \quad \underbrace{\frac{1}{2 \cdot 2 \cdot 3 \cdot a \cdot a \cdot a \cdot a}}_{}$$

Denominators in factored form.

The LCD is the product of the different factors in the denominators, where the power of each factor is the greatest number of times that it occurs in any single denominator.

The highest power of this factor
in any denominator

$$\text{LCD} = 2 \cdot 2 \cdot 3 \cdot 3 \cdot a \cdot a \cdot a \cdot a = 2^2 3^2 a^4$$

To Find the LCD of Rational Expressions

- Factor each denominator completely.

- Multiply these factors, using for the power of each factor the greatest number of times that it occurs in any of the denominators. The product of all these factors is the LCD.

EXAMPLE 4

Find the LCD of each group of rational expressions.

a. $\dfrac{1}{6}$ and $\dfrac{2}{n}$ **b.** $\dfrac{1}{10x}$ and $\dfrac{7}{12x^2}$

c. $\dfrac{3}{8p}, \dfrac{7}{10p^2q}$, and $\dfrac{1}{4pqr}$

Solution

a. We begin by factoring the denominators of $\dfrac{1}{6}$ and $\dfrac{2}{n}$:

Factor 6: $2 \cdot 3$

Factor n: n

No factor is repeated more than once in any denominator.
So the LCD is the product of the factors.

$$\text{LCD} = 2 \cdot 3 \cdot n = 6n$$

b. $\dfrac{1}{10x}$ and $\dfrac{7}{12x^2}$

Factor $10x$: $2 \cdot 5 \cdot x$
Factor $12x^2$: $2^2 \cdot 3 \cdot x^2$

The factors 2 and x each appear at most twice in any denominator.
The factors 3 and 5 each appear at most once in any denominator.

$$\text{LCD} = 2^2 \cdot 3 \cdot 5 \cdot x^2 = 60x^2$$

PRACTICE 4

For each group of rational expressions, determine the least common denominator.

a. $\dfrac{8}{y}$ and $\dfrac{x}{20}$

b. $\dfrac{1}{6t}$ and $\dfrac{4}{3t^2}$

c. $\dfrac{1}{5x}, \dfrac{7}{15xy^3}$, and $\dfrac{x+y}{2x^2}$

EXAMPLE 4 (continued)

c. $\dfrac{3}{8p}$, $\dfrac{7}{10p^2q}$, and $\dfrac{1}{4pqr}$

Factor $8p$: $2^3 \cdot p$

Factor $10p^2q$: $2 \cdot 5 \cdot p^2 \cdot q$

Factor $4pqr$: $2^2 \cdot p \cdot q \cdot r$

$LCD = 2^3 \cdot 5 \cdot p^2 \cdot q \cdot r = 40p^2qr$

In Example 4, we found the LCD of rational expressions whose denominators were monomials. Now, let's consider the case where the denominators may be trinomials or have binomial factors.

EXAMPLE 5

Find the LCD of each group of rational expressions.

a. $\dfrac{4x + 3}{3x^2 - 3x}$ and $\dfrac{x - 6}{x^2 - 2x + 1}$

b. $\dfrac{x + 4}{x + 7}$, $\dfrac{3}{x - 4}$, and $\dfrac{8x - 1}{x + 9}$

c. $\dfrac{2y}{x^2 - 2xy + y^2}$, $\dfrac{x}{x^2 - y^2}$, and $\dfrac{x - y}{x^2 + 2xy + y^2}$

Solution

a. First, we factor the denominators of $\dfrac{4x + 3}{3x^2 - 3x}$ and $\dfrac{x - 6}{x^2 - 2x + 1}$.

Factor $3x^2 - 3x$: $3x(x - 1)$

Factor $x^2 - 2x + 1$: $(x - 1)(x - 1) = (x - 1)^2$

The factor $3x$ appears at most once and the factor $(x - 1)$ appears at most twice in either denominator.

$$LCD = 3x(x - 1)^2$$

b. $\dfrac{x + 4}{x + 7}$, $\dfrac{3}{x - 4}$, and $\dfrac{8x - 1}{x + 9}$

Factor $x + 7$: $x + 7$

Factor $x - 4$: $x - 4$

Factor $x + 9$: $x + 9$

No factor appears more than once in any denominator.

$$LCD = (x + 7)(x - 4)(x + 9)$$

c. $\dfrac{2y}{x^2 - 2xy + y^2}$, $\dfrac{x}{x^2 - y^2}$, and $\dfrac{x - y}{x^2 + 2xy + y^2}$

Factor $x^2 - 2xy + y^2$: $(x - y)(x - y) = (x - y)^2$

Factor $x^2 - y^2$: $(x - y)(x + y)$

Factor $x^2 + 2xy + y^2$: $(x + y)(x + y) = (x + y)^2$

$LCD = (x - y)^2(x + y)^2$

Determine the LCD for each group of rational expressions.

a. $\dfrac{9n + 1}{2n^2 + 2n}$ and $\dfrac{n - 5}{n^2 + 2n + 1}$

b. $\dfrac{7p + 1}{p + 2}$, $\dfrac{5p}{p - 1}$, and $\dfrac{3p + 1}{p + 5}$

c. $\dfrac{s - t}{s^2 + 4st + 4t^2}$, $\dfrac{3t}{s^2 - 4t^2}$, and $\dfrac{6s}{s^2 - 4st + 4t^2}$

Adding and Subtracting Rational Expressions with Different Denominators

In order to add or subtract rational expressions with different denominators, we will need to change each expression to an equivalent rational expression whose denominator is the LCD.

EXAMPLE 6

Write the following rational expressions in terms of their LCD.

a. $\dfrac{3}{5x^2}$ and $\dfrac{x+1}{6xy}$ $\qquad$ **b.** $\dfrac{6x+1}{x^2-4}$ and $\dfrac{x}{x^2+5x+6}$

Solution

a. The LCD of $\dfrac{3}{5x^2}$ and $\dfrac{x+1}{6xy}$ is $30x^2y$. Since $5x^2 \cdot 6y = 30x^2y$, we multiply the numerator and denominator of $\dfrac{3}{5x^2}$ by $6y$ so that the denominator becomes the LCD.

$$\frac{3}{5x^2} = \frac{3 \cdot 6y}{5x^2 \cdot 6y} = \frac{18y}{30x^2y}$$

Note that multiplying the numerator and the denominator by $6y$ is the same as multiplying the expression by 1. So $\dfrac{3}{5x^2}$ and $\dfrac{18y}{30x^2y}$ are equivalent expressions.

Since $6xy \cdot 5x = 30x^2y$, we multiply the numerator and denominator of $\dfrac{x+1}{6xy}$ by $5x$ so that the denominator becomes the LCD.

$$\frac{x+1}{6xy} = \frac{(x+1) \cdot 5x}{6xy \cdot 5x} = \frac{5x(x+1)}{30x^2y}$$

b. To find the LCD, we begin by factoring the denominators of the expressions.

$$\text{Factor } x^2 - 4: \quad (x+2)(x-2)$$
$$\text{Factor } x^2 + 5x + 6: \quad (x+2)(x+3)$$

The LCD of $\dfrac{6x+1}{x^2-4}$ and $\dfrac{x}{x^2+5x+6}$ is $(x+2)(x-2)(x+3)$.

Since $(x^2-4)(x+3) = (x+2)(x-2)(x+3)$, we multiply the numerator and denominator of $\dfrac{6x+1}{x^2-4}$ by $(x+3)$ so that the denominator becomes the LCD.

$$\frac{6x+1}{x^2-4} = \frac{(6x+1)(x+3)}{(x+2)(x-2)(x+3)}$$

Since $(x^2+5x+6)(x-2) = (x+2)(x+3)(x-2)$, we multiply the numerator and denominator of $\dfrac{x}{x^2+5x+6}$ by $(x-2)$ so that the denominator becomes the LCD.

$$\frac{x}{x^2+5x+6} = \frac{x(x-2)}{(x+2)(x+3)(x-2)}$$

PRACTICE 6

Express each pair of rational expressions in terms of their LCD.

a. $\dfrac{2}{7p^3}$ and $\dfrac{p+3}{2p}$

b. $\dfrac{3y-2}{y^2-9}$ and $\dfrac{y}{y^2-6y+9}$

Now, let's look at how to add and subtract rational expressions with different denominators using the concept of equivalent rational expressions.

> ### To Add (or Subtract) Rational Expressions with Different Denominators
> - Find the LCD of the rational expressions.
> - Write each rational expression with a common denominator, usually the LCD.
> - Add (or subtract) the numerators.
> - Simplify, if possible.

EXAMPLE 7

Perform the indicated operation.

a. $\dfrac{1}{2n} - \dfrac{3}{5n}$ **b.** $\dfrac{2}{3x^2} + \dfrac{5}{9x}$

Solution

a. The LCD of the rational expressions is $10n$.

$\dfrac{1}{2n} - \dfrac{3}{5n} = \dfrac{1 \cdot 5}{2n \cdot 5} - \dfrac{3 \cdot 2}{5n \cdot 2}$ Write as equivalent rational expressions with the LCD as the denominator.

$= \dfrac{5}{10n} - \dfrac{6}{10n}$ Simplify.

$= \dfrac{-1}{10n}$ Subtract the numerators.

$= -\dfrac{1}{10n}$ Simplify.

b. The LCD of the rational expressions is $9x^2$.

$\dfrac{2}{3x^2} + \dfrac{5}{9x} = \dfrac{2 \cdot 3}{3x^2 \cdot 3} + \dfrac{5 \cdot x}{9x \cdot x}$ Write as equivalent rational expressions with the LCD as the denominator.

$= \dfrac{6}{9x^2} + \dfrac{5x}{9x^2}$ Simplify.

$= \dfrac{5x + 6}{9x^2}$ Add the numerators.

EXAMPLE 8

Combine.

a. $\dfrac{y + 5}{y - 2} - \dfrac{y - 3}{y}$

b. $\dfrac{7a + 5}{a - 1} - \dfrac{a}{1 - a}$

Solution

a. The LCD of the rational expression is $y(y - 2)$.

$$\frac{y + 5}{y - 2} - \frac{y - 3}{y}$$

$$= \frac{(y + 5) \cdot y}{(y - 2) \cdot y} - \frac{(y - 3) \cdot (y - 2)}{y \cdot (y - 2)} \qquad \text{Write in terms of the LCD.}$$

$$= \frac{y^2 + 5y}{y(y - 2)} - \frac{y^2 - 5y + 6}{y(y - 2)} \qquad \text{Multiply.}$$

$$= \frac{(y^2 + 5y) - (y^2 - 5y + 6)}{y(y - 2)} \qquad \text{Subtract the numerators.}$$

$$= \frac{y^2 + 5y - y^2 + 5y - 6}{y(y - 2)} \qquad \begin{array}{l}\text{Remove the parentheses in} \\ \text{the numerator.}\end{array}$$

$$= \frac{10y - 6}{y(y - 2)} \qquad \text{Simplify.}$$

$$= \frac{2(5y - 3)}{y(y - 2)} \qquad \text{Factor the numerator.}$$

b. The LCD of the rational expressions can be either $a - 1$ or $1 - a$, since $1 - a = -(a - 1)$. To get the LCD $(a - 1)$, we factor out -1 in the denominator of the expression $\dfrac{a}{1 - a}$.

$$\frac{7a + 5}{a - 1} - \frac{a}{1 - a}$$

$$= \frac{7a + 5}{a - 1} - \frac{a}{-(a - 1)} \qquad \text{Write } 1 - a \text{ as } -(a - 1).$$

$$= \frac{7a + 5}{a - 1} - \left(-\frac{a}{a - 1}\right) \qquad \text{Write } \frac{a}{-(a-1)} \text{ as } -\frac{a}{(a - 1)}.$$

$$= \frac{7a + 5}{a - 1} + \frac{a}{a - 1} \qquad \text{Simplify.}$$

$$= \frac{7a + 5 + a}{a - 1} \qquad \text{Add the numerators.}$$

$$= \frac{8a + 5}{a - 1} \qquad \text{Simplify.}$$

EXAMPLE 9

Combine.

a. $\dfrac{9}{2x + 4} + \dfrac{x}{x^2 - 4}$

b. $\dfrac{2}{x - 3} - \dfrac{x - 1}{9 - x^2}$

PRACTICE 9

Combine.

a. $\dfrac{1}{4x - 16} + \dfrac{2x}{x^2 - 16}$

b. $\dfrac{3x - 7}{x^2 - 1} + \dfrac{2}{1 - x}$

EXAMPLE 9 (continued)

Solution

a. $\dfrac{9}{2x+4} + \dfrac{x}{x^2-4}$

$$= \dfrac{9}{2(x+2)} + \dfrac{x}{(x+2)(x-2)} \qquad \text{Factor the denominators.}$$

$$= \dfrac{9 \cdot (x-2)}{2(x+2) \cdot (x-2)} + \dfrac{x \cdot 2}{(x+2)(x-2) \cdot 2} \qquad \begin{array}{l}\text{Write in terms of the}\\ \text{LCD } 2(x+2)(x-2).\end{array}$$

$$= \dfrac{9x-18}{2(x+2)(x-2)} + \dfrac{2x}{2(x+2)(x-2)} \qquad \text{Multiply.}$$

$$= \dfrac{11x-18}{2(x+2)(x-2)} \qquad \begin{array}{l}\text{Add the numerators and}\\ \text{combine like terms.}\end{array}$$

b. $\dfrac{2}{x-3} - \dfrac{x-1}{9-x^2}$

$$= \dfrac{2}{x-3} - \dfrac{x-1}{(3-x)(3+x)}$$

$$= \dfrac{2}{x-3} - \dfrac{x-1}{-(x-3)(x+3)} \qquad \text{Write } 3-x \text{ as } -(x-3).$$

$$= \dfrac{2}{x-3} - \left[-\dfrac{x-1}{(x-3)(x+3)} \right]$$

$$= \dfrac{2}{x-3} + \dfrac{x-1}{(x-3)(x+3)}$$

$$= \dfrac{2 \cdot (x+3)}{(x-3) \cdot (x+3)} + \dfrac{x-1}{(x-3)(x+3)} \qquad \begin{array}{l}\text{Write in terms of the}\\ \text{LCD } (x-3)(x+3).\end{array}$$

$$= \dfrac{2x+6}{(x-3)(x+3)} + \dfrac{x-1}{(x-3)(x+3)} \qquad \begin{array}{l}\text{Add the numerators and}\\ \text{combine like terms.}\end{array}$$

$$= \dfrac{(2x+6)+(x-1)}{(x-3)(x+3)}$$

$$= \dfrac{3x+5}{(x-3)(x+3)}$$

EXAMPLE 10

Perform the indicated operations.

a. $\dfrac{7n}{n^2+4n+3} - \dfrac{3n-2}{n^2+2n+1}$

b. $\dfrac{y}{y-1} - \dfrac{y}{3} - \dfrac{4}{y+2}$

Solution

a. $\dfrac{7n}{n^2+4n+3} - \dfrac{3n-2}{n^2+2n+1}$

$$= \dfrac{7n}{(n+3)(n+1)} - \dfrac{3n-2}{(n+1)^2} \qquad \text{Factor the denominators.}$$

$$\begin{aligned}= &\dfrac{7n \cdot (n+1)}{(n+3)(n+1) \cdot (n+1)}\\[6pt] &- \dfrac{(3n-2) \cdot (n+3)}{(n+1)^2 \cdot (n+3)}\end{aligned} \qquad \begin{array}{l}\text{Write in terms of the}\\ \text{LCD } (n+3)(n+1)^2.\end{array}$$

PRACTICE 10

Perform the indicated operation.

a. $\dfrac{y}{y^2+5y+6} - \dfrac{4y+1}{y^2+3y+2}$

b. $\dfrac{2x}{5} - \dfrac{1}{x+1} - \dfrac{x-1}{4x}$

$$= \frac{7n^2 + 7n}{(n + 3)(n + 1)^2} - \frac{3n^2 + 7n - 6}{(n + 3)(n + 1)^2} \qquad \text{Multiply.}$$

$$= \frac{(7n^2 + 7n) - (3n^2 + 7n - 6)}{(n + 3)(n + 1)^2} \qquad \text{Subtract the numerators.}$$

$$= \frac{7n^2 + 7n - 3n^2 - 7n + 6}{(n + 3)(n + 1)^2} \qquad \begin{array}{l}\text{Remove the parentheses} \\ \text{in the numerator.}\end{array}$$

$$= \frac{4n^2 + 6}{(n + 3)(n + 1)^2} \qquad \text{Combine like terms.}$$

$$= \frac{2(2n^2 + 3)}{(n + 3)(n + 1)^2} \qquad \text{Factor the numerator.}$$

b. The LCD is $3(y - 1)(y + 2)$.

$$\frac{y}{y - 1} - \frac{y}{3} - \frac{4}{y + 2}$$

$$= \frac{y \cdot 3(y + 2)}{(y - 1) \cdot 3(y + 2)} - \frac{y \cdot (y - 1)(y + 2)}{3 \cdot (y - 1)(y + 2)} - \frac{4 \cdot 3(y - 1)}{(y + 2) \cdot 3(y - 1)}$$

$$= \frac{3y^2 + 6y}{3(y - 1)(y + 2)} - \frac{y^3 + y^2 - 2y}{3(y - 1)(y + 2)} - \frac{12y - 12}{3(y - 1)(y + 2)}$$

$$= \frac{(3y^2 + 6y) - (y^3 + y^2 - 2y) - (12y - 12)}{3(y - 1)(y + 2)}$$

$$= \frac{3y^2 + 6y - y^3 - y^2 + 2y - 12y + 12}{3(y - 1)(y + 2)}$$

$$= \frac{-y^3 + 2y^2 - 4y + 12}{3(y - 1)(y + 2)}$$

EXAMPLE 11

A car gets c mpg in the city and d mpg more on the highway. Determine the amount of gas the car uses on a drive of x mi in the city and y mi on the highway, written as a single rational expression.

Solution   In the city, the car is driven for x mi at c mpg. So the amount of gas consumed is $\frac{x}{c}$ gal. On the highway, the car is driven y mi at $(c + d)$ mpg. Therefore $\frac{y}{c + d}$ gal are consumed. Altogether, the amount of gas used is:

$$\frac{x}{c} + \frac{y}{c + d} = \frac{x}{c} \cdot \frac{c + d}{c + d} + \frac{y}{c + d} \cdot \frac{c}{c} \qquad \text{The LCD is } c(c + d).$$

$$= \frac{x(c + d) + yc}{c(c + d)}$$

So the car uses $\dfrac{x(c + d) + yc}{c(c + d)}$ gal of gas.

PRACTICE 11

To find the percent change for the cost of an item, retailers can use the expression

$$100\left(\frac{C_1}{C_0} - 1\right),$$

where C_1 is the new cost and C_0 is the old cost. Write as a single rational expression.

Perform the indicated operation. Simplify, if possible.

1. $\dfrac{5a}{12} + \dfrac{11a}{12}$

2. $\dfrac{y}{5} + \dfrac{4y}{5}$

3. $\dfrac{5t}{3} - \dfrac{2t}{3}$

4. $\dfrac{2x}{15} - \dfrac{8x}{15}$

5. $\dfrac{10}{x} + \dfrac{1}{x}$

6. $\dfrac{4}{a} + \dfrac{3}{a}$

7. $\dfrac{6}{7y} - \dfrac{1}{7y}$

8. $\dfrac{-7}{8x} + \dfrac{3}{8x}$

9. $\dfrac{5x}{2y} + \dfrac{x}{2y}$

10. $\dfrac{4a}{3b} + \dfrac{8a}{3b}$

11. $\dfrac{2p}{5q} - \dfrac{3p}{5q}$

12. $\dfrac{x}{6y} - \dfrac{11x}{6y}$

13. $\dfrac{2}{x+1} + \dfrac{7}{x+1}$

14. $\dfrac{8}{p+q} + \dfrac{1}{q+p}$

15. $\dfrac{5}{x+2} - \dfrac{9}{2+x}$

16. $\dfrac{10}{r+s} - \dfrac{14}{s+r}$

17. $\dfrac{a}{a+3} + \dfrac{1}{a+3}$

18. $\dfrac{p}{p-1} + \dfrac{3}{p-1}$

19. $\dfrac{3x}{x-8} + \dfrac{2x+1}{x-8}$

20. $\dfrac{4p}{p-5} + \dfrac{p-2}{p-5}$

21. $\dfrac{7x+1}{5x+2} - \dfrac{3x}{5x+2}$

22. $\dfrac{2x}{2x-1} - \dfrac{x+1}{2x-1}$

23. $\dfrac{9x+17}{2x+5} - \dfrac{3x+2}{2x+5}$

24. $\dfrac{5a+1}{2a+3} - \dfrac{a-5}{2a+3}$

25. $\dfrac{-7+5n}{3n-1} + \dfrac{7n+3}{3n-1}$

26. $\dfrac{-5+8x}{5x-1} + \dfrac{4-3x}{5x-1}$

27. $\dfrac{x^2-1}{x^2-4x-2} - \dfrac{x^2-x+3}{x^2-4x-2}$

28. $\dfrac{9+3x-x^2}{x^2+x+1} + \dfrac{x^2-5}{x^2+x+1}$

29. $\dfrac{x}{x^2-3x+2} + \dfrac{2}{x^2-3x+2} + \dfrac{x^2-4x}{x^2-3x+2}$

30. $\dfrac{a}{a^2-5a+4} + \dfrac{2}{a^2-5a+4} + \dfrac{a^2-4a}{a^2-5a+4}$

31. $\dfrac{2x-1}{3x^2-x+2} + \dfrac{8}{3x^2-x+2} - \dfrac{3x}{3x^2-x+2}$

32. $\dfrac{4y-3}{y^2+7y+1} - \dfrac{y}{y^2+7y+1} + \dfrac{6-y}{y^2+7y+1}$

The following expressions represent denominators of rational expressions. Find their LCD.

33. $5(x+2)$ and $3(x+2)$

34. $9(4c-1)$ and $6(4c-1)$

35. $(p-3)(p+8)$ and $(p-3)(p-8)$

36. $(b+c)(5b)$ and $(b+c)(2b)$

37. $t, t+3$, and $t-3$

38. $s, 4s$, and $s+5$

39. $t^2+7t+10$ and t^2-25

40. n^2-1 and n^2+6n-7

41. $3s^2-11s+6$ and $3s^2+4s-4$

42. $2x^2+x-15$ and $-2x^2+9x-10$

Write each pair of rational expressions in terms of their LCD.

43. $\dfrac{1}{3x}$ and $\dfrac{5}{4x^2}$

44. $\dfrac{2}{5y^3}$ and $\dfrac{3}{y^2}$

45. $\dfrac{5}{2a^2}$ and $\dfrac{a-3}{7ab}$

46. $\dfrac{x+2}{4xy}$ and $\dfrac{7}{6x^2}$

47. $\dfrac{8}{n(n+1)}$ and $\dfrac{5}{(n+1)^2}$

48. $\dfrac{2}{c(c-3)^2}$ and $\dfrac{4}{(c-3)}$

49. $\dfrac{3n}{4n+4}$ and $\dfrac{2n}{n^2-1}$

50. $\dfrac{4y}{y^2-1}$ and $\dfrac{y}{2y+2}$

51. $\dfrac{2n}{n^2+6n+5}$ and $\dfrac{3n}{n^2+2n-15}$

52. $\dfrac{7t}{t^2-2t+1}$ and $\dfrac{4t}{t^2-5t+4}$

C *Perform the indicated operations. Simplify, if possible.*

53. $\dfrac{5}{3x}+\dfrac{1}{2x}$

54. $\dfrac{3}{4a}+\dfrac{2}{5a}$

55. $\dfrac{2}{3x^2}-\dfrac{5}{6x}$

56. $\dfrac{9}{7c^2}-\dfrac{3}{5c^3}$

57. $\dfrac{-2}{3x^2y}+\dfrac{4}{3xy^2}$

58. $\dfrac{6}{p^2q}+\dfrac{8}{pq^2}$

59. $\dfrac{1}{x+1}+\dfrac{1}{x-1}$

60. $\dfrac{2}{a+b}+\dfrac{2}{a-b}$

61. $\dfrac{p+6}{3}-\dfrac{2p+1}{7}$

62. $\dfrac{b+9}{5}-\dfrac{5-7b}{2}$

63. $x-\dfrac{10-4x}{2}$

64. $\dfrac{-2t+1}{8}-3t$

65. $\dfrac{3a+1}{6a}-\dfrac{a^2-2}{2a^2}$

66. $\dfrac{y^2-2}{2y^2}-\dfrac{2y-7}{4y}$

67. $\dfrac{a^2}{a-1}-\dfrac{1}{1-a}$

68. $\dfrac{6}{x-4}-\dfrac{x}{4-x}$

69. $\dfrac{4}{c-4}+\dfrac{c}{4-c}$

70. $\dfrac{a}{a-b}+\dfrac{b}{b-a}$

71. $\dfrac{x-5}{x+1}-\dfrac{x+2}{x}$

72. $\dfrac{p+4}{p}-\dfrac{p+3}{p-1}$

73. $\dfrac{4x-5}{x-4}+\dfrac{1-3x}{4-x}$

74. $\dfrac{5n}{1-n}+\dfrac{3n-2}{n-1}$

75. $\dfrac{5x}{x^2+x-2}+\dfrac{6}{x+2}$

76. $\dfrac{p}{p^2-3p+2}+\dfrac{4}{p-1}$

77. $\dfrac{4}{3n-9}-\dfrac{n}{n^2+2n-15}$

78. $\dfrac{-2x}{x^2+7x+12}+\dfrac{5}{4x+16}$

79. $\dfrac{2}{t+5}-\dfrac{t+6}{25-t^2}$

80. $\dfrac{4}{x-2}-\dfrac{x-1}{4-x^2}$

81. $\dfrac{4x}{x^2+2x+1}-\dfrac{2x+5}{x^2+4x+3}$

82. $\dfrac{y-1}{y^2-4y+4}+\dfrac{3y}{y^2-y-2}$

83. $\dfrac{2t-1}{2t^2+t-3}+\dfrac{2}{t-1}$

84. $\dfrac{4}{y-2}+\dfrac{3y-1}{y^2-6y+8}$

85. $\dfrac{4x}{x-1}+\dfrac{2}{3x}+\dfrac{x}{x^2-1}$

86. $\dfrac{c}{c+2}+\dfrac{3}{4c}+\dfrac{2c}{c^2-4}$

87. $\dfrac{5y}{3y-1}-\dfrac{3}{y-4}+\dfrac{y+1}{3y^2-13y+4}$

88. $\dfrac{6x-1}{2x+1}+\dfrac{x}{x-1}-\dfrac{4x}{2x^2-x-1}$

89. $\dfrac{a-1}{(a+3)^2} - \dfrac{2a-3}{a+3} - \dfrac{a}{4a+12}$

90. $\dfrac{y}{8y-16} + \dfrac{3y+4}{y-2} - \dfrac{y-3}{(y-2)^2}$

Mixed Practice

Write each pair of rational expressions in terms of their LCD.

91. $\dfrac{p+3}{6p^2}$ and $\dfrac{5}{8pq}$

92. $\dfrac{3}{y(y-2)}$ and $\dfrac{8}{(y-2)^2}$

Perform the indicated operations. Simplify, if possible.

93. $\dfrac{9}{xy^2} + \dfrac{6}{x^2y}$

94. $\dfrac{7}{y-3} - \dfrac{y}{3-y}$

95. $\dfrac{c-3}{c} - \dfrac{c-2}{c+1}$

96. $\dfrac{-2a}{a^2+7a+10} + \dfrac{4}{3a+15}$

97. $\dfrac{b+3}{b^2-2b-3} - \dfrac{4}{b^2-6b+9}$

98. $\dfrac{2x}{x^2-1} + \dfrac{x}{x+1} + \dfrac{2}{3x}$

99. $\dfrac{6}{m+n} - \dfrac{11}{n+m}$

100. $\dfrac{5x-2}{3x-1} - \dfrac{2x}{3x-1}$

101. $\dfrac{r}{r^2-r-6} + \dfrac{3}{r^2-r-6} + \dfrac{r^2-5r}{r^2-r-6}$

102. If $j^2 - 4$ and $j^2 + 2j - 8$ represent denominators of rational expressions, find their LCD.

Applications

Solve.

103. The position of an object thrown upward is given by the expression $vt + \dfrac{gt^2}{2}$. Here v is the initial velocity, t is the time, and g is a constant related to gravity. Write as a single rational expression.

104. Under certain conditions, the chances of an event happening can be represented by the rational expression $\dfrac{f}{t}$, whereas $\dfrac{t-f}{t}$ represents the chances of the event *not* happening. In this expression, f is the number of favorable outcomes and t is the total number of outcomes. What is the sum of these two expressions?

105. A bank pays an interest rate r compounded annually on all account balances. If a customer wanted the balance in an account to be \$1000 at the end of one year, she would need to have a current balance of $\dfrac{1000}{1+r}$ dollars. However, if she were willing to wait two years for the balance to reach \$1000, then her current balance would need to be only $\dfrac{1000}{(1+r)^2}$ dollars. Express the difference between the quantities $\dfrac{1000}{1+r}$ dollars and $\dfrac{1000}{(1+r)^2}$ dollars as a single rational expression.

106. An expression to find the length of base b of a trapezoid is $\dfrac{2A}{h} - a$. Another expression is $\dfrac{2A-ah}{h}$. Explain whether these two expressions are equivalent.

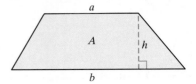

107. A trucker drove 20 mi at a speed r mph, and then returned at double the speed. How long did the whole trip take? Write the answer as a single rational expression.

108. A train traveled m mi at a speed of s mph. A bus following the same route traveled 5 mph slower. How much longer did the bus take than the train to make this trip? Write the answer as a single rational expression.

109. To maintain a checking account, a bank charges a customer \$3 per month and \$0.10 per check. For x checks, the cost per check is $\left(\dfrac{3}{x} + 0.1 \right)$ dollars. Write this cost as a single rational expression.

110. In baseball, an expression for *runners per inning* is $\dfrac{H}{I} + \dfrac{W}{I}$, where H represents the number of hits, W the number of walks, and I the number of innings pitched. Explain whether the expression $\dfrac{H + W}{2I}$ is equivalent to $\dfrac{H}{I} + \dfrac{W}{I}$.

• Check your answers on page A-24

MIND Stretchers

Groupwork

1. When we break a given rational expression into *partial fractions*, we are expressing it as the sum of two other rational expressions. For instance, we could write the rational expression $\dfrac{4x + 1}{x^2 + x}$ as $\dfrac{3}{x + 1} + \dfrac{1}{x}$.

a. Confirm that $\dfrac{4x + 1}{x^2 + x} = \dfrac{3}{x + 1} + \dfrac{1}{x}$.

b. Find the missing numerators so that $\dfrac{5x + 1}{x^2 - 1} = \dfrac{?}{x - 1} + \dfrac{?}{x + 1}$.

Critical Thinking

2. Find the following product.

$$\left(1 + \frac{1}{n}\right)\left(1 + \frac{1}{n + 1}\right)\left(1 + \frac{1}{n + 2}\right)\left(1 + \frac{1}{n + 3}\right) \cdots \left(1 + \frac{1}{n + 99}\right)\left(1 + \frac{1}{n + 100}\right)$$

Mathematical Reasoning

3. A formula sometimes used to add arithmetic fractions is:

$$\frac{p}{q} + \frac{r}{s} = \frac{ps + qr}{qs}$$

a. Working with a pair of arithmetic fractions of your choice, confirm that this formula gives the correct sum by comparing it to the answer derived using an alternative method.

b. Consider the left side of this formula as a sum of rational expressions. Explain why the right side of the formula is correct.

Cultural Note

The ancient Egyptians expressed most fractions as the sum of *unit fractions* (fractions whose numerators are 1). For instance, they would write the fraction $\frac{2}{5}$ as $\frac{1}{3} + \frac{1}{15}$. More generally, the rational expression $\frac{x+y}{xy}$ can be written as the sum of the unit fractions $\frac{1}{x}$ and $\frac{1}{y}$.

(*Source:* Jan Gullberg, *From the Birth of Numbers*, W. W. Norton, 1997)

In arithmetic, a *complex fraction* is a fraction that in turn contains one or more fractions in its numerator, denominator, or both:

$$\text{Main fraction line} \longrightarrow \cfrac{\dfrac{1}{3} \; \longleftarrow \text{ Fraction in the numerator}}{\dfrac{4}{5} \; \longleftarrow \text{ Fraction in the denominator}}$$

Such fractions can be put in standard form as the quotient of two integers by recalling that the main fraction line represents division:

$$\cfrac{\dfrac{1}{3}}{\dfrac{4}{5}} = \frac{1}{3} \div \frac{4}{5} = \frac{1}{3} \cdot \frac{5}{4} = \frac{5}{12}$$

A comparable expression in algebra is called *a complex rational expression* (or a *complex algebraic fraction*). Such an expression contains a rational expression in its numerator, denominator, or both. Some examples are shown below:

$$\frac{x^2 + 5x - 2}{\dfrac{3}{4}} \qquad \frac{7}{\dfrac{a + b}{2}} \qquad \cfrac{\dfrac{3n}{2} + \dfrac{1}{n}}{\dfrac{n - 4}{n^2}}$$

As in the case of an arithmetic complex fraction, a complex rational expression is usually simplified, that is, written as the quotient of two polynomials with no common factors.

We consider two methods of simplifying a complex rational expression: the *division method* and the *LCD method.*

The Division Method

In the division method, we begin by writing the numerator and denominator as single rational expressions, simplifying if necessary. Then, we write the complex rational expression as the numerator divided by the denominator.

For instance, suppose we want to simplify the complex rational expression below:

$$\cfrac{\dfrac{x}{2} + \dfrac{1}{2}}{\dfrac{x^2}{3} + \dfrac{x}{3}}$$

We first write the numerator and denominator as single rational expressions.

$$\cfrac{\dfrac{x}{2} + \dfrac{1}{2}}{\dfrac{x^2}{3} + \dfrac{x}{3}} = \cfrac{\dfrac{x + 1}{2}}{\dfrac{x^2 + x}{3}} \quad \longleftarrow \quad$$

Add the rational expressions in the numerator.

Add the rational expressions in the denominator.

$$= \frac{x + 1}{2} \div \frac{x^2 + x}{3} \qquad$$ Write the expression as the numerator divided by the denominator.

$$= \frac{x+1}{2} \cdot \frac{3}{x^2+x} \qquad \text{Take the reciprocal of the divisor and change the operation to multiplication.}$$

$$= \frac{\overset{1}{\cancel{x+1}}}{2} \cdot \frac{3}{x(\cancel{x+1})}_{1} \qquad \text{Factor and divide out the common factor.}$$

$$= \frac{3}{2x} \qquad \text{Simplify.}$$

So we can conclude that the original complex rational expression $\dfrac{\dfrac{x}{2} + \dfrac{1}{2}}{\dfrac{x^2}{3} + \dfrac{x}{3}}$ is equivalent to the rational expression $\dfrac{3}{2x}$.

To Simplify a Complex Rational Expression: Division Method

- Write the numerator and denominator as single rational expressions in simplified form.
- Write the expression as the numerator divided by the denominator.
- Divide.
- Simplify, if possible.

Let's use the division method in the following examples.

EXAMPLE 1

Simplify.

a. $\dfrac{\dfrac{2}{x}}{\dfrac{6}{x^3}}$

b. $\dfrac{y}{\dfrac{y}{2} + \dfrac{y^2}{3}}$

Solution

a. Neither the numerator nor the denominator of the complex rational expression can be simplified. So we write the expression as the numerator divided by the denominator.

$$\frac{\dfrac{2}{x}}{\dfrac{6}{x^3}} = \frac{2}{x} \div \frac{6}{x^3}$$

$$= \frac{2}{x} \cdot \frac{x^3}{6} \qquad \text{Take the reciprocal of the divisor and change the operation to multiplication.}$$

$$= \frac{\overset{1}{2}}{\underset{1}{\cancel{x}}} \cdot \frac{\overset{x^2}{\cancel{x^3}}}{\underset{3}{6}} \qquad \text{Divide out the common factors.}$$

$$= \frac{x^2}{3} \qquad \text{Simplify.}$$

PRACTICE 1

Simplify.

a. $\dfrac{\dfrac{3}{x^4}}{\dfrac{5}{x}}$

b. $\dfrac{2x}{\dfrac{x^2}{4} + \dfrac{x}{2}}$

b. Let's express the denominator as a single rational expression.

$$\frac{y}{\dfrac{y}{2} + \dfrac{y^2}{3}} = \frac{y}{\dfrac{y \cdot 3}{2 \cdot 3} + \dfrac{y^2 \cdot 2}{3 \cdot 2}}$$ Add the rational expressions in the denominator. The LCD is 6.

$$= \frac{y}{\dfrac{3y + 2y^2}{6}}$$

$$= y \div \frac{3y + 2y^2}{6}$$ Write the expression as the numerator divided by the denominator.

$$= y \cdot \frac{6}{3y + 2y^2}$$ Take the reciprocal of the divisor, and change the operation to multiplication.

$$= \overset{1}{\cancel{y}} \cdot \frac{6}{\underset{1}{\cancel{y}}(3 + 2y)}$$ Factor, and divide out the common factor.

$$= \frac{6}{3 + 2y}$$ Simplify.

EXAMPLE 2

Simplify: $\dfrac{1 + \dfrac{1}{x}}{1 - \dfrac{1}{x}}$

Solution We begin by simplifying the numerator and denominator so that each is a single rational expression.

$$\frac{1 + \dfrac{1}{x}}{1 - \dfrac{1}{x}} = \frac{\dfrac{x + 1}{x}}{\dfrac{x - 1}{x}}$$

$$= \frac{x + 1}{x} \div \frac{x - 1}{x}$$

$$= \frac{x + 1}{x} \cdot \frac{x}{x - 1}$$

$$= \frac{x + 1}{\underset{1}{\cancel{x}}} \cdot \frac{\overset{1}{\cancel{x}}}{x - 1}$$

$$= \frac{x + 1}{x - 1}$$

PRACTICE 2

Simplify: $\dfrac{2 - \dfrac{1}{n}}{2 + \dfrac{1}{n}}$

The LCD Method

Now, we consider the LCD method of simplifying a complex rational expression. In this method, we multiply the numerator and denominator of the complex rational expression by the LCD of all rational expressions that appear within it.

Let's simplify the complex rational expression $\dfrac{\dfrac{x}{2} + \dfrac{1}{2}}{\dfrac{x^2}{3} + \dfrac{x}{3}}$ given on the first page of this section using the LCD method. First, we must find the LCD of all the rational expressions in the numerator and denominator. The denominators within the complex rational expression are 2 and 3, so the LCD is 6.

$$\frac{\dfrac{x}{2} + \dfrac{1}{2}}{\dfrac{x^2}{3} + \dfrac{x}{3}} = \frac{\left(\dfrac{x}{2} + \dfrac{1}{2}\right) \cdot 6}{\left(\dfrac{x^2}{3} + \dfrac{x}{3}\right) \cdot 6}$$ Multiply the numerator and denominator by the LCD 6.

$$= \frac{\dfrac{x}{2} \cdot 6 + \dfrac{1}{2} \cdot 6}{\dfrac{x^2}{3} \cdot 6 + \dfrac{x}{3} \cdot 6}$$ Use the distributive property in the numerator and in the denominator.

$$= \frac{3x + 3}{2x^2 + 2x}$$ Simplify.

$$= \frac{3\overset{1}{\cancel{(x+1)}}}{2x\underset{1}{\cancel{(x+1)}}}$$ Factor and divide out the common factor.

$$= \frac{3}{2x}$$ Simplify.

Note that the division method and the LCD method result in the same answer.

To Simplify a Complex Rational Expression: LCD Method

- Find the LCD of all the rational expressions *within* both the numerator and denominator.

- Multiply the numerator and denominator of the complex rational expression by this LCD.

- Simplify, if possible.

EXAMPLE 3

Simplify.   **a.** $\dfrac{\dfrac{3}{y^2}}{\dfrac{1}{y}}$ **b.** $\dfrac{n^2}{\dfrac{1}{n} + \dfrac{2}{n^2}}$

Solution

a. $\dfrac{\dfrac{3}{y^2}}{\dfrac{1}{y}} = \dfrac{\dfrac{3}{y^2} \cdot y^2}{\dfrac{1}{y} \cdot y^2}$ Multiply by the LCD y^2.

$$= \frac{3}{y}$$ Simplify.

PRACTICE 3

Simplify.

a. $\dfrac{\dfrac{4}{x}}{\dfrac{2}{x^3}}$

b. $\dfrac{\dfrac{1}{y} + \dfrac{3}{y^2}}{2y}$

b. $\dfrac{n^2}{\dfrac{1}{n} + \dfrac{2}{n^2}} = \dfrac{n^2 \cdot n^2}{\left(\dfrac{1}{n} + \dfrac{2}{n^2}\right) \cdot n^2}$ Multiply by the LCD n^2.

$= \dfrac{n^2 \cdot n^2}{\dfrac{1}{n} \cdot n^2 + \dfrac{2}{n^2} \cdot n^2}$ Use the distributive property.

$= \dfrac{n^4}{n + 2}$ Simplify.

EXAMPLE 4

Simplify.

a. $\dfrac{3 + \dfrac{1}{x^2}}{3 - \dfrac{1}{x}}$ **b.** $\dfrac{\dfrac{1}{x^2} - \dfrac{1}{y^2}}{\dfrac{3}{x} - \dfrac{3}{y}}$

Solution

a. Multiply the numerator and denominator by the LCD x^2.

$$\dfrac{3 + \dfrac{1}{x^2}}{3 - \dfrac{1}{x}} = \dfrac{\left(3 + \dfrac{1}{x^2}\right) \cdot x^2}{\left(3 - \dfrac{1}{x}\right) \cdot x^2}$$

$$= \dfrac{3 \cdot x^2 + \dfrac{1}{x^2} \cdot x^2}{3 \cdot x^2 - \dfrac{1}{x} \cdot x^2}$$

$$= \dfrac{3x^2 + 1}{3x^2 - x}$$

b. Multiply the numerator and denominator by the LCD x^2y^2.

$$\dfrac{\dfrac{1}{x^2} - \dfrac{1}{y^2}}{\dfrac{3}{x} - \dfrac{3}{y}} = \dfrac{\left(\dfrac{1}{x^2} - \dfrac{1}{y^2}\right) \cdot x^2y^2}{\left(\dfrac{3}{x} - \dfrac{3}{y}\right) \cdot x^2y^2}$$

$$= \dfrac{\dfrac{1}{x^2}(x^2y^2) - \dfrac{1}{y^2}(x^2y^2)}{\dfrac{3}{x}(x^2y^2) - \dfrac{3}{y}(x^2y^2)}$$

$$= \dfrac{y^2 - x^2}{3xy^2 - 3x^2y} = \dfrac{\overset{1}{\cancel{(y - x)}}(y + x)}{3xy\underset{1}{\cancel{(y - x)}}}$$

$$= \dfrac{y + x}{3xy}$$

PRACTICE 4

Express in simplest terms.

a. $\dfrac{4 + \dfrac{1}{y}}{4 - \dfrac{1}{y^2}}$

b. $\dfrac{\dfrac{1}{2a^2} - \dfrac{1}{2b^2}}{\dfrac{5}{a} + \dfrac{5}{b}}$

EXAMPLE 5

A resistor is an electrical device such as a lightbulb that offers resistance to the flow of electricity. When two lightbulbs with resistance R_1 and R_2 are connected in a certain kind of circuit, the combined resistance is given by the following expression:

$$\dfrac{1}{\dfrac{1}{R_1} + \dfrac{1}{R_2}}$$

Simplify this expression for the combined resistance.

Solution

$$\frac{1}{\dfrac{1}{R_1} + \dfrac{1}{R_2}} = \frac{1 \cdot R_1 R_2}{\left(\dfrac{1}{R_1} + \dfrac{1}{R_2}\right) \cdot R_1 R_2}$$

$$= \frac{1 \cdot R_1 R_2}{\dfrac{1}{R_1} \cdot R_1 R_2 + \dfrac{1}{R_2} \cdot R_1 R_2} = \frac{R_1 R_2}{R_2 + R_1}$$

PRACTICE 5

The *harmonic mean* is a kind of average. The harmonic mean of the three numbers a, b, and c is given by the expression $\dfrac{3}{\dfrac{1}{a} + \dfrac{1}{b} + \dfrac{1}{c}}$.

Simplify this complex rational expression.

Ⓐ *Simplify.*

1. $\dfrac{\dfrac{x}{5}}{\dfrac{x^2}{10}}$

2. $\dfrac{\dfrac{3}{s^2}}{\dfrac{s^3}{2}}$

3. $\dfrac{\dfrac{a+1}{2}}{\dfrac{a-1}{2}}$

4. $\dfrac{\dfrac{n-1}{n}}{\dfrac{n+1}{2n}}$

5. $\dfrac{3+\dfrac{1}{x}}{3-\dfrac{1}{x^2}}$

6. $\dfrac{1+\dfrac{1}{y^2}}{8-\dfrac{1}{y}}$

7. $\dfrac{\dfrac{1}{3d}-\dfrac{1}{d^2}}{d-\dfrac{9}{d}}$

8. $\dfrac{a-\dfrac{4}{a}}{\dfrac{1}{a^2}+\dfrac{1}{2a}}$

9. $\dfrac{1-\dfrac{4y^2}{x^2}}{3+\dfrac{6y}{x}}$

10. $\dfrac{\dfrac{p}{q}+2}{\dfrac{p^2}{q^2}-4}$

11. $\dfrac{\dfrac{2}{y}-\dfrac{1}{5}}{\dfrac{5}{y}-1}$

12. $\dfrac{\dfrac{2}{t}-\dfrac{3}{t^2}}{10+\dfrac{2}{t}}$

13. $\dfrac{1+\dfrac{4}{x}+\dfrac{4}{x^2}}{1+\dfrac{5}{x}+\dfrac{6}{x^2}}$

14. $\dfrac{1-\dfrac{1}{a^2}}{4-\dfrac{5}{a}+\dfrac{1}{a^2}}$

15. $\dfrac{3+\dfrac{1}{y+1}}{5-\dfrac{1}{y+1}}$

16. $\dfrac{2+\dfrac{1}{b-1}}{2-\dfrac{1}{b-1}}$

17. $\dfrac{\dfrac{x}{4}-\dfrac{x}{8}}{\dfrac{2}{y^2}+\dfrac{2}{y}}$

18. $\dfrac{\dfrac{n}{2}+\dfrac{n}{3}}{\dfrac{1}{m}-\dfrac{1}{m^2}}$

19. $\dfrac{\dfrac{x}{x+1}-\dfrac{2}{x}}{\dfrac{x}{3}}$

20. $\dfrac{\dfrac{y}{5}}{\dfrac{y}{y+2}+\dfrac{3}{y}}$

Mixed Practice

Simplify.

21. $\dfrac{\dfrac{m+2}{3m}}{\dfrac{m-1}{m}}$

22. $\dfrac{3-\dfrac{1}{d-2}}{3+\dfrac{1}{d-2}}$

23. $\dfrac{\dfrac{4}{u}-\dfrac{2}{u+1}}{\dfrac{u}{2}}$

24. $\dfrac{x-\dfrac{9}{x}}{\dfrac{1}{3x}-\dfrac{1}{x^2}}$

25. $\dfrac{\dfrac{3}{y^2}+\dfrac{4}{y}}{6+\dfrac{3}{y}}$

26. $\dfrac{3-\dfrac{1}{b}-\dfrac{2}{b^2}}{1-\dfrac{1}{b^2}}$

Applications

Solve.

27. An expression from the study of electricity is

$$\frac{V}{\dfrac{1}{2R} + \dfrac{1}{2R + 2}},$$

where V represents voltage and R represents resistance. Simplify this expression.

28. At the beginning of the year, C dollars are invested in one business and D dollars in another. By the end of the year, the rate of return for the first investment was r and for the second, s. The value of the portfolio at the end of the year was therefore $C(1 + r) + D(1 + s)$. To show that the overall rate of return for the two investments was $\dfrac{Cr + Ds}{C + D}$, check that the expression

$$\frac{C(1 + r) + D(1 + s)}{1 + \dfrac{Cr + Ds}{C + D}}$$

when simplified is the total initial investment.

29. *The earned run average* (ERA) is a statistic used in baseball to represent the average number of earned runs that a pitcher allows. A pitcher's ERA can be calculated using the expression

$$\frac{E}{\dfrac{I}{9}},$$

where E stands for the number of earned runs a pitcher gave up after pitching I innings. Simplify this complex rational expression.

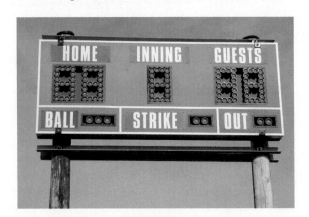

30. The expression

$$\frac{\dfrac{m}{c}}{1 - \dfrac{p^2}{c^2}}$$

is important in the design of airplanes. Write this expression in simplified form.

31. On a trip, an airplane flew at an average speed of a mph, returning on the same route at an average speed of b mph. The plane's average speed for the round trip is given by the following complex rational expression:

$$\frac{2}{\dfrac{1}{a} + \dfrac{1}{b}}$$

Simplify.

32. An object travels at speed s for a distance d and later travels at speed S for distance D. The following expression represents the average speed for the entire trip:

$$\frac{d + D}{\dfrac{d}{s} + \dfrac{D}{S}}$$

Simplify this expression.

33. The weight of an object decreases as its distance from the Earth's surface increases. Suppose an object weighs w kg at sea level. Then at a height of h km above sea level, it will weigh (in kilograms)

$$\frac{w}{\left(1 + \dfrac{h}{6400}\right)^2}.$$

Show that $\dfrac{w}{\left(1 + \dfrac{h}{6400}\right)^2}$ and $\dfrac{6400^2 w}{(6400 + h)^2}$ are equivalent expressions.

34. The sound from a car horn as it approaches an observer seems to change its frequency. This is an example of the Doppler effect, which is studied in physics. The observer will hear a sound with frequency

$$\frac{S}{\dfrac{S - v}{f}},$$

where f is the actual frequency of the sound, S is the speed of sound, and v is the speed of the approaching object. Simplify the expression.

• Check your answers on page A-24.

MINDStretchers

Research

1. The harmonic mean is related to musical harmony. Either in your college library or on the Web, investigate the relationship between music and the harmonic mean. Summarize your findings in a few sentences.

Writing

2. Select a complex rational expression of your choice and simplify it using (a) the division method and (b) the LCD method. Which method do you prefer? Explain why.

Groupwork

3. Find a complex rational expression with numerator $\dfrac{4}{y}$ that simplifies to $\dfrac{3}{x + 5}$.

OBJECTIVES

(A) To solve an equation involving rational expressions

(B) To solve applied problems involving rational equations

What Rational Equations Are and Why They Are Important

So far in this chapter, we have discussed rational *expressions*. Now, let's consider rational (or fractional) *equations*, that is, equations that contain one or more rational expressions. Here are some examples:

$$\frac{n}{2} - 3 = \frac{n}{5} \qquad \frac{1}{t} + \frac{1}{3} = \frac{1}{2} \qquad \frac{3}{x+1} - \frac{1}{8} = \frac{5}{x^2 - 1}$$

Many situations involving rates of work, motion, and proportions can be modeled by rational equations.

Solving Rational Equations

In solving rational equations, it is important not to confuse a rational expression with a rational equation.

$$\frac{4x}{3} + \frac{x}{3} \qquad\qquad \frac{4x}{3} + \frac{x}{3} = 5$$

Rational *expression*   Rational *equation*

The key to solving a rational equation is to *clear the equation of rational expressions*. We do this by first determining their *least common denominator*, and then by multiplying both sides of the equation by the LCD.

EXAMPLE 1

Solve and check: $\dfrac{x}{2} + \dfrac{x}{6} = \dfrac{2}{3}$

Solution The rational expressions in this equation are $\dfrac{x}{2}, \dfrac{x}{6}$, and $\dfrac{2}{3}$.

The denominators of these expressions are 2, 6, and 3, so the LCD is 6. To clear the equation of rational expressions, we multiply each side of the equation by 6.

$$\frac{x}{2} + \frac{x}{6} = \frac{2}{3}$$

$$6 \cdot \left(\frac{x}{2} + \frac{x}{6} \right) = 6 \cdot \frac{2}{3} \qquad \text{Multiply each side of the equation by the LCD.}$$

$$6 \cdot \frac{x}{2} + 6 \cdot \frac{x}{6} = 6 \cdot \frac{2}{3} \qquad \text{Use the distributive property.}$$

$$3x + x = 4 \qquad \text{Simplify.}$$

$$4x = 4 \qquad \text{Combine like terms.}$$

$$x = 1 \qquad \text{Divide each side by 4.}$$

PRACTICE 1

Solve and check: $\dfrac{y}{2} - \dfrac{y}{3} = \dfrac{1}{12}$

Check

$$\frac{x}{2} + \frac{x}{6} = \frac{2}{3}$$

$$\frac{1}{2} + \frac{1}{6} \stackrel{?}{=} \frac{2}{3} \qquad \text{Substitute 1 for } x.$$

$$\frac{3}{6} + \frac{1}{6} \stackrel{?}{=} \frac{2}{3}$$

$$\frac{4}{6} \stackrel{?}{=} \frac{2}{3}$$

$$\frac{2}{3} = \frac{2}{3} \qquad \text{True}$$

So the solution to the given equation is 1.

The following rule describes the general procedure for solving equations of this type:

To Solve a Rational Equation

- Find the LCD of all rational expressions in the equation.
- Multiply each side of the equation by the LCD.
- Solve the equation.
- Check the solution(s) in the original equation.

EXAMPLE 2

Solve and check: $\dfrac{4}{n} - \dfrac{n+1}{3} = 1$

Solution The LCD of the rational expressions in the equation is $3n$.

$$\frac{4}{n} - \frac{n+1}{3} = 1$$

$$3n \cdot \left(\frac{4}{n} - \frac{n+1}{3} \right) = 3n \cdot 1 \qquad \begin{array}{l}\text{Multiply each side of the equation} \\ \text{by the LCD.}\end{array}$$

$$3n \cdot \frac{4}{n} - 3n \cdot \frac{n+1}{3} = 3n \cdot 1 \qquad \begin{array}{l}\text{Distribute } 3n \text{ on the} \\ \text{left side of the equation.}\end{array}$$

$$\overset{1}{3n} \cdot \frac{4}{\underset{1}{n}} - \overset{1}{3n} \cdot \frac{n+1}{\underset{1}{3}} = 3n \cdot 1$$

$$12 - n(n+1) = 3n \qquad \text{Simplify.}$$

$$12 - n^2 - n = 3n \qquad \text{Use the distributive property.}$$

$$n^2 + n + 3n - 12 = 0$$

$$n^2 + 4n - 12 = 0 \qquad \text{Write in standard form.}$$

$$(n+6)(n-2) = 0 \qquad \text{Factor the left side of the equation.}$$

$$n+6 = 0 \quad \text{or} \quad n-2 = 0 \qquad \text{Set each factor equal to 0.}$$

$$n = -6 \qquad\qquad n = 2$$

PRACTICE 2

Solve and check: $\dfrac{x-2}{5} - 1 = -\dfrac{2}{x}$

EXAMPLE 2 (continued)

Check

Substitute -6 for n.

$$\frac{4}{n} - \frac{n+1}{3} = 1$$

$$\frac{4}{-6} - \frac{-6+1}{3} \stackrel{?}{=} 1$$

$$-\frac{2}{3} - \left(-\frac{5}{3}\right) \stackrel{?}{=} 1$$

$$-\frac{2}{3} + \frac{5}{3} \stackrel{?}{=} 1$$

$$\frac{3}{3} \stackrel{?}{=} 1$$

$$1 = 1 \qquad \text{True}$$

Substitute 2 for n.

$$\frac{4}{n} - \frac{n+1}{3} = 1$$

$$\frac{4}{2} - \frac{2+1}{3} \stackrel{?}{=} 1$$

$$2 - 1 \stackrel{?}{=} 1$$

$$1 = 1 \qquad \text{True}$$

Our check confirms that the solutions are -6 and 2.

EXAMPLE 3

Solve and check: $\dfrac{2}{x+3} + \dfrac{1}{x-3} = -\dfrac{6}{x^2-9}$

Solution First, we find the LCD of the rational expressions in this equation, which is $x^2 - 9$, or $(x+3)(x-3)$. Then, we muliply each side of the equation by the LCD.

$$\frac{2}{x+3} + \frac{1}{x-3} = -\frac{6}{x^2-9}$$

$$\frac{2}{x+3} + \frac{1}{x-3} = -\frac{6}{(x+3)(x-3)}$$

$$(x+3)(x-3)\cdot\frac{2}{x+3} + (x+3)(x-3)\cdot\frac{1}{x-3} = (x+3)(x-3)\cdot\frac{-6}{(x+3)(x-3)}$$

$$\overset{1}{\cancel{(x+3)}}(x-3)\cdot\frac{2}{\underset{1}{\cancel{x+3}}} + (x+3)\overset{1}{\cancel{(x-3)}}\cdot\frac{1}{\underset{1}{\cancel{x-3}}} = \overset{1}{\cancel{(x+3)}}\overset{1}{\cancel{(x-3)}}\cdot\frac{-6}{\underset{1}{\cancel{(x+3)}}\underset{1}{\cancel{(x-3)}}}$$

$$2(x-3) + (x+3) = -6$$
$$2x - 6 + x + 3 = -6$$
$$3x - 3 = -6$$
$$3x = -3$$
$$x = -1$$

Check

$$\frac{2}{x+3} + \frac{1}{x-3} = -\frac{6}{x^2-9}$$

$$\frac{2}{-1+3} + \frac{1}{-1-3} \stackrel{?}{=} -\frac{6}{(-1)^2-9} \qquad \text{Substitute } -1 \text{ for } x.$$

$$\frac{2}{2} + \frac{1}{-4} \stackrel{?}{=} -\frac{6}{-8} \qquad \text{Simplify.}$$

$$\frac{2}{2} + \left(-\frac{1}{4}\right) \stackrel{?}{=} -\left(-\frac{6}{8}\right)$$

$$1 - \frac{1}{4} \stackrel{?}{=} \frac{3}{4}$$

$$\frac{3}{4} = \frac{3}{4} \qquad \text{True}$$

So the solution is -1.

PRACTICE 3

Solve and check: $\dfrac{4}{y + 2} + \dfrac{2}{y - 1} = \dfrac{12}{y^2 + y - 2}$

When multiplying each side of a rational equation by a variable expression, the resulting equation may have a solution that does not satisfy the original equation. If such a result makes a denominator in the original equation 0, then the rational expression is undefined. These **extraneous solutions** are *not* solutions of the original equation. So in solving rational equations, it is particularly important to check all possible solutions.

EXAMPLE 4

Solve and check: $\dfrac{x^2}{x - 2} = \dfrac{4}{x - 2}$

Solution   The LCD of the rational expressions is $x - 2$.

$$\frac{x^2}{x - 2} = \frac{4}{x - 2}$$

$$(x - 2) \cdot \frac{x^2}{x - 2} = (x - 2) \cdot \frac{4}{x - 2} \qquad \text{Multiply each side of the equation by the LCD.}$$

$$\overset{1}{(x - 2)} \cdot \frac{x^2}{\underset{1}{x - 2}} = \overset{1}{(x - 2)} \cdot \frac{4}{\underset{1}{x - 2}}$$

$$x^2 = 4$$

$$x^2 - 4 = 0$$

$$(x + 2)(x - 2) = 0 \qquad \text{Factor the left side of the equation.}$$

$$x + 2 = 0 \quad \text{or} \quad x - 2 = 0 \qquad \text{Set each factor equal to 0.}$$

$$x = -2 \qquad\qquad x = 2 \qquad \text{Solve for } x.$$

Check

Substitute -2 for x.

$$\frac{x^2}{x - 2} = \frac{4}{x - 2}$$

$$\frac{(-2)^2}{-2 - 2} \overset{?}{=} \frac{4}{-2 - 2}$$

$$-\frac{4}{4} = -\frac{4}{4} \qquad \text{True}$$

Substitute 2 for x.

$$\frac{x^2}{x - 2} = \frac{4}{x - 2}$$

$$\frac{2^2}{2 - 2} \overset{?}{=} \frac{4}{2 - 2}$$

$$\frac{4}{0} = \frac{4}{0} \qquad \text{Undefined}$$

Since we get undefined fractions when we substitute 2 for x in the original equation, 2 is *not* a solution. So the solution is -2. Without solving, how could we have known that 2 is *not* a solution of the original equation? Explain.

PRACTICE 4

Solve and check: $x = \dfrac{9}{x + 3} + \dfrac{3x}{x + 3}$

Rational equations play an important role in **work problems**. In these problems, we typically want to compute how long it will take to complete a task.

The key to solving a work problem is to determine the *rate of work*, that is, the fraction of the task that is completed in one unit of time. For instance, if it takes a secretary 5 hr to type a report, then the secretary's rate of work—the fraction of the report typed in 1 hr— would be $\frac{1}{5}$. If it takes a painter 6 hr to paint a room, then $\frac{1}{6}$ of the room would be painted in an hour so that $\frac{1}{6}$ is the painter's rate of work.

Consider the following example.

EXAMPLE 5

Two company employees, one senior and the other junior, are responsible for carrying out a project. If the two employees had worked alone, the junior employee would have completed the project in 6 hr and the senior employee would have completed it in 4 hr. How long would it take the two employees to carry out the project working together?

Solution   Since this is a work problem, we can use the following equation to determine how long it will take the two employees to carry out the project working together:

Rate of work · Time worked = Part of the task completed

Using this equation, we can set up a table. Let t represent the time it takes the two employees to complete the project working together. Note here that the task to be completed is the project.

	Rate of Work · Time Worked =		Part of the Task Completed
Senior employee	$\frac{1}{4}$	t	$\frac{1}{4} \cdot t$
Junior employee	$\frac{1}{6}$	t	$\frac{1}{6} \cdot t$

Since the sum of the parts of the task completed must equal one complete task, we have:

$$\frac{1}{4} \cdot t + \frac{1}{6} \cdot t = 1$$

$$\frac{t}{4} + \frac{t}{6} = 1$$

To solve this equation, we multiply each side by the LCD 12.

$$12 \cdot \frac{t}{4} + 12 \cdot \frac{t}{6} = 12 \cdot 1$$

$$3t + 2t = 12$$

$$5t = 12$$

$$t = 2\tfrac{2}{5} = 2.4$$

So working together, it would take the two employees 2.4 hr (or 2 hr 24 min) to complete the project. We can confirm by checking, which we leave as an exercise.

PRACTICE 5

A town water tank has two pumps. Working alone, the less powerful pump can fill the tank in 10 hr, whereas the more powerful pump can fill it in 6 hr. How long will it take both pumps working together to fill the tank?

Another application of rational equations is **motion problems**. Recall that in these problems, an object moves at a constant rate r for time t and travels a distance d. These three quantities are related by the following formula:

$$d = rt$$

If we solve this equation for t, we get $t = \dfrac{d}{r}$, which is the formula that we apply in Example 6. This example involves the time a round trip took, which is the sum of the time going and the time returning.

EXAMPLE 6

A family on vacation drove 50 mi to a hotel, and then returned home following the same route. Because of lighter traffic, the family drove at twice the speed going to the hotel as compared to returning home. If the round trip took 3 hr, at what speed did the family return home?

Solution   We are looking for the family's speed returning home. Let's represent this unknown quantity by r. Since the family traveled twice as fast going to the hotel as returning home, their speed going to the hotel must have been $2r$. We use the formula $\dfrac{d}{r} = t$ to find the time traveled in each direction, and set up a table.

	Distance	÷ Rate	= Time
Going to the hotel	50	$2r$	$\dfrac{50}{2r}$
Returning home	50	r	$\dfrac{50}{r}$

Since it is given that the round trip took 3 hr in all, we write:

$$\frac{50}{2r} + \frac{50}{r} = 3$$

To solve this rational equation, we multiply each side of the equation by the LCD $2r$.

$$2r \cdot \frac{50}{2r} + 2r \cdot \frac{50}{r} = 2r \cdot 3$$

$$\overset{1}{2r} \cdot \frac{50}{2\overset{}{r}} + \overset{1}{2r} \cdot \frac{50}{\overset{}{r}} = 2r \cdot 3$$

$$50 + 100 = 6r$$

$$6r = 150$$

$$r = 25$$

So the family returned home at 25 mph. We can confirm this solution by checking.

PRACTICE 6

A business executive traveled 1800 mi by jet, continuing the trip an additional 300 mi on a propeller plane. The speed of the jet was 3 times that of the propeller plane. If the entire trip took 6 hr, what was the speed of the propeller plane?

Some rational equations are formulas or literal equations that relate two or more variables. Consider the following example.

EXAMPLE 7

The formula $\dfrac{1}{f} = \dfrac{1}{p} + \dfrac{1}{q}$ gives the focal length f of a lens, where p is the distance between the lens and an object and q is the distance between the image and the lens.

a. Solve this equation for f.

b. Use the formula found in part (a) to find the value of f when $p = 40$ cm and $q = 10$ cm.

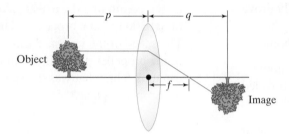

Solution

a. Our task is to solve the equation for f.

$$\frac{1}{f} = \frac{1}{p} + \frac{1}{q}$$

$$fpq \cdot \frac{1}{f} = fpq \cdot \left(\frac{1}{p} + \frac{1}{q}\right)$$

$$fpq \cdot \frac{1}{f} = fpq \cdot \frac{1}{p} + fpq \cdot \frac{1}{q}$$

$$f\!\!\!\!/pq \cdot \frac{1}{f\!\!\!\!/} = f p\!\!\!\!/q \cdot \frac{1}{p\!\!\!\!/} + f p q\!\!\!\!/ \cdot \frac{1}{q\!\!\!\!/}$$

$$pq = fq + fp$$

Since we are solving for f, we factor it out on the right side of the equation.

$$pq = f(q + p)$$

$$f = \frac{pq}{p + q}$$

b. Substituting 40 for p and 10 for q in the formula $f = \dfrac{pq}{p + q}$ we get:

$$f = \frac{40 \cdot 10}{40 + 10}$$

$$= \frac{400}{50}$$

$$= 8$$

So the focal length of the lens is 8 cm.

PRACTICE 7

Economists are interested in how the free market works. Suppose that a commodity sells at the price x dollars per unit. The number of units sold, called the *demand D,* can be modeled by the following equation:

$$D = \frac{500}{x} + 50$$

a. Solve for x in terms of D.

b. Find the value of x when the demand is 450 units.

7.5 Exercises

FOR
EXTRA
HELP

MyMathLab *Math* XL
          PRACTICE   WATCH   READ   REVIEW

A *Solve and check.*

1. $\dfrac{y}{2} + \dfrac{7}{10} = -\dfrac{4}{5}$

2. $\dfrac{x}{6} - \dfrac{1}{10} = \dfrac{2}{5}$

3. $\dfrac{1}{t} - \dfrac{7}{3} = -\dfrac{1}{3}$

4. $\dfrac{5}{p} + \dfrac{1}{3} = \dfrac{6}{p}$

5. $x + \dfrac{1}{x} = 2$

6. $y + \dfrac{2}{y} = 3$

7. $\dfrac{t+1}{2t-1} - \dfrac{5}{7} = 0$

8. $\dfrac{b-3}{3b-2} - \dfrac{4}{5} = 0$

9. $\dfrac{t-2}{3} = 4$

10. $\dfrac{4}{x+1} = 3$

11. $\dfrac{2n}{n+1} = \dfrac{2}{n+1} + 1$

12. $\dfrac{4}{s-3} - \dfrac{3s}{s-3} = 2$

13. $\dfrac{5x}{x+1} = \dfrac{x^2}{x+1} + 2$

14. $\dfrac{x^2}{x-2} = \dfrac{2x}{x-2} + 3$

15. $\dfrac{x}{x-3} - \dfrac{6}{x} = 1$

16. $\dfrac{2}{t} + \dfrac{t}{t+1} = 1$

17. $1 + \dfrac{4}{x^2} = \dfrac{4}{x}$

18. $\dfrac{1}{2x} + \dfrac{3}{x^2} = 1$

◉ 19. $\dfrac{2}{p+1} - \dfrac{1}{p-1} = \dfrac{2p}{p^2-1}$

20. $\dfrac{2}{y+2} - \dfrac{5}{2-y} = \dfrac{3y}{y^2-4}$

◉ 21. $\dfrac{3}{x} - \dfrac{1}{x+4} = \dfrac{5}{x^2+4x}$

22. $\dfrac{2}{y-2} + \dfrac{3}{y} = \dfrac{4}{-2y+y^2}$

23. $1 - \dfrac{6x}{(x-4)^2} = \dfrac{2x}{x-4}$

24. $\dfrac{4x}{x+1} - 2 = -\dfrac{3x}{(x+1)^2}$

25. $\dfrac{n+1}{n^2+2n-3} = \dfrac{n}{n+3} - \dfrac{1}{n-1}$

26. $\dfrac{7}{x-5} - \dfrac{5x+6}{x^2-3x-10} = \dfrac{x}{x+2}$

Mixed Practice

Solve and check.

27. $\dfrac{2}{b+3} = \dfrac{5b}{b^2-9} - \dfrac{3}{3-b}$

28. $\dfrac{3}{r} + \dfrac{1}{4} = \dfrac{12}{r}$

29. $\dfrac{6m}{m+1} - 3 = \dfrac{8m}{(m+1)^2}$

30. $\dfrac{a-3}{3(a-2)} - \dfrac{3}{8} = 0$

31. $2 - \dfrac{2x}{x+3} = \dfrac{6}{x+1}$

32. $\dfrac{8}{s} + \dfrac{s}{s+4} = 1$

Applications

B *Solve.*

33. Two brothers share a house. It would take the younger brother working alone 45 min to clean the attic, whereas it would take the older brother only 30 min. If the two brothers worked together, how long would it take them to clean the attic?

34. One crew can pave a road in 24 hr and a second crew can do the same job in 16 hr. How long would it take for the two crews working together to pave the road?

35. A clerical worker takes 4 times as long to finish a job as it does an executive secretary. Working together, it takes them 3 hr to finish the job. How long would it take the clerical worker, working alone, to finish the job?

36. One pipe can fill a tank in 1 min. A second pipe takes 2 min to fill the same tank. Working together, how long will it take both pipes to fill the tank?

37. A car traveled twice as fast on a dry road as it did making the return trip on a slippery road. The trip was 60 mi each way, and the round trip took 3 hr. What was the speed of the car on the dry road?

38. A cyclist rode uphill and then turned around and made the same trip downhill. The downhill speed was 3 times the uphill speed. If the trip each way was 18 mi and the entire trip lasted 2 hr, at what speed was the cyclist going uphill?

39. Steam exerts pressure on a pipe. In the formula

$$p = \frac{P}{LD},$$

p represents the pressure per square inch in the pipe, P stands for the total pressure in the pipe, L is the pipe's length, and D is the pipe's diameter. Solve this formula for D.

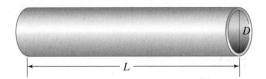

40. To approximate the appropriate dosage of medicine for a child, doctors sometimes use Young's rule:

$$C = \frac{aA}{a + 12},$$

where C represents the child's dosage, a stands for the child's age, and A is the adult dosage. Solve this formula for a.

41. The city of Miami has two main interstate highways, I-75 and I-95. Suppose that an evacuation was planned for a portion of the city due to an impending hurricane. If all traffic moves smoothly, assume that I-75 alone could handle the evacuation traffic in x hr and I-95 in $(x + 3)$ hr, where x depends on the size of the evacuation.
a. How long would it take to carry out the evacuation using both highways?
b. How long would the evacuation take if $x = 21$ hr?

42. While you were in a car, the first 10 megabytes (MB) of a 170 MB download to your cell phone was on a 3G network. The remainder of the download, which altogether took m min, was on a 4G network that downloads 8 times as fast.
a. How fast (in megabytes per minute) was the 4G network?

b. How fast was this network if $m = 100$ min?

• Check your answers on page A-24.

MINDStretchers

Groupwork

1. Working with a partner, show that there are no positive real numbers a and b for which the equation $\dfrac{1}{a} + \dfrac{1}{b} = \dfrac{1}{a + b}$ is true.

Mathematical Reasoning

2. Use your knowledge of systems of equations to solve the following system:

$$\textbf{(1)} \quad \frac{1}{x} + \frac{3}{y} = 1$$

$$\textbf{(2)} \quad -\frac{4}{x} + \frac{3}{y} = -3$$

Historical

3. A riddle from 1500 years ago deals with the age of the Greek mathematician Diophantus when he died. According to this riddle, "... *his boyhood lasted* $\dfrac{1}{6}$ *of his life; he married after* $\dfrac{1}{7}$ *more; his beard grew after* $\dfrac{1}{12}$ *more, and his son was born 5 years later; the son lived to half his father's age, and the father died 4 years after the son.*" Solve this riddle by determining Diophantus' age at his death. (*Source:* www.history.mcs.st-and.ac.uk)

Recall in Section 7.5 that to solve rational equations we multiplied every term by the LCD in order to clear the denominators. However, rational equations of a particular form, where one rational expression equals another, lend themselves to an alternative method for clearing denominators. Here are some examples of equations of that form:

$$\frac{x}{12} = \frac{2}{3} \qquad \frac{5}{6} = \frac{10}{n} \qquad \frac{1}{n+1} = \frac{3}{n-1}$$

Two rational expressions are equal to one another when their *cross products* are equal. So rather than multiplying by the LCD, we can simply *set the cross products equal*. Let's apply each method to the first equation.

LCD Method

$$\frac{x}{12} = \frac{2}{3}$$

$$12 \cdot \frac{x}{12} = 12 \cdot \frac{2}{3} \qquad \text{Multiply each side of the equation by the LCD 12.}$$

$$\overset{1}{\cancel{12}} \cdot \frac{x}{\cancel{12}} = \overset{4}{\cancel{12}} \cdot \frac{2}{\cancel{3}} \qquad \text{Divide out common factors.}$$

$$1 \cdot x = 4 \cdot 2 \qquad \text{Simplify.}$$

$$x = 8$$

Cross-Product Method

$$\frac{x}{12} = \frac{2}{3}$$

$$\frac{x}{12} \diagup\!\!\!\!\!\diagdown \frac{2}{3} \qquad \text{Cross products}$$

$$3 \cdot x = 12 \cdot 2 \qquad \text{Cross multiply.}$$

$$3x = 24 \qquad \text{Simplify.}$$

$$\frac{3x}{3} = \frac{24}{3} \qquad \text{Divide each side of the equation by 3.}$$

$$x = 8$$

So in both methods, our solution appears to be 8, but is it correct? We can check by substituting in the original equation:

$$\frac{x}{12} = \frac{2}{3}$$

$$\frac{8}{12} \overset{?}{=} \frac{2}{3}$$

$$\frac{2}{3} = \frac{2}{3} \qquad \text{True}$$

The example above suggests that when two rational expressions are equal to one another, we can solve the equation by either the LCD method or the cross-product method. In this section, we solve these types of equations using the cross-product method.

EXAMPLE 1

Solve: $\dfrac{5}{6} = \dfrac{10}{n}$

Solution We clear the denominators by cross multiplying.

$$\dfrac{5}{6} = \dfrac{10}{n} \qquad \text{Cross multiply.}$$

$$5n = 60$$

$$\dfrac{5n}{5} = \dfrac{60}{5} \qquad \text{Divide both sides by 5.}$$

$$n = 12$$

Check

$$\dfrac{5}{6} = \dfrac{10}{n}$$

$$\dfrac{5}{6} \stackrel{?}{=} \dfrac{10}{12} \qquad \text{Substitute 12 for } n.$$

$$\dfrac{5}{6} = \dfrac{5}{6} \qquad \text{True}$$

PRACTICE 1

Solve: $\dfrac{4}{5} = \dfrac{p}{10}$

Equations of the type shown in Example 1 commonly arise from **ratio and proportion problems**.

A *ratio* is a comparison of two numbers, expressed as a quotient. For instance, if we compare the numbers 5 and 8, their ratio would be 5 to 8, written as $\dfrac{5}{8}$ or 5 : 8. Here are some other ratios:

- A ratio of 6 women to 4 men, written as $\dfrac{6}{4}$, or in simplest terms, $\dfrac{3}{2}$

- A ratio of 230 congressmen voting for a bill to 180 congressmen voting against the same bill, written as $\dfrac{230}{180}$, or $\dfrac{23}{18}$

A ratio of quantities with different units is called a *rate*. Here are some examples:

- A basketball team's record of 8 wins and 2 losses, written as $\dfrac{8 \text{ wins}}{2 \text{ losses}}$

- A wage of $20 per hour, written as $\dfrac{20 \text{ dollars}}{1 \text{ hour}}$

When two ratios (or rates) are equal, we say that they are *in proportion*. For instance, the ratios $\dfrac{1}{2}$ and $\dfrac{4}{8}$ are in proportion. The equation $\dfrac{1}{2} = \dfrac{4}{8}$ is called *a proportion*.

DEFINITION

A **proportion** is a statement that two ratios, $\dfrac{a}{b}$ and $\dfrac{c}{d}$, are equal, written $\dfrac{a}{b} = \dfrac{c}{d}$, where $b \neq 0$ and $d \neq 0$.

Now, let's consider an example of *solving a proportion problem*. Suppose that you saved $500 in 4 mo. At this rate, how long would it take you to save for a computer that costs $750?

To answer this question, we can write a proportion in which the rates compare the amount of savings in a given amount of time. We want to find the amount of time corresponding to a savings of $750. Let's call this missing value x.

$$\text{Savings in dollars} \rightarrow \frac{500}{4} = \frac{750}{x} \leftarrow \text{Savings in dollars}$$
$$\text{Time in months} \rightarrow \phantom{\frac{500}{4}} \phantom{\frac{750}{x}} \leftarrow \text{Time in months}$$

We then set the cross products equal in order to find the missing value.

$$\frac{500}{4} = \frac{750}{x}$$
$$500 \cdot x = 4 \cdot 750 \qquad \text{Cross multiply.}$$
$$500x = 3000$$
$$x = \frac{3000}{500}$$
$$x = 6$$

So it will take you 6 months to save for the computer.

To Solve a Proportion

- Find the cross products, and set them equal.
- Solve the resulting equation.
- Check the solution in the original equation.

EXAMPLE 2

The actual length of the bedroom shown in the floor plan to the right is 15 ft. What is the actual width of the room?

$l = 2$ in.

$w = 1.5$ in.

Solution   The actual measurements and the floor plan measurements are in proportion.

$$\text{The actual width in feet} \rightarrow \frac{w}{15} = \frac{1.5}{2} \leftarrow \text{The floor plan width in inches}$$
$$\text{The actual length in feet} \rightarrow \phantom{\frac{w}{15}} \phantom{\frac{1.5}{2}} \leftarrow \text{The floor plan length in inches}$$
$$2w = (15)(1.5) \qquad \text{Cross multiply.}$$
$$\frac{2w}{2} = \frac{22.5}{2} \qquad \text{Divide each side of the equation by 2.}$$
$$w = 11.25$$

Check

$$\frac{w}{15} = \frac{1.5}{2}$$
$$\frac{11.25}{15} \stackrel{?}{=} \frac{1.5}{2}$$
$$0.75 = 0.75 \qquad \text{True}$$

Since our answer checks, we can conclude that the bedroom is actually 11.25 ft wide.

PRACTICE 2

It takes 80 lb of sodium hydroxide to neutralize 98 lb of sulfuric acid. At this rate, how many pounds of sodium hydroxide are needed to neutralize 49 lb of sulfuric acid?

Would we get the same answer if we had set up the proportion in Example 2 as $\dfrac{w}{1.5} = \dfrac{15}{2}$? Explain.

EXAMPLE 3

A 125-lb adult gets a dosage of 2 mL of a particular drug. At this rate, how much additional drug does a 175-lb adult require?

Solution   We begin by writing a proportion in which we use x to represent the amount of additional drug required.

$$\dfrac{2}{125} = \dfrac{2 + x}{175} \quad \begin{array}{l} \leftarrow \text{Amount of drug in milliliters} \\ \leftarrow \text{Weight in pounds} \end{array}$$

$$2 \cdot 175 = 125(2 + x) \qquad \text{Cross multiply.}$$

$$350 = 250 + 125x \qquad \text{Use the distributive property.}$$

$$125x = 100 \qquad \text{Combine like terms.}$$

$$x = \dfrac{100}{125} \qquad \text{Divide each side of the equation by 125.}$$

$$x = 0.8$$

Check

$$\dfrac{2}{125} = \dfrac{2 + x}{175}$$

$$\dfrac{2}{125} \stackrel{?}{=} \dfrac{2 + 0.8}{175} \qquad \text{Substitute 0.8 for } x.$$

$$\dfrac{2}{125} \stackrel{?}{=} \dfrac{2.8}{175} \qquad \text{Simplify.}$$

$$0.016 = 0.016 \qquad \text{True}$$

So an additional 0.8 mL of the drug is required for a 175-lb adult.

PRACTICE 3

A homeowner pays $900 a month on a $100,000 mortgage. If instead she had a $75,000 mortgage, how much less money would she be paying per month at the same interest rate?

Another application of ratio and proportion is the geometric topic of **similar triangles**. These are triangles with the same shape but not necessarily the same size.

In the diagram to the right, $\triangle ABC$ and $\triangle DEF$ are similar. These triangles have corresponding angles that have equal measures, that is:

$$m\angle A = m\angle D$$
$$m\angle B = m\angle E$$
$$m\angle C = m\angle F$$

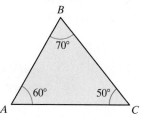

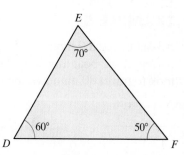

Note that in similar triangles, corresponding sides are opposite equal angles. Each side of $\triangle ABC$ corresponds to a side of $\triangle DEF$.

$$\overline{AB} \text{ corresponds to } \overline{DE}.$$
$$\overline{BC} \text{ corresponds to } \overline{EF}.$$
$$\overline{AC} \text{ corresponds to } \overline{DF}.$$

It can be shown that the lengths of the corresponding sides of similar triangles are in proportion:

$$\dfrac{AB}{DE} = \dfrac{BC}{EF} = \dfrac{AC}{DF} \quad \begin{array}{l} \leftarrow \triangle ABC \\ \leftarrow \triangle DEF \end{array}$$

We can use this relationship to find missing sides of similar triangles, as the following example illustrates:

EXAMPLE 4

In the following diagram, $\triangle PQR$ and $\triangle STU$ are similar. Find x, the length of side $\overline{SU}$.

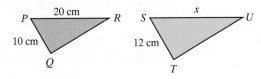

Solution   Since the triangles are similar, corresponding sides are in proportion. To solve for x, let's set up a proportion involving the lengths of sides $\overline{PQ}$, $\overline{ST}$, $\overline{PR}$, and $\overline{SU}$.

$$\frac{PQ}{ST} = \frac{PR}{SU}$$

$$\frac{10}{12} = \frac{20}{x} \qquad \text{Substitute the given values.}$$

$$10x = 240 \qquad \text{Cross multiply.}$$

$$x = \frac{240}{10}$$

$$x = 24$$

So the length of $\overline{SU}$ is 24 cm.

PRACTICE 4

Triangle ABC and triangle DEC are similar. Find x, the length of side $\overline{DE}$.

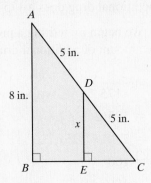

In Section 7.5, we discussed motion problems involving rational equations. Here, we continue the discussion of this type of problem. In Example 5, we again use the formula $t = \dfrac{d}{r}$ in order to set two travel times equal to one another.

EXAMPLE 5

A boat travels 60 mph in still water. Find the speed of the river's current if the boat traveled 70 mi up the river in the same time that it took to travel 80 mi down the river.

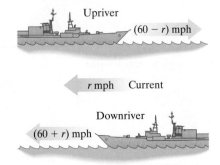

PRACTICE 5

The speed of the jet stream was 300 mph. When flying in the direction of the jet stream, a plane flies 1000 mi in the same time that it takes to fly 250 mi against the jet stream. Find the speed of the plane in still air.

Solution Let r represent the speed of the river's current. When traveling downriver, the speed of the current is added to that of the boat. When traveling upriver, the speed of the river is subtracted from that of the boat. Applying the formula $t = \dfrac{d}{r}$, we can set up the following table:

| | d $\div$ r $=$ t | | |
	Distance	Rate	Time
Upriver	70	$60 - r$	$\dfrac{70}{60 - r}$
Downriver	80	$60 + r$	$\dfrac{80}{60 + r}$

We are told that these two times are equal, so we write the following proportion and solve for r:

$$\frac{80}{60 + r} = \frac{70}{60 - r}$$

$$80(60 - r) = 70(60 + r) \qquad \text{Cross multiply.}$$

$$4800 - 80r = 4200 + 70r$$

$$600 = 150r$$

$$r = 4$$

The speed of the river's current was 4 mph. How would you check that this answer is correct?

Solve and check.

1. $\dfrac{x}{10} = \dfrac{4}{5}$

2. $\dfrac{2}{9} = \dfrac{x}{27}$

3. $\dfrac{n}{100} = \dfrac{4}{5}$

4. $\dfrac{22}{35} = \dfrac{44}{x}$

5. $\dfrac{8}{7} = \dfrac{s}{21}$

6. $\dfrac{1}{v} = \dfrac{10}{3}$

7. $\dfrac{8 + x}{12} = \dfrac{22}{36}$

8. $\dfrac{75}{20} = \dfrac{30}{p - 1}$

9. $\dfrac{y + 3}{14} = \dfrac{y}{7}$

10. $\dfrac{n - 6}{3} = \dfrac{n}{5}$

11. $\dfrac{x - 1}{8} = \dfrac{x + 1}{12}$

12. $\dfrac{3}{t - 2} = \dfrac{9}{t + 2}$

13. $\dfrac{x}{8} = \dfrac{2}{x}$

14. $\dfrac{4}{n} = \dfrac{n}{16}$

15. $\dfrac{2}{y} = \dfrac{y - 4}{16}$

16. $\dfrac{5}{x + 2} = \dfrac{x}{7}$

17. $\dfrac{a}{a + 3} = \dfrac{4}{5a}$

18. $\dfrac{1}{3n} = \dfrac{n}{n + 2}$

19. $\dfrac{y + 1}{y + 6} = \dfrac{y}{y + 6}$

20. $\dfrac{2x + 3}{x - 3} = \dfrac{3x}{x - 3}$

Mixed Practice

Solve and check.

21. $\dfrac{3}{r} = \dfrac{r - 4}{7}$

22. $\dfrac{a}{a + 3} = \dfrac{3}{2a}$

23. $\dfrac{4}{w - 3} = \dfrac{7}{w + 3}$

24. $\dfrac{7}{9} = \dfrac{v}{72}$

25. $\dfrac{7}{m} = \dfrac{63}{6}$

26. $\dfrac{45}{40} = \dfrac{18}{t + 1}$

Applications

Solve.

27. The owner of a small business is considering the purchase of a laser printer that can print 40 pages in 2 min. At this rate, how long would it take to print a 25-page report?

28. In a mayoral election, 48 out of every 100 voters voted for a certain candidate. How many of the 200,000 voters chose the candidate?

29. A trucker drives 60 mi on 8 gal of gas. At the same rate, how many gallons of gas will it take him to drive 120 mi?

30. A chauffeur took 2 hr longer to drive 275 mi than he took to drive 165 mi. If his speed was the same on both trips, at what speed was he driving?

31. A runner traveled 3 mi in the same time that a cyclist traveled 10 mi. The speed of the cyclist was 14 mph greater than that of the runner. What was the cyclist's speed?

32. A train travels 225 mi in the same time that a bus travels 200 mi. If the speed of the train is 5 mph greater than that of the bus, find the speed of the bus.

33. A train goes 30 mph faster than a bus. The train travels 400 mi in the same time as the bus goes 250 mi. What are their speeds?

34. A pilot flies 600 mi with a tailwind of 20 mph. Against the wind, he flies only 500 mi in the same amount of time. What is the speed of the plane in still air?

35. $\triangle ABC$ and $\triangle DEF$ are similar. Find y, the length of side $\overline{AB}$.

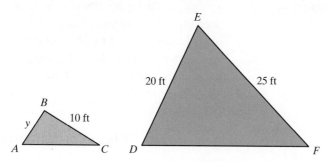

36. In the following map of Kansas, the triangle formed by Salina, Hays, and Great Bend is similar to the triangle formed by Dodge City, Wichita, and Great Bend. Find the distance between Dodge City and Great Bend.

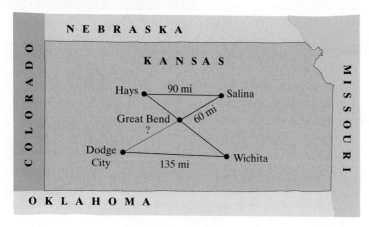

37. When a 6-foot-tall man is 8 ft away from a tree, the tip of his shadow and the tip of the tree's shadow meet 4 ft behind him. The two right triangles shown are similar. Find the height of the tree.

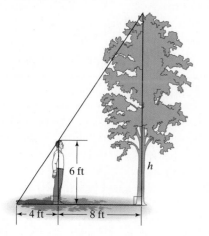

38. Similar triangles can be used to find how far a ship is from the shore. Find the value of x in the diagram.

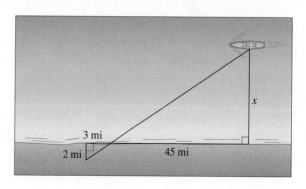

39. There are 20 guests at a party. If 4 more women were to arrive at the party, then 50% of the guests would be women. How many women are there at the party?

40. At a local bank, the simple interest rate on a savings account is 2% higher than the interest rate for a checking account. After one year, a deposit in the checking account would earn interest of $60, whereas the same deposit would earn interest of $120 in the savings account. Find the interest rate on each account.

• Check your answers on page A-25.

MIND*Stretchers*

Research

1. The mathematician Eratosthenes headed the great library at Alexandria in Egypt more than 2000 years ago. In addition to his work on prime numbers, Eratosthenes used the method of ratio and proportion to approximate the circumference of the Earth. Either in your college library or on the Web, investigate how he did this and how successful he was. Summarize your findings.

Mathematical Reasoning

2. If $\dfrac{a}{b} = \dfrac{c}{d}$, can you show that $\dfrac{a+b}{b} = \dfrac{c+d}{d}$? Explain.

Groupwork

3. Working with a partner, make up an equation where one rational expression equals another.

   a. Solve the equation using the LCD method.
   b. Solve the same equation using the cross-product method.
   c. Which method do you prefer? Explain why.

Concept/Skill	Description	Example
[7.1] **Rational expression**	An algebraic expression $\frac{P}{Q}$ that can be written as the quotient of two polynomials, P and Q, where $Q \neq 0$.	$\frac{2x}{x+5}$, where $x + 5 \neq 0$
[7.1] **Equivalent rational expressions**	Rational expressions that have the same value, no matter what value replaces the variable.	$\frac{2}{3}$ and $\frac{2x}{3x}$, where $x \neq 0$
[7.1] **To find an equivalent rational expression**	• Multiply the numerator and denominator of $\frac{P}{Q}$ by the same polynomial R. $$\frac{P}{Q} = \frac{PR}{QR},$$ where $Q \neq 0$ and $R \neq 0$.	$\frac{3}{5y+1} = \frac{3 \cdot y}{(5y+1) \cdot y} = \frac{3y}{5y^2 + y}$
[7.1] **To simplify a rational expression**	• Factor the numerator and denominator. • Divide out any common factors in the numerator and denominator.	$\frac{3n^2 - 3}{n^2 + n - 2}$ $$= \frac{3(\overset{1}{\cancel{n-1}})(n+1)}{(\underset{1}{\cancel{n-1}})(n+2)} = \frac{3(n+1)}{n+2}$$
[7.2] **To multiply rational expressions**	• Factor the numerators and denominators. • Divide the numerators and denominators by all common factors. • Multiply the remaining factors in the numerators and the remaining factors in the denominators.	$\frac{6x}{x^2 - 4x} \cdot \frac{x^2 - 16}{3x}$ $$= \frac{6x}{x(x-4)} \cdot \frac{(x+4)(x-4)}{3x}$$ $$= \frac{\overset{2}{\cancel{6x}}^{\,1}}{\underset{1}{x}\underset{1}{(\cancel{x-4})}} \cdot \frac{(x+4)\overset{1}{(\cancel{x-4})}}{\underset{1}{3x}}$$ $$= \frac{2(x+4)}{x}$$
[7.2] **To divide rational expressions**	• Take the reciprocal of the divisor, and change the operation to multiplication. • Follow the rule for multiplying rational expressions.	$\frac{x^2 - x - 6}{3x - 9} \div \frac{4x + 8}{5x}$ $$= \frac{x^2 - x - 6}{3x - 9} \cdot \frac{5x}{4x + 8}$$ $$= \frac{(\overset{1}{\cancel{x+2}})(\overset{1}{\cancel{x-3}})}{3(\underset{1}{\cancel{x-3}})} \cdot \frac{5x}{4(\underset{1}{\cancel{x+2}})}$$ $$= \frac{5x}{12}$$

continued

Concept/Skill	Description	Example
[7.3] **To add (or subtract) rational expressions with the same denominator**	• Add (or subtract) the numerators and keep the same denominator. • Simplify, if possible.	$\dfrac{3n-5}{n+1}+\dfrac{n+6}{n+1}$ $=\dfrac{(3n-5)+(n+6)}{n+1}$ $=\dfrac{3n-5+n+6}{n+1}=\dfrac{4n+1}{n+1}$ $\dfrac{2n^2+4n+6}{n-5}-\dfrac{n^2+10n+1}{n-5}$ $=\dfrac{(2n^2+4n+6)-(n^2+10n+1)}{n-5}$ $=\dfrac{2n^2+4n+6-n^2-10n-1}{n-5}$ $=\dfrac{n^2-6n+5}{n-5}$ $=\dfrac{(n-1)(n\overset{1}{-5})}{(n\underset{1}{-5})}=n-1$
[7.3] **To find the LCD of rational expressions**	• Factor each denominator completely. • Multiply these factors, using for the power of each factor the greatest number of times that it occurs in any of the denominators. The product of all these factors is the LCD.	$\dfrac{x}{x^2-4x+3}$ and $\dfrac{x+5}{2x-2}$ Factor x^2-4x+3: $(x-1)(x-3)$ Factor $2x-2$: $2(x-1)$ LCD $=2(x-1)(x-3)$
[7.3] **To add (or subtract) rational expressions with different denominators**	• Find the LCD of the rational expressions. • Write each rational expression with a common denominator, usually the LCD. • Add (or subtract) the numerators. • Simplify, if possible.	$\dfrac{n}{n^2+3n+2}+\dfrac{3}{2n+2}$ $=\dfrac{n}{(n+1)(n+2)}+\dfrac{3}{2(n+1)}$ $=\dfrac{n\cdot 2}{(n+1)(n+2)\cdot 2}$ $\quad+\dfrac{3\cdot(n+2)}{2(n+1)\cdot(n+2)}$ $=\dfrac{2n}{2(n+1)(n+2)}+\dfrac{3n+6}{2(n+1)(n+2)}$ $=\dfrac{5n+6}{2(n+1)(n+2)}$ $\dfrac{5y}{y^2-9}-\dfrac{2}{y+3}$ $=\dfrac{5y}{(y+3)(y-3)}-\dfrac{2}{y+3}$ $=\dfrac{5y}{(y+3)(y-3)}-\dfrac{2\cdot(y-3)}{(y+3)\cdot(y-3)}$ $=\dfrac{5y}{(y+3)(y-3)}-\dfrac{2y-6}{(y+3)(y-3)}$ $=\dfrac{5y-(2y-6)}{(y+3)(y-3)}$ $=\dfrac{3y+6}{(y+3)(y-3)}=\dfrac{3(y+2)}{(y+3)(y-3)}$

Concept/Skill	Description	Example
[7.4] **Complex rational expression (complex algebraic fraction)**	A rational expression that contains rational expressions in its numerator, denominator, or both.	$$\dfrac{x+1}{\frac{2}{3}},\ \dfrac{\frac{a-b}{4}}{5},$$ $$\dfrac{\frac{6n}{5}-\frac{1}{n}}{\frac{n+3}{n^2}}$$
[7.4] **To simplify a complex rational expression: division method**	• Write the numerator and denominator as single rational expressions in simplified form. • Write the expression as the numerator divided by the denominator. • Divide. • Simplify, if possible.	$$\dfrac{\frac{x}{4}}{\frac{x^2}{3}+\frac{x}{2}}=\dfrac{\frac{x}{4}}{\frac{x^2}{3}\cdot\frac{2}{2}+\frac{x}{2}\cdot\frac{3}{3}}$$ $$=\dfrac{\frac{x}{4}}{\frac{2x^2+3x}{6}}$$ $$=\frac{x}{4}\div\frac{2x^2+3x}{6}$$ $$=\frac{x}{4}\cdot\frac{6}{2x^2+3x}$$ $$=\frac{\overset{1}{x}}{\underset{2}{4}}\cdot\frac{\overset{3}{6}}{\underset{1}{x(2x+3)}}$$ $$=\frac{3}{2(2x+3)}$$
[7.4] **To simplify a complex rational expression: LCD method**	• Find the LCD of all the rational expressions *within* both the numerator and denominator. • Multiply the numerator and denominator of the complex rational expression by this LCD. • Simplify, if possible.	$$\dfrac{1-\frac{1}{y}}{1-\frac{1}{y^2}}=\dfrac{\left(1-\frac{1}{y}\right)\cdot y^2}{\left(1-\frac{1}{y^2}\right)\cdot y^2}$$ $$=\dfrac{1\cdot y^2-\frac{1}{\cancel{y}}\cdot\overset{y}{\cancel{y^2}}}{1\cdot y^2-\frac{1}{\cancel{y^2}}\cdot\overset{1}{\cancel{y^2}}}$$ $$=\frac{y^2-y}{y^2-1}$$ $$=\dfrac{y(\cancel{y-1})}{(y+1)\underset{1}{(\cancel{y-1})}}$$ $$=\frac{y}{y+1}$$

continued

Concept/Skill	Description	Example
[7.5] **To solve a rational equation**	• Find the LCD of all rational expressions in the equation. • Multiply each side of the equation by the LCD. • Solve the equation. • Check your solution(s) in the original equation.	$$\frac{1}{x+3} + \frac{1}{x-3} = \frac{2}{x^2-9}$$ $$\frac{1}{x+3} + \frac{1}{x-3} = \frac{2}{(x+3)(x-3)}$$ $$(x+3)(x-3) \cdot \frac{1}{x+3}$$ $$+ (x+3)(x-3)\frac{1}{x-3}$$ $$= (x+3)(x-3) \cdot \frac{2}{(x+3)(x-3)}$$ $$x - 3 + x + 3 = 2$$ $$2x = 2$$ $$x = 1$$ **Check** Substitute 1 for x. $$\frac{1}{x+3} + \frac{1}{x-3} = \frac{2}{x^2-9}$$ $$\frac{1}{1+3} + \frac{1}{1-3} \overset{?}{=} \frac{2}{1^2-9}$$ $$\frac{1}{4} - \frac{1}{2} \overset{?}{=} -\frac{2}{8}$$ $$-\frac{1}{4} = -\frac{1}{4} \quad \text{True}$$
[7.6] **Proportion**	A statement that two ratios, $\frac{a}{b}$ and $\frac{c}{d}$, are equal, written $\frac{a}{b} = \frac{c}{d}$, where $b \neq 0$ and $d \neq 0$.	The equation $\frac{3}{4} = \frac{10}{x}$ is a proportion, where $\frac{3}{4}$ and $\frac{10}{x}$ are ratios.
[7.6] **To solve a proportion**	• Find the cross products, and set them equal. • Solve the resulting equation. • Check the solution in the original equation.	$$\frac{8}{x-3} = \frac{6}{x+4}$$ $$8(x+4) = 6(x-3)$$ $$8x + 32 = 6x - 18$$ $$2x = -50$$ $$x = -25$$ **Check** Substitute -25 for x. $$\frac{8}{x-3} = \frac{6}{x+4}$$ $$\frac{8}{-25-3} \overset{?}{=} \frac{6}{-25+4}$$ $$\frac{8}{-28} \overset{?}{=} \frac{6}{-21}$$ $$-\frac{2}{7} = -\frac{2}{7} \quad \text{True}$$

Say Why

Fill in each blank.

1. The rational expression $\dfrac{2x-3}{x+4}$ _____ undefined
   is/is not
   when $x = \dfrac{3}{2}$ because _____
   _____.

2. The expression $\dfrac{a^2(3-a)}{a(5+a)}$ _____ in simplest form
   is/is not
   because _____
   _____.

3. The LCD of $\dfrac{5}{6x^3y}$ and $\dfrac{1}{2xy^2z}$ _____ $6x^3y^2z$
   is/is not
   because _____
   _____.

4. The expression $\dfrac{2}{\frac{x}{y}}$ _____ a complex rational
   is/is not
   expression because _____
   _____.

5. The solution $a = 3$ _____ an extraneous solution
   is/is not
   to the equation $1 + \dfrac{9}{a^2} = \dfrac{6}{a}$ because _____
   _____.

6. The proportion $\dfrac{8}{10} = \dfrac{20}{25}$ _____ true because
   is/is not
   _____
   _____.

[7.1]

7. Identify the values for which the given rational expression is undefined.

   a. $\dfrac{4}{x+1}$ b. $\dfrac{6x+12}{x^2-x-6}$

8. Determine whether each pair of rational expressions is equivalent.

   a. $\dfrac{2x}{y} \overset{?}{=} \dfrac{10x^2y}{5xy^2}$ b. $\dfrac{x^2-9}{x^2+6x+9} \overset{?}{=} \dfrac{x-3}{x+3}$

Simplify.

9. $\dfrac{12m}{20m^2}$

10. $\dfrac{15n-18}{9n+6}$

11. $\dfrac{x^2+2x-8}{4-x^2}$

12. $\dfrac{2x^2-3x-20}{3x^2-13x+4}$

[7.2]   *Perform the indicated operation.*

13. $\dfrac{10mn}{3p^2} \cdot \dfrac{9np}{5m^2}$

14. $\dfrac{y-5}{4y+6} \cdot \dfrac{6y+9}{3y-15}$

15. $\dfrac{x+6}{x^2+x-30} \cdot \dfrac{x^2-10x+25}{2x+5}$

16. $\dfrac{2a^2-2a-4}{4-a^2} \cdot \dfrac{2a^2+a-6}{4a^2-2a-6}$

17. $\dfrac{x^2y}{2x} \div xy^2$

18. $\dfrac{5m+10}{2m-20} \div \dfrac{7m+14}{14m-20}$

19. $\dfrac{5y^2}{x^2-36} \div \dfrac{25xy-25y}{x^2-7x+6}$

20. $\dfrac{2x^2+x-1}{x^2+8x+7} \div \dfrac{6x^2+x-2}{x^2+14x+49}$

Write each pair of rational expressions in terms of their LCD.

21. $\dfrac{1}{5x}$ and $\dfrac{3}{20x^2}$

22. $\dfrac{4}{n-1}$ and $\dfrac{n}{n+4}$

23. $\dfrac{1}{3x+9}$ and $\dfrac{x}{x^2+4x+3}$

24. $\dfrac{2}{3x^2-5x-2}$ and $\dfrac{1}{4-x^2}$

Perform the indicated operation.

25. $\dfrac{3t+1}{2t}+\dfrac{t-1}{2t}$

26. $\dfrac{5y}{y+7}-\dfrac{y-28}{y+7}$

27. $\dfrac{5y+4}{4y^2-2y}-\dfrac{2}{2y-1}$

28. $\dfrac{n}{3n+15}+\dfrac{n-2}{n^2+5n}$

29. $\dfrac{4}{x-3}-\dfrac{4x+1}{9-x^2}$

30. $\dfrac{y+3}{4-y^2}+\dfrac{1}{2-y}$

31. $\dfrac{2}{m+1}+\dfrac{6m-2}{m^2-2m-3}$

32. $\dfrac{3x-2}{x^2-x-12}-\dfrac{x+3}{x-4}$

33. $\dfrac{2x}{x^2+4x+4}-\dfrac{x-1}{x^2-2x-8}$

34. $\dfrac{n+4}{2n^2-3n+1}+\dfrac{n+1}{2n^2+5n-3}$

[7.4] *Simplify.*

35. $\dfrac{\dfrac{x}{2}}{\dfrac{3x^2}{7}}$

36. $\dfrac{1-\dfrac{9}{y}}{1-\dfrac{81}{y^2}}$

37. $\dfrac{\dfrac{1}{x}+\dfrac{1}{y}}{\dfrac{1}{2x}+\dfrac{1}{2y}}$

38. $\dfrac{4-\dfrac{3}{x}-\dfrac{1}{x^2}}{2-\dfrac{5}{x}+\dfrac{3}{x^2}}$

[7.5] *Solve and check.*

39. $\dfrac{2x}{x-4}=5-\dfrac{1}{x-4}$

40. $\dfrac{y+1}{y}+\dfrac{1}{2y}=4$

41. $\dfrac{5}{2x}+\dfrac{3}{x+1}=\dfrac{7}{x}$

42. $\dfrac{y-2}{y-4}=\dfrac{1}{y+2}+\dfrac{y+3}{y^2-2y-8}$

43. $\dfrac{x}{x+2}-\dfrac{2}{2-x}=\dfrac{x+6}{x^2-4}$

44. $\dfrac{3}{n^2-5n+4}-\dfrac{1}{n^2-4n+3}=\dfrac{n-3}{n^2-7n+12}$

[7.6] *Solve and check.*

45. $\dfrac{8}{5}=\dfrac{72}{x}$

46. $\dfrac{28}{x+3}=\dfrac{7}{9}$

47. $\dfrac{5}{3+y}=\dfrac{3}{7y+1}$

48. $\dfrac{11}{x-2}=\dfrac{x+7}{2}$

Solve.

49. A company found that the cost per booklet for printing x booklets can be represented by $\left(0.72 + \dfrac{200}{x}\right)$ dollars.

Write this cost as a single rational expression.

50. Four friends decide to split the cost of renting a car equally. They discover that if they let one more friend share in the rental, the cost for each of the original four friends will be reduced by $10. What is the total cost of the car rental?

51. A hiker walks a distance d mi at a speed of r mph, and then returns on the same path at a speed of s mph. The hiker's average speed for the entire trip can be represented by

$$\frac{2d}{\dfrac{d}{r} + \dfrac{d}{s}}.$$

Simplify this expression.

52. With the water running at full force, it takes 10 min to fill a bathtub. It then takes 15 min for the bathtub to drain. If by mistake the water is running at full force while the tub is draining, how long will it take the tub to fill?

53. A family on vacation drove 400 mi at two different speeds, 50 mph and 60 mph. The total driving time was 7 hr. How many miles did the family drive at 50 mph?

54. One student takes x hr to design a Web page. Another student takes an hour longer. What part of the job will be finished in an hour if the two students work together?

55. The director of a college cafeteria knows that it cost $31,000 to serve 15,000 meals in September. She expects to serve about 20,000 meals in October. How much should she anticipate spending on the October meals?

56. Find a rational expression equal to the sum of the reciprocals of three consecutive integers, starting with n.

• Check your answers on page A-25.

CHAPTER 7   Posttest

EXTRA
HELP

Test Prep
VIDEOS

in MyMathLab, and on YouTube (search "Akst Introductory Alg" and click on "Channels").

To see if you have mastered the topics in this chapter, take this test.

1. Identify the values for which the rational expression $\dfrac{3x}{x - 8}$ is undefined.

2. Show that $\dfrac{y - 3}{y}$ is equivalent to $-\dfrac{3y - y^2}{y^2}$.

Simplify.

3. $\dfrac{15a^3b}{12ab^2}$

4. $\dfrac{x^2 - 4x}{xy - 4y}$

5. $\dfrac{3b^2 - 27}{b^2 - 4b - 21}$

6. $\dfrac{\dfrac{3}{x^2} - \dfrac{1}{x}}{\dfrac{9}{x^2} - 1}$

7. Write $\dfrac{4n - 1}{n^2 + 6n - 16}, \dfrac{2}{n + 8}$, and $\dfrac{n}{4n - 8}$ in terms of their LCD.

Perform the indicated operation.

8. $\dfrac{7x - 10}{x + 6} - \dfrac{5x - 22}{x + 6}$

9. $\dfrac{3}{2y - 8} + \dfrac{2}{4y^2 - 16y}$

10. $\dfrac{5}{d - 3} - \dfrac{d - 4}{d^2 - d - 6}$

11. $\dfrac{5}{2x^2 - 3x - 2} - \dfrac{x}{4 - x^2}$

12. $\dfrac{n + 1}{3n - 18} \cdot \dfrac{n - 6}{6n^3 - 6n}$

13. $\dfrac{a^2 - 25}{a^2 - 2a - 24} \div \dfrac{a^2 + a - 30}{a^2 - 36}$

14. $\dfrac{x^2 + 6x + 8}{x^2 + x - 2} \div \dfrac{x + 4}{2x^2 + 12x + 16}$

Solve and check.

15. $\dfrac{1}{y - 5} + \dfrac{y + 4}{25 - y^2} = \dfrac{1}{y + 5}$

16. $\dfrac{2y}{y - 4} - 2 = \dfrac{4}{y + 5}$

17. $\dfrac{x}{x + 6} = \dfrac{1}{x + 2}$

Solve.

18. Large files, which account for up to half of Internet traffic, can be downloaded to a computer from different sources and at different speeds at the same time. Suppose that there are three sources that contribute to the downloading of a large file. Working alone, one source would complete the downloading of the file in $2m$ min, the second in $3m$ min, and the third in $5m$ min. Working from the three sources together, how long would it take to download the file? (*Sources:* bittorrent.com and wikipedia.org)

19. A company owns two mail processing machines. The newer machine works twice as fast as the older one. Together, the two machines process 1000 pieces of mail in 20 min. How long does it take each machine, working alone, to process 1000 pieces of mail?

20. In the diagram shown, the heights and shadows of the woman and of the tree are in proportion. Find the height of the tree in meters.

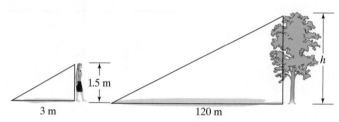

3 m   1.5 m   120 m   h

• Check your answers on page A-25.

Cumulative Review Exercises

To help you review, solve the following.

1. Replace ▨ with $>$, $<$, or $=$ to make a true statement:
$-|-4.6|$ ▨ -4.7

2. Simplify: $y - [2y - 3(y - 1)]$

3. Solve the formula $A = \dfrac{1}{2}h(b + B)$ for h.

4. Solve $4n - 5(n + 2) < -7$. Graph the solution on the number line.

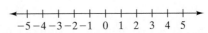

5. Find the equation of the line that is perpendicular to the graph of $y = \dfrac{1}{3}x$ and that passes through the point $(2, -4)$.

6. Graph the inequality $y \geq 2$ on the coordinate plane.

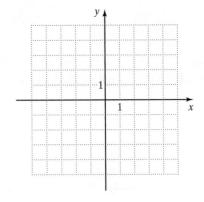

7. Solve by substitution, and then check.
$$x + y = 8$$
$$y = 2x - 1$$

8. Remove parentheses and simplify:
$(3x - 5) + (x^2 - 8x + 11)$

9. Simplify: $(3k - 6l)^2$

10. Factor, if possible: $6x^3 - 3x^2 - 3x$

11. Solve: $(n + 5)(n - 2) = 0$

12. Factor: $100y^2 - 81$

13. Simplify: $\dfrac{2 - \dfrac{6y}{x}}{1 - \dfrac{9y^2}{x^2}}$.

14. Solve and check: $\dfrac{c}{c + 6} = \dfrac{1}{c + 2}$

15. In round numbers, the population of the United States doubled between the years 1900 and 1950, and again between 1950 and 2000. If p represents the population in 1900, express in exponential form the population in 2000. (*Source:* census.gov)

16. A broker invested twice as much money in Stock A as in Stock B. Stock A had a return of 6% and Stock B a return of 4%. The total return was $1200. How much was invested in each stock?

17. The fastest train in the world is not France's TGV or the Japanese Bullet Train but, rather, the Pudong International Airport Express in Shanghai, China. This train travels at a maximum speed of 268 mph. At this rate, how far, to the nearest mile, would it travel in 2 min? (*Source:* beijing-visitor.com)

18. The U.S. annual per capita consumption of fruit in recent years can be modeled by the equation $y = 0.6x + 100$, where x is the number of years after 1970 and y is the number of pounds of fruit consumed per person.
(*Source: The World Almanac and Book of Facts 2011*)

   a. On a coordinate plane, label the axes, and then plot the graph of this equation.

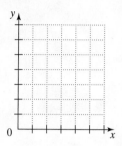

   b. Identify the slope of this line and explain its significance according to this model.

   c. Identify the y-intercept and explain its significance.

19. Two buses following the same route leave from the same station at different times. The Reston bus leaves 2 hr later than the Arlington bus. The Arlington bus is traveling at 40 mph and the Reston bus is traveling at 60 mph. How long will it take the Reston bus to overtake the Arlington bus?

20. A country's balance of payments can be found by subtracting its imports from its exports. If x is the number of years after 2000, the total exports of the United States for a particular year can be approximated by the polynomial $-5.6x^3 + 82.7x^2 - 227.6x + 1111.1$, and the total U.S. imports can be approximated by $-10.3x^3 + 132.8x^2 - 295.8x + 1494.8$ (both in billions of dollars). Find a polynomial to model the U.S. balance of payments. (*Source:* bea.gov)

• Check your answers on page A-25.

Radical Expressions and Equations

Square Roots and Lighting

A theatrical lighting designer focuses the audience's attention on what is lit: actors, costumes, props, and scenery. In planning a production for the stage, the designer must take into account how much illumination a lighting fixture will deliver from a particular distance. An actor further from the fixture seems more dimly lit than the one who is closer.

In general, the distance d associated with an illumination of I on a person or object is given by the equation

$$d = \sqrt{\frac{k}{I}},$$

where k is a constant that depends on the fixture, d is in feet, and I is in foot-candles. It follows from this formula that if a lighting designer wants to increase the illumination on someone by a factor of four, the distance to the light source must be divided by the square root of 4, that is, by 2.

(*Source:* J. Michael Gillette, *Theatrical Design and Production*, McGraw-Hill, 1999)

Assume all variables and radicands represent nonnegative real numbers.

Simplify.

1. $\sqrt{81}$

2. $-\sqrt{27}$

3. $\sqrt{45a^2}$

4. $\sqrt{\dfrac{x}{64}}$

5. $6\sqrt{2} + \sqrt{2} - 3\sqrt{2}$

6. $\sqrt{12} + 2\sqrt{75}$

7. $\sqrt{9x^3} - 4x\sqrt{x} + x\sqrt{36x}$

Multiply or divide. Simplify, if possible.

8. $\sqrt{6} \cdot \sqrt{3}$

9. $\sqrt{2xy} \cdot \sqrt{10xy^3}$

10. $\dfrac{\sqrt{30}}{\sqrt{5}}$

11. $\sqrt{n}(\sqrt{n} + 2)$

12. $(\sqrt{3} - 1)(\sqrt{3} + 4)$

Simplify.

13. $\sqrt{\dfrac{5x}{6}}$

14. $\dfrac{\sqrt{40x^3}}{\sqrt{2x}}$

15. Rationalize the denominator:
$$\dfrac{8 + \sqrt{7}}{\sqrt{2}}$$

Solve and check.

16. $\sqrt{x} - 1 = 5$

17. $y = \sqrt{4y - 3}$

Solve.

18. The velocity of a car (in meters per second) that starts from rest with a constant acceleration of a (in meters per second^2) can be found using the expression $\sqrt{2as}$, where s is the distance traveled (in meters). If the acceleration of the car is 2 m/sec^2, find its velocity after it has traveled 100 m.

19. An Olympic gymnast is practicing her floor exercise routine. The floor is a square whose sides measure 12 m. In the routine, she uses the diagonal of the square surface to complete a tumbling sequence. Find the distance the gymnast covers in the tumbling sequence. Express this distance as both a radical in simplest form and as a decimal rounded to the nearest tenth of a meter.

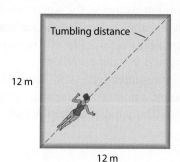

20. Traffic accident investigators use the lengths of skid marks to determine the minimum speed a car was traveling before it skids to a stop. The minimum speed S (in miles per hour) can be approximated by the formula
$$S = \sqrt{30fL},$$
where f is the drag factor (or coefficient of friction) for the road surface and L is the length (in feet) of the skid marks. Solve this formula for L in terms of S and f.

• Check your answers on page A-25.

What Radicals Are and Why They Are Important

So far, we have considered two types of algebraic expressions, namely polynomials and rational expressions. In this chapter, we extend the discussion to a third type, called *radical expressions*. **Radical expressions**, such as $4\sqrt{y}$ or $5 - \sqrt{x}$, are algebraic expressions that involve *square roots*. Note that in this text we use the terms *radical* and *square root* interchangeably.

When we find a square root of a number, we ask ourselves: What number multiplied by itself gives the original number? This question arises naturally in many situations, notably in geometry. For instance, recall the formula for the area A of a square with side s.

The formula $A = s^2$ states that to find the area of a square, we square the length of a side. If we know the value of A and wish to determine the value of s, we find the square root of A, that is, $s = \sqrt{A}$.

In this chapter, first we discuss evaluating and simplifying radical expressions, as well as carrying out the usual operations on them—adding, subtracting, multiplying, and dividing. Then, we consider how to solve equations involving radical expressions.

Introduction to Radicals

Let's begin with the definition of square root.

Actually, every positive number has *two* square roots, one positive and the other negative. For example, the square roots of 4 are $+2$ and -2 because $(+2)^2 = 4$ and $(-2)^2 = 4$. The symbol $\sqrt{}$, called the **radical sign**, stands for the positive or *principal square root,* but is commonly referred to as "**the** square root." For instance, $\sqrt{9}$ is read "the square root of 9", or "radical 9", and represents $+3$, or 3. By contrast, the negative square root of 9, namely, -3, is represented by $-\sqrt{9}$. Throughout the remainder of this text, when we speak of the square root of a number we mean its principal square root.

The number under the radical sign is called the **radicand.**

$$\nearrow \sqrt{n} \nwarrow$$
The radical sign · · · The radicand

We read $\sqrt{n}$ as "the square root of n."

This chapter focuses on radicals with a nonnegative radicand, such as $\sqrt{0}$ and $\sqrt{3}$. We will not discuss radicals with a negative radicand, say, $\sqrt{-4}$. Such a radical is not a real number, since no real number when squared results in a negative number.

Evaluating Radicals

Let's continue our discussion of radicals by finding the value of radicals with radicands that are whole numbers. A whole number is said to be a **perfect square** if it is the square of another whole number. For instance, 9 is a perfect square, since $9 = 3^2$. On the other hand, 8 is not a perfect square, since it is not the square of any whole number. Perfect squares play a special role in finding the value of square roots.

The square root of a perfect square is always a whole number. For instance, $\sqrt{9} = 3$, since $3^2 = 9$.

EXAMPLE 1

Find the value of the following radical expressions.

a. $\sqrt{16}$ b. $-3\sqrt{25}$

Solution

a. $\sqrt{16} = 4$ since $4^2 = 16$.   b. $-3\sqrt{25} = -3 \cdot 5 = -15$

PRACTICE 1

Evaluate the following radical expressions.

a. $\sqrt{4}$

b. $-5\sqrt{49}$

TIP In evaluating radicals, remember that squaring, doubling, and halving a number are *not* the same as finding its square root.

The square root of a whole number that is not a perfect square is an *irrational number.* Recall from Section 1.1 that irrational numbers have decimal representations which neither terminate nor repeat. For instance, we have seen that $\sqrt{2}$ and $\sqrt{3}$ are irrational numbers.

$$\sqrt{2} = 1.4142\ldots, \text{ which is approximately } 1.414.$$
$$\sqrt{3} = 1.7320\ldots, \text{ which is approximately } 1.732.$$

A calculator can be used to approximate the value of these square roots by decimals rounded to a specific place value.

EXAMPLE 2

Evaluate $\sqrt{7}$, rounded to the nearest thousandth.

Solution   Since the radicand 7 is not a perfect square, its square root is an irrational number. Using a calculator, we see that $\sqrt{7} = 2.6457\ldots$. Rounding to the nearest thousandth gives us 2.646.

PRACTICE 2

Find the value of $\sqrt{10}$, rounded to the nearest thousandth.

Simplifying Radicals

The operations of squaring and taking a square root undo each other. They are opposite operations in the same way that adding a number is the opposite of subtracting that number. The following two properties of radicals stem from this relationship.

Squaring a Square Root

For any nonnegative real number a,

$$(\sqrt{a})^2 = a.$$

In words, this property states that when we take the square root of a nonnegative number and then square the result, we get the original number. For instance, $(\sqrt{9})^2 = 9$ and $(\sqrt{8})^2 = 8$.

Taking the Square Root of a Square

For any nonnegative real number a,

$$\sqrt{a^2} = a.$$

In words, this property states that when we square a nonnegative number and then take the square root, the result is the original number. For instance, $\sqrt{3^2} = 3$.

We will use both of these properties repeatedly throughout this chapter. Do you see the difference between the two properties? Explain.

EXAMPLE 3

Simplify each radical expression.

a. $(\sqrt{25})^2$ **b.** $(\sqrt{2})^2$ **c.** $\sqrt{4^2}$ **d.** $\sqrt{9^2}$

Solution

a. $(\sqrt{25})^2 = 25$ Use the property of squaring a square root.

b. $(\sqrt{2})^2 = 2$

c. $\sqrt{4^2} = 4$ Use the property of taking the square root of a square.

d. $\sqrt{9^2} = 9$

PRACTICE 3

Simplify.

a. $(\sqrt{6})^2$

b. $(\sqrt{5})^2$

c. $\sqrt{7^2}$

d. $\sqrt{1^2}$

Some radicands contain variables. Consider, for instance, the expression $\sqrt{x}$. For negative values of x, the radicand is negative and so $\sqrt{x}$ is not a real number. *For purposes of simplicity, we will assume throughout the remainder of this text that all radicands and all variables are nonnegative.*

Let's look at some radicals that are perfect squares involving variables. For instance, the radicands for $\sqrt{x^8}$, $\sqrt{49y^6}$, and $\sqrt{25a^2b^4}$ are all perfect squares.

$$\sqrt{x^8} = \sqrt{(x^4)^2} = x^4$$
$$\sqrt{49y^6} = \sqrt{(7y^3)^2} = 7y^3$$
$$\sqrt{25a^2b^4} = \sqrt{(5ab^2)^2} = 5ab^2$$

Note that the exponent of a variable in any perfect square is always an even number, such as 8 or 10. When finding the square root of an *even* power, we can simply divide the power by 2. For instance,

$$\sqrt{x^8} = x^4 \leftarrow \frac{8}{2} = 4$$

EXAMPLE 4

Simplify each radical expression.

a. $\sqrt{n^6}$ **b.** $-\sqrt{16x^8}$ **c.** $\sqrt{100a^4b^2}$

Solution

a. The radicand n^6 is a perfect square, so we can use the property of taking the square root of a square:

$$\sqrt{n^6} = \sqrt{(n^3)^2} = n^3$$

b. $-\sqrt{16x^8} = -\sqrt{(4x^4)^2} = -4x^4$

c. $\sqrt{100a^4b^2} = \sqrt{(10a^2b)^2} = 10a^2b$

PRACTICE 4

Simplify.

a. $\sqrt{x^4}$

b. $\sqrt{64t^{10}}$

c. $-\sqrt{121x^2y^2}$

Just as we simplify fractions to express them in lowest terms, so we put radical expressions *in simplified form.* In this way, we can recognize that expressions that appear to be different are really the same. Also, simplified radical expressions are often easier to work with.

But what exactly does it mean to simplify a radical? It is easier to say when a radical expression is *not* simplified than to say when it *is* simplified. *A radical is not considered to be in simplified form if its radicand is divisible by a perfect square.* For instance, the radical $\sqrt{8}$ is not in simplified form because the perfect square 4 is a factor of the radicand 8. To simplify such a radical, we apply the *product rule of radicals*.

The Product Rule of Radicals

If a and b are any nonnegative real numbers, then

$$\sqrt{ab} = \sqrt{a} \cdot \sqrt{b}.$$

In words, this rule states that the square root of a product is the product of the square roots.

To check that the product rule works in a particular example, we consider the radical $\sqrt{4 \cdot 9}$.

$$\sqrt{4 \cdot 9} \overset{?}{=} \sqrt{4} \cdot \sqrt{9}$$
$$\sqrt{36} \overset{?}{=} 2 \cdot 3$$
$$6 = 2 \cdot 3 \qquad \text{True}$$

So $\sqrt{4 \cdot 9} = \sqrt{4} \cdot \sqrt{9}$, as the product rule implies.

Now, let's use this rule to simplify $\sqrt{8}$.

$$\sqrt{8} = \sqrt{4 \cdot 2} \qquad \text{Factor out the perfect square 4 from the radicand.}$$
$$= \sqrt{4} \cdot \sqrt{2} \qquad \text{Use the product rule of radicals.}$$
$$= 2\sqrt{2} \qquad \sqrt{4} = 2.$$

So $2\sqrt{2}$ is the simplified form of $\sqrt{8}$.

Some radicals contain radicands that have more than one possible perfect square factor. To simplify these radicals, we generally factor out the *largest* perfect square factor of the radicand, as shown in Example 5(a).

EXAMPLE 5

Simplify.

a. $\sqrt{48}$ **b.** $-4\sqrt{27}$ **c.** $\dfrac{\sqrt{24}}{6}$

Solution

a. $\sqrt{48} = \sqrt{16 \cdot 3}$ Factor out the perfect square 16.
$\phantom{\sqrt{48}} = \sqrt{16} \cdot \sqrt{3}$ Use the product rule of radicals.
$\phantom{\sqrt{48}} = 4\sqrt{3}$ Take the square root of the perfect square.

b. $-4\sqrt{27} = -4\sqrt{9 \cdot 3}$ Factor out the perfect square 9.
$\phantom{-4\sqrt{27}} = -4\sqrt{9} \cdot \sqrt{3}$ Use the product rule of radicals.
$\phantom{-4\sqrt{27}} = -4 \cdot 3\sqrt{3}$ Take the square root of the perfect square.
$\phantom{-4\sqrt{27}} = -12\sqrt{3}$ Simplify.

PRACTICE 5

Express in simplified form.

a. $\sqrt{72}$

b. $2\sqrt{40}$

c. $\dfrac{\sqrt{75}}{-15}$

EXAMPLE 5 (continued)

c. $\dfrac{\sqrt{24}}{6} = \dfrac{\sqrt{4 \cdot 6}}{6}$ Factor out the perfect square 4.

$\quad = \dfrac{\sqrt{4} \cdot \sqrt{6}}{6}$ Use the product rule of radicals.

$\quad = \dfrac{2\sqrt{6}}{6}$ Take the square root of the perfect square.

$\quad = \dfrac{\overset{1}{2\sqrt{6}}}{\underset{3}{6}}$ Divide out the common factor.

$\quad = \dfrac{\sqrt{6}}{3}$ Simplify.

Note that in Example 5(c), we cannot divide out the 6 in the numerator with the 6 in the denominator, since one is under a radical sign and the other is not.

> **TIP** Be careful when dividing out common factors in radical expressions.
>
> $$\dfrac{\sqrt{3}}{3} \neq \dfrac{\sqrt{\cancel{3}}}{\cancel{3}} \quad \text{but} \quad \dfrac{3\sqrt{3}}{3} = \dfrac{\cancel{3}\sqrt{3}}{\cancel{3}}$$

Now, let's apply the product rule to radical expressions involving variables. The key here is to factor out the largest possible even power of each variable. Note that these problems test our knowledge of exponents as well as of radicals.

EXAMPLE 6

Simplify.

a. $\sqrt{y^5}$ b. $-\sqrt{20a^6}$ c. $\sqrt{60x^5y}$

Solution

a. $\sqrt{y^5} = \sqrt{y^4 \cdot y}$ Factor out the perfect square y^4, using the product rule of exponents.

$\quad = \sqrt{y^4} \cdot \sqrt{y}$ Use the product rule of radicals.

$\quad = y^2\sqrt{y}$ Take the square root of the perfect square.

b. $-\sqrt{20a^6} = -\sqrt{4 \cdot 5 \cdot a^6}$ Factor out the perfect squares 4 and a^6.

$\quad = -\sqrt{4} \cdot \sqrt{a^6} \cdot \sqrt{5}$ Use the product rule of radicals.

$\quad = -2a^3\sqrt{5}$ Take the square root of the perfect squares.

c. $\sqrt{60x^5y} = \sqrt{4 \cdot 15 \cdot x^4 \cdot x \cdot y}$

$\quad = \sqrt{4} \cdot \sqrt{x^4} \cdot \sqrt{15xy}$

$\quad = 2x^2\sqrt{15xy}$

PRACTICE 6

Express in simplified form.

a. $\sqrt{x^3}$

b. $\sqrt{18n^4}$

c. $-\sqrt{50ab^2}$

Some radical expressions involve quotients. If the radicand is a perfect square, we can take the square root directly as shown in the following example.

EXAMPLE 7

Find the square root. **a.** $\sqrt{\dfrac{1}{4}}$ **b.** $\sqrt{\dfrac{x^2}{49}}$ **c.** $\sqrt{\dfrac{a^2}{b^8}}$

Solution

a. $\sqrt{\dfrac{1}{4}} = \dfrac{1}{2}$ since $\left(\dfrac{1}{2}\right)^2 = \dfrac{1}{4}$.   **b.** $\sqrt{\dfrac{x^2}{49}} = \dfrac{x}{7}$ since $\left(\dfrac{x}{7}\right)^2 = \dfrac{x^2}{49}$.

c. $\sqrt{\dfrac{a^2}{b^8}} = \dfrac{a}{b^4}$ since $\left(\dfrac{a}{b^4}\right)^2 = \dfrac{a^2}{b^8}$.

PRACTICE 7

Find the square root.

a. $\sqrt{\dfrac{1}{16}}$ **b.** $\sqrt{\dfrac{y^2}{4}}$ **c.** $\sqrt{\dfrac{x^4}{y^6}}$

Some radical expressions contain quotients that are not perfect squares. To simplify these radicals, we use the following rule:

The Quotient Rule of Radicals

If a is a nonnegative real number and b is a positive real number, then

$$\sqrt{\dfrac{a}{b}} = \dfrac{\sqrt{a}}{\sqrt{b}}.$$

In words, this rule states that the square root of a quotient is the quotient of the square roots.

To check that the quotient rule works in a particular example, let's consider the radical $\sqrt{\dfrac{4}{9}}$.

$$\sqrt{\dfrac{4}{9}} \stackrel{?}{=} \dfrac{\sqrt{4}}{\sqrt{9}}$$

$$\sqrt{\left(\dfrac{2}{3}\right)^2} \stackrel{?}{=} \dfrac{2}{3}$$

$$\dfrac{2}{3} = \dfrac{2}{3} \qquad \text{True}$$

So $\sqrt{\dfrac{4}{9}} = \dfrac{\sqrt{4}}{\sqrt{9}}$, which is in agreement with the quotient rule.

EXAMPLE 8

Simplify.

a. $\sqrt{\dfrac{7}{4}}$ **b.** $\sqrt{\dfrac{3x}{25}}$ **c.** $\sqrt{\dfrac{2a^5b^{10}}{9}}$

Solution

a. $\sqrt{\dfrac{7}{4}} = \dfrac{\sqrt{7}}{\sqrt{4}}$ Use the quotient rule of radicals.

$\phantom{a.\ \sqrt{\dfrac{7}{4}}} = \dfrac{\sqrt{7}}{2}$ Take the square root of the perfect square.

b. $\sqrt{\dfrac{3x}{25}} = \dfrac{\sqrt{3x}}{\sqrt{25}}$ Use the quotient rule of radicals.

$\phantom{b.\ \sqrt{\dfrac{3x}{25}}} = \dfrac{\sqrt{3x}}{5}$ Take the square root of the perfect square.

PRACTICE 8

Simplify.

a. $\sqrt{\dfrac{3}{16}}$

b. $\sqrt{\dfrac{2y}{49}}$

c. $\sqrt{\dfrac{5x^7y^2}{4}}$

EXAMPLE 8 (continued)

c. $\sqrt{\dfrac{2a^5b^{10}}{9}} = \dfrac{\sqrt{2a^5b^{10}}}{\sqrt{9}}$ Use the quotient rule of radicals.

$\qquad = \dfrac{\sqrt{a^4b^{10}} \cdot \sqrt{2a}}{\sqrt{9}}$ Use the product rule of radicals.

$\qquad = \dfrac{a^2b^5\sqrt{2a}}{3}$ Take the square root of each perfect square.

Now, let's see if we can use our knowledge of radicals in solving some applied problems. The following problem deals with the Pythagorean theorem, a topic that we discussed in Section 6.5. Recall that this theorem describes the relationship between the lengths of the three sides of a right triangle, which is expressed as $c^2 = a^2 + b^2$. We can rewrite the theorem as

$$c = \sqrt{a^2 + b^2},$$

where c is the length of the hypotenuse, and a and b are the lengths of the other two sides.

> **TIP** In evaluating a radical such as $\sqrt{a^2 + b^2}$ with arithmetic operations in the radicand, parentheses are understood around the radicand. So in evaluating $\sqrt{a^2 + b^2}$, we compute $a^2 + b^2$, take the square root, and then simplify, if possible.

EXAMPLE 9

A flat panel television is rectangular in shape, with dimensions 25 in. by 15 in. Find the length of the panel's diagonal, both as a radical and as a decimal rounded to the nearest inch.

Solution   Let x represent the length of the diagonal of the panel. The lengths of the other two sides of the triangle shown are 15 in. and 25 in. Using the Pythagorean theorem, we get:

$$x = \sqrt{a^2 + b^2}$$
$$= \sqrt{15^2 + 25^2}$$
$$= \sqrt{225 + 625}$$
$$= \sqrt{850}$$
$$= \sqrt{25 \cdot 34}$$
$$= \sqrt{25} \cdot \sqrt{34}$$
$$= 5\sqrt{34}$$

So the length of the diagonal of the panel is $5\sqrt{34}$ in., or approximately 29 in.

PRACTICE 9

Two Jeeps try to pull a car out of the snow, tugging at a 90° angle to each other.

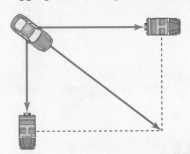

If one Jeep exerts a force of 600 lb and the other a force of 800 lb, the magnitude of the resulting force on the car (in pounds) can be found by computing $\sqrt{600^2 + 800^2}$. Calculate this force.

Mathematically Speaking

Fill in each blank with the most appropriate term or phrase from the given list.

odd	even	square root
product	divide by 2	square
quotient	positive square root	take the square root
double	irrational	

1. A(n) _____ of a nonnegative real number a is a number that when squared is a.

2. The principal square root of 16, written $\sqrt{16}$, means the _____ of 16.

3. The square root of a whole number that is not a perfect square is a(n) _____ number.

4. If we take the square root of a nonnegative number and then _____ the result, we get the original number.

5. When we square a nonnegative number and then _____, the result is the original number.

6. The exponent of the variable in any perfect square is always a(n) _____ number.

7. The square root of a product is the _____ of the square roots.

8. The square root of a quotient is the _____ of the square roots.

A *Find the value of each radical. When a radicand is not a perfect square, use a calculator to evaluate, rounding to the nearest thousandth.*

9. $\sqrt{36}$

10. $\sqrt{25}$

11. $\sqrt{1}$

12. $\sqrt{9}$

13. $-\sqrt{100}$

14. $-\sqrt{64}$

15. $3\sqrt{49}$

16. $2\sqrt{81}$

17. $\sqrt{5}$

18. $\sqrt{6}$

19. $3\sqrt{2}$

20. $4\sqrt{3}$

B *Simplify.*

21. $(\sqrt{16})^2$

22. $(\sqrt{5})^2$

23. $(\sqrt{11})^2$

24. $(\sqrt{10})^2$

25. $(\sqrt{5x})^2$

26. $(\sqrt{8y})^2$

27. $\sqrt{2^2}$

28. $\sqrt{3^2}$

29. $\sqrt{9^2}$

30. $\sqrt{14^2}$

31. $\sqrt{n^8}$

32. $\sqrt{s^2}$

33. $\sqrt{49y^2}$

34. $\sqrt{16t^4}$

35. $\sqrt{9x^4}$

36. $\sqrt{100y^6}$

37. $\sqrt{25x^2y^{10}}$

38. $\sqrt{49p^{12}q^4}$

Simplify each expression, factoring out any perfect square.

39. $\sqrt{32}$

40. $\sqrt{75}$

41. $-\sqrt{108}$

42. $-\sqrt{98}$

43. $6\sqrt{27}$

44. $3\sqrt{20}$

45. $\dfrac{\sqrt{48}}{12}$

46. $\dfrac{\sqrt{50}}{10}$

47. $\sqrt{11x^2}$

48. $\sqrt{7n^4}$

49. $\sqrt{n^5}$

50. $\sqrt{t^7}$

51. $\sqrt{20x^3}$

52. $\sqrt{32y^9}$

53. $-\sqrt{12p^2q}$

54. $-\sqrt{24a^3b^4}$

55. $9\sqrt{10x^3y^4}$

56. $2\sqrt{25a^7b^6}$

Simplify.

57. $\sqrt{\dfrac{4}{25}}$

58. $\sqrt{\dfrac{1}{36}}$

59. $-\sqrt{\dfrac{1}{4}}$

60. $-\sqrt{\dfrac{49}{9}}$

61. $\sqrt{\dfrac{81}{n^6}}$

62. $\sqrt{\dfrac{121}{x^8}}$

63. $\sqrt{\dfrac{x^4}{y^2}}$

64. $\sqrt{\dfrac{a^2}{b^6}}$

65. $-\sqrt{\dfrac{3}{4}}$

66. $-\sqrt{\dfrac{2}{9}}$

67. $\sqrt{\dfrac{5n}{16}}$

68. $\sqrt{\dfrac{3t}{25}}$

69. $\sqrt{\dfrac{3x^2y^6}{4}}$

70. $\sqrt{\dfrac{5a^4b^2}{9}}$

71. $-\sqrt{\dfrac{27x^6y}{16}}$

72. $-\sqrt{\dfrac{32ab^8}{25}}$

Mixed Practice

Simplify.

73. $\sqrt{\dfrac{36}{y^6}}$

74. $\sqrt{\dfrac{6m^4n^2}{49}}$

Find the value of each radical. When a radicand is not a perfect square, use a calculator to evaluate, rounding to the nearest thousandth.

75. $-\sqrt{144}$

76. $2\sqrt{8}$

Simplify each expression, factoring out any perfect square.

77. $5\sqrt{12}$

78. $\sqrt{11t^5}$

79. $-\sqrt{40a^4b^3}$

Simplify.

80. $\left(\sqrt{27x}\right)^2$

81. $\sqrt{r^6}$

82. $\sqrt{64m^8}$

Applications

Solve.

83. The *geometric mean m* of two numbers a and b is the square root of their product. The geometric mean can be thought of as a kind of average.

    a. Express this relationship as a formula.

    b. Find the geometric mean of 2 and 8.

84. To approximate the maximum speed s (in knots) of a sailboat, sailors multiply 1.3 by the square root of the length of the boat's waterline w (in feet).

|← Waterline →|

    a. Express this relationship as a formula.

    b. Use this formula to approximate the maximum speed, to the nearest knot, of a sailboat with a 16-foot waterline.

85. When an object is dropped from a height h (in meters), it takes approximately $\sqrt{\dfrac{h}{5}}$ sec to reach the ground.

  a. Find the time it takes for an object dropped from a height of 20 m to reach the ground.

  b. If the same object is dropped from double the height, will it take twice as long to reach the ground? Explain.

86. The expression $2\pi\sqrt{\dfrac{L}{32}}$ approximates the time (in seconds) that a pendulum of length L (in feet) takes to make one complete swing back and forth.

  a. How long does it take the pendulum shown to make one complete swing? Use 3.14 for π, and round to the nearest second.

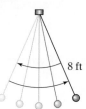

  b. How long would it take a pendulum 4 times the length of that in part (a) to make a complete swing? Is this answer 4 times the answer to part (a)?

87. Under certain conditions, the expression $2\sqrt{5L}$ can be used to approximate the speed of a car (in miles per hour) that has left a skid mark of length L (in feet). If an investigating officer arrives at the scene of an accident where these conditions apply and finds a skid mark 180 ft long, at what speed was the car traveling at the time of the accident?

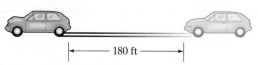

88. Firefighters can approximate the speed S at which water leaves a typical nozzle by using the formula

$$S = 12\sqrt{P},$$

where the pressure P is measured in pounds per square inch (psi) and the speed is in feet per second. Find the speed of the water if the pressure is 50 psi. Express your answer both as a simplified radical and as a decimal rounded to the nearest whole number. (*Source:* Jim Cottrell, "Fire Stream Physics," *Firefighter's News,* 1995)

89. A surveyor wishes to find the distance between the towns B and C across from one another on a lake, as shown in the illustration. Find this distance expressed both as a radical in simplified form and as a decimal rounded to the nearest mile.

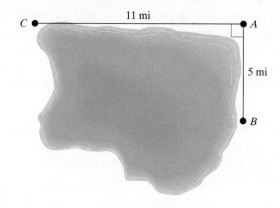

90. A plane is about to land at an airport. The flight path of the plane's descent is shown in the following diagram. What is the distance from the plane to the airport? Express this distance both as a radical and as a decimal rounded to the nearest multiple of 100 ft.

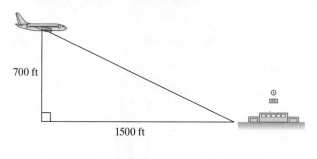

• Check your answers on page A-25.

MINDStretchers

Research

1. Using your college library or the Web, investigate whether there is a connection between the radical symbol and the symbol for a drugstore prescription.

   Describe the results of your research.

Writing

2. Consider the following statement: *Just as with adding and subtracting, the operations of **squaring** and **taking a square root** are opposites.* In a few sentences, support or refute this statement.

Investigation

3. Consider the statement: For any positive numbers, the larger the number, the larger its square root.

   a. Use a calculator either to support this assertion with many examples or to give a counterexample. Sum up your findings.

   b. If you supported this assertion, do your examples prove the assertion true? If you gave a counterexample, does the counterexample prove the assertion false? Explain.

Cultural Note

The square root is related to one of the most revolutionary experiments in the history of science. Working some 500 years ago, Galileo Galilei discovered that contrary to ancient authority, the speed of falling objects, which he had dropped from the Leaning Tower of Pisa, was proportional to the square root of the height from which they were dropped. Because of such experiments, Galileo is considered to be the founder of modern physics.

(*Source:* Jeanne Bendick, *Along Came Galileo*, Beautiful Feat Books, 1999)

Sums and differences of many radicals, in contrast to other kinds of numbers such as fractions and decimals, cannot be simplified. For instance, there is no way to combine $\sqrt{2}$ and $\sqrt{3}$, so we cannot simplify the expression $\sqrt{2} + \sqrt{3}$. Similarly, we cannot simplify the expression $\sqrt{5} - 1$. However, we can approximate the value of these expressions in decimal form using a calculator.

Some other radicals *can* be combined. In this section, we discuss such radicals.

Adding and Subtracting Like Radicals

When adding or subtracting *like* radicals, we can simplify the result.

> **DEFINITIONS**
>
> **Like radicals** are radical expressions that have the same radicand. **Unlike radicals** are radical expressions with different radicands.

For instance, $4\sqrt{2}$ and $3\sqrt{2}$ are like radicals. By contrast, the radicals $7\sqrt{2}$ and $\sqrt{3}$ are unlike.

We use the distributive property to add or subtract like radicals, just as we do for adding or subtracting like terms.

Adding Like Terms

$$4x + 3x = (4 + 3)x = 7x$$

Like terms

Adding Like Radicals

$$4\sqrt{2} + 3\sqrt{2} = (4 + 3)\sqrt{2} = 7\sqrt{2}$$

Like radicals

Subtracting Like Terms

$$4x - 3x = (4 - 3)x = x$$

Like terms

Subtracting Like Radicals

$$4\sqrt{2} - 3\sqrt{2} = (4 - 3)\sqrt{2} = \sqrt{2}$$

Like radicals

EXAMPLE 1

Add or subtract.

a. $5\sqrt{3} + 2\sqrt{3}$

b. $7\sqrt{x} - 2\sqrt{x} - 4\sqrt{x}$

c. $4\sqrt{2y + 1} - \sqrt{2y + 1}$

d. $5\sqrt{3} + 3\sqrt{5}$

Solution

a. $5\sqrt{3} + 2\sqrt{3} = (5 + 2)\sqrt{3}$ Combine the coefficients using the distributive property.

$\qquad = 7\sqrt{3}$ Simplify.

b. $7\sqrt{x} - 2\sqrt{x} - 4\sqrt{x} = (7 - 2 - 4)\sqrt{x}$ Use the distributive property.

$\qquad = 1\sqrt{x}$ Simplify.

$\qquad = \sqrt{x}$

c. $4\sqrt{2y + 1} - \sqrt{2y + 1} = (4 - 1)\sqrt{2y + 1}$

$\qquad = 3\sqrt{2y + 1}$

d. $5\sqrt{3} + 3\sqrt{5}$ cannot be simplified because the radicals are not like.

PRACTICE 1

Add or subtract.

a. $8\sqrt{5} - 2\sqrt{5}$

b. $3\sqrt{n} + \sqrt{n} + 7\sqrt{n}$

c. $10\sqrt{t^2 - 3} + \sqrt{t^2 - 3}$

d. $4\sqrt{6} - 2\sqrt{2}$

Adding and Subtracting Unlike Radicals

Some *unlike* radicals become *like* when they are simplified. When this happens, they can be combined.

EXAMPLE 2

Combine. Simplify, if possible.

a. $\sqrt{12} + \sqrt{27}$ **b.** $7\sqrt{50} - \sqrt{72} + \sqrt{32}$

c. $-\sqrt{4x} + 3\sqrt{x}$ **d.** $a\sqrt{a^2b} - 5b\sqrt{b}$

Solution

a. $\sqrt{12} + \sqrt{27} = \sqrt{4 \cdot 3} + \sqrt{9 \cdot 3}$ Factor out perfect squares.

$\qquad\qquad\quad = \sqrt{4} \cdot \sqrt{3} + \sqrt{9} \cdot \sqrt{3}$ Use the product rule of radicals.

$\qquad\qquad\quad = 2\sqrt{3} + 3\sqrt{3}$ Take the square root of the perfect squares.

$\qquad\qquad\quad = (2 + 3)\sqrt{3}$ Use the distributive property.

$\qquad\qquad\quad = 5\sqrt{3}$ Simplify.

b. $7\sqrt{50} - \sqrt{72} + \sqrt{32}$

$= 7\sqrt{25 \cdot 2} - \sqrt{36 \cdot 2} + \sqrt{16 \cdot 2}$ Factor out perfect squares.

$= 7 \cdot \sqrt{25} \cdot \sqrt{2} - \sqrt{36} \cdot \sqrt{2} + \sqrt{16} \cdot \sqrt{2}$ Use the product rule of radicals.

$= 7 \cdot 5\sqrt{2} - 6\sqrt{2} + 4\sqrt{2}$ Take the square roots of the perfect squares.

$= 35\sqrt{2} - 6\sqrt{2} + 4\sqrt{2}$ Simplify.

$= (35 - 6 + 4)\sqrt{2}$ Use the distributive property.

$= 33\sqrt{2}$ Simplify.

c. $-\sqrt{4x} + 3\sqrt{x} = -2\sqrt{x} + 3\sqrt{x} = (-2 + 3)\sqrt{x} = \sqrt{x}$

d. $a\sqrt{a^2b} - 5b\sqrt{b} = a \cdot a\sqrt{b} - 5b\sqrt{b}$

$\qquad\qquad\quad = a^2\sqrt{b} - 5b\sqrt{b}$

$\qquad\qquad\quad = (a^2 - 5b)\sqrt{b}$

PRACTICE 2

Combine. Simplify, if possible.

a. $\sqrt{50} + \sqrt{98}$

b. $\sqrt{12} + 2\sqrt{75} - 6\sqrt{27}$

c. $-3\sqrt{16t} + \sqrt{9t}$

d. $\sqrt{25ab^4} + 7b^2\sqrt{a}$

EXAMPLE 3

It takes $\sqrt{20}$ sec for an object to fall from the top of a 320-ft building, and $\sqrt{5}$ sec for the object to fall from the top of an 80-ft building. How much longer will it take the object to fall from the taller building? Express this quantity both in radical form and as a decimal rounded to the nearest second.

Solution   We need to find the difference between $\sqrt{20}$ sec and $\sqrt{5}$ sec.

$$
\begin{aligned}
\sqrt{20} - \sqrt{5} &= \sqrt{4 \cdot 5} - \sqrt{5} \\
&= \sqrt{4} \cdot \sqrt{5} - \sqrt{5} \\
&= 2\sqrt{5} - \sqrt{5} \\
&= \sqrt{5}
\end{aligned}
$$

So it takes the object $\sqrt{5}$ sec, or approximately 2 sec, longer to fall from the taller building.

PRACTICE 3

A young couple wants to build a square cabin with an area of 200 m² on the corner of a square plot of land with area 1800 m². Find the length of the front yard shown in the illustration. Express this length as a decimal rounded to the nearest meter.

←Front yard→

Mathematically Speaking

Fill in each blank with the most appropriate term or phrase from the given list.

distributive property	common	unlike
cannot	associative property of addition	can
like		

1. Radical expressions that have the same radicand are called _____ radicals.

2. The _____ is used to add or subtract like radicals.

3. Radicals that are _____, even after being simplified, cannot be combined.

4. If we rewrite $2\sqrt{12} + 3\sqrt{75}$ as $4\sqrt{3} + 15\sqrt{3}$, we _____ combine the terms.

Combine and simplify, if possible.

5. $5\sqrt{7} + 3\sqrt{7}$

6. $7\sqrt{11} + \sqrt{11}$

7. $3\sqrt{2} - 8\sqrt{2}$

8. $\sqrt{3} - 10\sqrt{3}$

9. $6\sqrt{3} - 3\sqrt{6}$

10. $2\sqrt{7} + 7\sqrt{2}$

11. $-5\sqrt{11} - 10\sqrt{11} + 2\sqrt{11}$

12. $-4\sqrt{5} + 4\sqrt{5} + 2\sqrt{5}$

13. $7t\sqrt{3} + 2t\sqrt{3}$

14. $3y\sqrt{2} + 5y\sqrt{2}$

15. $13\sqrt{x} + 10\sqrt{x}$

16. $2\sqrt{3k} + 9\sqrt{3k}$

17. $2\sqrt{x} - 3\sqrt{y}$

18. $\sqrt{m} - 2\sqrt{n}$

19. $6\sqrt{x+1} - \sqrt{x+1}$

20. $\sqrt{2m-1} - 7\sqrt{2m-1}$

21. $\sqrt{8} - \sqrt{32}$

22. $\sqrt{12} - \sqrt{27}$

23. $\sqrt{50} + \sqrt{72}$

24. $\sqrt{48} + \sqrt{12}$

25. $-\sqrt{12} + 5\sqrt{3}$

26. $3\sqrt{2} - \sqrt{18}$

27. $6\sqrt{75} - 2\sqrt{12}$

28. $5\sqrt{72} - 4\sqrt{50}$

29. $5\sqrt{8} - 3\sqrt{12} + \sqrt{2}$

30. $2\sqrt{3} + \sqrt{20} + \sqrt{5}$

31. $2\sqrt{16y} + 3\sqrt{4y}$

32. $2\sqrt{9x} + 8\sqrt{4x}$

33. $\sqrt{9x} - \sqrt{16x^3}$

34. $2\sqrt{25a^3} - \sqrt{a}$

35. $\sqrt{25p} + \sqrt{64p} + \sqrt{p}$

36. $\sqrt{49a} - 2\sqrt{a} + 9\sqrt{a}$

37. $2\sqrt{16x} - \sqrt{x} - 3\sqrt{4x}$

38. $\sqrt{81y} + 6\sqrt{y} - 8\sqrt{25y}$

39. $-5x\sqrt{2x^3y^4} + x\sqrt{2x^5y^2}$

40. $-2\sqrt{9ab^5} - 3a\sqrt{4a^3b}$

Mixed Practice

Combine and simplify, if possible.

41. $\sqrt{27} - \sqrt{3} + \sqrt{6}$

42. $5\sqrt{7} + 7\sqrt{5}$

43. $6\sqrt{7p} - 2\sqrt{7p}$

44. $\sqrt{b} - 3\sqrt{36b^3}$

45. $\sqrt{75} + \sqrt{48}$

46. $8\sqrt{2} - \sqrt{72}$

Applications

B *Solve. Express the answer as a radical in simplified form.*

47. For the isosceles triangle shown, find:

    a. the lengths of the missing sides.

    b. the perimeter.

48. The size of a television set is commonly given by the length of the screen's diagonal. For the two sets pictured, find:

    a. the lengths of their diagonals.

    b. the difference of the diagonal lengths.

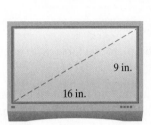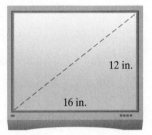

49. Two square tiles are pictured below:

    Area = 90 in²   Area = 40 in²

    a. Determine the side lengths of the two tiles.

    b. How much longer is a side of the larger tile than the smaller tile?

50. The accompanying map shows the route that a college recruiter takes in calling on colleges each week. She starts at home (A), visits colleges in towns B, C, and D in that order, and then returns home.

    a. How long is the road from A to B?

    b. What is the length of the recruiter's weekly trip?

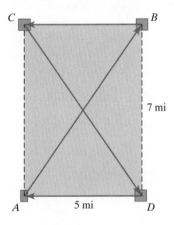

51. A manufacturer supplies machine parts to a retailer. The number n of machine parts and the total price P (in dollars) are related by the following equation:

$$P = 9\sqrt{n}$$

How much more will the manufacturer charge the retailer for 4000 machine parts than for 1000 machine parts?

52. It takes $\dfrac{\sqrt{50 - h}}{4}$ sec for an object to fall from a height of 50 ft to a height of h ft above the ground. From a height of 50 ft, how much longer does it take to drop to a height of 10 ft above the ground than to drop to a height of 40 ft above the ground?

• Check your answers on page A-25.

Groupwork

1. Working with a partner, choose several arbitrary nonnegative values for the variables a and b in the two left columns of the following table:

a	b	$a + b$	$\sqrt{a}$	$\sqrt{b}$	$\sqrt{a} + \sqrt{b}$	$\sqrt{a + b}$

   a. Using a calculator, complete the table with each entry rounded to the nearest thousandth.

   b. Compare the entries in the two rightmost columns. Are the entries in one column consistently larger than the entries in the other column? State a conjecture based on this observation.

Patterns

2. Consider the isosceles right triangle ABC shown below:

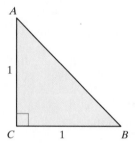

On side $\overline{AB}$, build right triangle ABD with $AD = 1$.

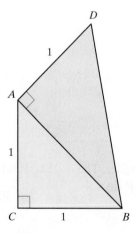

   a. Find the lengths of $\overline{AB}$ and $\overline{DB}$.
   b. How much longer is $\overline{DB}$ than $\overline{AB}$?

c. Now build another right triangle on $\overline{DB}$ with a leg of length 1. What is the length of $\overline{EB}$?

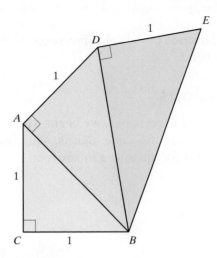

d. Describe the pattern that you observe in this *spiral* of triangles.

Writing

3. Recall that the radicals $\sqrt{2}$ and $5\sqrt{2}$ are called *like*, whereas the radicals $7\sqrt{2}$ and $\sqrt{3}$ are called *unlike*. List two other kinds of numbers where a distinction is made between like and unlike. Why are these distinctions made?

8.3 Multiplication and Division of Radical Expressions

The last section covered the addition and subtraction of radical expressions. In this section, we discuss how to find their products and quotients.

A To multiply radical expressions

Multiplying Radical Expressions

B To divide radical expressions

In considering how to find the product of radical expressions, we begin with the simplest case—multiplying two radical expressions that are the same, that is, squaring a radical expression. Recall the property used in Section 8.1 for squaring a square root: $(\sqrt{a})^2 = a$.

C To rationalize the denominator in a radical expression

D To solve applied problems involving the multiplication or division of radical expressions

EXAMPLE 1

Simplify. **a.** $\sqrt{3} \cdot \sqrt{3}$ **b.** $(\sqrt{5x})^2$ **c.** $(2\sqrt{n+1})^2$

Solution Here, we use the property of squaring a square root.

a. $\sqrt{3} \cdot \sqrt{3} = (\sqrt{3})^2 = 3$ **b.** $(\sqrt{5x})^2 = 5x$

c. $(2\sqrt{n+1})^2 = 2^2 \cdot (\sqrt{n+1})^2$ Use the rule of raising a product to a power.

$$= 4(n+1)$$
$$= 4n + 4$$

PRACTICE 1

Simplify:

a. $\sqrt{5} \cdot \sqrt{5}$

b. $(\sqrt{2y^3})^2$

c. $(3\sqrt{t+1})^2$

Now, let's look at multiplying two radical expressions that are different from one another. Here we apply the product rule of radicals, discussed in Section 8.1. Note that we can write this rule as

$$\sqrt{a} \cdot \sqrt{b} = \sqrt{ab},$$

where a and b are nonnegative numbers. In words, this rule states that the product of square roots is the square root of the product.

EXAMPLE 2

Multiply. Simplify, if possible.

a. $\sqrt{3} \cdot \sqrt{5}$ **b.** $(-5\sqrt{10})(6\sqrt{2})$ **c.** $\sqrt{12n^3} \cdot \sqrt{3n}$

Solution Here, we use the product rule of radicals before simplifying.

a. $\sqrt{3} \cdot \sqrt{5} = \sqrt{3 \cdot 5} = \sqrt{15}$

b. $(-5\sqrt{10})(6\sqrt{2}) = -5 \cdot 6 \cdot \sqrt{10 \cdot 2}$
$$= -30\sqrt{20}$$
$$= -30\sqrt{4 \cdot 5}$$
$$= -30 \cdot 2\sqrt{5}$$
$$= -60\sqrt{5}$$

c. $\sqrt{12n^3} \cdot \sqrt{3n} = \sqrt{12n^3 \cdot 3n}$
$$= \sqrt{36n^4}$$
$$= 6n^2$$

PRACTICE 2

Find the product. Simplify, if possible.

a. $\sqrt{7} \cdot \sqrt{10}$

b. $(9\sqrt{6})(-4\sqrt{3})$

c. $\sqrt{8y} \cdot \sqrt{2y^5}$

Next, we consider the multiplication of radical expressions that may contain more than one term. Just as with the multiplication of polynomials, the key here is to use the distributive property.

EXAMPLE 3

Find the product. Simplify, if possible.

a. $\sqrt{8}(2\sqrt{3} + \sqrt{2})$ **b.** $\sqrt{x}(5\sqrt{y} - 2)$

Solution

a. $\sqrt{8}(2\sqrt{3} + \sqrt{2})$

$= \sqrt{8}(2\sqrt{3}) + \sqrt{8}(\sqrt{2})$ Use the distributive property.

$= 2\sqrt{24} + \sqrt{16}$ Use the product rule of radicals.

$= 2\sqrt{4 \cdot 6} + 4$ Factor out a perfect square from the radicand 24.

$= 2 \cdot 2\sqrt{6} + 4$ Use the property of taking the square root of a square.

$= 4\sqrt{6} + 4$ Simplify.

b. $\sqrt{x}(5\sqrt{y} - 2)$

$= \sqrt{x}(5\sqrt{y}) - \sqrt{x}(2)$ Use the distributive property.

$= 5\sqrt{xy} - 2\sqrt{x}$ Use the product rule of radicals.

PRACTICE 3

Multiply. Simplify, if possible.

a. $\sqrt{6}(3\sqrt{3} - \sqrt{8})$

b. $\sqrt{a}(\sqrt{b} + 3)$

EXAMPLE 4

Find the product and simplify.

a. $(4\sqrt{2} - 1)(7\sqrt{2} + 3)$ **b.** $(2\sqrt{a} + 4)(\sqrt{a} - 1)$

Solution

a. $(4\sqrt{2} - 1)(7\sqrt{2} + 3)$

$= (4\sqrt{2} - 1)(7\sqrt{2} + 3)$

$= (4\sqrt{2})(7\sqrt{2}) + (4\sqrt{2})3$ Use the FOIL method.
$\quad + (-1)(7\sqrt{2}) + (-1)(3)$

$= 4 \cdot 7 \cdot 2 + 4 \cdot 3\sqrt{2} - 7\sqrt{2} - 3$ Use the property of squaring a square root.

$= 56 + 12\sqrt{2} - 7\sqrt{2} - 3$ Simplify.

$= 53 + 5\sqrt{2}$ Combine like terms.

b. $(2\sqrt{a} + 4)(\sqrt{a} - 1)$

$= (2\sqrt{a})(\sqrt{a}) + (2\sqrt{a})(-1) + 4\sqrt{a} + 4(-1)$ Use the FOIL method.

$= 2a - 2\sqrt{a} + 4\sqrt{a} - 4$

$= 2a + 2\sqrt{a} - 4$

PRACTICE 4

Find the product and simplify.

a. $(2\sqrt{3} + 4)(\sqrt{3} - 1)$

b. $(\sqrt{x} + 2)(3\sqrt{x} - 2)$

Recall from Section 5.6 the formula for multiplying the sum and difference of the same two terms.

$$(a + b)(a - b) = a^2 - b^2$$

In the following example, the binomial factors are radical expressions.

EXAMPLE 5

Find the product and simplify.

a. $(\sqrt{10} + 5)(\sqrt{10} - 5)$ **b.** $(\sqrt{x} - \sqrt{y})(\sqrt{x} + \sqrt{y})$

Solution   We apply the formula for the product of the sum and difference of two terms.

a. $(\sqrt{10} + 5)(\sqrt{10} - 5) = (\sqrt{10})^2 - (5)^2$
$$= 10 - 25$$
$$= -15$$

b. $(\sqrt{x} - \sqrt{y})(\sqrt{x} + \sqrt{y}) = (\sqrt{x})^2 - (\sqrt{y})^2 = x - y$

PRACTICE 5

Multiply and simplify:

a. $(\sqrt{7} - 3)(\sqrt{7} + 3)$

b. $(\sqrt{p} + \sqrt{q})(\sqrt{p} - \sqrt{q})$

Note that in Example 5, the products contain no radical sign. How would you explain why the radical signs drop out?

Recall from Section 5.6 that when squaring a binomial, the following formulas apply:

$$(a + b)^2 = a^2 + 2ab + b^2 \qquad \text{The square of a sum}$$
$$(a - b)^2 = a^2 - 2ab + b^2 \qquad \text{The square of a difference}$$

In the next example, the binomials being squared contain radicals.

EXAMPLE 6

Simplify.

a. $(\sqrt{3} + 5x)^2$ **b.** $(\sqrt{p} - 2)^2$

Solution

a. $(\sqrt{3} + 5x)^2 = (\sqrt{3})^2 + 2(\sqrt{3})(5x) + (5x)^2$ Use the
$$= (\sqrt{3})^2 + 2(5)(\sqrt{3})(x) + (5)^2x^2$$ formula for
$$= 3 + 10x\sqrt{3} + 25x^2$$ squaring a
$$= 25x^2 + 10x\sqrt{3} + 3$$ binomial sum.

b. $(\sqrt{p} - 2)^2 = (\sqrt{p})^2 - 2(\sqrt{p})(2) + 2^2$ Use the formula for
$$= p - 2(\sqrt{p})(2) + 4$$ squaring a binomial
$$= p - 4\sqrt{p} + 4$$ difference.

PRACTICE 6

Simplify.

a. $(\sqrt{2} + b)^2$

b. $(\sqrt{x} - 6)^2$

EXAMPLE 7

In an *equilateral* triangle, all three sides are equal in length. The height h of this kind of triangle is given by the expression $\dfrac{s\sqrt{3}}{2}$, where s is the length of each side of the triangle. What is the height of the triangle shown?

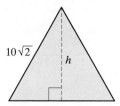

Solution   In the triangle shown, $s = 10\sqrt{2}$. So we get:

$$h = \frac{s\sqrt{3}}{2}$$

$$= \frac{10\sqrt{2} \cdot \sqrt{3}}{2}$$

$$= \frac{10\sqrt{6}}{2} = 5\sqrt{6}$$

PRACTICE 7

The number of 24-hour days that it takes a planet in our solar system to revolve once around the Sun is approximated by the expression $0.2\,(\sqrt{R})^3$, where R is the average distance of the planet from the Sun (in millions of kilometers). For the planet Mercury, the average distance is about 60 million km. How many 24-hour days does it take Mercury to revolve around the Sun, to the nearest whole day?

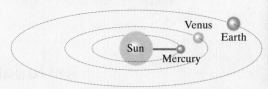

Dividing Radical Expressions

Now, let's consider dividing radical expressions. Here, we use the quotient rule of radicals discussed in Section 8.1. Note that we can write this rule as

$$\frac{\sqrt{a}}{\sqrt{b}} = \sqrt{\frac{a}{b}},$$

where a is nonnegative and b is positive. This rule allows us, in dividing two square roots, to bring the radicands under a single radical sign. We can then divide the radicands. This is particularly useful if their quotient happens to be a perfect square.

EXAMPLE 8

Find the quotient and simplify.

a. $\dfrac{\sqrt{26}}{\sqrt{2}}$ **b.** $\dfrac{\sqrt{16y^7}}{\sqrt{y}}$ **c.** $\dfrac{5\sqrt{10t}}{\sqrt{90t^3}}$

Solution   Here, we use the quotient rule of radicals before simplifying.

a. $\dfrac{\sqrt{26}}{\sqrt{2}} = \sqrt{\dfrac{26}{2}} = \sqrt{13}$ **b.** $\dfrac{\sqrt{16y^7}}{\sqrt{y}} = \sqrt{\dfrac{16y^7}{y}} = \sqrt{16y^6} = 4y^3$

c. $\dfrac{5\sqrt{10t}}{\sqrt{90t^3}} = 5\sqrt{\dfrac{10t}{90t^3}}$ Use the quotient rule of radicals.

$\qquad = 5\sqrt{\dfrac{1}{9t^2}}$ Simplify the radicand.

$\qquad = 5 \cdot \dfrac{1}{3t}$ Use the property of taking the square root of a square.

$\qquad = \dfrac{5}{3t}$

PRACTICE 8

Divide and simplify.

a. $\dfrac{\sqrt{21}}{\sqrt{3}}$

b. $\dfrac{\sqrt{4x^5}}{\sqrt{x}}$

c. $\dfrac{\sqrt{2y}}{10\sqrt{8y^5}}$

When the radicand of a radical expression is a fraction, the expression is not considered to be simplified. To simplify, we can apply the quotient rule of radicals.

Radicands with perfect squares in the denominator lend themselves to this approach.

EXAMPLE 9

Simplify.   **a.** $\sqrt{\dfrac{x^3}{y^8}}$ **b.** $\sqrt{\dfrac{9p^4}{q^{10}}}$ **c.** $\sqrt{\dfrac{7x^2}{4}}$

Solution   Using the quotient rule of radicals, we get:

a. $\sqrt{\dfrac{x^3}{y^8}} = \dfrac{\sqrt{x^2 \cdot x}}{\sqrt{y^8}} = \dfrac{x\sqrt{x}}{y^4}$ **b.** $\sqrt{\dfrac{9p^4}{q^{10}}} = \dfrac{\sqrt{9p^4}}{\sqrt{q^{10}}} = \dfrac{3p^2}{q^5}$

c. $\sqrt{\dfrac{7x^2}{4}} = \dfrac{\sqrt{7x^2}}{\sqrt{4}} = \dfrac{x\sqrt{7}}{2}$

PRACTICE 9

Simplify.

a. $\sqrt{\dfrac{m^5}{n^4}}$ **b.** $\sqrt{\dfrac{y^6}{25x^2}}$

c. $\sqrt{\dfrac{5a^2}{9}}$

Rationalizing the Denominator

Some radical expressions are written as fractions with radicals in their denominators:

$$\frac{5}{\sqrt{2}} \qquad \frac{3x}{\sqrt{x-1}} \qquad \frac{v}{\sqrt{1-m^2}}$$

This type of expression, which can be difficult to evaluate without a calculator, is not considered to be simplified. However, we can always *rationalize the denominator* of such an expression, that is, rewrite the expression in an equivalent form that contains no radical in its denominator. To do this, we multiply the numerator and denominator by a square root that will make the radicand in the denominator a perfect square.

EXAMPLE 10

Rationalize the denominator.

a. $\dfrac{3}{\sqrt{5}}$ **b.** $\dfrac{\sqrt{6}}{\sqrt{x}}$ **c.** $\dfrac{\sqrt{25y^6}}{\sqrt{8}}$

Solution   To rationalize the denominators, we multiply the numerator and denominator by a square root that makes the radicand in the denominator a perfect square.

a. $\dfrac{3}{\sqrt{5}} = \dfrac{3}{\sqrt{5}} \cdot \dfrac{\sqrt{5}}{\sqrt{5}}$ Multiply the numerator and denominator by $\sqrt{5}$.

$= \dfrac{3\sqrt{5}}{\sqrt{5^2}}$

$= \dfrac{3\sqrt{5}}{5}$

b. $\dfrac{\sqrt{6}}{\sqrt{x}} = \dfrac{\sqrt{6}}{\sqrt{x}} \cdot \dfrac{\sqrt{x}}{\sqrt{x}} = \dfrac{\sqrt{6x}}{x}$

PRACTICE 10

Rationalize the denominator.

a. $\dfrac{1}{\sqrt{2}}$

b. $\dfrac{\sqrt{5}}{\sqrt{s}}$

c. $\dfrac{\sqrt{49r^4}}{\sqrt{12}}$

c. $\dfrac{\sqrt{25y^6}}{\sqrt{8}} = \dfrac{5y^3}{\sqrt{8}}$ Take the square root in the numerator.

$$= \dfrac{5y^3}{\sqrt{8}} \cdot \dfrac{\sqrt{8}}{\sqrt{8}}$$

$$= \dfrac{5y^3\sqrt{8}}{8}$$

$$= \dfrac{5y^3(2\sqrt{2})}{8}$$

$$= \dfrac{5y^3\sqrt{2}}{4}$$

When the radicand of a radical expression is a fraction, we can simplify the radical by first applying the quotient rule and then rationalizing the denominator.

EXAMPLE 11

Simplify. **a.** $\sqrt{\dfrac{3}{5}}$ **b.** $\sqrt{\dfrac{x}{12}}$

Solution

a. $\sqrt{\dfrac{3}{5}} = \dfrac{\sqrt{3}}{\sqrt{5}} = \dfrac{\sqrt{3}}{\sqrt{5}} \cdot \dfrac{\sqrt{5}}{\sqrt{5}} = \dfrac{\sqrt{15}}{5}$

b. $\sqrt{\dfrac{x}{12}} = \dfrac{\sqrt{x}}{\sqrt{12}}$

$$= \dfrac{\sqrt{x}}{\sqrt{12}} \cdot \dfrac{\sqrt{3}}{\sqrt{3}}$$

$$= \dfrac{\sqrt{3x}}{\sqrt{36}}$$

$$= \dfrac{\sqrt{3x}}{6}$$

PRACTICE 11

Simplify.

a. $\sqrt{\dfrac{1}{6}}$

b. $\sqrt{\dfrac{n}{20}}$

To review, a radical expression is considered simplified if the following conditions are all met:

	Simplified	Not Simplified
• The radicand has no factor that is a perfect square.	$5\sqrt{2}$	$\sqrt{50} = \sqrt{25 \cdot 2}$
• There are no fractions under the radical sign.	$\dfrac{\sqrt{3}}{2}$	$\sqrt{\dfrac{3}{4}}$
• There are no radicals in the denominator.	$\dfrac{\sqrt{6}}{3}$	$\dfrac{2}{\sqrt{6}}$

Some radical expressions are in the form of fractions, with more than one term in the numerator. We may need to rationalize the denominator of such an expression.

EXAMPLE 12

Rationalize the denominator.

a. $\dfrac{\sqrt{2} + 3}{\sqrt{5}}$ b. $\dfrac{\sqrt{x} - 1}{\sqrt{y}}$

Solution

a. $\dfrac{\sqrt{2} + 3}{\sqrt{5}} = \dfrac{(\sqrt{2} + 3)}{\sqrt{5}} \cdot \dfrac{\sqrt{5}}{\sqrt{5}}$

$= \dfrac{(\sqrt{2} + 3)\sqrt{5}}{5}$

$= \dfrac{(\sqrt{2})(\sqrt{5}) + 3\sqrt{5}}{5}$

$= \dfrac{\sqrt{10} + 3\sqrt{5}}{5}$

b. $\dfrac{\sqrt{x} - 1}{\sqrt{y}} = \dfrac{\sqrt{x} - 1}{\sqrt{y}} \cdot \dfrac{\sqrt{y}}{\sqrt{y}}$

$= \dfrac{(\sqrt{x} - 1)\sqrt{y}}{y}$

$= \dfrac{(\sqrt{x})(\sqrt{y}) - \sqrt{y}}{y}$

$= \dfrac{\sqrt{xy} - \sqrt{y}}{y}$

PRACTICE 12

Rationalize the denominator.

a. $\dfrac{\sqrt{5} - 1}{\sqrt{3}}$

b. $\dfrac{\sqrt{c} + 2}{\sqrt{b}}$

Thus far, we have rationalized denominators containing a single radical term. But suppose that there are two radical terms in a denominator. In this case, the key is to identify the *conjugate* of the denominator. The expressions $a + b$ and $a - b$ are called conjugates of one another. Note that conjugates come in pairs, since they are the sum and difference of the same two terms.

Recall that in Example 5(a) we multiplied two radical expressions, $\sqrt{10} + 5$ and $\sqrt{10} - 5$, which are conjugates of one another. We used the formula $(a + b)(a - b) = a^2 - b^2$, and then observed that the product -15 contains no radical sign. The elimination of radical signs suggests a procedure for rationalizing a denominator with two terms, namely, *multiplying both the numerator and the denominator by the conjugate of the denominator.*

EXAMPLE 13

Rationalize the denominator.

a. $\dfrac{4}{1 + \sqrt{3}}$

b. $\dfrac{x}{\sqrt{y} - \sqrt{2}}$

PRACTICE 13

Rationalize the denominator.

a. $\dfrac{8}{3 - \sqrt{2}}$

b. $\dfrac{a}{\sqrt{b} + \sqrt{5}}$

Solution

a. $\dfrac{4}{1+\sqrt{3}} = \dfrac{4}{1+\sqrt{3}} \cdot \dfrac{1-\sqrt{3}}{1-\sqrt{3}}$

Multiply the numerator and the denominator by the conjugate of the denominator.

$$= \dfrac{4(1-\sqrt{3})}{(1+\sqrt{3})(1-\sqrt{3})}$$

$$= \dfrac{4-4\sqrt{3}}{1^2-(\sqrt{3})^2}$$

Use the formula for findng the product of the sum and difference of two terms.

$$= \dfrac{4-4\sqrt{3}}{1-3}$$

$$= \dfrac{2(2-2\sqrt{3})}{-2}$$

$$= \dfrac{\overset{1}{\cancel{2}}(2-2\sqrt{3})}{\underset{1}{\cancel{-2}}}$$

$$= -2+2\sqrt{3}$$

b. $\dfrac{x}{\sqrt{y}-\sqrt{2}} = \dfrac{x}{\sqrt{y}-\sqrt{2}} \cdot \dfrac{\sqrt{y}+\sqrt{2}}{\sqrt{y}+\sqrt{2}}$

Multiply the numerator and the denominator by the conjugate of the denominator.

$$= \dfrac{x(\sqrt{y}+\sqrt{2})}{(\sqrt{y}-\sqrt{2})(\sqrt{y}+\sqrt{2})}$$

$$= \dfrac{x\sqrt{y}+x\sqrt{2}}{y-2}$$

Use the formula for finding the product of the sum and difference of two terms.

In Examples 13(a) and (b), note that when we multiplied the numerator and denominator by the conjugate of the denominator, the radical sign is eliminated in the denominator, as expected.

EXAMPLE 14

The velocity v (in kilometers per second) of a meteor streaking toward the Earth can be modeled by the expression $\dfrac{450}{\sqrt{d}}$, where d is its distance from the center of the Earth (in kilometers). Find the velocity of a meteor when it is 20,000 km from the Earth's center, written as a radical in simplified form. Using a calculator, also express this velocity as a decimal rounded to the nearest kilometer per second.

Solution We substitute 20,000 for d in $v = \dfrac{450}{\sqrt{d}}$.

$$v = \dfrac{450}{\sqrt{20{,}000}} = \dfrac{450}{\sqrt{10{,}000 \cdot 2}} = \dfrac{450}{\sqrt{10{,}000} \cdot \sqrt{2}}$$

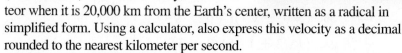

$$= \dfrac{\overset{9}{\cancel{450}}}{\underset{2}{\cancel{100}}\sqrt{2}} = \dfrac{9}{2\sqrt{2}} = \dfrac{9}{2\sqrt{2}} \cdot \dfrac{\sqrt{2}}{\sqrt{2}} = \dfrac{9\sqrt{2}}{2 \cdot 2} = \dfrac{9\sqrt{2}}{4}$$

So the velocity of the meteor is $\dfrac{9\sqrt{2}}{4}$ km/sec, or approximately 3 km/sec.

PRACTICE 14

The formula $P = \dfrac{590}{\sqrt{t}}$ approximates the pulse rate (in beats per minute) for an adult who is t in. tall. Find the pulse rate of an adult 72 in. tall, written as a radical expression in simplified form. Using a calculator, also express this pulse rate rounded to the nearest whole number of beats per minute.

Mathematically Speaking

Fill in each blank with the most appropriate term or phrase from the given list.

> simplify
>
> take the square roots before dividing
>
> distributive property
>
> contain more than one term
>
> associative property of multiplication
>
> are different from one another
>
> bring the radicands under a single radical sign
>
> rationalize

1. To multiply two radical expressions that _____, we use the product rule $\sqrt{a} \cdot \sqrt{b} = \sqrt{ab}$.

2. The key to multiplying radical expressions that contain more than one term is to use the _____.

3. The quotient rule of radicals, $\dfrac{\sqrt{a}}{\sqrt{b}} = \sqrt{\dfrac{a}{b}}$, where $a \geq 0$ and $b > 0$, allows us to _____.

4. To _____ the denominator of an expression means to rewrite the expression in an equivalent form that contains no radical in its denominator.

Multiply. Simplify, if possible.

5. $\sqrt{21} \cdot \sqrt{21}$

6. $\sqrt{17} \cdot \sqrt{17}$

7. $\sqrt{15} \cdot \sqrt{15}$

8. $\sqrt{13} \cdot \sqrt{13}$

9. $(\sqrt{3n})^2$

10. $(\sqrt{4x})^2$

11. $(\sqrt{5y})^2$

12. $(\sqrt{7a})^2$

13. $(4\sqrt{x-1})^2$

14. $(3\sqrt{y+1})^2$

15. $(2\sqrt{t+5})^2$

16. $(5\sqrt{n-3})^2$

17. $\sqrt{18} \cdot \sqrt{3}$

18. $\sqrt{6} \cdot \sqrt{15}$

19. $(-2\sqrt{5})(7\sqrt{10})$

20. $(3\sqrt{24})(-5\sqrt{2})$

21. $\sqrt{8x^3} \cdot \sqrt{2x}$

22. $\sqrt{3x} \cdot \sqrt{12x}$

23. $\sqrt{3r} \cdot \sqrt{5r}$

24. $\sqrt{2n} \cdot \sqrt{7n}$

25. $\sqrt{2x} \cdot \sqrt{5} \cdot \sqrt{10y}$

26. $\sqrt{6y} \cdot \sqrt{7xy} \cdot \sqrt{2x}$

27. $\sqrt{3}(\sqrt{3} - 1)$

28. $\sqrt{2}(\sqrt{2} + 10)$

29. $\sqrt{x}(\sqrt{x} - 7)$

30. $\sqrt{y}(8 + \sqrt{y})$

31. $\sqrt{a}(4\sqrt{b} + 1)$

32. $\sqrt{p}(3\sqrt{q} - 5)$

33. $(\sqrt{5} + 3)(\sqrt{5} + 2)$

34. $(4 - \sqrt{7})(3 + \sqrt{7})$

35. $(\sqrt{7} - \sqrt{9})(\sqrt{7} + \sqrt{9})$

36. $(\sqrt{6} + \sqrt{4})(\sqrt{6} - \sqrt{4})$

37. $(8\sqrt{3} + 1)(5\sqrt{3} - 2)$

38. $(4\sqrt{2} + 1)(5\sqrt{2} - 3)$

39. $(\sqrt{n} + 5)(3\sqrt{n} - 1)$

40. $(2 - 5\sqrt{t})(4 + \sqrt{t})$

41. $(6 - \sqrt{3})(6 + \sqrt{3})$

42. $(\sqrt{2} - 1)(\sqrt{2} + 1)$

43. $(5 + 2\sqrt{3})(5 - 2\sqrt{3})$

44. $(4\sqrt{10} - 3)(4\sqrt{10} + 3)$

45. $(\sqrt{x} + 2)(\sqrt{x} - 2)$

46. $(5 - \sqrt{y})(5 + \sqrt{y})$

47. $(\sqrt{a} + \sqrt{b})(\sqrt{a} - \sqrt{b})$

48. $(\sqrt{m} - \sqrt{n})(\sqrt{m} + \sqrt{n})$

49. $(\sqrt{3x} - \sqrt{y})(\sqrt{3x} + \sqrt{y})$

50. $(\sqrt{q} + \sqrt{5p})(\sqrt{q} - \sqrt{5p})$

51. $(\sqrt{2} - x)^2$

52. $(4y + \sqrt{3})^2$

53. $(\sqrt{x} - 1)^2$

54. $(2 + \sqrt{n})^2$

3 *Find the quotient and simplify.*

55. $\dfrac{\sqrt{15}}{\sqrt{3}}$

56. $\dfrac{\sqrt{10}}{\sqrt{2}}$

⊙ 57. $\dfrac{\sqrt{5}}{\sqrt{125}}$

58. $\dfrac{\sqrt{6}}{\sqrt{24}}$

59. $\dfrac{\sqrt{4a^3}}{\sqrt{a}}$

60. $\dfrac{\sqrt{9d^7}}{\sqrt{d^3}}$

⊙ 61. $\dfrac{4\sqrt{5y}}{\sqrt{45y^5}}$

62. $\dfrac{\sqrt{3x^4}}{5\sqrt{48x^8}}$

⊙ 63. $\sqrt{\dfrac{a^4}{b^6}}$

64. $\sqrt{\dfrac{m^8}{n^2}}$

65. $\sqrt{\dfrac{16x^{12}}{y^8}}$

66. $\sqrt{\dfrac{49p^6}{q^{10}}}$

⊙ 67. $\sqrt{\dfrac{5x^{10}}{36}}$

68. $\sqrt{\dfrac{7y^2}{100}}$

Simplify.

⊙ 69. $\dfrac{2}{\sqrt{3}}$

70. $\dfrac{1}{\sqrt{2}}$

71. $\dfrac{\sqrt{5}}{\sqrt{y}}$

72. $\dfrac{\sqrt{7}}{\sqrt{a}}$

73. $\sqrt{\dfrac{2}{11}}$

74. $\sqrt{\dfrac{3}{7}}$

⊙ 75. $\sqrt{\dfrac{x^2}{5}}$

76. $\sqrt{\dfrac{n^4}{3}}$

77. $\sqrt{\dfrac{t}{50}}$

78. $\sqrt{\dfrac{y}{32}}$

79. $\sqrt{\dfrac{a}{2}}$

80. $\sqrt{\dfrac{6y}{5}}$

Rationalize the denominator.

81. $\dfrac{\sqrt{5} + 2}{\sqrt{3}}$

82. $\dfrac{6 - \sqrt{2}}{\sqrt{10}}$

83. $\dfrac{\sqrt{n} - 1}{\sqrt{m}}$

84. $\dfrac{4 - \sqrt{a}}{\sqrt{b}}$

⊙ 85. $\dfrac{15}{4 + \sqrt{6}}$

86. $\dfrac{8}{\sqrt{3} + 1}$

87. $\dfrac{11}{4 - \sqrt{5}}$

88. $\dfrac{10}{\sqrt{7} - 2}$

89. $\dfrac{4}{\sqrt{5} - \sqrt{3}}$

90. $\dfrac{5}{\sqrt{6} + \sqrt{2}}$

91. $\dfrac{a}{\sqrt{b} - \sqrt{3}}$

92. $\dfrac{x}{\sqrt{5} + \sqrt{y}}$

Mixed Practice

Rationalize the denominator.

93. $\dfrac{\sqrt{c} - 6}{\sqrt{d}}$

94. $\dfrac{32}{5 - \sqrt{17}}$

Find the quotient and simplify, if possible.

95. $\dfrac{\sqrt{81a^5}}{\sqrt{a}}$

96. $\dfrac{\sqrt{5b^3}}{4\sqrt{45b^7}}$

97. $\sqrt{\dfrac{64s^8}{t^{12}}}$

Simplify.

98. $\dfrac{\sqrt{15}}{\sqrt{y}}$

99. $\sqrt{\dfrac{c}{72}}$

Multiply, and then simplify.

100. $\sqrt{20k} \cdot \sqrt{5k^3}$

101. $\sqrt{r}(1 + 3\sqrt{r})$

102. $(3\sqrt{11} - 4)(3\sqrt{11} + 4)$

103. $(\sqrt{7m} - \sqrt{n})(\sqrt{7m} + \sqrt{n})$

104. $(8x + \sqrt{6})^2$

Applications

Solve.

105. Variance (V) and standard deviation (s) are two measures that statisticians use for studying data sets. These two measures are related by the formula $s = \sqrt{V}$. If one data set has double the variance of another, how many times the standard deviation of the second data set is that of the first?

106. The expression $\sqrt{2gh}$ can be used to determine the velocity of a free-falling object (in feet per second), where $g = 32$ ft/sec^2, and the object has fallen h ft. If a ball has fallen 60 ft, is the ball's velocity double that of a ball that has fallen 30 ft?

107. The distance d between two points (x_1, y_1) and (x_2, y_2) on the coordinate plane is given by the formula
$$d = \sqrt{(x_2 - x_1)^2 + (y_2 - y_1)^2}.$$
Check that point $(6, 10)$ is twice as far from the origin as the point $(3, 5)$.

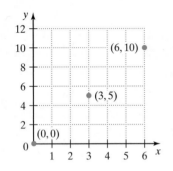

108. What is the area of the cross section of the pyramid pictured?

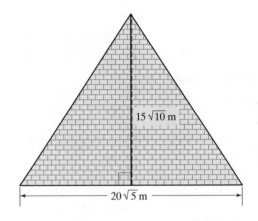

109. A hailstone will take t sec to drop d ft, where
$$t = \sqrt{\frac{d}{16}}.$$
How much time will it take a hailstone to drop 500 ft?

110. Chemists study the motion of gas molecules, called *diffusion*. The rates of diffusion (r_1 and r_2) of two gases are related by the formula
$$\frac{r_1}{r_2} = \frac{\sqrt{m_1}}{\sqrt{m_2}},$$
where m_1 and m_2 are the masses of the gas molecules. Find the ratio $\frac{r_1}{r_2}$, if $m_1 = 44$ units and $m_2 = 4$ units. Express the answer as a simplified radical.

111. An expression for finding the radius r of a cylindrical subwoofer is $\sqrt{\dfrac{V}{\pi h}}$, where V is the volume of the subwoofer, and h is its height. Simplify this expression, rationalizing the denominator.

112. Suppose that an investment of P dollars grows at a fixed annual rate of return r. If at the end of two years the investment is worth A dollars, then the rate of return can be found from the formula

$$r = \frac{\sqrt{A} - \sqrt{P}}{\sqrt{P}}.$$

Rewrite this expression, rationalizing the denominator.

• Check your answers on page A-26.

MINDStretchers

Writing

1. Explain why each of the following radical expressions is *not* simplified.

    a. $\sqrt{x^2 - 2ax + a^2}$ **b.** $\sqrt{\dfrac{x}{y}}$ **c.** $\dfrac{p}{\sqrt{q}}$

Mathematical Reasoning

2. Many quadratic equations have solutions that involve radical expressions. Determine whether $x = 1 - \sqrt{2}$ is a solution of the equation $x^2 - 2x = 1$.

Groupwork

3. Heron's formula states that the area of any triangle is equal to $\sqrt{s(s - a)(s - b)(s - c)}$, where the side lengths of the triangle are a, b, and c, and s is *half* the sum of the three side lengths, that is, $s = \dfrac{a + b + c}{2}$.

    a. Working with a partner, apply Heron's formula to three triangles of your choice, filling in the four left columns in the following table.

Side Length a	Side Length b	Side Length c	Area of the Triangle	Area of the Triangle with Doubled Sides

    b. For each of these triangles, form a new triangle whose sides are double the length of the sides of the original triangle. Use Heron's formula to compute the area of these new triangles and enter the results in the right column of the previous table.

    c. In your three examples, does doubling all the side lengths of a triangle double its area?

Finally, let's turn our attention to solving *radical equations*.

DEFINITION
A **radical equation** is an equation with a variable in one or more radicands.

Some examples of radical equations are:

$$\sqrt{3x} = 18 \qquad \sqrt{2n} = \sqrt{3n + 1} \qquad y - 2 = \sqrt{y} + 1$$

Can you explain why the equation $x + 3 = \sqrt{5}$ is not a radical equation?

The key to solving a radical equation is to find an equivalent equation with no square root. To do this, we can use the *squaring property of equality*.

The Squaring Property of Equality

For any real numbers a and b, if $a = b$, then $a^2 = b^2$.

In words, this property states that if two numbers are equal, then their squares are equal.

In solving a radical equation, we begin by *isolating the radical*. Applying the squaring property of equality, we then square each side of the equation. Squaring eliminates the radical and makes the equation easier to solve. However, the resulting equation may have solutions that are not solutions to the original equation. Recall that such "solutions" are called extraneous solutions. So when solving a radical equation, it is particularly important to check all possible solutions in the original equation.

EXAMPLE 1

Solve and check: $\sqrt{x} - 2 = 5$

Solution

$$\sqrt{x} - 2 = 5$$
$$\sqrt{x} - 2 + 2 = 5 + 2 \qquad \text{Add 2 to each side of the equation.}$$
$$\sqrt{x} = 7 \qquad \text{Simplify.}$$
$$(\sqrt{x})^2 = (7)^2 \qquad \text{Use the squaring property of equality.}$$
$$x = 49$$

Check $\sqrt{x} - 2 = 5$

$$\sqrt{49} - 2 \overset{?}{=} 5 \qquad \text{Substitute 49 for } x \text{ in the original equation.}$$
$$7 - 2 \overset{?}{=} 5$$
$$5 = 5 \qquad \text{True}$$

So 49 is the solution.

PRACTICE 1

Solve and check: $\sqrt{y} + 3 = 7$

EXAMPLE 2

Solve and check: $\sqrt{x+1} + 3 = 0$

Solution
$$\sqrt{x+1} + 3 = 0$$
$$\sqrt{x+1} = -3$$
$$(\sqrt{x+1})^2 = (-3)^2$$
$$x + 1 = 9$$
$$x = 8$$

Check
$$\sqrt{x+1} + 3 = 0$$
$$\sqrt{8+1} + 3 \overset{?}{=} 0 \qquad \text{Substitute 8 for } x \text{ in the original equation.}$$
$$\sqrt{9} + 3 \overset{?}{=} 0$$
$$3 + 3 \overset{?}{=} 0$$
$$6 \neq 0 \qquad \text{False}$$

Our check fails, so 8 is an extraneous solution, that is, it is *not* a solution to the original equation. Since 8 is the only possible solution, the equation has no solution.

PRACTICE 2

Solve and check: $\sqrt{2t-5} + 7 = 0$

Some radical equations involve more than one square root, as in the following example:

EXAMPLE 3

Solve and check: $\sqrt{2n+1} = \sqrt{5n-2}$

Solution   In this equation, the radicals are already isolated, so we begin by using the squaring property of equality to square each side of the equation.
$$\sqrt{2n+1} = \sqrt{5n-2}$$
$$(\sqrt{2n+1})^2 = (\sqrt{5n-2})^2$$
$$2n + 1 = 5n - 2$$
$$1 = 3n - 2$$
$$3 = 3n$$
$$n = 1$$

Check
$$\sqrt{2n+1} = \sqrt{5n-2}$$
$$\sqrt{2\cdot1+1} \overset{?}{=} \sqrt{5\cdot1-2} \qquad \begin{array}{l}\text{Substitute 1 for } n \text{ in the}\\ \text{original equation.}\end{array}$$
$$\sqrt{3} = \sqrt{3} \qquad \text{True}$$

So 1 is the solution.

PRACTICE 3

Solve and check:
$$\sqrt{4x+7} = \sqrt{6x-11}$$

These examples suggest the following rule:

To Solve a Radical Equation

- Isolate a term with a radical.
- Square each side of the equation.
- Where possible, combine like terms.
- Solve the resulting equation.
- Check the possible solution(s) in the original equation.

Some radical equations are equivalent to quadratic equations with more than one solution.

EXAMPLE 4

Solve and check: $1 + \sqrt{1 - x} = x$

Solution

$$1 + \sqrt{1 - x} = x$$

$$\sqrt{1 - x} = x - 1 \qquad \text{Isolate the radical.}$$

$$(\sqrt{1 - x})^2 = (x - 1)^2 \qquad \text{Square each side of the equation.}$$

$$1 - x = x^2 - 2x + 1$$

$$0 = x^2 - x$$

$$x(x - 1) = 0 \qquad \text{Factor.}$$

$$x = 0 \quad \text{or} \quad x - 1 = 0 \qquad \text{Set each factor equal to 0.}$$

$$x = 1$$

Check

Substitute 0 for x.

$$1 + \sqrt{1 - x} = x$$

$$1 + \sqrt{1 - 0} \stackrel{?}{=} 0$$

$$1 + 1 \stackrel{?}{=} 0$$

$$2 \neq 0 \qquad \text{False}$$

Substitute 1 for x.

$$1 + \sqrt{1 - x} = x$$

$$1 + \sqrt{1 - 1} \stackrel{?}{=} 1$$

$$1 + 0 \stackrel{?}{=} 1$$

$$1 = 1 \qquad \text{True}$$

We see that $x = 1$ is a solution, whereas $x = 0$ is not. So the only solution to the original equation is 1.

EXAMPLE 5

Solve and check: $2\sqrt{x + 6} = \sqrt{x^2 + 19}$

Solution

$$2\sqrt{x + 6} = \sqrt{x^2 + 19}$$

$$(2\sqrt{x + 6})^2 = (\sqrt{x^2 + 19})^2 \qquad \text{Square each side of the equation.}$$

$$4(x + 6) = x^2 + 19$$

$$4x + 24 = x^2 + 19$$

$$0 = x^2 - 4x - 5$$

$$(x + 1)(x - 5) = 0$$

$$x + 1 = 0 \quad \text{or} \quad x - 5 = 0$$

$$x = -1 \qquad x = 5$$

Check

Substitute -1 for x.

$$2\sqrt{x + 6} = \sqrt{x^2 + 19}$$

$$2\sqrt{(-1) + 6} \stackrel{?}{=} \sqrt{(-1)^2 + 19}$$

$$2\sqrt{5} \stackrel{?}{=} \sqrt{20}$$

$$2\sqrt{5} = 2\sqrt{5} \qquad \text{True}$$

Substitute 5 for x.

$$2\sqrt{x + 6} = \sqrt{x^2 + 19}$$

$$2\sqrt{5 + 6} \stackrel{?}{=} \sqrt{(5^2) + 19}$$

$$2\sqrt{11} \stackrel{?}{=} \sqrt{44}$$

$$2\sqrt{11} = 2\sqrt{11} \qquad \text{True}$$

So both -1 and 5 are solutions.

EXAMPLE 6

The approximate time t (in seconds) that it takes an object to fall freely a distance d (in feet) is given by the formula

$$t = \sqrt{\frac{d}{16}}.$$

If a skydiver had 3 sec of free fall, from what height did he leap?

Solution

$$t = \sqrt{\frac{d}{16}}$$

$$3 = \sqrt{\frac{d}{16}}$$

$$(3)^2 = \left(\sqrt{\frac{d}{16}}\right)^2$$

$$9 = \frac{d}{16}$$

$$144 = d, \quad \text{or} \quad d = 144$$

Check

$$t = \sqrt{\frac{d}{16}}$$

$$3 \stackrel{?}{=} \sqrt{\frac{144}{16}}$$

$$3 \stackrel{?}{=} \sqrt{9}$$

$$3 = 3 \quad \text{True}$$

So the skydiver must have leaped from a height of 144 ft.

PRACTICE 6

The sharper a road turns, the lower the speed a car can safely travel on the road without skidding. A formula used to find the maximum safe speed on a road is

$$s = \sqrt{2.5r},$$

where r is the radius of the road's curve (in feet) and s is the maximum safe speed of the car (in miles per hour). Find the radius r that will permit a maximum safe speed of 50 mph.

Recall from Section 2.4 that formulas state a relationship between two or more variables. In some cases, these formulas contain radicals.

EXAMPLE 7

To escape a planet's gravity (in meters per second^2), a spacecraft must achieve an initial velocity v (in meters per second) given by the formula

$$v = \sqrt{2gR},$$

where g is the planet's force of gravity and R is the planet's radius (in meters). Solve this formula for R.

Solution

$$v = \sqrt{2gR}$$

$$v^2 = 2gR$$

$$\frac{v^2}{2g} = R$$

$$R = \frac{v^2}{2g}$$

PRACTICE 7

The distance d (in miles) that a passenger can see from an airplane at altitude h (in feet) on a clear day can be approximated using the formula

$$d = \sqrt{\frac{3h}{2}}.$$

Solve this formula for h.

Mathematically Speaking

Fill in each blank with the most appropriate term or phrase from the given list.

rational equation	extraneous solution	their squares
radical solution	their square roots	square
isolate	radical equation	

1. A(n) _____ is an equation in which a variable appears in one or more radicands.

2. The squaring property of equality states that if two numbers are equal, then _____ are equal.

3. In solving a radical equation, the first step is to _____ a term with a radical.

4. A possible solution of a radical equation that does not make the equation true is called a(n) _____.

Solve and check.

5. $\sqrt{x} = 3$

6. $\sqrt{n} = 7$

7. $\sqrt{2x} = 8$

8. $\sqrt{3x} = 9$

9. $\sqrt{x} + 6 = 0$

10. $\sqrt{4s} + 1 = 0$

11. $\sqrt{a} - 4 = 4$

12. $\sqrt{y} - 10 = -1$

13. $\sqrt{x + 3} = 3$

14. $\sqrt{n - 2} = 7$

15. $\sqrt{3t + 7} = 4$

16. $\sqrt{2t + 1} = 1$

17. $\sqrt{9t - 14} = \sqrt{2t}$

18. $\sqrt{3y + 2} = \sqrt{5y}$

19. $\sqrt{4x + 1} = \sqrt{2x + 7}$

20. $\sqrt{7x - 5} = \sqrt{4x + 13}$

21. $-2\sqrt{n - 2} = \sqrt{3n + 4}$

22. $\sqrt{5y + 2} = -3\sqrt{y - 2}$

23. $\sqrt{y - 1} + 4 = 6$

24. $1 + \sqrt{n + 4} = 11$

25. $3 - \sqrt{3x + 1} = 2$

26. $12 - \sqrt{2y - 1} = 7$

⊙ 27. $\sqrt{x + 6} - x = 4$

28. $\sqrt{n + 8} - n = 2$

⊙ 29. $7 + \sqrt{2x + 9} = x + 4$

30. $\sqrt{4y^2 + 5} = y + 4$

31. $7\sqrt{v} + v = -10$

32. $x + 6\sqrt{x} = -8$

33. $n - 3\sqrt{n + 2} = -4$

34. $y - 2\sqrt{y + 5} = -2$

35. $5\sqrt{y - 6} = \sqrt{y^2 - 14}$

36. $\sqrt{x^2 - 9} = 4\sqrt{x - 3}$

37. $\sqrt{2n^2 - 7} + 3 = n$

38. $\sqrt{4 - 3x} + 10 = x + 8$

39. $\sqrt{4x + 13} - 2x = -1$

40. $\sqrt{3x + 7} + 5 = 3x$

Mixed Practice

Solve and check.

41. $a - 4\sqrt{a + 5} = -8$

42. $\sqrt{2x^2 - 2x + 4} = x + 1$

43. $\sqrt{t} - 11 = -3$

44. $13 - \sqrt{3m - 5} = 9$

45. $\sqrt{7y + 8} = 6$

46. $\sqrt{9a - 13} = \sqrt{5a + 7}$

47. $\sqrt{9 - 4x} = x - 3$

48. $\sqrt{x^2 + 2} = 3\sqrt{x - 2}$

Applications

Solve.

49. An electrical appliance with resistance R (in ohms) draws current I (in amps) and has power P (in watts). These quantities are related by the following formula:

$$I = \sqrt{\frac{P}{R}}$$

If an appliance has a resistance of 25 ohms and draws 10 amps, find its power.

50. For a sphere, the formula $r = \sqrt{\dfrac{S}{4\pi}}$ relates the radius r and the surface area S of the sphere. Find the surface area of a sphere with radius 8 in. Write your answer in terms of π.

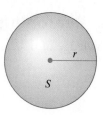

51. The period of a spring is the time it takes for the spring to stretch from one position, down, and then back again to its original position. The formula $T = 2\pi\sqrt{\dfrac{m}{k}}$, known as Hooke's law, expresses the period T (in seconds) in terms of the mass m (in grams) bobbing on the spring, where k is a constant. If $k = 8$, what mass will produce a period of 2 sec?

52. The formula for the diagonal of a rectangle d in terms of its length l and width w is:

$$d = \sqrt{l^2 + w^2}$$

Find the length of a rectangle if its width is 12 in. and its diagonal is 20 in.

53. The age of a tree t (in years) can be approximated using the formula

$$t = \frac{\sqrt{\dfrac{A}{\pi}} - b}{g},$$

where g is the width of each growth ring (in inches), b is the bark width (in inches), and A is the area of the tree's cross-section (in square inches). (*Source:* National Climatic Data Center.)

a. Solve this formula for A.

b. Using this model, find the cross-section area of a 16-year-old tree with bark width $\dfrac{2}{3}$ in. and growth ring width $\dfrac{1}{3}$ in.

54. An investment pays an annual rate of return r on a principal of P dollars. The value V of the investment (in dollars) after 2 yr is related to the rate of return and the principal as follows:

$$r = \frac{\sqrt{V}}{\sqrt{P}} - 1$$

a. Solve this formula for V.

b. If the principal was \$100 and the rate of return is 0.05, what is the value of the investment?

• Check your answers on page A-26.

Groupwork

1. In general, $\sqrt{a} + \sqrt{b} \neq \sqrt{a + b}$. Working with a partner, determine if there are any positive numbers a and b for which $\sqrt{a} + \sqrt{b} = \sqrt{a + b}$. Justify your answer.

Mathematical Reasoning

2. Using your knowledge of radical equations, show how you would solve the following equation.

$$\sqrt{x - 16} + \sqrt{x + 11} = 9$$

Technology

3. On a coordinate plane, graph the radical equation $y = \sqrt{x}$

   a. by using a table.

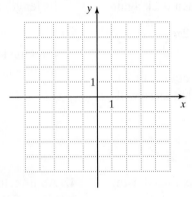

   b. by using a graphing calculator or graphing software.

Concept/Skill	Description	Example
[8.1] **Square root (of a nonnegative real number a)**	A number that when squared is a.	2 and -2 are square roots of 4 because $(2)^2 = 4$ and $(-2)^2 = 4$.
[8.1] **Radicand**	The number under a radical sign.	The radicand of $\sqrt{5}$ is 5.
[8.1] **Perfect square**	A whole number that is the square of another whole number.	9 is a perfect square because $$9 = 3^2.$$
[8.1] **Squaring a square root**	For any nonnegative real number a, $$(\sqrt{a})^2 = a.$$	$(\sqrt{8})^2 = 8$
[8.1] **Taking the square root of a square**	For any nonnegative real number a, $$\sqrt{a^2} = a.$$	$\sqrt{3^2} = 3$
[8.1] **The product rule of radicals**	If a and b are any nonnegative real numbers, then $$\sqrt{ab} = \sqrt{a} \cdot \sqrt{b}.$$	$\sqrt{4 \cdot 9} = \sqrt{4} \cdot \sqrt{9}$
[8.1] **The quotient rule of radicals**	If a is a nonnegative real number and b is a positive real number, then $$\sqrt{\frac{a}{b}} = \frac{\sqrt{a}}{\sqrt{b}}.$$	$\sqrt{\frac{4}{9}} = \frac{\sqrt{4}}{\sqrt{9}}$
[8.2] **Like radicals**	Radical expressions that have the same radicand.	$\sqrt{2}$ and $5\sqrt{2}$ are *like* radicals.
[8.2] **Unlike radicals**	Radical expressions with different radicands.	$7\sqrt{2}$ and $\sqrt{3}$ are *unlike* radicals.
[8.2] **To add or subtract *like* radicals**	• Use the distributive property. • Then, simplify.	$4\sqrt{2} + 3\sqrt{2} = (4+3)\sqrt{2} = 7\sqrt{2}$
[8.2] **To add or subtract *unlike* radicals**	• Simplify the unlike radicals, if possible. • If like radicals, then add or subtract.	$\sqrt{12} + \sqrt{27} = \sqrt{4 \cdot 3} + \sqrt{9 \cdot 3}$ $= 2\sqrt{3} + 3\sqrt{3}$ $= 5\sqrt{3}$
[8.3] **To multiply radicals**	• Apply the product rule of radicals. • Simplify, if possible.	$(\sqrt{10})(2\sqrt{2}) = 2 \cdot \sqrt{10 \cdot 2}$ $= 2\sqrt{20}$ $= 2 \cdot 2\sqrt{5}$ $= 4\sqrt{5}$
[8.3] **To divide radicals**	• Apply the quotient rule of radicals. • Simplify, if possible.	$\frac{\sqrt{20x^4}}{\sqrt{5x^2}} = \sqrt{\frac{20x^4}{5x^2}}$ $= \sqrt{4x^2}$ $= 2x$

continued

Concept/Skill	Description	Example
[8.3] To rationalize a denominator	• If the denominator is a radical term, then multiply the numerator and denominator by the denominator. • If the denominator is a binomial with a radical term, then multiply the numerator and denominator by the conjugate of the denominator.	$\dfrac{\sqrt{x}}{\sqrt{8}} = \dfrac{\sqrt{x}}{\sqrt{8}} \cdot \dfrac{\sqrt{8}}{\sqrt{8}}$ $= \dfrac{\sqrt{8x}}{\sqrt{8^2}}$ $= \dfrac{\sqrt{4 \cdot 2x}}{8}$ $= \dfrac{2\sqrt{2x}}{8} = \dfrac{\sqrt{2x}}{4}$ $\dfrac{4}{1-\sqrt{3}} = \dfrac{4}{1-\sqrt{3}} \cdot \dfrac{1+\sqrt{3}}{1+\sqrt{3}}$ $= \dfrac{4(1+\sqrt{3})}{(1-\sqrt{3})(1+\sqrt{3})}$ $= \dfrac{4(1+\sqrt{3})}{1^2 - (\sqrt{3})^2}$ $= \dfrac{4(1+\sqrt{3})}{1-3}$ $= \dfrac{4(1+\sqrt{3})}{-2}$ $= -2(1+\sqrt{3})$ $= -2 - 2\sqrt{3}$
[8.4] Radical equation	An equation with a variable in one or more radicands.	$\sqrt{3x} = 18$
[8.4] The squaring property of equality	For any real numbers a and b, if $a = b$, then $a^2 = b^2$.	If $\sqrt{x} = 7$, then $(\sqrt{x})^2 = (7)^2$.
[8.4] To solve a radical equation	• Isolate a term with a radical. • Square each side of the equation. • Where possible, combine like terms. • Solve the resulting equation. • Check the possible solution(s) in the original equation.	$\sqrt{x} + 1 = 5$ $\sqrt{x} = 4$ $(\sqrt{x})^2 = 4^2$ $x = 16$ **Check** $\sqrt{x} + 1 = 5$ $\sqrt{16} + 1 \overset{?}{=} 5$ $4 + 1 \overset{?}{=} 5$ $5 = 5 \quad$ True

Say Why
Fill in each blank.

1. The number $\sqrt{6}$ _____ an irrational number
   $\frac{}{\text{is/is not}}$
   because _____
   _____ .

2. The square of $\sqrt{27}$ _____ 27 because
   $\frac{}{\text{is/is not}}$
   _____
   _____ .

3. The radical $\sqrt{72}$ _____ in simplified form
   $\frac{}{\text{is/is not}}$
   because _____
   _____ .

4. The expression $\frac{\sqrt{5}}{\sqrt{3}}$ _____ simplified
   $\frac{}{\text{is/is not}}$
   because _____
   _____ .

5. The equation $x + \sqrt{27} = 4$ _____ a radical
   $\frac{}{\text{is/is not}}$
   equation because _____
   _____ .

6. The squaring property of equality _____ used to
   $\frac{}{\text{is/is not}}$
   solve a radical equation because _____
   _____ .

Assume that all variables and radicands represent nonnegative real numbers.

[8.1] *Simplify.*

7. $-\sqrt{49}$

8. $\sqrt{6^2}$

9. $(\sqrt{7x})^2$

10. $\sqrt{28}$

11. $-3\sqrt{18}$

12. $\sqrt{32x^3}$

13. $\sqrt{\dfrac{9}{25}}$

14. $-\sqrt{\dfrac{3t}{16}}$

15. $\sqrt{\dfrac{144}{x^{100}}}$

16. $2\sqrt{25a^5b^3}$

[8.2] *Combine.*

17. $2\sqrt{5} + \sqrt{5}$

18. $\sqrt{n} + 7\sqrt{n}$

19. $4x\sqrt{3} - 3x\sqrt{3}$

20. $\sqrt{27} - 2\sqrt{75}$

21. $x\sqrt{4x} + \sqrt{9x^3}$

22. $\sqrt{50a} - 3\sqrt{8a} + 8\sqrt{2a}$

[8.3] *Multiply. Simplify, if possible.*

23. $\sqrt{5} \cdot \sqrt{3}$

24. $\sqrt{8n} \cdot \sqrt{2n}$

25. $\sqrt{5a^2b^3} \cdot \sqrt{10ab^3}$

26. $\sqrt{x}(\sqrt{x} - 4)$

27. $(\sqrt{7} - 1)(\sqrt{7} + 1)$

28. $(\sqrt{y} + 1)(2\sqrt{y} - 3)$

29. $(\sqrt{y} + 5)^2$

Divide and simplify.

30. $\dfrac{\sqrt{54}}{\sqrt{6}}$

31. $\dfrac{\sqrt{3}}{\sqrt{12}}$

32. $\dfrac{\sqrt{48x}}{\sqrt{3x}}$

33. $\dfrac{\sqrt{24a^6}}{\sqrt{2a^3}}$

Rationalize the denominator.

34. $\dfrac{2}{\sqrt{11}}$

35. $\dfrac{\sqrt{3x^2}}{\sqrt{6x}}$

36. $\dfrac{4\sqrt{8} + \sqrt{2}}{\sqrt{2}}$

37. $\dfrac{10}{\sqrt{7} - 1}$

[8.4]   *Solve and check.*

38. $\sqrt{n} - 3 = 5$

39. $\sqrt{2x + 1} = 4$

40. $\sqrt{4n - 5} = \sqrt{n + 10}$

41. $x - \sqrt{3x + 1} = 3$

Mixed Applications

Solve.

42. The length of one side of a square city block is $50\sqrt{2}$ ft. Find the area of the block.

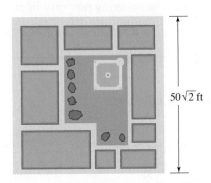

$50\sqrt{2}$ ft

43. The formula for the distance that an astronaut can see on any planet is $d = \sqrt{\dfrac{rh}{2640}}$, where h is his height (in feet) and r is the radius of the planet (in miles). If the astronaut is 6 ft tall and the radius of Mars is about 2100 mi, how far can he see on Mars? Round the answer to the nearest mile.

44. The length of the diagonal of the metal box shown in the figure below is modeled by the expression $\sqrt{l^2 + w^2 + h^2}$, where l, w, and h represent the length, width, and height of the box, respectively. If the box is 5 in. long, 4 in. wide, and 3 in. high, will a screwdriver 8 in. long fit diagonally in the box?

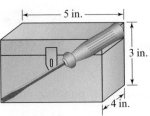

5 in.

3 in.

4 in.

45. Find, in radical form, the perimeter of the kite shown below.

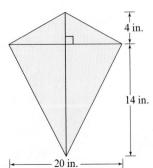

4 in.

14 in.

20 in.

46. A formula for the radius of a sphere r is

$$r = \sqrt{\dfrac{S}{4\pi}},$$

where S is the surface area of the sphere. Write this formula, rationalizing the denominator.

47. Scientists who study lakes are concerned with their shape. These scientists have grouped lakes according to the concept of *shoreline development*, a measure of how closely a lake resembles a circle. Shoreline development D is defined by the formula

$$D = \frac{L}{2\sqrt{\pi A}},$$

where L is the length of the lake's shoreline and A is the area of the lake. (*Source:* David G. Frey, Ed., *Limnology in North America*)

a. Check that if a lake is perfectly circular, the shoreline development is equal to 1.

b. Rationalize the denominator in the formula.

c. Goslute Lake, which existed in the American southwest some 50 million years ago, had a shoreline development equal to 1.58. Solve for its area in terms of the length of its shoreline.

48. For the box shown with volume V, the length of a side x of the square base is $\sqrt{\dfrac{V}{15}}$. Simplify this expression.

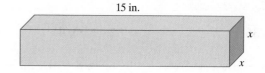

15 in.

• Check your answers on page A-26.

CHAPTER 8 Posttest

EXTRA HELP

Test Prep VIDEOS   in MyMathLab, and on YouTube (search "Akst Introductory Alg" and click on "Channels").

To see if you have mastered the topics in this chapter, take this test.

Assume all variables and radicands represent nonnegative real numbers.

Simplify.

1. $2\sqrt{36}$

2. $-3\sqrt{45}$

3. $\sqrt{32x^3y}$

4. $\sqrt{\dfrac{5n}{16}}$

5. $\sqrt{x} - 3\sqrt{x} + 5\sqrt{x}$

6. $\sqrt{8} - 4\sqrt{50} + \sqrt{18}$

7. $t\sqrt{4t} + 2\sqrt{16t^3}$

Multiply or divide. Simplify, if possible.

8. $(\sqrt{12})(\sqrt{2})$

9. $(\sqrt{5x^3y})(\sqrt{5x^2y})$

10. $\dfrac{\sqrt{75}}{\sqrt{3}}$

11. $\sqrt{y}(5\sqrt{y} - 1)$

12. $(\sqrt{3} + 4)^2$

Simplify.

13. $\sqrt{\dfrac{2p}{5}}$

14. $\dfrac{\sqrt{48x^4}}{\sqrt{2x}}$

15. Rationalize the denominator: $\dfrac{\sqrt{3} - 2\sqrt{6}}{\sqrt{3}}$

Solve and check.

16. $\sqrt{5x + 1} - 2 = 4$

17. $\sqrt{x + 11} = \sqrt{7x - 1}$

18. One of several expressions for the *windchill temperature* (WCT) is

$$91 + 0.08(3.7\sqrt{V} + 6 - 0.3V)(T - 91),$$

where *T* is the temperature (in degrees Fahrenheit) and *V* is the wind speed (in miles per hour). Find the WCT, rounded to the nearest whole number, if the temperature is 42°F and the wind speed is 16 mph.

19. Find the length of the ladder shown, expressed as a simplified radical.

24 ft

6 ft

20. On a clear day, the distance *d* (in miles) that a lookout on a ship can see to the horizon from height *h* (in feet) can be approximated by the formula $d = 1.2\sqrt{h}$. How high must a lookout climb to see a ship 6 mi away?

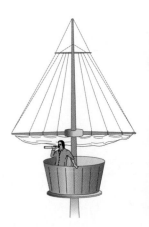

• Check your answers on page A-26.

To help you review, solve the following.

1. Simplify: $\dfrac{2}{3} + \left(-\dfrac{4}{5}\right)$

2. Evaluate the formula $A = P(1 + rt)$, if $P = \$1500$, $r = 3\%$ per year, and $t = 2$ yr.

3. Solve and check: $-\dfrac{2c}{9} = -10$

4. Solve and check: $c = -2(c - 1)$

5. Compute the slope m of the line that passes through the points $(4, 0)$ and $(3, 5)$. Plot these points on the coordinate plane and draw the line.

6. Consider the equation $4x - 6y = 12$.
   a. Find its slope and x- and y-intercepts.

   b. Its graph is which of the following types? ╱ ╲ — |

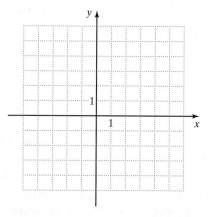

7. For the system of equations graphed, determine the number of solutions.

8. Find the product: $(4x^2 - 3)(4x - 1)$

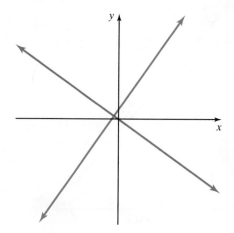

9. Factor: $y^2 - 12y + 32$

10. Factor: $3x^3 + 24x^2 + 48x$

11. Add: $\dfrac{9}{c - 2} + \dfrac{c - 3}{c^2 - c - 2}$

12. Solve and check: $\dfrac{2}{m} = \dfrac{m - 3}{9}$

13. Solve and check: $\sqrt{n + 6} + 7 = 9$

14. Solve and check: $11 - \sqrt{x - 4} = 8$

15. The speed of sound in air s is modeled by the formula

$$s = 0.6t + 331,$$

where the speed is in meters per second and the temperature t is in degrees Celsius. At what temperature is this speed equal to 343 m/sec? (*Source:* Peter J. Nolan, *Fundamentals of College Physics*)

16. The average distance from the Sun is approximately 1.5×10^8 km to the Earth and about 6×10^9 km to the dwarf planet Pluto. The distance to Pluto is how many times the distance to the Earth, expressed in scientific notation? (*Source:* Jeffrey Bennett et al., *The Cosmic Perspective*)

17. A country's balance of payments can be found by subtracting its imports from its exports. If x is the number of years since 2000, the total exports of the United States can be approximated by the polynomial $-6x^3 + 83x^2 - 228x + 1111$, and the total U.S. imports can be approximated by $-10x^3 + 133x^2 - 296x + 1495$ (both polynomials in billions of dollars). Use this model to approximate the U.S. balance of payments. (*Source:* bea.gov)

18. The first week, a student with a part-time job made $150 in x hr. The second week, he made $175 working 1 more hour than the previous week. What was the ratio of the student's average pay per hour in the second week as compared to the first week?

19. At a studio that creates 3-D animation movies, a computer graphics server would take 72 hr working by itself to process the most intensive action sequences in a single animated movie frame. By contrast, a new desktop computer, although much more expensive, would take only 36 hr. (*Source:* tomshardware.com)

a. How many hours would it take for a computer graphics server and a new desktop computer working together to process one of these sequences?

b. One night, the studio used 5000 servers and 2000 desktop computers to process the animation. How many hours did it take to process a sequence?

20. Two hikers start a trip by walking due west from a campsite, as shown below. They then turn due north and walk to a waterfall. What is the distance d from the camp to the waterfall expressed in radical form? Also express this distance as a decimal, rounded to the nearest hundred meters.

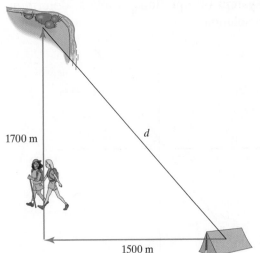

1700 m

d

1500 m

• Check your answers on page A-26.

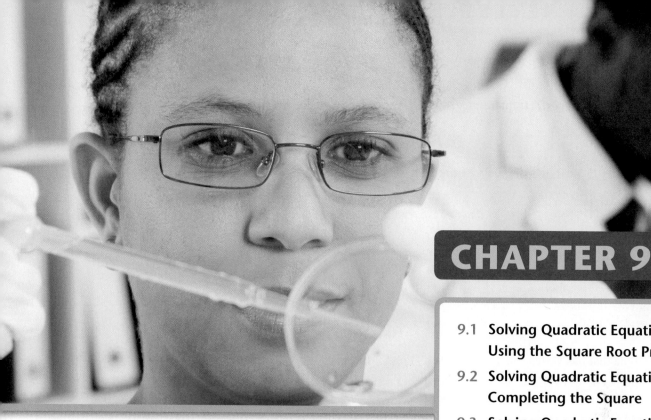

CHAPTER 9

9.1 **Solving Quadratic Equations by Using the Square Root Property**

9.2 **Solving Quadratic Equations by Completing the Square**

9.3 **Solving Quadratic Equations by Using the Quadratic Formula**

9.4 **Graphing Quadratic Equations in Two Variables**

Quadratic Equations

Using Quadratic Equations to Test New Drugs

In testing new medicines or food additives, scientists try to determine their effect on the body by collecting and analyzing data. A method for analyzing data is to picture the data by plotting points on a coordinate plane.

For instance, the graph to the upper right illustrates the effect of various kinds of insulin on diabetic patients. It shows that some kinds of insulin have a longer-lasting effect than others.

The *dose-effect curves* shown to the lower right deal with raising poultry. The curves show how the weight of turkeys changes when they are given a food additive called methionine. This substance, in the appropriate dosage, is thought to enhance turkey size and health.

Notice the shape of these graphs. Such curves can sometimes be approximated by *parabolas*, that is, by curves that correspond to quadratic equations. Scientists can then use these quadratic equations to summarize the observed data and to make predictions.

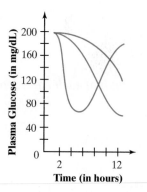

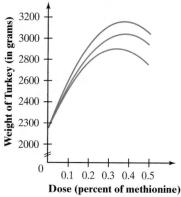

(*Sources:* Bertram G. Katzung, *Basic and Clinical Pharmacology,* 1995; Donald A. Berry, *Statistical Methodology in the Pharmaceutical Sciences,* 1990; David H. Tedeschi and Ralph E. Tedeschi, *Importance of Fundamental Principles in Drug Evaluation,* 1968)

1. Solve $3x^2 = 54$ using the square root property of equality.

2. Solve $A = 4\pi r^2$ for r. Assume that all variables are positive.

3. Fill in the blank to make a perfect square trinomial: $x^2 + 12x + [\ ]$.

4. Solve $3x^2 - 12x + 6 = 0$ by completing the square.

5. Solve $-2x^2 + 2x + 5 = 0$ by using the quadratic formula.

Solve.

6. $(x - 6)^2 = 25$

7. $n^2 + 7n + 12 = 0$

8. $y^2 - 6y = -4$

9. $4x^2 + 9 = 21$

10. $5n^2 - 10n - 4 = 0$

11. $(x + 2)(x + 2) = 18$

12. Sketch a parabola that has vertex $(0, 0)$ and passes through the points $(1, -1), (-1, -1), (2, -4),$ and $(-2, -4)$.

13. Find the vertex and axis of symmetry of the graph of $y = x^2 - 6x + 7$.

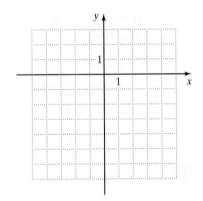

Sketch the graph of each equation.

14. $y = -x^2$

15. $y = x^2 - 2x + 1$

16. $y = 4x - 2x^2$

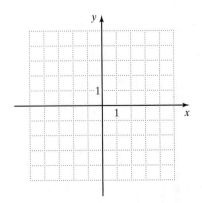

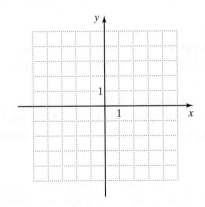

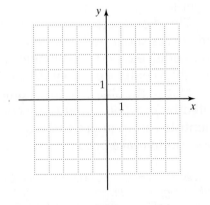

Solve.

17. The final velocity (in meters per second) of an object starting from rest that is accelerated at a constant rate is given by $v^2 = 2as$, where s is the distance traveled (in meters) during the acceleration a. Solve this equation for v.

18. A box manufacturer makes open boxes from 30 in. by 36 in. pieces of cardboard by cutting squares of equal size from the corners and turning up the sides. According to the manufacturing specifications, each box is to have a base with an area of 720 in^2. What size squares should be cut from the corners of the piece of cardboard in order to meet the box specifications?

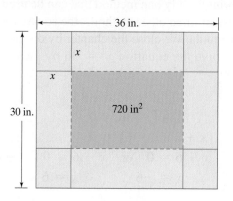

19. A Cessna airplane traveled 216 mi against a 15-mph headwind and 420 mi with a 25-mph tailwind. If the total flying time was 4 hr, what is the speed of the airplane in still air?

20. A rhythmic gymnast throws a ball straight up from a height of 5 ft with an initial velocity of 48 ft/sec. The equation $h = -16t^2 + 48t + 5$ models the ball's height h (in feet) after t sec. What is the maximum height the ball reaches?

• Check your answers on page A-26.

OBJECTIVES

A To solve a quadratic equation using the square root property

B To solve a literal equation or formula using the square root property

C To solve applied problems involving quadratic equations

Recall from Section 6.5 that a *quadratic equation* is an equation that can be written in the standard form $ax^2 + bx + c = 0$, where a, b, and c are real numbers and $a \neq 0$. In that section, we saw how quadratic equations can be used to solve problems related to physics, business, sports, and other fields. We solved such equations by factoring.

Factoring is only one method that can be used to solve quadratic equations. In this chapter, we consider three other methods: the square root property, completing the square, and the quadratic formula. We close the chapter by discussing how to graph quadratic equations.

Let's solve the equation $x^2 = 36$ by factoring, which is the method we discussed in Section 6.5.

$$x^2 = 36$$
$$x^2 - 36 = 0 \qquad \text{Write the equation in standard form.}$$
$$(x + 6)(x - 6) = 0 \qquad \text{Factor.}$$
$$x + 6 = 0 \quad \text{or} \quad x - 6 = 0 \qquad \text{Set each factor equal to 0.}$$
$$x = -6 \qquad\qquad x = 6$$

The solutions are -6 and 6, which we can write as ± 6 (read "plus or minus 6").

The solutions ± 6 can be expressed as $\pm \sqrt{36}$. This suggests that we can also solve the equation $x^2 = 36$ simply by taking the square root of both sides.

The Square Root Property of Equality

If n is a nonnegative number and $x^2 = n$, then $x = \pm \sqrt{n}$, that is, $x = \sqrt{n}$ or $x = -\sqrt{n}$.

This property allows us to take the square root of each side of an equation, resulting in an equivalent equation, provided that each radicand is nonnegative.

Let's use this property to solve quadratic equations of the form $x^2 = n$ and $ax^2 = n$. Consider these examples.

EXAMPLE 1

Solve by taking square roots.

a. $x^2 = 49$ **b.** $3x^2 = 24$ **c.** $9 + 16t^2 = 0$

Solution

a. $x^2 = 49$

$\quad x = \pm \sqrt{49} \qquad$ Use the square root property of equality.

$\quad x = \pm 7 \qquad$ Simplify.

Check

Substitute 7 for x.

$$x^2 = 49$$
$$(7)^2 \overset{?}{=} 49$$
$$49 = 49 \qquad \text{True}$$

Substitute -7 for x.

$$x^2 = 49$$
$$(-7)^2 \overset{?}{=} 49$$
$$49 = 49 \qquad \text{True}$$

The solutions are 7 and -7.

PRACTICE 1

Use the square root property of equality to solve.

a. $x^2 = 100$

b. $2x^2 = 36$

c. $25 - 81t^2 = 0$

b. In this example, we must first divide each side of the equation by the coefficient of the x-squared term before using the square root property of equality.

$$3x^2 = 24$$
$$x^2 = 8 \qquad \text{Divide each side by 3.}$$
$$x = \pm\sqrt{8} \qquad \text{Use the square root property of equality.}$$
$$x = \pm 2\sqrt{2} \qquad \text{Simplify.}$$

The solutions are $2\sqrt{2}$ and $-2\sqrt{2}$.

c. $9 + 16t^2 = 0$

$$16t^2 = -9 \qquad \text{Subtract 9 from each side of the equation.}$$
$$t^2 = -\frac{9}{16} \qquad \text{Divide each side by 16.}$$

Recall from Section 8.1 that a radical with a negative radicand is not a real number. So there are no real-number solutions of the equation $9 + 16t^2 = 0$.

Now, let's look at how we use the square root property of equality to solve quadratic equations of the form $(ax + b)^2 = d$ and $a(x + c)^2 = d$, that is, equations containing the square of a binomial.

EXAMPLE 2

Solve using the square root property of equality.

a. $(y - 1)^2 = 25$ **b.** $4(x + 4)^2 = 8$ **c.** $(3x + 1)^2 = 18$

Solution

a. $(y - 1)^2 = 25$

$$y - 1 = \pm\sqrt{25} \qquad \text{Use the square root property of equality.}$$
$$y - 1 = \pm 5 \qquad \text{Simplify.}$$
$$y - 1 = 5 \quad \text{or} \quad y - 1 = -5$$
$$y = 6 \qquad\qquad y = -4$$

The solutions are 6 and -4.

b. $4(x + 4)^2 = 8$

$$(x + 4)^2 = 2 \qquad \text{Divide each side by 4.}$$
$$x + 4 = \pm\sqrt{2} \qquad \text{Use the square root property of equality.}$$
$$x + 4 = \sqrt{2} \quad \text{or} \quad x + 4 = -\sqrt{2}$$
$$x = -4 + \sqrt{2} \qquad\qquad x = -4 - \sqrt{2}$$

The solutions are $-4 + \sqrt{2}$ and $-4 - \sqrt{2}$.

PRACTICE 2

Solve using the square root property.

a. $(n + 2)^2 = 9$

b. $3(x - 4)^2 = 15$

c. $(4m - 1)^2 = 54$

EXAMPLE 2 (continued)

c. $(3x + 1)^2 = 18$

$\qquad 3x + 1 = \pm\sqrt{18}$ Use the square root property of equality.

$\qquad 3x + 1 = \pm3\sqrt{2}$

$3x + 1 = 3\sqrt{2}$ or $3x + 1 = -3\sqrt{2}$

$\qquad 3x = -1 + 3\sqrt{2} \qquad\qquad 3x = -1 - 3\sqrt{2}$

$\qquad x = \dfrac{-1 + 3\sqrt{2}}{3} \qquad\qquad x = \dfrac{-1 - 3\sqrt{2}}{3}$

The solutions are $\dfrac{-1 + 3\sqrt{2}}{3}$ and $\dfrac{-1 - 3\sqrt{2}}{3}$.

In Example 2(c), note that since 3 is not a common factor of both terms in the numerator, we cannot cancel the 3 in the numerator with the 3 in the denominator.

EXAMPLE 3

Solve: $x - 5 = \dfrac{3}{x - 5}$

Solution

$x - 5 = \dfrac{3}{x - 5}$

$(x - 5)^2 = 3$ Multiply each side by $(x - 5)$.

$\quad x - 5 = \pm\sqrt{3}$ Use the square root property of equality.

$x - 5 = \sqrt{3}$ or $x - 5 = -\sqrt{3}$

$\qquad x = 5 + \sqrt{3} \qquad\qquad x = 5 - \sqrt{3}$

The solutions are $5 + \sqrt{3}$ and $5 - \sqrt{3}$.

PRACTICE 3

Solve: $x + 4 = \dfrac{2}{x + 4}$

Recall that we solved linear literal equations in Section 2.4. Now, let's apply the square root property to solving quadratic literal equations. We assume that all variables and radicands are positive.

EXAMPLE 4

Solve for the indicated variable.

a. $V = \pi r^2 h$ for r (Volume formula for a cylinder)

b. $c^2 = a^2 + b^2$ for b (Pythagorean theorem)

Solution

a. Solving for r, we get:

$\qquad V = \pi r^2 h$

$\qquad \dfrac{V}{\pi h} = r^2$ Divide each side by πh.

$\pm\sqrt{\dfrac{V}{\pi h}} = r$ Use the square root property of equality.

$\qquad r = \pm\sqrt{\dfrac{V}{\pi h}} = \dfrac{\pm\sqrt{\pi h V}}{\pi h}$

PRACTICE 4

Solve for the indicated variable.

a. $E = mc^2$ for c (c = speed of light)

b. $c^2 = a^2 + b^2$ for a

Since the length of the radius is positive, we accept only the positive square root: $r = \sqrt{\dfrac{V}{\pi h}}$

b. Solving for b, we get:

$$c^2 = a^2 + b^2$$
$$c^2 - a^2 = b^2 \qquad \text{Add } -a^2 \text{ to each side.}$$
$$\pm\sqrt{c^2 - a^2} = b \qquad \text{Use the square root property.}$$

So $b = \pm\sqrt{c^2 - a^2}$. Since b represents a side of a right triangle and its length is positive, we accept only the positive square root, namely,

$$b = \sqrt{c^2 - a^2}.$$

Can the expression $\pm\sqrt{c^2 - a^2}$ in Example 4(b) be simplified? Explain.

EXAMPLE 5

A circular table top and a square table top have the same area. (Recall that the area of a square with side s is given by the formula $A = s^2$, whereas $A = \pi r^2$ represents the area of a circle with radius r.)

a. Express s in terms of r.

b. Approximating π by 3.14, find, to the nearest inch, the side of the square table if the radius of the circular table is 20 in.

Solution

a. Since the square and circular areas are equal, $s^2 = \pi r^2$. To express s in terms of r, we need to solve this equation for s.

$$s^2 = \pi r^2$$
$$s = \pm\sqrt{\pi}\, r$$

Since the side is positive in length, the solution is $s = \sqrt{\pi}\, r$, or $s = r\sqrt{\pi}$.

b. Next, we substitute 20 for r and simplify.

$$s = \sqrt{\pi}(20)$$
$$\approx \sqrt{3.14}(20)$$
$$\approx 35.44$$

So the side of the square table is approximately 35 in. long.

PRACTICE 5

A nuclear power plant on the Japanese coastline suffered a meltdown because of an earthquake and a tsunami. As a result, the government ordered a mandatory evacuation in a semicircular region centered around the power plant. (*Source:* latimes.com)

a. The area A of a semicircle with radius r is given by $A = \dfrac{1}{2}\pi r^2$.

Solve this equation for r.

b. If the area of the region around the power plant was approximately 225 mi², find, to the nearest mile, the radius of the evacuated region.

EXAMPLE 6

If P dollars are invested at interest rate r (in decimal form) compounded annually, then at the end of 2 yr the amount will have grown to $A = P(1 + r)^2$.

a. Solve this equation for rate r.

b. What is the interest rate needed for an investment of $1000 to grow to $1210 in 2 yr?

Solution

a. First, we solve the equation $A = P(1 + r)^2$ for r.

$$A = P(1 + r)^2$$
$$P(1 + r)^2 = A$$
$$(1 + r)^2 = \frac{A}{P}$$
$$1 + r = \pm\sqrt{\frac{A}{P}}$$
$$r = -1 \pm \sqrt{\frac{A}{P}}$$

Since the interest rate is positive, we conclude that $r = -1 + \sqrt{\frac{A}{P}}$.

b. Substituting $1000 for P and $1210 for A, we get:

$$r = -1 + \sqrt{\frac{A}{P}}$$
$$= -1 + \sqrt{\frac{1210}{1000}}$$
$$= -1 + \sqrt{\frac{121}{100}}$$
$$= -1 + \frac{11}{10}$$
$$= 0.1$$

So the interest rate must be 10%.

PRACTICE 6

Loggers use *board feet* as a unit to measure the volume of timber. The formula

$$B = (d - 4)^2$$

gives an estimate of the number of board feet B in a 16-foot log, where d is the diameter of the log in inches and $d > 4$ in.

a. Solve the formula for diameter d.

b. What is the diameter of a log containing 16 board feet?

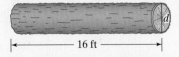

|← 16 ft →|

Would the value of r in Example 6 have changed if in the original equation we had substituted first and then solved for r? Explain.

Mathematically Speaking

Fill in each blank with the most appropriate term or phrase from the given list.

divide both sides by a	the coefficient of x^2	positive or negative square root of n
square of a binomial	square root of a binomial	take the square root of both sides
principal square root of n	binomial equation	
quadratic equation		

1. A(n) _____ is an equation that can be written in the standard form $ax^2 + bx + c = 0$, where a, b, and c are real numbers and $a \neq 0$.

2. The square root property of equality states that if p is a nonnegative number and $x^2 = n$, then x is equal to the _____.

3. Before using the square root property of equality to solve quadratic equations of the form $ax^2 = n$, we must _____.

4. We can use the square root property of equality to solve quadratic equations of the form $(ax + b)^2 = c$ or $a(x + c)^2 = d$, that is, equations containing the _____.

Ⓐ *Solve by using the square root property.*

5. $y^2 = 9$

6. $x^2 = 16$

7. $p^2 = 2$

8. $x^2 = 5$

9. $n^2 = \dfrac{1}{4}$

10. $t^2 = \dfrac{16}{25}$

11. $5t^2 = 20$

12. $2n^2 = 72$

13. $4x^2 = 28$

14. $7n^2 = 21$

15. $3s^2 = 36$

16. $6x^2 = 48$

17. $\dfrac{1}{4}p^2 = 20$

18. $\dfrac{1}{3}s^2 = 7$

19. $5 + t^2 = 11$

20. $x^2 - 9 = 3$

21. $x^2 - 8 = 9$

22. $p^2 - 10 = 1$

23. $4n^2 - 9 = 0$

24. $9x^2 - 25 = 0$

25. $4x^2 + 18 = 8$

26. $9n^2 - 5 = -7$

27. $15 - 16x^2 = 3$

28. $6 - 18y^2 = 2$

29. $(n + 2)^2 = 9$

30. $(c - 5)^2 = 16$

31. $(x - 7)^2 = 49$

32. $(y + 9)^2 = 81$

33. $(x + 6)^2 = 5$

34. $(x - 2)^2 = 2$

35. $(5 - s)^2 = \dfrac{9}{16}$

36. $(1 - n)^2 = \dfrac{1}{4}$

37. $(y - 4)^2 = 9$

38. $(x + 2)^2 = 25$

39. $2(p - 5)^2 = 6$

40. $3(t + 4)^2 = 21$

41. $5(x + 1)^2 = 40$

42. $4(n - 3)^2 = 48$

43. $(3x + 1)^2 = 4$

44. $(2t - 3)^2 = 9$

45. $(4y + 5)^2 = 3$

46. $(8a - 1)^2 = 2$

47. $(2x - 7)^2 = 20$

48. $(3x + 4)^2 = 27$

49. $\left(\dfrac{1}{2}x - 5\right)^2 = 10$ **50.** $\left(\dfrac{1}{3}t + 1\right)^2 = 8$ **51.** $(x + 1)^2 + 49 = 0$ **52.** $(m - 5)^2 + 100 = 0$

53. $a + 8 = \dfrac{5}{a + 8}$ **54.** $x - 1 = \dfrac{2}{x - 1}$ **55.** $x - 3 = \dfrac{24}{x - 3}$ **56.** $n + 6 = \dfrac{27}{n + 6}$

57. $(y - 1)(y + 1) = 4$ **58.** $(n - 3)(n + 3) = -1$

Solve for the indicated variable.

59. $ax^2 - b = 0$ for x **60.** $h = 16t^2$ for t ⊙ **61.** $K = \dfrac{4\pi^2}{v^2 r}$ for v **62.** $W = i^2 Rt$ for i

63. $\dfrac{x^2}{16} - \dfrac{y^2}{25} = 1$ for y **64.** $E = \dfrac{1}{2}mv^2$ for v

Mixed Practice

Solve by using the square root property.

65. $(5y - 6)^2 = 44$ **66.** $p^2 - 2 = 30$

67. $5 - 75m^2 = -4$ **68.** $(y - 4)^2 = 11$

69. $5(x + 2)^2 = 60$ **70.** $6t^2 = 42$

71. $a - 4 = \dfrac{63}{a - 4}$ **72.** Solve for x: $\dfrac{x^2}{4} + \dfrac{y^2}{9} = 1$

Applications

Solve.

73. Two campus security officers communicate with one another using walkie-talkies. After meeting, one of the officers walks north at 45 ft/min and the other walks east at 60 ft/min. To the nearest minute, for how long can they hear one another if the walkie-talkies have a maximum range of 300 ft?

74. The length of a rectangular floor is twice the width. The area of the floor is 32 ft². What are the dimensions of the room?

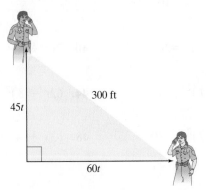

300 ft

45t

60t

75. A candy company wants to reduce the cost of its product. The candy is cylindrical in shape, with height h and with radius r. Management wants to reduce the size of the candies by 5%, while keeping the same shape. (Reminder: The formula for the volume of a cylinder is $V = \pi r^2 h$.)

a. If the company reduces only the height of the candy, what must the new height H be?

b. If the company reduces only the radius of the candy, what must the new radius R be?

76. A formula for the volume of the cone shown below is
$$V = \frac{1}{3}\pi r^2 h.$$

a. Solve the formula for r.

b. Find the radius of the cone, rounded to the nearest tenth of an inch. Use $\pi \approx 3.14$.

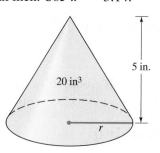

5 in.

20 in³

r

77. The graph of the equation $x^2 + y^2 = r^2$, where $r > 0$, is a circle whose center is the origin on the coordinate plane, as shown in the figure. Solve this equation for y.

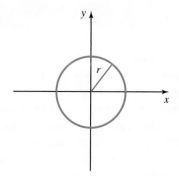

78. If a businessman deposits $900 in an account with interest rate r compounded annually, the amount of money A in the account after 2 yr is given by the following equation:
$$A = 900(1 + r)^2$$

Solve this equation for r.

79. An object d ft from a point source of light receives I foot-candles of light, where
$$I = \frac{4050}{d^2}.$$

Solve this formula for the distance.

80. The owner of a pizzeria wants 20% of the surface of a pizza to be the crust around the edge (see diagram below). If the radius of the pizza (including the crust) is r, find the appropriate width c of the pizza crust.

• Check your answers on page A-27.

MINDStretchers

Mathematical Reasoning

1. Consider the inequality $x^2 - 9 \le 0$.

   a. Identify three positive and three negative numbers that satisfy this inequality.

   b. Solve the inequality.

Groupwork

2. Working in a small group, solve the following equations:

   a. $x^4 = 16$ **b.** $t^3 = 9t$ **c.** $(n^2 - 1)^2 = 16$ **d.** $(y^2 + 5)^2 = 9$

Critical Thinking

3. Consider the equation $x^2 = a$. Solve for x. How many real solutions does this equation have? Explain.

Cultural Note

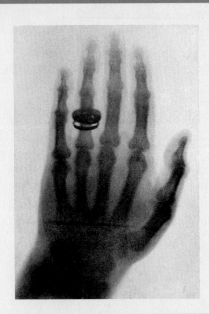

The photograph to the left is one of the first taken by the German physicist Wilhelm Roentgen in 1895 after he discovered strange rays that he called *X-rays* since they were a mysterious phenomenon. Roentgen's discovery allowed doctors to see inside the body for the first time without surgery and earned him the first Nobel Prize awarded for physics.

The practice of using an *x* and other letters from the end of the alphabet to represent mathematical unknowns goes back to the seventeenth-century French mathematician René Descartes, who made major contributions to the development of algebra.

(*Source:* Florian Cajori, *A History of Mathematical Notations*, The Open Court Publishing Company, 1929)

In this section, we solve quadratic equations by **completing the square**. This method combines two previous approaches for solving quadratic equations, namely, taking square roots and factoring perfect square trinomials.

For instance, consider the equation $x^2 + 10x = 2$. This equation is different from those of the form solved in the previous section because the expression $x^2 + 10x$ is not a perfect square.

To use the square root property, we need to write the form of the left-hand side of the equation as a perfect square trinomial. Recall that in Section 5.6 we discussed perfect squares such as the following:

		The coefficient of the quadratic term is 1.		The coefficient of the linear term		The constant term
Perfect square		↓		↓		↓
↓						
$(x + 1)^2$	$=$	x^2	$+$	$2x$	$+$	1
$(x + 2)^2$	$=$	x^2	$+$	$4x$	$+$	4
$(x + 3)^2$	$=$	x^2	$+$	$6x$	$+$	9
$(x + 4)^2$	$=$	x^2	$+$	$8x$	$+$	16

• • •

Note that in each perfect square trinomial, the coefficient of the quadratic term is 1, and the constant term is the square of one-half the coefficient of the linear term. For example, in $x^2 + 8x + 16$, we see that:

$$16 = \left(\frac{1}{2} \cdot 8\right)^2$$

More generally, when we square the binomial $x + c$ we get:

$$(x + c)^2 = x^2 + 2cx + c^2$$

where

$$\underset{\substack{\uparrow \\ \text{The constant} \\ \text{term}}}{c^2} = \underset{\substack{\uparrow \\ \text{The coefficent of} \\ \text{the linear term}}}{\left(\frac{1}{2} \cdot 2c\right)^2}$$

Observing this pattern leads us to a method of finding the value of the constant term that makes the general quadratic trinomial $x^2 + bx + c$ a perfect square.

To Complete the Square for the Expression $x^2 + bx$

- Find one-half of the coefficient of the linear term x: $\frac{1}{2}b$

- Square this value: $\left(\frac{1}{2}b\right)^2$

- Add the resulting value to the expression.

EXAMPLE 1

Fill in the blank to make each trinomial a perfect square.

a. $y^2 + 6y +$ ▢   **b.** $x^2 - 3x +$ ▢

Solution

a. $y^2 + 6y +$ ▢

The coefficient of the linear term y is 6.

$$\frac{1}{2}(6) = 3 \qquad \text{Find one-half the coefficient of } y.$$

$$3^2 = 9 \qquad \text{Square 3.}$$

The constant term to be added to $y^2 + 6y$ is 9. So we get the perfect square trinomial $y^2 + 6y + 9$, which is equal to $(y + 3)^2$.

b. $x^2 - 3x +$ ▢

The coefficient of the linear term x is -3.

$$\frac{1}{2}(-3) = -\frac{3}{2}$$

$$\left(-\frac{3}{2}\right)^2 = \frac{9}{4}$$

The constant term to be added to $x^2 - 3x$ is $\frac{9}{4}$. So we get:

$x^2 - 3x + \frac{9}{4}$, which is equal to $\left(x - \frac{3}{2}\right)^2$.

PRACTICE 1

Fill in the blank to make each trinomial a perfect square.

a. $x^2 - 12x +$ ▢

b. $n^2 + 5n +$ ▢

Now, let's apply the method of completing the square to solve a quadratic equation of the form $ax^2 + bx + c = 0$, where $a = 1$. Consider, for instance, the equation $x^2 + 8x = 5$. We begin by completing the square on the left side of the equation $x^2 + 8x = 5$.

$$x^2 + 8x = 5 \qquad \text{Take one-half the coefficient of the linear term } x: \frac{1}{2}(8) = 4.$$
$$\text{Then, square it: } 4^2 = 16.$$

We add 16 to the left side of the equation to complete the square, but to maintain equality we must also add 16 to the right side of the equation.

$$x^2 + 8x + 16 = 5 + 16$$
$$x^2 + 8x + 16 = 21$$

Next, we write the left side of the equation $x^2 + 8x + 16 = 21$ as the square of a binomial and then, solve the equation.

$$(x + 4)^2 = 21$$
$$x + 4 = \pm\sqrt{21} \qquad \text{Use the square root property of equality.}$$
$$x + 4 = \sqrt{21} \qquad \text{or} \quad x + 4 = -\sqrt{21}$$
$$x = -4 + \sqrt{21} \qquad\qquad x = -4 - \sqrt{21}$$

So the solutions are $-4 + \sqrt{21}$ and $-4 - \sqrt{21}$.

Let's look at some additional examples of solving quadratic equations by completing the square.

EXAMPLE 2

Solve by completing the square.

a. $y^2 - 6y = 7$ **b.** $x^2 - 5x + 3 = 0$

Solution

a.
$$y^2 - 6y = 7$$
$$y^2 - 6y + 9 = 7 + 9$$

> Take one-half the coefficient of the linear term: $\frac{1}{2}(-6) = -3$. Then, square it: $(-3)^2 = 9$. Add 9 to each side of the equation.

$$y^2 - 6y + 9 = 16$$
$$(y - 3)^2 = 16$$

> Write $y^2 - 6y + 9$ as the square of a binomial.

$$y - 3 = \pm 4$$

> Use the square root property of equality.

$$y - 3 = 4 \quad \text{or} \quad y - 3 = -4$$
$$y = 7 \qquad\qquad y = -1$$

The solutions of $y^2 - 6y = 7$ are 7 and -1. How would you check these solutions?

b. Before completing the square, we move all variable terms to one side of the equation and the constant term to the other side.

$$x^2 - 5x + 3 = 0$$
$$x^2 - 5x = -3$$

> Add -3 to each side.

$$x^2 - 5x + \frac{25}{4} = -3 + \frac{25}{4}$$

> Take one-half the coefficient of the linear term: $\frac{1}{2}(-5) = -\frac{5}{2}$. Then, square it: $\left(-\frac{5}{2}\right)^2 = \frac{25}{4}$. Add $\frac{25}{4}$ to each side.

$$x^2 - 5x + \frac{25}{4} = \frac{13}{4}$$

> Simplify.

$$\left(x - \frac{5}{2}\right)^2 = \frac{13}{4}$$

> Write $x^2 - 5x + \frac{25}{4}$ as the square of a binomial.

$$x - \frac{5}{2} = \pm\sqrt{\frac{13}{4}}$$

> Use the square root property.

$$x - \frac{5}{2} = \sqrt{\frac{13}{4}} \qquad \text{or} \qquad x - \frac{5}{2} = -\sqrt{\frac{13}{4}}$$

$$x = \frac{5}{2} + \frac{\sqrt{13}}{2} \qquad\qquad x = \frac{5}{2} - \frac{\sqrt{13}}{2}$$

$$x = \frac{5 + \sqrt{13}}{2} \qquad\qquad x = \frac{5 - \sqrt{13}}{2}$$

The solutions are $\dfrac{5 + \sqrt{13}}{2}$ and $\dfrac{5 - \sqrt{13}}{2}$.

PRACTICE 2

Solve by completing the square.

a. $y^2 + 4y = 21$

b. $n^2 + 7n + 5 = 0$

Next, we consider quadratic equations of the form $ax^2 + bx + c = 0$, where $a \neq 1$. Before completing the square, we must divide each side of the equation by a, the coefficient of the second-degree (or quadratic) term.

EXAMPLE 3

Solve by completing the square.

a. $2y^2 - 4y = 14$ **b.** $5m^2 - 10m + 4 = 0$

Solution

a.
$$2y^2 - 4y = 14$$
$$y^2 - 2y = 7 \qquad \text{Divide each side of the equation by 2, the coefficient of the second-degree term.}$$

$$y^2 - 2y + 1 = 7 + 1 \qquad \text{Complete the square: add } \left(\frac{-2}{2}\right)^2 = 1 \text{ to each side.}$$
$$(y - 1)^2 = 8$$
$$y - 1 = \pm\sqrt{8}$$
$$y - 1 = \sqrt{8} \qquad \text{or} \quad y - 1 = -\sqrt{8}$$
$$y = 1 + 2\sqrt{2} \qquad\qquad y = 1 - 2\sqrt{2}$$

The solutions are $1 + 2\sqrt{2}$ and $1 - 2\sqrt{2}$.

b. $5m^2 - 10m + 4 = 0$
$$m^2 - 2m + \frac{4}{5} = 0 \qquad \text{Divide each side by 5, the coefficient of the second-degree term.}$$

$$m^2 - 2m = -\frac{4}{5} \qquad \text{Subtract } \frac{4}{5} \text{ from each side.}$$

$$m^2 - 2m + 1 = -\frac{4}{5} + 1 \qquad \text{Complete the square: add } \left(\frac{-2}{2}\right)^2 = 1 \text{ to each side.}$$

$$(m - 1)^2 = \frac{1}{5}$$

$$m - 1 = \pm\sqrt{\frac{1}{5}}$$

$$m - 1 = +\sqrt{\frac{1}{5}} \quad \text{or} \quad m - 1 = -\sqrt{\frac{1}{5}}$$

$$m = 1 + \sqrt{\frac{1}{5}} \qquad\qquad m = 1 - \sqrt{\frac{1}{5}}$$

$$m = 1 + \frac{\sqrt{5}}{5} \qquad\qquad m = 1 - \frac{\sqrt{5}}{5}$$

The solutions are $1 + \frac{\sqrt{5}}{5}$ and $1 - \frac{\sqrt{5}}{5}$.

PRACTICE 3

Solve by completing the square.

a. $2y^2 - 16y = 8$

b. $4x^2 - 4x - 1 = 0$

The preceding examples suggest the following general method for solving a quadratic equation by completing the square.

To Solve the Quadratic Equation $ax^2 + bx + c = 0$ by Completing the Square

- If $a \neq 1$, then divide the equation by a, the coefficient of the second-degree term. If $a = 1$, then proceed to the next step.

- Move all terms with variables to one side of the equals sign and all constants to the other side.

- Take one-half the coefficient of the linear term. Then, square it and add this square to each side of the equation.

- Factor the side of the equation containing the variable terms, writing it as the square of a binomial.

- Use the square root property of equality.

- Solve the resulting equations.

EXAMPLE 4

Solve $4n^2 - 12n + 5 = 0$ by completing the square.

Solution

$$4n^2 - 12n + 5 = 0$$

$$n^2 - 3n + \frac{5}{4} = 0 \qquad \text{Divide each side by 4.}$$

$$n^2 - 3n = -\frac{5}{4}$$

$$n^2 - 3n + \frac{9}{4} = -\frac{5}{4} + \frac{9}{4} \qquad \text{Complete the square.}$$

$$\left(n - \frac{3}{2}\right)^2 = 1$$

$$n - \frac{3}{2} = \pm 1$$

$$n = \frac{3}{2} + 1 \quad \text{or} \quad n = \frac{3}{2} - 1$$

$$n = \frac{5}{2} \qquad\qquad n = \frac{1}{2}$$

The solutions are $\frac{5}{2}$ and $\frac{1}{2}$.

PRACTICE 4

Solve $9x^2 - 9x - 4 = 0$ by completing the square.

EXAMPLE 5

The rectangular room pictured has an area of 280 ft² and a perimeter of 68 ft.

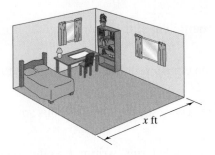

a. Express all four sides of the room in terms of x.

b. Find the dimensions of the room.

Solution

a. First, we must find the lengths of the four sides in terms of x. The area of the rectangle, which is given to be 280 ft², is the product of the length and the width. Since one side of the rectangle is x, the adjacent side is $\dfrac{280}{x}$ ft. So the dimensions of the rectangle are as pictured:

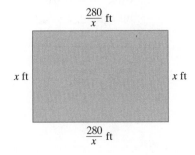

b. To find the dimensions of the room, we recall that the perimeter of the room is 68 ft. The perimeter is the sum of the side lengths:

$$2x + 2\left(\frac{280}{x}\right) = 68$$

$$x + \frac{280}{x} = 34$$

$$x^2 + 280 = 34x$$

$$x^2 - 34x = -280$$

$$x^2 - 34x + 289 = -280 + 289$$

$$(x - 17)^2 = 9$$

$$x - 17 = \pm 3$$

$$x = 17 \pm 3$$

$$x = 20 \quad \text{or} \quad x = 14$$

Note that if one side is 20, then the adjacent side, represented by $\dfrac{280}{x}$, is 14. And if one side is 14, then the adjacent side is 20.

So the room is 20 ft by 14 ft.

PRACTICE 5

In ping-pong, the length of the top of the ping-pong table is 1 ft less than twice the width. The area of the ping-pong table is 36 ft².

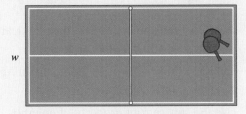

a. Write an expression for the length of the ping-pong table in terms of the width w.

b. Find the length and width of the top of the table.

A   *Fill in the blank to make each trinomial a perfect square.*

1. $x^2 + 6x + \boxed{}$ **2.** $t^2 + 4t + \boxed{}$ **3.** $x^2 - 2x + \boxed{}$ **4.** $n^2 - 10n + \boxed{}$

5. $x^2 + 5x + \boxed{}$ **6.** $x^2 - 3x + \boxed{}$ **7.** $y^2 - y + \boxed{}$ **8.** $n^2 + 9n + \boxed{}$

Solve by completing the square.

9. $x^2 + 4x = 0$ **10.** $p^2 + 6p = 0$ **11.** $b^2 - 10b = -1$ **12.** $x^2 - 20x = -50$

13. $y^2 + 14y - 15 = 0$ **14.** $t^2 + 10t + 21 = 0$ **15.** $x^2 - 6x - 4 = 0$ **16.** $x^2 - 4x - 2 = 0$

⊙ 17. $n^2 + 8n + 20 = 0$ **18.** $x^2 - 2x + 2 = 0$ **19.** $x^2 - x - 3 = 0$ **20.** $y^2 + 7y + 8 = 0$

21. $2y^2 + 8y = 24$ **22.** $3x^2 + 12x = -12$ **23.** $6x^2 + 12x - 5 = 0$ **24.** $5a^2 + 10a - 3 = 0$

25. $4n^2 - 20n + 7 = 0$ **26.** $9y^2 - 18y - 7 = 0$ **27.** $4n^2 - 3n - 4 = 0$ **28.** $2x^2 - x - 2 = 0$

29. $2x^2 - 4x + 8 = 0$ **30.** $5x^2 - 20x = -30$ **31.** $(x - 3)(x + 1) = 1$ **32.** $(d + 2)(d + 3) = 8$

Mixed Practice

Solve by completing the square.

33. $2x^2 + x - 4 = 0$ **34.** $b^2 + 12b + 27 = 0$

35. $n^2 - 6n + 10 = 0$ **36.** $2x^2 - 4x - 38 = 0$

Fill in the blank to make each trinomial a perfect square.

37. $t^2 + 7t + \boxed{}$ **38.** $a^2 - 8a + \boxed{}$

Applications

B   *Solve by completing the square.*

39. For the duration of an experiment, a biologist discovers that the number of insects N in a tank after d days is approximated by the model $N = d^2 + 12d + 6$. After how many days are there 51 insects?

40. An electronic company's revenue R (in dollars) is the product of the number of machines n that it sells and the price of a machine $(100 - n)$. This relationship can be expressed as:

$$R = n(100 - n)$$

How many machines must be sold for the revenue to be equal to $1600?

41. If fireworks on the ground are shot straight upward at a speed of 32 ft/sec, then the height of the fireworks h (in feet) after t sec can be represented by the expression $-16t^2 + 32t$.

  a. At what time will the fireworks be 16 ft above the ground?

  b. At what time (to the nearest tenth of a second) will the fireworks be 8 ft above the ground?

42. Consider the rectangle shown below with the measurements given in feet.

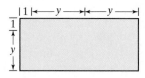

  a. Write an expression for the width and length of the rectangle.

  b. Write an expression for the area of the rectangle.

  c. If the area of the rectangle is 25 ft^2, find the value of y to the nearest tenth of a foot.

43. Two friends leave a party, one driving north and the other east. They are 10 mi apart when one of them is 2 mi farther from the party than the other. At that time, how far were they each from the party?

44. A commuter walks to and from work, which is 2 mi from her apartment. Going home, she walks 1 mph slower than going to work. If it takes her 1 h 10 min for the round trip, at what speed does she walk to work?

45. An office has two copying machines. Working alone, one machine takes 3 min longer to do a particular job than the other machine. When the machines are working together, it takes 4 min to do the job. To the nearest minute, how long would it take the faster machine to do the job alone?

46. Two painters are working on a house. If the first painter were doing the job by himself, it would take him 2 hr longer to paint the house than if the second painter were doing the job by himself. Working together, it takes the painters 5 hr to complete the job. To the nearest hour, how long would it take each painter to do the job working alone?

• Check your answers on page A-27.

Critical Thinking

1. A circular graph with radius r and centered around the point (h, k) can be represented by the equation $(x - h)^2 + (y - k)^2 = r^2$. Find the center and the radius of the circle represented by $x^2 - 4x + y^2 + 6y = -12$.

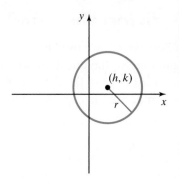

Mathematical Reasoning

2. We can represent the polynomial $x^2 + 10x$ geometrically by the area enclosed in the following diagram. In a few sentences, explain how we can use this diagram to "complete the square."

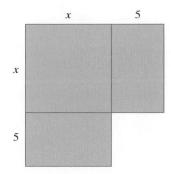

Groupwork

3. Working with a partner, consider the equation $x^2 + bx + c = 0$. The claim is made that this equation will have no real solution if $c > \dfrac{b^2}{4}$.

   a. Choose some values of b and of c to check this claim. Record your findings.

   b. Prove or disprove the claim by completing the square.

Deriving the Quadratic Formula

In Section 9.2, we used the method of completing the square each time we solved a quadratic equation. So the question arises: Why not use this method and solve the general equation $ax^2 + bx + c = 0$, where $a \neq 0$, once and for all? By solving the general equation for x, we can derive the *quadratic formula*:

$$ax^2 + bx + c = 0$$

$$x^2 + \frac{b}{a}x + \frac{c}{a} = 0 \qquad \text{Divide each side by } a.$$

$$x^2 + \frac{b}{a}x = -\frac{c}{a} \qquad \text{Add } -\frac{c}{a} \text{ to each side.}$$

$$x^2 + \frac{b}{a}x + \frac{b^2}{4a^2} = -\frac{c}{a} + \frac{b^2}{4a^2} \qquad \text{Complete the square: } \left(\frac{1}{2}\cdot\frac{b}{a}\right)^2 = \frac{b^2}{4a^2}; \text{ and add } \frac{b^2}{4a^2} \text{ to each side.}$$

$$\left(x + \frac{b}{2a}\right)^2 = \frac{b^2 - 4ac}{4a^2} \qquad \text{Write the left side as the square of a binomial, and write the right side as a single fraction.}$$

$$x + \frac{b}{2a} = \pm\sqrt{\frac{b^2 - 4ac}{4a^2}} \qquad \text{Assuming that the radicand is nonnegative, use the square root property of equality.}$$

$$x = -\frac{b}{2a} \pm \frac{\sqrt{b^2 - 4ac}}{2a} \qquad \text{Solve for } x \text{ by adding } -\frac{b}{2a} \text{ to each side.}$$

$$x = \frac{-b \pm \sqrt{b^2 - 4ac}}{2a} \qquad \text{Combine fractions.}$$

So the solutions of $ax^2 + bx + c = 0$ are $\dfrac{-b + \sqrt{b^2 - 4ac}}{2a}$ and $\dfrac{-b - \sqrt{b^2 - 4ac}}{2a}$.

This result gives us the *quadratic formula,* which can be used to solve any quadratic equation written in standard form.

The Quadratic Formula

If $ax^2 + bx + c = 0$, where a, b, and c are real numbers and $a \neq 0$, then

$$x = \frac{-b \pm \sqrt{b^2 - 4ac}}{2a}.$$

Solving Quadratic Equations Using the Quadratic Formula

The quadratic formula allows us to solve any quadratic equation of the form $ax^2 + bx + c = 0$, where $a \neq 0$. For instance, consider the equation $x^2 + 3x = 10$. To solve, we first write the equation in standard form: $ax^2 + bx + c = 0$.

$$x^2 + 3x = 10$$
$$x^2 + 3x - 10 = 0$$

In this equation $a = 1$, $b = 3$, and $c = -10$. So we substitute these values for a, b, and c in the formula, and then simplify.

$$x = \frac{-b \pm \sqrt{b^2 - 4ac}}{2a}$$

$$= \frac{-(3) \pm \sqrt{(3)^2 - 4(1)(-10)}}{2(1)} \qquad \text{Substitute 1 for } a, \text{ 3 for } b, \text{ and } -10 \text{ for } c.$$

$$= \frac{-3 \pm \sqrt{9 + 40}}{2} \qquad \text{Simplify.}$$

$$= \frac{-3 \pm \sqrt{49}}{2} \qquad \text{Simplify.}$$

$$= \frac{-3 \pm 7}{2} \qquad \text{Take the square root.}$$

$$x = \frac{-3 + 7}{2} \quad \text{or} \quad x = \frac{-3 - 7}{2}$$

$$= \frac{4}{2} = 2 \qquad\qquad = \frac{-10}{2} = -5$$

So the solutions are 2 and -5.

Note that the equation $x^2 + 3x = 10$ can also be solved by factoring.

$$x^2 + 3x - 10 = 0$$
$$(x + 5)(x - 2) = 0$$
$$x + 5 = 0 \quad \text{or} \quad x - 2 = 0$$
$$x = -5 \qquad\qquad x = 2$$

So solving by factoring yields the same solutions as those we found by using the quadratic formula.

Can all equations that are solved by the quadratic formula also be solved by factoring? Explain.

To Solve a Quadratic Equation $ax^2 + bx + c = 0$ Using the Quadratic Formula

- Write the equation in standard form, if necessary.
- Identify the coefficients a, b, and c.
- Substitute values for a, b, and c in the formula $x = \dfrac{-b \pm \sqrt{b^2 - 4ac}}{2a}$.
- Simplify.

In showing how to apply the quadratic formula, first we consider quadratic equations of the form $ax^2 + bx + c = 0$, where $a = 1$.

EXAMPLE 1

Solve by using the quadratic formula.

a. $m^2 - 9m + 18 = 0$

b. $x^2 + 3x = -3$

PRACTICE 1

Solve by using the quadratic formula.

a. $n^2 - n = 0$

b. $m^2 + m = -1$

EXAMPLE 1 (continued)

Solution

a. $m^2 - 9m + 18 = 0$

This equation is already in standard form, so $a = 1$, $b = -9$, and $c = 18$. We substitute these values in the quadratic formula.

$$m = \frac{-b \pm \sqrt{b^2 - 4ac}}{2a}$$

$$= \frac{-(-9) \pm \sqrt{(-9)^2 - 4(1)(18)}}{2(1)}$$

$$= \frac{9 \pm \sqrt{81 - 72}}{2}$$

$$= \frac{9 \pm \sqrt{9}}{2}$$

$$= \frac{9 \pm 3}{2}$$

$$m = \frac{9 + 3}{2} \quad \text{or} \quad m = \frac{9 - 3}{2}$$

$$= \frac{12}{2} \qquad\qquad = \frac{6}{2}$$

$$= 6 \qquad\qquad = 3$$

The solutions are 6 and 3.

b. $x^2 + 3x = -3$

First, let's write $x^2 + 3x = -3$ in standard form:
$x^2 + 3x + 3 = 0$. We see that $a = 1$, $b = 3$, and $c = 3$. We substitute these values in the formula.

$$x = \frac{-b \pm \sqrt{b^2 - 4ac}}{2a}$$

$$= \frac{-(3) \pm \sqrt{(3)^2 - 4(1)(3)}}{2(1)}$$

$$= \frac{-3 \pm \sqrt{9 - 12}}{2}$$

$$= \frac{-3 \pm \sqrt{-3}}{2}$$

In Section 9.1, we saw that a radical with a negative radicand is not a real number. Here note that the radicand is negative. So there are no real solutions of the equation $x^2 + 3x = -3$.

TIP Before using the quadratic formula, make sure that the equation you are solving is in standard form.

Next, let's consider quadratic equations of the form $ax^2 + bx + c = 0$, where $a \neq 1$.

EXAMPLE 2

Use the quadratic formula to solve.

a. $4t^2 = 12t - 1$ **b.** $2y^2 + 3y = 1$

Solution

a.
$$4t^2 = 12t - 1$$
$$4t^2 - 12t + 1 = 0 \qquad \text{Write the equation in standard form.}$$

$$t = \frac{-b \pm \sqrt{b^2 - 4ac}}{2a}$$

$$= \frac{-(-12) \pm \sqrt{(-12)^2 - 4(4)(1)}}{2(4)} \qquad \begin{array}{l} \text{Substitute 4 for } a, \\ -12 \text{ for } b, \text{ and} \\ 1 \text{ for } c. \end{array}$$

$$= \frac{12 \pm \sqrt{144 - 16}}{8}$$

$$= \frac{12 \pm \sqrt{128}}{8}$$

$$x = \frac{12 + \sqrt{128}}{8} \quad \text{or} \quad x = \frac{12 - \sqrt{128}}{8}$$

$$= \frac{12 + 8\sqrt{2}}{8} \qquad\qquad = \frac{12 - 8\sqrt{2}}{8}$$

$$= \frac{4(3 + 2\sqrt{2})}{8} \qquad\quad = \frac{4(3 - 2\sqrt{2})}{8}$$

$$= \frac{3 + 2\sqrt{2}}{2} \qquad\qquad = \frac{3 - 2\sqrt{2}}{2}$$

The solutions are $\dfrac{3 + 2\sqrt{2}}{2}$ and $\dfrac{3 - 2\sqrt{2}}{2}$.

b.
$$2y^2 + 3y = 1$$
$$2y^2 + 3y - 1 = 0 \qquad \text{Write the equation in standard form.}$$

$$y = \frac{-b \pm \sqrt{b^2 - 4ac}}{2a}$$

$$= \frac{-3 \pm \sqrt{(3)^2 - 4(2)(-1)}}{2(2)} \qquad \begin{array}{l} \text{Substitute 2 for } a, 3 \\ \text{for } b, \text{ and } -1 \text{ for } c. \end{array}$$

$$= \frac{-3 \pm \sqrt{9 + 8}}{4}$$

$$= \frac{-3 \pm \sqrt{17}}{4}$$

The solutions are $\dfrac{-3 + \sqrt{17}}{4}$ and $\dfrac{-3 - \sqrt{17}}{4}$.

PRACTICE 2

Use the quadratic formula to solve.

a. $5s^2 - 8s = 3$

b. $3m^2 = 7 - 3m$

Sometimes in solving a quadratic equation, we need to approximate the solutions using a calculator. This is often true when solving applied problems.

EXAMPLE 3

Solve $2t^2 + 4t = 7$. Round the solutions to the nearest tenth.

Solution

$$2t^2 + 4t = 7$$

$$2t^2 + 4t - 7 = 0 \qquad \text{Write the equation in standard form.}$$

$$t = \frac{-b \pm \sqrt{b^2 - 4ac}}{2a} \qquad \text{Use the quadratic formula.}$$

$$= \frac{-4 \pm \sqrt{(4)^2 - 4(2)(-7)}}{2(2)} \qquad \text{Substitute 2 for } a, 4 \text{ for } b, \text{ and } -7 \text{ for } c.$$

$$= \frac{-4 \pm \sqrt{16 + 56}}{4}$$

$$= \frac{-4 \pm \sqrt{72}}{4}$$

$$= \frac{-4 \pm 6\sqrt{2}}{4}$$

$$t = \frac{-4 + 6\sqrt{2}}{4} \quad \text{or} \quad t = \frac{-4 - 6\sqrt{2}}{4}$$

$$= \frac{-2 + 3\sqrt{2}}{2} \qquad\qquad = \frac{-2 - 3\sqrt{2}}{2}$$

$$\approx 1.121 \qquad\qquad\qquad \approx -3.121$$

The solutions are approximately 1.1 and −3.1.

PRACTICE 3

Solve $3x^2 = 6x + 4$. Round the solutions to the nearest tenth.

EXAMPLE 4

An experienced clerk can complete a data-entry project in 2 hr less than a new clerk can working alone. If the clerks work together, it would take 3 hr to complete the data-entry project. How long would it take each clerk to complete the data-entry project working alone? Round your answer to the nearest tenth of an hour.

Solution If it takes the experienced clerk n hours to complete the project working alone, then she works at the rate $\frac{1}{n}$ of the project per hour. If it takes the new clerk $(n + 2)$ hours working alone, then he works at the rate $\frac{1}{n + 2}$ of the project per hour.

Let's make a table using the fact that the product of rate and time worked is the amount of work completed.

	Rate of Work	Time Worked	Part of the Project Completed
Experienced Clerk	$\frac{1}{n}$	3	$\frac{3}{n}$
New Clerk	$\frac{1}{n + 2}$	3	$\frac{3}{n + 2}$

PRACTICE 4

Working together, two pumps can empty a tank in 3 hr. Working alone, one pump takes 5 hr more than the other to carry out this task. Using a calculator, determine how long, rounded to the nearest tenth of an hour, it takes each pump working alone to empty the tank.

Because the total amount of work completed in 3 hr is 1 project, we can write the equation $\dfrac{3}{n} + \dfrac{3}{n+2} = 1$. Now, we solve for n.

$$\frac{3}{n} + \frac{3}{n+2} = 1$$

$$n(n+2)\left(\frac{3}{n} + \frac{3}{n+2}\right) = n(n+2) \cdot 1$$

$$\overset{1}{\cancel{n}}(n+2)\frac{3}{\underset{1}{\cancel{n}}} + n\cancel{(n+2)}\frac{3}{\cancel{n+2}} = n(n+2)$$

$$3(n+2) + 3n = n(n+2)$$

$$3n + 6 + 3n = n^2 + 2n$$

$$6n + 6 = n^2 + 2n$$

$$0 = n^2 - 4n - 6, \text{ or } n^2 - 4n - 6 = 0$$

Using the quadratic formula to solve for n, we get:

$$n = \frac{-(-4) \pm \sqrt{(-4)^2 - 4(1)(-6)}}{2(1)}$$

$$= \frac{4 \pm \sqrt{16 + 24}}{2}$$

$$= \frac{4 \pm \sqrt{40}}{2}$$

$$= \frac{4 \pm 2\sqrt{10}}{2}$$

$$= \frac{2(2 \pm \sqrt{10})}{2}$$

$$= 2 \pm \sqrt{10}$$

$$n = 2 + \sqrt{10} \quad \text{or} \quad n = 2 - \sqrt{10}$$
$$\approx 5.162 \qquad\qquad\quad \approx -1.162$$

Because the amount of time it takes the experienced clerk to complete the project must be positive, we take n to be approximately 5.2. So working alone, it takes the experienced clerk about 5.2 hr to complete the data entry project. Because the new clerk's time working alone is 2 hr more than the experienced clerk's time, the new clerk's time is about 7.2 hr, rounded to the nearest tenth of an hour.

In this chapter, we have used four methods of solving quadratic equations in the form $ax^2 + bx + c = 0$. These methods are factoring, the square root property, the quadratic formula, and completing the square. The following table lists when to use each method:

Method	When to Use Method
Factoring	• Use when the polynomial can be factored.
The Square Root Property	• Use when $b = 0$.
The Quadratic Formula	• Use when the first two methods do not apply.
Completing the Square	• Use only when specified. This method is easier to use when a is 1 and b is even.

Write each quadratic equation in standard form. Then, identify the values of a, b, and c.

Quadratic Equation	Standard Form	$a =$	$b =$	$c =$
1. $2x^2 + 9x - 1 = 0$				
2. $x^2 - 3x + 4 = 0$				
3. $-x^2 + 3x = 8$				
4. $x^2 + 4 = -x$				
5. $x^2 + 2x = 3x - 8$				
6. $5y + 7 = 2y^2 + 1$				
7. $\frac{1}{3}y^2 - \frac{1}{2}y = -\frac{1}{4}$				
8. $-1.04 + 0.001x + 0.002x^2 = 0$				
9. $(2x + 1)(3x - 5) = 10$				
10. $(4x - 9)(x) = 2$				

Solve by using the quadratic formula.

11. $t^2 + 2t - 3 = 0$ 12. $x^2 + 8x + 7 = 0$ 13. $n^2 + 3n + 6 = 0$ 14. $x^2 - 5x + 7 = 0$

15. $7 + 6x - x^2 = 0$ 16. $1 - x - x^2 = 0$ 17. $y^2 - 4y = 1$ 18. $x^2 + 6x = -9$

19. $3p^2 - 4p + 1 = 0$ 20. $5p^2 + p - 3 = 0$ 21. $4x^2 - 6x = -1$ 22. $6y^2 + 8y = 2$

23. $2n^2 = 7n + 30$ 24. $4t^2 = -4t + 9$ 25. $x^2 + 3x + 2 = 0$ 26. $x^2 - 6x + 5 = 0$

27. $3 + x^2 + 4x = 0$ 28. $2y^2 + 15 = 11y$ 29. $3p^2 = 2p$ 30. $x^2 = -3x$

31. $4n^2 + 3n - 1 = 0$ 32. $s^2 - 7s + 2 = 0$ 33. $t^2 + 5 = 7t$ 34. $y^2 = 4y + 2$

35. $(n + 7)(n - 8) = 5$ 36. $(w - 3)(w - 2) = 12$ 37. $5(x - 1) = x(x + 2)$ 38. $x(x + 1) = -2(x + 5)$

39. $\frac{x^2}{3} - x = -\frac{1}{2}$ 40. $\frac{3}{n - 2} - \frac{1}{n - 1} = 2$

Mixed Practice

Solve by using the quadratic formula.

41. $4m + 5 + m^2 = 0$ 42. $4k^2 + 6k = 2$

43. $-6 + s^2 = s$ 44. $(n + 3)(n - 5) = -10$

Write each quadratic equation in standard form. Then, identify the values of a, b, and c.

Quadratic Equation	Standard Form	$a =$	$b =$	$c =$
45. $\frac{2}{5}d^2 - \frac{4}{5}d = \frac{1}{5}$				
46. $(3x + 2)(2x - 1) = 4$				

Applications

Solve by using the quadratic formula. Use a calculator, if necessary.

47. The sum of the consecutive whole numbers from 1 through n is given by the formula $S = \dfrac{n(n + 1)}{2}$. If this sum is 91, what is n?

48. The formula $d = \dfrac{n(n - 3)}{2}$ is used to find the number of diagonals in a polygon with n sides. How many sides does a polygon with 20 diagonals have?

49. A lab coordinator and a lab technician work in a mathematics laboratory. Working alone, the lab technician can set up the lab in 20 min less time than the lab coordinator. Together, they can set up the lab in 40 min. How long, to the nearest minute, does it take each of them to set up the lab working alone?

50. A college copy center has two photocopying machines. The older copier takes 2 hr longer than the newer copier to do a job. Together, the two copiers would take 5 hr to do the job. If the older machine breaks down, how long to the nearest hour will it take the newer machine to do the job?

51. A car travels 20 mph faster than a bicycle. The car travels 40 mi in 4 hr less than it takes the bicycle. Find the speed of the car, rounded to the nearest mile per hour.

52. Researchers investigating a new medicine found that the percent p (in decimal form) of the medicine in the blood after t sec can be modeled by $p = 0.08t - 0.01t^2$. After how many seconds, to the nearest tenth of a second, does the blood contain 10% of the medicine?

53. On a recent date, the Fahrenheit temperature y in Anaheim, California, between 11:00 A.M. and 6 P.M. could be modeled by

$$y = -0.4x^2 + 2.7x + 59.4,$$

where x represents the number of hours after 11:00 A.M. At approximately what times was the temperature $61°$ F?

54. The equation $y = 0.1x^2 + 2.4x + 291$ models the United States population y (in millions) between 2003 and 2010, where x represents the number of years since 2003. According to this model, in which year was the population approximately 305 million? (*Source:* cms.gov)

• Check your answers on page A-27.

MINDStretchers

Mathematical Reasoning

1. The radicand in the quadratic formula, $b^2 - 4ac$, is called the *discriminant*. Consider the following equations:
$$x^2 - 8x + 16 = 0 \qquad x^2 + 2x - 8 = 0 \qquad x^2 - 3x + 5 = 0$$

   a. Compute the discriminant for each equation.

   b. How many real solutions does an equation with a positive discriminant have?
   An equation with a zero discriminant?
   An equation with a negative discriminant?

Research

2. The equation $x^3 - x^2 + 4x = 3$ is said to be a *cubic equation*, because the highest power of the variable is 3. Investigate on the Web or in your college library whether mathematicians have discovered a *cubic formula* for solving cubic equations, just as the quadratic formula solves quadratic equations. Summarize your findings in a short paragraph.

Patterns

3. Find the solutions of each quadratic equation. Then, fill in the table.

Equation	Sum of Solutions	Product of Solutions
$x^2 + 5x + 6 = 0$		
$x^2 - 4x - 12 = 0$		
$2x^2 + 3x + 1 = 0$		
$3x^2 - x - 2 = 0$		

Consider the quadratic equation $ax^2 + bx + c = 0$. Based on the results in the table, identify a relationship between the coefficients of the terms of a quadratic equation and the sum and product of its solutions.

Recall from Chapter 3 our discussion of linear equations in two variables. There we graphed equations of the form $y = mx + b$. In this section, we graph quadratic equations in two variables. These equations are of the form $y = ax^2 + bx + c$, where $a \neq 0$. Examples of quadratic equations in two variables are:

$$y = x^2 \qquad y = 5 - 2x^2 \qquad y = x^2 - 4x + 3$$

Equations of this type, when graphed, have varied real-world applications. For example, they may provide insight into how best to achieve major business goals, such as maximizing profit or minimizing cost.

Graphing Quadratic Equations in Two Variables

Just as with the graphing of linear equations, we can graph quadratic equations by choosing values of one variable and then computing the corresponding values of the other variable to make a table of values.

For instance, let's consider the graph of the equation $y = x^2$. First, we select values for x, and then find the corresponding values for y.

x	$y = x^2$	(x, y)
-2	$y = (-2)^2 = 4$	$(-2, 4)$
-1	$y = (-1)^2 = 1$	$(-1, 1)$
0	$y = (0)^2 = 0$	$(0, 0)$
1	$y = (1)^2 = 1$	$(1, 1)$
2	$y = (2)^2 = 4$	$(2, 4)$

Next, we plot these points on a coordinate plane, and then draw a smooth curve through them. The result is the graph of $y = x^2$.

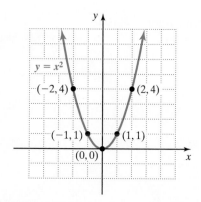

The graph of an equation of the form $y = ax^2 + bx + c$ is called a **parabola**. A parabola has the following characteristics:

- The graph is U-shaped, opening either upward or downward.
- The highest or lowest point of a parabola is called the **vertex** (or *turning point*) of the parabola.
- The graph has an axis of symmetry, which is a vertical line that passes through the vertex. This **axis of symmetry** divides the parabola into two parts that are mirror images of one another.
- The curve goes on indefinitely.

Consider the following graphs:

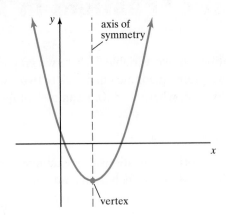

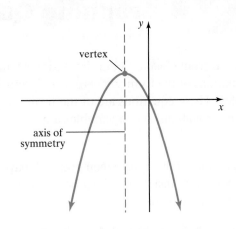

The parabola on the left opens upward, with the minimum y-value at the vertex. By contrast, the parabola on the right opens downward, with the maximum y-value at the vertex.

EXAMPLE 1

Graph: $y = 2x^2$

Solution First, we select values for x, and then find the corresponding values for y.

x	$y = 2x^2$	(x, y)
-2	$y = 2(-2)^2 = 2 \cdot 4 = 8$	$(-2, 8)$
-1	$y = 2(-1)^2 = 2 \cdot 1 = 2$	$(-1, 2)$
0	$y = 2(0)^2 = 2 \cdot 0 = 0$	$(0, 0)$
1	$y = 2(1)^2 = 2 \cdot 1 = 2$	$(1, 2)$
2	$y = 2(2)^2 = 2 \cdot 4 = 8$	$(2, 8)$

Next, we plot the points on a coordinate plane, and then draw a smooth curve through them. The result is the graph of the parabola $y = 2x^2$.

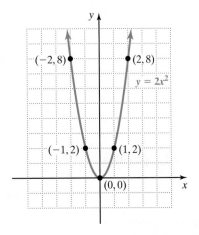

PRACTICE 1

Graph: $y = -2x^2$

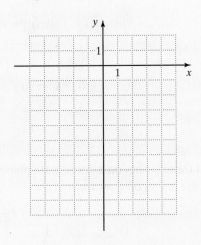

What are the differences and similarities between the graphs of $y = 2x^2$ and $y = -2x^2$?

Note that in Example 1, the vertex is the point $(0, 0)$ and the axis of symmetry is the y-axis. This is always true for the graph of any quadratic equation of the form $y = ax^2$.

Now, let's look at the graphs of equations that are of the more general form $y = ax^2 + bx + c$. Although we can always make a table of x- and y-values to graph a quadratic equation, we consider two other methods of graphing quadratic equations. One method is *graphing the vertex and several points on either side of the vertex*. The other method is *graphing the vertex and the x- and y-intercepts*.

Recall that the vertex is the point on the parabola where the curve turns and that it lies on the axis of symmetry.

The Equation of the Axis of Symmetry of a Parabola

For a parabola given by the quadratic equation $y = ax^2 + bx + c$, the equation of the axis of symmetry is $x = -\dfrac{b}{2a}$.

Since the axis of symmetry is a vertical line, its equation is of the form $x =$ a constant. From our discussion of vertical lines, we know that every point on the axis of symmetry, including the vertex, has x-coordinate $-\dfrac{b}{2a}$. We can use this fact to find the coordinates of the vertex as follows:

To Find the Vertex of a Parabola

For a parabola given by the quadratic equation $y = ax^2 + bx + c$,

- The x-coordinate of the vertex is $-\dfrac{b}{2a}$.

- The y-coordinate of the vertex is found by substituting $-\dfrac{b}{2a}$ for x into the equation and then computing y.

Now, let's use the vertex and several points on either side of the vertex to graph a quadratic equation.

EXAMPLE 2

Graph: $y = 5 - 2x^2$

Solution   Writing the equation in standard form, $y = -2x^2 + 0x + 5$, we see that $a = -2$ and $b = 0$. The x-coordinate of the vertex is:

$$-\frac{b}{2a} = \frac{0}{2(-2)} = 0$$

Substituting 0 for x into the equation $y = 5 - 2x^2$, we find the y-coordinate of the vertex:

$$y = 5 - 2x^2 = 5 - 2(\mathbf{0})^2 = 5$$

So the vertex is $(0, 5)$.

PRACTICE 2

Graph: $y = -3x^2 + 1$

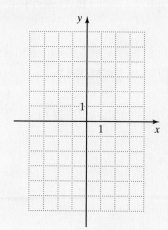

EXAMPLE 2 (continued)

Next, we find points on either side of the vertex.

x	$y = 5 - 2x^2$	(x, y)
-2	$y = 5 - 2x^2 = 5 - 2(-2)^2 = -3$	$(-2, -3)$
-1	$y = 5 - 2x^2 = 5 - 2(-1)^2 = 3$	$(-1, 3)$
0	$y = 5 - 2x^2 = 5 - 2(0)^2 = 5$	$(0, 5)$ ← vertex
1	$y = 5 - 2x^2 = 5 - 2(1)^2 = 3$	$(1, 3)$
2	$y = 5 - 2x^2 = 5 - 2(2)^2 = -3$	$(2, -3)$

Now, we plot the points and draw a smooth curve through them. The result is the graph of the parabola $y = 5 - 2x^2$. Note that the y-axis, with equation $x = 0$, is the axis of symmetry for the parabola.

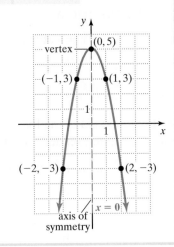

In Example 1, we saw that for $y = 2x^2$, where a is positive, the parabola opens upward. On the other hand, in Example 2, the graph of $y = 5 - 2x^2$, where a is negative, opens downward. Knowing the sign of a is useful in graphing $y = ax^2 + bx + c$.

> **TIP** When a is positive, the parabola opens upward. When a is negative, the parabola opens downward.

Finally, we consider the graphing of quadratic equations of the form $y = ax^2 + bx + c$ using both the vertex and the x- and y-intercepts.

EXAMPLE 3

Consider $y = x^2 - 4x + 3$.

a. Sketch its graph.　　**b.** Describe the graph.

Solution

a. To graph $y = x^2 - 4x + 3$, first we find the vertex.
The x-coordinate of the vertex is:

$$-\frac{b}{2a} = -\frac{-4}{2(1)} = \frac{4}{2} = 2$$

PRACTICE 3

Consider $y = -x^2 + 2x + 3$.

a. Sketch its graph.

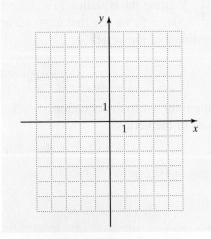

Substituting 2 for x, we get:

$$y = x^2 - 4x + 3 = (2)^2 - 4(2) + 3 = 4 - 8 + 3 = -1$$

So the vertex of the parabola is $(2, -1)$, and the axis of symmetry is $x = 2$.

Next, we find the x- and y-intercepts. Recall that to find the x-intercept of a graph, we let $y = 0$ and solve for x.

$$y = x^2 - 4x + 3$$
$$0 = x^2 - 4x + 3$$
$$0 = (x - 3)(x - 1)$$
$$x - 3 = 0 \quad \text{or} \quad x - 1 = 0$$
$$x = 3 \qquad\qquad x = 1$$

So the x-intercepts are $(3, 0)$ and $(1, 0)$.

Similarly, to find the y-intercept, we let $x = 0$ and solve for y.

$$y = x^2 - 4x + 3$$
$$= (0)^2 - 4(0) + 3 = 3$$

The y-intercept is therefore $(0, 3)$.

Finally, we plot the vertex, the x-intercepts, and the y-intercept on a coordinate plane. Then, we draw a smooth curve through the points. The result is the graph of the equation $y = x^2 - 4x + 3$.

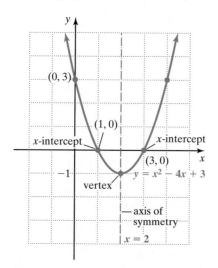

b. Describe the graph.

b. The parabola $y = x^2 - 4x + 3$ opens upward since $a = 1$ is positive. The curve turns at the vertex $(2, -1)$, the lowest point of the graph. The equation of the axis of symmetry is $x = 2$.

Can you explain the relationship between the y-intercept of a parabola and the constant term of the corresponding quadratic equation in the form $y = ax^2 + bx + c$?

EXAMPLE 4

A Roman candle is shot straight upward from the ground with an initial velocity of 96 ft/sec. After t sec, its distance s in feet above the ground is given by the equation $s = 96t - 16t^2$. Graph the equation. Find the maximum height reached by the Roman candle.

Solution Let's graph the equation $s = 96t - 16t^2$ by using the vertex and t- and s-intercepts. First, we find the vertex. The t-coordinate of the vertex is:

$$-\frac{b}{2a} = -\frac{96}{2(-16)} = 3$$

Substituting 3 for t, we get the value of s as follows:

$$s = 96t - 16t^2 = 96(3) - 16(3)^2 = 288 - 144 = 144$$

So the vertex has coordinates $(3, 144)$. Note that the line of symmetry is $t = 3$.

Next, we find the t- and s-intercepts of the graph. To find the t-intercept, we let $s = 0$ and solve for t.

$$s = 96t - 16t^2$$
$$0 = 96t - 16t^2$$
$$0 = 16t(6 - t)$$
$$16t = 0 \quad \text{or} \quad 6 - t = 0$$
$$t = 0 \qquad\qquad t = 6$$

So the t-intercepts are $(0, 0)$ and $(6, 0)$. Similarly, to find the s-intercept, we let $t = 0$ and solve for s.

$$s = 96t - 16t^2$$
$$s = 96(0) - 16(0)^2 = 0$$

The s-intercept is therefore $(0, 0)$.

Finally, we plot the vertex and the t- and s-intercepts on a coordinate plane to draw a smooth curve through the points.

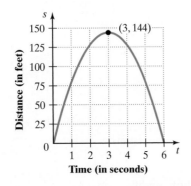

Since the maximum s-value is 144, the maximum height reached by the Roman candle is 144 ft. Notice that the distance above the ground increases between 0 and 3 sec and then decreases between 3 and 6 sec after the Roman candle is shot.

PRACTICE 4

The total profit or loss P in dollars made by a street vendor is given by the equation $P = x^2 - 10x$, where x is the number of items sold. Graph this equation. Find the number of items sold that results in the most money lost.

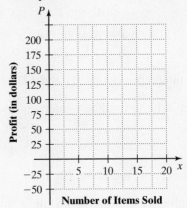

Note that in Example 4, the graph of $s = 96t - 16t^2$ does not go below the t-axis. Why is this the case? Explain.

Mathematically Speaking

Fill in each blank with the most appropriate term or phrase from the given list.

positive	vertex	negative
axis of symmetry	equation of the axis of	graph of an equation
equation	symmetry	

1. The _____ of the form $y = ax^2 + bx + c$ is called a parabola.

2. The highest or lowest point of a parabola is said to be its _____.

3. The vertical line passing through a parabola's vertex is called its _____.

4. For the graph of $y = ax^2 + bx + c$, the _____ is $x = -\dfrac{b}{2a}$.

5. The parabola corresponding to the equation $y = ax^2 + bx + c$ opens upward when a is _____.

6. The parabola corresponding to the equation $y = ax^2 + bx + c$ opens downward when a is _____.

Graph a parabola with the following characteristics.

7. Passes through the points $(-2, 5)$, $(-1, 2)$, $(0, 1)$, $(1, 2)$, and $(2, 5)$.

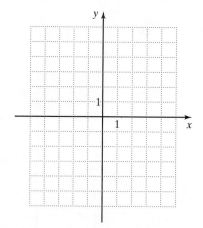

8. Passes through the points $(-2, -3)$, $(-1, 0)$, $(0, 1)$, $(1, 0)$, and $(2, -3)$.

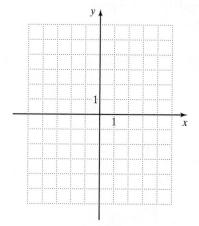

9. Has vertex $(0, 0)$ and also passes through the points $(-2, 4)$, $(-1, 1)$, $(1, 1)$, and $(2, 4)$.

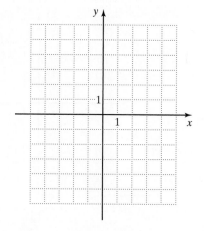

10. Has vertex $(0, 3)$ and also passes through the points $(2, -1)$, $(-2, -1)$, $(3, -6)$, and $(-3, -6)$.

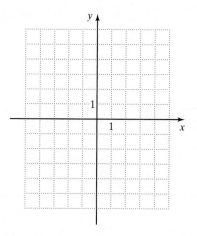

11. Has vertex $(2, -1)$, y-intercept $(0, 3)$, and x-intercepts $(1, 0)$ and $(3, 0)$.

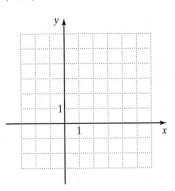

12. Has vertex $(-3, -1)$, y-intercept $(0, 8)$, and passes through the point $(-6, 8)$.

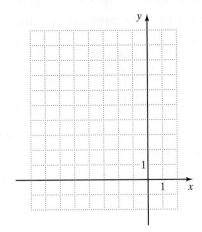

Find the axis of symmetry and the vertex of the graph of each equation.

13. $y = \dfrac{1}{2}x^2$

14. $y = \dfrac{1}{4}x^2$

15. $y = 5 - 4x + x^2$

16. $y = 2x^2 - 8x - 1$

Sketch the graph of each equation.

17. $y = 3x^2$

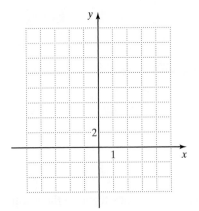

18. $y = 4x^2$

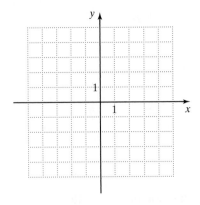

19. $y = -3x^2$

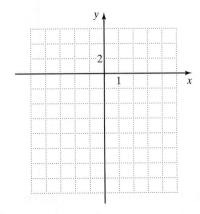

20. $y = -5x^2$

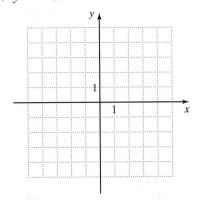

21. $y = 4 - x^2$

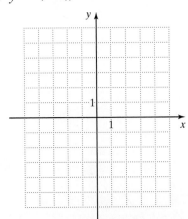

22. $y = 3 - x^2$

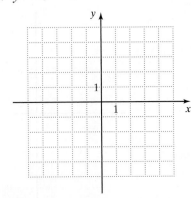

23. $y = x^2 - 9$

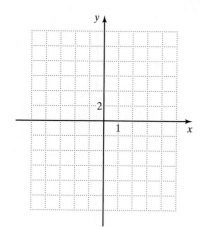

24. $y = x^2 - 4$

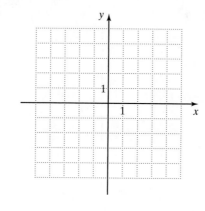

25. $y = x^2 - 2x$

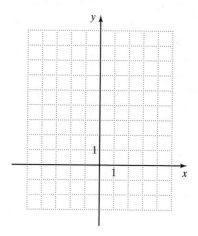

26. $y = x^2 - 4x$

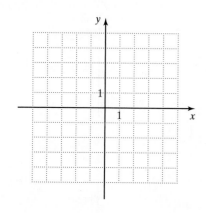

27. $y = -x^2 + 4x$

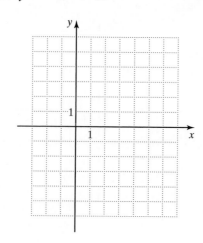

28. $y = -x^2 + 2x$

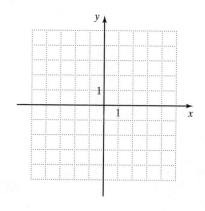

29. $y = -x^2 + 3x + 2$

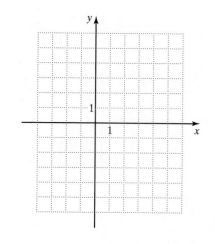

30. $y = -x^2 + 5x - 4$

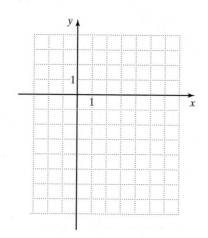

31. $y = x^2 - x - 6$

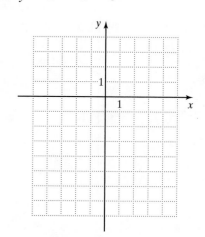

32. $y = x^2 - 4x + 4$

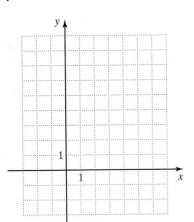

33. $y = 3x^2 - 4x + 1$

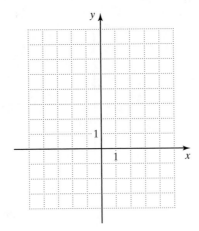

34. $y = 2x^2 + 3x - 2$

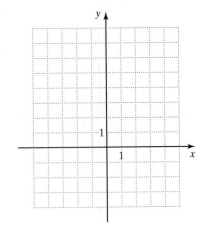

35. $y = (x + 2)^2$

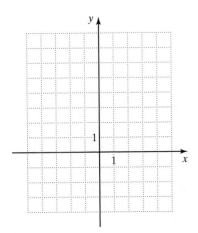

36. $y = (x + 3)^2$

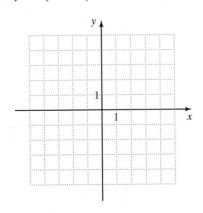

37. $y = (x - 2)^2$

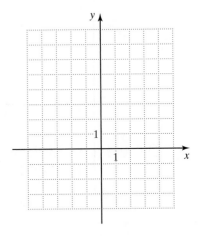

38. $y = (x - 3)^2$

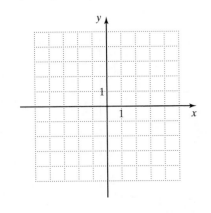

39. $y = 2 - 3x^2$

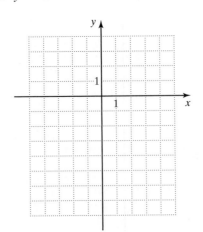

40. $y = 3x - 2x^2$

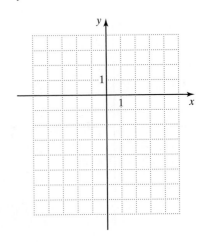

41. $y = -\dfrac{1}{2}x^2 - 5$

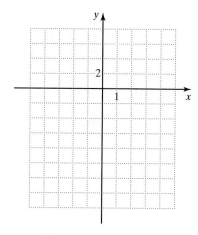

42. $y = -\dfrac{1}{4}x^2 - 4$

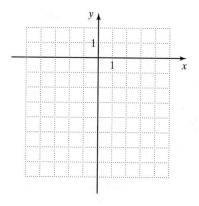

43. $y = \dfrac{3}{4}x^2 + 2$

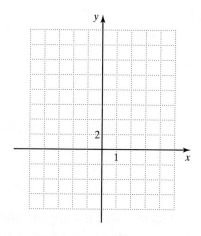

44. $y = \dfrac{1}{2}x^2 + 1$

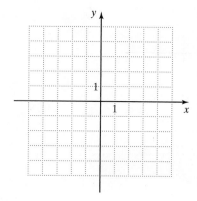

Mixed Practice

Solve.

45. Graph a parabola that has vertex $(0, -5)$ and also passes through the points $(-2, -1), (-1, -4), (1, -4), (2, -1)$.

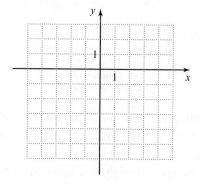

46. Find the axis of symmetry and the vertex of the graph of the equation $y = 4 - 3x^2 + 6x$.

Sketch the graph of each equation.

47. $y = 2x^2 - 3x + 1$

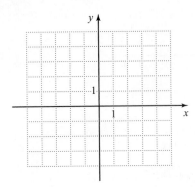

48. $y = -x^2 + 2x$

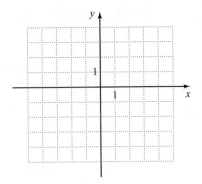

49. $y = -x^2 - 3x + 2$

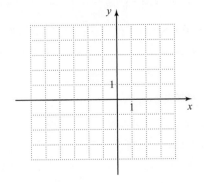

50. $y = \frac{1}{4}x^2 - 2$

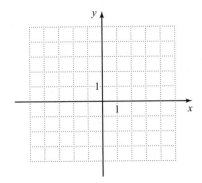

Applications

Solve.

51. A company's annual payroll (in millions of dollars) in various years is modeled by $0.4x^2 - 2.4x + 23.9$, where x represents the number of years after 2005.

a. Graph the following equation:
$y = 0.4x^2 - 2.4x + 23.9$

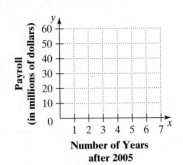

b. According to this model, in what year was the payroll minimized?

c. In a sentence, describe the trend in the team's payroll.

52. The manager of a coat factory found that the unit cost y (in dollars) of producing x coats can be approximated using the expression $0.002x^2 - 2.7x + 1000$.

a. Graph the equation $y = 0.002x^2 - 2.7x + 1000$.

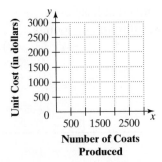

b. According to this model, the unit cost of producing coats will be lowest if how many coats are produced?

c. Describe how the unit cost varies with the number of coats produced.

53. The daily profit y of a company varies with the number of products x it sells per day. The daily profit can be approximated by the expression $-0.1x^2 + 10x$.

a. Graph the equation $y = -0.1x^2 + 10x$.

b. How many products must the company sell per day to maximize the profit?

c. Describe how the profit changes as sales increase.

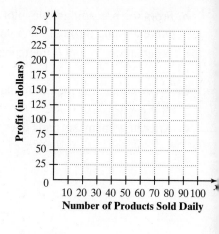

Profit (in dollars)

Number of Products Sold Daily

54. A child throws a ball straight upward from a height of 6 ft with an initial velocity of 80 ft/sec. The ball's height h ft above the ground after t sec is modeled by the following equation:

$$h = 6 + 80t - 16t^2$$

a. Graph this equation.

b. Will the ball reach a height of 4 ft above the ground?

c. Describe how the height of the ball changes over time.

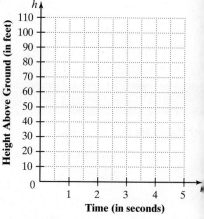

Height Above Ground (in feet)

Time (in seconds)

55. The manager of a store determines that the store can sell x items at a price of $100 - \frac{1}{4}x$ dollars per item.

a. Write a formula for the total revenue R that these sales generate.

b. Using the formula in part (a), complete a table of the revenue generated by the sale of 0, 100, 200, 300, and 400 items.

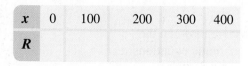

x	0	100	200	300	400
R					

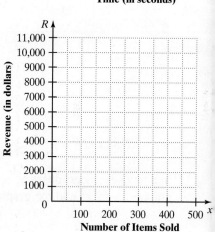

Revenue (in dollars)

Number of Items Sold

c. Graph the formula.

d. How many items should the store sell to maximize the revenue?

e. What is that maximum revenue?

56. The formula $h = 147t - 4.9t^2$ can be used to predict the height in meters of a small rocket launched straight upward from the ground with an initial velocity of 147 m/sec. The number of seconds after launch is represented by t.

a. Using the given formula, complete the table.

t	0	5	10	15	20	25	30
h							

b. From the preceding table, plot the ordered pairs (t, h). Draw a smooth curve passing through the points.

c. What are the t- and h-intercepts of the graph?

d. Approximate the maximum height reached by the rocket.

e. After how many seconds does the rocket hit the ground?

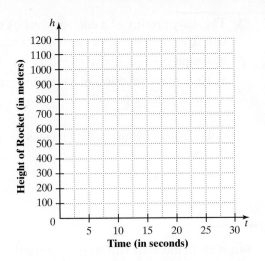

• Check your answers on page A-28.

MINDStretchers

Patterns

1. Consider the graph of $y = x^2$.

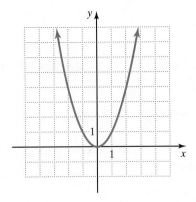

In a few sentences, describe any relationship that you observe between the graph of $y = x^2$ and the graph of each of the following quadratic equations.

a. $y = (x + 1)^2$ **b.** $y = (x - 1)^2$

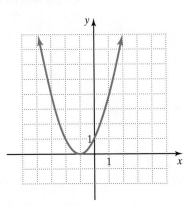   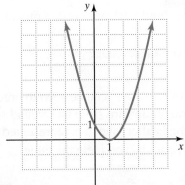

c. $y = x^2 + 1$

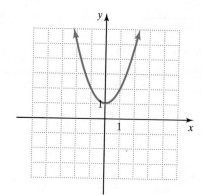

d. $y = x^2 - 1$

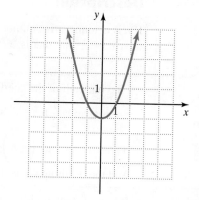

Research

2. Using the Web or your college library, find five situations in which parabolas are important in the real world. Write a sentence about each example.

Mathematical Reasoning

3. For the three graphs shown to the right, identify the graph(s) that fit each description below. Each graph is of the form $y = ax^2 + bx + c$.

   a. $a > 0$

   b. $a < 0$

   c. The graph with the *greatest* value of $|a|$

   d. The graph with the *least* value of $|a|$

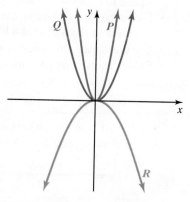

Concept/Skill	Description	Example
[9.1] **The square root property of equality**	If n is a nonnegative number and $x^2 = n$, then $x = \pm\sqrt{n}$, that is, $x = \sqrt{n}$ or $x = -\sqrt{n}$.	$x^2 = 15$ $x = \sqrt{15}$ or $x = -\sqrt{15}$
[9.2] **To complete the square for the expression** $x^2 + bx$	• Find one-half of the coefficient of the linear term x: $\dfrac{1}{2}b$ • Square this value: $\left(\dfrac{1}{2}b\right)^2$ • Add the resulting value to the expression.	$x^2 + 8x + c$: $\dfrac{1}{2}(8) = 4$ $4^2 = 16$ $x^2 + 8x + 16 = (x + 4)^2$
[9.2] **To solve the quadratic equation** $ax^2 + bx + c = 0$ **by completing the square**	• If $a \neq 1$, then divide the equation by a, the coefficient of the second-degree term. If $a = 1$, then proceed to the next step. • Move all terms with variables to one side of the equals sign and all constants to the other side. • Take one-half of the coefficient of the linear term. Then, square it and add this square to both sides of the equation. • Factor the side containing the variables, writing it as the square of a binomial. • Use the square root property of equality. • Solve the resulting equations.	$3x^2 + 18x + 12 = 0$ $x^2 + 6x + 4 = 0$ $\quad x^2 + 6x = -4$ $\dfrac{1}{2}(6) = 3$ $3^2 = 9$ $x^2 + 6x + 9 = -4 + 9$ $\quad (x + 3)^2 = 5$ $\quad\quad x + 3 = \pm\sqrt{5}$ $\quad\quad\quad x = -3 \pm \sqrt{5}$ Solutions: $x = -3 + \sqrt{5}$ and $x = -3 - \sqrt{5}$
[9.3] **The quadratic formula**	If $ax^2 + bx + c = 0$, where a, b, and c are real numbers and $a \neq 0$, then $$x = \frac{-b \pm \sqrt{b^2 - 4ac}}{2a}.$$	$2x^2 + 4x - 7 = 0$, where $a = 2$, $b = 4$, and $c = -7$

Concept/Skill	Description	Example
[9.3] **To solve a quadratic equation** $ax^2 + bx + c = 0$ **using the quadratic formula**	• Write the equation in standard form, if necessary. • Identify the coefficients a, b, and c. • Substitute values for a, b, and c in the formula $$x = \frac{-b \pm \sqrt{b^2 - 4ac}}{2a}.$$ • Simplify.	$3x^2 = 6x + 4$ $3x^2 - 6x - 4 = 0$ $a = 3$, $b = -6$, and $c = -4$ $$x = \frac{-b \pm \sqrt{b^2 - 4ac}}{2a}$$ $$= \frac{-(-6) \pm \sqrt{(-6)^2 - 4(3)(-4)}}{2(3)}$$ $$= \frac{6 \pm \sqrt{36 + 48}}{6}$$ $$= \frac{6 \pm \sqrt{84}}{6}$$ $$= \frac{6 \pm 2\sqrt{21}}{6}$$ $$x = \frac{3 \pm \sqrt{21}}{3}$$ Solutions: $$x = \frac{3 + \sqrt{21}}{3} \text{ and } x = \frac{3 - \sqrt{21}}{3}$$
[9.4] **Parabola**	The graph of an equation of the form $y = ax^2 + bx + c$.	
[9.4] **Vertex of a parabola**	The highest (maximum y-value) or lowest (minimum y-value) point of a parabola.	

continued

Concept/Skill	Description	Example
[9.4] **The axis of symmetry**	The axis of symmetry is a vertical line that passes through the vertex of a parabola. For a parabola given by the quadratic equation $y = ax^2 + bx + c$, the equation of the axis of symmetry is $x = -\dfrac{b}{2a}$.	For $y = x^2 - 4x + 3$, the equation of the axis of symmetry, where $a = 1$ and $b = -4$, is $$x = -\frac{b}{2a} = -\frac{(-4)}{2(1)} = 2.$$
[9.4] **To find the vertex of a parabola**	For a parabola given by the quadratic equation $y = ax^2 + bx + c$: • The x-coordinate of the vertex is $-\dfrac{b}{2a}$. • The y-coordinate of the vertex is found by substituting $-\dfrac{b}{2a}$ for x into the equation and computing y.	For $y = x^2 + 2x - 3$, $a = 1$, $b = 2$, and $c = -3$. $$x = -\frac{b}{2a} = -\frac{2}{2(1)} = -1$$ $$y = x^2 + 2x - 3$$ $$y = (-1)^2 + 2(-1) - 3 = -4$$ Vertex: $(-1, -4)$

CHAPTER 9   Review Exercises

Say Why
Fill in each blank.

1. If $a^2 = 14$, then a _____ equal to $\sqrt{14}$ or $-\sqrt{14}$
   is/is not
   because _____
   _____.

2. The expression $x^2 + 2ax + a^2$ _____ a perfect
   is/is not
   square because _____
   _____.

3. The graph of $y = 2x^2 - 3x + 1$ _____ a parabola
   is/is not
   because _____
   _____.

4. The axis of symmetry of the graph of
   $y = -4x^2 + 5x - 3$ _____ a vertical line because
   is/is not
   _____.

5. The graph of the equation $y = -4x^2 + 5x - 3$ opens
   _____ because
   upward/downward
   _____.

6. The parabola $y = 3x^2$ has a _____
   mininum/maximum
   value because _____
   _____.

[9.1]   *Solve by using the square root property.*

7. $y^2 = 24$ 8. $4x^2 = 12$ 9. $x^2 + 5 = 0$ 10. $(2n - 5)^2 = 18$

Solve for the indicated variable.

11. $A = \pi r^2$ for r (Area of a circle)   12. $d = \frac{1}{2}gt^2$ for t ($t = $ time)

[9.2]   *Fill in the blank to complete the square.*

13. $x^2 - 10x + $ 14. $x^2 + 7x + $

Solve by completing the square.

15. $n^2 - 6n = 27$ 16. $y^2 + 3y = 4$ 17. $2x^2 + 8x = 4$ 18. $4y^2 = 4y + 1$

[9.3]   *Solve by using the quadratic formula.*

19. $y^2 - 2y - 1 = 0$ 20. $-x^2 = 8x + 1$ 21. $2n^2 + n = 5$ 22. $(x + 3)(x - 2) = -10$

23. $y^2 = -5y$ 24. $\frac{2}{x} + \frac{1}{x + 3} = 1$

Graph a parabola with the following characteristics.

25. Passes through the points
$(0, -3), (1, -4), (-1, 0), (2, -3),$ and $(3, 0)$.

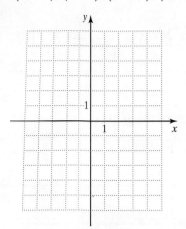

26. Has vertex $(-1, -4)$, y-intercept $(0, -3)$, and x-intercepts $(-3, 0)$, and $(1, 0)$.

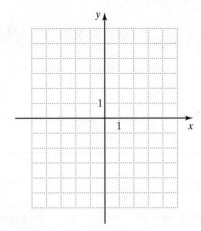

For each equation, find the axis of symmetry and the vertex of the corresponding parabola.

27. $y = 4x + x^2$

28. $y = -x^2 - 2x + 3$

Sketch the graph of each equation.

29. $y = \frac{1}{2}x^2$

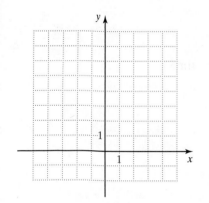

30. $y = x^2 - 5x - 6$

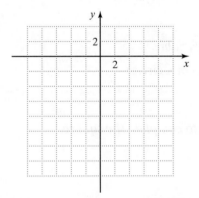

31. $y = -x^2 - 2x$

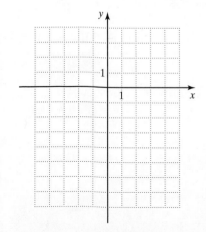

32. $y = 8 - 2x - 3x^2$

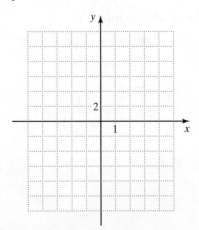

Mixed Applications

Solve. Use a calculator, if necessary.

33. Suppose a map company wants to make a globe with a surface area of 615 in². Use the formula $A = 4\pi r^2$, where A is the surface area, r is the radius of the sphere, and π is approximately 3.14. Find the radius to the nearest tenth of an inch.

34. A rug company uses the equation $P = 90x - x^2$ to determine its profit P for selling x rugs. If the company had a profit of $2000, how many rugs did it sell?

35. The length of a rectangular banner is 5 ft more than twice its width. To the nearest tenth of a foot, find the dimensions if the banner has an area of 20 ft².

36. The formula $2D = n(n - 3)$ gives the number of diagonals D in a polygon of n sides. An architect designs a building with 14 diagonals. How many sides does the building have?

37. A travel agency offers a vacation package to the Bahamas at a discount rate. It determined that the amount of profit P per person can be modeled by the equation $P = 40n - n^2$, where n is the number of persons.

   a. Graph $P = 40n - n^2$.

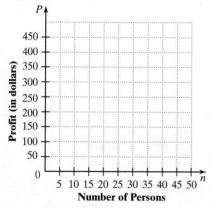

   b. What is the agency's profit per person if 16 people go to the Bahamas?

   c. If the profit was $351 per person, how many people went to the Bahamas?

   d. What is the maximum profit per person that the agency can make?

38. A ball is thrown straight up into the air from a height of 6 ft with an initial velocity of 40 ft/sec. The height h in feet after t sec is modeled by the equation $h = -16t^2 + 40t + 6$.

   a. Graph the equation.

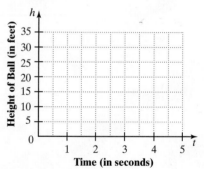

   b. In how many seconds does the ball reach its maximum height?

   c. What is the maximum height of the ball?

An inexperienced Web designer takes 8 days longer to develop a particular website than an experienced designer. Working together, the two designers can develop the site in 3 days. How long would it take each, working alone, to develop the site?

40. Because of traffic, a commuter drives home at an average speed that is 10 mph slower than her speed driving to work. The distance between her home and work is 15 mi. If it takes her 1 hr, round trip, what is her average speed driving to work to the nearest mile per hour?

• Check your answers on page A-29.

CHAPTER 9   Posttest

EXTRA HELP

Test Prep VIDEOS   in MyMathLab, and on YouTube (search "AkstIntr Alg" and click on "Channels").

To see if you have mastered the topics in this chapter, take this test.

1. Solve $4(x + 1)^2 = 32$ using the square root property of equality.

2. Solve $V = \dfrac{1}{3}\pi r^2 h$ for r. Assume that all variables are positive.

3. Fill in the blank to make the trinomial a perfect square: $x^2 - x + \underline{\quad}$.

4. Solve $2y^2 - 6y - 2 = 0$ by completing the square.

5. Solve $2y^2 + 3y - 4 = 0$ using the quadratic formula.

Solve.

6. $6x^2 = 72$

7. $x^2 + 6x = 40$

8. $25(y - 4)^2 = 49$

9. $4x^2 - 8x - 3 = 0$

10. $(3x - 4)(x + 4) = -12$

11. $\dfrac{3}{n - 1} + \dfrac{5}{n + 1} = 1$

12. Sketch a parabola that has vertex $(-1, 1)$ and passes through the points $(-3, -3)$, $(-2, 0)$, $(0, 0)$, and $(1, -3)$.

13. Find the vertex and the axis of symmetry of the graph of $y = x^2 + 3x + 4$.

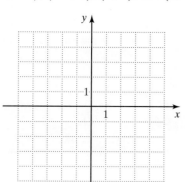

Sketch the graph of each equation.

14. $y = x^2 - 2x - 3$

15. $y = 2x^2 + x - 15$

16. $y = (2 - x)^2$

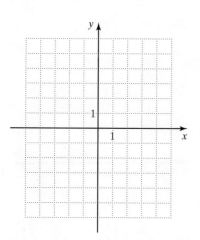

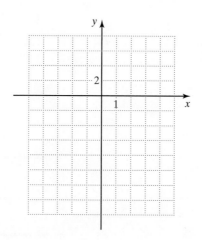

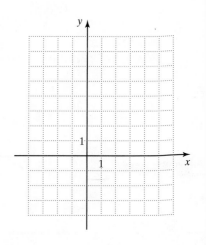

The period of a pendulum is the time it takes for it to swing back and forth. The formula $l = 0.81t^2$ relates the length l (in feet) of a pendulum to the time t (in seconds) that the pendulum takes to swing back and forth. Solve this formula for t.

18. German federal highways, known as *autobahns*, have no mandatory speed limit. The equation $d = 0.0056s^2 + 0.14s$ models the distance d (in meters) that a car traveling at a speed of s km/hr needs to stop. Determine, to the nearest whole number, the maximum speed of a car that can stop within 75 m on an autobahn. (*Source:* wikipedia.org)

19. The recent U.S. average per capita personal income y (in dollars) can be roughly modeled by the quadratic expression $10x^2 - 860x + 14{,}560$, where x is the number of years since 1900. Using this model, in what year in the future would we expect the per capita income to reach $75,000? (*Source: The New York Times Almanac, 2011*)

20. The equation $h = 80t - 16t^2$ models the height h (in feet) an object propelled straight up from the ground with an initial velocity of 80 ft/sec reaches in t sec. What is the maximum height the object reaches?

• Check your answers on page A-29.

To help you review, solve the following.

1. Evaluate: $(3x + y)^2 + z$ if $x = -2, y = 5$, and $z = -9$

2. Simplify: $7 + 9[11 - 4(3x + 6 - x)]$

3. Solve the equation $ax - by = c$ for y. Then, find the value of y when $a = 2, b = 5, c = 6$, and $x = 1$.

4. Graph: $2x - 3y < 6$

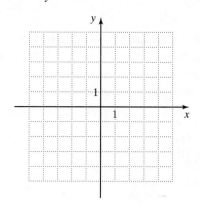

5. Solve: $\begin{aligned} -4x + 3y &= 13 \\ 3x + 4y &= 9 \end{aligned}$

6. Multiply: $(3n - 2)(5 + 6n)$

7. Divide: $\dfrac{4x^3}{y^2 - 4} \div \dfrac{6xy - 18x}{y^2 - y - 6}$

8. Factor: $9c^2d^2 + 6c^3d + 3cd^3$

9. Solve: $3x^2 - 19x - 14 = 0$

10. Simplify: $\dfrac{\dfrac{1}{4x} - \dfrac{1}{x^2}}{x - \dfrac{16}{x}}$

11. Solve: $\dfrac{c}{c + 6} = \dfrac{1}{c + 2}$

12. Simplify: $\dfrac{\sqrt{x^3y^4}}{\sqrt{xy}}$

13. Solve: $x^2 - 8x + 4 = 0$

14. Solve: $3x^2 + 5x = 2$

15. In 1862, the U.S. Congress enacted the nation's first income tax, at the rate of 3%. At this rate, what was the income of a taxpayer whose tax amounted to $22.50? (*Source: The Seattle Times*)

16. A multimedia company started with 2 employees. In 6 months, the company had 7 employees. The number of employees increased at a steady rate.

  a. Write a linear equation that models the relationship between the number of employees n and number of months t since the company started.

  b. Graph the equation in Quadrant I.

  c. Explain the relationship between the equation in part (a) and the slope of the graph in part (b).

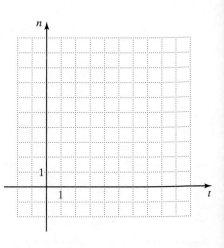

...ain bacteria population triples in size every day. ...ppose a sample starts with 400 bacteria. The expression $400 \cdot 3^x$ models the number of bacteria in the sample after x days. Evaluate the expression for $x = 0$, 1, and 2.

18. An average-sized audio file is 0.004 gigabytes (GB), and an average-sized TV episode video file is 0.350 GB. A certain portable multimedia player has 80 GB of storage. Express as an inequality the possible number of average-sized audio a and video files b that can be stored on this device. (*Source:* apple.com)

19. The speed of sound in air is approximately 300 m/sec. The sound from a firecracker exploding in an empty field expands outward in a growing circle. (*Source:* Everest, *Master Handbook of Acoustics*)

a. How far from the firecracker can its sound be heard in t sec?

b. In how many seconds, to the nearest second, will the sound of the firecracker be heard over an area of a square kilometer?

20. A sandbag is dropped from a hot-air balloon that is 128 ft above the ground. The height of the sandbag above the ground is given by the equation $h = -16t^2 + 128$, where h is the height in feet and t is the time in seconds.

a. Graph the equation.

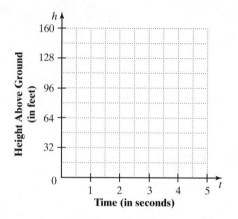

b. How far did the sandbag fall from second 0 to second 1? Explain.

• Check your answers on page A-30.

Appendix A.1

Table of Symbols

$+$	add
$-$	subtract
$\cdot, \times, (a)(b), 2y$	multiply
$\dfrac{a}{b}, \div, x+1\overline{)x^2-1}$	divide
x^n	x raised to the power n
$(\ \ )$	parentheses (a grouping symbol)
$[\ \ ]$	brackets (a grouping symbol)
π	pi (a number approximately equal to $\dfrac{22}{7}$ or 3.14)
$-a$	the opposite, or additive inverse, of a
$\dfrac{1}{a}$	the reciprocal, or multiplicative inverse, of a
$=$	is equal to
$\approx$	is approximately equal to
$\neq$	is not equal to
$<$	is less than
$\leq$	is less than or equal to
$>$	is greater than
$\geq$	is greater than or equal to
(x, y)	an ordered pair whose first coordinate is x and whose second coordinate is y
$\circ$	degree (for angles)
$\sqrt{a}$	the principal square root of a
$\lvert a \rvert$	the absolute value of a

Appendix A.2

Factoring the Sum of Cubes and the Difference of Cubes

As we discussed in Section 6.4, there are formulas that make it easier to factor polynomials having a special form. Two formulas not covered in that section deal with the *sum of cubes* and the *difference of cubes*. These formulas provide a shortcut for factoring a binomial that is in the form of the cube of one term plus or minus the cube of another term. To state the formulas, we let the terms be a and b, with corresponding cubes a^3 and b^3.

Factoring the Sum of Cubes
$$a^3 + b^3 = (a + b)(a^2 - ab + b^2)$$

Factoring the Difference of Cubes
$$a^3 - b^3 = (a - b)(a^2 + ab + b^2)$$

TIP   In these two formulas, the sign of b^3 on the left side of the equation is
- the *same* as that of b in the binomial factor on the right.
- the *opposite* of that of the middle term of the trinomial on the right.

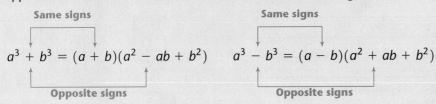

Same signs

$$a^3 + b^3 = (a + b)(a^2 - ab + b^2) \qquad a^3 - b^3 = (a - b)(a^2 + ab + b^2)$$

Opposite signs   Opposite signs

EXAMPLE 1

Indicate whether each binomial is a sum or difference of cubes.

a. $x^3 - 8$ **b.** $x^3 + y^3$ **c.** $27p^3 - q^6$ **d.** $x^2 + x^3$

Solution

a. In the expression $x^3 - 8$, both x^3 and 8 (or 2^3) are perfect cubes, where x corresponds to a and 2 to b.
$$x^3 - 8 = x^3 - 2^3$$
$$\qquad\qquad a^3 \quad - \quad b^3$$
So $x^3 - 8$ is a *difference of cubes*.
b. In $x^3 + y^3$, both x^3 and y^3 are *perfect cubes* and correspond to a^3 and b^3, respectively in our formula. So $x^3 + y^3$ is a *sum of cubes*.
c. For $27p^3 - q^6$, both $27p^3$ and q^6 are perfect cubes because $27p^3 = (3p)^3$ and $q^6 = (q^2)^3$. So $27p^3 - q^6$ is a *difference of cubes*.
d. In $x^2 + x^3$, the first term x^2 is not a perfect cube. So the expression is *neither a sum of cubes nor a difference of cubes*.

PRACTICE 1

Indicate whether each binomial is a sum or difference of cubes.

a. $n^3 - 1$

b. $p^3 + q^4$

c. $x^3 + y^9$

d. $8x^6 - x^3$

EXAMPLE 2

Factor:

a. $x^3 - y^3$

b. $y^3 + 8$

Solution

a. The expression $x^3 - y^3$ is a difference of cubes, so we can apply that formula.

$$x^3 - y^3 = (x)^3 - (y)^3 = (x - y)(x^2 + xy + y^2)$$

$$a^3 \quad - \quad b^3 \qquad (a - b)(a^2 + ab + b^2)$$

So the factorization of $x^3 - y^3$ is $(x - y)(x^2 + xy + y^2)$. How can you check that this factorization is correct?

b. The expression $y^3 + 8$ is a sum of cubes.

$$y^3 + 8 = (y)^3 + (2)^3 = (y + 2)[y^2 - (y)(2) + (2)^2]$$

$$a^3 \quad + \quad b^3 \qquad (a + b)(a^2 - ab + b^2)$$

We conclude that the factorization of $y^3 + 8$ is $(y + 2)(y^2 - 2y + 4)$.

PRACTICE 2

Factor:

a. $r^3 + s^3$

b. $y^3 - 64$

EXAMPLE 3

Factor:

a. $x^3 + 27$

b. $2x^3 + 2$

c. $64n^3 - n^6$

Solution

a. Since $27 = 3^3$, $x^3 + 27$ is a sum of cubes.

$$x^3 + 27 = x^3 + 3^3 \qquad \text{Use the sum-of-cubes formula.}$$
$$= (x + 3)(x^2 - x \cdot 3 + 3^2)$$
$$= (x + 3)(x^2 - 3x + 9)$$

So the factorization of $x^3 + 27$ is $(x + 3)(x^2 - 3x + 9)$.

b. We note that the two terms $2x^3$ and 2 of the binomial $2x^3 + 2$ have 2 as the greatest common factor.

$$2x^3 + 2 = 2(x^3 + 1) \qquad \text{Factor out the GCF.}$$
$$= 2(x + 1)(x^2 - x + 1) \qquad \text{Use the sum of cubes formula.}$$

So the factorization of $2x^3 + 2$ is $2(x + 1)(x^2 - x + 1)$.

c. We see that the two terms of the binomial $64n^3 - n^6$ have n^3 as the greatest common factor.

$$64n^3 - n^6 = n^3(64 - n^3) \qquad \text{Factor out the GCF.}$$
$$= n^3(4 - n)(16 + 4n + n^2) \qquad \text{Use the difference-of-cubes formula.}$$

So the factorization of $64n^3 - n^6$ is $n^3(4 - n)(16 + 4n + n^2)$.

PRACTICE 3

Factor:

a. $125 - x^3$

b. $5n^3 - 40$

c. $2n + 54n^4$

Factor, if possible.

1. $y^3 - 1$

2. $t^3 - 8$

3. $p^3 + q^3$

4. $p^2 + q^2$

5. $x^3 + y^6$

6. $a^3 - b^3$

7. $n^3 + 8$

8. $x^3 + 27$

9. $8 - x^5$

10. $8 - x^6$

11. $27n^3 + 1$

12. $8 + n^4$

13. $5x^3 - 5$

14. $3x^3 - 24$

15. $2p^3 + 54$

16. $128n^3 + 2n^6$

17. $x^6 - 125x^3$

18. $8x^3 + x^6$

• Check your answers on page A-30.

Appendix A.3

Introduction to Graphing Calculators

This appendix covers the basic graphing features of a graphing calculator (or graphing software) used in this text. Note that the keystrokes and screens presented here and in the calculator inserts found throughout the text may be different from those on your graphing calculator. Refer to your user's manual for specific information and instructions on accessing and using the features of your particular model.

Graphing Equations

To graph an equation, it must be entered into the graphing calculator's *equation editor* in "$y =$" form. On many graphing calculators, the equation editor can be displayed by pressing the $\boxed{Y =}$ key. Note that you may enter more than one equation in the equation editor.

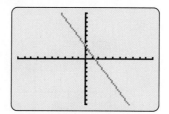

Equation Editor

For example, to graph the equation $4x + 2y = 6$, you must first solve the equation for y, getting $y = -2x + 3$. Then, enter the expression $-2x + 3$ to the right of $\backslash\mathbf{Y1} =$ in the equation editor. Finally, press the $\boxed{\textbf{GRAPH}}$ key to display the graph on a coordinate plane.

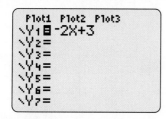

Note that most graphing calculators have two separate keys to distinguish between subtraction and negation. The $\boxed{=}$ key is used for subtraction and the $\boxed{(-)}$ key is used for negation.

You must be careful when entering equations containing fractions. To ensure that the graphing calculator interprets the input correctly, you may need to enclose all or part of the fraction in parentheses. For instance, to graph the equation $y = \dfrac{1}{2}x$, enter the expression $(1/2)x$ to the right of $\backslash\mathbf{Y1} =$ in the equation editor. Then, press $\boxed{\textbf{GRAPH}}$ to display the graph of the equation.

673

Viewing Windows

The screen on which a graph is displayed is called the *viewing window* and it represents a portion of a coordinate plane. The viewing window is defined by the following values:

Xmin: the minimum x value displayed on the x-axis

Xmax: the maximum x value displayed on the x-axis

Xscl: the distance between adjacent tick marks on the x-axis

Ymin: the minimum y value displayed on the y-axis

Ymax: the maximum y value displayed on the y-axis

Yscl: the distance between adjacent tick marks on the y-axis

You can set the viewing window by entering the minimum and maximum values and the scales for the axes in the *window editor*. The window editor can be displayed by pressing the **WINDOW** key. The screen on the left shows the window editor and displays the settings for the corresponding *standard viewing window* shown on the right.

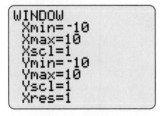

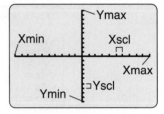

Window Editor   **Viewing Window**

The graph of an equation can be easily misinterpreted if an inappropriate viewing window is selected. For instance, compare the graph of the quadratic equation $y = x^2 - 15x + 14$, as shown in the following three viewing windows:

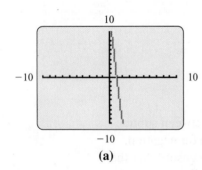

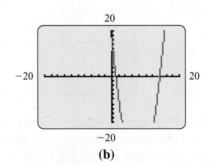

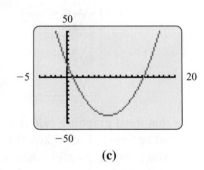

(a)   (b)   (c)

Although each of these viewing windows displays the graph of the equation, the viewing window in (c) is best because it shows all the key features of the graph, that is, the x- and y-intercepts and the vertex. Selecting an appropriate viewing window may require some practice, but familiarity with key features of the graphs of the equations discussed in the text will facilitate this process.

Other graphing calculator features, such as **TRACE**, **ZOOM**, and **INTERSECT**, are discussed in calculator inserts throughout the text.

Answers

CHAPTER R

Chapter R   Pretest, p. 2

1. 7^3 **2.** 400 **3.** 8 **4.** 16 **5.** 1, 2, 3, 4, 6, 12 **6.** $2^2 \cdot 5$ **7.** $\frac{2}{3}$
8. $\frac{2}{5}$ **9.** $7\frac{1}{2}$ **10.** $\frac{4}{27}$ **11.** $\frac{8}{3}$, or $2\frac{2}{3}$ **12.** Nine and thirteen
thousandths **13.** 3.1 **14.** 20.3103 **15.** 5.66 **16.** 37.843 **17.** 23.5
18. 0.18 **19.** 0.0031 **20.** 0.07 **21.** \$2 million **22.** $\frac{1}{25}$
23. two times **24.** 10% **25.** 0.95

Section R.1   Practices, pp. 3–5

1, *p. 3:* 2^5 **2,** *p. 3:* 784 **3,** *p. 4:* 10^9 **4,** *p. 4:* 16 **5,** *p. 5:* 1
6, *p. 5:* More

Section R.2   Practices, pp. 6–7

1, *p. 6:* 50 **2,** *p. 7:* 180 **3,** *p. 7:* 24 hr

Section R.3   Practices, pp. 8–15

1, *p. 8:* $\frac{29}{9}$ **2,** *p. 9:* $2\frac{2}{3}$ **3,** *p. 9:* $\frac{4}{5}$ **4,** *p. 10:* $\frac{2}{3}$ **5,** *p. 10:* $\frac{2}{5}$
6, *p. 11:* $\frac{5}{3}$, or $1\frac{2}{3}$ **7,** *p. 11:* $\frac{3}{10}$ **8,** *p. 12:* $8\frac{1}{8}$ **9,** *p. 12:* $5\frac{1}{6}$ **10,** *p. 13:* $\frac{3}{8}$
11, *p. 13:* $\frac{7}{22}$ **12,** *p. 14:* $\frac{63}{8}$, or $7\frac{7}{8}$ **13,** *p. 14:* 6 **14,** *p. 15:* $\frac{8}{5}$, or $1\frac{3}{5}$
15, *p. 15:* 160 million

Section R.4   Practices, pp. 17–21

1, *p. 17:* **a.** $\frac{1}{2}$ **b.** $2\frac{73}{1000}$ **2,** *p. 17:* Four and three thousandths
3, *p. 18:* 748.08 **4,** *p. 18:* 42.092 **5,** *p. 19:* 1.179 **6,** *p. 19:* 9.835
7, *p. 20:* 327,000 **8,** *p. 20:* 18.04 **9,** *p. 21:* 0.00086 **10,** *p. 21:* 2.9

Section R.5   Practices, pp. 22–23

1, *p. 22:* $\frac{7}{100}$ **2,** *p. 22:* 0.05 **3,** *p. 23:* 2.5% **4,** *p. 23:* 25%
5, *p. 23:* 0.40, or 0.4

Exercises R, pp. 24–26

1. 6^5 **2.** 7^8 **3.** $2^2 \cdot 10^3$ **4.** $5^4 \cdot 4^3$ **5.** 25,000 **6.** 576 **7.** 3 **8.** 0
9. 14 **10.** 37 **11.** 4 **12.** 2 **13.** 1, 2, 3, 5, 6, 10, 15, 25, 50, 75, 150
14. 1, 3, 19, 57 **15.** Prime **16.** Prime **17.** Composite
18. Composite **19.** $2 \cdot 3 \cdot 7$ **20.** $2 \cdot 3^3$ **21.** $2^4 \cdot 3$ **22.** $2^2 \cdot 5^2$
23. 24 **24.** 60 **25.** 72 **26.** 84 **27.** $\frac{19}{5}$ **28.** $\frac{73}{10}$ **29.** $5\frac{3}{4}$ **30.** $3\frac{4}{9}$
31. $\frac{1}{2}$ **32.** $\frac{2}{3}$ **33.** $5\frac{1}{2}$ **34.** $6\frac{2}{7}$ **35.** $\frac{5}{9}$ **36.** 1 **37.** $\frac{1}{4}$ **38.** 1
39. $\frac{34}{35}$ **40.** $\frac{25}{18}$, or $1\frac{7}{18}$ **41.** $\frac{1}{4}$ **42.** $\frac{7}{20}$ **43.** $6\frac{1}{2}$ **44.** $7\frac{2}{5}$ **45.** $10\frac{3}{5}$
46. 8 **47.** $3\frac{1}{3}$ **48.** $\frac{1}{9}$ **49.** $1\frac{4}{5}$ **50.** $2\frac{1}{2}$ **51.** $6\frac{1}{2}$ **52.** $1\frac{1}{5}$ **53.** $2\frac{7}{8}$
54. $\frac{7}{12}$ **55.** $\frac{2}{15}$ **56.** $\frac{2}{3}$ **57.** 14 **58.** $\frac{110}{3}$, or $36\frac{2}{3}$ **59.** $\frac{91}{6}$, or $15\frac{1}{6}$
60. 3 **61.** $\frac{17}{3}$, or $5\frac{2}{3}$ **62.** $\frac{5}{3}$, or $1\frac{2}{3}$ **63.** $\frac{1}{9}$ **64.** $\frac{1}{40}$ **65.** $\frac{24}{7}$, or $3\frac{3}{7}$
66. $\frac{9}{5}$, or $1\frac{4}{5}$ **67.** $\frac{1}{2}$ **68.** $\frac{102}{5}$, or $20\frac{2}{5}$ **69.** $\frac{7}{8}$ **70.** $2\frac{3}{500}$
71. Hundredths **72.** Tenths **73.** Seventy-two hundredths **74.** Five
hundredths **75.** Three and nine thousandths **76.** Twelve and two hundred thirty-five thousandths **77.** 7.3 **78.** 9.5 **79.** 4.39 **80.** 8.69
81. 18.11 **82.** 24.13 **83.** 1.873 **84.** 0.9 **85.** 2.912 **86.** 3.744
87. 2710 **88.** 530 **89.** 0.0015 **90.** 3.19 **91.** 5 **92.** 20 **93.** 7.35
94. 8.96 **95.** 14.5 **96.** 33 **97.** $\frac{3}{4}$ **98.** $\frac{1}{25}$ **99.** $1\frac{3}{50}$ **100.** $2\frac{1}{2}$
101. 0.06 **102.** 0.08 **103.** 1.5 **104.** 1.8 **105.** 31% **106.** 5%
107. 1.45% **108.** 1.48% **109.** 10% **110.** 37.5% **111.** 80%
112. 175% **113.** 73°F **114.** 8:10 A.M. **115.** $\frac{3}{4}$ mi **116.** 7 lb **117.** \$18
118. 1.66 m **119.** \$1.30 **120.** \$6.70 **121.** $\frac{7}{50}$ **122.** 10^6

Chapter R   Posttest, p. 27

1. 512 **2.** 37 **3.** 1, 2, 4, 5, 10, 20 **4.** $\frac{13}{4}$ **5.** $\frac{5}{18}$ **6.** $1\frac{1}{2}$ **7.** $12\frac{1}{24}$
8. $\frac{13}{90}$ **9.** $3\frac{19}{20}$ **10.** $\frac{3}{5}$ **11.** 2 **12.** $2\frac{3}{16}$ **13.** Two and three hundred
ninety-six thousandths **14.** 16.202 **15.** 6.99 **16.** 44.678 **17.** 2070
18. 0.0005 **19.** 0.125; 12.5% **20.** 70%; $\frac{7}{10}$ **21.** 6 times **22.** 1.9
23. 625.3 m² **24.** 0.2 **25.** 15 times

CHAPTER 1

Chapter 1   Pretest, p. 29

1. +\$2000 **2.** yes **3.** 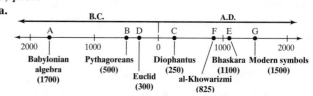 **4.** 5
5. $\frac{2}{3}$ **6.** < **7.** −13 **8.** −2 **9.** −26 **10.** $\frac{4}{1}$, or 4 **11.** 9
12. Ten less than the product of three and n (Answers may vary.)
13. -6^4 **14.** −8 **15.** $4n - 7$ **16.** $14x - 10$ **17.** 6051 m
18. $\frac{L}{3}$, or $\frac{1}{3}L$ **19.** 30 cm **20.** The third round

Section 1.1   Practices, pp. 31–37

1, *p. 31:* −5°F **2,** *p. 32:* **3,** *p. 33:*

a. 41 **b.** $\frac{8}{9}$ **c.** −1.7 **d.** $\frac{2}{5}$ **4,** *p. 34:* **a.** $\frac{1}{2}$ **b.** 0 **c.** 9 **d.** −3 **5,** *p. 35:* **a.** True
b. False **c.** True **d.** True **e.** False **6,** *p. 35:*

3, $-\frac{1}{2}$, −1.6, and −2.4 **7,** *p. 36:* The Caspian Sea; $-92 < -52 < 0$
8, *p. 37:*
a.

	B.C.				A.D.	
A		B D	C		F E	G
2000	1000		0		1000	2000

Babylonian algebra (1700)   Pythagoreans (500)   Euclid (300)   Diophantus (250)   al-Khowarizmi (825)   Bhaskara (1100)   Modern symbols (1500)

b. A, B, D, C, F, E, and G

Exercises 1.1, pp. 38–42

1. natural numbers **3.** negative numbers **5.** origin **7.** neither terminate nor **9.** −5 km **11.** −22.5°C **13.** −\$160

		Whole Numbers	Integers	Rational Numbers	Real Numbers
15.	−7		✓	✓	✓
17.	$3\frac{1}{6}$			✓	✓
19.	10	✓	✓	✓	✓

21.
23.
25.
27.

29. 3 **31.** −15 **33.** 3.5 **35.** 4 **37.** $\frac{2}{3}$ **39.** 4.6 **41.** $-\frac{1}{2}$
43. 4 and −4 **45.** Impossible; absolute value is always positive or zero.
47. True **49.** False **51.** True **53.** > **55.** > **57.** = **59.** =
61. < **63.**

−1.5 −$\frac{1}{2}$ 0   3$\frac{1}{2}$; 3$\frac{1}{2}$, 0, −$\frac{1}{2}$, −1.5

−4 −3 −2 −1 0 1 2 3 4

65.

−3 −2 −1   1 2   ; 2, 1, −1, −2, −3 **67.** −$53

−4 −3 −2 −1 0 1 2 3 4

69. Rational numbers and real numbers **71.** −1.5 **73.** False **75.** <
77. Today **79.** −129°F **81. a.** Sirius **b.** −13
c.

Beta
Sagittae

−15   −10   −5   0   +5   +10   +15

83. a. Aristotle **b.** Attila **c.** Socrates **d.** Tiger Woods

Section 1.2   Practices, pp. 43–47

1, *p. 43:* −1

−4 −3 −2 −1 0 1 2 3 4
     End   Start

2, *p. 43:* 0;

−5 −4 −3 −2 −1 0 1 2 3 4 5
  Start   End

3, *p. 44:* 0.5;

−2   −1   0 0.5 1   2
            Start End

4, *p. 45:* −31 **5,** *p. 45:* 3.5 **6,** *p. 45:* 0 **7,** *p. 46:* $-\frac{5}{18}$
8, *p. 46:* −10 **9,** *p. 47:* $36.50 **10,** *p. 47:* −3.588

Exercises 1.2, pp. 48–50

1. negative **3.** opposites **5.** larger **7.** commutative property
of addition **9.** 1 **11.** 0 **13.** −8 **15.** Additive inverse property
17. Commutative property of addition **19.** Additive identity property
21. Associative property of addition **23.** Additive identity property
25. 23 **27.** −5 **29.** −12 **31.** −80 **33.** −30 **35.** 0 **37.** 4.3
39. 1.4 **41.** −10.5 **43.** −5.7 **45.** −16.3 **47.** −1 **49.** $-\frac{2}{5}$ **51.** $\frac{2}{5}$
53. $\frac{5}{6}$ **55.** −88 **57.** 0 **59.** −7 **61.** 10.48 **63.** −12 **65.** 1.245
67. Associative property of addition **69.** −7 **71.** −15 **73.** $-1\frac{1}{3}$
75. 5° above 0° (+5) **77.** Lost $16,000(−$16,000) **79.** The Pittsburgh Steelers by four points **81.** Yes; he will have $283.26 in the account.
83. 31,300 ft

Section 1.3   Practices, pp. 52–54

1, *p. 52:* 5 **2,** *p. 53:* 3 **3,** *p. 53:* −4 **4,** *p. 53:* −15.7 **5,** *p. 53:* 28
6, *p. 54:* 17 **7,** *p. 54:* −3 **8,** *p. 54:* About 870 years older

Exercises 1.3, pp. 55–56

1. 17 **3.** −31 **5.** −44 **7.** 71 **9.** −35 **11.** −32 **13.** 32
15. −45 **17.** −62 **19.** 44 **21.** 1000 **23.** −1.42 **25.** −7.8
27. 0 **29.** 10.3 **31.** $-1\frac{1}{6}$ **33.** $-17\frac{1}{4}$ **35.** $6\frac{1}{10}$ **37.** 12
39. −7 **41.** −18 **43.** −21 **45.** −26.487 **47.** 5 **49.** 48
51. −38 **53.** $1\frac{7}{12}$ **55.** 14 **57.** 28 centuries **59.** 9000 ft
61. $276,039 **63.** 10,336 ft **65. a.** Krypton: 4°; neon: 3°;
bromine: 66° **b.** bromine **c.** bromine

Section 1.4   Practices, pp. 58–62

1, *p. 58:* 100 **2,** *p. 59:* −15 **3,** *p. 60:* **a.** 8 **b.** $-\frac{5}{27}$ **c.** 0.12 **d.** −4.75
e. 0 **f.** $\frac{2}{3}$ **4,** *p. 60:* 64 **5,** *p. 61:* −240 **6,** *p. 61:* −28 **7,** *p. 62:* −30
8, *p. 62:* −10 **9,** *p. 62:* 4 **10,** *p. 62:* About −5.9; the rock is moving
downward at a velocity of 5.9 ft/sec.

Exercises 1.4, pp. 63–66

1. positive **3.** multiply two numbers in either order **5.** any number
and zero **7.** odd **9.** Commutative property of multiplication
11. Associative property of multiplication **13.** Multiplicative identity
property **15.** Multiplication property of zero **17.** −12 **19.** 21

21. −3 **23.** $-\frac{4}{27}$ **25.** $-\frac{16}{27}$ **27.** 0.9 **29.** −60 **31.** 120 **33.** 0
35. 144 **37.** 120 **39.** $-\frac{1}{27}$ **41.** 0.98 **43.** −23 **45.** 27 **47.** −3
49. −5 **51.** 6 **53.** −20 **55. a.** 5 **b.** 1 **c.** −3 **d.** −7 **e.** −11
57. 3.64 **59.** −120 **61. a.** −6 **b.** −1 **c.** 4 **d.** 9 **e.** 14 **63.** 4
65. $10 (won $10) **67.** The team scored 2 more points than its opponents (+2). **69.** −15 in. (dropped 15 in.) **71. a.** −135 calories
b. +106 calories **c.** −852 calories

Section 1.5   Practices, pp. 67–71

1, *p. 67:* **a.** −8 **b.** 7 **c.** $-\frac{1}{2}$ **d.** −0.7 **e.** 60 **2,** *p. 69:* **a.** $-\frac{1}{5}$
b. −8 **c.** $\frac{3}{4}$ **d.** $-\frac{5}{8}$ **3,** *p. 70:* **a.** $-\frac{4}{3}$, or $-1\frac{1}{3}$ **b.** 25 **4,** *p. 70:* **a.** −12
b. −3 **5,** *p. 71:* −125 minutes (Down 125 minutes)

Exercises 1.5, p. 72

1. absolute values **3.** different signs **5.** undefined **7.** multiplicative
inverse **9. a.** −2 **b.** $\frac{1}{5}$ **c.** $-\frac{4}{3}$ **d.** $\frac{5}{16}$ **e.** −1 **11.** 8 **13.** −9
15. 0 **17.** 25 **19.** 25 **21.** 4 **23.** 5 **25.** $-\frac{1}{8}$, or −0.125 **27.** $\frac{1}{2}$, or
0.5 **29.** $-\frac{6}{5}$ **31.** −32 **33.** $-\frac{1}{8}$ **35.** −0.5 **37.** −20 **39.** −8
41. 10 **43.** 2.54 **45.** −1.17 **47.** −16 **49.** 1 **51.** 2 **53.** $\frac{1}{2}$
55. 4 **57.** 14 **59.** $-\frac{3}{2}$ **61.** 0.12 **63.** 5 **65.** 6 **67.** +200 customers (an increase of 200 customers per year) **69.** −26 **71.** A decrease
of about 4736 people per year (−4735.5) **73.** An average loss of 4 yd
(−4 yd) **75.** Expenses averaged $6000 per month (−$6000).
77. Yes.

Section 1.6   Practices, pp. 76–82

1, *p. 76:* **a.** 3 **b.** 1 **2,** *p. 77:* Answers may vary. **a.** $\frac{1}{3}$ of p **b.** the difference between 9 and x **c.** s divided by −8 **d.** n plus −6 **e.** the product of $\frac{3}{8}$
and m **3,** *p. 78:* **a.** twice x minus the product of 3 and y **b.** 4 plus $3m$
c. 5 times the difference between a and b **d.** the difference between r and
s divided by the sum of r and s **4,** *p. 78:* **a.** $\frac{1}{6}n$ **b.** $n + (−5)$
c. $m − (−4)$ **d.** $\frac{100}{x}$ **e.** $−2y$ **5,** *p. 79:* **a.** $m + (−n)$ **b.** $5y − 11$
c. $\frac{m + n}{mn}$ **d.** $−6(x + y)$ **6,** *p. 79:* $A = \frac{1}{2}h(b + B)$ **7,** *p. 80:* **a.** −36
b. 324 **8,** *p. 81:* **a.** $2^4(−5)^2$ **b.** $(−6)^4(8)^2$ **9,** *p. 81:* **a.** $−x^5$
b. $2m^3n^4$ **10,** *p. 82:* The population after 10 hr was $243x$, or 3^5x.

Exercises 1.6, pp. 83–85

1. variable **3.** algebric expression **5.** the quotient of x and 7
7. base **9.** 1 **11.** 2 **13.** 3 **15.** 3 plus t **17.** 4 less than x
19. 7 times r **21.** the quotient of a and 4 **23.** the product of $\frac{4}{5}$ and w
25. the sum of negative 3 and z **27.** twice n plus 1 **29.** 4 times the
quantity x minus y **31.** 1 minus 3 times x **33.** the product of a and b
divided by the sum of a and b **35.** twice x minus 5 times y **37.** $x + 5$
39. $d − 4$ **41.** $−6a$ **43.** $y + (−15)$ **45.** $\frac{1}{8}k$ **47.** $\frac{m}{n}$ **49.** $a − 2b$
51. $4z + 5$ **53.** $12(x − y)$ **55.** $\frac{b}{a − b}$ **57.** −9 **59.** −432
61. $(−2)^3(4)^2$ **63.** $6^2(−3)^3$ **65.** $(2)^4(−1)^2$ **67.** $3n^3$ **69.** $−4a^3b^2$
71. $−y^3$ **73.** $10a^3b^2c$ **75.** $−x^2y^3$ **77.** 8 times the quantity w minus y
79. the product of s and r divided by the difference between r and s
81. $a^2(−b)^2$ **83.** −36 **85.** $x + 2y$ **87.** $(90 + x + y)°$
89. $\frac{30,000}{p}$ dollars **91.** $(t + x)$ dollars **93.** $(2^3 \cdot 5000)$ dollars
95. s^2 **97.** $500(20 − n)$ dollars **99.** $(ab − cd)$ ft^2

Section 1.7   Practices, pp. 87–89

1, *p. 87:* **a.** 15 **b.** 30 **2,** *p. 87:* **a.** 18 **b.** 16 **c.** 52 **d.** −2 **3,** *p. 88:*
a. $\frac{7}{3}$ **b.** $\frac{3}{4}$ **c.** 81 **d.** −81 **4,** *p. 88:* 39.7 gal **5,** *p. 89:* $F = \frac{9}{5}C + 32$
6, *p. 89:* The distance d is 80 mi. **7,** *p. 89:* **a.** $K = C + 273$ **b.** K is 267.

Exercises 1.7, pp. 90–93

1. −2 **3.** 16 **5.** −32 **7.** −7 **9.** 0 **11.** 12 **13.** 5 **15.** 56
17. −7 **19.** 15 **21.** −14.5 **23.** −16 **25.** 15

27.

x	0	1	2	-1	-2
$2x + 5$	5	7	9	3	1

29.

y	0	1	2	3	4
$y - 0.5$	-0.5	0.5	1.5	2.5	3.5

31.

x	0	2	4	-2	-4
$-\frac{1}{2}x$	0	-1	-2	1	2

33.

n	2	4	6	-2	-4
$\frac{n}{2}$	1	2	3	-1	-2

35.

g	0	1	2	-1	-2
$-g^2$	0	-1	-4	-1	-4

37.

a	0	1	2	-1	-2
$a^2 + 2a - 2$	-2	1	6	-3	-2

39. $-20°$ **41.** \$2200 **43.** $7\frac{1}{2}$ ft **45.** -3 **47.** 314 m **49.** 20 mg
51. 13.5 cm^2 **53.** 3140 in^3 **55.** 1
57.

x	0	1	2	-1	-2
$-2x + 4$	4	2	0	6	8

59. -11 **61.** 10.5 in^2 **63.** $A = \frac{a + b + c}{3}$ **65.** $a^2 + b^2 = c^2$
67. $E = mc^2$ **69.** $l = 0.4w + 25$ **71.** The object falls 64 ft.
73. a. $m = \frac{100(s - c)}{c}$ **b.** 40% **75. a.** $C = 9f + 4(c + p)$
b. 53 calories

Section 1.8 Practices, pp. 95–99

1, p. 95: a. coefficients: 1 and -3; terms: m and $-3m$; like **b.** coefficient: 5; terms: $5x$ and 7; unlike **c.** coefficients: 2 and -3; terms: $2x^2y$ and $-3xy^2$; unlike **d.** coefficients: 1, 2, and -4; terms: m, $2m$, and $-4m$; like
2, p. 96: a. $-40r - 10s$ **b.** $5w + w$ **c.** $3g - 9h$ **d.** $1.5y + 3$
3, p. 96: a. $6x$ **b.** $-6y$ **c.** $-2a + b$ **d.** 0 **4, p. 97: a.** $-2y^2$ **b.** Cannot be simplified **c.** $3xy^2$ **5, p. 97:** $3y - 10$ **6, p. 97:** $-2a + 3b$
7, p. 98: $4y - 1$ **8, p. 98:** $-2y - 18$ **9, p. 98:** $-10y + 13$
10, p. 99: $5c + 12(c - 40)$; $(17c - 480)$ dollars

Exercises 1.8, pp. 100–102

1. coefficient **3.** distributive property **5.** negative **7.** 7 **9.** -5
11. 1 **13.** -1 **15.** -0.1 **17.** $\frac{2}{3}$ **19.** 2; -5 **21.** Terms: $2a$ and $-a$; like **23.** Terms: $5p$ and 3; unlike **25.** Terms: $4x^2$ and $-6x^2$; like
27. Terms: $-20n$ and $-3n$; like **29.** $-7x + 7y$ **31.** $a - 10a$
33. $-0.5r - 1.5$ **35.** $10x$ **37.** $-11n$ **39.** $14a$ **41.** $2y + 2$
43. 0 **45.** Cannot be simplified **47.** $4r^2t^2$ **49.** Cannot be simplified
51. $2x + 2$ **53.** $11x - 1$ **55.** $-3y + 10$ **57.** $4x - 9$ **59.** $-n + 39$
61. $5x + 1$ **63.** $-3x + 13$ **65.** $-3a - 14$ **67.** Cannot be simplified
69. Terms; $3a$ and $3a^2$; unlike **71.** 1 **73.** $3y + 7$ **75.** $x + x + 40$; $(2x + 40)°$ **77.** $d + 2(d + 4)$; $(3d + 8)$ dollars
79. $n + (n + 1) + (n + 2)$; $3n + 3$
81. $1c + 0b - 0.25(54 - b - c)$; $1.25c + 0.25b - 13.5$

Chapter 1 Review Exercises, pp. 106–110

1. is; possible answer: it can be expressed as the quotient of two integers, $\frac{1}{2}$ **2.** is; possible answer: their sum is 0 **3.** are not; possible answer: their product is not 1 **4.** is not; possible answer: the expression to be squared is $6a$ **5.** is; possible answer: it can be expressed as 5 to the third power **6.** are not; possible answer: they have different exponents of the variable p and different exponents of the variable q **7.** $+3$ mi **8.** $-\$160$
9.
10.
11.
12. **13.** 4 **14.** -6.5 **15.** $-\frac{2}{3}$
16. 0.7 **17.** 4 **18.** 0 **19.** 2.6 **20.** $\frac{5}{9}$ **21.** True **22.** False **23.** -5
24. -4 **25.** Commutative property of addition **26.** Additive identity property **27.** Associative property of addition **28.** Additive inverse property **29.** 0 **30.** 2 **31.** 2 **32.** -15 **33.** -5 **34.** -85
35. -8.1 **36.** 15.3 **37.** 9 **38.** -11 **39.** -55 **40.** -3
41. -27 **42.** 27 **43.** 16 **44.** -5 **45.** -1.42 **46.** $-8\frac{5}{8}$ **47.** 5
48. 10 **49.** Commutative property of multiplication **50.** Associative property of multiplication **51.** Multiplicative identity property
52. Multiplication property of zero **53.** -10 **54.** -21 **55.** -5400
56. 2400 **57.** 27 **58.** $-\frac{1}{4}$ **59.** 6000 **60.** 36 **61.** -23 **62.** 14
63. 26 **64.** 38 **65.** -12 **66.** 6 **67.** $-\frac{3}{2}$ **68.** $\frac{1}{8}$ **69.** 3 **70.** -6
71. $-2\frac{1}{5}$ **72.** $-\frac{6}{5} = -1\frac{1}{5}$ **73.** 32 **74.** 2 **75.** -1 **76.** -2 **77.** 13
78. -20 **79.** 3 **80.** 2 **81.** 1 **82.** 4 For Exs. 83–86, answers may vary.
83. the sum of negative 6 and w **84.** the product of negative $\frac{1}{3}$ and x
85. 6 more than negative 3 times n **86.** 5 times the quantity p minus q
87. $x - 10$ **88.** $\frac{1}{2}s$ **89.** $\frac{p}{q}$ **90.** $R - 2V$ **91.** $6(4n - 2)$
92. $\frac{-4a}{5b + c}$ **93.** $(-3)^4$ **94.** $(-5)^3 3^2$ **95.** $4x^3$ **96.** $-5a^2b^3c$
97. 29 **98.** $-\frac{20}{9}$ **99.** -20 **100.** 60 **101.** 1 **102.** 1
103. $-5x + 5y$ **104.** $14x - 2y$ **105.** $-2x^2$ **106.** r^2t^2
107. $2a - 9$ **108.** $-3x - 2$ **109.** $-4x - 15$ **110.** $a - 17$
111. $+\$700$ **112.** $-\$7000$ **113.** $-\$2.00$ **114.** Exothermic $(+3°C)$
115. $(I - 0.5h)$ degrees **116.** 0.97 per 100; 0.97% **117.** 281 m
118. Tue: $-\$0.09$, Wed: $-\$0.04$, Thu: $+\$0.24$ Fri: $+\$0.32$ **119.** $6w$
120. Twice as bright **121.** $3^3 \cdot 10$ bacteria **122.** 71 ft **123.** $0.0013P$
124. $[0.05x + 0.07(600 - x)]$ dollars **125.** Approximately 406 B.C.
126. The first loss is 3 times the second loss. **127.** 20 amperes
128. The account is overdrawn by $\$10 (-10)$. **129.** $(x + 12y)$ dollars
130. $(f + 4s)$ students

Chapter 1 Posttest, p. 111

1. $-10,000$ **2.** Yes **3.** **4.** -7
5. 3.5 **6.** True **7.** 7 **8.** 3 **9.** 10 **10.** $\frac{1}{12}$ **11.** -50 **12.** $x + 2y$
13. $(-5)^3$ **14.** -5 **15.** $7y + 4$ **16.** $2t + 3$ **17.** $1.05d$ dollars
18. An improvement of \$70,000 **19.** Yes; the team kept the ball.
20. $2,604,000d$

CHAPTER 2

Chapter 2 Pretest, p. 113

1. No **2.** Subtract 2 (or add -2). **3.** $y = -5$ **4.** $n = -8$
5. $x = -9$ **6.** $x = -\frac{1}{2}$ **7.** $y = 11$ **8.** $x = 3$ **9.** $n = -2$
10. $x = \frac{5}{3}$ **11.** $v = w + 5u$ **12.** 25% **13.** 20 **14.** 60
15. **16.** $x > 0$;
17. 12 min **18.** 20 centerpieces **19.** $m = \frac{2E}{v^2}$ **20.** Option A is a better deal if the member uses the gym more than 15 hours per month $(x > 15)$.

Section 2.1 Practices: pp. 115–119

1, p. 115: No, 4 is not a solution. **2, p. 115:** Yes, -8 is a solution.
3, p. 116: $y = 5$ **4, p. 117:** $n = -17$ **5, p. 117:** $s = -\frac{1}{2}$
6, p. 118: $x = 0.1$ **7, p. 119:** An increase of 12.3°C

Exercises 2.1, pp. 120–123

1. equation **3.** equivalent equations **5.** addition property of equality
7. a. True **b.** False **c.** True **d.** True **9.** Subtract 4 (or add −4).
11. Subtract 3.5 (or add −3.5). **13.** Add 1. **15. Add** $2\frac{1}{5}$. **17.** $y = -23$
19. $t = 0$ **21.** $a = -12$ **23.** $z = -6$ **25.** $x = -42$ **27.** $t = 6$
29. $r = 6$ **31.** $n = 13$ **33.** $x = -1$ **35.** $y = -3\frac{1}{2}$ **37.** $m = 2.9$
39. $t = -3.6$ **41.** $a = -5$ **43.** $m = -\frac{1}{2}$ **45.** $y = 1.88$ **47.** $x + 2 = 12$; $x = 10$ **49.** $n - 4 = 21; n = 25$ **51.** $x + (-3) = -1; x = 2$
53. $n + 7 = 11; n = 4$ **55.** d **57.** a **59.** Subtract $\frac{2}{3}$ (or add $-\frac{2}{3}$).
61. a. False **b.** True **c.** False **d.** True **63.** Possible answer: $x - 2.5 = -3.8; x = -1.3$ **65.** $m = -23$ **67.** $n = -4\frac{1}{3}$
69. $x + 10 = 44; x = 34$ mph **71.** $x + 190 = 370; x = 180$ calories
73. $h - 170 = 215; h = 385$ m **75.** $x + 118.5 = 180; x = 61.5°$

Section 2.2 Practices: pp. 124–128

1, p. 124: $y = 63$ **2, p. 125:** $y = 9$ **3, p. 125:** $x = -10$ **4, p. 126:** $z = 13$ **5, p. 126:** $y = -14$ **6, p. 127:** The total bill was $758.
7, p. 130: It will take about 2.3 hr.

Exercises 2.2, pp. 129–131

1. Multiply by 3. **3.** Divide by −5. **5.** Divide by −2.2. **7.** Multiply by $\frac{4}{3}$.
9. Multiply by $-\frac{2}{5}$. **11.** $x = -5$ **13.** $n = 18$ **15.** $a = 4.8$
17. $x = -0.5$ **19.** $c = -7$ **21.** $r = -22$ **23.** $x = 12$ **25.** $y = -\frac{5}{2}$
27. $n = 8$ **29.** $c = -3$ **31.** $x = 2.88$ **33.** $a = -2$ **35.** $y = \frac{2}{3}$
37. $x = -2.30$ **39.** $x = -6.82$ **41.** $-4x = 56; x = -14$ **43.** $\frac{n}{0.2} = 1.1$; $n = 0.22$ **45.** $\frac{x}{3.5} = 30; x = 105$ **47.** $\frac{1}{6}x = 2\frac{4}{5}; x = \frac{84}{5}$ **49.** c **51.** a
53. $x = -4$ **55.** $a = -24$ **57.** $\frac{x}{5} = 2; x = 10$ **59.** Divide by −5.2.
61. $0.02x = 10.5; x = 525$ yr **63.** $70r = 3348; r \approx 48$ mph
65. $0.05c = 20; c = 400$ copies **67.** $\frac{2}{3}x = 800{,}000; x = \$1{,}200{,}000$
69. $\frac{1}{5}d = 1000; d = 5000$ m **71.** $7.50t = 187.50; t = 25$ hr
73. $12x = 10{,}020; x = \$835$ per month

Section 2.3 Practices: pp. 132–140

1, p. 132: $y = 4$ **2, p. 133:** $c = 45$ **3, p. 133:** $b = -3$ **4, p. 134:** $n = \frac{4}{3}$ **5, p. 134:** $t = 3$ **6, p. 135:** $f = -\frac{3}{4}$ **7, p. 135:** $w = 4$
8, p. 136: $z = -5$ **9, p. 137:** $t = -4$ **10, p. 137:** $y = -2$ **11, p. 138:** The car will have a value of $6500 in 5 yr. **12, p. 138:** The express train will catch up with the local train in 2.5 hr, or $2\frac{1}{2}$ hr. **13, p. 139:** 75 mi
14, p. 140: 20 mph

Exercises 2.3, pp. 141–143

1. $x = 3$ **3.** $t = -2$ **5.** $m = -5$ **7.** $n = 12$ **9.** $x = -75$ **11.** $t = 2$
13. $b = -19$ **15.** $x = 39$ **17.** $r = -50$ **19.** $y = -2$ **21.** $z = -6$
23. $a = -7$ **25.** $t = 0$ **27.** $y = -1$ **29.** $r = \frac{10}{3}$ **31.** $x = -7$
33. $y = 2$ **35.** $a = 14$ **37.** $t = \frac{3}{2}$ **39.** $y = 0$ **41.** $z = 2$
43. $m = -2$ **45.** $y = 1.32$ **47.** $n = 0.27$ **49.** a **51.** d **53.** $x = -4$
55. $z = -2$ **57.** $y = 4.06$ **59.** $45 + 135x = 1260; x = 9$. The student is carrying 9 credits. **61.** $x + 2x = 3690; x = 1230$. One candidate received 1230 votes; the other candidate received 2460 votes.
63. $3 + 2(t - 1) = 9; t = 4$. The car was parked in the garage for 4 hr.
65. $0.02x + 0.01(5000 - x) = 85; x = 3500$. 3500 large postcards and 1500 small postcards can be printed. **67.** $24(t + \frac{1}{3}) = 36t; t = \frac{2}{3}$. It took $\frac{2}{3}$ hr, or 40 min, to catch the bus. **69.** $27r + 27(r + 2) = 432$; $r = 7$. One snail is crawling at a rate of 7 cm/min, the other is crawling at a rate of 9 cm/min. **71.** $2r + 2(r + 4) = 212; r = 51$. The speed of the slower truck is 51 mph.

Section 2.4 Practices: pp. 145–148

1, p. 145: $p = 1 - q$ **2, p. 146:** $r = \frac{t + s}{3}$ **3, p. 146:** $x = \frac{5ac}{4}$
4, p. 146: $x = \frac{y - b}{m}$ **5, p.147: a.** $r = \frac{A - P}{Pt}$ **b.** $r = 0.025$, or 2.5%
6, p.147: a. $h = \frac{2A}{b}$ **b.** $h = 14$ in. **7, p. 148: a.** $A = \frac{1}{2}h(b + B)$
b. $b = \frac{2A - hB}{h}$ **c.** $b = 5$ cm

Exercises 2.4, pp. 149–151

1. literal equation **3.** algebraic expression **5.** $y = x - 10$
7. $d = c + 4$ **9.** $d = \frac{-3y}{a}$ **11.** $n = 4p$ **13.** $z = \frac{2a}{xy}$ **15.** $x = \frac{7 - y}{3}$ **17.** $y = \frac{12 - 3x}{4}$ **19.** $y = 4t$ **21.** $b = \frac{p - r}{5}$
23. $l = \frac{2m - h}{4}$ **25.** $r = \frac{d}{t}$ **27.** $b = P - a - c$ **29.** $d = \frac{C}{\pi}$
31. $R = \frac{P}{I^2}$ **33.** $a = 3A - b - c$ **35.** $a = S - dn + d$
37. a. $r = \frac{I}{Pt}$ **b.** $r = 0.03$ or 3% per year **39. a.** $b = \frac{A}{h}$ **b.** $b = 5$ m
41. $h = \frac{V}{\pi r^2}$ **43.** $b = \frac{5m}{2ac}$ **45.** $z = \frac{3 - 4w}{9}$ **47.** $R = P + C; R = \$2500$
49. a. $K = \frac{V}{T}$ **b.** $V = KT$ **51. a.** $C = \frac{W}{150} \cdot A$ **b.** $A = \frac{150C}{W}$
53. a. $m = \frac{t}{5}$ **b.** $t = 5m$ **c.** The thunder will be heard in 12.5 sec ($t = 12.5$).
55. a. $C = 2\pi r$ **b.** $r = \frac{C}{2\pi}$ **c.** $r \approx 0.8$ ft

Section 2.5 Practices: pp. 152–157

1, p. 152: 20 **2, p. 152:** About $45 billion **3, p. 153:** 10.35 m
4, p. 153: $551 billion **5, p. 154:** $87\frac{1}{2}\%$ **6, p. 154:** 32% of the presidents had been vice president. **7, p. 155:** The number of nursing homes decreased by 2.4%. **8, p. 155:** The stock index dropped more in 1929. **9, p. 156:** 6.5% ($r = 0.065$) **10, p. 156:** She invested $7000 in a mutual fund and $14,000 in bonds. **11, p. 157:** 2 g

Exercises 2.5, pp. 158–161

1. base **3.** times **5.** 6 **7.** 23 **9.** 2.87 kg **11.** $40 **13.** 4
15. 32 in² **17.** $120 **19.** $19,500 **21.** 1.75 **23.** 4600 m
25. $62\frac{1}{2}\%$, or 62.5% **27.** $33\frac{1}{3}\%$ **29.** 125% **31.** 10%
33. $62\frac{1}{2}\%$.or 62.5% **35.** $140 **37.** 40% **39.** 20 **41.** 0.035
43. $3\frac{1}{3}\%$ **45.** 15 oz **47.** 175% **49.** 120 **51.** 9.6 **53.** 105
55. 50 million more eligible voters **57.** 18.75% **59.** There are 32 employees. **61.** 20% **63.** The workforce was 18.75 million people.
65. 54 tables **67.** No **69.** $100 **71.** $250 **73.** $20,000 was invested at 8% and $14,000 was invested at 10%. **75.** $10,000 was invested at 5%. **77.** $\frac{1}{3}$ cup additional olive oil **79.** 6 oz

Section 2.6 Practices, pp. 162–169

1, p. 162: No, 4 is not a solution.
2, p. 163: [number line, -5 to 5]
3, p. 163: [number line, -5 to 5]
4, p. 163: [number line, -5 to 5]
5, p. 165: $n > -1$; [number line, -5 to 5]
6, p. 165: $x \le 5\frac{1}{2}$; [number line, -2 to 8]
7, p. 166: $x > -7$; [number line, -8 to 2]
8, p. 167: $x \le 3$; [number line, -5 to 5]
9, p. 167: $x < -5$; [number line, -11 to -1]
10, p. 167: $x > -5$; [number line, -8 to 2]
11, p. 168: $z \le -5$ **12, p. 168:** $x < 2$
13, p. 168: $(x + 3) + (x + 2) + x \ge 14; x \ge 3$; The perimeter will be greater than or equal to 14 in. for any value of x greater than or equal to 3.
14, p. 169: $15(8.5) + 7.5t \ge 300; t \ge 23$. She should work at least 23 hr on the second job.

Exercises 2.6, pp. 170–174

1. inequality **3.** open **5.** unchanged **7.** negative
9. a. False **b.** True **c.** False **d.** False
11. [number line, -2 to 8]
13. [number line, -8 to 2]

15.
17.
19.
21.
23.
25.
27. $v < -7$;
29. $y > 0$;
31. $y \le 3.5$;
33. $v \le 2$;
35. $2 \ge x$, or $x \le 2$;
37. $a < -3$;
39. $y < -2$;
41. $x \ge 0$;
43. $a \le -4$;
45. $-9 \ge n$, or $n \le -9$;
47. $n > 3$ **49.** $x \le 6$ **51.** $y < -7$ **53.** $n \ge 13$ **55.** $m \ge 7$
57. $x > 7$ **59.** $z < 0$ **61.** $x \le 0.25$ **63.** $x \le -3$ **65.** $y > -3$
67. $x \le 0.4$ **69.** $x < \frac{1}{2}$ **71.** $n \ge 4.5$ **73.** $x < -6$ **75.** $y < -625$
77. d **79.** d **81.**
83. a. False **b.** True **c.** True **d.** True
85. $m \le -6$; **87.** $a \le \frac{16}{3}$
89. $x < 5$ **91.** $\frac{81 + 85 + 91 + x}{4} > 85$; $x > 83$. The student must score above 83. **93.** $\frac{250 + 250 + 150 + 130 + 180 + x}{6} \ge 200$; $x \ge 240$. The store must make at least $240 in sales. **95.** $0.50 + 0.10x \ge 2$; $x \ge 15$. Each call lasts at least 15 min. **97.** $0.03h > 1000 + 0.025h$; $h > 200,000$. She should accept the deal on a house that she sells for more than $200,000.
99. $200 - 2.5x < 180$; $x > 8$. He will weigh less than 180 lb after 8 mo.
101. $6 \le p \le 18$

Chapter 2 Review Exercises, pp. 177–179

1. is not; Possible answer: when we substitute -7 for x we do not get a true statement **2.** is; Possible answer: it can be written as $-\frac{3}{4}x + 0 = 2$, which is in the form $ax + b = c$ **3.** is; Possible answer: the addition property of equality allows us to add -5 to both sides of the equation **4.** are; Possible answer: multiplying each side of the first equation by 2 gives us the second equation **5.** is not; Possible answer: $8 - 6.5$ is not greater than 1.5 **6.** is not; Possible answer: if we divide each side of an inequality by a negative number, we reverse the inequality's direction **7.** No. **8.** 0 is a solution. **9.** $x = -9$ **10.** $t = -2$ **11.** $a = -14$
12. $n = 11$ **13.** $y = 7.9$ **14.** $r = 15.2$ **15.** $x = -6$ **16.** $z = -10$
17. $x = -10$ **18.** $d = -3$ **19.** $y = 4$ **20.** $x = -3$ **21.** $n = 41$
22. $r = -150$ **23.** $t = -9$ **24.** $y = -12$ **25.** $x = 3$ **26.** $t = -9$
27. $a = -14$ **28.** $r = 54$ **29.** $y = 9$ **30.** $t = 1$ **31.** $x = 6$
32. $y = -3$ **33.** $z = 3$ **34.** $n = 5$ **35.** $c = -\frac{2}{3}$ **36.** $p = \frac{5}{2}$
37. $x = -5$ **38.** $x = -\frac{2}{3}$ **39.** $n = 0$ **40.** $x = -\frac{17}{6}$ **41.** $x = \frac{8}{5}$
42. $x = -1$ **43.** $a = 2c + 5b$ **44.** $a = \frac{bn}{2}$ **45.** $a = \frac{P-b}{2}$
46. $h = \frac{3V}{B}$ **47.** 40 **48.** 4 **49.** 160% **50.** $62\frac{1}{2}\%$, or 62.5%

51. 25.5 **52.** $70 **53.**
54.
55.
56.
57. $n \ge -2$;
58. $y > 5$;
59. $t \ge 0$;
60. $y \le 2$;
61. $x \ge 2$;
62. $n < 5$; **63.** 16,000 Btu
64. 60° **65.** 140 guests **66.** 5 sides **67.** One candidate received 11,925 votes; the other received 27,285 votes. **68. a.** $C = 2 + 16y$ **b.** $y = \frac{C - 2}{16}$ **69.** The trucks will meet 4 hr after departure. **70.** 24 students **71.** 1000 mi **72.** Seaver received 99% of the votes cast.
73. 35% **74.** 10:15 P.M. **75.** $30 **76.** Van Buren's electoral vote count dropped 65%. **77.** 2 L **78.** 5 pt **a.** $p = 2.2k$ **b.** $k = \frac{p}{2.2}$
80. 4000 books

Chapter 2 Posttest, p. 180
1. -2 is not a solution. **2.** $x = -9$ **3.** $n = -6$ **4.** $y = 11$
5. $y = 8$ **6.** $x = 3$ **7.** $s = 2$ **8.** $x = 2$ **9.** $a = 1$ **10.** $x = \frac{5}{6}$
11. $p = t - 5n$ **12.** 20 **13.** 200% **14.**
15. $z \ge -3$; **16.** 11 mi
17. $L = \frac{S + 21}{3}$ **18.** 46,000,000 operations and procedures
19. 10 mph and 12 mph **20.** The monthly cost of Plan A exceeds the monthly cost of Plan B if more than 75 min of calls are made outside the network.

Chapter 2 Cumulative Review, p. 181
1. -6 yd **2.** **3.** 2 **4.** True **5.** 1 **6.** 2.4
7. 5 **8.** 5050 **9.** $-11x + 36$ **10.** $a = -20$ **11.** $x = 36$
12. $m = 4$ **13.** $m = -\frac{5p}{2n}$ **14.** 32 **15.** -2 lb **16.** 3^4 **17.** 2.5 hr
18. a. $A = 50 + 25(t - 1)$ **b.** $t = \frac{A - 25}{25}$ **c.** 4 hr **19.** $3w + d$
20. Greater than $925,000

CHAPTER 3

Chapter 3 Pretest, pp. 183–185
1. **2.** IV **3.** $m = \frac{1}{2}$

4. $\overleftrightarrow{AB}$ is parallel to $\overleftrightarrow{CD}$, since they have the same slope: -2. **5.** The slope of $\overleftrightarrow{PQ}$ is -1 and the slope of $\overleftrightarrow{RS}$ is 1. $\overleftrightarrow{PQ}$ is perpendicular to $\overleftrightarrow{RS}$, since the product of their slopes is -1. **6.** x-intercept: $(3, 0)$;

y-intercept: $(0, 4)$ **7.**

x	4	7	$\frac{5}{2}$	2
y	3	9	0	-1

8. 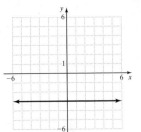 **9.**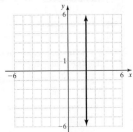

10. 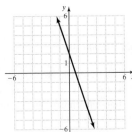 **11.**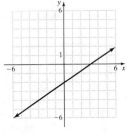

12. Slope is 2; y-intercept is $(0, -5)$ **13.** $y = 5x - 8$
14. Slope-intercept form: $y = 2x + 8$; point-slope form:
$y - 8 = 2(x - 0)$ **15.** Slope-intercept form: $y = x - 3$;
Point-slope form: $y - 1 = (x - 4)$

16.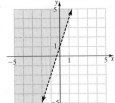

17. The slope of the line is positive. As the population of a state
increases, the number of representatives in Congress from that state
increases. **18.** Variety A grows faster.
19. a. $c = 2.5x$ **b.**
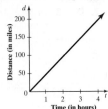
c. The slope of the line is 2.5. It represents the cost of renting a movie.
20. a. $d = 50t$ **b.**
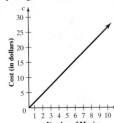
c. The slope of the graph is 50. It represents the speed the sales represen-
tative is driving.

Section 3.1   Practices, pp. 188–191
1, p. 188:   **2, p. 189: a.** II **b.** IV **c.** III **d.** I
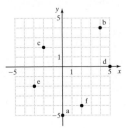

3, p. 189:   **4, p. 191:** The value of the car
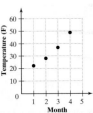

decreases as the number of years increases.   **5, p. 191:** From A to B and
B to C, the line segments slant up to the right, indicating that the runner's
heartbeats per minute increase over this period of time. From C to D, the
line segment slants downward, to the right, indicating that the runner's
heartbeats per minute decrease. Possible scenario: The runner starts out
warming up by jogging slowly for a certain length of time (A to B), then
the runner jogs more quickly for some time (B to C), and finally, the runner jogs more slowly (C to D), resting at D.

Exercises 3.1, pp. 192–196
1. origin **3.** ordered pair **5.** below
7.   **9.**
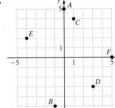 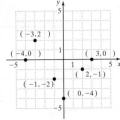

11. III **13.** II **15.** IV **17.** I
19.   **21.** II **23.** I
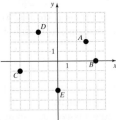

25. a. $A(20, 40)$, $B(52, 90)$, $C(76, 80)$, and $D(90, 28)$ **b.** Students
A, B, and C scored higher in English than in mathematics.

27.   **29. a.**
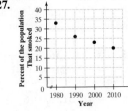 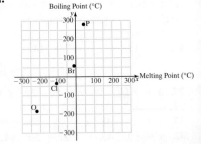

b. The *y*-coordinate is larger. The pattern shows that for each substance its boiling point is higher than its melting point.   **31.** The number of senators from a state (2) is the same regardless of the size of the state's population.   **33.** The graph in (a) could describe this motion. As the child moves away from the wall, the distance from the wall increases (line segment slants upward to the right). When the child stands still, the distance from the wall does not change (horizontal line segment). Finally, as the child moves toward the wall, the child's distance from the wall decreases (line segment slants downward to the right).

Section 3.2   Practices, pp. 200–209

1, p. 200: $m = \frac{1}{3}$ **2, p. 201:** $m = -\frac{6}{5}$ **3, p. 201:**

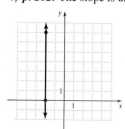   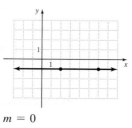   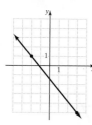

$m = 0$

4, p. 202: The slope is undefined.   **5, p. 203:** Slope of $\overleftrightarrow{PQ}$: $-\frac{2}{3}$; slope of $\overleftrightarrow{RS}$: $-\frac{1}{2}$ **6, p. 203:** Scenario A is most desirable. The slope of the line is negative, which indicates a decrease in the number of people ill over time.

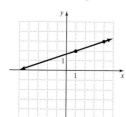

7, p. 204:   **8, p. 204: a.**   **b.** $m = -560$

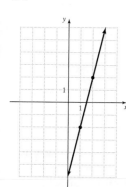   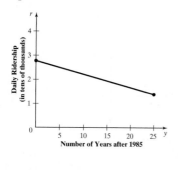

c. The slope indicates that the daily ridership decreased by 560 per year.
9, p. 206: The slope of $\overleftrightarrow{EF}$ is $-\frac{5}{4}$ and the slope of $\overleftrightarrow{GH}$ is $-\frac{5}{4}$. Since their slopes are equal, the lines are parallel.   **10, p. 206: a.** Yes, the lines are parallel since their slopes are both $\frac{15}{4}$.   **b.** Yes; the lines on the graph appear to be parallel.   **c.** The salaries increased at the same rate.   **d.** The starting salary of the multimedia designer was about $57,000.
11, p. 208: The slope of $\overleftrightarrow{AB}$ is 2 and the slope of $\overleftrightarrow{AC}$ is $\frac{1}{2}$. Since the product of their slopes is not equal to -1, the lines are not perpendicular.
12, p. 209: The slope of the diagonal from $(0, 0)$ to $(6, 6)$ is 1. The slope of the diagonal from $(0, 6)$ to $(6, 0)$ is -1. Since the product of the slopes is -1, the diagonals of the square are perpendicular.

Exercises 3.2, pp. 210–218

1. rate of change   **3.** negative   **5.** horizontal   **7.** parallel
9. $\frac{3}{4}$ **11.** undefined

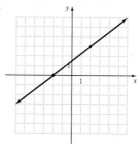   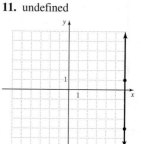

13. $-\frac{2}{5}$ **15.** 0

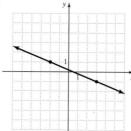   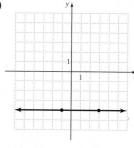

17. -7

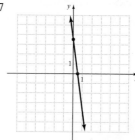

19. The slope of $\overleftrightarrow{AB}$ is -1. The slope of $\overleftrightarrow{CD}$ is 2.   **21.** Positive slope; neither   **23.** Negative slope; neither   **25.** Undefined slope; vertical
27. Zero slope; horizontal   **29.**

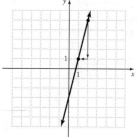

31.   **33.**

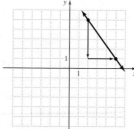   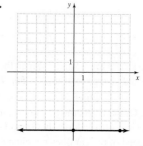

35.

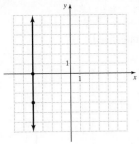

37. a. $\overleftrightarrow{PQ}$: $m = 4$; $\overleftrightarrow{RS}$: $m = 4$; the lines are parallel. **b.** $\overleftrightarrow{PQ}$: $m = -\frac{3}{2}$; $\overleftrightarrow{RS}$: $m = \frac{2}{3}$; the lines are perpendicular. **39.** Zero slope; horizontal

41.

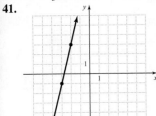

43. a. $\overleftrightarrow{AB}$: $m = -\frac{1}{2}$; $\overleftrightarrow{CD}$: $m = 2$; the lines are perpendicular.
b. $\overleftrightarrow{AB}$: $m = -\frac{4}{3}$; $\overleftrightarrow{CD}$: $m = -\frac{4}{3}$; the lines are parallel.
45. Undef.

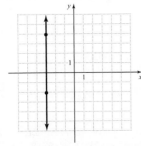

47. a. The slope is positive. **b.** A positive slope indicates that as the temperature of the gas increases, the pressure in the tube increases.
49. a. Motorcycle A **b.** Motorcycle B **c.** The slope is the change in distance over time, or the average speed of the motorcycles. **51.** The slope of each line is 1. Since the slopes of the lines are equal, the garbage deposits at the landfills are growing at the same rate. **53. a.**

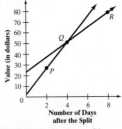

b. The slopes are 12 and 7. Since the slopes are not equal, the rate of increase did change over time. **55.** The product of the slopes of the two lines is $-5 \cdot \frac{1}{3}$, which is not equal to -1, so $\overleftrightarrow{AD}$ is not the shortest route.
57. a. Graph II; As the car travels, its distance increases with time. This implies a positive slope. **b.** Graph I; The car is set for a constant speed. The speed of the car does not change over time. This implies a 0 slope.

Section 3.3   Practices, pp. 221–230

1, *p. 221:*

x	0	5	−3	$-\frac{1}{2}$	−2
y	1	11	−5	0	−3

2, *p. 224:*

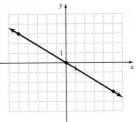

3, *p. 224:* a.

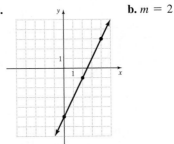

b. $m = 2$

4, *p. 226:* *x*-intercept: $(4, 0)$;
y-intercept: $(0, -2)$;

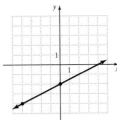

5, *p. 227:*

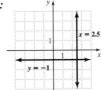

6, *p. 228:* **a.** $C = 0.03s + 40$
b.

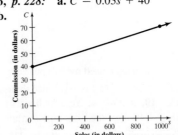
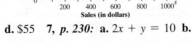

c. $m = 0.03$; for every sale, the commission increases by 0.03 times the value of the sale.

d. \$55 **7, *p. 230:* a.** $2x + y = 10$ **b.**

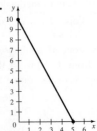

c. For each year the athlete is paid $2 million, the number of years that she could be paid $1 million decreases by 2 years. **d.** The *x*-intercept is the number of years of the contract if she was paid $2 million in each year of the contract. The *y*-intercept is the number of years of the contract if she was paid $1 million in each year of the contract.

Exercises 3.3, pp. 231–242

1. solution   **3.** three points   **5.** *y*-intercept

7.

x	4	7	$\frac{8}{3}$
y	4	13	0

9.

x	3.5	6	$-\frac{1}{10}$	$\frac{8}{5}$
y	-17.5	-30	$\frac{1}{2}$	-8

11.

x	0	-4	8	4
y	3	6	-3	0

13.

x	3	6	-3	0
y	0	1	-2	-1

15.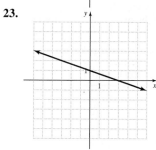

17.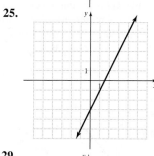

19.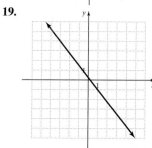

21.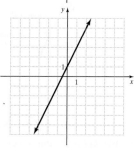

23.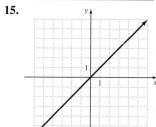

25.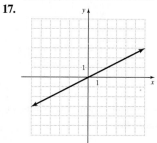

27.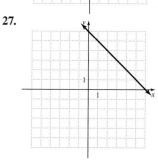

29.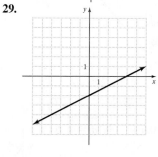

31. 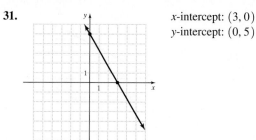 *x*-intercept: $(3, 0)$
y-intercept: $(0, 5)$

33. 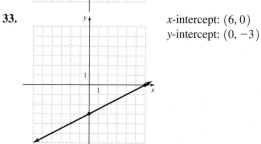 *x*-intercept: $(6, 0)$
y-intercept: $(0, -3)$

35. 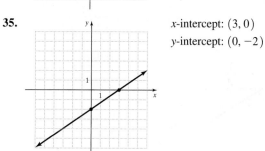 *x*-intercept: $(3, 0)$
y-intercept: $(0, -2)$

37. 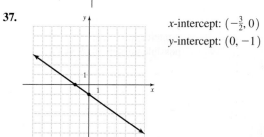 *x*-intercept: $\left(-\frac{3}{2}, 0\right)$
y-intercept: $(0, -1)$

39. 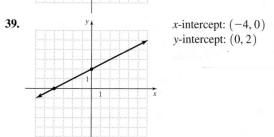 *x*-intercept: $(-4, 0)$
y-intercept: $(0, 2)$

41. 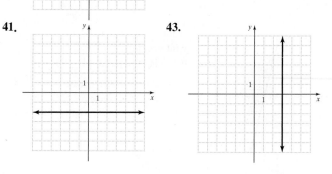 **43.**

45.

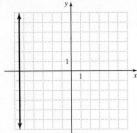

47.

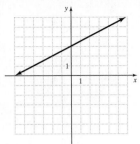

49.

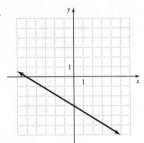

51.

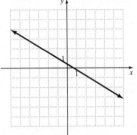

53.

x	-3	$\frac{5}{2}$	8	1
y	12	1	-10	4

55.

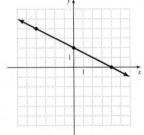

x-intercept: $(4, 0)$
y-intercept: $(0, 2)$
possible third point: $(-4, 4)$

57.

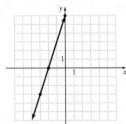

x-intercept: $(-2, 0)$
y-intercept: $(0, 6)$
possible third point: $(-3, -3)$

59.

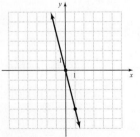

x-intercept: $(0, 0)$
y-intercept: $(0, 0)$
possible third point: $(1, -4)$

61.

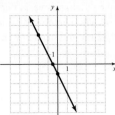

x-intercept: $\left(-\frac{1}{2}, 0\right)$
y-intercept: $(0, -1)$
possible third point: $(-2, 3)$

63.

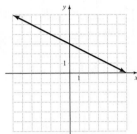

65. a.

t	0	0.5	1	1.5	2
v	10	-6	-22	-38	-54

A positive value of v means that the object is moving upward.
A negative value of v means that the object is moving
downward. **b.**

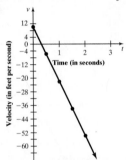

c. The v-intercept is the initial velocity of the object.
d. The t-intercept represents the time when the object changes from an
upward motion to a downward motion.

67. a. $P = 100m + 500$ **b.**

m	1	2	3
P	600	700	800

c.

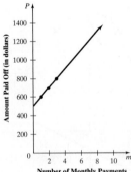

69. a. $0.05n + 0.1d = 2$ **b.**

c. Only nonnegative integer values make sense, since you cannot have fractions of a nickel or a dime.

71. a. $F = 5d + 40$

b.
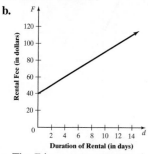

c. The F-intercept represents the fixed cost of renting the laptop for 0 days.

Section 3.4   Practices: pp. 243–249

1, p. 243: Slope is -2; y-intercept is $(0, 3)$.   **2, p. 244:** Slope is -1; y-intercept is $(0, 0)$.   **3, p. 244:** $y = \frac{3}{2}x - 2$ **4, p. 244:** $y = 4x + 6$ **5, p. 245:** $y = 1x + 2$, or $y = x + 2$ **6, p. 245:** $y = -2x - 1$ **7, p. 245:** $y = -\frac{1}{2}x - 2$ **8, p. 245:** $w = 45 - 3t$

9, p. 246:

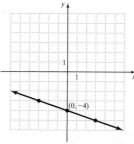

10, p. 247: $y - 0 = 2(x - 7)$ **11, p. 248:** $y - 7 = 1(x - 7)$ or $y - 0 = 1(x - 0)$ **12, p. 248:** Point-slope form: $w - 27 = 5(b - 4)$; slope- intercept form: $w = 5b + 7$ **13, p. 249:**

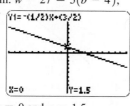

The displayed coordinates of the y-intercept are $x = 0$ and $y = 1.5$.

Exercises 3.4, pp. 250–257

1. Slope-intercept   **3.** y-intercept   **5.** For $y = 3x - 5$: 3, $(0, -5)$, $\diagup$, $(\frac{5}{3}, 0)$; for $y = -2x$: -2, $(0, 0)$, $\diagdown$, $(0, 0)$; for $y = 0.7x + 3.5$: 0.7, $(0, 3.5)$, $\diagup$, $(-5, 0)$; for $y = \frac{3}{4}x - \frac{1}{2}$: $\frac{3}{4}$, $(0, -\frac{1}{2})\diagup$, $(\frac{2}{3}, 0)$; for $6x + 3y = 12$: -2, $(0, 4)$; $\diagdown$, $(2, 0)$; for $y = -5$: 0, $(0, -5)$, —, no x-intercept; for $x = -2$: undefined, no y-intercept, $|$, $(-2, 0)$

7. Slope: -1; y-intercept: $(0, 2)$ **9.** Slope: 3; y-intercept: $(0, -4)$ **11.** $y = x - 10$ **13.** $y = -\frac{1}{10}x + 1$ **15.** $y = -\frac{3}{2}x + \frac{1}{4}$ **17.** $y = \frac{2}{5}x - 2$ **19.** $y = 3x + 14$ **21.** b   **23.** a

25.
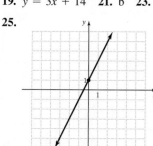

27.
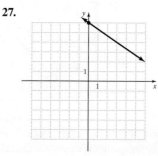

29.
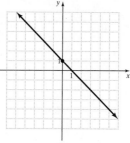

31.
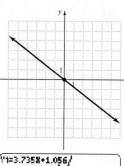

33.
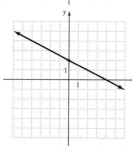

35.
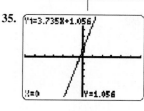

37. $y = 3x + 7$ **39.** $y = 5x - 20$ **41.** $y = -\frac{1}{2}x + 4$ **43.** $y = -x + 3$ **45.** $y = \frac{1}{6}x - \frac{29}{6}$ **47.** $x = -3$ **49.** $y = -6$ **51.** $y = \frac{3}{4}x + 3$ **53.** $y = -x - 2$ **55.** $y = 2$ **57.** $x = -2.5$ **59.** For $y = -7x + 2$: -7, $(0, 2)$, $\diagdown$, $(\frac{2}{7}, 0)$; for $y = 4x$: 4, $(0, 0)$, $\diagup$, $(0, 0)$; for $y = 2.5x + 10$: 2.5, $(0, 10)$, $\diagup$, $(-4, 0)$; for $y = \frac{2}{3}x - \frac{1}{4}$: $\frac{2}{3}$, $(0, -\frac{1}{4})$, $\diagup$, $(\frac{3}{8}, 0)$; for $5x + 4y = 20$: $-\frac{5}{4}$, $(0, 5)$, $\diagdown$, $(4, 0)$; for $x = 9$: undefined, no y-intercept, $|$, $(9, 0)$; for $y = -3.2$: 0, $(0, -3.2)$, $-$, no x-intercept

61. $y = 4x - 5$ **63.** d   **65.** $y = \frac{1}{2}x - 1$

67.

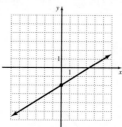

69. $y = -x + 5$

71. a. $\frac{9}{5}$ **b.** $F = \frac{9}{5}C + 32$ **c.** Water boils at 100°C.

73. a. $y - 4500 = 6(x - 500)$ **b.** $y = 6x + 1500$ **c.** The y-intercept represents the monthly flat fee the utility company charges its residential customers.

75. a. $I = 0.03S + 1500$ **b.**

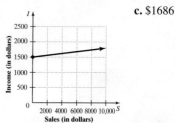

c. $1686

77. $P = \frac{1}{33}d + 1$ **79.** $L = \frac{5}{6}F + 10$

Section 3.5   Practices: pp. 260–263

1, p. 260: No, $(1, 3)$ is not a solution to the inequality.

2, p. 261:

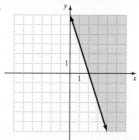

3, p. 262:

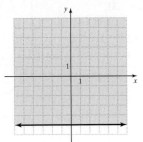

4, p. 262:

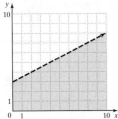

5, p. 263: a. $d + g \leq 3000$ **b.**

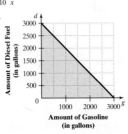

c. The d-intercept represents the maximum amount of diesel fuel the refinery can produce if no gasoline is produced. The g-intercept represents the maximum amount of gasoline the refinery can produce if no diesel fuel is produced.

Exercises 3.5, pp. 264–271

1. Half-plane **3.** Graph **5.** Broken **7.** No, not a solution **9.** Yes, a solution **11.** No, not a solution

13.

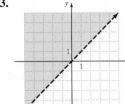

15.

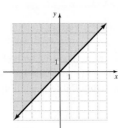

17.

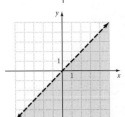

19. d **21.** b

23.

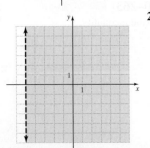

25.

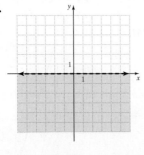

27.

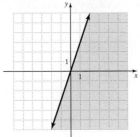

29.

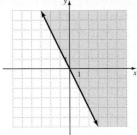

31.

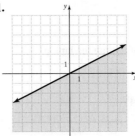

33.

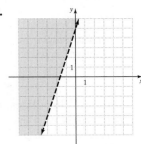

35.

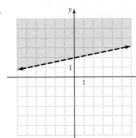

37.

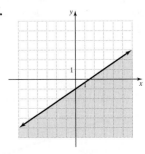

39.

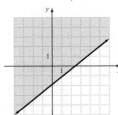

41. No, not a solution **43.** b

45.

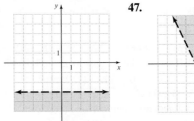

47.

49. a. $h < \frac{1}{4}i$ **b.**

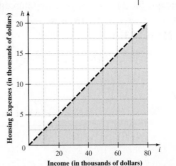

c. Choice of point may vary. Possible point: $(20, 2.5)$. The guideline holds since the inequality is true when the values are substituted into the original inequality.

51. a. $x + y \geq 200$ **b.**

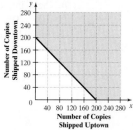

c. Choice of point may vary. Possible point: $(200, 60)$. Check:
$200 + 60 \overset{?}{\geq} 200, 260 \geq 200$, True. At least 200 copies are shipped.

53. a. $30x + 75y \geq 1500$ or $2x + 5y \geq 100$

b.

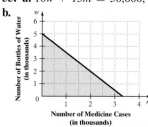

c. Since the point $(20, 20)$ lies in the solution region, selling 20 small and 20 large gift baskets will generate the desired revenue.

55. a. $10w + 15m \leq 50,000$, or $2w + 3m \leq 10,000$

b.

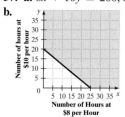

c. Answers may vary. Possible answers: 1500 cases of medicine and 1000 bottles of water: $(1.5, 1)$; 300 cases of medicine and 500 bottles of water: $(0.3, 0.5)$; 1200 cases of medicine and 2000 bottles of water: $(1.2, 2)$

57. a. $8x + 10y \geq 200$, or $4x + 5y \geq 100$

b.

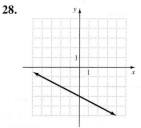

c. Answers may vary. Possible answers: 20 hr at the job paying \$8 per hour and 10 hr at the job paying \$10 per hour; 10 hr at the job paying \$8 per hour and 15 hr at the job paying \$10 per hour

59.

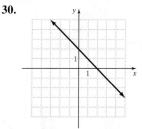

Chapter 3   Review Exercises: p. 277

1. are not; Possible answer: their y-coordinates have different signs   **2.** does not; Possible answer: the equation is false when 3 is substituted for x and -1 for y **3.** is not; Possible answer: the coefficient of x is negative   **4.** is; Possible answer: the coefficient of x is 0   **5.** is; Possible answer: $x = -3$ when $y = 0$ **6.** is; Possible answer: its change in x-values is 0, and division by 0 is undefined
7. is not; Possible answer: the product of their slopes is 1, not -1
8. is; Possible answer: substituting $x = 0$ and $y = 0$ into the inequality makes it true

9.

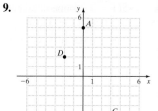

10.

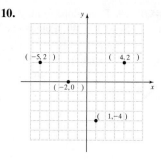

11. IV   **12.** III

13. 5;

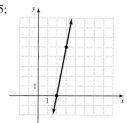

14. 0;

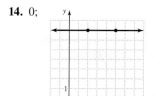

15.

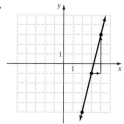

16.

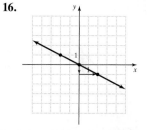

17. Positive slope   **18.** Undefined slope   **19.** Negative slope   **20.** Zero slope   **21.** Parallel   **22.** Perpendicular   **23.** $(30, 0)$ **24.** $(0, 50)$

25.

x	0	1	$\frac{5}{2}$	3
y	-5	-3	0	1

26.

x	2	5	-4	8
y	1	-2	7	-5

27.

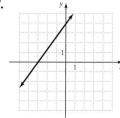

28.

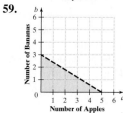

29.

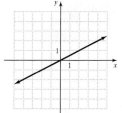

30.

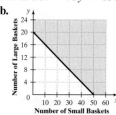

31. $y = x - 10$ **32.** $y = -\frac{1}{2}x - \frac{1}{2}$ **33.** $4, (0, -16), \nearrow, (4, 0)$
34. $-\frac{1}{3}, (0, 0), \searrow, (0, 0)$ **35.** The slope of a line perpendicular to this line is -2.   **36.** The slope of a line parallel to this line is 3.   **37.** Point-slope form: $y - 5 = -(x - 3)$; slope-intercept form: $y = -x + 8$

38. $y = 0$ **39.** Point-slope form: $y - 5 = -5(x - 1)$; slope-intercept form: $y = -5x + 10$ **40.** Point-slope form: $y - 1 = \frac{1}{5}(x - 3)$; slope-intercept form: $y = \frac{1}{5}x + \frac{2}{5}$ **41.** $y = -\frac{3}{2}x + 3$ **42.** $y = \frac{3}{2}x - 3$
43. No, it is not a solution. **44.** Yes, it is a solution.

45.
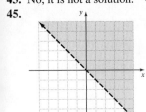

46.
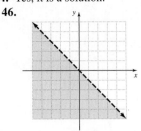

47.
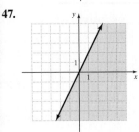

48.
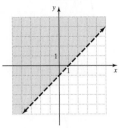

49. a.
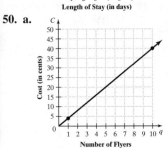

b. The A-intercept is $(0, 0)$. The A-intercept means that the cost for renting a room for 0 days is \$0.

50. a.
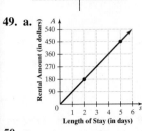

b. The slope of the line is 4. The slope represents the rate the print shop charges for each flyer, which is 4 cents.

51. The graph in part (a) could describe this motion. As the man drives toward the town, the distance between the man and the town decreases implying a negative slope. When the man stops, the distance between the man and the town remains the same, as indicated by the horizontal line segment. When the man drives toward the town again, the distance again decreases, implying a negative slope.
52. In the first part of the flight, the airplane takes off and ascends to a particular altitude (line segment slanting up to the right), then it flies at that same altitude during the second and longest part of the flight (horizontal line segment), and finally in the last part of the flight, it descends and lands (line segment slanting down to the right).
53. a. $i = 0.09s + 20{,}000$ **b.**
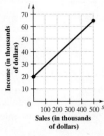

54. a.
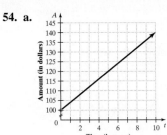

b. The A-intercept of the graph is $(0, 100)$. The A-intercept represents the initial balance in the bank account.

55. a.
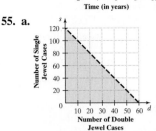

b. Answers may vary. Possible answer: 20 double jewel cases and 30 single jewel cases.

56. a.
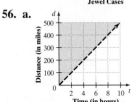

b. Choice of point may vary. Possible answer: $(2, 110)$; the coordinates mean that you caught up and passed your friend if you covered a distance of 110 mi in 2 hr.

Chapter 3   Posttest: pp. 284–286

1.
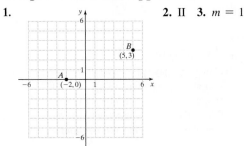

2. II **3.** $m = 1$

4. The graphs are parallel. The slope of $y = 3x + 1$ is 3 and the slope of $y = 3x - 2$ is 3. Since the slopes of the two lines are equal, their graphs are parallel. **5.** The slope of $\overleftrightarrow{AB}$ is $\frac{7}{2}$. The slope of $\overleftrightarrow{CD}$ is $-\frac{2}{7}$. $\overleftrightarrow{AB}$ is perpendicular to $\overleftrightarrow{CD}$, since the product of their slopes is -1. **6.** x-intercept: $(-5, 0)$, y-intercept: $(0, 2)$ **7.** The slope is positive. As the distance driven increases, the rental cost increases. **8.** Yes, the points do lie on the same line. The line containing $(0, 0)$ and $(-2, -4)$ is $y = 2x$. The line containing $(0, 0)$ and $(1, 2)$ is $y = 2x$.

9.

x	-3	5	$\frac{1}{3}$	1
y	10	-14	0	-2

10.
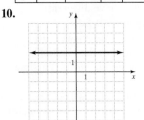

11.
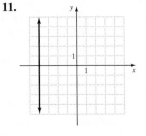

12.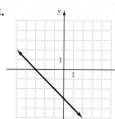

13.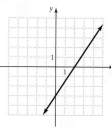

14.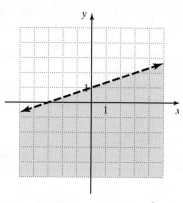

14. Slope: 3; y-intercept: $(0, 1)$ **15.** $y = 2x - 5$ **16.** $y = -x - 3$
17. Point-slope form: $y - 5 = \frac{3}{7}(x - 3)$; slope-intercept form: $y = \frac{3}{7}x + \frac{26}{7}$
18.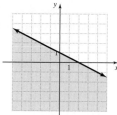

15. Point-slope form: $y - 62 = -\frac{8}{5}(x - 10)$; slope-intercept form:
$y = -\frac{8}{5}x + 78$ **16.** $l = \frac{P - 2w}{2}$ **17.** \$4 trillion   **18.** On the average,
males gain weight over time until age 40, when they begin to lose weight.
19. In the summer   **20. a.** $y = 6x$
b.

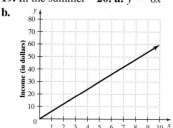

19. 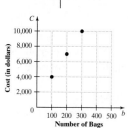 $C = 1000 + 30b$

20. $y \geq 124{,}000 + 8000x$

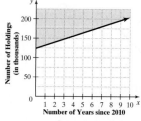

c. x-intercept: $(0, 0)$, y-intercept: $(0, 0)$

CHAPTER 4

Chapter 4   Pretest, pp. 290–291
1. a. Not a solution **b.** A solution **c.** Not a solution   **2.** One solution
3. $(-3, 1)$ **4.** $(5, 2)$

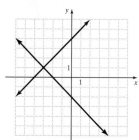

5. Infinitely many solutions,   **6.** No solution
namely all points on the line

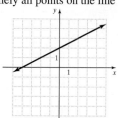 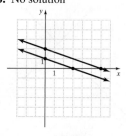

Chapter 3   Cumulative Review: pp. 287–288
1. $>$ **2.** -4.3 **3.** $\frac{1}{4}$ **4.** -24 **5.** 22 **6.** $8x - 1$ **7.** -1.3 **8.** -4
9. 17 **10.** $x > 2$
$-2\ -1\ \ 0\ \ 1\ \ 2\ \ 3\ \ 4\ \ 5\ \ 6\ \ 7\ \ 8$ **11. a.** II **b.** IV

12. 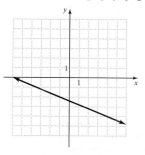   **13.** $y = -\frac{3}{4}x + \frac{25}{4}$

7. $(-5, -6)$ **8.** $(2, 1)$ **9.** $a = -5, b = 1$ **10.** $(1, -3)$ **11.** Infinitely many solutions, namely all ordered pairs that satisfy both equations
12. $(-4, 1)$ **13.** $(-1, 4)$ **14.** $n = -15, m = -7$ **15.** No solution
16. $(2, 0)$ **17.** The college awarded 2476 bachelor's degrees and 619 associate's degrees. **18.** Fifty \$5 tickets were printed. **19.** \$80,000 was invested in the fund at 5% interest, and \$120,000 was invested in the fund at 6%. **20.** The speed of the boat was 6 mph, and the speed of the current was 0.5 mph.

Section 4.1 Practices, pp. 293–301

1, *p. 293*: a. Yes, it is a solution of the system. **b.** No, it is not a solution of the system. **2, *p. 295*: a.** One solution **b.** Infinitely many solutions
c. No solution **3, *p. 296*:** $(3, -1)$

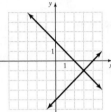

4, *p. 298*: No solution

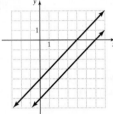

5, *p. 298*: Infinitely many solutions, namely all points on the line

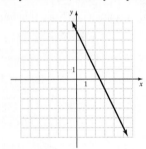

6, *p. 299*: a. $\begin{cases} m + v = 1150 \\ v = m - 100 \end{cases}$

b.

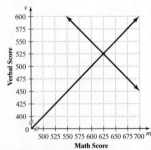

c. $(625, 525)$ **d.** The point of intersection indicates that she got a score of 625 on her test of math skills and a score of 525 on her test of verbal skills. **7, *p. 300*: a.** $y = 1.50x + 450$ **b.** $y = 3x$

c.

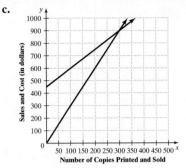

d. The break-even point is $(300, 900)$. So when 300 newsletters are printed and sold, the cost of printing the newsletter and the income from sales will be the same, \$900.

8, *p. 301*:

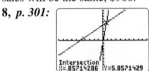

The approximate solution is $(0.857, 5.857)$.

Exercises 4.1, pp. 302–313

1. system of equations **3.** are parallel **5. a.** Not a solution; **b.** Not a solution; **c.** A solution **7. a.** Not a solution; **b.** Not a solution; **c.** A solution **9. a.** III **b.** IV **c.** II **d.** I

11. $(3, 1)$

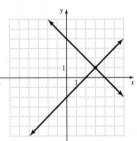

13. $(0, 4)$

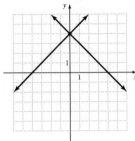

15. $(1, 5)$

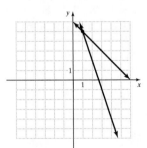

17. $(0, 1)$

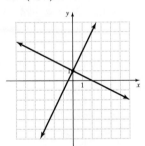

19. $(-2, -1)$

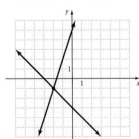

21. Infinitely many solutions, namely all points on the line

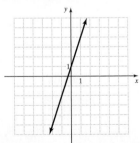

23. No solution

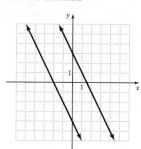

25. Infinitely many solutions, namely all points on the line

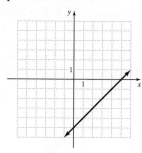

27. No solution

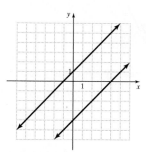

29. $(-2, -2)$

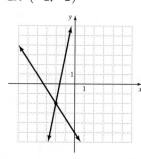

31. No solution

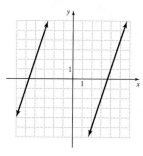

33. $(-1, 2)$

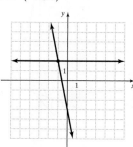

35. $(0, -2)$

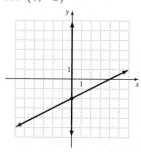

37. $(-6, -4)$

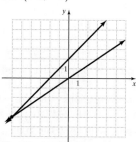

39. Infinitely many solutions, namely all points on the line

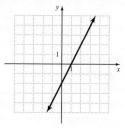

41. $(3, -2)$

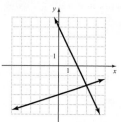

43. $(2, 2)$

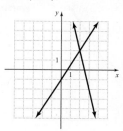

45. b   **47. a.** $x + y = 57{,}000$
$$y = x + 3000$$

b.

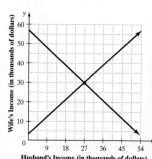

c. The husband made \$27,000 and the wife made \$30,000.

49. a. $y = 40x + 75$ (Mike)
$$y = 30x + 100$$ (Sally)

b.

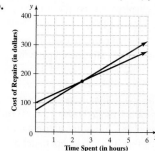

c. The plumbers would charge the same amount for 2.5 hr of work.   **d.** Sally charges less.

51.

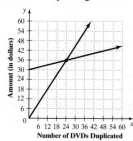

The break-even point for duplicating DVDs is $(24, 36)$.

53.

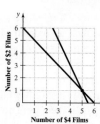

Five \$4 films were rented.

55.

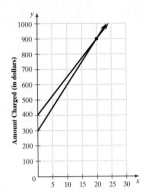

The health clubs charge the same amount ($900) for 20 mo.

Section 4.2 Practices, pp. 314–320

1, p. 314: $(4, -3)$ **2, p. 315:** $m = \frac{32}{7}, n = -\frac{5}{7}$ **3, p. 316:** $(0, -1)$
4, p. 316: No solution **5, p. 317:** Infinitely many solutions, namely all ordered pairs that satisfy both equations **6, p. 317: a.** $c = 35n + 20$ (TV Deal), $c = 25n + 30$ (Movie Deal) **b.** $n = 1$ and $c = 55$
c. The cost is the same ($55) for both cable deals if you sign up for one month. **7, p. 318:** 12.5 oz of the 20% copper alloy and 2.5 oz of the 50% copper alloy are required. **8, p. 320:** $80,000 in McDonald's and $40,000 in Wendy's

Exercises 4.2, pp. 321–322

1. $(3, 7)$ **3.** $(-4, -3)$ **5.** $(-12, 4)$ **7.** $(2, 1)$ **9.** $(0, 0)$ **11.** $(-1, 2)$
13. $(6, -6)$ **15.** No solution **17.** $(\frac{7}{2}, -\frac{1}{2})$ **19.** No solution
21. Infinitely many solutions, namely all ordered pairs that satisfy both equations **23.** Infinitely many solutions, namely all ordered pairs that satisfy both equations **25.** $p = 3, q = 5$ **27.** $s = -2, t = 1$
29. $(-3, 6)$ **31.** No solution **33.** $(-\frac{3}{2}, \frac{1}{2})$ **35. a.** $c = 1.25m + 3$, $c = 1.50m + 2$ **b.** $m = 4, c = 8$. The solution indicates that both companies charge the same amount ($8) for a 4-mi taxi ride. **37.** 80 full-price tickets were sold. **39.** She can combine 2.5 L of the antiseptic that is 30% alcohol with 7.5 L of the antiseptic that is 70% alcohol to get the desired concentration. **41.** There were 3 women in one department and 72 women in the other department. **43.** $4000 was loaned at 6%, and $1000 was loaned at 7%. **45.** $23,000 was invested at 7% and $17,000 was invested at 9%.

Section 4.3 Practices, pp. 324–329

1, p. 324: $(-2, 8)$ **2, p. 325:** $(2, -5)$ **3, p. 325:** $(0, 6)$
4, p. 326: $(2, -2)$ **5, p. 327:** $(10, 8)$ **6, p. 327:** Infinitely many solutions, namely all ordered pairs that satisfy both equations
7, p. 328: The whale's speed in calm water is 30 mph and the speed of the current is 10 mph. **8, p. 329:** 100 adults and 75 students attended the game.

Exercises 4.3, pp. 331–333

1. $(5, -2)$ **3.** $(-\frac{3}{2}, -\frac{5}{2})$ **5.** $p = -3, q = -16$ **7.** $(-1, 0)$
9. No solution **11.** Infinitely many solutions, namely all ordered pairs that satisfy both equations **13.** $(-2, \frac{1}{2})$ **15.** $s = 0, d = -2$
17. $(7, 4)$ **19.** $(-1, 2)$ **21.** $p = \frac{9}{2}, q = -\frac{11}{2}$ **23.** $(-3.5, 2.5)$
25. $(2, -2)$ **27.** $(\frac{8}{3}, -\frac{4}{3})$ **29.** No solution **31.** $a = -3, b = -2$
33. Infinitely many solutions, namely all ordered pairs that satisfy both equations **35.** $(-\frac{3}{4}, \frac{1}{4})$ **37.** The speed of the pass if there were no wind would be 13 yd per sec. **39.** The zoo collected 83 full-price admissions and 140 half-price admissions. **41.** There were 738 Lords and 650 Members of Parliament. **43.** It takes the computer 3 nanoseconds to carry out one sum and 4 nanoseconds to carry out one product. **45.** The rate for full-page ads is $950 and for half-page ads is $645.

Chapter 4 Review Exercises, pp. 337–341

1. is not; Possible answer: the ordered pair does not satisfy both equations in the system **2.** does not have; Possible answer: there are no points of intersection **3.** has; Possible answer: all points on the line are solutions **4.** does; Possible answer: the coefficients of the x-terms are opposites **5.** No, it is not a solution of the system. **6. a.** No solution **b.** One solution **c.** Infinitely many solutions **7. a.** III **b.** IV **c.** II
d. I **8.** $(1, 5)$ **9.** No solution

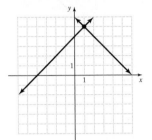 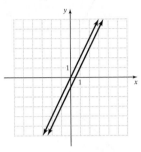

10. Infinitely many solutions, namely all ordered pairs that satisfy both equations

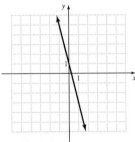

11. $(0, 5)$

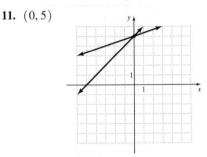

12. $(-1, 4)$ **13.** $a = 2, b = 2$ **14.** No solution **15.** Infinitely many solutions, namely all ordered pairs that satisfy both equations
16. $(4, -3)$ **17.** Infinitely many solutions, namely all ordered pairs that satisfy both equations **18.** $(6, -3)$ **19.** $(2, -5)$
20. a. $y = 0.50x + 1750, y = 5.50x$ **b.** The student must type 350 pages in order to break even. **21. a.** $s = 10h, s = 8h + 50$ **b.** 25 hr **22.** The area of the screen is 6720 ft^2. **23.** The tennis court is 36 ft wide and 78 ft long. **24.** The coin box contained 200 nickels and 150 dimes. **25.** One train travels at a rate of 60 mph and the other travels at a rate of 65 mph. **26.** The team made 818 two-point baskets and 267 three-point baskets. **27.** The pharmacist should mix 150 mL of the 30% solution and 50 mL of the 10% solution. **28.** 1400 L of the 50% solution and 600 L of water are needed to fill the tank. **29.** The client put $40,000 in municipal bonds and $10,000 in corporate stocks. **30.** $7000 was invested in the high-risk fund, and $3000 was invested in the low-risk fund. **31.** The speed of the slower plane was 400 mph. **32.** The bird flies at a speed of 21 mph in still air and the speed of the wind was 5 mph. **33. a.** 57 senators **b.** 43 senators **34. a.** The two metals will be the same temperature after 12 min. **b.** The iron will be 14° colder than the copper after 14 min.

Chapter 4 Posttest, pp. 342–343

1. a. A solution **b.** Not a solution **c.** Not a solution **2.** The system has no solution. **3.** $(3, 0)$

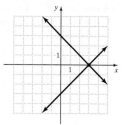

4. The system has no solution.

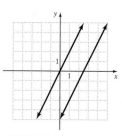

5. $(-1, 2)$

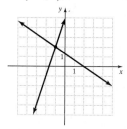

6. An infinite number of solutions, namely all points on the line

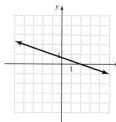

7. $(-4, 1)$ **8.** $(1, 3)$ **9.** $u = \frac{3}{4}, v = \frac{17}{4}$ **10.** $(2, -5)$
11. No solution **12.** $p = 0, q = 0.5$ **13.** $(-2, -7)$
14. Infinitely many solutions, namely all ordered pairs that satisfy both equations **15.** $(\frac{1}{7}, -\frac{6}{7})$ **16.** $(3, -2)$ **17.** The winning candidate got 4204 votes. **18.** One serving of turkey and two servings of salmon **19.** 1 gal of the 20% iodine solution and 3 gal of the 60% iodine solution **20.** The speed of the wind was 20 mph and the speed of the plane in still air is 150 mph.

Chapter 4 Cumulative Review, pp. 344–346

1. True **2.** -26 **3.** 6 **4.** $5x - 3y - 2$ **5.** No, 5 is not a solution.
6. $x = -3$ **7.** 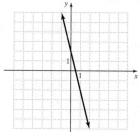 ; $x > 1$
8. The slope is $\frac{3}{4}$. **9.** Slope: $m = -\frac{1}{2}$; y-intercept: $(0, 2)$
10.

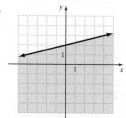

11. 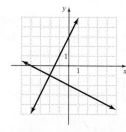 **12.** $(-2, -1)$

---

13. $a = 3, b = 2$ **14.** $(-4, 1)$ **15.** 0.0765 g **16.** The two companies charge the same amount ($95) for a one-day rental if the car is driven 300 mi. **17.** The plane can fly 1554 mph $(S = 1554)$. **18.** 3 L of plasma and 2 L of cells

19.

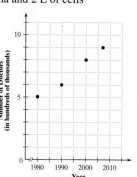

20. About 106,400 mi²

CHAPTER 5

Chapter 5 Pretest, p. 348

1. x^9 **2.** y^4 **3.** -3 **4.** $16x^8y^6$ **5.** $\frac{a^3}{b^{15}}$ **6.** $\frac{x^2}{25y^8}$
7. a. $6x^4, 5x^3, x^2, -7x,$ and 8 **b.** 6, 5, 1, -7, and 8 **c.** 4 **d.** 8
8. $3n^2 + n + 2$ **9.** $x^2 + x - 4$ **10.** $-6a^2 + 9a^2b + 3ab - 4b^2$
11. $3x^4 - 12x^3 + 27x^2$ **12.** $2n^3 + 7n^2 - 3n - 18$ **13.** $4x^2 - 3x - 27$
14. $9y^2 - 49$ **15.** $25 - 20n + 4n^2$ **16.** $t^2 - 2t - 5$
17. $4x + 5$ **18.** 200 mol of hydrogen will contain 1.2×10^{26} molecules. **19.** $100 + 200r + 100r^2$ **20.** 8 billion people

Section 5.1 Practices, pp. 349–355

1, p. 349: a. 10,000 **b.** $-\frac{1}{32}$ **c.** y^6 **d.** $-y$ **2, p. 350: a.** 1 **b.** 1
c. 1 **d.** -1 **e.** 1 **3, p. 351: a.** 10^{12} **b.** $(-4)^6$ **c.** n^{10} **d.** y^5
e. We cannot apply the product rule because the bases are not the same. **4, p. 352: a.** 7^5 **b.** $(-9)^1$, or -9 **c.** s^0, or 1 **d.** r^7
e. We cannot apply the quotient rule because the bases are not the same. **5, p. 352: a.** y^9 **b.** x^5y^6 **c.** a^2 **6, p. 353: a.** $\frac{1}{9^2} = \frac{1}{81}$
b. $\frac{1}{n^5}$ **c.** $-\frac{1}{3y}$ **d.** $\frac{1}{5^3} = \frac{1}{125}$ **7, p. 354: a.** $\frac{s}{8}$ **b.** $\frac{3}{x}$ **c.** $\frac{1}{r^6}$ **d.** $\frac{3^2}{g^5} = \frac{9}{g^5}$
e. x^3 **8, p. 354: a.** a^3 **b.** $\frac{2x^2}{5}$ **c.** $\frac{r^3s}{2}$ **9, p. 355: a.** 2^{23} **b.** 2^{30}

Exercises 5.1, pp. 356–359

1. added **3.** 0 power **5.** 125 **7.** $\frac{9}{16}$ **9.** 0.16 **11.** -0.25 **13.** -8
15. 81 **17.** $-\frac{1}{8}$ **19.** x^4 **21.** pq **23.** 1 **25.** -1 **27.** 10^{11} **29.** 4^8
31. a^6 **33.** Cannot be simplified **35.** n^7 **37.** Cannot be simplified
39. 8^2 **41.** 5^3 **43.** y^1, or y **45.** a^6 **47.** Cannot be simplified
49. x^0, or 1 **51.** y^6 **53.** p^7q^5 **55.** $x^3y^2z^3$ **57.** a^1, or a **59.** x^2
61. $\frac{1}{5}$ **63.** $\frac{1}{x}$ **65.** $-\frac{1}{3a}$ **67.** $\frac{1}{2^4}$ **69.** $-\frac{1}{3^4}$ **71.** $\frac{8}{n^3}$ **73.** $\frac{1}{x^2}$
75. $-\frac{x}{3^2}$ **77.** $\frac{y^3}{x^2}$ **79.** $\frac{q}{r}$ **81.** $\frac{4y^2}{x}$ **83.** $\frac{1}{p^5}$ **85.** p^3 **87.** $\frac{1}{a}$ **89.** $2n^4$
91. p^4q **93.** $\frac{1}{t^5}$ **95.** x^7 **97.** a^1, or a **99.** $\frac{b^3}{a^3}$ **101.** s^3t^6 **103.** $-\frac{8}{27}$
105. Cannot be simplified **107.** $\frac{1}{y^6}$ **109.** $\frac{1}{q^5}$ **111.** x^4y^3
113. a. $35 \cdot 2^5 = 1120$ people were ill on the sixth day of the epidemic; $35 \cdot 2^9 = 17,920$ people were ill on the tenth day. **b.** The number of people ill on the tenth day was 2^4, or 16, times as great as the number ill on the sixth day. **115.** $60 \times (0.95)^{11}$ ppm **117. a.** Volume of small box: $8x^3$, Volume of large box: $125x^3$ **b.** The volume of the larger box is $\frac{125}{8}$ times the volume of the small box.

Section 5.2 Practices, pp. 360–368

1, p. 360: a. $2^6 = 64$ **b.** $\frac{1}{7^3} = \frac{1}{343}$ **c.** q^8 **d.** $\frac{-1}{p^{15}} = -\frac{1}{p^{15}}$
2, p. 361: a. $49a^2$ **b.** $-64x^3$ **c.** $-64x^3$ **3, p. 361: a.** $36a^{18}$
b. $q^{16}r^{20}$ **c.** $-2a^3b^{21}$ **d.** $\frac{49}{a^2c^{10}}$ **4, p. 362: a.** $\frac{y^2}{9}$ **b.** $\frac{u^{10}}{v^{10}}$ **c.** $\frac{y^2}{9}$
d. $\frac{100a^{10}}{9b^4c^2}$ **e.** $125x^3y^6$ **5, p. 363: a.** $\frac{x^2}{25}$ **b.** $\frac{1}{4u}$ **c.** $\frac{b^6}{a^{10}}$ **6, p. 364:** 253.9

7, _p. 364_: 0.0000000043 **8, _p. 365_:** 8×10^{12} **9, _p. 365_:** 7.1×10^{-11}
10, _p. 365_: a. 2.464×10^2 **b.** 4×10^{11} **11, _p. 366_:** 9×10^{14}
12, _p. 367_: 1×10^{-3}, or 0.001 **13, _p. 367_:** 2.5E$^-$17 (Answers may vary.) **14, _p. 368_:** 7.3E$^-$10 (Answers may vary.)
15, _p. 368_: 4.6×10^8, or 460,000,000

Exercises 5.2, pp. 369–373

1. factors **3.** raise each factor to that power **5.** raise the reciprocal of the quotient **7.** left **9.** 2^8, or 256 **11.** 5^4, or 625 **13.** 10^{10}, or 10,000,000,000 **15.** $\frac{1}{4^4}$, or $\frac{1}{256}$ **17.** x^{24} **19.** y^8 **21.** $\frac{1}{x^6}$ **23.** n^4
25. $64x^3$ **27.** $64y^2$ **29.** $-64n^{15}$ **31.** $64y^8$ **33.** $\frac{1}{9a^2}$ **35.** $\frac{1}{p^7q^7}$
37. $r^{12}t^6$ **39.** $4p^{10}q^2$ **41.** $-2m^{12}n^{24}$ **43.** $-\frac{64m^{15}}{n^{30}}$ **45.** $\frac{1}{a^{12}b^8}$
47. $\frac{16y^6}{x^4}$ **49.** $\frac{125}{b^3}$ **51.** $\frac{c^2}{b^2}$ **53.** $-\frac{a^7}{b^7}$ **55.** $\frac{a^6}{27}$ **57.** $-\frac{p^{15}}{q^{10}}$ **59.** $\frac{4}{a}$ **61.** $\frac{8x^{15}}{y^6}$
63. $\frac{1}{p^5q^5}$ **65.** $81x^4y^{12}$ **67.** $\frac{v^4}{16u^4}$ **69.** $\frac{y^4z^{16}}{16x^8}$ **71.** $\frac{t^{12}}{r^{10}}$ **73.** $\frac{b^6}{8a^{12}}$
75. 317,000,000 **77.** 0.000001 **79.** 6,200,000 **81.** 0.00004025
83. 4.2×10^8 **85.** 3.5×10^{-6} **87.** 2.17×10^{11} **89.** 7.31×10^{-9}
91.

Standard Notation	Scientific Notation (written)	Scientific Notation (displayed on a calculator)
975,000,000	9.75×10^8	9.75E8
487,000,000	4.87×10^8	4.87E8
0.0000000001652	1.652×10^{-10}	1.652E$-$10
0.000000067	6.7×10^{-8}	6.7E$-$8
0.0000000000001	1×10^{-13}	1E$-$13
3,281,000,000	3.281×10^9	3.281E9

93. 9×10^7 **95.** 2.075×10^{-4} **97.** 1.68×10^1 **99.** 1.25×10^{10}
101. 3×10^2 **103.** 3×10^8 **105.** 0.0003067
107.

Standard Notation	Scientific Notation (written)	Scientific Notation (displayed on a calculator)
428,000,000,000	4.28×10^{11}	4.28E11
3,240,000	3.24×10^6	3.24E6
0.000005224	5.224×10^{-6}	5.224E$-$6
0.000000057	5.7×10^{-8}	5.7E$-$8
0.000000682	6.82×10^{-7}	6.82E$-$7
48,360,000	4.836×10^7	4.836E7

109. $\frac{9b^{10}}{a^4}$ **111.** $-\frac{1}{64y^3}$ **113.** $-\frac{16y^{12}}{x^8z^4}$ **115.** 7×10^{-8}
117. The larger volume is 8 times the smaller volume.
119. 4,000,000,000 bytes and 17,000,000,000 bytes
121. 7×10^{-7} m **123.** 2×10^{11} cells
125. 0.00000000000000000000000017 g **127.** 7,600,000,000
129. 1.6×10^{13} red blood cells **131. a.** 1.86×10^5 mi/sec
b. About 8.495×10^8 sec, or about 27 yr

Section 5.3 Practices, pp. 375–379

1, _p. 375_: (a) Terms: $-10x^2$, $4x$, and 20 (b) Coefficients: -10, 4, and 20 **2, _p. 375_:**

Polynomial	Monomial	Binomial	Trinomial
$5x^2 + 2x + 9$			✓
$-4x^2$	✓		
$12p - 1$		✓	

3, _p. 376_: a. Degree 1 **b.** Degree 2 **c.** Degree 3 **d.** Degree 0
4, _p. 376_:

Polynomial	Constant Term	Leading Term	Leading Coefficient
$-3x^7 + 9$	9	$-3x^7$	-3
x^5	0	x^5	1
$x^4 - 7x - 1$	-1	x^4	1
$3x + 5x^3 + 20$	20	$5x^3$	5

5, _p. 377_: a. $9x^5 - 7x^4 + 9x^2 - 8x - 6$ **b.** $7x^5 + x^3 - 3x^2 + 8$
6, _p. 378_: $x^2 + 6x + 20$ **7, _p. 378_: a.** -1 **b.** 19 **8, _p. 378_:** 356 ft
9, _p. 379_: 60°F

Exercises 5.3, pp. 380–383

1. monomial **3.** power **5.** leading term **7.** constant term
9. Polynomial **11.** Not a polynomial **13.** Polynomial
15. Not a polynomial **17.** (a) Terms: $5x^4$, $-2x^3$, and 1; (b) coefficients: 5, -2, and 1
19.

Polynomial	Monomial	Binomial	Trinomial
$5x - 1$		✓	
$-5a^2$	✓		
$-6a + 3$		✓	
$x^3 + 4x^2 + 2$			✓

21. $-4x^3 + 3x^2 - 2x + 8$; degree 3 **23.** $-3y + 2$; degree 1
25. $-p^4 + 3p^3 + 4p^2 - p + 10$; degree 4 **27.** $-4y^5 - y^3 - 2y + 2$; degree 5 **29.** $5a^2 - a$; degree 2 **31.** $3p^3$; degree 3
33.

Polynomial	Constant Term	Leading Term	Leading Coefficient
$-x^7 + 2$	2	$-x^7$	-1
$2x - 30$	-30	$2x$	2
$-5x + 1 + x^2$	1	x^2	1
$7x^3 - 2x - 3$	-3	$7x^3$	7

35. $10x^3 - 7x^2 + 10x + 6$ **37.** $r^3 + 3r^2 - 8r + 14$
39. $3x^2 - 5x - 6$ **41.** $2n^3 + 14n^2 + 20n + 10$ **43.** $0x^2$
45. $0x$ **47.** 11; -17 **49.** 37; 79 **51.** 13.73849; -4.74751
53. $5a^4 - a^3 + 2a - 4$; degree 4 **55.** For $4x - 20$: -20, $4x$, 4; for $-7x^6 + 9$: 9, $-7x^6$, -7; for $3x + 2 + x^2$: 2, x^2, 1; for $6x - x^5 + 11$: 11, $-x^5$, -1 **57.** Not a polynomial **59.** -1.89375; 83.02657 **61.** $0x^3$ and $0x^0$, or 0 **63.** A polynomial in x; degree 16
65. 120 ft **67.** \$50 **69.** 700 radio stations

Section 5.4 Practices, pp. 385–389

1, _p. 385_: $9x^2 + 3x - 43$ **2, _p. 385_:** $10p^2 + pq - 10q^2$ **3, _p. 386_:** $11n^2 + 2n - 3$ **4, _p. 386_:** $8p^3 + 2p^2q - 2pq^2 - 2q^3 + 25$
5, _p. 387_: a. $3r + 3s$ **b.** $-3q$ **6, _p. 388_:** $-3x^2 - 13x$
7, _p. 388_: $-5x^2 + 32x - 26$ **8, _p. 388_:** $-p^2 - 11pq + 17q^2$
9, _p. 389_: $(-0.04x + 1.22)$ million more attendees

Exercises 5.4, pp. 390–392

1. $2x^2 + 8x + 2$ **3.** $5n^3 + 9n$ **5.** $2p^2 + 3p - 1$
7. $11x^2 - 3xy + 2y^2$ **9.** $3p^3 + 2p^2q - 3pq^2 - 3q^3 + 5$
11. $10x^2 + 17x - 5$ **13.** $5x^3 + x^2 + 9x + 2$ **15.** $-3a^3 + 6a^2 + 7ab^2$
17. $-x^2 - 2x + 11$ **19.** $-2x^3 + 9x^2 - 13x + 11$ **21.** $x^2 - 2x - 9$
23. $5x^2 - 5y^2 + 4$ **25.** $-5p^2 + 7p + 6$ **27.** $t^3 - 12t^2 + 8$

29. $3r^3 - 17r^2s - 2$ **31.** $-x - r$ **33.** $2p - 3q - r$
35. $3y^2 + 3y + 4$ **37.** $m^3 - 12m + 16$ **39.** $2x^3 - 5x^2 - 10x + 9$
41. $9x^2 + 4x + 7$ **43.** $-2x + 12$ **45.** $8x^2y^2 - 12xy - 14$
47. $2m^2 - 5m - 9$ **49.** $5x^3 + 3x^2 + 2x - 12$ **51.** $5m - 8$
53. $-2x^2 + 6x + 3$ **55.** $5p^2 + 11p - 7$ **57.** **a.** $6.28r^2 + 6.28rh$
b. $12.56r^2$ or $12.56h^2$ **59.** $(-28x^3 - 61x^2 - 41x + 12,543)$ thousand more barrels per day   **61.** **a.** $(0.8x^3 + 6.8x^2 - 20.6x + 21.9)$ millions of dollars more for the Giants   **b.** \$43 million

Section 5.5   Practices, pp. 394–399

1, p. 394: $40x^5$ **2, p. 394:** $-350a^4b^5$ **3, p. 394:** $25x^2y^4$
4, p. 395: $70s^3 - 21s$ **5, p. 395:** $12m^6n^7 - 4m^4n^4 - 2m^3n^3$
6, p. 395: $-14s^5 + 34s^4 + 22s^3 + s^2$ **7, p. 396:** $2a^2 + a - 3$
8, p. 397: $16x^2 - 2x - 3$ **9, p. 397:** $14m^2 + 5mn - n^2$
10, p. 398: $n^3 - 3n^2 - 6n + 8$ **11, p. 398:** $8n^3 + 15n^2 + n + 6$
12, p. 399: $24x^4 + 56x^3 + 27x^2 + 60x - 7$ **13, p. 399:**
$p^3 - 3p^2q + 3pq^2 - q^3$ **14, p. 399:** $P + 2Pr + Pr^2$

Exercises 5.5, pp. 400–402

1. $-24x^2$ **3.** $-9t^5$ **5.** $20x^6$ **7.** $-70x^8$ **9.** $16p^3q^3r^2$ **11.** $64x^2$
13. $\frac{1}{8}t^{12}$ **15.** $-350a^4$ **17.** $-24a^4b^3c$ **19.** $7x^2 - 5x$ **21.** $45t^2 + 5t^3$
23. $24a^5 - 42a^4$ **25.** $12x^3 - 8x^2$ **27.** $x^5 - 2x^4 + 4x^3$
29. $15x^3 + 25x^2 + 30x$ **31.** $-45x^3 + 27x^2 + 63x$
33. $6x^5 + 24x^4 - 6x^3 - 6x^2$ **35.** $28pq - 4p^3$ **37.** $-3p^4q^2 + 9pq^3$
39. $6a^6b^5 + 20a^3b^8$ **41.** $-6x^2 + 26x$ **43.** $8x^3 - 16x^2 + 7x$
45. $-23x^3 + 35x^2 - 45x$ **47.** $7x^4y - 8x^3y - 18x^2y^2$
49. $-45a^5b^4 + 51a^3b^6$ **51.** $y^2 + 5y + 6$ **53.** $x^2 - 8x + 15$
55. $a^2 - 4$ **57.** $2w^2 - w - 21$ **59.** $-10y^2 + 17y - 3$
61. $20p^2 - 18p + 4$ **63.** $u^2 - v^2$ **65.** $-2p^2 + 3pq - q^2$
67. $3a^2 - 7ab + 2b^2$ **69.** $4pq + 3p - 32q - 24$
71. $x^3 - 6x^2 + 10x - 3$ **73.** $2x^3 + 5x^2 - 13x + 5$
75. $a^3 - b^3$ **77.** $3x^3 - 6x^2 - 105x$ **79.** $12n^3 - 3n$
81. $8m^4n - 6m^3n + 8m^2n^2$ **83.** $40s^7$ **85.** $-36y^3 - 28y^2 + 32y$
87. $-24x^{11}$ **89.** $-4a^3b + 2a^2b^4$ **91.** $x^3 - 4x^2 + 5x - 2$
93. $(x^2 + 30x)$ mm^2 **95.** $(-3000x^2 + 55,000x + 1,500,000)$
dollars   **97.** **a.** $(5000 + 15,000r + 15,000r^2 + 5000r^3)$ dollars
b. $(5000r + 10,000r^2 + 5000r^3)$ dollars   **c.** \$605

Section 5.6   Practices, pp. 404–407

1, p. 404: $p^2 + 20p + 100$ **2, p. 405:** $s^2 + 2st + t^2$
3, p. 405: $16p^2 + 40pq + 25q^2$ **4, p. 406:** $25x^2 - 20x + 4$
5, p. 406: $u^2 - 2uv + v^2$ **6, p. 406:** $4x^2 - 36xy + 81y^2$
7, p. 407: $t^2 - 100$ **8, p. 407:** **a.** $r^2 - s^2$ **b.** $64s^2 - 9t^2$
9, p. 407: $100 - 49k^4$ **10, p. 407:** $S^2 - s^2$

Exercises 5.6, pp. 408–410

1. plus   **3.** negative   **5.** $y^2 + 4y + 4$ **7.** $x^2 + 8x + 16$
9. $x^2 - 22x + 121$ **11.** $36 - 12n + n^2$ **13.** $x^2 + 2xy + y^2$
15. $9x^2 + 6x + 1$ **17.** $16n^2 - 40n + 25$ **19.** $81x^2 + 36x + 4$
21. $a^2 + a + \frac{1}{4}$ **23.** $64b^2 + 16bc + c^2$ **25.** $25x^2 - 20xy + 4y^2$
27. $x^2 - 6xy + 9y^2$ **29.** $16x^6 + 8x^3y^4 + y^8$ **31.** $a^2 - 1$
33. $16x^2 - 9$ **35.** $9y^2 - 100$ **37.** $m^2 - \frac{1}{4}$ **39.** $n^2 - 0.09$
41. $16a^2 - b^2$ **43.** $9x^2 - 4y^2$ **45.** $1 - 25n^2$ **47.** $x^3 - 25x$
49. $5n^4 + 70n^3 + 245n^2$ **51.** $n^4 - m^8$ **53.** $a^4 - b^4$
55. $36n^2 + 48n + 16$ **57.** $16p^2 - 81$ **59.** $64 - 16a + a^2$
61. $-32x^4 + 48x^3 - 18x^2$ **63.** $x^2 + y^2 - 10x - 2y + 26$
65. $A + \frac{AP}{50} + \frac{AP^2}{10,000}$ **67.** $\frac{3m^2 - 2am - 2bm - 2cm + a^2 + b^2 + c^2}{2}$
69. **a.** $4x$ **b.** Yes **c.** The area of the square is y^2 larger than the area of the rectangle.

Section 5.7   Practices, pp. 411–416

1, p. 411: $-4n^5$ **2, p. 411:** $-4pr^3$ **3, p. 412:** $3x^2 - 2x$
4, p. 412: $-7x^5 - 5x^2 + 4$ **5, p. 412:** $-a^6b^3 + \frac{ab}{5} - 3$

6, p. 414: $2x + 3$ **7, p. 415:** $x^2 + 2x + 3 + \frac{2}{3x + 1}$ **8, p. 415:** $3s - 5$
9, p. 416: $n^2 - 4n - 3 + \frac{-13}{4n - 3}$ **10, p. 416:** **a.** The future value of the investment after 1 yr is $10(1 + r)^1$, or $10(1 + r)$. The future value of the investment after 2 yr is $10(1 + r)^2$. **b.** $10r + 10$ and $10r^2 + 20r + 10$ **c.** The future value of the investment after 2 yr is $(r + 1)$ times as great as the future value of the investment after 1 yr.

Exercises 5.7, pp. 417–419

1. coefficients   **3.** remainder   **5.** $2x^2$ **7.** $-4a^7$ **9.** $\frac{4x}{3}$
11. $4q^2$ **13.** $3u^2v^2$ **15.** $-\frac{15ab^2}{7}$ **17.** $-\frac{3u^3v^2}{2}$ **19.** $3n + 5$
21. $2b^3 - 1$ **23.** $-6a - 4$ **25.** $3 - 2x^2$ **27.** $2a^2 - 3a + 5$
29. $-\frac{n^2}{5} + 2n + 1$ **31.** $\frac{5a}{2} + \frac{b^2}{2}$ **33.** $-4xy^2 + 3 + y$
35. $2q^2 - pq^2 + \frac{3p^2}{2}$ **37.** $x - 7$ **39.** $7x - 2$ **41.** $3x - 1$
43. $5x + 3$ **45.** $7x + 2$ **47.** $x + \frac{5}{x + 2}$ **49.** $2x + 1$
51. $2x - 3 + \frac{-2}{4x + 3}$ **53.** $x^2 - 6x + 5$ **55.** $2x^2 - x - 3 + \frac{7}{3x - 4}$
57. $5x + 20 + \frac{78}{x - 4}$ **59.** $2x^2 + 3x + 4 + \frac{15}{2x - 3}$ **61.** $x^2 - 3x + 9$
63. $-7r$ **65.** $-\frac{2}{3}ab$ **67.** $-3n^3 + 8n + \frac{1}{3}$ **69.** $2x + 1$
71. $x^2 - 1 + \frac{3}{4x + 1}$ **73.** $-4m^2 + 9$ **75.** **a.** 1100 words
b. $\left(\dfrac{9x^2 - 340x + 3600}{x}\right)$, or $\left(9x - 340 + \dfrac{3600}{x}\right)$ words per point
c. 110 words per point   **77.** **a.** $\frac{d}{r} = t$, or $t = \frac{d}{r}$ **b.** It takes $(t^2 - 7t + 14)$ hr.   **79.** $(x^2 + 16x + 150)$ dollars per person

Chapter 5   Review Exercises, pp. 425–428

1. is not; Possible answer: the product rule of exponents says to add the exponents when multiplying powers of the same base   **2.** is; Possible answer: the quotient rule says to subtract the exponent in the denominator from the exponent in the numerator   **3.** cannot; Possible answer: a simplified expression does not have any negative exponents   **4.** is not; Possible answer: 0.24 is less than 1   **5.** is not; Possible answer: the power of the variable is negative   **6.** does; Possible answer: it can be written as a coefficient of x^0.   **7.** $-x^3$ **8.** -1 **9.** n^{11}
10. x^7 **11.** n^3 **12.** p^3 **13.** y^7 **14.** a^3b^3 **15.** y **16.** n^2
17. $\frac{1}{5x}$ **18.** $-\frac{3}{n^2}$ **19.** $\frac{v^4}{64}$ **20.** y^4 **21.** $\frac{1}{x}$ **22.** $\frac{y^3}{5}$ **23.** a^{10}
24. $\frac{1}{t^6}$ **25.** $\frac{1}{x^2y}$ **26.** x^2y **27.** $10^8 = 100,000,000$ **28.** $-x^9$
29. $4x^6$ **30.** $-64m^{15}n^3$ **31.** $\frac{3}{x^{12}}$ **32.** $\frac{b^8}{a^6}$ **33.** $\frac{x^4}{81}$ **34.** $\frac{a^2}{b^6}$ **35.** $\frac{y^6}{x^6}$
36. $x^{10}y^5$ **37.** $\frac{16a^6}{b^8c^2}$ **38.** $\frac{v^4}{49u^{10}w^2}$ **39.** 37,000,000,000
40. 1,630,000,000   **41.** 0.00005022   **42.** 0.00000000006   **43.** 1.2×10^{12}
44. 4.27×10^8 **45.** 4×10^{-14} **46.** 5.6×10^{-7} **47.** 5.88×10^9
48. 6.3×10^3 **49.** 6×10^6 **50.** 6×10^{-10} **51.** Polynomial
52. Not a polynomial   **53.** Trinomial   **54.** Binomial
55. $-3y^3 + y^2 + 8y - 1$; degree 3, leading term: $-3y^3$, leading coefficient: -3 **56.** $n^4 - 7n^3 - 6n^2 + n$; degree 4, leading term: n^4, leading coefficient: 1   **57.** $-x^3 + x^2 + 2x + 13$
58. $3n^3 + 4n^2 - 6n + 4$ **59.** $12; 0$ **60.** $0; -16$
61. $x^2 - x + 13$ **62.** $-2y^3 - y^2 - y - 5$ **63.** $a^2 - 2ab - 3b^2$
64. $5s^3t + s^2t + 9s^2 - 6st - 3t^2$ **65.** $2x^2 - 8x - 8$
66. $-n^3 + 3n^2 + n$ **67.** $5y^4 - 5y^3 + 2y^2 - 6y - 3$
68. $2x^3 + 8x^2 - 12x + 3$ **69.** $4t^2 + 4t$ **70.** $-2x - y$
71. $2y^2 - 5y + 2$ **72.** $-5x^2 + 9$ **73.** $-6x^5$ **74.** $-144a^3b^5$
75. $8x^2y^2 - 10xy^3$ **76.** $-5x^4 + 15x^3 - 5x^2$ **77.** $n^2 + 10n + 21$
78. $3x^2 + 9x - 54$ **79.** $8x^2 - 6x + 1$ **80.** $9a^2 + 3ab - 2b^2$
81. $2x^4 + 6x^3 - 5x^2 - 13x + 6$ **82.** $y^3 - 9y^2 + 15y - 2$
83. $-6y^2 + 13y$ **84.** $-x^3 + 6x^2 - 6x$ **85.** $a^2 - 2a + 1$
86. $s^2 + 8s + 16$ **87.** $4x^2 + 20x + 25$ **88.** $9 - 24t + 16t^2$
89. $25a^2 - 20ab + 4b^2$ **90.** $u^4 + 2u^2v^2 + v^4$ **91.** $m^2 - 16$
92. $36 - n^2$ **93.** $49n^2 - 1$ **94.** $4x^2 - y^2$ **95.** $16a^2 - 9b^2$
96. $x^3 - 100x$ **97.** $-48t^4 + 120t^3 - 75t^2$ **98.** $p^4 - 2p^2q^2 + q^4$
99. $3x^2$ **100.** $-2a^2b^3c$ **101.** $6x^2 - 2$ **102.** $5x^3 + 3x^2 - 2x - 1$

103. $3x - 7$ **104.** $x^2 - 2x - 1 + \frac{12}{2x - 1}$ **105.** 1.39×10^{10} yr
106. 6,240,000,000,000,000,000 eV **107.** 3×10^{-5} m
108. 3,700,000 mi^2 **109.** There will be 36 handshakes.
110. The object is 480.4 m above the ground. **111.** 77°F
112. 1700 two-year colleges **113.** **a.** $(3b^2 - 16b)$ m^2
b. 3115 m^2 **114.** **a.** $(13x + 280)$ dollars per person
b. $300 **115.** **a.** lwh in^3 **b.** $12^3 lwh$ in^3, or $1728 lwh$ in^3
116. $V = \pi r^3$

Chapter 5 Posttest, p. 429

1. x^7 **2.** n^6 **3.** $\frac{7}{a}$ **4.** $-27x^6 y^3$ **5.** $\frac{x^8}{y^{12}}$ **6.** $\frac{y^3}{27x^6}$
7. **a.** $-x^3, 2x^2, 9x, -1$ **b.** $-1, 2, 9,$ and -1 **c.** 3 **d.** -1
8. $2y^2 - y + 5$ **9.** $-x^2 + x - 9$ **10.** $3x^2 y^2 - 4x^2$
11. $10m^3 n^3 - 20m^2 n^3 + 2m^2 n^4$ **12.** $y^4 - 3y^3 + 2y^2 + 4y - 4$
13. $6x^2 + 19x - 7$ **14.** $49 - 4n^2$ **15.** $4m^2 - 12m + 9$
16. $-4s^2 - 5s + 9$ **17.** $t^2 - t - 1 + \frac{4}{3t - 2}$ **18.** 1×10^{-7} m
19. First house: $(1500x + 140{,}000)$ dollars; second house:
$(800x + 90{,}000)$ dollars **20.** The account balance is $1060.90.

Chapter 5 Cumulative Review, pp. 430–431

1. 7.36 **2.** 61 **3.** $2y + 18$ **4.** 1 **5.** $m = \frac{y - b}{x}$ **6.** $x \leq -5$
7. $m = -\frac{1}{2}$ **8.** Slope: $\frac{2}{3}$; y-intercept: $(0, -2)$

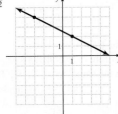

9.

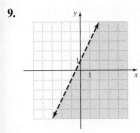

10. $(-1, -2)$

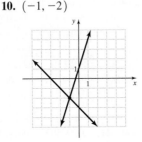

11. $(2, -2)$ **12.** $m^2 - 16m + 16$ **13.** $-16a^2 + 49$
14. $-3x + 1$ **15.** **a.** $x + 4, x + 8,$ and $x + 12$
b. Yes; $1980 - 1972 = 8$, which is a multiple of 4.
16. **a.** $0.20b$ **b.** $c = 1.20b$ **17.** **a.** $C = 125d + 300$
b.

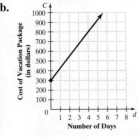

c. The C-intercept represents the cost of the vacation package with
no hotel stay. **18.** **a.** $c = 1000m + 1500$ (small one-bedroom)
$c = 900m + 1800$ (large studio) **b.** $m = 3$ and $c = 4500$
c. The cost ($4500) is the same for renting the apartments for 3 mo.
19. 9×10^{13} kg·m^2/sec^2 **20.** $(24\pi x^2 + 72\pi x + 54\pi)$ in^2

CHAPTER 6

Chapter 6 Pretest, p. 433

1. $18a$ **2.** $4p(q + 4)$ **3.** $5xy(2x - x^2 y^2 + y)$ **4.** $(x + 2)(3x + 2)$
5. $(n - 3)(n - 8)$ **6.** $(a - 3)(a + 7)$ **7.** $3y(y - 1)(y - 3)$, or
$3y(1 - y)(3 - y)$ **8.** $(5a - 4b)(a + 2b)$ **9.** $-2(3n + 1)(2n - 7)$
10. $(2x - 7)^2$ **11.** $(5n + 3)(5n - 3)$ **12.** $y(x + 2y)(x - 2y)$
13. $(y^3 - 4)(y^3 - 5)$ **14.** $0, 6$ **15.** $\frac{2}{3}, -1$ **16.** $3, -5$
17. $h = \frac{A - 2lw}{2l + 2w}$ **18.** $-(16t + 1)(t - 4)$ ft **19.** $(S + 15)(S - 15)$ ft^2
20. The length of the screen is 32 in. and the height is 24 in.

Section 6.1 Practices, pp. 435–439

1, p. 435: 24 **2, p. 435:** a **3, p. 436:** $6xy^2$ **4, p. 436:** $2y^2(5 + 4y^3)$
5, p. 437: $7a(3ab - 2)$ **6, p. 437:** $2ab^2(4a - 3b)$
7, p. 437: $12(2a^2 - 4a + 1)$ **8, p. 437:** $a = \frac{s^2}{b + c}$
9, p. 438: $(y - 3)(4 + y)$, or $(y - 3)(y + 4)$
10, p. 438: $(x - 1)(3y - 2)$ **11, p. 438:** $(4 - 3x)(1 + 2x)$
12, p. 439: $(b - 5)(a + 4)$ **13, p. 439:** $(y - z)(5 - y)$
14, p. 439: $t(v_0 + \frac{1}{2}at)$

Exercises 6.1, pp. 440–442

1. factoring **3.** greatest common factor (GCF) **5.** 27 **7.** x^3 **9.** $4b$
11. $4y^3$ **13.** $3a^2 b^2$ **15.** $3x - 1$ **17.** $x(x + 7)$ **19.** $3(x + 2)$
21. $8(3x^2 + 1)$ **23.** $9(3m - n)$ **25.** $x(7x - 2)$ **27.** $b^2(5 - 6b)$
29. $5x(2x^2 - 3)$ **31.** $ab(ab + 1)$ **33.** $xy(6y + 7x)$ **35.** $9pq(3q - 2p)$
37. $2x^3 y(1 + 6y^3)$ **39.** $3(c^3 + 2c^2 + 4)$ **41.** $b^2(9b^2 - 3b + 1)$
43. $2m^2(m^2 + 5m - 3)$ **45.** $b^2(5b^3 - 3b + 2)$ **47.** $5x(3x^3 - 2x^2 - 5)$
49. $4ab(a + 2ab - 3)$ **51.** $3cd(3cd + 4c^2 + d^2)$ **53.** $(x - 1)(x + 3)$
55. $(a - 1)(5a - 3)$ **57.** $(s + 7)(r - 2)$ **59.** $(x - y)(a - b)$
61. $(y + 2)(3x - 1)$ **63.** $(b - 1)(b - 5)$ **65.** $(y - 1)(y + 5)$
67. $(t - 3)(1 + t)$ **69.** $(b - 7)(9a - 2)$ **71.** $(r + 3)(s + t)$
73. $(x + 6)(y - 4)$ **75.** $(5x - 3z)(3y + 4z)$ **77.** $(z + 4)(2x + 5y)$
79. $P = \frac{TM}{C + L}$ **81.** $l = \frac{S - 2wh}{2w + 2h}$ **83.** $8p(2p^2 + 3)$ **85.** $12rs(4s - 5r)$
87. $6(7j^2 - 1)$ **89.** $(s - 3)(t - 7)$ **91.** $(y - 4)(3x - 5)$ **93.** $4a^2 b$
95. $m(v_2 - v_1)$ **97.** $0.5n(n - 1)$ **99.** $\frac{1}{2}n(n - 3)$ **101.** $n = \frac{P - D}{C + T}$
103. $50(14n + 18s - 5e)$ dollars

Section 6.2 Practices, pp. 445–451

1, p. 445: $(x + 1)(x + 4)$, or $(x + 4)(x + 1)$
2, p. 446: $(y - 4)(y - 5)$, or $(y - 5)(y - 4)$
3, p. 446: Prime polynomial **4, p. 447:** $(y - 4)(y - 8)$
5, p. 447: $(p - q)(p - 3q)$, or $(p - 3q)(p - q)$
6, p. 448: $(x - 2)(x + 3)$ **7, p. 448:** $(x + 2)(x - 23)$
8, p. 449: $(y - 4)(y + 6)$ **9, p. 449:** $(a + 3b)(a - 8b)$
10, p. 450: $y(y + 1)(y - 10)$ **11, p. 450:** $8x(x - 1)(x + 2)$
12, p. 450: $-(x - 1)(x + 11)$ **13, p. 451:** $-16(t + 1)(t - 3)$

Exercises 6.2, pp. 452–454

1. not factorable **3.** opposite signs **5.** f **7.** e **9.** b
11. $x - 5$ **13.** $x + 4$ **15.** $x - 1$ **17.** $(x + 2)(x + 4)$
19. $(x - 1)(x + 6)$ **21.** Prime polynomial **23.** $(x - 2)(x + 4)$
25. $(x - 1)(x - 3)$ **27.** $(y - 4)(y - 8)$ **29.** $(t + 1)(t - 5)$
31. Prime polynomial **33.** $(y - 4)(y - 5)$ **35.** $(y + 3)(y - 16)$
37. $(b + 4)(b + 7)$ **39.** $(x - 7)(x - 7)$ **41.** $-(y + 5)(y - 10)$
43. $(x - 8)(x - 8)$ **45.** $(x - 2)(x - 8)$ **47.** $(w - 3)(w - 27)$
49. $(p - q)(p - 7q)$ **51.** $(p + q)(p - 5q)$ **53.** $(m - 5n)(m - 7n)$
55. $(x + y)(x + 8y)$ **57.** $5(x + 2)(x - 3)$ **59.** $2(x - 2)(x + 7)$
61. $6(t - 1)(t - 2)$ **63.** $3(x + 2)(x + 4)$ **65.** $y(y - 2)(y + 5)$
67. $a(a + 3)(a + 5)$ **69.** $t^2(t - 2)(t - 12)$ **71.** $4a(a - 1)(a - 2)$
73. $2x(x + 3)(x + 5)$ **75.** $4x(x - 3)(x - 4)$ **77.** $2s(s - 4)(s + 7)$
79. $2c^2(c - 5)(c + 7)$ **81.** $ax(x - 2)(x - 16)$ **83.** Prime polynomial
85. $5x^2(x - 5)(x + 2)$ **87.** $-(w + 4)(w - 10)$ **89.** $6(m + 2)(m - 3)$
91. $(t - 5)(t - 12)$ **93.** b **95.** $(p + q)(p + q) = 1$, or $(p + q)^2 = 1$

97. $16(t-2)(t+5)$ **99. a.** $(6x^2 + 24x + 18)$ in.^{2} **b.** $6(x+1)(x+3)$
101. $\frac{57}{400,000}(x-3000)(x+1000)$

Section 6.3 Practices, pp. 457–461
1, p. 457: $(5x+4)(x+2)$ **2, p. 457:** $(6x-7)(x-3)$
3, p. 458: $(7y-2)(y+7)$ **4, p. 459:** $(2x-5)(x+2)$
5, p. 459: $3x(3x+1)(2x-3)$ **6, p. 460:** $3(6c-5d)(2c+d)$
7, p. 460: $(2x+1)(x-4)$ **8, p. 461:** $x(2x-5)(2x-7)$

Exercises 6.3, pp. 462–464
1. e **3.** b **5.** c **7.** $3x+1$ **9.** $x-3$ **11.** $x-3$
13. $(3x+5)(x+1)$ **15.** $(2y-1)(y-5)$ **17.** $(3x+2)(x+4)$
19. Prime polynomial **21.** $(6y+5)(y-1)$ **23.** $(2y-7)(y-2)$
25. $(3a+2)(3a-8)$ **27.** $(4x-1)(x-3)$ **29.** $(3y+2)(4y+3)$
31. $(2m-3)(m-7)$ **33.** $-(3a-1)(2a+3)$ **35.** $(8y-11)(y+2)$
37. Prime polynomial **39.** $(8a+1)(a+8)$ **41.** $(3x-1)(2x+9)$
43. $(4y-3)(2y-5)$ **45.** $2(7y-5)(y-2)$ **47.** $4(7a-1)(a+1)$
49. $-2(3b+1)(b-7)$ **51.** $2y(3y+2)(2y+7)$
53. $2a^2(7a-5)(a-2)$ **55.** $xy(2x+3)(x+5)$
57. $2ab(3b-1)(b-7)$ **59.** $(5c-d)(4c-d)$ **61.** $(2x+y)(x-3y)$
63. $(4a-b)(2a-b)$ **65.** $3(3x+2y)(2x-y)$
67. $2(4c-5d)(2c-3d)$ **69.** $3(3u+v)(3u+v)$
71. $3x(7x-3y)(2x+3y)$ **73.** $-5x^2y(3x+y)(2x-3y)$
75. $a(5x+2y)(x-6y)$ **77.** $(m-1)(3m-2)$ **79.** $(4x-3)(2x+1)$
81. $2x(x+3)(7x+1)$ **83.** $(2m-3n)(4m-3n)$ **85.** $(r-1)(7r-2)$
87. d **89.** $-(5t-4)(t+5)$ **91. a.** $(2x^2+25x+72)$ ft^2
b. $(2x+9)(x+8)$ ft^2 **93.** $(2n-5)(2n-1)$; since the difference of
the factors is $(2n-1)-(2n-5) = 2n-1-2n+5 = 4$, the fac-
tors represent two integers that differ by 4 no matter what integer n represents.

Section 6.4 Practices, pp. 465–469
1, p. 465: a. The trinomial is a perfect square. **b.** The trinomial is not
a perfect square. **c.** The trinomial is a perfect square. **d.** The trinomial
is not a perfect square. **e.** The trinomial is a perfect square.
2, p. 466: $(n+10)^2$ **3, p. 467:** $(t-2)^2$ **4, p. 467:** $(5c-4d)^2$
5, p. 467: $(x^2+4)^2$ **6, p. 468: a.** The binomial is a difference of
squares. **b.** The binomial is not a difference of squares. **c.** The binomial is
not a difference of squares. **d.** The binomial is a difference of squares.
7, p. 468: $(y+11)(y-11)$ **8, p. 469:** $(3x+5y)(3x-5y)$
9, p. 469: $(8x^4+9y)(8x^4-9y)$ **10, p. 469:** $16(4+t)(4-t)$

Exercises 6.4, pp. 470–472
1. perfect square trinomial **3.** difference of squares **5.** Perfect square
trinomial **7.** Neither **9.** Perfect square trinomial **11.** Difference of
squares **13.** Difference of squares **15.** Perfect square trinomial
17. Neither **19.** Perfect square trinomial **21.** Neither **23.** $(x-6)^2$
25. $(y+10)^2$ **27.** $(a-2)^2$ **29.** Prime polynomial **31.** $(2a-9)^2$
33. $(7x+2)^2$ **35.** $(6-5x)^2$ **37.** $(m+13n)^2$ **39.** $(15a-b)^2$
41. $(y^2+1)^2$ **43.** $6(x+1)^2$ **45.** $3m(3m-2)^2$ **47.** $4s^2t(t+10)^2$
49. $(m+8)(m-8)$ **51.** $(y+9)(y-9)$ **53.** $(12+x)(12-x)$
55. $(10m+9)(10m-9)$ **57.** Prime polynomial
59. $(1+3x)(1-3x)$ **61.** $(x+2y)(x-2y)$
63. $(10x+3y)(10x-3y)$ **65.** $3k(k+7)(k-7)$
67. $4y^2(y+3)(y-3)$ **69.** $3x^2y(3+y)(3-y)$
71. $2(ab+7)(ab-7)$ **73.** $(16+r^2)(4+r)(4-r)$
75. $5(x^2+4y^2)(x+2y)(x-2y)$ **77.** $(c-d)(x+2)(x-2)$
79. $(x-y)(4+a)(4-a)$ **81.** $(3c+8d)^2$ **83.** $(a+15b^2)(a-15b^2)$
85. $6(3a^2b-1)^2$ **87.** $(9+w^2)(3-w)(3+w)$ **89.** $(6u+5)^2$
91. Difference of squares **93.** $4\pi(r_1+r_2)(r_1-r_2)$ **95.** $4x(1-x)^2$
97. $k(v_2+v_1)(v_2-v_1)$

Section 6.5 Practices, pp. 473–478
1, p. 473: a. Linear **b.** Quadratic **2, p. 474:** $\frac{1}{3}, -5$ **3, p. 475:** $0, -6$
4, p. 475: $-\frac{1}{4}, 3$ **5, p. 476:** $1, -5$ **6, p. 477:** The dimensions of the

frame should be 8 in. by 10 in. **7, p. 478:** The scooter going north has
traveled 12 mi.

Exercises 6.5, pp. 479–481
1. quadratic equation **3.** in standard form **5.** Quadratic **7.** Linear
9. Quadratic **11.** $-3, 4$ **13.** 1 **15.** $0, -\frac{5}{3}$ **17.** $-\frac{1}{2}, 5$ **19.** $-\frac{3}{2}, \frac{3}{2}$
21. $0, \frac{2}{3}$ **23.** $0, 2$ **25.** $0, \frac{1}{5}$ **27.** $-2, -3$ **29.** $7, -8$ **31.** $-\frac{1}{2}, 3$
33. $\frac{2}{3}, -\frac{1}{2}$ **35.** $\frac{1}{6}$ **37.** $-11, 11$ **39.** $\frac{3}{2}$ **41.** $\frac{1}{2}$ **43.** $\frac{1}{3}, -2$ **45.** $-2, 3$
47. $3, 4$ **49.** $2, -4$ **51.** $-\frac{1}{3}, -1$ **53.** $-\frac{1}{2}, -1$ **55.** $0, -5$ **57.** $-\frac{1}{2}, \frac{1}{2}$
59. $-\frac{1}{2}, \frac{1}{2}$ **61.** -3 **63.** $-3, 4$ **65.** $-2, 3$ **67.** $4, -5$ **69.** $-\frac{1}{3}, -4$
71. $2, 4$ **73.** $1, 2$ **75.** $\frac{1}{4}, -3$ **77.** $\frac{1}{2}, -3$ **79.** $\frac{1}{5}, -4$ **81.** 8 **83.** $0, \frac{5}{3}$
85. $0, -\frac{1}{2}$ **87.** There were 15 teams in the league. **89.** One car traveled
6 mi and the other traveled 8 mi. **91.** 200 ft **93.** The average annual
rate of return is 25%. **95. a.** $\frac{2}{105}(105-x)(105+x)$ **b.** It spans from
-105 yd to 105 yd, and so is 210 yd wide.

Chapter 6 Review Exercises, pp. 485–486
1. is; Possible answer: it cannot be factored **2.** is not; Possible answer:
$(2x^2-x-3)$ can be factored further (it is not a prime polynomial)
3. is; Possible answer: it can be expressed as $(x-2y)^2$ **4.** can;
Possible answer: the binomial is a difference of squares **5.** 12
6. $3m^2$ **7.** $3(x-2y)$ **8.** $2pq(8p^2q+9p-2q)$ **9.** $(n-1)(1+n)$
10. $(x-5)(b-2)$ **11.** $r = \frac{d}{t_1+t_2}$ **12.** $x = \frac{c-y}{a-b}$
13. Prime polynomial **14.** Prime polynomial **15.** $(y+6)(y+7)$
16. $(m-2n)(m-5n)$ **17.** $-2(x-2)(x+6)$
18. $3x(x+y)(x-5y)$ **19.** $(3x-1)(x+2)$ **20.** $(5n+3)(n+2)$
21. Prime polynomial **22.** $(3x+4)(2x-3)$
23. $(2a-7b)(a+5b)$ **24.** $-(2a-3)(2a-5)$
25. $3y(3y-1)(y+7)$ **26.** $q(2p+q)(p-2q)$ **27.** $(b-3)^2$
28. $(8+x)(8-x)$ **29.** $(5y-2)^2$ **30.** $(3a+4b)^2$
31. $(9p+10q)(9p-10q)$ **32.** $(2x^4-7)^2$
33. $3(4x^2+y^2)(2x+y)(2x-y)$ **34.** $(x-1)(x+3)(x-3)$
35. $-2, 1$ **36.** $0, 4$ **37.** $0, -6$ **38.** $-\frac{1}{2}$ **39.** $2, 8$ **40.** $-\frac{2}{3}, 1$
41. $-\frac{5}{2}, 1$ **42.** $3, -4$ **43.** $aL(t_2-t_1)$ **44.** $(t_2-t_1)[v-16(t_2+t_1)]$
45. The distance between the two intersections is 2500 ft.
46. The length of the horizontal diagonal of the kite is 32 in.
47. The rocket will reach a height of 18 ft above the launch in $\frac{1}{4}$ sec
and $\frac{9}{2}$ sec, or in 0.25 sec and 4.5 sec. **48.** $h = \frac{2A}{b+B}$

Chapter 6 Posttest, p. 487
1. $3x^2$ **2.** $2y(x-7)$ **3.** $2p(2p-3q)(2p-q)$ **4.** $(a-b)(x-y)$
5. $(n+3)(n-16)$ **6.** $(x+2)(x-4)$ **7.** $-5x(x+1)(x-4)$
8. $(4x-3y)(x+4y)$ **9.** $-3(2x-3)^2$ **10.** $(3x+5y)^2$
11. $(11+2x)(11-2x)$ **12.** $(pq+1)(pq-1)$
13. $(y+2)^2(y-2)^2$ **14.** $-8, 1$ **15.** $\frac{1}{3}, -2$ **16.** $-\frac{3}{2}, 2$
17. $mg(y_2-y_1)$ **18.** 8 m **19.** $2x(2x+55)$ ft^2
20. The weight reaches the ground in $\frac{3}{4}$, or 0.75 sec.

Chapter 6 Cumulative Review, pp. 488–489
1. -3 **2.** $8a+7b+2$ **3.** 8 **4.** 12.5%
5. 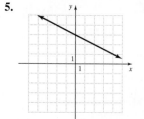 **6.**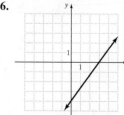

7. No solution **8.** $(-1, 7)$ **9.** n **10.** -31 **11.** $21x^2+x-2$
12. $(3x-7)(x+2)$ **13.** $(4x-3)^2$ **14.** $-3, \frac{1}{4}$ **15.** $-\$0.15$
16. The charge is $(f+8c)$ dollars. **17.** The company must sell 50 shirts
per day in order to break even. **18.** Positive **19.** 1.5×10^8 km
20. $2\pi r(r+h)$

Chapter 7   Pretest, p. 491

1. The expression is undefined when $x = -6$.
2. $\frac{n-2}{3} = \frac{(n-2)\cdot n}{3 \cdot n} = \frac{n^2 - 2n}{3n}$ **3.** $\frac{4x}{y^2}$ **4.** $4a$ **5.** $-\frac{w}{w+6}$ **6.** $\frac{10}{n^2}$
7. $\frac{6(a-2)}{9a(a-2)}$ and $\frac{a}{9a(a-2)}$ **8.** $\frac{5y-2}{y+1}$ **9.** $\frac{15x+2}{12x^2}$ **10.** $\frac{2c+12}{(c-3)(c+3)}$
11. $\frac{3x-1}{(x-1)^2(x+1)}$ **12.** $\frac{4a^2}{3n^2}$ **13.** $\frac{y-4}{5y(y+4)}$ **14.** $\frac{1}{x+4}$ **15.** 5   **16.** $-\frac{2}{3}$
17. $\frac{3}{2}$ and 3   **18.** $\frac{2x + 1500}{x}$ dollars   **19.** The average speed during the
first part of the trip was 65 mph and the average speed during the second
part was 55 mph.   **20.** It takes the photocopier $2\frac{1}{2}$, or 2.5, minutes to
make 30 copies.

Section 7.1   Practices, pp. 493–498

1, p. 493: a. The expression is undefined when n is equal to 4.   **b.** The
expression is undefined when n is equal to -3 or 3.   **2, p. 494: a.** The
expressions are equivalent.   **b.** The expressions are equivalent.
3, p. 495: a. $\frac{2n^2}{m}$ **b.** $-\frac{3x^2}{2}$ **4, p. 496: a.** $\frac{y-4}{2y+3}$ **b.** The expression
cannot be simplified.   **c.** $\frac{1}{3}$ **5, p. 496: a.** $\frac{t-x}{z-3g}$ **b.** $\frac{4(n-1)}{n+2}$ **c.** $\frac{y-1}{y+3}$
6, p. 497: a. -1 **b.** -3 **c.** $-\frac{3}{n+5}$ **d.** $-\frac{3s+2}{s+1}$ **7, p. 498: a.** $\frac{r}{2}$
b. The expression is undefined when $r = 0$.

Exercises 7.1, pp. 499–501

1. rational expression   **3.** equivalent   **5.** $x = 0$ **7.** $y = 2$
9. $x = -2$ **11.** $n = \frac{1}{2}$ **13.** $x = -1$ or $x = 1$ **15.** $x = -4$ or $x = 5$
17. The expressions are equivalent.   **19.** The expressions are equivalent.
21. The expressions are not equivalent.   **23.** The expressions are equivalent.
25. The expressions are not equivalent.   **27.** $\frac{5a^3}{6}$ **29.** $\frac{1}{4x^3}$ **31.** $\frac{3t}{2s^2}$
33. $8a^3b^3$ **35.** $\frac{p}{5q}$ **37.** $\frac{5}{4}$ **39.** $-\frac{8x}{3}$ **41.** $x - 2$ **43.** $\frac{x+1}{2x+3}$ **45.** $\frac{a}{b}$
47. $\frac{2}{3}$ **49.** $t - 1$ **51.** -1 **53.** $-\frac{1}{2}$ **55.** $\frac{b-4}{b+4}$ **57.** $\frac{x+3}{5}$
59. $\frac{a(a+7)}{a-12}$ **61.** $\frac{t+1}{t+2}$ **63.** $-\frac{4d+3}{4d-3}$ **65.** $-\frac{2s}{2s-3}$ **67.** $\frac{2x+1}{2x-1}$
69. $\frac{2y-1}{y(2y+3)}$ **71.** $\frac{2ab}{2b+a}$ **73.** $\frac{m+7n}{2(m+6n)}$ **75.** $\frac{2y^2}{3x}$ **77.** $\frac{3(x-1)}{x+3}$
79. -1 **81.** $\frac{2x+3}{3x}$ **83.** $p = \frac{3}{2}$ **85.** The expressions are equivalent.
87. C in.   **89. a.** $(x^2 + 7x + 10)$ ft^2 **b.** $(x^2 + 2x)$ ft^2
c. $\frac{x^2 + 7x + 10}{x^2 + 2x}, \frac{x+5}{x}$ **d.** $\frac{13}{8}$ **91. a.** $\frac{\pi r_1^2}{\pi r_2^2 - \pi r_1^2}; \frac{r_1^2}{r_2^2 - r_1^2}$ **b.** $\frac{\pi r_2^2 - \pi r_1^2}{\pi r_3^2 - \pi r_2^2}; \frac{r_2^2 - r_1^2}{r_3^2 - r_2^2}$

Section 7.2   Practices, pp. 503–507

1, p. 503: a. $\frac{2n}{7m}$ **b.** $\frac{p}{15q}$ **2, p. 504: a.** $\frac{3}{2t}$ **b.** $\frac{2}{6g-1}$
3, p. 504: a. $2(y+7)(y-2)$ **b.** $-\frac{x+5}{x+1}$ **4, p. 506: a.** $\frac{3a}{2b}$
b. $\frac{4p^2}{q}$ **5, p. 506: a.** $\frac{(x+3)(x+3)}{(x-10)(x+1)}$, or $\frac{(x+3)^2}{(x-10)(x+1)}$ **b.** $\frac{1}{3}$
c. $\frac{y(y+3)}{5(y+2)(y-1)}$ **6, p. 507:** $\frac{Wv^2}{gr}$

Exercises 7.2, pp. 508–510

1. $\frac{1}{4t}$ **3.** $\frac{6}{ab}$ **5.** $\frac{10}{3x^9}$ **7.** $-\frac{14}{x}$ **9.** $\frac{5}{x^3}$ **11.** $\frac{8n-3}{n}$ **13.** $\frac{4}{5}$ **15.** $-\frac{1}{2}$
17. $\frac{3(x+2y)}{4}$ **19.** $\frac{p^2-1}{p^2-4}$ **21.** $\frac{n+3}{n+1}$ **23.** $\frac{2y-1}{2y-5}$ **25.** $\frac{16}{x^4}$ **27.** $3x$
29. $\frac{1}{2}$ **31.** 2   **33.** $\frac{10}{x}$ **35.** $-\frac{3}{t^2}$ **37.** $\frac{6y^2}{x^4}$ **39.** $\frac{(c+3)(c-7)}{(c-5)(c+9)}$ **41.** $\frac{5}{12}$
43. $\frac{1}{2(1-x)}$ **45.** $-p$ **47.** $\frac{x+y}{5x}$ **49.** $-\frac{1}{t+2}$ **51.** $\frac{x+5}{x+6}$
53. $\frac{3(p+2)(p-4)}{(p+5)^2}$ **55.** $-\frac{1}{r^2}$ **57.** $\frac{2}{3-x}$ **59.** $-\frac{1}{6p^2}$ **61.** $\frac{3(y+3)}{4}$
63. $\frac{3x(x-1)}{3x+1}$ **65.** $\frac{A-p}{pr}$ **67.** $\frac{pqB}{10,000}$ dollars   **69.** $\frac{V^2r}{(R+r)^2}$
71. a. $\frac{4r}{3h}$ **b.** $\frac{2r}{h+r}$ **c.** $\frac{2(h+r)}{3h}$

Section 7.3   Practices, pp. 512–521

1, p. 512: a. $\frac{10}{y+2}$ **b.** $\frac{5r}{s}$ **c.** 6   **d.** $3n-1$ **2, p. 513: a.** $\frac{5}{v}$ **b.** $\frac{3t}{5}$
c. $-\frac{p}{3q}$ **3, p. 513: a.** 2   **b.** $2x$ **c.** $-\frac{3}{x-5}$ **4, p. 515: a.** LCD $= 20y$
b. LCD $= 6t^2$ **c.** LCD $= 30x^2y^3$ **5, p. 516: a.** LCD $= 2n(n+1)^2$
b. LCD $= (p+2)(p-1)(p+5)$ **c.** LCD $= (s+2t)^2(s-2t)^2$

6, p. 517: a. $\frac{4}{14p^3}$ and $\frac{7p^2(p+3)}{14p^3}$ **b.** $\frac{(3y-2)(y-3)}{(y+3)(y-3)^2}$ and $\frac{y(y+3)}{(y+3)(y-3)^2}$
7, p. 518: a. $\frac{11}{12p}$ **b.** $\frac{3y-2}{15y^2}$ **8, p. 518: a.** $\frac{9x+6}{x(x+3)} = \frac{3(3x+2)}{x(x+3)}$
b. $\frac{2x-5}{x-1}$ **9, p. 519: a.** $\frac{9x+4}{4(x-4)(x+4)}$ **b.** $\frac{x-9}{(x+1)(x-1)}$
10, p. 520: a. $\frac{-3y^2 - 12y - 3}{(y+3)(y+2)(y+1)} = \frac{-3(y^2 + 4y + 1)}{(y+3)(y+2)(y+1)}$
b. $\frac{8x^3 + 3x^2 - 20x + 5}{20x(x+1)}$ **11, p. 521:** $\frac{100(C_1 - C_0)}{C_0}$

Exercises 7.3, pp. 522–525

1. $\frac{4a}{3}$ **3.** t **5.** $\frac{11}{x}$ **7.** $\frac{5}{7y}$ **9.** $\frac{3x}{y}$ **11.** $-\frac{p}{5q}$ **13.** $\frac{9}{x+1}$
15. $-\frac{4}{x+2}$ **17.** $\frac{a+1}{a+3}$ **19.** $\frac{5x+1}{x-8}$ **21.** $\frac{4x+1}{5x+2}$ **23.** 3
25. 4   **27.** $\frac{x-4}{x^2 - 4x - 2}$ **29.** 1   **31.** $\frac{-x+7}{3x^2 - x + 2}$
33. LCD $= 15(x+2)$ **35.** LCD $= (p-3)(p+8)(p-8)$
37. LCD $= t(t+3)(t-3)$ **39.** LCD $= (t+2)(t+5)(t-5)$
41. LCD $= (3s-2)(s-3)(s+2)$ **43.** $\frac{4x}{12x^2}$ and $\frac{15}{12x^2}$
45. $\frac{35b}{14a^2b}$ and $\frac{2a(a-3)}{14a^2b}$ **47.** $\frac{8(n+1)}{n(n+1)^2}$ and $\frac{5n}{n(n+1)^2}$
49. $\frac{3n(n-1)}{4(n+1)(n-1)}$ and $\frac{8n}{4(n+1)(n-1)}$
51. $\frac{2n(n-3)}{(n+1)(n+5)(n-3)}$ and $\frac{3n(n+1)}{(n+1)(n+5)(n-3)}$
53. $\frac{13}{6x}$ **55.** $\frac{4-5x}{6x^2}$ **57.** $\frac{-2y+4x}{3x^2y^2}$, or $-\frac{2(y-2x)}{3x^2y^2}$ **59.** $\frac{2x}{(x+1)(x-1)}$
61. $\frac{p+39}{21}$ **63.** $3x-5$ **65.** $\frac{a+6}{6a^2}$ **67.** $\frac{a^2+1}{a-1}$ **69.** -1
71. $\frac{-8x-2}{x(x+1)}$, or $-\frac{2(4x+1)}{x(x+1)}$ **73.** $\frac{7x-6}{x-4}$ **75.** $\frac{11x-6}{(x-1)(x+2)}$
77. $\frac{n+20}{3(n-3)(n+5)}$ **79.** $\frac{3t-4}{(t+5)(t-5)}$ **81.** $\frac{2x^2+5x-5}{(x+1)^2(x+3)}$
83. $\frac{6t+5}{(2t+3)(t-1)}$ **85.** $\frac{12x^3 + 17x^2 - 2}{3x(x-1)(x+1)}$ **87.** $\frac{5y^2 - 28y + 4}{(3y-1)(y-4)}$
89. $\frac{-9a^2 - 11a + 32}{4(a+3)^2}$ **91.** $\frac{4q(p+3)}{24p^2q}$ and $\frac{15p}{24p^2q}$ **93.** $\frac{3(3x+2y)}{x^2y^2}$
95. $-\frac{3}{c(c+1)}$ **97.** $\frac{b^2 - 4b - 13}{(b-3)^2(b+1)}$ **99.** $-\frac{5}{m+n}$ **101.** $\frac{r-1}{r+2}$
103. $\frac{2vt + gt^2}{2}$ **105.** $\frac{1000r}{(1+r)^2}$ dollars   **107.** The trip took $\frac{30}{r}$ hr.
109. $\frac{3+0.1x}{x}$ dollars

Section 7.4   Practices, pp. 528–532

1, p. 528: a. $\frac{3}{5x^3}$ **b.** $\frac{8}{x+2}$ **2, p. 529:** $\frac{2n-1}{2n+1}$ **3, p. 530: a.** $2x^2$
b. $\frac{y+3}{2y^3}$ **4, p. 531: a.** $\frac{4y^2 + y}{4y^2 - 1}$ **b.** $\frac{b-a}{10ab}$ **5, p. 532:** $\frac{3abc}{bc + ac + ab}$

Exercises 7.4, pp. 533–535

1. $\frac{2}{x}$ **3.** $\frac{a+1}{a-1}$ **5.** $\frac{x(3x+1)}{3x^2 - 1}$ **7.** $\frac{1}{3d(d+3)}$ **9.** $\frac{x-2y}{3x}$ **11.** $\frac{10-y}{5(5-y)}$
13. $\frac{x+2}{x+3}$ **15.** $\frac{3y+4}{5y+4}$ **17.** $\frac{xy^2}{16(y+1)}$ **19.** $\frac{3(x^2 - 2x - 2)}{x^2(x+1)}$ **21.** $\frac{m+2}{3(m-1)}$
23. $\frac{4(u+2)}{u^2(u+1)}$ **25.** $\frac{4y+3}{3y(2y+1)}$ **27.** $\frac{2VR(R+1)}{2R+1}$ **29.** $\frac{9E}{I}$ **31.** $\frac{2ab}{a+b}$ mph
33. $\frac{w}{(1+\frac{h}{6400})^2} = \frac{w}{(\frac{6400+h}{6400})^2} = \frac{w}{\frac{(6400+h)^2}{6400^2}} = \frac{6400^2 w}{(6400+h)^2}$

Section 7.5   Practices, pp. 536–542

1, p. 536: $\frac{1}{2}$ **2, p. 537:** 2, 5   **3, p. 539:** 2   **4, p. 539:** 3
5, p. 540: Working together, it will take both pumps $3\frac{3}{4}$ hr
(or 3 hr 45 min) to fill the tank.   **6, p. 541:** The speed of the propeller
plane was 150 mph.   **7, p. 542: a.** $x = \frac{500}{D-50}$ **b.** $1.25 per unit

Exercises 7.5, pp. 543–544

1. -3 **3.** $\frac{1}{2}$ **5.** 1   **7.** 4   **9.** 14   **11.** 3   **13.** 1, 2   **15.** 6   **17.** 2
19. -3 **21.** $-\frac{7}{2}$ **23.** 2, -8 **25.** 4, -1 **27.** $-\frac{5}{2}$ **29.** $-\frac{1}{3}$, 3   **31.** No
solution   **33.** It will take them 18 min to clean the attic.   **35.** It would
take the clerical worker 15 hr to finish the job working alone.   **37.** The
speed on the dry road was 60 mph.   **39.** $D = \frac{P}{Lp}$ **41. a.** The evacuation
would take $\frac{x^2 + 3x}{2x + 3}$ hr.   **b.** It would take 11 hr 12 min.

Section 7.6 Practices, pp. 547–551

1, *p. 547:* 8 **2**, *p. 548:* 40 lb of sodium hydroxide are needed to neutralize 49 lb of sulfuric acid. **3**, *p. 549:* She would be paying $225 less if she had a $75,000 mortgage at the same interest rate. **4**, *p. 550:* The length of $\overline{DE}$ is 4 in. **5**, *p. 550:* The speed of the plane in still air is 500 mph.

Exercises 7.6, pp. 552–554

1. 8 **3.** 80 **5.** 24 **7.** $-\frac{2}{3}$ **9.** 3 **11.** 5 **13.** $-4, 4$ **15.** $8, -4$ **17.** $-\frac{6}{5}, 2$ **19.** No solution **21.** $7, -3$ **23.** 11 **25.** $\frac{2}{3}$ **27.** It would take $1\frac{1}{4}$ min (or 1 min 15 sec) to print a 25-page report. **29.** It will take 16 gal of gas to drive 120 mi. **31.** The cyclist's speed was 20 mph. **33.** The speed of the bus is 50 mph and the speed of the train is 80 mph. **35.** $\overline{AB} = 8$ ft **37.** The height of the tree is 18 ft. **39.** There are 8 women at the party.

Chapter 7 Review Exercises, pp. 559–561

1. is not; Possible answer: the denominator is not 0 when $\frac{3}{2}$ is substituted for x **2.** is not; possible answer: the expression can be simplified by dividing out the common factor a **3.** is; possible answer: it is the product of the highest power of each factor in either denominator **4.** is; possible answer: it contains a rational expression in its denominator **5.** is not; possible answer: $a = 3$ is a solution of the original equation **6.** is; possible answer: the cross products, $8 \cdot 25$ and $10 \cdot 20$, are equal **7. a.** $x = -1$ **b.** $x = 3$ and $x = -2$ **8. a.** Equivalent **b.** Equivalent
9. $\frac{3}{5m}$ **10.** $\frac{5n - 6}{3n + 2}$ **11.** $-\frac{x + 4}{x + 2}$ **12.** $\frac{2x + 5}{3x - 1}$ **13.** $\frac{6n^2}{pm}$ **14.** $\frac{1}{2}$
15. $\frac{x - 5}{2x + 5}$ **16.** -1 **17.** $\frac{1}{2y}$ **18.** $\frac{5(7m - 10)}{7(m - 10)}$ **19.** $\frac{y}{5(x + 6)}$ **20.** $\frac{x + 7}{3x + 2}$
21. $\frac{4x}{20x^2}$ and $\frac{3}{20x^2}$ **22.** $\frac{4(n + 4)}{(n - 1)(n + 4)}$ and $\frac{n(n - 1)}{(n - 1)(n + 4)}$
23. $\frac{x + 1}{3(x + 3)(x + 1)}$ and $\frac{3x}{3(x + 3)(x + 1)}$
24. $\frac{2(x + 2)}{(3x + 1)(x - 2)(x + 2)}$ and $-\frac{3x + 1}{(3x + 1)(x - 2)(x + 2)}$
25. 2 **26.** 4 **27.** $\frac{y + 4}{2y(2y - 1)}$ **28.** $\frac{n^2 + 3n - 6}{3n(n + 5)}$ **29.** $\frac{8x + 13}{(x - 3)(x + 3)}$
30. $\frac{2y + 5}{(2 + y)(2 - y)}$ **31.** $\frac{8m - 8}{(m + 1)(m - 3)}$, or $\frac{8(m - 1)}{(m + 1)(m - 3)}$
32. $\frac{-x^2 - 3x - 11}{(x + 3)(x - 4)}$ **33.** $\frac{x^2 - 9x + 2}{(x + 2)^2(x - 4)}$ **34.** $\frac{2n^2 + 7n + 11}{(2n - 1)(n - 1)(n + 3)}$
35. $\frac{7}{6x}$ **36.** $\frac{y}{y + 9}$ **37.** 2 **38.** $\frac{4x + 1}{2x - 3}$ **39.** 7 **40.** $\frac{1}{2}$ **41.** -3
42. $3, -1$ **43.** -144. 2 **45.** 45 **46.** 33 **47.** $\frac{1}{8}$ **48.** $4, -9$
49. $\frac{0.72x + 200}{x}$ dollars **50.** The total cost of the car rental is $200.
51. $\frac{2rs}{s + r}$ **52.** It will take 30 min to fill the tub. **53.** The family drove 100 miles at 50 mph. **54.** $\frac{2x + 1}{x(x + 1)}$ of the job will be done in an hour.
55. She should anticipate spending about $41,333. **56.** $\frac{3n^2 + 6n + 2}{n(n + 1)(n + 2)}$

Chapter 7 Posttest, pp. 562–563

1. The expression is undefined when $x = 8$.
2. $-\frac{3y - y^2}{y^2} = -\frac{y(3 - y)}{y^2} = \frac{\overset{1}{\cancel{y}}(3 - y)}{\overset{y}{\cancel{y^2}}} = \frac{-(3 - y)}{y} = \frac{y - 3}{y}$ **3.** $\frac{5a^2}{4b}$
4. $\frac{x}{y}$ **5.** $\frac{3(b - 3)}{b - 7}$ **6.** $\frac{1}{x + 3}$ **7.** $\frac{4(4n - 1)}{4(n + 8)(n - 2)}$, $\frac{8(n - 2)}{4(n + 8)(n - 2)}$, and $\frac{n(n + 8)}{4(n + 8)(n - 2)}$ **8.** 2 **9.** $\frac{3y + 1}{2y(y - 4)}$ **10.** $\frac{4d + 14}{(d - 3)(d + 2)} = \frac{2(2d + 7)}{(d - 3)(d + 2)}$
11. $\frac{2x^2 + 6x + 10}{(2x + 1)(x - 2)(x + 2)}$, or $\frac{2(x^2 + 3x + 5)}{(2x + 1)(x - 2)(x + 2)}$ **12.** $\frac{1}{18n(n - 1)}$
13. $\frac{a + 5}{a + 4}$ **14.** $\frac{2(x + 4)(x + 2)}{x - 1}$ **15.** 6 **16.** -14 **17.** $2, -3$
18. It would take $\frac{30m}{31}$ min to download the file. **19.** Working alone, the newer machine can process 1000 pieces of mail in 30 min and the older machine can process 1000 pieces of mail in 60 min. **20.** The height of the tree is 60 m.

Chapter 7 Cumulative Review, pp. 564–565

1. $>$ **2.** $2y - 3$ **3.** $h = \frac{2A}{b + B}$
4. $n > -3$; ⟵———◯——————⟶
  $-5\ -4\ -3\ -2\ -1\ 0\ 1\ 2\ 3\ 4\ 5$
5. $y = -3x + 2$

6.
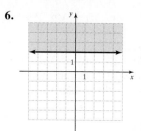

7. $(3, 5)$ **8.** $x^2 - 5x + 6$
9. $9k^2 - 36kl + 36l^2$
10. $3x(2x + 1)(x - 1)$

11. $-5, 2$ **12.** $(10y + 9)(10y - 9)$ **13.** $\frac{2x}{x + 3y}$ **14.** $-3, 2$
15. $2^2 p$ **16.** The investments were $15,000 in Stock A and $7500 in Stock B. **17.** 9 mi

18. a.

[graph: Per Capita Consumption of Fruit (in pounds), y-axis 0 to 140; Number of Years after 1970, x-axis 5 10 15 20 25 30]

b. 0.6; the annual per capita consumption of fruit increased by 0.6 lb per year **c.** The y-intercept is $(0, 100)$; 100 represents the number of pounds of fruit consumed per capita in 1970 **19.** It will take the Reston bus 4 hr to overtake the Arlington bus.
20. $(4.7x^3 - 50.1x^2 + 68.2x - 383.7)$ billions of dollars

CHAPTER 8

Chapter 8 Pretest, p. 567

1. 9 **2.** $-3\sqrt{3}$ **3.** $3a\sqrt{5}$ **4.** $\frac{\sqrt{x}}{8}$ **5.** $4\sqrt{2}$ **6.** $12\sqrt{3}$ **7.** $5x\sqrt{x}$
8. $3\sqrt{2}$ **9.** $2xy^2\sqrt{5}$ **10.** $\sqrt[6]{6}$ **11.** $n + 2\sqrt{n}$ **12.** $-1 + 3\sqrt{3}$
13. $\frac{\sqrt{30x}}{6}$ **14.** $2x\sqrt{5}$ **15.** $\frac{8\sqrt{2} + \sqrt{14}}{2}$ **16.** 36 **17.** 1, 3 **18.** The velocity of the car is 20 m/sec **19.** The gymnast covers $12\sqrt{2}$ m, or approximately 17.0 m in the tumbling sequence. **20.** $L = \frac{s^2}{30f}$

Section 8.1 Practices, pp. 569–574

1, *p. 569:* **a.** 2 **b.** -35 **2**, *p. 569:* 3.162 **3**, *p. 570:* **a.** 6 **b.** 5 **c.** 7 **d.** 1 **4**, *p. 570:* **a.** x^2 **b.** $8t^5$ **c.** $-11xy$ **5**, *p. 571:* **a.** $6\sqrt{2}$ **b.** $4\sqrt{10}$ **c.** $-\frac{\sqrt{3}}{3}$ **6**, *p. 572:* **a.** $x\sqrt{x}$ **b.** $3n^2\sqrt{2}$ **c.** $-5b\sqrt{2a}$
7, *p. 573:* **a.** $\frac{1}{4}$ **b.** $\frac{y}{y^3}$ **c.** $\frac{x^2}{y^3}$ **8**, *p. 573:* **a.** $\frac{\sqrt{3}}{4}$ **b.** $\frac{\sqrt{2y}}{7}$ **c.** $\frac{x^3y\sqrt{5x}}{2}$
9, *p. 574:* 1000 lb

Exercises 8.1, pp. 575–577

1. square root **3.** irrational **5.** take the square root **7.** product **9.** 6
11. 1 **13.** -10 **15.** 21 **17.** 2.236 **19.** 4.243 **21.** 16 **23.** 11
25. $5x$ **27.** 2 **29.** 9 **31.** n^4 **33.** $7y$ **35.** $3x^2$ **37.** $5xy^5$
39. $4\sqrt{2}$ **41.** $-6\sqrt{3}$ **43.** $18\sqrt{3}$ **45.** $\frac{\sqrt{3}}{2}$ **47.** $x\sqrt{11}$
49. $n^2\sqrt{n}$ **51.** $2x\sqrt{5x}$ **53.** $-2p\sqrt{3q}$ **55.** $9xy^2\sqrt{10x}$ **57.** $\frac{2}{5}$
59. $-\frac{1}{2}$ **61.** $\frac{9}{n^3}$ **63.** $\frac{x^2}{y}$ **65.** $-\frac{\sqrt{3}}{2}$ **67.** $\frac{\sqrt{5n}}{4}$ **69.** $\frac{xy^3\sqrt{3}}{2}$ **71.** $-\frac{3x^3\sqrt{3y}}{4}$
73. $\frac{6}{y^3}$ **75.** -12 **77.** $10\sqrt{3}$ **79.** $-2a^2b\sqrt{10b}$ **81.** r^3
83. a. $m = \sqrt{a \cdot b}$ **b.** $m = 4$ **85. a.** It takes the object 2 sec to reach the ground. **b.** No; $\sqrt{\frac{40}{5}} = \sqrt{8} = 2\sqrt{2}$, which is not equal to $2 \cdot 2$. **87.** The car was traveling at a speed of 60 mph at the time of the accident. **89.** The distance between the towns is $\sqrt{146}$ mi, or about 12 mi.

Section 8.2 Practices, pp. 579–581

1, *p. 579:* **a.** $6\sqrt{5}$ **b.** $11\sqrt{n}$ **c.** $11\sqrt{t^2 - 3}$ **d.** Cannot be simplified because the radicals are not like. **2**, *p. 580:* **a.** $12\sqrt{2}$ **b.** $-6\sqrt{3}$ **c.** $-9\sqrt{t}$ **d.** $12b^2\sqrt{a}$ **3**, *p. 581:* The length of the front yard is $20\sqrt{2}$ m, or approximately 28 m.

Exercises 8.2, pp. 582–583

1. like **3.** unlike **5.** $8\sqrt{7}$ **7.** $-5\sqrt{2}$ **9.** Cannot be combined
11. $-13\sqrt{11}$ **13.** $9t\sqrt{3}$ **15.** $23\sqrt{x}$ **17.** Cannot be combined
19. $5\sqrt{x + 1}$ **21.** $-2\sqrt{2}$ **23.** $11\sqrt{2}$ **25.** $3\sqrt{3}$ **27.** $26\sqrt{3}$

29. $11\sqrt{2} - 6\sqrt{3}$ 31. $14\sqrt{y}$ 33. $(3-4x)\sqrt{x}$ 35. $14\sqrt{p}$ 37. $\sqrt{x}$
39. $(-5x^2y^2 + x^3y)\sqrt{2x}$, or $(x-5y)x^2y\sqrt{2x}$ 41. $2\sqrt{3} + \sqrt{6}$
43. $4\sqrt{7p}$ 45. $9\sqrt{3}$ 47. **a.** Each missing side measures $\sqrt{61}$ units.
b. The perimeter of the triangle is $10 + 2\sqrt{61}$ units. 49. **a.** $3\sqrt{10}$ in.
and $2\sqrt{10}$ in. **b.** The side of the larger tile is $\sqrt{10}$ in. longer. 51. The
manufacturer will charge $90\sqrt{10}$ more dollars for 4000 machine parts
than for 1000 machine parts.

Section 8.3 Practices, pp. 586–593

1, *p. 586:* **a.** 5 **b.** $2y^3$ **c.** $9t + 9$ 2, *p. 586:* **a.** $\sqrt{70}$ **b.** $-108\sqrt{2}$
c. $4y^3$ 3, *p. 587:* **a.** $9\sqrt{2} - 4\sqrt{3}$ **b.** $\sqrt{ab} + 3\sqrt{a}$
4, *p. 587:* **a.** $2 + 2\sqrt{3}$ **b.** $3x + 4\sqrt{x} - 4$ 5, *p. 588:* **a.** -2
b. $p - q$ 6, *p. 588:* **a.** $2 + 2b\sqrt{2} + b^2$, or $b^2 + 2b\sqrt{2} + 2$
b. $x - 12\sqrt{x} + 36$ 7, *p. 589:* It takes Mercury about 93 twenty-four-
hour days to revolve around the Sun. 8, *p. 589:* **a.** $\sqrt{7}$ **b.** $2x^2$ **c.** $\frac{1}{20y^2}$
9, *p. 590:* **a.** $\frac{m^2\sqrt{m}}{n^2}$ **b.** $\frac{y^3}{5x}$ **c.** $\frac{a\sqrt{5}}{3}$ 10, *p. 590:* **a.** $\frac{\sqrt{2}}{2}$ **b.** $\frac{\sqrt{5s}}{s}$ **c.** $\frac{7r^2\sqrt{3}}{6}$
11, *p. 591:* **a.** $\frac{\sqrt{6}}{6}$ **b.** $\frac{\sqrt{5n}}{10}$ 12, *p. 592:* **a.** $\frac{\sqrt{15} - \sqrt{3}}{3}$ **b.** $\frac{\sqrt{bc} + 2\sqrt{b}}{b}$
13, *p. 592:* **a.** $\frac{24 + 8\sqrt{2}}{7}$ **b.** $\frac{a\sqrt{b} - a\sqrt{5}}{b - 5}$ 14, *p. 593:* $\frac{295\sqrt{2}}{6}$ beats per
minute, or approximately 70 beats per minute

Exercises 8.3, pp. 594–597

1. are different from one another 3. bring the radicands under a single
radical sign 5. 21 7. 15 9. $3n$ 11. $5y$ 13. $16x - 16$
15. $4t + 20$ 17. $3\sqrt{6}$ 19. $-70\sqrt{2}$ 21. $4x^2$ 23. $r\sqrt{15}$
25. $10\sqrt{xy}$ 27. $3 - \sqrt{3}$ 29. $x - 7\sqrt{x}$ 31. $4\sqrt{ab} + \sqrt{a}$
33. $11 + 5\sqrt{5}$ 35. -2 37. $118 - 11\sqrt{3}$ 39. $3n + 14\sqrt{n} - 5$
41. 33 43. 13 45. $x - 4$ 47. $a - b$ 49. $3x - y$
51. $2 - 2x\sqrt{2} + x^2$ 53. $x - 2\sqrt{x} + 1$ 55. $\sqrt{5}$ 57. $\frac{1}{5}$ 59. $2a$
61. $\frac{4}{3y^2}$ 63. $\frac{a^2}{b^3}$ 65. $\frac{4x^6}{y^4}$ 67. $\frac{x^5\sqrt{5}}{6}$ 69. $\frac{2\sqrt{3}}{3}$ 71. $\frac{\sqrt{5y}}{y}$ 73. $\frac{\sqrt{22}}{11}$
75. $\frac{x\sqrt{5}}{5}$ 77. $\frac{\sqrt{2t}}{2}$ 79. $\frac{\sqrt{2a}}{2}$ 81. $\frac{\sqrt{15} + 2\sqrt{3}}{3}$ 83. $\frac{\sqrt{mn} - \sqrt{m}}{m}$
85. $\frac{12 - 3\sqrt{6}}{2}$ 87. $4 + \sqrt{5}$ 89. $2\sqrt{5} + 2\sqrt{3}$ 91. $\frac{a\sqrt{b} + a\sqrt{3}}{b - 3}$
93. $\frac{\sqrt{cd} - 6\sqrt{d}}{d}$ 95. $9a^2$ 97. $\frac{8s^4}{t^6}$ 99. $\frac{\sqrt{2c}}{12}$ 101. $\sqrt{r} + 3r$
103. $7m - n$ 105. The standard deviation of the first data set is $\sqrt{2}$,
or approximately 1.4, times that of the second. 107. The distance from
$(0,0)$ to $(3,5)$ is $\sqrt{34}$, and the distance from $(0,0)$ to $(6,10)$ is $2\sqrt{34}$.
109. It will take the hailstone $\frac{5\sqrt{5}}{2}$ sec, or about 5.6 sec to drop 500 ft.
111. $\frac{\sqrt{\pi h V}}{\pi h}$

Section 8.4 Practices, pp. 598–604

1, *p. 598:* 16 2, *p. 599:* No solution 3, *p. 599:* 9 4, *p. 600:* 2, 3
5, *p. 600:* 4, 5 6, *p. 601:* A radius of 1000 ft will permit a maximum
safe speed of 50 mph. 7, *p. 601:* $h = \frac{2}{3}d^2$

Exercises 8.4, pp. 602–603

1. radical equation 3. isolate 5. 9 7. 32 9. No solution 11. 64
13. 6 15. 3 17. 2 19. 3 21. No solution 23. 5 25. 0
27. -2 29. 8 31. No solution 33. $-1, 2$ 35. 8, 17
37. No solution 39. 3 41. 4, -4 43. 64 45. 4 47. No solution
49. Its power is 2500 watts. 51. A mass of $\frac{8}{\pi^2}$ g will produce a period of
2 sec. 53. **a.** $A = \pi(gt + b)^2$ **b.** $A = 36\pi$, or approximately 113 in^2

Chapter 8 Review Exercises, pp. 607–609

1. is; Possible answer: it is the square root of a number that is not a perfect
square 2. is; Possible answer: the square of the square root of a nonnegative number is that number 3. is not; Possible answer: the radicand 72
is divisible by the perfect square 36 4. is not; Possible answer: a radical
is in the denominator 5. is not; Possible answer: the radicand does not
contain a variable 6. is; Possible answer: squaring can eliminate the
radical resulting in a simpler equation 7. -7 8. 6 9. $7x$ 10. $2\sqrt{7}$

11. $-9\sqrt{2}$ 12. $4x\sqrt{2x}$ 13. $\frac{3}{5}$ 14. $-\frac{\sqrt{3t}}{4}$ 15. $\frac{12}{x^{50}}$ 16. $10a^2b\sqrt{ab}$
17. $3\sqrt{5}$ 18. $8\sqrt{n}$ 19. $x\sqrt{3}$ 20. $-7\sqrt{3}$ 21. $5x\sqrt{x}$ 22. $7\sqrt{2a}$
23. $\sqrt{15}$ 24. $4n$ 25. $5ab^3\sqrt{2a}$ 26. $x - 4\sqrt{x}$ 27. 6
28. $2y - \sqrt{y} - 3$ 29. $y + 10\sqrt{y} + 25$ 30. 3 31. $\frac{1}{2}$ 32. 4
33. $2a\sqrt{3a}$ 34. $\frac{2\sqrt{11}}{11}$ 35. $\frac{\sqrt{2x}}{2}$ 36. 9 37. $\frac{5\sqrt{7} + 5}{3}$ 38. 64
39. $\frac{15}{2}$ 40. 5 41. 8 42. The area of the city block is 5000 ft^2.
43. The astronaut can see approximately 2 mi. 44. No; an 8-in.
screwdriver will not fit diagonally in the box. 45. $(4\sqrt{29} + 4\sqrt{74})$ in.
46. $r = \frac{\sqrt{S\pi}}{2\pi}$ 47. **a.** $D = \frac{2\pi r}{2\sqrt{\pi(\pi r^2)}} = \frac{2\pi r}{2\pi r} = 1$ **b.** $D = \frac{L\sqrt{\pi A}}{2\pi A}$
c. $A = \frac{L^2}{4\pi(1.58)^2}$ 48. $\frac{\sqrt{15V}}{15}$ in.

Chapter 8 Posttest, p. 610

1. 12 2. $-9\sqrt{5}$ 3. $4x\sqrt{2xy}$ 4. $\frac{\sqrt{5n}}{4}$ 5. $3\sqrt{x}$ 6. $-15\sqrt{2}$
7. $10t\sqrt{t}$ 8. $2\sqrt{6}$ 9. $5x^2y\sqrt{x}$ 10. 5 11. $5y - \sqrt{y}$
12. $19 + 8\sqrt{3}$ 13. $\frac{\sqrt{10p}}{5}$ 14. $2x\sqrt{6x}$ 15. $1 - 2\sqrt{2}$ 16. 7 17. 2
18. The windchill temperature is 28°F. 19. The length of the ladder is
$6\sqrt{17}$ ft. 20. A lookout must climb 25 ft to see a ship 6 mi away.

Chapter 8 Cumulative Review, pp. 611–612

1. $-\frac{2}{15}$ 2. \$1590 3. $c = 45$ 4. $\frac{2}{3}$
5.

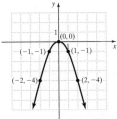

Slope: -5

6. **a.** $m = \frac{2}{3}$; $(3, 0)$; $(0, -2)$ **b.** / 7. The system has one
solution. 8. $16x^3 - 4x^2 - 12x + 3$ 9. $(y - 4)(y - 8)$
10. $3x(x + 4)^2$ 11. $\frac{10c + 6}{(c-2)(c+1)}$, or $\frac{2(5c + 3)}{(c-2)(c+1)}$ 12. $-3, 6$
13. -2 14. 13 15. The speed is 343 m/sec when the temperature is
20°C. 16. 4×10^1 17. $(4x^3 - 50x^2 + 68x - 384)$ billion dollars
18. $\frac{7x}{6(x+1)}$ 19. **a.** Working together, the server and the computer
would take 24 hr. **b.** It took 0.008 hr to process the sequence. 20. The
distance from the camp to the waterfall is $100\sqrt{514}$ m, or approximately
2300 m.

CHAPTER 9

Chapter 9 Pretest, pp. 614–615

1. $3\sqrt{2}, -3\sqrt{2}$ 2. $r = \frac{1}{2}\sqrt{\frac{A}{\pi}} = \frac{\sqrt{A\pi}}{2\pi}$ 3. 36 4. $2 + \sqrt{2}, 2 - \sqrt{2}$
5. $\frac{1 + \sqrt{11}}{2}, \frac{1 - \sqrt{11}}{2}$ 6. 11, 1 7. $-3, -4$ 8. $3 + \sqrt{5}, 3 - \sqrt{5}$
9. $\sqrt{3}, -\sqrt{3}$ 10. $\frac{5 + 3\sqrt{5}}{5}, \frac{5 - 3\sqrt{5}}{5}$ 11. $-2 + 3\sqrt{2}, -2 - 3\sqrt{2}$
12.

13. vertex: $(3, -2)$; axis of symmetry: $x = 3$

14.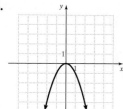

15.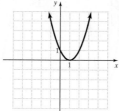

16.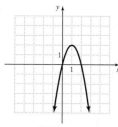

17. $v = \sqrt{2as}$ **18.** 3-in. by 3-in. squares should be cut from the corners to meet the specifications. **19.** The speed of the plane in still air is 150 mph. **20.** The ball reaches a maximum height of 41 ft.

Section 9.1   Practices, pp. 616–620

1, p. 616: a. $10, -10$ **b.** $3\sqrt{2}, -3\sqrt{2}$ **c.** $\frac{5}{9}, -\frac{5}{9}$ **2, p. 617: a.** $1, -5$ **b.** $4 + \sqrt{5}, 4 - \sqrt{5}$ **c.** $\frac{1 + 3\sqrt{6}}{4}, \frac{1 - 3\sqrt{6}}{4}$ **3, p. 618:** $-4 + \sqrt{2}$, $-4 - \sqrt{2}$ **4, p. 618: a.** $c = \sqrt{\frac{E}{m}} = \frac{\sqrt{Em}}{m}$ **b.** $a = \sqrt{c^2 - b^2}$
5, p. 619: a. $r = \frac{\sqrt{2A}}{\pi}$ **b.** The radius was about 12 mi.
6, p. 620: a. $d = 4 + \sqrt{B}$ **b.** 8 in.

Exercises 9.1, pp. 621–623
1. quadratic equation **3.** divide both sides by a **5.** $3, -3$
7. $\sqrt{2}, -\sqrt{2}$ **9.** $\frac{1}{2}, -\frac{1}{2}$ **11.** $2, -2$ **13.** $\sqrt{7}, -\sqrt{7}$ **15.** $2\sqrt{3}, -2\sqrt{3}$
17. $4\sqrt{5}, -4\sqrt{5}$ **19.** $\sqrt{6}, -\sqrt{6}$ **21.** $\sqrt{17}, -\sqrt{17}$ **23.** $\frac{3}{2}, -\frac{3}{2}$
25. No real solution **27.** $\frac{\sqrt{3}}{2}, -\frac{\sqrt{3}}{2}$ **29.** $1, -5$ **31.** $14, 0$
33. $-6 + \sqrt{5}, -6 - \sqrt{5}$ **35.** $\frac{17}{4}, \frac{23}{4}$ **37.** $7, 1$ **39.** $5 + \sqrt{3}, 5 - \sqrt{3}$
41. $-1 + 2\sqrt{2}, -1 - 2\sqrt{2}$ **43.** $\frac{1}{3}, -1$ **45.** $\frac{-5 + \sqrt{3}}{4}, \frac{-5 - \sqrt{3}}{4}$
47. $\frac{7 + 2\sqrt{5}}{2}, \frac{7 - 2\sqrt{5}}{2}$ **49.** $10 + 2\sqrt{10}, 10 - 2\sqrt{10}$
51. No real solution **53.** $-8 + \sqrt{5}, -8 - \sqrt{5}$
55. $3 + 2\sqrt{6}, 3 - 2\sqrt{6}$ **57.** $\sqrt{5}, -\sqrt{5}$ **59.** $x = \pm\sqrt{\frac{b}{a}}$ **61.** $v = \pm\frac{2\pi}{\sqrt{Kr}}$
63. $y = \pm\frac{5\sqrt{x^2 - 16}}{4}$ **65.** $\frac{6 + 2\sqrt{11}}{5}, \frac{6 - 2\sqrt{11}}{5}$ **67.** $\frac{\sqrt{3}}{5}, -\frac{\sqrt{3}}{5}$
69. $-2 + 2\sqrt{3}, -2 - 2\sqrt{3}$ **71.** $4 + 3\sqrt{7}, 4 - 3\sqrt{7}$ **73.** They can hear each other for 4 min. **75. a.** $H = 0.95h$ **b.** The new radius is approximately $0.97r$. **77.** $y = \pm\sqrt{r^2 - x^2}$ **79.** $d = 45\sqrt{\frac{2}{l}} = \frac{45\sqrt{2l}}{l}$

Section 9.2   Practices, pp. 626–630
1, p. 626: a. 36 **b.** $\frac{25}{4}$ **2, p. 627: a.** $3, -7$ **b.** $\frac{-7 + \sqrt{29}}{2}, \frac{-7 - \sqrt{29}}{2}$
3, p. 628: a. $4 + 2\sqrt{5}, 4 - 2\sqrt{5}$ **b.** $\frac{1 + \sqrt{2}}{2}, \frac{1 - \sqrt{2}}{2}$ **4, p. 629:** $\frac{4}{3}, -\frac{1}{3}$
5, p. 630: a. The length of the table is $2w - 1$. **b.** The length is 8 ft and the width is 4.5 ft.

Exercises 9.2, pp. 631–632
1. 9 **3.** 1 **5.** $\frac{25}{4}$ **7.** $\frac{1}{4}$ **9.** $0, -4$ **11.** $5 + 2\sqrt{6}, 5 - 2\sqrt{6}$
13. $1, -15$ **15.** $3 + \sqrt{13}, 3 - \sqrt{13}$ **17.** No real solution
19. $\frac{1 + \sqrt{13}}{2}, \frac{1 - \sqrt{13}}{2}$ **21.** $2, -6$ **23.** $-1 + \frac{\sqrt{66}}{6}, -1 - \frac{\sqrt{66}}{6}$
25. $\frac{5 + 3\sqrt{2}}{2}, \frac{5 - 3\sqrt{2}}{2}$ **27.** $\frac{3 + \sqrt{73}}{8}, \frac{3 - \sqrt{73}}{8}$ **29.** No real solution
31. $1 + \sqrt{5}, 1 - \sqrt{5}$ **33.** $\frac{-1 + \sqrt{33}}{4}, \frac{-1 - \sqrt{33}}{4}$ **35.** No real solution
37. $\frac{49}{4}$ **39.** There are 51 insects after 3 days. **41. a.** The fireworks will be 16 ft above the ground 1 sec after being shot into the air. **b.** They will

be 8 ft above the ground in approximately 0.3 sec and again in 1.7 sec.
43. One friend was 6 mi from the party and the other was 8 mi from the party. **45.** It would take the faster machine about 7 min to do the job alone.

Section 9.3   Practices, pp. 635–638
1, p. 635: a. $1, 0$ **b.** No real solution **2, p. 637: a.** $\frac{4 + \sqrt{31}}{5}, \frac{4 - \sqrt{31}}{5}$
b. $\frac{-3 + \sqrt{93}}{6}, \frac{-3 - \sqrt{93}}{6}$ **3, p. 638:** $2.5, -0.5$ **4, p. 638:** Working alone, it takes one pump about 4.4 hr and the other about 9.4 hr to empty the tank.

Exercises 9.3, pp. 640–641

	Standard Form	$a =$	$b =$	$c =$
1.	$2x^2 + 9x - 1 = 0$	2	9	-1
3.	$-x^2 + 3x - 8 = 0$	-1	3	-8
5.	$x^2 - x + 8 = 0$	1	-1	8
7.	$\frac{1}{3}y^2 - \frac{1}{2}y + \frac{1}{4} = 0$	$\frac{1}{3}$	$-\frac{1}{2}$	$\frac{1}{4}$
9.	$6x^2 - 7x - 15 = 0$	6	-7	-15

11. $1, -3$ **13.** No real solution **15.** $-1, 7$ **17.** $2 + \sqrt{5}, 2 - \sqrt{5}$
19. $1, \frac{1}{3}$ **21.** $\frac{3 + \sqrt{5}}{4}, \frac{3 - \sqrt{5}}{4}$ **23.** $6, -\frac{5}{2}$ **25.** $-1, -2$ **27.** $-1, -3$
29. $0, \frac{2}{3}$ **31.** $-1, \frac{1}{4}$ **33.** $\frac{7 + \sqrt{29}}{2}, \frac{7 - \sqrt{29}}{2}$ **35.** $\frac{1 + 7\sqrt{5}}{2}, \frac{1 - 7\sqrt{5}}{2}$
37. No real solution **39.** $\frac{3 + \sqrt{3}}{2}, \frac{3 - \sqrt{3}}{2}$ **41.** No real solution
43. $-2, 3$ **45.** $\frac{2}{5}d^2 - \frac{4}{5}d - \frac{1}{5} = 0; a = \frac{2}{5}, b = -\frac{4}{5}, c = -\frac{1}{5}$
47. n is 13. **49.** Working alone, it takes the lab coordinator about 91 min to set up the lab, and it takes the lab technician about 71 min to set up the lab. **51.** The speed of the car is approximately 27 mph. **53.** At about 11:40 A.M. and 5 P.M.

Section 9.4   Practices, pp. 644–648
1, p. 644: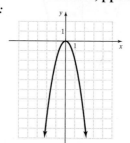

2, p. 645: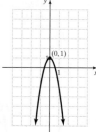

3, p. 646: a.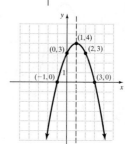

b. The graph of the parabola opens downward since $a = -1$ is negative. The curve turns at the vertex $(1, 4)$, the highest point of the graph. The equation of the axis of symmetry is $x = 1$.

4, p. 648: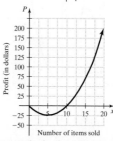

Selling 5 items results in the most money lost ($25).

Exercises 9.4, pp. 649–656

1. graph of an equation   **3.** axis of symmetry   **5.** positive

7.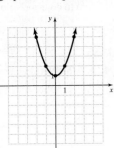

9.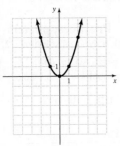

11.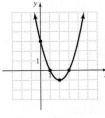

13. Axis of symmetry: $x = 0$; vertex: $(0, 0)$ **15.** Axis of symmetry: $x = 2$; vertex: $(2, 1)$

17.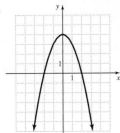

19.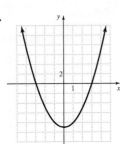

21.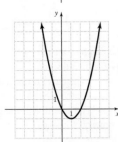

23.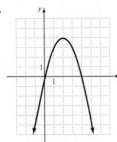

25.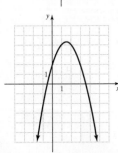

27.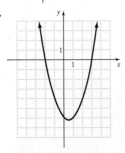

29.

31.

33.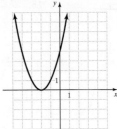

35.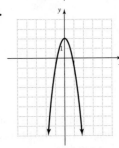

37.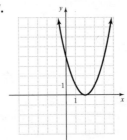

39.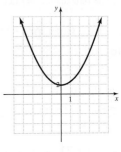

41.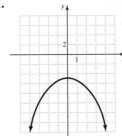

43.

45.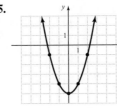

47.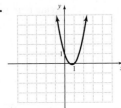

49.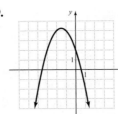

51. a.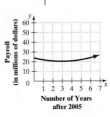

b. In the year 2008   **c.** The payroll decreases from 2005 to 2008 and increases thereafter.

53. a.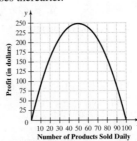

b. The company must sell 50 products each day to maximize the profit.   **c.** The profit increases when 0 to 50 products are sold, but then decreases for sales of more than 50 products.

55. a. $R = x(100 - \frac{1}{4}x) = 100x - \frac{1}{4}x^2$

b.

x	0	100	200	300	400
R	0	7500	10,000	7500	0

c.

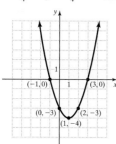

d. The store should sell 200 items to maximize the revenue.   **e.** The maximum revenue is $10,000.

Chapter 9   Review Exercises, pp. 661–664

1. is; Possible answer: of the square root property of equality   **2.** is; Possible answer: the constant term a^2 is the square of half the coefficient of the linear term $2a$ **3.** is; Possible answer: the equation has the form $y = ax^2 + bx + c$, where $a \neq 0$ **4.** is; Possible answer: its equation is $x = \frac{5}{8}$ **5.** downward; Possible answer: the coefficient of x^2 is negative   **6.** minimum; Possible answer: the coefficient of the quadratic term is positive   **7.** $2\sqrt{6}, -2\sqrt{6}$ **8.** $\sqrt{3}, -\sqrt{3}$ **9.** No real solution
10. $\frac{5 + 3\sqrt{2}}{2}, \frac{5 - 3\sqrt{2}}{2}$ **11.** $r = \sqrt{\frac{A}{\pi}}$ **12.** $t = \sqrt{\frac{2d}{g}} = \frac{\sqrt{2dg}}{g}$ **13.** 25
14. $\frac{49}{4}$ **15.** $9, -3$ **16.** $1, -4$ **17.** $-2 + \sqrt{6}, -2 - \sqrt{6}$
18. $\frac{1 + \sqrt{2}}{2}, \frac{1 - \sqrt{2}}{2}$ **19.** $1 + \sqrt{2}, 1 - \sqrt{2}$ **20.** $-4 + \sqrt{15}, -4 - \sqrt{15}$
21. $\frac{-1 + \sqrt{41}}{4}, \frac{-1 - \sqrt{41}}{4}$ **22.** No real solution   **23.** $-5, 0$ **24.** $\sqrt{6}, -\sqrt{6}$
25.

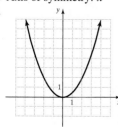

26.

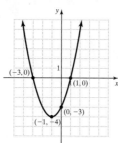

27. Axis of symmetry: $x = -2$; vertex: $(-2, -4)$
28. Axis of symmetry: $x = -1$; vertex: $(-1, 4)$
29.

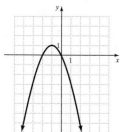

30.

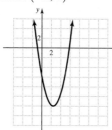

31.

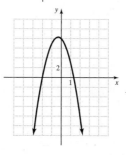

32.

(graph)

33. The radius of the sphere is approximately 7.0 in.   **34.** The company had sold either 40 rugs or 50 rugs since the profit is $2000 for both.   **35.** The banner is approximately 2.2 ft by 9.3 ft.   **36.** The building has 7 sides.
37. a.

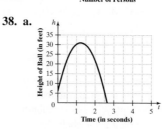

b. The profit per person if 16 people go is $384.   **c.** Either 13 people or 27 people went to the Bahamas since the profit is $351 for both.   **d.** The maximum profit per person that the agency can make is $400 (when 20 people go to the Bahamas).

38. a.

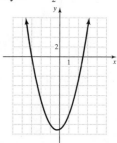

b. The ball reaches its maximum height in $\frac{5}{4}$, or 1.25, sec.   **c.** The maximum height of the ball is 31 ft.   **39.** Working alone, it would take the experienced Web designer 4 days to develop the website and it would take the inexperienced designer 12 days to develop the website.
40. Her average speed driving to work is approximately 36 mph.

Chapter 9   Posttest, pp. 665–666

1. $-1 + 2\sqrt{2}, -1 - 2\sqrt{2}$ **2.** $r = \sqrt{\frac{3V}{\pi h}} = \frac{3\sqrt{\pi h V}}{\pi h}$ **3.** $\frac{1}{4}$
4. $\frac{3 + \sqrt{13}}{2}, \frac{3 - \sqrt{13}}{2}$ **5.** $\frac{-3 + \sqrt{41}}{4}, \frac{-3 - \sqrt{41}}{4}$ **6.** $2\sqrt{3}, -2\sqrt{3}$ **7.** $4, -10$
8. $\frac{27}{5}, \frac{13}{5}$ **9.** $\frac{2 + \sqrt{7}}{2}, \frac{2 - \sqrt{7}}{2}$ **10.** $\frac{-4 + 2\sqrt{7}}{3}, \frac{-4 - 2\sqrt{7}}{3}$
11. $4 + \sqrt{15}, 4 - \sqrt{15}$ **12.**

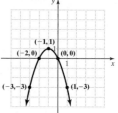

13. Vertex: $\left(-\frac{3}{2}, \frac{7}{4}\right)$; axis of symmetry $x = -\frac{3}{2}$
14.

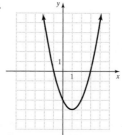

15.

(graph)

16.

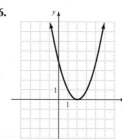

17. $t = \frac{\sqrt{l}}{0.9}$ **18.** The maximum speed is 104 km/hr.   **19.** The year 2032
20. The object reaches a maximum height of 100 ft (when $t = 2.5$ sec).

Chapter 9 Cumulative Review, pp. 667–668

1. -8 **2.** $-72x - 110$ **3.** $n = \frac{ax - c}{6}; -\frac{4}{5}$

4.

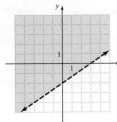

5. $(-1, 3)$ **6.** $18n^2 + 3n - 10$ **7.** $\frac{2x^2}{3(y - 2)}$ **8.** $3cd(2c + d)(c + d)$

9. $-\frac{2}{3}, 7$ **10.** $\frac{1}{4x(x + 4)}$ **11.** $-3, 2$ **12.** $xy\sqrt{y}$ **13.** $4 \pm 2\sqrt{3}$

14. $-2, \frac{1}{3}$ **15.** \$750

16. a. $n = \frac{5}{6}t + 2$

b.

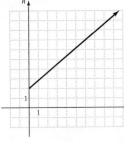

c. The slope of the graph, $\frac{5}{6}$, represents the rate of increase in the number of employees (5 employees every 6 months). In the equation it is the coefficient of t.

17. For $x = 0$, the number of bacteria is 400; for $x = 1$, the number of bacteria is 1200; for $x = 2$, the number of bacteria is 3600.

18. $0.004a + 0.35b \leq 80$ **19. a.** $300t$ meters away   **b.** 2 sec

20. a.

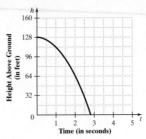

b. The sandbag has fallen 16 ft. At $t = 0$, the sandbag is at a height of 128 ft above the ground, and at 1 sec the sandbag has fallen to a height of 112 ft above the ground. So it has fallen a total of $(128 - 112)$ ft, or 16 ft.

APPENDIX A.2

Appendix A.2   Practices, pp. 670–671

1, p. 670: a. The binomial is a difference of cubes.   **b.** The binomial is neither a sum of cubes nor a difference of cubes.   **c.** The binomial is a sum of cubes.   **d.** The binomial is a difference of cubes.

2, p. 671: a. $(r + s)(r^2 - rs + s^2)$ **b.** $(y - 4)(y^2 + 4y + 16)$

3, p. 671: a. $(5 - x)(25 + 5x + x^2)$ **b.** $5(n - 2)(n^2 + 2n + 4)$ **c.** $2n(1 + 3n)(1 - 3n + 9n^2)$

Exercises A.2, p. 672

1. $(y - 1)(y^2 + y + 1)$ **3.** $(p + q)(p^2 - pq + q^2)$

5. $(x + y^2)(x^2 - xy^2 + y^4)$ **7.** $(n + 2)(n^2 - 2n + 4)$

9. Prime polynomial.   **11.** $(3n + 1)(9n^2 - 3n + 1)$

13. $5(x - 1)(x^2 + x + 1)$ **15.** $2(p + 3)(p^2 - 3p + 9)$

17. $x^3(x - 5)(x^2 + 5x + 25)$

Glossary

The numbers in brackets following each glossary term represent the section in which that term is discussed.

absolute value [1.1] The absolute value of a number is its distance from 0 on the number line.

addition property of equality [2.1] This property states that when adding any real number to each side of an equation, the result is an equivalent equation.

additive identity property [1.2] For any real number a, $a + 0 = a$ and $0 + a = a$.

additive inverse property [1.2] For any real number a, there is exactly one real number $-a$ such that $a + (-a) = 0$ and $(-a) + a = 0$.

additive inverses [1.2] Two numbers that have a sum of 0.

algebraic expression [1.6] An expression in which constants and variables are combined using standard arithmetic operations.

associative property of addition [1.2] The associative property of addition states that when adding three numbers, the sum is the same regardless of how they are grouped. For any three numbers a, b, and c,
$$(a + b) + c = a + (b + c).$$

associative property of multiplication [1.4] The associative property of multiplication states that when multiplying three numbers, the product is the same regardless of how they are grouped. For any three numbers a, b, and c, $(a \cdot b) \cdot c = a \cdot (b \cdot c)$.

average (or mean) [R.1] The average (or mean) of a set of numbers is the sum of those numbers divided by however many numbers are in the set.

axis of symmetry [9.4] The vertical line passing through the vertex of a parabola. The axis of symmetry divides the parabola into two parts that are mirror images of one another.

base (exponent) [1.6, 5.1] The base is the number that is a repeated factor when written with an exponent.

binomial [5.3] A binomial is a polynomial with two terms.

break-even point [4.1] The point at which the income for a business equals its expenses.

coefficient [1.8, 5.3] The numerical factor of each variable term; for example, in the expression $5x$, 5 is called the coefficient.

common factor [6.1] A common factor of two or more integers is an integer that is a factor of each integer.

common multiple [R.2] A number that is a multiple of two or more numbers is called a common multiple.

commutative property of addition [1.2] The commutative property of addition states that the sum of two numbers is the same regardless of order. For any two real numbers a and b, $a + b = b + a$.

commutative property of multiplication [1.4] The commutative property of multiplication states that the product of two numbers is the same regardless of order. For any two real numbers a and b, $a \cdot b = b \cdot a$.

completing the square [9.2] A method for solving quadratic equations that involves taking square roots and factoring perfect square trinomials.

complex rational expression (complex algebraic fraction) [7.4] An expression that contains a rational expression in its numerator, denominator, or both.

composite number [R.2] A whole number that has more than two factors.

conjugate [8.3] The conjugate of $a + b$ is $a - b$.

constant [1.6] A known quantity whose value does not change.

constant term [5.3] In a polynomial, the constant term is the term of degree 0.

coordinate plane [3.1] The flat surface on which we draw graphs.

coordinates [3.1] A pair of numbers in a given order that corresponds to a location on the coordinate plane.

decimal [R.4] A decimal has a whole number part, which precedes the decimal point, and a fractional part, which follows the decimal point.

decimal places [R.4] The decimal places are the places to the right of the decimal point.

degree of a monomial [5.3] The degree of a monomial is the power of the variable in the monomial.

degree of a polynomial [5.3] The degree of a polynomial is the highest degree of any of its terms.

demand curve [4.1] The demand curve illustrates that when the price of an item increases, the quantity of items sold declines. It is commonly approximated by a straight line with a negative slope.

denominator [R.3] The number below the fraction line in a fraction is called the denominator. It represents the number of parts into which the whole is divided.

difference of squares [6.4] An expression in the form $a^2 - b^2$, which can be factored as $(a + b)(a - b)$.

distributive property [1.8] The distributive property states that a number times the sum of two quantities is equal to the number times one quantity plus the number times the other quantity. For all real numbers, a, b, and c, $a \cdot (b + c) = a \cdot b + a \cdot c$.

elimination (or addition) method [4.3] A method used to solve a system of equations that is based on a closely related property of equality that states:

If $a = b$ and $c = d$, then $a + c = b + d$.

equation [2.1] An equation is a mathematical statement that two expressions are equal.

equilibrium [4.1] The point of intersection of the supply curve and the demand curve where the quantity supplied is equal to the quantity demanded.

equivalent equations [2.1] Equivalent equations are equations that have the same solution.

equivalent fractions [R.3] Equivalent fractions are fractions that represent the same value.

equivalent rational expressions [7.1] Rational expressions that have the same value, no matter what value replaces the variable.

exponent (or power) [R.1, 1.6, 5.1] An exponent (or power) is a number that indicates how many times another number is multiplied by itself.

exponential notation [1.6] Exponential notation is a shorthand method for representing repeated multiplication of the same factor.

extraneous solutions [7.5, 8.4] Extraneous solutions are *not* solutions of the original equation.

factored completely [6.2] A polynomial is factored completely when it is expressed as the product of a monomial and one or more prime polynomials.

factoring by grouping [6.1] When trying to factor a polynomial that has four terms, it may be possible to group pairs of terms in such a way that a common binomial factor can be found. This method is called factoring by grouping.

factors [R.2] In a multiplication problem involving two or more whole numbers, the whole numbers that are multiplied are called factors.

FOIL method [5.5] A method for multiplying two binomials. Multiply First terms, Outside terms, Inside terms, and Last terms, then combine like terms.

formula [1.7] A formula is an equation that indicates how variables are related to one another.

fraction [R.3] A fraction can mean a part of a whole or the quotient of two whole numbers.

fraction line (or fraction bar) [R.3] The fraction line separates the numerator from the denominator and stands for the phrase *out of* or *divided by*.

graph [3.3] The graph of a linear equation in two variables consists of all points whose coordinates make the equation true.

graphing method [4.1] A method of solving a linear system of equations in which we graph the equations that make up the system and any point of intersection is a solution of the system.

greatest common factor (GCF) [6.1] The greatest common factor (GCF) of two or more integers is the greatest integer that is a factor of each integer.

greatest common factor (GCF) of two or more monomials [6.1] The greatest common factor (GCF) of two or more monomials is the product of the greatest common factor of the coefficients and for each variable, the variable to the lowest power to which it is raised in any of the monomials.

horizontal line [3.2] When the slope of a line is 0, its graph is a horizontal line.

improper fraction [R.3] An improper fraction is a fraction whose numerator is larger than or equal to its denominator.

inequality [2.6] An inequality is any mathematical statement containing $<$, $\leq$, $>$, $\geq$, or $\neq$.

inequality symbols [1.1] The symbols $\neq$, $<$, $\leq$, $>$, and $\geq$, which are used to compare numbers.

integers [1.1] The integers are the numbers $\ldots, -4, -3, -3, -1, 0, +1, +2, +3, +4, \ldots$ continuing indefinitely in both directions.

irrational numbers [1.1] Real numbers that cannot be written as the quotient of two integers.

leading coefficient [5.3] The leading coefficient is the coefficient of the leading term in a polynomial.

leading term [5.3] The leading term of a polynomial is the term in the polynomial with the highest degree.

least common denominator (LCD) [R.3] The least common denominator (LCD) for any set of fractions is the least common multiple of their denominators.

least common multiple (LCM) [R.2] The least common multiple (LCM) of two or more numbers is the smallest nonzero number that is a multiple of each number.

like fractions [R.3] Like fractions are fractions with the same denominator.

like radicals [8.2] Like radicals are radical expressions that have the same radicand.

like terms [1.8] Like terms are terms that have the same variables with the same exponents.

linear equation in one variable [2.1] An equation that can be written in the form $ax + b = c$, where a, b, and c are real numbers and $a \neq 0$.

linear equation in two variables [3.3] A linear equation in two variables, x and y, is an equation that can be written in the *general form* $Ax + By = C$, where A, B, and C are real numbers and A and B are not both 0.

linear inequality in two variables [3.5] A linear inequality in two variables is an inequality that can be written in the form $Ax + By < C$, where A, B, and C are real numbers and A and B are not both 0. The inequality symbol can be $<, >, \leq$, or $\geq$.

literal equation [2.4] A literal equation is an equation involving two or more variables.

mean (or average) [R.1] The mean (or average) of a set of numbers is the sum of those numbers divided by however many numbers are in the set.

mixed number [R.3] A mixed number consists of a whole number and a proper fraction.

monomial [5.3] A monomial is an expression that is the product of a real number and variables raised to nonnegative integer powers.

multiples [R.2] The multiples of a number are the products of that number and the whole numbers.

multiplication property of equality [2.2] This property states that when multiplying each side of an equation by any nonzero real number, the result is an equivalent equation.

multiplication property of zero [1.4] The multiplication property of zero states that the product of any number and 0 is 0. For any real number a, $a \cdot 0 = 0$ and $0 \cdot a = 0$.

multiplicative identity property [1.4] The multiplicative identity property states that the product of any number and 1 is the original number. For any real number a, $a \cdot 1 = a$ and $1 \cdot a = a$.

multiplicative inverse property [1.5] The multiplicative inverse property states that the product of a number and its multiplicative inverse is 1. For any nonzero real number a, $a \cdot \frac{1}{a} = 1$ and $\frac{1}{a} \cdot a = 1$.

multiplicative inverses (or reciprocals) [1.5] Two nonzero numbers that have a product of 1.

natural numbers [1.1] The natural numbers are $1, 2, 3, 4, 5, 6, \ldots$.

negative number [1.1] A negative number is a number smaller than 0.

negative slope [3.2] On a graph, the slope of a line that slants downward from left to right.

numerator [R.3] The number above the fraction line in a fraction is called the numerator. It tells us how many parts of the whole the fraction contains.

opposites [1.1] Two real numbers that are the same distance from 0 on the number line but on opposite sides of 0 are called opposites. For any real number n, its opposite is $-n$.

ordered pair [3.1] A pair of numbers that represents a point in the coordinate plane.

order of operations [R.1] The order of operations is a rule we agree to follow when a mathematical expression involves more than one mathematical operation.

origin [1.1, 3.1] On the number line, the point at 0; in the coordinate plane, the point where the axes intersect, $(0, 0)$.

parabola [9.4] The U-shaped graph of an equation of the form $y = ax^2 + bx + c$ that opens either upward or downward.

parallel lines [3.2] Two nonvertical lines are parallel if and only if their slopes are equal. That is, if the slopes are m_1 and m_2, then $m_1 = m_2$.

percent [R.5] A percent is a ratio or fraction with a denominator of 100. A number written with the % sign means "divided by 100."

percent decrease [2.5] In a percent problem, if the quantity is decreasing, it is called a percent decrease.

percent increase [2.5] In a percent problem, if the quantity is increasing, it is called a percent increase.

perfect square [8.1] A whole number is said to be a perfect square if it is the square of another whole number.

perfect square trinomial [6.4] A trinomial that can be factored as the square of a binomial, for example, $a^2 + 2ab + b^2 = (a + b)^2$ and $a^2 - 2ab + b^2 = (a - b)^2$.

perpendicular lines [3.2] Two nonvertical lines are perpendicular if and only if the product of their slopes is -1. That is, if the slopes are m_1 and m_2, then $m_1 \cdot m_2 = -1$.

point-slope form [3.4] The point-slope form of a linear equation is written as $y - y_1 = m(x - x_1)$, where x_1, y_1, and m are constants. In this form, m is the slope and (x_1, y_1) is a point that lies on the graph of the equation.

polynomial [5.3] A polynomial is an algebraic expression with one or more monomials added or subtracted.

positive number [1.1] A positive number is a number larger than 0.

positive slope [3.2] On a graph, the slope of a line that slants upward from left to right.

power (or exponent) [R.1, 1.6, 5.1] A power (or exponent) is a number that indicates how many times another number is multiplied by itself.

prime factorization [R.2] Prime factorization is the process of writing a whole number as a product of its prime factors.

prime number [R.2] A prime number is a whole number that has exactly two factors, namely itself and 1.

prime polynomial [6.2] A polynomial that is not factorable.

principal square root [8.1] The square root of a number that is nonnegative.

proper fraction [R.3] A proper fraction is a fraction whose numerator is smaller than its denominator.

proportion [7.6] A proportion is a statement that two ratios, a/b and c/d, are equal, written $a/b = c/d$, where $b \neq 0$ and $d \neq 0$.

Pythagorean theorem [6.5] The Pythagorean theorem states that for every right triangle, the sum of the squares of the lengths of the legs equals the square of the length of the hypotenuse: $a^2 + b^2 = c^2$, where a and b are the lengths of the legs and c is the length of the hypotenuse.

quadrant [3.1] One of four regions of a coordinate plane separated by the x- and y-axes.

quadratic equation (second-degree equation) [6.5, 9.1] An equation that can be written in the form $ax^2 + bx + c = 0$, where $a, b,$ and c are real numbers and $a \neq 0$.

quadratic formula [9.3] The quadratic formula states that if $ax^2 + bx + c = 0$, where $a, b,$ and c are real numbers and $a \neq 0$, then $x = (-b \pm \sqrt{b^2 - 4ac})/2a$.

radical equation [8.4] A radical equation is an equation with a variable in one or more radicands.

radical expressions [8.1] Radical expressions are algebraic expressions that involve square roots.

radical sign [8.1] The symbol $\sqrt{}$.

radicand [8.1] The number under the radical sign.

rate of change [3.2] Slope can be interpreted as a rate of change. It indicates how fast the quantity is changing and if the quantity being graphed increases or decreases.

ratio [7.6] A ratio is a comparison of two numbers, expressed as a quotient.

rational equation [7.5] An equation that contains one or more rational expressions.

rational expression [7.1] A rational expression P/Q is an algebraic expression that can be written as the quotient of two polynomials, P and Q, where $Q \neq 0$.

rational numbers [1.1] Numbers that can be written in the form a/b, where a and b are integers and $b \neq 0$.

rationalize the denominator [8.3] To rewrite an expression in an equivalent form that contains no radical in its denominator.

real numbers [1.1] Numbers that can be represented as points on a number line.

reciprocal (or multiplicative inverse) [R.3, 1.5] Two non-zero numbers that have a product of 1. For example, the reciprocal of the fraction $\frac{2}{3}$ is $\frac{3}{2}$.

reduced to lowest terms (or simplified form) [R.3, 7.1] A fraction is said to be reduced to lowest terms when the only common factor of its numerator and its denominator is 1.

regression line [3.1] A straight line that is closest to passing through the points on a graph.

scientific notation [5.2] A number is in scientific notation if it is written in the form $a \times 10^n$, where n is an integer and a is greater than or equal to 1 but less than 10 ($1 \leq a < 10$).

second-degree equation (quadratic equation) [6.5] An equation that can be written in the form $ax^2 + bx + c = 0$, where $a, b,$ and c are real numbers and $a \neq 0$.

similar triangles [7.6] Triangles with the same shape but not necessarily the same size.

simplest form (or reduced to lowest terms) [R.3, 7.1] A fraction is said to be in simplest form when the only common factor of its numerator and its denominator is 1.

slope [3.2] The ratio of the change in y-values to the change in x-values along a line. The slope m of a line passing through the points (x_1, y_1) and (x_2, y_2) is defined to be $m = (y_2 - y_1)/(x_2 - x_1)$, where $x_1 \neq x_2$.

slope-intercept form [3.4] A linear equation is in slope-intercept form if it is written as $y = mx + b$, where m and b are constants. In this form, m is the slope and $(0, b)$ is the y-intercept of the graph of the equation.

solution of an equation [2.1] A solution of an equation is a value of the variable that makes the equation a true statement.

solution of an equation in two variables [3.3] A solution of an equation in two variables is an ordered pair of numbers that when substituted for the variables makes the equation true.

solution of an inequality [2.6] A solution of an inequality is any value of the variable that makes the inequality true.

solution of a system of two linear equations [4.1] A solution of a system of two linear equations in two variables is an ordered pair of numbers that makes both equations in the system true.

solution of an inequality in two variables [3.5] A solution of an inequality in two variables is an ordered pair of numbers that when substituted for the variables makes the inequality a true statement.

solve an inequality [2.6] To solve an inequality is to find all of its solutions.

square root [8.1] The square root of a nonnegative real number a is a number that when squared is a.

square root property of equality [9.1] If n is a nonnegative number and $x^2 = n$, then $x = \pm\sqrt{n}$; that is $x = \sqrt{n}$ or $x = -\sqrt{n}$.

substitution method [4.2] A method for solving a system of equations in which one linear equation is solved for one of the variables and then the result is substituted into the other equation.

supply curve [4.1] The supply curve illustrates that as selling prices increase, wholesalers are inclined to make more goods available to retailers. It is approximated by a straight line with positive slope.

system of equations [4.1] A system of equations is a group of two or more equations solved simultaneously.

term [1.6] A term is a number, a variable, or the product or quotient of numbers and variables.

trinomial [5.3] A trinomial is a polynomial with three terms.

unit fraction [7.3] A fraction whose numerator is 1.

unlike fractions [R.3] Unlike fractions are fractions with different denominators.

unlike radicals [8.2] Unlike radicals are radical expressions with different radicands.

unlike terms [1.8] Unlike terms are terms that do not have the same variables with the same exponents.

variable [1.6] An unknown quantity represented by any letter or symbol or a quantity that can change in value.

vertex [9.4] The highest or lowest point of a parabola is called the vertex.

vertical line [3.2] When the slope of a line is undefined, its graph is a vertical line.

whole numbers [1.1] The whole numbers consist of 0 and the natural numbers: $0, 1, 2, 3, 4, 5, \ldots$.

x-axis [3.1] The horizontal number line in the coordinate plane.

x-coordinate [3.1] The first number in an ordered pair that represents a horizontal distance in the coordinate plane.

x-intercept [3.3] The x-intercept of a line is the point where the graph crosses the x-axis.

y-axis [3.1] The vertical number line in the coordinate plane.

y-coordinate [3.1] The second number in an ordered pair that represents a vertical distance in the coordinate plane.

y-intercept [3.3] The y-intercept of a line is the point where the graph crosses the y-axis.

zero-product property [6.5] The zero-product property states that if the product of two factors is 0, then either one or both of the factors must be 0.

Index